Connecticut Wildlife

Connecticut

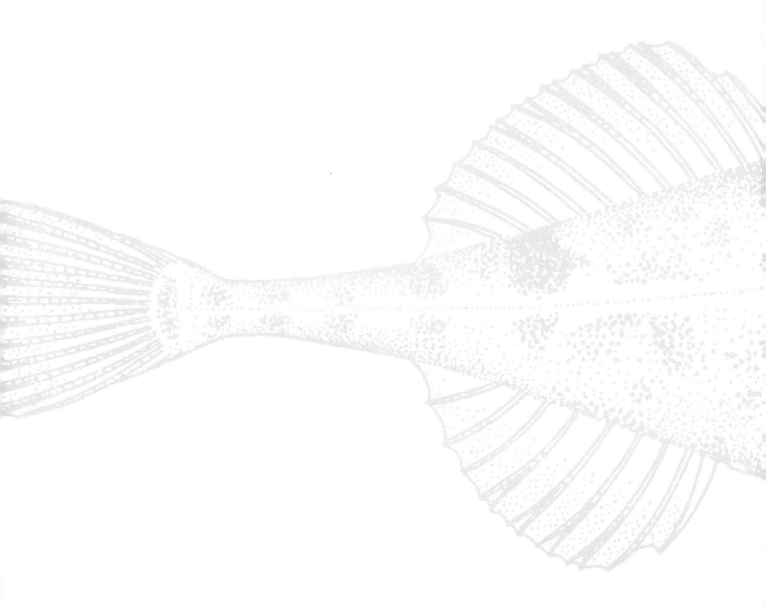

University Press of New England • *Hanover and London*

Wildlife

Biodiversity, Natural History, and Conservation

Geoffrey A. Hammerson

Published by University Press of New England,
One Court Street, Lebanon, NH 03766
www.upne.com
© 2004 by Geoffrey A. Hammerson
Printed in the United States of America
5 4 3 2 1

All rights reserved. No part of this book may be reproduced in any form or by any electronic or mechanical means, including storage and retrieval systems, without permission in writing from the publisher, except by a reviewer, who may quote brief passages in a review. Members of educational institutions and organizations wishing to photocopy any of the work for classroom use, or authors and publishers who would like to obtain permission for any of the material in the work, should contact Permissions, University Press of New England, One Court Street, Lebanon, NH 03766.

Library of Congress Cataloging-in-Publication Data
Hammerson, Geoffrey A.
Connecticut wildlife : biodiversity, natural history, and conservation / Geoffrey A. Hammerson.
   p. cm.
Includes bibliographical references (p.   ).
ISBN 1-58465-369-8 (pbk. : alk. paper)
1. Animals—Connecticut. 2. Plants—Connecticut.
3. Biotic communities—Connecticut. I. Title.
QL166H36 2004
791.9746—dc22   2003027342

Photographs and drawings © 2004 Geoffrey A. Hammerson unless otherwise credited. Additional photographs © 2004 Stephen P. Broker, Paul Fusco, or Larry Master.

For critical funding that supported the publication of this book the author thanks

the Connecticut Association of Conservation and Inland Wetlands Commissions,

Connecticut River Watershed Council, and

The Rockfall Foundation.

For Sandy, with love

Contents

Acknowledgments — xi
Introduction — xiii
A Note on the Text — xv

Chapter 1. The Landscape — 1
  Geology — 1
  Bird's-Eye View — 1
  Topography and Coastline — 1
  Post-Glacial Changes — 2
  Human Impact — 6

Chapter 2. The Seasons — 9
  Climate — 9
  Seasonal Change — 10
  Long-term Change — 10

Chapter 3. Coastal Waters and Wetlands: Estuarine Ecosystems — 11
  Long Island Sound—Physical Environment — 11
  Open-Water Biota — 12
  Rocky Shores — 15
  Muddy and Sandy Shores and Shallows — 20
  Salt Marshes — 25
  Brackish Marshes — 31
  Stress Gradients in Estuarine Ecosystems — 33
  Coastal Forests — 33
  Conservation — 33

Chapter 4. Streams and Associated Wetlands: Riverine Ecosystems — 40
  Open Waters and Aquatic Beds — 43
  Riverine and Floodplain Marshes — 47
  Floodplain Forests — 51
  Beaver-Influenced Wetlands — 53
  Conservation — 54
  The Connecticut River — 56

Chapter 5. Lakes and Ponds: Lacustrine Ecosystems — 65
  Physical Characteristics — 65
  Vegetation — 67
  Fauna — 68
  Conservation — 69

Chapter 6. Inland Wetlands: Palustrine Ecosystems — 71
  Freshwater Marshes and Meadows — 71
  Shrub Swamps — 73
  Forested Wetlands — 75
  Peatlands: Bogs and Fens — 78
  Vernal Pools — 82
  Wetland Indicators — 86
  Plants and the Wetland Environment — 87
  Conservation — 91

Chapter 7. Uplands: Terrestrial Ecosystems — 95
  Deciduous Forests — 95
  Coniferous and Mixed Forests and Woodlands — 110
  Forest Edges and Shrubby Old Fields — 113
  Fields and Grasslands — 115
  Suburban Yards — 117
  Cities — 117
  Conservation — 117

Chapter 8. Algae, Fungi, and Lichens — 122
  Algae — 122
  Fungi and Lichens — 125

Chapter 9. Plants — 128
  Nonvascular Plants — 129
  Vascular Plants — 130
  Ecology of Trees, Shrubs, and Vines in Connecticut — 135
  Ecology of Early Spring Wildflowers — 175
  Autumn Wildflowers — 178
  Plants in Winter — 180

Parasitic Plants　181
Understanding Orchids　183
Non-native Invasive Plants　184

Chapter 10. Sponges through Ectoprocts　186
Sponges　186
Jellyfishes, Hydroids, and Sea Anemones　187
Comb Jellies　190
Flatworms　191
Ribbon Worms　191
Nematodes　191
Horsehair Worms　192
Ectoprocts　192

Chapter 11. Segmented Worms　194
Polychaetes　194
Earthworms and Relatives　195
Leeches　195

Chapter 12. Mollusks　197
Snails and Other Gastropods　197
Clams, Mussels, and Oysters　204
Squids and Octopuses　210

Chapter 13. Chelicerates　211
Horseshoe Crab　211
Sea Spiders　211
Spiders　212
Harvestmen　214
Ticks and Mites　214

Chapter 14. Insects, Centipedes, and Millipedes　217
Flightless Insect Relatives　217
Mayflies　218
Dragonflies and Damselflies　218
Stoneflies　221
Walkingsticks　222
Grasshoppers, Crickets, Katydids, and Relatives　222
Earwigs　226
Mantids and Cockroaches　226
Termites　227
True Bugs　227
Cicadas, Aphids, and Relatives　229
Dobsonflies, Antlions, Lacewings, and Relatives　232
Beetles　233
Wasps, Bees, and Ants　239
Caddisflies　247
Butterflies and Moths　248
Common Scorpionflies　262
True Flies　262
Fleas　265

Insects in Winter　266
Galls and Gall Makers　267
Centipedes and Millipedes　271

Chapter 15. Crustaceans　273
Fairy Shrimps and Clam Shrimps　273
Cladocerans　273
Copepods　274
Ostracods　274
Barnacles　275
Isopods　275
Amphipods　276
Crabs and Relatives　277

Chapter 16. Echinoderms and Tunicates　283
Echinoderms　283
Tunicates　284

Chapter 17. Fishes　286
Lampreys　286
Cartilaginous Fishes　287
Ray-finned Fishes　288
Habitat　304
Food and Feeding　304
Reproduction　304
Fishes in Winter　305
Conservation　305

Chapter 18. Amphibians　307
Salamanders　307
Frogs and Toads　310
Habitats　313
Reproduction　314
Amphibians in Winter　314
Conservation　315

Chapter 19. Reptiles　317
Turtles　317
Lizards　320
Snakes　321
Habitats　324
Reproduction and Activity Calendar　324
Reptiles in Winter　325
Conservation　327

Chapter 20. Birds　329
Loons, Grebes, Storm Petrels, Gannets, and Cormorants　330
Wading Birds and Vultures　331
Swans, Geese, and Ducks　333
Eagles, Hawks, and Falcons　336

Pheasants, Grouse, Turkeys, and Quail	339	Great Apes	383
Rails, Gallinules, and Coots	340	Rabbits and Hares	383
Shorebirds	341	Rodents	384
Gulls, Terns, and Skimmers	343	Whales and Porpoises	389
Doves, Parakeets, and Cuckoos	344	Carnivores	389
Owls, Nighthawks, and Nightjars	345	Hoofed Mammals	394
Swifts, Hummingbirds, and Kingfishers	347	Habitats	396
Woodpeckers	347	Food and Feeding	396
Flycatchers	348	Reproduction	398
Songbirds	349	Mammals in Winter	399
Seasonal Pattern of Bird Diversity	362	Conservation	403
Habitats	363
Food and Feeding	363
Reproduction	367
Migration	369
Birds in Winter	371
Conservation	372

Chapter 21. Mammals   379
    Opossums   379
    Shrews and Moles   380
    Bats   382

Chapter 22. A Naturalist's Calendar   405

Appendix A. Endangered, Threatened, and Special Concern Species in Connecticut   431

Appendix B. Conservation Supplement   433

Bibliography   435

Index   451

Acknowledgments

The learning process that led to the development of this book was facilitated, stimulated, shared, and made superbly enjoyable by colleagues, friends, and especially several hundred students in the Graduate Liberal Studies Program (GLSP) and Biology Department at Wesleyan University. Our collective observations and experiences during innumerable field trips over two decades form the heart and soul of this book. Steve Broker, former associate director of the GLSP, deserves special thanks for initially bringing me on board at Wesleyan and for his role in developing a curriculum of field-oriented courses in biology and ecology. His successor, Greg McHone, and GLSP director Barbara MacEachern, were constant sources of support and encouragement over many years. My daughter Kay joined me for many snake hunts and night prowls during the early years of this project, before she was old enough to know better. In recent years Sandra Frimel faithfully aided and abetted my sometimes dubious activities in the field and at home, good-naturedly condoning my desire to examine road kills (sometimes even delivering them to me!), or patiently understanding whenever I suddenly got up in the middle of a conversation to investigate a mysterious sound or something fluttering outside the window. Larry Master and NatureServe speeded up my progress by allowing me to devote a little work time to this project. I thank Paul Fusco, Steve Broker, and Larry Master for their excellent photographic contributions. Dale Schweitzer and Jay Cordeiro helped me with some invertebrate identifications. Jay Cordeiro kindly updated my mollusk information. I'm deeply grateful to the editorial and production staff at University Press of New England—Mike Burton, Ellen Wicklum, and especially Jessica Stevens—for their wise counsel, professional expertise, and endless good spirit. Additionally, I am indebted to the generations of dedicated field biologists who artfully transformed their hard work and thoughtful studies into useful publications that became the foundation for this book.

Introduction

> Wildlife: "the native fauna and flora of a particular region"
> —*Oxford English Dictionary*

I'm somewhat of a newcomer to Connecticut. My pioneering ancestors left the Midwest and crossed the Great Plains, Rocky Mountains, and Great Basin desert to seek gold in California's Sierra Nevada in the early 1850s. They didn't find fortune there but did decide to stay in the Golden State, where, several generations later, I was born, raised, and attended college. I always regarded myself as a Californian and thought of New England as almost a foreign land. Recently, however, I was surprised to learn that some of my earlier relatives departed England and arrived as teenagers at the Plymouth Colony in Massachusetts in 1635. By the mid-1700s, my ancestors were residents of what is now Connecticut. Some fought (on the winning side!) in the Revolutionary War.

So my move to Connecticut in 1984 was in a sense a home-coming, though I didn't know it at the time. I felt apprehensive about leaving my life-long home in the West, but my arrival in Connecticut was at the same time a source of excitement, a chance to observe the plant and animal life of a region that was almost totally new to me—a savory experience for a naturalist.

Since 1984, I've tried to learn as much as possible about the natural world of Connecticut's hills, valleys, wetlands, shores, and waters. My experiences included many fine days spent studying woodland wildflowers and butterflies, birding along coastal shores, tracking mammals in snowy woods and fields, or paddling lakes, rivers, and Long Island Sound. Less comfortable but no less enjoyable were several wetland surveys in drenching rains and many summer explorations that involved crawling into sauna-like swamp thickets. Sometimes I used high-tech methods to probe the hidden lives of secretive animals, but more often I simply walked slowly and watched quietly what was happening around me, or paid curious attention when small wild guests entered my home or appeared outside lighted windows.

A few years ago, I decided to summarize what I've learned in a single book that would be useful to the state's students, teachers, naturalists, and conservationists, or to anyone who might simply be curious about nature as seen in their own backyard or local park. My field notes provided a wealth of information. To broaden my knowledge beyond my own experiences, I examined hundreds of pertinent books and research papers by other scientists. This book provides a fairly detailed overview of the state's biodiversity, natural history, and conservation issues, presented in a way that I hope will be useful to any interested person, including both scientists and the general public.

I emphasize that I consider "wildlife" to include all organisms, plant, animal, and otherwise, living wild in nature, not just vertebrates, or charismatic groups, or those with obvious economic value. And I expand that further to include ecological patterns, processes, and interactions as well. It's all wild life.

It would be impossible to include in a single volume of reasonable size ecological information on all of the biological diversity included even in the small region covered by this book. Similarly, I could not describe all of Connecticut's multitudinous conservation actions and programs. So this book is limited, but it does cover extensively the state's mammals, birds, reptiles, and amphibians, and the vast majority of fishes, plus many of the more observable, characteristic, or interesting invertebrates, as well as the major ecological patterns and processes and charactcristic plants and animals of each of the state's major ecosystems. Vertebrate species generally are discussed individually, but for most invertebrates, of which there are an overwhelming number of species, I provide generalized information that applies to selected groups or brief or in-depth discussions for selected members of a particular group. My treatment of plants includes information on their distribution and prominence among the state's major ecosystems, and I also summarize ecological characteristics of many of the state's trees, shrubs, and vines.

My focus throughout is on such questions as: What are the basic characteristics of the major ecosystems? Which plants and animals occur in the region? What are their major ecological associations, and how do they affect other organisms? What daily, seasonal, annual, and long-term changes are evident? When, where, and how do they produce offspring? Which species and ecosystems are in need of conservation attention? The specific kinds of information covered are not necessarily the same for each group of organisms.

This large book is most definitely not a *field* guide. In fact, I did not include much information on identification. My intent was that this book complement available identification guides that generally provide little in the way of ecological information or information specific to Connecticut. Although this book is not an identification guide, the numerous illustrations should help you to identify many of the organisms I discuss.

This book does not have to be read in any particular sequence. I don't recommend that you try reading the book straight through from front to back. Instead, I suggest that you use it as a reference work. You may wish to look up information on particular habitats or species before or after you go for a walk or out in your boat. Or simply read any section that piques your interest. Many of the organisms mentioned in the ecosystem sections in the first several chapters are discussed in more detail in later chapters that cover plants, animals, and other groups.

Please consider two important points when using this book. First, our knowledge of the natural world is actually quite rudimentary. We know very little about the lives of most wild organisms, and most processes at work in nature are barely understood. Second, the living world is dynamic, constantly changing. Most of today's patterns and processes are in a state of flux; this is a basic characteristic of ecosystems, but often it is not recognized. So this book should not be considered anything more than a progress report on ecosystems and organisms about which we know little and which will not be the same tomorrow as they are today. Because organisms and events in the natural world are incompletely known and highly variable, you can expect to observe deviations from the generalized patterns described in this book.

Progress in understanding the natural world is a two-part process that involves careful observation and recording those observations in a lasting form, such as in a field journal or through labeled drawings, photographs, or specimens. Memories of things we have experienced in the field certainly enrich our lives, but memory is an unreliable means to objective understanding. To facilitate your nature education beyond what you gain from this book, I encourage you to keep a tangible record of your field observations. One of the pleasures of writing this book was to come back from the field and compare my new knowledge to what I had previously learned and written about a topic. If you learn the environmental requirements of plants and look for the tell-tale signs of animals, you may be able to decipher the history, present condition and major ecological influences, and probable future of the places you visit—an enjoyable and challenging puzzle!

Humans have been a primary factor in shaping the present environment and will continue to affect the state's land and waters. There are few places where human influence cannot be readily seen. This statement applies not only to the northeastern United States but also seems to be the case throughout the world—just look out the window on your next cross-country flight and try to find a large area where human activity is not evident! Unfortunately, human impacts tend strongly toward biological impoverishment. My colleague David Armstrong aptly expressed it this way: "It is sobering to realize that what is natural history today may be just plain history tomorrow." I hope that this book fosters an increased appreciation of the region's natural biological diversity and a desire to protect it. Whether we recognize it or not, ecosystems that have all of their component parts and processes intact are a primary element in determining the quality of our lives. Damaged ecosystems inevitably result in economic hardship, and they rob us of high-quality opportunities for enjoyment of some of the basic pleasures of life. Beautiful natural areas, which for many of us are absolutely essential spiritual resources, have disappeared at an alarming rate. Happily, there is cause for optimism about the preservation of nature in the region, reflected in the many active conservationists who live in Connecticut and the important protective measures they have established. I hope this book encourages you to join us, if you haven't already.

A Note on the Text

Throughout this book (but not in figure captions), species that are not native to Connecticut are indicated by an asterisk, unless the text makes it clear that the species is an exotic. An asterisk in parentheses (*) denotes genera represented in Connecticut by both native and introduced species.

Scientific names are provided in the main entry for a species in chapters 8 through 21 and elsewhere for clarification or if the species has no English name. When a scientific name is not provided in a particular discussion (as in chapters 3 through 7), you can find it with the English name in the index or in the main entry for the species in chapters 8 through 21. Scientific names generally are from the following sources: plants, Magee and Ahles (1999); marine invertebrates, Weiss (1995); mollusks, Turgeon et al. (1998); insect families, Arnett (2000); various other animal groups, NatureServe (www.natureserve.org).

As a scientist, I prefer to use scientific units of measurement, but for the convenience of most users of this book I express most quantitative measurements in more familiar units (e.g., degrees Fahrenheit, inches, miles, pounds). Sometimes I use millimeters for very small measurements. Here are some conversions: ¼ inch = 6.4 millimeters (mm), 1 inch = 2.54 centimeters (cm) or 25.4 mm, 1 foot = 30.5 cm, 1 yard = 0.9 meters (m), 1 mile = 1.6 kilometers, 1 acre = 0.4 hectares or 4,047 m^2, 1 square mile = 640 acres = 2.59 km^2, 1 pound = 0.45 kilograms.

I use terms such as abundant, common, uncommon, rare, and the like, to convey a subjective indication of both the absolute abundance of a species and the likelihood that one might observe it in the appropriate season and habitat. I have not attempted to define these descriptors quantitatively because absolute abundance of most species is unknown or highly variable, and detectability depends on observer skill, effort, and methods. Where appropriate and possible, the subjective abundance/detectability descriptors are used in conjunction with locational or temporal terms to indicate geographical and seasonal variations (e.g., locally common; rare inland but common along the coast; rare in winter).

Appendix A includes a brief discussion of official state conservation designations (Endangered, Threatened, and Special Concern) as defined under the Connecticut Endangered Species Act of 1989 (Public Act 89-224). Because these statuses are reviewed periodically and may change over time, I did not include them in this book. Check the Department of Environmental Protection website (http://dep.state.ct.us) for the most up-to-date information on state statuses.

Connecticut Wildlife

Chapter 1

The Landscape

Connecticut, a small state on the east-central coast of North America, encompasses the northern end of the Appalachian Mountain chain and lies north of the vast Atlantic Coastal Plain of the mid-Atlantic and southeastern United States (Fig. 1.1). The state is bordered by New York to the west and south, Massachusetts to the north, and Rhode Island to the east. With an area of approximately 4,845 square miles, only Rhode Island and Delaware are smaller. The human population (3,290,000) averages roughly one person per acre, but in most areas density is either much higher or lower than this.

Geology

Despite its small size, Connecticut encompasses a surprisingly diverse array of environments, in large part reflecting the state's varied geological foundations. The geological landscape consists of hills of hard, crystalline rocks in the Western and Eastern Uplands (Fig. 1.2), divided by a central lowland zone of softer sedimentary shales and sandstones of Triassic age in the Central Valley, with an area of marble valleys (consisting of Cambrian and Ordovician limestones mostly metamorphosed into marble) along the Housatonic River system in western Connecticut (Figs. 1.3 and 1.4).

The central Connecticut Valley lowland is interrupted here and there by several north-south-trending "trap-rock" ridges (Figs. 1.5 and 1.6). These areas include the Metacomet Ridge and specific features such as Saltonstall Ridge, Totoket Mountain, Higby Mountain, Beseck Mountain, Lamentation Mountain, the Hanging Hills, Avon Mountain, Talcott Mountain, Penwood Mountain, and West Suffield Mountain. The features generally consist of the exposed, westward-facing edge of a tilted layer of basalt. The basalt flowed onto the surface, cooled, and was later covered by sedimentary layers (brownstone). The west edge of each ridge, below a basalt cliff, usually consists of steep talus weathered from the exposed, up-tilted edge of the basalt. The steep cliffs and talus slopes exist because the hard basalt is more resistant to erosion than the softer brownstone. Joining these ridges are other notable basalt features, including West Rock, East Rock, and Sleeping Giant in the south and Manitook Mountain, Barndoor Hills, the Hedgehog, the Sugarloaf, and Onion Mountain to the north. These represent basalt dikes and sills that cooled underground and were later exposed by erosion.

Glacial till and stratified glacial and alluvial deposits of sand and gravel cover the bedrock in the Western and Eastern Uplands. In the Connecticut River valley north of Rocky Hill, surface deposits include alluvial sand, silt, clay, gravel, and organic material with interposed silt or clay—remnants of a huge ice-age lake (Glacial Lake Hitchcock). Lake-bottom deposits of fine sand, silt, and clay exist in the state's southern lowlands.

Bird's-Eye View

Through the window of an airplane aloft, extensive forests of deciduous trees, patches of coniferous forest and fields, and the sparkling surfaces of thousands of lakes and ponds and a network of 8,400 miles of streams give Connecticut in summer the appearance of a lumpy green shag carpet, threadbare here and there and strewn with reflective shards and slivers. In winter, the carpet is a threadbare brown and gray, revealing the rocky nature of much of the landscape and, when snow-whitened, emphasizing the many trapezoidal fields that have been cut into the forest. Satellite images of the state clearly reveal densely urbanized areas, primarily in the southwestern and central portions of the state, and the much larger areas of sparsely populated forest (Fig. 1.7).

Topography and Coastline

For the most part, Connecticut consists of moderate hills, with some fairly large, relatively flat areas along the coast and in the Central Valley. Most of the state, particularly the

Fig. 1.1. Satellite image of coastal northeastern North America from Maine to North Carolina, showing Hurricane Erin over the Atlantic Ocean, 11 September 2001.

southern and central regions, is less than 500 feet above sea level, though hills of 500 to 1,000 feet are common (Fig. 1.8). The highest hills, mostly reaching 1,000 to 1,500 feet but extending above 1,500 feet in a few places, are in the northwestern part of the state. The highest elevation in Connecticut is 2,380 feet above sea level on the south slope of Mount Frissel in the extreme northwestern corner of the state. Elevations in the uplands of northeastern and west-central Connecticut are mostly 500 to 1,000 feet.

The 253-mile coastline of Connecticut, bordering the large estuary of Long Island Sound, is irregular and bordered in places by small islands. Some islands, such as the Norwalk Islands, consist of masses of sand, gravel, and boulders left by glaciers (Fig. 1.9). Others, including the Thimble Islands, are partially submerged mounds of solid bedrock (Fig. 1.10). Adding tidal river and island frontage in coastal towns to the mainland coast of Long Island Sound, the state's coastal water frontage totals about 550 miles.

To the south, Long Island, a large land mass visible on clear days from the Connecticut shore, shelters the coast from the brunt of the Atlantic Ocean's heaviest surf. The warm plume of the northern end of the Gulf Stream flows northward in the western Atlantic Ocean to the east.

Post-Glacial Changes

The present landscape of Connecticut is a young one. Only 20,000 years ago, all of this region was buried under a thick sheet of ice. Easily observable evidence of glaciation includes ice-polished bedrock surfaces bearing north-south gouges made by rocks imbedded in the flowing ice; long, streamlined drumlin hills that formed under moving ice; and huge rocks set down in odd locations by a melting glacier.

Sea level was about 400 feet lower during the Ice Age, the water being locked up in glaciers, so the coast of northeastern North America was more than 100 miles seaward of its

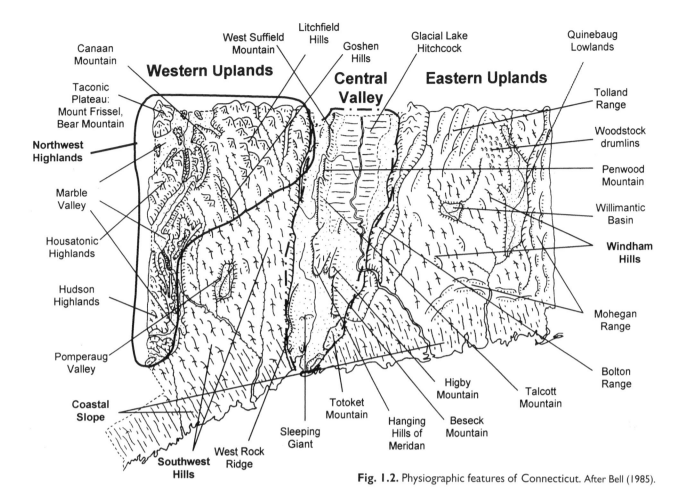

Fig. 1.2. Physiographic features of Connecticut. After Bell (1985).

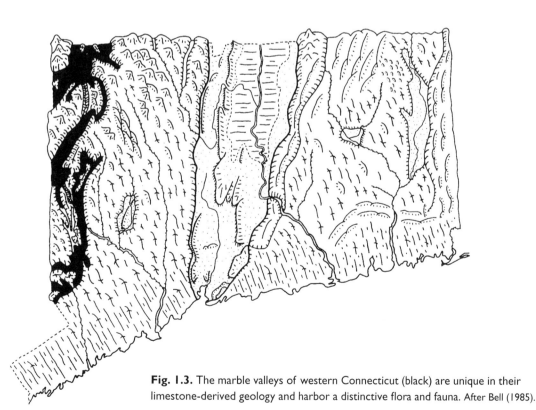

Fig. 1.3. The marble valleys of western Connecticut (black) are unique in their limestone-derived geology and harbor a distinctive flora and fauna. After Bell (1985).

present location, in areas now submerged by the Atlantic Ocean. Subsequent warming of the climate resulted in melting of the ice (exposing all of Connecticut by about 15,500 years ago), a rise in sea level, and the development of a tundra-covered proto-Connecticut, a situation that lasted a few thousand years. A large glacial lake existed in north-central Connecticut south to Rocky Hill for several thousand years. Piles of glacially transported debris (moraines), now represented by the Captain Islands, Norwalk Islands, Falkner Island, and, from Madison eastward, less conspicuous inland deposits elsewhere, were left behind as the southern edge of the glacier melted northward (Fig. 1.11). These terminal moraines represent locations where melting of the ice kept pace with its southward flow, depositing debris as at the end of a conveyor belt. The moraines extend westward into New Jersey and eastward to Cape Cod, Martha's Vineyard, and Nantucket, and are represented on Long Island by two long west-east ridges (Harbor Hill Moraine to the north, Ronkonkoma Moraine to the south). Long Island Sound originated during post-glacial time (about 15,000 years ago) when the rising sea began to flood the basin formerly occupied by a large freshwater lake (glacial Lake Connecticut) that formed after ice recession. Glacial lakes Hitchcock and Middletown, which

Fig. 1.4. Extensive areas of marble, historically mined for included deposits of iron ore, are unique to western Connecticut. Stephen P. Broker.

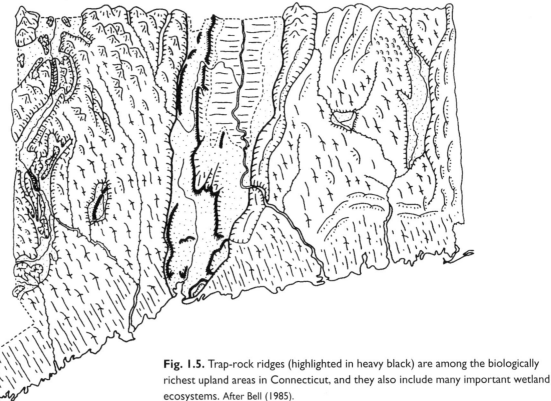

Fig. 1.5. Trap-rock ridges (highlighted in heavy black) are among the biologically richest upland areas in Connecticut, and they also include many important wetland ecosystems. After Bell (1985).

occupied most of north-central Connecticut, drained about 13,500 to 15,500 years ago. Sea level rose to within 130 feet of its present level by about 9,000 years ago and was 25 feet below today's level by 4,000 years ago.

The tundra landscape that dominated recently deglaciated Connecticut was followed by a conifer-dominated forest as woody vegetation became more common. Spruce-fir forest covered southern Connecticut about 10,000 to 14,000 years ago, followed by a period when pine was common. By about 10,000 years ago, colonization by hardwoods resulted in a widespread mixed forest, and by 8,000 years ago, a forest dominated by oaks, hickories, chestnuts, maples, birches, and hemlocks developed, though chestnut was a relatively late arrival, attaining regional importance about 2,000 years ago. Hemlocks declined in northeastern North America about 5,000 years ago, then increased a few thousand years ago, perhaps as the climate became cooler and more moist. By a few thousand years ago, the species composition of Connecticut forests was much as it is today, except that American chestnut was then more abundant. This scenario is based on an analysis of pollen grains found in the layered deposits at the bottom of boggy lakes. The

Fig. 1.6. Basalt cliffs and talus slopes of Connecticut's trap-rock ridges provide dramatic contrasts with surrounding lowlands. Stephen P. Broker.

Fig. 1.7. Satellite image of Connecticut (southwest corner not included), showing cleared and urbanized areas (lightest tones in central area), forest (middle tones), and large bodies of water (black).

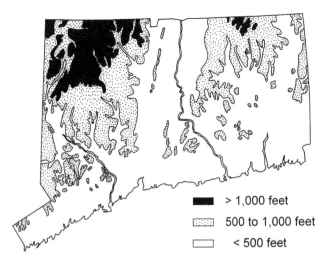

Fig. 1.8. Connecticut topography (elevation above sea level). Modified from Joseph D. Zeranski and Thomas R. Baptist. 1990. *Connecticut Birds* (Hanover: University Press of New England).

Fig. 1.10. The bedrock foundation of the Thimble Islands is evident in this photograph. Stephen P. Broker.

specific sequence of vegetation development reflected the differing climate tolerances of each species, impacts of established vegetation on other plants, and differential rates of dispersal of plant seeds by birds and mammals.

Human Impact

Significant human alteration of the Connecticut landscape began about 10,000 years ago with the arrival of the first colonists. These pioneers were not the English but rather Native American descendents of people who earlier crossed from Asia to North America when those continents were connected or in closer proximity during glacial periods when sea level was lower. Native Americans affected the environment through their use of fire, which created locally deforested areas and open forests with a food-rich understory of blueberries and huckleberries, and through fishing, shellfish gathering, and hunting of caribou and other mammals and birds. Cultivation of food plants and collection of nuts, seeds, and fruits also affected local ecosystems.

Later settlement by Europeans, beginning in the 1630s, resulted in extensive clearing of the forests for agriculture and fuel for industry, such that by the early to mid-1800s three-fourths of the state was deforested. Accompanying this was extensive erosion, drainage of wetlands, water pollution, and decimation of native wildlife populations.

The trend since the late 1800s has been toward reforestation (60 percent of the state is now forested), resulting from abandonment of agriculture with increasing industrialization. This is evidenced today by the many stone walls, marking old field and pasture borders, that occur in the woods (Fig. 1.12). Recently, however, deforestation and fragmentation, and loss of old fields and pastures, have begun again as residential development expands into thousands of small forest clearings and former fields.

Attempts to improve water and air quality have produced mixed results. The installation of sewage treatment plants has resulted in better water quality, though some rivers remain impoverished by residual industrial pollution. Some fishes are too toxic to eat safely, and contaminant-related beach closures are still all too common. Air pollution is a nagging regional problem. Photochemical air pollutants and nitrogen compounds resulting from emissions in the New York City metropolitan area are carried by prevailing

Fig. 1.9. The Norwalk Islands, as shown in this 1991 composite aerial photograph, consist of material deposited by Pleistocene glaciers and reworked by more recent currents and storms.

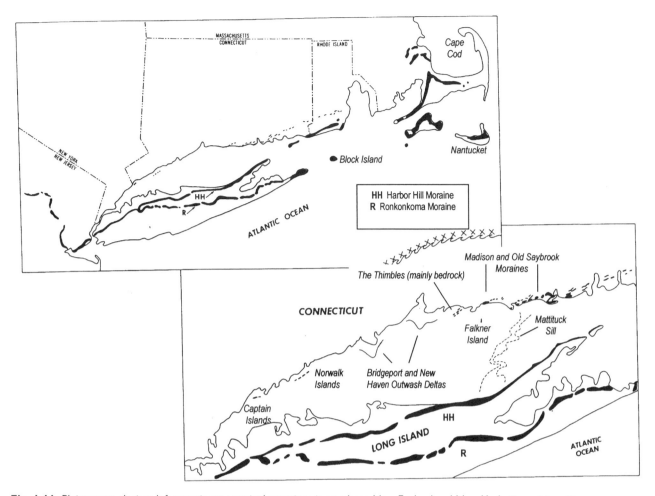

Fig. 1.11. Pleistocene glaciers left prominent terminal moraines in southern New England and New York. Coastal Area Management Program (1977).

Fig. 1.12. Testimony to an historical landscape, stone walls that marked the boundaries of former fields and pastures are a common feature of Connecticut forests. Most of these were constructed in the early 1800s to contain sheep.

winds northeastward through Connecticut and as far as northeastern Massachusetts. As a result, southwestern Connecticut has exceptionally high ozone concentrations and atmospheric deposition of nitrogen.

Some impacts, however, have been overwhelmingly positive. Connecticut's residents have established 30 state forests with almost 140,000 acres, more than 90 state parks encompassing some 30,000 acres, and 36 state wildlife management areas totaling nearly 17,500 acres. Most of these provide important wildlife habitat. These state lands complement the coastal Stewart B. McKinney National Wildlife Refuge plus properties of more than 100 local land trusts and numerous other preserves that support significant wildlife populations.

These are just a sampling of human impacts on Connecticut's environment. Additional and more detailed information is included throughout this book.

Chapter 2

The Seasons

In temperate regions such as Connecticut, dramatic seasonal changes characterize the life and ecology of virtually all organisms. These changes are related to seasonal changes in climate and daylight. Through the process of evolution, plant and animal life cycles, especially the production of offspring, often are timed to coincide with the most favorable environmental conditions. So the timing of life cycles generally varies from place to place, and to some extent from year to year, as does the climate.

Climate

The climate of Connecticut varies with elevation and proximity to the coast (Fig. 2.1). Due to the influence of the large body of water of Long Island Sound, coastal regions of the state are slower to cool in fall and slower to warm in spring than are inland areas. Likewise, compared to comparable elevations in inland locations, the coast is cooler in summer (an effect of afternoon breezes off the Sound) and warmer in winter. However, these patterns are not without exceptions. For example, mean monthly temperature at Norwalk on the coast and in the hills of Middletown in the Central Valley are almost identical.

The season's first frosts may occur as early as late September inland and late October along the coast, though this is quite variable. In most of the state, frosts at night are occasional in October and become regular in November. Freezing temperatures generally end in mid-May in northern Connecticut, late April in the Middletown area, and in mid-April in coastal areas. In some years, ice may not form on pond surfaces until late November or early December.

Cloudy skies are a regular feature of Connecticut's weather. On an annual basis, the percentage of possible sunshine averages only about 55 to 60 percent, with late summer and early fall tending to have the highest frequency of clear days.

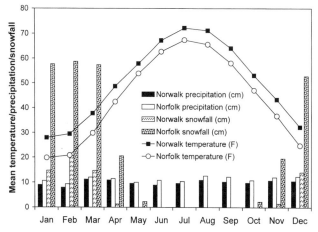

Fig. 2.1. Climate of two areas in Connecticut. Norwalk (elevation 37 feet) is on the coast and Norfolk (1,337 feet) is in the Northwest Highlands.

Monthly precipitation averages about 3 to 5 inches throughout the year. Most snowfall occurs from December through March, though significant storms sometimes occur in April as well. The highlands of northwestern Connecticut receive much more snowfall than other areas of the state (Fig. 2.1).

Average conditions as just described are somewhat misleading. The so-called "normal" temperature and precipitation levels mentioned by weather reporters are mere averages of conditions that normally are quite variable. Environmental conditions often change greatly over short periods of time or from year to year during the same season. For example, in central Connecticut, the spring (April–June) drought of 1999 was severest of the century (total of 5.67 inches of precipitation for the three months), yet the 10.74 inches of rain that fell in September of the same year made that month the wettest September of the century with the exception of 1938, when a major hurricane hit the state and produced 14.18 inches. The spring of 2001 lacked significant rainfall from late March through late May and caused extensive death of new leaves on many trees. In 1999 in central Connecticut, the hot late spring to early summer (June–July) was equaled during the 1900s only by the years 1949 and 1912, but July of 2000 was the coolest in more than 100 years. The fall of 2001 was exceptionally dry. And in recent years, snowfall has been erratic, ranging from very little to record-breaking amounts (as in the winter of 1993–1994). And occasional late summer hurricanes, usually in August or September, yield some of the state's most severe weather variations (e.g., Great New England Hurricane, 21 September 1938; Hurricane Carol, 31 August 1954; Hurricane Gloria, 27 September 1985). In the natural world, environmental extremes generally are more ecologically influential than averages.

Seasonal Change

Spring and fall are particularly dynamic seasons. Both the daily amount of daylight and the mean temperature change much more quickly in spring and fall than in summer and winter (Fig. 2.2). Similarly, the daily number of minutes of daylight increases and decreases much more near the equinoxes in spring and fall than at the summer and winter solstices. Changes in the daily amount of light and dark, registered by nervous systems of animals and the phytochrome pigments of plants, cause biochemical responses that affect the organisms' developmental processes, reproduction, and behavior.

Seasonal change occurs primarily because the Earth is tilted (23.5 degrees from vertical) on its axis of rotation as it moves in an elliptical orbit, traveling about 18.5 miles per second at a distance of about 91 to 95 million miles from the sun (the Earth is closer to the Sun in winter than in summer). The tilt of the Earth is one of our world's most important ecological factors, affecting the proportions of night and day, climate, and consequently the life cycles of most plants and animals. During Connecticut's summer, the northern half of the globe is tilted toward the Sun and the weather is relatively warm and the days are long. In winter, the northern hemisphere is angled away from the Sun. The low angle of the winter sun and the brief amount of

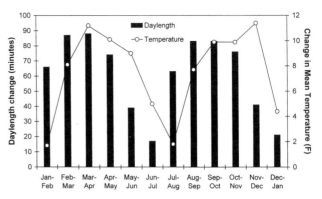

Fig. 2.2. Monthly change in daylight and temperature in Connecticut is greatest in spring and fall and minimal in summer and winter. Daylength is calculated as the number of minutes between sunrise and sunset.

daylight result in substantially less warming. In Connecticut, sunset occurs about 15.2 hours after sunrise at the summer solstice but only 9.2 hours after sunrise at the winter solstice. Night and day are the same length, about 12 hours, at the vernal and autumnal equinoxes, when the Sun rises due east and sets due west. The Earth's tilt also causes climate-influencing movements of air masses in the troposphere (the lower-most atmospheric layer).

Long-term Change

Connecticut's climate is changing toward warmer and wetter conditions. For example, the temperature in Storrs increased from an average of 45.8°F in 1892 through 1921 to 48.2°F during 1966 through 1995. Annual precipitation increased in some locations by 20 percent. These changes, which are projected to continue, may lead to significant changes in Connecticut's ecosystems as sea level rises along coastal shores, streamflows change, and various vegetation types expand or contract. Human health could suffer as well. As the climate becomes warmer and wetter, pollution problems may intensify and populations of microbes and disease-carrying insects could increase.

Chapter 3

Coastal Waters and Wetlands: Estuarine Ecosystems

The natural world can be viewed as a mosaic of ecosystems (or communities or habitats), each of which has relatively consistent physical and biotic conditions and distinctive ecological processes, joined by transitional areas where these conditions and processes change more strongly than within an ecosystem. Some of the transitions are abrupt, as where steep slopes of a ravine meet a stream, whereas others are more gradual, such as where a wide, slow river meets the sea. These gradual transitions generally require subjective or arbitrary definitions of where one system ends and another begins. Each ecosystem nevertheless has characteristics that distinguish it from other ecosystems, and ecosystems provide a useful framework for describing the state's many environmental variations.

The ecosystems or major habitat types of Connecticut can be grouped into five broad categories: estuarine (salt water and associated wetlands), riverine, lacustrine (lakes), palustrine (inland wetlands), and upland or terrestrial. Each of these includes easily recognizable subdivisions. For example, estuarine systems include bays, coastal river mouths, salt marshes, rocky intertidal areas, and other habitats.

In this section I first discuss deep-water and other perennially subtidal portions of the Long Island Sound ecosystem. Then I look at intertidal ecosystems—areas that are submerged at high tide and exposed at low tide.

Long Island Sound—Physical Environment

A "sound" is a long, wide ocean inlet. Long Island Sound is a large, mostly enclosed body of water between lands in Connecticut and Long Island (New York) (Fig. 3.1). It is approximately 110 miles long and up to about 21 miles wide. Encompassing 1,310 square miles, it is one of the largest estuaries in the United States. It extends eastward from the eastern margin of the New York City metropolitan area to its connection with the Atlantic Ocean at "the Race," an area of strong currents near Plum Island and Fishers Island (both in New York).

Long Island Sound is relatively shallow. Water depth averages about 65 to 80 feet, reaching a maximum of about 320 feet in the vicinity of the Race at the eastern mouth of the Sound. Within Connecticut (about 55 percent of the Sound is in New York), most of the Sound is less than 50 feet deep; the maximum depth of about 250 feet is 3.5 miles south of the coast at Waterford. Maximum depth between Hammonasset Beach State Park and Falkner Island is about 50 feet. Rock ledges, reefs, islands, and shoals are common along the Connecticut side of the Sound. The bottom of the Sound is primarily mud or sandy mud, with limited areas of deeper water dominated by sand (e.g., from the Connecticut River southwestward).

Salinity of Long Island Sound is less than that of pure sea water—about 25 to 29 parts per thousand in the central part of the Sound. The most saline waters are near the open water of the Atlantic Ocean at the Sound's eastern end. Salinity is lowest in spring, when the inflow of water from the Connecticut River, the major freshwater supply (70 percent of the total), is highest. High salinity water is denser than low salinity water, so salinity often is lower at the surface than at greater depths.

Water temperature varies greatly with season and depth. Temperature at a depth of 6 to 7 feet is about 32 to 41°F (0 to 4°C) in winter and 68 to 73°F (20 to 23°C) in late summer. The Sound's waters usually are well mixed and of largely uniform temperature from top to bottom except in mid- to late summer when the upper layer becomes significantly warmer. Often you can detect the top of the distinctly cool subsurface layer by swimming out from shore to deep water and extending your feet straight down.

Tidal range, the vertical difference between the level of high tide and low tide, is much larger in western Long Island Sound than it is nearer the Atlantic Ocean to the east. For example, the average tidal range is about 7.4 feet in Greenwich in the western part of the Sound, 4.9 feet at Madison and 3.5 feet at Old Saybrook on the central coast, and 2.7 feet

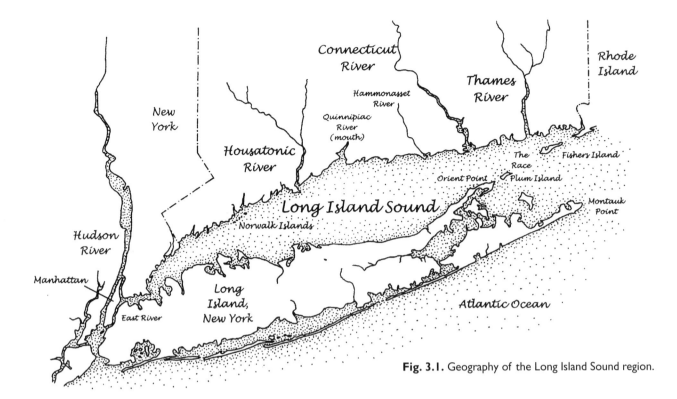

Fig. 3.1. Geography of the Long Island Sound region.

at Stonington in the east. These differences result from peculiarities in the movement of water that result from the shape and depth of the Sound's basin. It takes about 2.5 hours for a particular high or low tide to sweep along the entire length of the Connecticut coast and about 2.7 hours to move from the mouth of the Connecticut River to Middletown. Low tides and high tides alternate at intervals of approximately 6.2 hours.

The highest and lowest tides, called spring tides, occur around the time of the new and full moons. During the new moon, the Sun and Moon are aligned on the same side of Earth. In this orientation, the separate gravitational forces of the Sun and Moon on the ocean are in the same direction and thus are additive. The tidal range is also large during the full moon when the Moon, Earth, and Sun are aligned but the Moon and Sun are on opposite sides of Earth, such that these gravitational forces act in concert with centripetal forces. Centripetal force, arising from the rotation of Earth on its axis, plays a role in the tides by causing a tidal bulge (elevation of sea level) on the side of Earth opposite the Moon, where the gravitational force of the Moon is less than centrifugal force. The lowest high tides and highest low tides (neap tides) occur during the first and third quarters of the Moon (when the face of the Moon is half illuminated), when the Moon and Sun are at right angles with respect to the Earth. In this situation, the separate gravitational forces of the Moon and Sun pull in different directions, minimizing the tidal range. Distance of the Moon from Earth varies within each month, and tides are hightest when the Moon is closest to Earth. All of these factors result in ongoing tidal changes in any particular location. For example, in a particular month, daily tidal range at Old Saybrook might vary from 2 to 6 feet. Tides play a huge role in the lives of coastal organisms.

The surface of Long Island Sound tends to be relatively placid, with just small waves and often no whitecaps or just small ones. But strong winds generate large waves along the Connecticut shore, and waves of house-smashing size may accompany hurricanes, and strong tidal currents are a constant feature of the Sound's eastern mouth.

Open-Water Biota

Phytoplankton

The most abundant organisms in the open waters of Long Island Sound are microscopic phytoplankton (Fig. 3.2). Phytoplankton populations include mainly chlorophyll-containing diatoms and dinoflagellates (or dinomastigotes), represented in the Sound by probably more than 200 species, of which about 50 species are relatively abundant. Diatoms, which have silica shells and may exist as chains of cylindrical disks held together by chitinous threads or marginal tubules, drift passively in the water. Dinoflagellates have two whiplike flagella that allow them to swim.

Phytoplankton abundance varies seasonally and from year to year, influenced by variations in light, temperature,

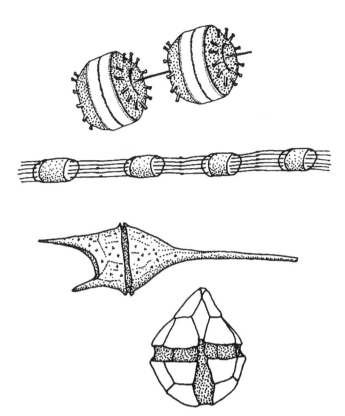

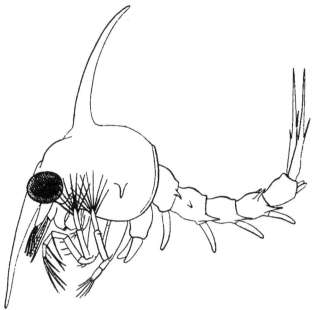

Fig. 3.3. The immature stages of many estuarine invertebrates, such as the rock crab larva shown here, are seasonally important zooplankton. From Verrill and Smith (1874).

Fig. 3.2. Examples of phytoplankton in Long Island Sound: diatoms, *Thalassiosira nordenskioldii* (top, part of chain as seen by electron microscope, some fine filamentous connectors not shown) and *Skeletonema costatum* (second from top, part of a chain as seen by light microscope), and dinoflagellates, *Ceratium lineatum* (third from top) and *Scrippsiella trochoidea* (bottom) (flagella not shown). Not drawn to scale.

salinity, and nutrient availability. Nutrient availability depends on supplies carried into the Sound by rivers or transported from bottom sediments into the water column by currents, turbulence, and upwelling.

Diatoms generally attain peak abundance in winter in the western and central basins of Long Island Sound and in late spring in the eastern Sound. Total phytoplankton abundance in winter may exceed 40 million cells per liter of water. Dinoflagellates tend to be most numerous in late spring and early summer.

Phytoplankton use sunlight and carbon dioxide to produce organic material and oxygen, and they are an important food source for zooplankton and other small filter-feeding animals. Indeed, phytoplankton are the primary foundation of the living ecosystem of Long Island Sound.

Zooplankton

Zooplankton are diverse microscopic or barely visible animals that live in the water column. Examples of estuarine zooplankton include copepod crustaceans and immature stages of fishes, barnacles, shrimp, crabs, polychaete worms, mollusks, and sea stars (Fig. 3.3). These subsist on phytoplankton and are ecologically crucial as prey for larger animals, such as small schooling fishes that are the primary food source for most large commercially or recreationally important fishes. The zooplankton in Long Island Sound are mainly species of small size that serve best as food for young fishes and efficient filter feeders. The Sound's zooplankton population is important for resident fishes, the young of various species that enter the Sound to feed, and also for the offspring of migratory fishes that enter the Sound to spawn.

Zooplankton abundance in Long Island Sound peaks in spring and summer, when a cubic meter (35 cubic feet) of water might contain over 200,000 copepods and larval forms of other invertebrates. Copepods (Fig. 15.3) dominate the zooplankton community and tend to be present throughout the year, whereas peak abundance of the larval stages of bottom-dwelling or attached invertebrates corresponds with a pulse of reproduction in late spring and early summer. A few species of cladocerans, another group of small crustaceans, are seasonally fairly abundant.

Macroinvertebrates

Moon jellies, lion's-mane jellyfishes, sea nettles, and comb jellies are slow-moving and sometimes very abundant

Open-Water Biota | 13

predators in the open waters of Long Island Sound. These gelatinous animals make a strong ecological impact by consuming tremendous numbers of invertebrate larvae and other zooplankton.

Information on other invertebrates, such as those inhabiting intertidal soft sediments or associated with rocky shores, is included in following sections in this chapter on "Rocky Shores" and "Muddy and Sandy Shores and Shallows."

Fishes

Long Island Sound is home to a diverse and productive fish fauna. Recent research by the Fisheries Division of the Connecticut Department of Environmental Protection yielded a wealth of information of the Sound's populations of finfish, lobster, and squid. Intensive trawl sampling throughout the Sound from 1984 to 1994 involved 2,859 trawl-net tows that yielded 83 fish species. Not included in this total are small shoreline species such as killifishes and sticklebacks that generally are not captured by trawling. Of the 83 species, 25 were very rare (each represented by fewer than 10 netted individuals).

These surveys revealed that species richness is highest in the central part of the Sound, which yielded an average of 13 species per tow. Overall fish abundance is highest in late summer and early fall, and the western and central parts of the Sound have higher fish abundance than does the eastern portion of the Sound. Abundance of bottom-oriented fishes peaks in spring and declines through summer, whereas pelagic fishes increase from spring to early fall.

Trawl samples of Long Island Sound's fish fauna in spring are dominated numerically by winter flounder and windowpane flounder, with little skate, red hake, scup, and fourspot flounder also abundant among bottom fishes. Atlantic herring is the most common pelagic fish in spring.

In summer trawl samples, scup, winter flounder, and windowpane flounder continue to be numerous on or near the bottom, while the pelagic butterfish is most abundant of all. During this time, most cold-water bottom fish move into deeper water and eventually out of the Sound as temperatures continue to increase. Meanwhile, warm-water migratory species such as bluefish, butterfish, weakfish, and scup move into the Sound. In late summer and early fall, butterfish and scup are by far the most numerous species, in large part due to reproduction that yields vast numbers of juveniles. By November, cooling temperatures cause warm-water migrants to move out of the Sound and overall fish abundance greatly declines. Butterfish, windowpane flounder, winter flounder, Atlantic herring, scup, and little skate are most plentiful in fall trawl samples.

No Connecticut port on Long Island Sound has nationally important levels of commercial fishery landings, but Montauk, New York, just outside Long Island Sound, ranks among the top 50 U.S. ports (14.3 million pounds valued at $13.1 million in 2001). New Bedford (Massachusetts) and Point Judith (Rhode Island) ranked number 1 and 16, respectively, in value of commercial landings in 2001.

Long Island Sound supports the complete life history of resident fishes, such as the tautog, winter flounder, windowpane, striped searobin, and many others. Some migratory fishes, such as scup and butterfish, use the Sound for spawning, feeding, and nursery habitat. Other migrants, including summer flounder and bluefish, spawn outside Long Island Sound but make important use of the Sound as summer feeding habitat. Some of these, such as red hake and silver hake, are represented in the Sound primarily by juveniles. In late summer and fall, the Sound hosts and feeds good numbers of hungry juveniles of various northward-roaming subtropical fishes, including filefish, crevalle jack, yellow jack, moonfish, rough scad, Spanish mackerel, and others.

The Sound's fishes exhibit a wide range of ecology and behavior. Flounders and skates often take small crustaceans and worms on or near soft bottoms. Grubbies ambush small invertebrates among algae near the bottom. Tautogs pluck and crush with their strong jaws mollusks and crustaceans attached to submerged rocks and piles. Moving throughout the water column are vast numbers of schooling fishes, such as menhaden, which consume phytoplankton and zooplankton, and butterfish, whose diet includes a wide variety of invertebrates and small fishes. In deeper water, intense feeding by schools of menhaden significantly reduces populations of large plankton. Bluefish and striped bass are important predators on schooling fishes.

Birds

Birds of many kinds feast on the bounty of the open waters of Long Island Sound throughout the year. In the colder months, the Sound's avifauna includes common loons, red-throated loons, horned grebes, pied-billed grebes, northern gannets (regular but rare), great cormorants, great black-backed gulls, herring gulls, ring-billed gulls, Bonaparte's gulls, mute swans, Canada geese, brants, mallards, black ducks, American wigeons, greater scaups, canvasbacks, common goldeneyes, buffleheads, white-winged scoters, long-tailed ducks, and red-breasted mergansers. In summer, the usual birds of open waters are double-crested cormorants, great black-backed gulls, herring gulls, ring-billed gulls, laughing gulls, common terns, roseate terns, least

Fig. 3.4. Small schooling fishes such as sand lance are important foods for least terns, as well as other water birds and larger fishes in Long Island Sound. Paul J. Fusco/Connecticut DEP Wildlife Division.

terns, ospreys, mute swans, and relatively few other waterfowl. Additional water birds that are associated primarily with salt marshes and tidal shores are mentioned in the appropriate following sections of this book.

Most of the birds, such as loons, grebes, cormorants, gannets, mergansers, gulls, and terns, forage on fishes, particularly small schooling species, whereas ospreys depend on larger fishes (Fig. 3.4). Grebes, gulls, and sometimes other birds also prey on invertebrates. Herring gulls often take shellfish and dead fishes. In subtidal waters, various dabbling ducks, geese, and swans mostly graze on algae and eelgrass, whereas diving ducks eat benthic mollusks and crustaceans as well as vegetable material. All these waterfowl use open waters not only for feeding but also as a place to rest and avoid land predators.

Mammals

Harbor seals are the only mammals (other than humans) present in the Sound in ecologically significant numbers. Seals are present primarily during the colder months and feed mostly on fishes, crustaceans, and squid. Harbor porpoises have a similar diet but are quite scarce in Long Island Sound, as are other sea-going mammals.

Rocky Shores

Rocky shores exist intermittently along the entire Connecticut coast and contribute uniquely to the state's biodiversity (Fig. 3.5). Some rocky shores are largely exposed bedrock whereas others consist of piles of boulders deposited by glaciers.

Fig. 3.5. Intertidal rocky shores are limited in extent but support many species not found elsewhere. These rocks, exposed at low tide, support thick growths of rockweed, knotted wrack, and Irish moss.

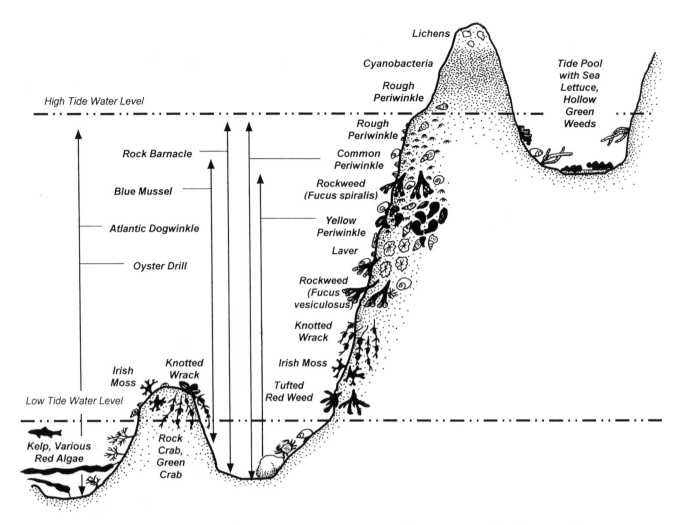

Fig. 3.6. Idealized intertidal zonation on a rocky shore. Particular sites may vary from this pattern. Not drawn to scale.

Life on rocky shores is arranged in zones corresponding with how often and long the organisms are doused or submerged in seawater (Fig. 3.6). The following sections review some of those patterns and the processes that cause them.

Algae and Lichens

Patches of lichens (*Verrucaria erichsenii*) and coatings of blackish cyanobacteria (*Calothrix*) color the highest rocks that receive a regular input of salt spray or an infrequent wetting by wave splash. Here inundation is too infrequent to support typical estuarine life forms and conditions are too salty for growth of upland plants.

At lower elevations, in the main intertidal zone, algae or seaweeds are an important component of rocky shore life. These algae, along with the planktonic and rock-coating diatoms, serve as food for the animals and form the basis of the rocky intertidal food web.

Seaweeds in calm, shallow, rocky intertidal waters include mermaid's hair, hollow green weeds, sea lettuce, and green fleece*, all green algae. Hollow green weeds and sea lettuce often tolerate pollution and low salinity. Crustose red and pink algae and *Ulothrix* (dark green filaments) can be found in tide pools. Banded red algae grow over a broad range of depths.

Rockweeds and knotted wrack, both brown algae, and thin, sheetlike laver, a red alga, often blanket intertidal rocks (Fig. 3.5). Knotted wrack frequents calm sites and is replaced by rockweeds in more surf-exposed areas; these two species form dense algal thickets that shelter animals from high temperatures and desiccation during low tide. Filamentous brown algae and red *Polysiphonia* tubed weeds grow attached to rockweeds and knotted wrack. Irish moss, a red alga, is common in backwash areas of intertidal rocks, generally below the level occupied by rockweed.

Areas that are well submerged even at low tide often support red algae such as branched, straplike dulse, and brown algae, including the long blades of kelp and whiplike smooth cord weed. Finely branched red algae occur in deepest areas supporting seaweeds; you may find dislodged examples floating in the water. Diatom films often cover intertidal rocks and make exploration a slippery affair.

Algae in the upper and mid-tidal levels tolerate high levels of water loss during low tide, then quickly rehydrate when resubmerged. They tolerate freezing temperatures in winter by limiting ice formation to extracellular areas. The relatively low upper limit of Irish moss may be set by its susceptibility to desiccation and/or freezing.

Further information on estuarine algae is included in Chapter 8.

Intertidal Invertebrates and Their Interactions

Barnacles, Mussels, and Algae. Filter-feeding barnacles (little gray barnacle and, lower, northern rock barnacle) and blue mussels are ubiquitous in the intertidal zone (Fig. 3.7). They feed on planktonic diatoms suspended in the water. The barnacles (particularly the little gray), being exceptionally tolerant of dry conditions, extend higher than the mussels. Of course, they must be submerged periodically if they are to feed, grow, and reproduce, so they do not extend much beyond the average high-tide level. Northern rock barnacles do poorly on sun-heated rocks but outcompete little gray barnacles at lower levels where physical stresses are reduced. In some situations, predatory green crabs* may reduce survival of northern rock barnacles at the upper border of their tidal range.

Fig. 3.7. Intertidal rocky shores support dense populations of attached organisms. Here filter-feeding barnacles dominate the upper and middle levels while rockweed, sea lettuce, and filter-feeding blue mussels compete for space at the lower level, where predatory oyster drills attack the mussels.

When exposed to air at low tide, barnacles, mollusks, and other upper and mid-intertidal invertebrates seal themselves closed. If conditions get really warm, blue mussels periodically open the shell slightly to allow gas exchange. But many upper intertidal invertebrates simply seal up, tolerate the adverse conditions, and wait for the next tide. They often use sheltering crevices and overhangs and thereby reduce stress from sun exposure. All of these high-intertidal organisms extend higher on the rocks on open coasts with much wave action than on sheltered shorelines where the ameliorating effect of splashing waves is reduced.

Attached organisms on rocky shores create favorable habitat for other species. For example, the rough surface of barnacles facilitates colonization by mussels, which eventually may overgrow the barnacles. Mussel beds in turn provide habitat for additional species that live in the spaces among these mollusks.

Mussels outcompete algae in the high intertidal zone, and they often form dense aggregations. Mussel dominance is facilitated by the scarcity of mussel predators in this zone. High densities of mussels have to compete for food, but they benefit from close living by being more resistant to heat, desiccation, wave action, and predators, and they probably find it easier to exchange gametes.

At lower intertidal levels, predators (see Predatory Snails, Crabs, and Sea Stars following) reduce mussel and barnacle densities, and algae may dominate, though some parts of the lower intertidal zone and adjoining shallow subtidal areas may be covered with encrusting or clone-forming invertebrates, such as tunicates, bryozoans, hydroids, and sponges. These largely filter-feeding species are intolerant of the physical stresses that characterize most of the intertidal zone and often dominate rocky subtidal areas where conditions are more benign.

Grazers. Conspicuous intertidal grazers include common and yellow periwinkles (the latter mainly on rockweeds in the lower intertidal) and sea urchins (mainly in subtidal areas). In the splash zone at the upper intertidal zone, rough periwinkles, which graze on diatoms and algae, sometimes are numerous in cracks and other semiprotected crevices. These snails are joined by various small scavenging and diatom- and algae-grazing isopods and amphipods.

Periwinkles affect intertidal communities not only by grazing on diatoms but also by influencing the species composition, distribution, and abundance of seaweeds (Fig. 3.8). Periwinkles graze on ephemeral, early successional algae such as sea lettuce, hollow green weeds, and laver and in doing so facilitate the establishment of later successional species, particularly rockweed. Sea lettuce and hollow green weeds have no defensive chemicals or structures and are

Rocky Shores | 17

Fig. 3.8. Common (upper left), rough (upper right), and yellow (bottom) periwinkles are important grazers on rocky shores.

Fig. 3.9. Two intertidal predators, Atlantic dogwinkle (top two) and Atlantic oyster drill (middle two) shown with a threeline mudsnail (bottom).

particularly vulnerable to grazers. In some situations, periwinkle grazing can inhibit the establishment of rockweed, as can competition with Irish moss (rockweed can colonize rocks and completely replace Irish moss if the latter is removed). Experiments in which grazing snails were removed from and prevented from re-entering exclosures clearly show the increased algal growth within the exclosure, contrasting sharply with the denuded areas outside the exclosures. By inhibiting the establishment of certain seaweeds, periwinkles open up space for the establishment of barnacles and mussels. Periwinkles may favor the persistence of already-established rockweeds (and species that use rockweed for shelter) by grazing on algae attached to the rockweed blades. This may benefit rockweed by decreasing the forces it experiences during storms.

Grazing by sea urchins in subtidal areas also can affect the distribution of algae. Factors that impact urchins, such as harvest by humans or disease, thus influence algae. For example, in southeastern Canada, disease-related urchin mortality associated with warming ocean temperatures resulted in development of kelp beds in areas formerly dominated by algal crusts that had formed in response to heavy urchin grazing. In deep water, urchin grazing on kelps facilitates development of horse mussel beds and algal crusts. Spaces among the mussels serve as refuges that help protect the urchins from their predators.

Amphipods and fishes may also affect subtidal algal populations and associated communities. By selectively grazing brown algae, grazing amphipods may allow dominance by red algae. But where omnivorous fishes that eat both amphipods and red and green algae are abundant, the advantage may shift back to brown algae. Such intricacies abound in nature.

Predatory Snails, Crabs, and Sea Stars. Green crabs*, Asian shore crabs*, rock crabs (mainly subtidal), and mud crabs are the dominant crustaceans of rocky shores. Green crabs, native to Europe, have been present in northeastern North America for over 200 years. Asian shore crabs, native to the western Pacific Ocean, became very common along Connecticut's rocky shores during the 1990s and have caused declines in populations of green crab and likely other intertidal species. American lobsters are bottom-dwelling predators and scavengers throughout the subtidal waters of the Sound.

The activities of predatory Atlantic dogwinkles, oyster drills, sea stars, and crabs may reduce barnacle and mussel populations and result in increased seaweed cover (Fig. 3.9). The crabs also may reduce populations of grazing snails and thus allow fast-growing green algae to flourish. Stressful temperature and moisture conditions generally keep these predators out of the upper intertidal zone.

Vertebrates

Fishes. Rocky intertidal areas in Connecticut do not have any fishes that are highly specialized for this habitat, but tautogs, which inhabit rocky subtidal areas, may move into the intertidal zone at high tide and feed on attached shellfishes, sometimes opening up spaces for colonization by algae or invertebrates. Of course, a large assortment of other mobile species may forage along rocky shores at high

Fig. 3.10. Purple sandpipers commonly forage for small invertebrates on rocky shores. Paul J. Fusco/Connecticut DEP Wildlife Division.

Fig. 3.11. Organisms on low-gradient shores such as this experience harsh physical conditions when exposed at low tide.

tide, and large tide pools sometimes contain mummichogs, striped killifishes, grubbies, and other small, shallow-water fishes. These fishes often eat free-swimming crustaceans.

Birds and Mammals. Birds use rocky intertidal habitats much more than do mammals, and amphibians and reptiles are absent. Swimming double-crested cormorants and red-breasted mergansers sometimes come close to shore and dive underwater to feed on the various small fishes in submerged rocky intertidal areas. Herring gulls, great black-backed gulls, least sandpipers, spotted sandpipers, semipalmated sandpipers, semipalmated plovers, ruddy turnstones, American oystercatchers, and various other invertebrate-eating shorebirds scour this zone for food at low tide. Cormorants, gulls, and terns often rest on prominent rocks exposed at low tide or otherwise isolated by water. Foraging sandpipers, plovers, and turnstones are especially prevalent on low-gradient cobble-boulder shores during their late-summer migration. Purple sandpipers hunt for invertebrates on rocky shores and jetties in winter (Fig. 3.10).

These avian predators, especially numerous, ubiquitous, and ever-present herring gulls that feed intensively on rocky shores, likely have a large direct impact on the relative abundance of intertidal invertebrates and probably cause shifts in intertidal communities by reducing populations of influential predators and herbivores. However, these impacts are poorly documented. What changes have been wrought by gull populations that are now artificially large due to increased food availability at dumps and other human sources? Does feeding on trash by gulls reduce intertidal feeding enough to offset the impact of the gulls' increased populations?

Birds can also affect vegetation on rocky coasts and thereby influence the composition and dynamics of intertidal communities. For example, intense grazing by large numbers of mute swans* may reduce algal cover and favor communities of attached animals.

Raccoons, occasional opossums and skunks, and Norway rats* feed among rocks at low tide, but we know little about mammalian influences on rocky shore communities.

Effects of Storms and Random Events

The preceding paragraphs only touch on the myriad and complex ecological relationships of rocky shores. Competitive and predatory interactions and community composition may vary greatly, depending on exposure to wave action, presence of pooled water at low tide, random events, and other factors. For example, storms impact the pattern of organisms on rocky shores. Direct wave shock and battering by wave-carried logs and rocks periodically remove attached seaweeds, barnacles, mussels, and other animals, opening up space for whatever potential colonists happen to be available in the water.

Low-Gradient Rocky Shores

Some gently sloping intertidal shores are dominated by cobbles in a matrix of gravel and sand (Fig. 3.11). These are harsh environments, subject to high levels of heat, desiccation, ice scouring, and human activity. The higher parts of cobbly shores may be quite barren. Lower, at mid-tidal levels, sea lettuce, hollow green weeds, rockweed, and algal crusts (*Ralfsia, Hildenbrandia*) compete for space with abundant barnacles and grazing hordes of common periwinkles*. Flocks of migrating brant graze on sea lettuce and grub in patches of smooth cordgrass that grow on low-gradient shores of cobble, gravel, and sand. Irish moss and green fleece* can be abundant on low intertidal cobbles, unless excluded by periwinkles.

Muddy and Sandy Shores and Shallows

Muddy and sandy shores provide a striking contrast with nearby rocky shores. Mud and sand dominate the floor of Long Island Sound, and soft-sediment tidal flats occur along or at the mouths of tidal rivers, such as at Greenwich Cove, outside the Saugatuck River, at Bridgeport Harbor, near the mouth of the Housatonic River, and around Griswold Point at the mouth of the Connecticut River. Soft-sediment biotas vary somewhat with water depth. The following material focuses on intertidal and shallow subtidal areas.

Fig. 3.12. Tidal flats support abundant burrowing animals. Here, exposed at low tide, clumps of blue mussels are visible, and quahogs, softshell clams, and various worms are burrowed in the muddy sand.

Tidal and Shallow Subtidal Flats

Estuarine tidal flats of mud, sand, and organic matter may look barren, unproductive, and uninviting, but these soft sediments harbor a surprising diversity and abundance of animal life, ranging from burrowing clams and polychaete worms to roving snails and crabs (Fig. 3.12). If you wish to explore soft mudflats and their biota up close without miring yourself, try wearing snow shoes or an old pair of cross-country skis. Some flats are firm enough to walk on unaided.

The relative amount of sand in the sediments is an important ecological factor. Diffusion of oxygen into fine, muddy sediments with high organic content occurs too slowly to offset microbial respiration, so mud bottoms may be anaerobic just beneath the surface, except where animal burrows facilitate water movement and allow deeper oxygen penetration. In mudflats, most life is concentrated in the upper, oxygenated portion of the sediments, especially at lower tidal levels where physical stresses are relatively mild.

In contrast, sandy sediments are relatively porous and readily flushed with oxygen-containing water to as much as several inches from the surface. Nevertheless, the burrowing fauna of sand flats generally is less productive and has lower diversity. Evidently, the burrowing lifestyle works better in mud, and the animals are able to deal with low-oxygen mud by pumping water through their burrows.

Mudflat animals subsist by grazing, filtering, gathering, and ingesting foods ranging from diatoms that live on and near the mud surface to organic debris and plankton delivered by the tides. Twice-daily food deliveries with each high tide allow many burrowing mudflat animals to stay in one place and devote a good deal of their body structures and energy to food extraction.

Epifauna and Burrowing Invertebrates. Researchers studying subtidal muddy and sand-shell bottoms of central Long Island Sound at depths of 20 to 100 feet documented densities of up to nearly 50,000 benthic invertebrates per square meter (10.8 square feet) of bottom, including some 140 invertebrate species, dominated by polychaete worms, amphipods, and mollusks. Three representative individual samples included 6,742 to 28,472 individuals of 29 to 42 species per square meter. A blind tanaid (*Leptognathia caeca*), an eyeless amphipod with pincerlike claws on the first pair of legs, was represented by more than 20,000 individuals in one of these samples.

These studies found that sand-shell bottoms harbor many more epifaunal species (those living on or just above the bottom) than do mud bottoms. Combining both types of substrate, the number of epifaunal species is roughly equivalent to the number of infaunal species, which are suspension- and deposit-feeders that live freely in the substrate or buried in tubes.

Opossum shrimp and sand shrimp are the most numerous of all epifauna in Long Island Sound. Other dominant species on sand-shell bottoms include common sea star, hydroids, longwrist hermit crab, four-eyed amphipods, and tube-dwelling polychaete worms (*Ampharete acutifrons*). Threeline mudsnail, Atlantic rock crab, Say mud crab, tube-dwelling amphipods (*Corophium*), and caprellid amphipods (skeleton shrimp) are additional important species. Patches of blue mussels sometimes also exist on sandy-silty bottoms in shallow subtidal water devoid of rocks. The mussels attach to shells and to each other but are easily lifted from the substrate.

Mud bottoms in the Sound often support a common painted worm/file yoldia/Atlantic nutclam community that also includes common sea star, opossum shrimp, sand shrimp, channeled barrel-bubble (a snail), threeline mudsnail, cone worm, and amethyst gem clam. Painted worms are pale polychaetes with a conspicuous red blood vessel running the length of the body.

On soft bottoms with high organic content, such as those typical of major industrial harbor areas, four-eyed amphipods are abundant epifaunal filter feeders, and burrowing, deposit-feeding polychaete worms (e.g., *Mediomastus ambiseta, Streblospio benedicti*) and burrowing, filter-feeding dwarf surfclams are common. Soft bottoms of less industrial harbors support burrowing, deposit-feeding polychaetes (e.g., *Leitoscoloplos, Streblospio benedicti*), as well as predatory polychaetes (*Eteone*) and filter-feeding gem clams.

The burrowing and feeding activity of invertebrates facilitates nutrient exchange between sediments and the water. Some burrowers, such as the clams, pump huge volumes of water through their systems, remove plankton and other organic matter, and release nutrient-rich wastes back into the water. Ampeliscid (four-eyed) amphipods filter feed, often from parchmentlike tubes, on muddy or sandy bottoms. Opossum shrimp are free-swimming filter feeders. Such suspension feeders dominate coarser sediments.

Deposit feeders such as certain polychaetes (painted worm, cone worm), clams (file yoldia, Atlantic nutclam), and sea cucumbers ingest sediments and digest the included organic material. The smallest invertebrates move among sediment grains and feed on bacteria or protozoa. On the surface, mudsnails and shrimps, sometimes in vast numbers, graze deposits rich in diatoms, microorganisms, and tiny algae. Deposit feeders tend to dominate finer sediments.

At the large end of the size range of bottom-dwelling invertebrates are American lobsters. These burrowing predators and scavengers live in both rocky and soft-bottomed subtidal waters of the Sound.

Aquatic Predators. The invertebrate populations of tidal and subtidal bottoms of mud or sand attract substantial populations of predatory fishes, crabs, snails, and sea stars. Atlantic silversides eat crustaceans; striped killifishes, mummichogs, and grubbies prey on various small invertebrates; and sheepshead minnows consume detritus and crustaceans. These fishes often inhabit very shallow water in the intertidal zone and retreat to adjacent subtidal areas at low tide. Many species of juvenile fishes feed heavily on copepods, opossum shrimps, and sand shrimps. Fourspine sticklebacks select tiny invertebrates and fish larvae among clumps of algae in quiet tidal lagoons. Common sea stars commonly make up a large proportion of the epifaunal biomass and prey on a wide assortment of invertebrates, especially mollusks. On subtidal bottoms and in intertidal areas at high tide, little skates, winter skates, flounders, spot, ribbon worms, and horseshoe crabs prey on worms, crustaceans, and mollusks. Bottom-feeding horseshoe crabs are especially evident during the spring spawning season and sometimes become stranded ashore on sandy flats during low tide. Predatory crabs (especially green crabs*) and snails may be common on tidal flats, where they feast on clams and smaller snails. All of these predators reduce the densities of sediment-dwelling prey.

Birds. The worms, crustaceans, mollusks, and fishes make tidal mudflats critical feeding habitats for large numbers of migrating shorebirds, including short-billed dowitchers, least sandpipers, semipalmated sandpipers, dunlins, greater yellowlegs, lesser yellowlegs, willets, black-bellied plovers, and semipalmated plovers. American black ducks and various other ducks feed on algae or invertebrates on submerged mudflats. Mute swans* graze attached algae. Great blue herons, green herons, great egrets, snowy egrets, and other wading birds stalk fishes in shallow tidal waters (Fig. 3.13).

Sand-Cobble Mosaics. Sandy intertidal areas with patches of stones or cobble, which provide attachment sites or cover, support augmented numbers of invertebrate species, including common periwinkles*, rough periwinkles, blue mussels, slippersnails, barnacles, hermit crabs, green crabs*, and Asian shore crabs*.

Eelgrass Meadows

Eelgrass beds or meadows occur on shallow, muddy, subtidal bottoms of coastal waters and are best developed in the eastern part of state, especially in small bays at the mouths of the Niantic, Thames, and Mystic rivers, and exist as far west as the Hammonasset River. Eelgrass is a locally important

Fig. 3.13. Submerged tidal flats are important feeding and resting areas for snowy egrets and many other birds and fishes. Paul J. Fusco/Connecticut DEP Wildlife Division.

Fig. 3.14. Sandy beaches are important places for humans and wildlife.

solar-energy-capturing part of the subtidal ecosystem. Eelgrass is a flowering perennial vascular plant, not a seaweed, and, despite its name, it is not a grass (it does serve as a development habitat for young American eels).

Eelgrass spreads primarily through vegetative reproduction (sprouts from spreading rhizomes), and its root system stabilizes sediments and helps prevent erosion. It does sometimes colonize new areas or bolster local populations by means of seed production and dispersal of detached, buoyant flowering shoots. Because it requires high light intensities, eelgrass is restricted to shallow water, especially if the water is turbid.

Eelgrass beds provide food, cover, favorable attachment sites, and critical nursery habitat for a diverse assemblage of small animals, including attached sponges, hydroids, anemones, entoprocts, ectoprocts, limpets, polychaetes, bay scallops, and tunicates to mobile worms, snails, isopods, amphipods, crabs, shrimps, fishes, and others species, many of which support sport and commercial fisheries and shellfisheries. The brant, a sea-going goose, is a major eelgrass grazer. Thus eelgrass beds are ecologically and economically important.

Eelgrass populations were decimated by a fungal disease (*Labyrinthula macrocystis*) in the 1930s. This led to erosion of sediments and declines in the populations of eelgrass-dependent animals such as bay scallops and brant. Eventually eelgrass populations slowly recovered. Theoretically, recovery should be further facilitated if improved water quality continues to foster growth in oyster populations. The water-filtering activity of oysters increases water clarity and thus benefits light-dependent eelgrass productivity. However, water quality problems have caused recent declines in eelgrass in eastern Long Island Sound.

Sandy Beaches and Dunes

Sandy beaches, our popular summer playgrounds, are scattered along the entire coast but amount to less than 20 percent of the total coastline. Sandy beaches in Connecticut are narrow, relatively steep, and generally have small, quiet waves (Fig. 3.14). Big, surf-pounded beaches are better developed along the ocean shores of Rhode Island, New York, and Massachusetts.

Coastal sand dunes are not actually part of the estuarine ecosystem, but I discuss these upland areas here because spatially and ecologically they are closely linked with estuarine shores and processes.

Narrow low dunes covered mostly by grass and other herbaceous vegetation occur on the inland side of sandy beaches (Fig. 3.15). Fingerlike sand spits with small dunes point toward river mouths in several areas (Fig. 3.16).

Fig. 3.15. Beach grass dominates the crest of a low dune along the Connecticut coast.

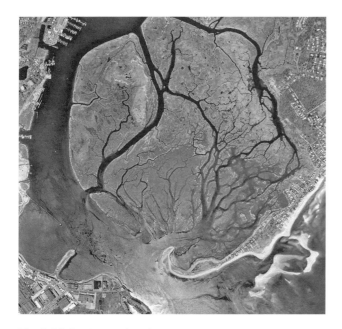

Fig. 3.16. Sand spits often form at river mouths and help provide conditions suitable for the development of salt marshes. In this aerial photograph, at Milford Point at the mouth of the Housatonic River, a spit with low dunes is bordered to the northwest by a tidal lagoon and the Nells Island marsh.

Fig. 3.17. Salt marsh peat (dark area at center), formerly protected on the landward side of sand spits, may end up in Long Island Sound as sea level rises and sand spits migrate northward.

Dune sand is derived from wind-blown and storm-transported beach sand, which in turn comes from bottom sediments of Long Island Sound, sediments transported in rivers, and eroded headlands. In Connecticut, dunes are not as well developed as they are in Rhode Island, on Cape Cod, or on barrier islands along the south side of Long Island where shores are fully exposed to dune-making processes—heavy surf and strong winds of the open ocean.

The dune environment is a severe one. Plants and animals on dunes experience many physical challenges, including strong winds that may dry out the plants, break them, and batter organisms with sand; hot sun and high temperatures; an unstable substrate of shifting sand that does not hold water very well; and constant deposition of salty aerosols.

Many dunes and beaches in Connecticut are on the move, migrating landward (north) due to the effect of the rising sea level, which in turn reflects a global warming trend resulting in widespread melting and recession of the world's glaciers. In the lower Connecticut River, sea level has risen about 9 inches per century over the past 300 years, resulting in the landward movement of coastal sand spits. These spits have overridden the salt marshes that formerly occurred behind them, and thick beds of marsh peat are exposed on the seaward side of these spits (Fig. 3.17). Some of these peat deposits, including one at Griswold Point at the mouth of the Connecticut River, have been dated at several hundred years old, suggesting that it may have taken that long for the spit to override the marsh. However, storms that wash over and sometimes breach the spits and their low dunes can greatly accelerate sand movement (Fig. 3.18). Dramatic changes in the coastal zone related to rising sea level may occur quite rapidly. It is prudent for us to acknowledge and plan for the possible consequences of global warming, even in the short-term. With some 20,000 commercial and residential buildings in Connecticut coastal areas that are at high risk of flooding during storms, the potential for major economic damage is large. Development of low-lying areas adjacent to the coast invites disaster!

Vegetation. The intertidal portion of sandy beaches has little or no vegetation, though hollow green weeds, sea lettuce, and rockweed are common seaweeds in areas with stones or cobble, and these and other seaweeds are common as beach-cast debris. Sea-rocket, Russian thistle*, spearscale, lamb's-quarters*, seaside spurge, cocklebur*, and (locally) seabeach sandwort occur in the upper parts of the beach where tides seldom reach.

The primary native dune plant is American beach grass (Fig. 3.15), a plant that thrives in (and depends on) areas of blowing sand. Rhizomes of beach grass extend into new deposits of wind-blown sand, then leafy sprouts arise. By reducing wind speed, beach grass causes airborne sand to fall to the ground. This builds up the dune, and the grass colonizes and stabilizes the newly deposited sand. Sea-rocket, beach pea, and seaside goldenrod often grow at the seaward margin of beach grass.

Other characteristic dune plants include a variable mixture of native and exotic weedy species, such as dusty miller*, seabeach orach, spearscale, seaside spurge, common

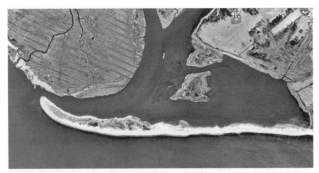

Fig. 3.18. Sand spits are dynamic habitats. Compare aerial photographs of the sand spit at Griswold Point in 1986 (upper) with the same area in 1995 (lower). The breach began in 1993 and has expanded since then.

cocklebur*, beach plum, bayberry, red-cedar, salt-spray rose*, Asian bittersweet*, shining sumac, Virginia creeper, ailanthus*, Russian thistle*, field mustard*, pepper-grass, false indigo*, beach pinweed, goosefoot *(Chenopodium)*, curly dock*, poison ivy, freshwater cordgrass, switchgrass, barnyard grass*, crabgrass*, sedges (*Carex silicea*), seaside germander, sea-blight, hedge-bindweed, climbing nightshade*, butter-and-eggs*, lettuce(*), lance-leaved tickseed*, common evening-primrose, soapwort*, and common mullein*. Beach plum stands out dramatically when it flowers prolifically in early May, as does seaside goldenrod in September and October.

Some sandy coastal areas in southern Connecticut support beach-heather, a low, mound- or mat-forming evergreen shrub. This rare plant, more common farther east in Rhode Island and on Cape Cod, is most conspicuous when flowering in late July to August. Golden-heather, a relative, is even rarer in the same habitat. Seabeach sandwort is another dune and upper beach specialist with a localized distribution.

Various mechanisms enable plants to cope with the harsh dune environment. For example, lengthwise curling of beach grass leaves such that the stomates (microscopic air valves) are not exposed to the wind reduces water loss. Seaside goldenrod has succulent leaves that serve as water storage structures. Beach pea roots contain nitrogen-fixing bacteria that help the plant cope with nutrient-poor sand.

Animals. Despite their small size, Connecticut's beaches and dunes support a unique array of wildlife not found in other habitats. Sandy upper beaches and dunes host myriad small animals, plus a few larger ones, only a few of which are mentioned here. Buried in the sand and hidden in daytime under beach-cast seaweed and other debris are numerous amphipod crustacean scavengers known as beach hoppers and beach fleas. They emerge into the open at night. In early fall you may find the small young-of-the-year with the larger adults. The larvae of seaweed flies (Diptera: Coelopidae) and various other flies develop in masses of beach-stranded seaweed. These flies function as decomposers of the algae and are a food resource for shorebirds. Scavenging earwigs sometimes occur under debris near the usual high-tide level. Sand crabs (not present in all areas) and digger amphipods are burrowers that feed on small bits of organic material on wave-swept beaches.

Cryptically colored sand locusts (Fig. 14.19) feed on beach and dune plants. Swarms of monarch butterflies, carpenter bees, wasps, and other insects sip nectar on prolifically flowering seaside goldenrods in September and early October. Tiger beetles and robber flies prey on other insects.

Piping plovers and least terns nest on upper beaches and low dunes where vegetation is sparse. The plovers move down onto intertidal beaches to feed on amphipods, small worms, and other invertebrates. These invertebrates attract small flocks of migrant and wintering sanderlings that run and feed along the water's edge. Sanderlings and other shorebirds extract the "meat" from beach-cast slippersnails, as do ring-billed gulls (Fig. 3.19). Migrating flocks of sandpipers, plovers, and other shorebirds often rest on sandy beaches and dunes. Song sparrows and northern mockingbirds occur in and near dune thickets, which also attract large numbers of migrating songbirds. Thousands of migrating tree swallows descend on fruiting bayberries in September and October. Horned larks and snow buntings are

Fig. 3.19. Sanderlings and ruddy turnstones feed opportunistically on beach-cast slippersnails in winter.

regular cold-season, seed-eating visitors to coastal beaches and sparsely vegetated dunes. Herring gulls, ring-billed gulls, great black-backed gulls, and laughing gulls are ubiquitous beach scavengers and predators.

Along with people, raccoons, opossums, striped skunks, and red foxes are common mammalian beach patrollers. Most of their activity consists of scavenging, with seasonal attacks on the eggs and young of nesting plovers and terns. On a smaller scale, shrews (Blarina, Sorex, and Cryptotis) hunt among dune plants for their insect prey. Dune thickets usually shelter plant-munching eastern cottontails*.

Salt Marshes

Like sandy beaches, dunes, and rocky intertidal areas, salt marshes are rare in Connecticut in the sense that they exist only in a very narrow zone along the coast (Fig. 3.20). Yet they have tremendous ecological and economic value. These flat coastal wetlands, dominated by grasses, occur especially at stream mouths, generally in protected areas (such as behind a sandy barrier spit) where sediments have been deposited to at least mid-tide level and salinities are greater than about 15 parts per thousand (ppt) (sediment deposits in the intertidal zone below the mid-tide level exist as tidal mudflats and are largely devoid of vegetation).

Most of the following information on the ecology of salt marshes in southern New England derives from research conducted by Mark Bertness, Paul Fell, William Niering, Scott Warren, and their colleagues and students at Brown University and Connecticut College.

Fig. 3.20. The pools in this salt marsh contain mummichogs and sheepshead minnows. The vegetation is mostly smooth cordgrass, with marsh-elder in the foreground.

History and Changes

Salt marshes became established in New England over the last 3,000 to 4,000 years, especially within the past 2,000 years, after the rate of the post-glacial rise in sea level decreased, allowing sediment deposition to keep up with the rise in sea level, which made it possible for marsh vegetation to develop. Beginning a few hundred years ago, the rate of relative sea-level rise in southern New England increased again, to approximately 9 to 10 inches per century. Studies in eastern Connecticut indicate that the accretion rate of high marsh peat has kept up with this rise in some locations, resulting in stable vegetation (salt-meadow cordgrass, black rush), but in other areas accretion has not kept up and high marsh communities have changed to wetter, more open communities of stunted smooth cordgrass and forbs. Continued rapid sea-level rise in New England may cause similar changes or even losses of salt marshes.

Marshes to Visit

You can get a good look at some of the largest salt marshes in Connecticut at Barn Island Wildlife Management Area, Bluff Point State Park, Hammonasset Beach State Park, East River, and Wheeler Wildlife Management Area at Milford Point. Brackish marshes dominate the lower Connecticut River system, though the marshes closest to Long Island Sound often have been referred to as salt marshes. A former area of salt marsh at Sherwood Island State Park was buried under dredged material in the mid-1950s—a sad and careless loss of a valuable resource. In many areas, you can explore these communities on foot without getting very wet, especially at low tide. A small boat will greatly facilitate your study of the rich life along salt marsh edges bordering tidal streams.

Vegetation Patterns

Salt marshes are highly productive ecosystems but, compared to freshwater marshes, relatively few species of plants thrive in these salty habitats. The vast bulk of the vegetation is made up of only four species, and extensive searching in any given marsh will yield fewer than two dozen species.

The distributions of salt marsh plants largely reflect the duration and frequency of tidal flooding (Fig. 3.21). High marshes may be flooded only during the highest (spring) tides that occur twice each month, whereas low marshes flood twice each day in accordance with the normal tidal rhythm. Marsh plants vary in their tolerance to inundation

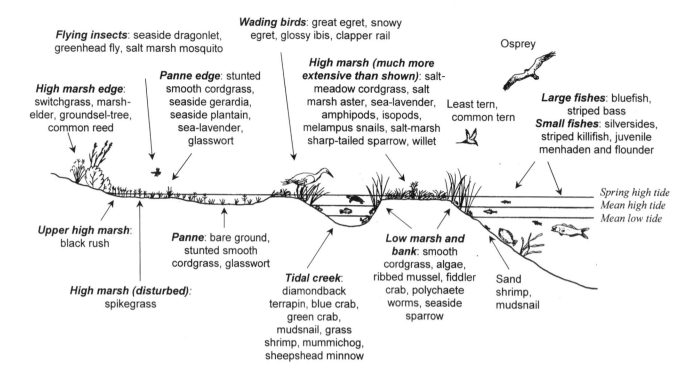

Fig. 3.21. Upper: An idealized Connecticut salt marsh. The location of plant species is strongly influenced by small differences in elevation above sea level. Not evident in the illustration is that the rich animal life in coastal marshes is based on a foundation of marsh plant detritus, associated decomposers, and microalgae. Not drawn to scale, and not all listed organisms are illustrated.

Lower: Diagrammatic summary of the roles of positive and negative interactions on the structure of Connecticut salt marsh plant communities. From Bertness and Leonard (1997). Used with permission of the Ecological Society of America.

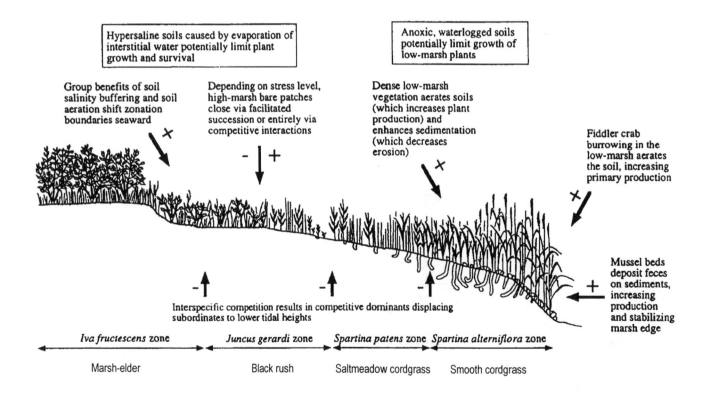

Fig. 3.22. Smooth cordgrass dominates the low marsh beside a tidal creek. Here the plants are exposed at low tide. High tide level is indicated by the change in color (lower part of plants is stained by muddy water).

Fig. 3.23. Smooth cordgrass (coarse grass at left, bordering water) of the low marsh meets salt-meadow cordgrass and spikegrass of the high marsh (finer grasses at right). Competitively dominant plants restrict smooth cordgrass to lower elevations whereas the superior competitors in the high marsh are unable to withstand the stressful physical conditions of the low marsh.

and so are distributed in zones roughly corresponding to elevation above sea level and thus susceptibility to flooding. Disturbance events such as storms that wash in and deposit rafts of debris that kill underlying vegetation, and competitive interactions, described below, also are important influences on plant distribution and abundance, particularly in high marsh.

The tall form of smooth cordgrass dominates the lowest areas adjacent to channels, mosquito ditches, or open water (Fig. 3.22). Such low marsh communities are most extensive east to the Connecticut River. Stems of smooth cordgrass contain air-conducting tubes that provide a good supply of oxygen to the roots, which as a result can grow in anoxic or low-oxygen mud that excludes most other plants. Smooth cordgrass is physiologically capable of growing throughout the better-drained high marsh where conditions are physiologically less stressful, but it is excluded from that zone as a result of strong competition with the dense root systems of high marsh plants. Sea lettuce and rockweed often grow near the bases of smooth cordgrass in tidal creeks.

High marshes generally are dominated by just two grass species and a rush: salt-meadow cordgrass, spikegrass, and black rush (Fig. 3.23). Salt-meadow cordgrass is excluded from low marsh because it is unable to oxygenate its roots adequately in anoxic marsh mud. Spikegrass, a poor competitor relative to the cordgrass species, is especially typical of areas recently denuded of live vegetation, such as may result from ice action in winter or burial by rafted mats of dead vegetation. The underground runners of spikegrass

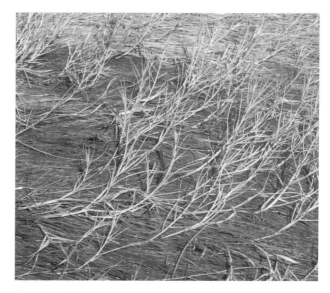

Fig. 3.24. Spikegrass colonizes a mat of dead marsh vegetation.

quickly colonize these areas (Fig. 3.24). That the lack of root competition is critical, and not primarily barrenness of the ground, is illustrated by the fact that spikegrass also readily spreads into dead but undecomposed mats of salt-meadow cordgrass. Salt-meadow cordgrass also may colonize barren places in the marsh.

Above the grassy zone of the high marsh is a zone of black rush, darkest of the common marsh plants. Black rush displaces spikegrass and competitively retricts salt-meadow cordgrass to lower marsh levels. But black rush is not a good colonist of areas that become hypersaline through evaporation when the sun beats down on unshaded bare patches. However, as a superior competitor, it is able to take over these areas once the grasses have colonized, shaded the substrate, and reduced substrate salinity through a decrease in evaporation. Black rush starts growth early, in May, before marsh grasses, but the plants are dead or dying by the end of August.

Depressions that trap water, and disturbance-caused bare areas, are subject to increased salinity through evaporation. These areas often support the short form of smooth cordgrass. Sea-lavender, annual and perennial salt-marsh asters, seaside gerardia, glassworts (especially in disturbed areas, which are colonized by seedlings), seaside arrow-grass, and seaside plantain also occur in these depressions, or pannes. (Attractive sea-lavender flowers are often picked for use in decorations and floral arrangements, but excessive harvesting can result in poor seedling recruitment and local population declines.) Glassworts, stunted smooth cordgrass, and spikegrass are among the plants that initially colonize disturbed substrates in restored marshes where common reed has been removed.

A shrub, marsh-elder, forms a band just above the high marsh (Fig. 3.25). Marsh-elder is absent from the lower elevations of the marsh because it is physiologically unable to tolerate prolonged flooded soil conditions, and its seedlings cannot effectively compete with the marsh's dense turf of perennial grasses and rushes. But marsh-elder seedlings do find places to become established, thanks to the passive effects of the adult plants, which sometimes trap plant debris transported by the highest tides. Mats of this debris kill underlying herbaceous vegetation, eventually creating a bare spot lacking competitors. Bare spots out in the open marsh generally become too saline for marsh-elder, due to water evaporation in the sun, but marsh-elder seedlings can survive in bare spots near adult marsh-elders because the larger plants, as well as dense stands of seedlings, shade the ground and prevent hypersalinization.

In general, many of the dominant salt marsh plants are restricted from lower elevations (which are more subject to flooding) by the harsh physical conditions whereas their

Fig. 3.25. A band of shrubby marsh-elder (center and right) borders the upper edge of many salt marshes.

upper elevational limits are strongly influenced by competitive interactions. Some plants that are not good competitors nevertheless persist by being able to colonize and establish themselves quickly in devegetated areas, such as where mats of floating debris have smothered and killed underlying vegetation or where sheets of ice in the marsh have floated and moved laterally during storms, ripping out the plants frozen into them.

Bordering salt marshes, just above marsh-elder, characteristic plants include groundsel-tree (a shrub, usually less common than marsh-elder), switchgrass, bayberry (a shrub), poison ivy, marshmallow*, common reed, seaside goldenrod, and various "weedy" species, as well as other plants typical of coastal dunes (see preceding material). In some areas, exotic Morrow's honeysuckle* and autumn-olive* are displacing native groundsel-tree. Plants such as three-square, freshwater cordgrass, red fescue, and marsh fern indicate places where salt marshes receive freshwater runoff from adjacent uplands.

The major salt marsh plants in Connecticut die back in winter and emerge in spring from underground rhizomes. Much of the resources accumulated by marsh plants in the warmer months are transferred to the rhizomes before winter, but dead plant material plays an important role in the marsh ecosystem. Energy and nutrients flow from marsh vegetation to predatory fishes, birds, and mammals mostly through a detritus-dependent array of decomposers. Direct grazing of live foliage by herbivorous animals does occur but is a less important pathway.

Plant Responses to Salty Soils

Salt marsh plants deal in various ways with the salty conditions that kill most other plants. In salt-sensitive plants, high salinity reduces the water uptake ability of the roots, which maintain low sodium concentrations, may expose plants to toxic concentrations of sodium and chloride ions, and may lead to deficiency in calcium or other ions. Salt-resistant plants are often termed halophytes.

Some plants (e.g., orach, *Atriplex*) accumulate and tolerate salt in their roots and are thus able to extract water from saline soil by osmosis. Succulent plants such as glasswort have leaves that swell by absorbing water, and this may prevent the concentration of absorbed salts from increasing too much. Smooth cordgrass and sea-lavender actively secrete excess salt via glands on the leaf surfaces. Some members of the goosefoot family (Chenopodiaceae) sequester salt in salt hairs or bladder cells on the leaf epidermis. Certain salt-tolerant plants absorb sodium into the roots, transport it to the shoot via the transpiration stream, and reroute it to the roots again for excretion.

Fauna

Invertebrates. Insects and spiders can be abundant in salt marshes in summer. Abundant seashore springtails feed on detritus and carrion and take refuge in tiny, submerged, air-filled spaces at high tide. Various homopteran insects suck the juices from salt marsh plants. Ground crickets and meadow katydids graze on high marsh plants. Beetles are represented by scavengers and detritus eaters in mud or masses of drifted vegetation, herbivorous species on various plants, and larval and adult predators on other insects. Praying mantises (mantids) perch on marsh plants and lash out with their raptorial front limbs to capture other insects. Several species of free-ranging and orb-weaving spiders also prey on insects and tiny mites. Hard to overlook are blood-sucking salt marsh mosquitoes and greenhead flies. Larval greenheads are predators in marsh mud. Seaside dragonlets, the only North American dragonfly that breeds in salt marshes, and (in late summer) migrating common green darners and black saddlebags, may feed on these and other flying insects. Seaside goldenrod and salt-marsh aster flowers attract a wide range of bees, wasps, syrphid flies, butterflies, and other insects in late summer and early fall.

If you stoop down and search carefully at the base of high-marsh grasses you'll likely find numerous eastern melampus snails grazing on the microscopic periphyton that coats the mud surface. These air-breathers often perch on grass stems above the water when the marsh is flooded. The

Fig. 3.26. Invertebrates, fishes, and birds are often abundant in tidal creeks in salt marshes.

golden ambersnail is another tiny air-breathing snail in upper edges of salt marshes and in brackish marshes. Marsh detritus often supports large populations of scavenging amphipods (*Orchestia grillus*) and isopods (*Philoscia vittata*).

Tidal creeks flowing through coastal marshes, and estuarine waters adjacent to salt marshes, are home to a rich invertebrate fauna (Fig. 3.26). Blue crabs are seasonally common in tidal creeks; they move seasonally between salt water and brackish water. Blue crabs are common enough in the lower river to support a modest recreational shellfishery. Saltwater and brackish-water creeks and marsh pools are inhabited by vast numbers of other crustaceans, including amphipods, isopods, grass shrimps, sand shrimps, mud crabs, and green crabs*, as well as mudsnails on subtidal and intertidal muds. Grass shrimp often swarm in vast numbers in shallow water along edges of tidal creeks. Water boatmen (*Trichocorixa*) can be abundant in summer in drainage ditches and marsh pools, and tiny midge larvae also occur in tidal creeks. You may also find striped anemones, clam worms, and rough periwinkles in marsh pools and channels.

Atlantic marsh fiddlers often are common on the muddy banks of salt marsh creeks. Fiddler crab burrows often turn these banks into "Swiss cheese" and, in conjunction with the flushing action of the rising and falling of tide, help aerate roots of smooth cordgrass and promote its vigorous growth (Fig. 3.27). Fiddler crabs scoop up sediments and feed on the organic material in it. Atlantic sand fiddlers replace marsh fiddlers in sandy sites.

In some areas, lower mud banks support large concentrations of scavenging mudsnails. Mudsnails have a heavy shell that may protect them from many potential predators and, accordingly, they are nonsecretive and often tremendously abundant.

Fig. 3.27. The banks of tidal creeks often are riddled with fiddler crab burrows. A fiddler crab is visible at center.

Ribbed-mussels attach to the bases of smooth cordgrass plants along tidal channels. They feed on bits of organic matter that they filter from the water. Rockweed often attaches to shells of ribbed-mussels. Herring gulls often pluck the mussels and drop them on rocks, pavement, or roofs, then fly down to feast on the meat within the broken shell.

In summary, most salt marsh crustaceans and mollusks are nourished by microscopic phytoplankton, algae, and microorganisms attached to rocks and shells or growing on the surface of marsh peat and mud, animal carrion, or decaying marsh plants.

Fishes. Tidal creeks in salt marshes are abundantly populated with small fishes, particularly mummichogs, plus striped killifishes, sheepshead minnows, spottail shiners, silversides, fourspine sticklebacks, American shad, young winter flounders, and others, with species composition varying with location, season, and salinity.

Schools of these fishes move into the marshes with the rising tide. Sheepshead minnows eat plant material and detritus, and the other fishes feed on marsh detritus and/or invertebrates that in turn depend on decaying marsh vegetation.

Mummichogs deposit eggs in the high marsh during spring tides (higher than average high tides that coincide with the new and full moons). These fishes feed on tiny eastern melampus snails, amphipods, and isopods in the high marsh (when high water makes them accessible). Even small ditches through salt marshes often harbor abundant mummichogs and sheepshead minnows. In early summer you can find juvenile mummichogs in small, isolated pools in the high marsh.

All of these small schooling fishes are critically important food resources for a wide assortment of sport and commercial fishes and crabs, as well as herons, egrets, and terns.

Reptiles. Diamondback terrapins—wary crustacean and mollusk eaters—are locally common in salty and brackish tidal creeks along the coast east to at least the Connecticut River. Sometimes omnivorous snapping turtles occur in this habitat. Other reptiles and amphibians, physiologically intolerant of salty water, generally are absent, though common garter snakes sometimes venture from uplands into the edges of high salt marshes.

Birds. Salt marshes are important bird habitats. Clapper rails, willets, saltmarsh sharp-tailed sparrows, and seaside sparrows nest only in salt marshes and the most saline brackish marshes. Additional breeding birds in salt marshes and their edges include mute swans*, American black ducks, mallards, Virginia rails, marsh wrens, common yellowthroats, song sparrows, swamp sparrows, and red-winged blackbirds. Ospreys nest on the many platforms erected for them in salt marshes, especially at the mouth of the Connecticut River. Great blue herons, great egrets, snowy egrets, black-crowned night-herons, glossy ibises, ospreys, yellowlegs, dunlins, several other kinds of migrant shorebirds, herring gulls, great black-backed gulls, least terns, common terns, tree swallows, bank swallows, barn swallows, common grackles, and many other less common birds take advantage of the rich food resources of Connecticut salt marshes. In fall, flocks of horned larks feed on sea-lavender fruits along the edges of salt marshes.

Mammals. The most common mammals of salt marshes are muskrats, raccoons, meadow voles, and shrews. Muskrats and voles eat living marsh vegetation. Grass-eating meadow voles are well hidden under grass cover in their runways in high salt marshes, but occasionally they fall prey to northern harriers, short-eared owls, and other predatory animals. Raccoons and shrews prey mainly on small animals, especially invertebrates, as do muskrats on occasion.

Brackish Marshes

Brackish, or semi-salty, marshes exist in a relatively narrow zone mainly along the lower parts of tidal streams, where fresh water draining from inland areas mixes with salt water of Long Island Sound (Fig. 3.28). The major area is along the lower Connecticut River. Brackish marshes of lesser extent are associated with several other rivers, such as the lower Housatonic River, lower Quinnipiac River, West River, East River, Branford River, and Thames River.

Brackish marshes are at most only about half as salty as sea water. Areas with salinities of about 0.5 to 18 ppt and water less than 6 feet deep at high tide generally support the tall, dense vegetation typical of these marshes. In early spring when streamflow is greatest, these areas are completely freshwater. Low flows in late summer and early fall may allow brackish water to move up to 15 miles or so upstream in the Connecticut River. Studies by Nels Barrett, Ken Metzler, Ron Rozsa, and others have elucidated the following vegetation patterns in Connecticut's brackish marshes.

Vegetation Patterns

As in salt marshes, plant distributions largely reflect water salinity and the duration and frequency of tidal flooding.

Fig. 3.28. Narrowleaf cattail, common reed, cordgrasses, rose-mallow, and wild rice are among the most conspicuous emergent plants along tidal creeks in brackish marshes. Here big cordgrass dominates.

Fig. 3.29. Rose-mallow, common in brackish marshes, grows here among narrowleaf cattails.

Examples of emergent plants in brackish marshes include narrowleaf cattail (almost pure stands in more saline areas), common reed, smooth cordgrass, big cordgrass, rose-mallow (Fig. 3.29), salt-marsh hemp, creeping bentgrass, common three-square, Olney three-square, wild rice, salt-marsh bulrush, soft-stem bulrush, spikerush, lilaeopsis, water pimpernel, bur-marigold, mudwort, arrowheads, salt-marsh fleabane, and water parsnip. The plants that dominate vary with salinity level. For example, closest to open water, cordgrass occurs in areas of higher salinity whereas three-square, arrowheads, and wild rice typify fresher waters.

Common reed is an invasive species that has been spreading and replacing other marsh vegetation along the lower Connecticut River (Fig. 3.30). The increase in common reed began in the late 1960s, and since then the area occupied has increased by about 1 to 2 percent per year. For further information on common reed, see the section on Marsh and Cove Degradation in "Conservation" in this chapter.

Along the Connecticut River at Great Meadow in Essex, an area transitional between freshwater and brackish water, paddlers exploring the low marsh (that is, the area submerged

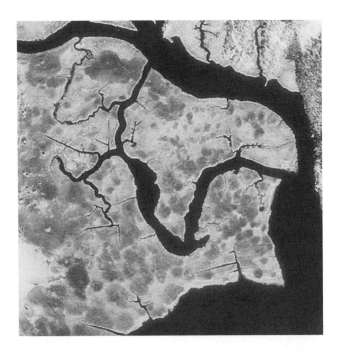

Fig. 3.30. Expanding patches of common reed are clearly visible as dark, circular, or irregular areas in this 1996 aerial photograph of brackish marsh at Lord Cove along the lower Connecticut River.

Fig. 3.31. Vast schools of juvenile menhaden filter-feed along the margins of brackish and salt marshes in late summer and early fall.

the longest during each tide) may find common three-square, water smartweed, salt-marsh hemp, and sweetflag. If you get out of your boat and venture into the marsh's slightly higher elevations, you're likely to find hybrid cattail, common reed, rose-mallow, arrow-arum, spikerush, purple loosestrife*, bur-marigold, mermaid-weed, marsh fern, and tussock sedge. If you approach this marsh from the upland side, look for freshwater cordgrass and switchgrass.

Meadows along the upland side of salt marshes that receive freshwater runoff may support a mixture of plants that can be found in salt marshes, brackish marshes, and freshwater marshes. So may former salt marshes in which tidal flow has been restricted, such as by ill-designed road crossings.

The submerged aquatic vegetation of brackish subtidal creeks, pools, and coves usually includes widgeon-grass and horned pondweed, plus sea lettuce and hollow green algae. In slightly saline waters, water-celery, waterweed, various pondweeds (e.g., curly*, ribbonleaf, and others), hornwort, Eurasian water-milfoil*, and mud-plantain can be common.

Fauna

Among the most conspicuous invertebrates of brackish marshes are red-jointed fiddlers, which are common in the less saline parts of creeks and coves bordering brackish marshes. They eat detritus and associated microorganisms. Invertebrates in the saltier sections of brackish marshes include some of those typical of salt marshes (see preceding section), whereas those at the upstream limits of brackish marsh include a small subset of those in freshwater river marshes (see Chapter 4).

The fish fauna includes menhaden (Fig. 3.31), silversides, banded killifishes, mummichogs, and many other riverine species, plus many of those discussed in the preceding salt marsh section. Populations of year-round residents are seasonally augmented by adults and juveniles of migratory anadromous fishes (e.g., shad and other herrings, sea lampreys, and American eels) that pass through brackish waters while moving between Long Island Sound and freshwater spawning or nursery areas in rivers.

The reptiles most frequently observed in brackish marshes are diamondback terrapins, snapping turtles, and painted turtles; few others are ever seen. These subsist largely on crustaceans, mollusks, and carrion.

Typical breeding birds include a mixture of fresh marsh and salt marsh species, such as least bitterns, mute swans*, Canada geese (an introduced population), American black ducks, mallards, clapper rails, king rails, Virginia rails, marsh wrens, common yellowthroats, sharp-tailed sparrows, seaside sparrows, song sparrows, swamp sparrows, and red-winged blackbirds. Common or characteristic summer nonbreeders include double-crested cormorants, great blue herons, snowy egrets, green herons, herring gulls, great black-backed gulls, ospreys, belted kingfishers, tree swallows, bank swallows, barn swallows, and common grackles. The bitterns, herons, egrets, cormorants, kingfishers, and ospreys are supported by abundant fish populations. Waterfowl feed on the leaves, stems, tubers, and seeds of various water plants and, seasonally, on insects and other invertebrates as well. Rails, sparrows, blackbirds, grackles, and wrens depend on abundant invertebrates and seeds of wetland plants. Swallows swoop through the air for flying insects.

Not surprisingly, muskrats and raccoons, ubiquitous wetland species, are the most common mammals. Herbivorous muskrats subsist on bulrushes, cattails, and many other wetland plants. Omnivorous raccoons prowl marsh edges and feed on available fruits, invertebrates, and fishes. Minks and river otters in low numbers hunt for fishes and large invertebrates in some brackish marshes.

Stress Gradients in Estuarine Ecosystems

Consideration of the stresses faced by seashore organisms helps illuminate the processes shaping salt marshes, rocky shores, and other intertidal communities (see Fig. 3.21). For basically terrestrial organisms such as flowering plants (e.g., marsh grasses), physical stresses increase with decreasing elevation (i.e., more frequent and longer flooding by the tides), whereas biotic stresses, such as competition, increase with increasing elevation (e.g., competition from the many upland and facultative wetland plants fringing the wetland habitat). In contrast, these stress gradients run in the opposite direction for marine aquatic organisms, which face many competitors and predators in subtidal areas and challenging temperature and moisture conditions at higher tidal levels.

Coastal Forests

Coastal forests are not actually part of the estuarine ecosystem, but it is convenient to discuss them here because they often immediately border salt marshes or intertidal shores. These forests (Fig. 3.32; also left background of Fig. 3.26) are dominated by deciduous trees and sometimes thick tangles of vines and bushes, often including black oak, white oak, scarlet oak, pignut hickory, mockernut hickory, bitternut hickory, sassafras, black gum, basswood, red maple, black cherry, red-cedar, shadbush, gray birch, black birch, bigtooth aspen, sweet pepperbush, catbriar, poison ivy, brambles, chokeberry, rose, hawthorn, Virginia creeper, winterberry, highbush blueberry, lowbush blueberry, huckleberry, arrowwood, smooth sumac, and hazelnut. American hornbeam, pink azalea, and witch-hazel join these a bit farther inland. Asian bittersweet, Morrow's honeysuckle, common buckthorn, wineberry, crab-apple, autumn-olive, multiflora rose, privet, and Norway maple are frequent exotics. Maples and birches may be scarce in coastal forest patches surrounded by salt marsh.

Coastal forests have no special adaptations to severe coastal storms, and hurricanes usually knock down many trees and kill foliage with prolonged blasts of salt spray. Sublethal effects of salt spray include a reduction in photosynthesis and reduced leaf growth. Rising sea level and re-

Fig. 3.32. Coastal forests include a high diversity of deciduous trees, shrubs, and vines.

sulting soil erosion are gradually removing the seaward edge of some coastal forests.

Animal life is essentially the same as in most inland deciduous forests (see Chapter 7). However, forests on some coastal islands are unique in supporting nesting populations of wading birds such as black-crowned and yellow-crowned night-herons, snowy egrets, great egrets, and little blue herons. Coastal forests and thickets also attract large numbers of migratory songbirds that have just completed or are getting ready for a crossing of Long Island Sound.

Conservation

More than 8 million people live in the drainage basin of Long Island Sound, so it is not surprising that humans have had a major impact on the ecosystem. Fortunately, one of the recent impacts is positive—conservation action. A major initiative for the conservation of the Sound is the Long Island Sound Estuary Program (LISEP), a research, management, and pollution abatement project funded by the National Estuary Program, which was established by Section 320 of the amended Clean Water Act of 1987. The LISEP is a cooperative effort involving researchers, federal and state regulators, user groups, and other concerned organizations and individuals working together to protect and improve the health of the Sound by implementing a Comprehensive Conservation and Management Plan (CCMP). This plan, finalized in 1994, identifies the major pollution problems affecting the Sound and proposes solutions. Much of the following discussion is based on the CCMP.

To date, the LISEP has found that the eastern portion of Long Island Sound is in relatively good condition, but some pollution problems have developed in the western Sound and the more industrialized harbors. In the following sections, I summarize the problems and potential solutions identified by the CCMP, plus additional conservation concerns that need to be addressed. If successful, the CCMP will ensure that the waters, shellfish, finfish, birds, and habitats of the Sound will continue to provide us with pleasant boating and valuable esthetic, recreational, educational, and commercial opportunities.

Hypoxia

Hypoxia—low dissolved oxygen (DO) level in the water—is a major problem in the western portion of Long Island Sound. Seasonal hypoxia has occurred there since at least the early 1970s. A major hypoxia-caused die-off of fishes and shellfish in 1987 raised concern over a "dying" Long Island Sound.

Hypoxia results from excessive nitrogen levels combined with normal temperature gradients in the water. Excessive nitrogen, acting as a fertilizer, promotes excessive growth or "blooms" of algae or cyanobacteria. The nitrogen comes from discharges from dozens of sewage treatment plants (discharging over 1 billion gallons of treated effluent into the Sound *every day*), stormwater runoff, acid rain (e.g., from nitrogen oxides in auto exhaust emissions), and the breakdown of organic matter. Atmospheric sources of nitrogen recently were confirmed as an important contributor to total nitrogen in riverine exports to coastal and estuarine ecosystems. Small, shallow streams can remove a significant amount of nitrogen from the water through the denitrification activity of stream-bottom bacteria, but large rivers, which receive large nitrogen inputs from sewage treatment plants and in which proportionally far less water is in contact with the stream bed, deliver most of the nitrogen they carry to the Sound.

During an algal bloom, vast numbers of algae die, fall to the bottom, and decompose. Decomposition by microorganisms and the normal needs of oxygen-consuming estuarine life draw oxygen from the deeper water. In summer, the Sun warms the surface water, but the deeper water remains cool. The density difference between the warm water and cool water prevents the oxygenated surface water from mixing with the oxygen-depleted (hypoxic) deeper water. Generally a minimum of about 3 milligrams of oxygen per liter of water (mg/l) is needed for healthy marine life. Summer dissolved oxygen levels fall below this critical threshold in the western and central parts of the Sound, with the lowest readings in the western basin. Bottom waters may fluctuate into and out of hypoxia throughout the summer. Hypoxic conditions tend to be coupled with high inputs of nutrients after heavy rains.

Organisms in hypoxic water may suffer increased mortality, stunted growth, or poor reproduction. Some hardy, mobile animals may be able to escape to better conditions, but sedentary benthic organisms and those more sensitive to changes in the environment suffer the consequences of hypoxia. Severe reductions in fish catches and die-offs of fishes such as Atlantic silversides, menhaden, and winter flounder, crabs, and lobsters have occurred in waters depleted in dissolved oxygen. Only by reducing the amount of nitrogen coming into the Sound can we hope to lessen the frequency and severity of hypoxia and its harmful consequences.

Nitrogen also has other impacts. In some areas, eelgrass beds have declined apparently as a result of increased nitrogen levels and reduced light transmission associated with increased water turbidity (which may result from decreased populations of clams and other filter feeders) and dense blooms of phytoplankton. In some locations, increased nitrogen levels may have caused replacement of eelgrass by the more nitrogen-tolerant sea lettuce.

More on Harmful Algal Blooms

Dinoflagellates of the genus *Alexandrium* contribute to harmful algal blooms in the northeastern United States. These organisms can lie dormant for years as cysts on the sea floor. The cysts eventually germinate, reproduce, and become food for clams, mussels, oysters, and scallops. Neurotoxins produced by the dinoflagellates may build up in the tissues of the shellfish. Humans that eat these shellfish may develop paralytic shellfish poisoning.

A new contributor to harmful algal blooms in the region appeared in the mid-1980s in bays at the east end of Long Island in New York and in Narragansett Bay in Rhode Island. The organism is an extremely small alga, new to science and recently named *Aureococcus anophagefferens*, but conveniently dubbed "brown tide" (in contrast to red tides caused by dinoflagellataes). It has reappeared in Long Island bays during most summers since its original discovery. It seems only a matter of time before it appears in the Connecticut portion of Long Island Sound.

The brown tide organism does its damage by reproducing in vast numbers and thus blocking out sunlight and thereby killing rich eelgrass communities and by monopolizing nutrients and outcompeting other algae. Zooplankton seem to be unable to consume *Aureococcus*, which may be too small, unpalatable, toxic, or otherwise unavailable. The decline in zooplankton certainly reduces food resources for fishes, which themselves feed many birds. Clams, mussels, scallops, and oysters also suffer from loss of their normal phytoplankton diet during brown tides, and they soon become emaciated and not worthy of harvest. Two consecutive years of blooms can totally eliminate bay scallops, which live less than two years.

Brown tide blooms do not appear to be associated with excessive levels of nitrogen or phosphorus in the water. Large inputs of iron and citric acid are possible culprits. Iron may come from deep water wells on Long Island and get washed into the bays as runoff. Citric acid, now often used as a replacement for phosphorus in household detergents, can hold iron in solution and make it available to the algae. In some bays, inadequate water circulation may be a contributing factor.

Toxins

A hazardous array of toxic contaminants reaches Long Island Sound through the numerous streams that drain the watershed. Primary sources of these contaminants include industry, sewage treatment plants, power plants, motor vehicles, agriculture, and runoff from commercial and residential areas. Airborne pollutants such as sulfur, lead, and nitrogen, which are emitted from smokestacks and automobiles, fall with rain to the ground and water surface. Pavement in urban areas exacerbates the problem by preventing water from filtering into the soil, increasing both the volume of runoff and the amount of pollution it carries. These toxic pollutants have harmed Connecticut's natural resources. For example, formerly vast oyster beds along the coast of Long Island Sound were virtually exterminated by river-borne water pollutants.

Today, potentially problematic contaminants in Long Island Sound include metals (cadmium, chromium, copper, lead, mercury, and zinc), chlorinated hydrocarbons (polychlorinated biphenyls [PCBs] and various pesticides, including DDT, DDE, chlordane, and dieldrin), and polynuclear aromatic hydrocarbons (e.g., fluorene, naphthalene). Although most of the pesticides are no longer in general use in the United States, they are persistent in the environment. Toxin levels in sediments of western Long Island Sound are relatively high, but the impacts of this on organisms are not fully understood. Relatively high concentrations of organic toxins have been found in blue mussels, and unhealthy PCB levels have been recorded in American eel, bluefish, striped bass, lobster hepatopancreas ("tomalley"), resulting in consumption advisories (www.state.ct.us/dph). Sublethal concentrations of toxins are known to cause anatomical, physiological, and reproductive abnormalities in fishes and shellfishes. Sediments in New Haven Harbor, the state's busiest commerical harbor, are widely contaminated with gasoline and fuel oil, which combined with low dissolved oxygen from excessive nutrients, are known to affect detrimentally the growth and survival of fishes (yet, surprisingly, abundance and growth rates of young winter flounder is much higher in New Haven Harbor than at the mouth of the relatively pristine Connecticut River!).

In response to low-probability transmission of West Nile virus and eastern equine encephalitus to humans by mosquitoes, salt marshes along the entire coast recently have been treated with larvicide. Larviciding is generally thought to be "safe" (except to mosquito larvae), but its impacts on nontarget species in natural ecosystems are not well documented. Some have suggested that there may be a connection between the application of large amounts of mosquito larvicide and the temporally associated die-off of lobsters in Long Island Sound. For further information on this topic, see the sections on lobsters and mosquitoes.

Efforts are being made to reduce toxin levels through more stringent restrictions on discharges, reductions in contaminant levels in stormwater runoff, and management practices that contain runoff and pollutants on land.

Pathogens

Certain pathogenic bacteria, viruses, and protozoa cause illness in humans and may affect fishes and shellfishes and thus harm commercial and recreational fisheries. These pathogens enter Long Island Sound in improperly treated or untreated sewage, through sewer systems, from ineffective septic tanks, and in discharges from boats. When sewage treatment plants malfunction or are overwhelmed by huge flows from sewers that carry both sewage and stormwater runoff during large rainfall events, untreated sewage and associated pathogens are released into rivers. These contamination events can result in the closure of beaches to swimming and shellfishing areas to harvesting. Sometimes the shellfish can be salvaged by moving them to clean water for self-cleansing and later harvest. Fortunately, upgrading of sewers and treatment facilities has led to reduced pathogen impacts on the Sound in recent years.

In 1998 and 1999, a die-off of lobsters in Long Island

Sound, especially in the western portion where fall landings declined over 90 percent, was associated with the presence of a protozoan parasite (*Paramoeba* sp.). Researchers are attempting to understand why the protozoan became a problem. Perhaps altered physical or chemical conditions in the sound reduced lobster health and made them more vulnerable to the parasite. Meanwhile, on 26 January 2000, the U.S. Secretary of Commerce officially declared a commercial fishery failure in the Long Island Sound lobster fishery.

In the late 1990s, oysters were affected by parasitic diseases known as Dermo and MSX. Increasing coastal water temperatures may have played a role in allowing these diseases to spread into Long Island Sound, where cold winters formerly kept them at bay.

Trash

Unsightly debris floating in the water and washed up on shores is one of the most obvious and obnoxious of the many human impacts on the Long Island Sound ecosystem. This debris is not only ugly but also leads to a vicious cycle of further degradation because people often see little reason in being fastidious in an already littered environment. Trash can pose a hazard to wildlife through ingestion or entanglement. Storm drains and combined sewer overflows are the major source of shore-deposited trash. Rainwater washes litter off streets, through sewers, and into coastal waters. Offshore dumping by commercial fleets and careless littering by shoreside visitors and recreational boaters are other sources of debris. Conscientious litter control on land and at sea will help solve the problem.

Non-native Invasive Sea Life

Exotic species such as those in ballast water released in port by commercial cargo ships, if they became established in Long Island Sound, could have potentially disastrous impacts on local fisheries and shellfisheries. A basic safeguard against such accidental introductions is to require that ships exchange ballast water out at sea such that direct translocations of organisms from one coastal site to another are avoided (species of the open sea are not likely to become established in an estuary and vice versa). That most Connecticut shorelines are now dominated by non-native crustaceans and mollusks such as the green crab (from Europe), Asian shore crab (from eastern Asia), and common periwinkle (from Europe) indicates that the danger is not just theoretical—coastal organisms from other continents can become well established in Long Island Sound. Likewise, exotic pathogens have been found in ballast water and may pose a threat not only to sea life but also to human health. "Guilty until proven innocent" is the only sensible attitude with regard to questions about the importation of non-native species.

Fisheries

Each year, some 350,000 marine anglers average about 1.4 million recreational fishing trips in Connecticut, with a recent annual harvest of 2 to 4 million fish. Additionally, approximately 1,000 commercial fishers fish or land fish in Connecticut. Until recently, the Long Island Sound lobster population supported a valuable commercial shellfishery, surpassed in dollar value only by oyster landings, as well as popular recreational opportunities.

Overfishing in recent decades has adversely affected some fisheries, such as surfclam, lobster, bluefish, and winter flounder. Effective management of commercial and recreational fisheries is needed to ensure that populations and harvest are sustainable in the long term.

Extensive human predation on fish and shellfish populations has directly altered estuarine community structure by changing the relative abundance of species (some species are harvested in huge numbers, while others are left alone). Indirect effects, though difficult to confirm, must be extensive. For example, reducing populations of predatory fishes and lobsters may allow prey populations to increase, which may augment food resources and populations of other predators and may allow the increased prey populations to exert a greater predatory and competitive impact on other species. These impacts have not been adequately studied.

Entanglement in commercial and recreational fishing gear results in mortality of sea turtles in Long Island Sound and adjacent estuarine and marine waters.

Shoreline Modifications

Breakwaters, groins, jetties, and sea walls have been erected in attempts to protect harbors and stop local erosion, but these structures have actually led to increased erosion and loss of some beaches because they interrupt sediment transport along the shore and interfere with natural beach-forming processes that replenish beaches. In some areas, groins and jetties also indirectly affect dunes by starving them of the beach sand from which they originate. Using a groin to capture sand in one place simply "robs" sand from elsewhere. Localized efforts to keep beaches as they are, and intensive development near beaches, do not address issues related to rising sea level and usually fail to recognize that beach dynamics operate on a landscape (large) scale. Effective beach conservation requires big-picture, long-term planning. Beaches need lots of room to respond to changing sea conditions, and we should recognize that attempts at shoreline fortification and establishment of "permanent" residential and commercial areas and property boundaries near dynamic beaches ultimately will result in both environmental and economic losses.

Marsh and Cove Degradation

About half of the state's coastal marshes have been destroyed by dredging or filling. Virtually all remaining salt marshes have been degraded by ditches that were cut into marshes, mainly between 1916 and 1940, in attempts to control mosquito populations. This not only failed to rid the marshes of mosquitoes but also reduced the value of the marshes as habitat for fishes, waterbirds, marsh sparrows, and other wildlife, and in some cases it created conditions favorable for invasion and expansion of common reed *(Phragmites)*.

Common reed is a grass that grows in dense monocultures and replaces species-rich natural marshes (Fig. 3.30). At the mouth of the Connecticut River, researchers from Connecticut College found that a few invasive reed clones became established between 1940 and 1960, then a large-scale increase occurred beginning in the late 1960s and early 1970s. In areas with the lowest salinity, clones appeared randomly throughout the marshes and have rapidly expanded radially outward. In areas with greater salt content, invasion began along creek edges and disturbed upland edges and expanded into the marshes. The studies indicated that high ground had more phragmites than did areas flooded longer by the tides.

We should note here that common reed is a native plant that has been present in Connecticut marshes for a few thousand years. However, the invasive form that grows as dense monocultures is an exotic form from Europe.

Common reed stands that replace cordgrass-rush communities have relatively low value to certain wildlife such as wading birds, ducks, and shorebirds that are excluded by the tall thick growth. Lori Benoit and Robert Askins found that salt-marsh species such as willets, seaside sparrows, and saltmarsh sharp-tailed sparrows not only disappear from reed-dominated areas but also become less frequent inhabitants of the small patches of non-reed marsh that remain among stands of invasive common reed. Additionally, muskrats appear to make less use of common reed than of the marsh plants its replaces, and plants that thrive in areas denuded by muskrats probably decline when common reed takes over. Populations of submerged aquatic flowering plants and macroalgae may be reduced in creeks bordering common reed.

Not all parts of the ecosystem are greatly impacted by common-reed invasion and establishment. Near the mouth of the Connecticut River, researchers from Connecticut College found that macroinvertebrate populations were not greatly changed when reed increased; the invertebrates appeared to be sustained by the large amount of leaf detritus that common reed produces. The fish faunas of creeks in reed marshes and in reed-free marshes showed no notable differences, and both types of marshes provided comparable food resources for fishes. However, fish biomass appeared to be greater in creeks free of reed. Some birds, such as red-winged blackbirds, marsh wrens, and swamp sparrows, use common reed as well as cattail marshes. In mid- to late summer, common reed stands are heavily used as night roosts by vast numbers of tree swallows.

Successful control of common reed often is facilitated by late summer applications of herbicide (e.g., Rodeo), followed by spring cutting with an amphibious mulching machine, combined with restoration of tidal flow. Treatment by herbicide application and mowing requires follow-up herbicide applications in subsequent years to kill survivng plants. A single mowing alone is not effective control. The Nature Conservancy and DEP recently cooperated in ongoing efforts to control common reed along the lower Connecticut River (see Fig. 4.23).

Some salt marshes and brackish marshes have been degraded through partial diking of the wetlands along road and railroad crossings, which restricts tidal flow, and by the installation of gates that further inhibit tidal flooding. These alterations reduce the salinity of marshes on the upland side of the structures and facilitate colonization by common reed.

Today, many marshes are being rehabilitated by the creation of deep pools and shallow pannes (mudflats) combined with renewal of tidal flushing (Fig. 3.33). These changes improve the habitat for young fishes and waterbirds, facilitate nutrient exchange between the marsh and adjacent open waters, allow mosquito-eating fishes access to marsh-dwelling mosquito larvae, and inhibit the growth of common reed. Marshes restored by re-establishing tidal flushing show increased plant and animal diversity.

Tidal coves also need restoration. Many small estuaries along the coast have been degraded as a result of land development that led to increased inputs of organic material, nutrients, and runoff into coves. This causes algal blooms and odors and clogs the basins with shellfish-smothering organic muds. Efforts have been made to restore some of the

Fig. 3.33. Removal of common reed and renewal of tidal flooding has made this restored marsh a much better wildlife habitat, supporting ducks, wading birds, and shorebirds where formerly there were few or none.

coves (such as Alewife Cove in southeastern Connecticut) by increasing tidal flushing, through dredging and using jetties to prevent sediment accumulation at cove mouths.

Public Access

Another problem that needs to be resolved is public access to the coastline, which is highly restricted by private landownership. Some areas of public shoreline (shores below mean high tide) have been posted in selfish (and illegal) attempts to restrict access. Keeping people away from shores and coastal waters fosters disinterest, which stifles efforts to protect and enhance the valuable natural resources of Long Island Sound. However, a balance is needed such that increasing recreational activity does not place coastal resources at risk from excessive human disturbance (for example, beach walkers and their roaming dogs often disturb coastal birds attempting to rest, nest, or care for their young). Some upper beaches and dunes in Connecticut are unnaturally barren due to heavy summertime foot traffic, which not only stresses and eliminates native plants but also creates conditions favorable for erosion and/or colonization by exotic plants, which are now common in most coastal dune areas.

Public access also leads to a seldom-discussed form of environmental degradation—noise pollution. Boaters in high-powered watercraft with inadequate sound mufflers commonly blast the ears of everyone within miles of their selfish excursions. We need better regulation of noise-making vessels.

Projects supported by the Long Island Sound Fund (see next section) have made significant improvements in public access in recent years. The recently published *Connecticut Coastal Access Guide,* available from the Connecticut Department of Environmental Protection, shows the locations and features of nearly 200 coastal access points. The Maritime Center in Norwalk and Project Oceanology at Avery Point in Groton are among the organizations that offer boat-based field trips to see first-hand the diverse sea life of Long Island Sound.

Support the Sound

Since passage of the Tidal Wetlands Act in 1969, over 1,600 acres of Connecticut's tidal wetlands at more than 50 sites have been restored. You can easily see examples of marsh rehabilitation at Hammonasset Beach State Park. But certain marshes are still under pressure from marina developers wishing to serve boat owners.

Fortunately, strict federal and state regulations now protect tidal wetlands from most abuses. Erection of structures, dredging and filling in the coastal zone, and various other activities in tidal, coastal, and navigable waters are regulated under Connecticut General Statutes (CGS Section 22a), including the Connecticut Tidal Wetlands Act and Connecticut Coastal Management Act, and Section 401 of the federal Clean Water Act. For information on regulations and permits needed for activities in coastal areas, contact the Office of Long Island Sound Programs (http://dep.state.ct.us/olisp).

You can help support efforts to clean up, protect, and restore Long Island Sound by purchasing a "Preserve the Sound" license plate (call 1-800-CT-SOUND). Proceeds go into the Long Island Sound Fund administered by the Department of Environmental Protection. The fund supports estuarine fisheries and wildlife research, restoration of tidal wetlands and embayments, habitat protection, research grants, and public access and education.

The U.S. Environmental Protection Agency has established offices in Connecticut and New York to oversee implementation of the LISEP and to provide a vehicle for public involvement in the protection of the Sound. For more information, contact the U.S. EPA Long Island Sound Office (www.epa.gov/region01/eco/lis).

The DEP's Connecticut Coastal Management Program,

initiated in 1980, works to balance protection of fragile coastal resources with sustainable economic uses of the shoreline. The program is administered by the DEP Office of Long Island Sound Programs.

Listen to the Sound 2000: A Citizens' Agenda for Long Island Sound, produced by National Audubon Society, Save the Sound, and the Regional Plan Association, is a document that presents a plan for establishing a reserve and public access system for the Sound, summarizes hearings held in Connecticut and New York, and lists sites that were recommended for inclusion in the system, with comments on their attributes and restoration needs (see http://ny.audubon.org/lis.html).

In December 2002, administrators of the U.S. Environmental Protection Agency and commissioners of the Connecticut Department of Environmental Protection and New York Department of Environmental Conservation signed the Long Island Sound 2003 Agreement. The agreement sets forth an ambitious plan to restore the health of Long Island Sound by 2014. It builds on the Long Island Sound Study's Comprehensive Conservation and Management Plan that was approved in 1994.

Chapter 4

Streams and Associated Wetlands: Riverine Ecosystems

We often view rivers as beginning in mountainous or hilly headwaters and ending at the ocean, or at their junction with another river, but in a sense rivers have no beginning and never end. They can be regarded as particularly vivid segments in the continuous movement of water molecules from one place to another and in their transformation from one form to another. Consider a water molecule now careening down a river rapid. Its recent history may have found it frozen on a snowy hillside, coursing through the kidney of an American shad, or rising in the vaporous mass of a thunderhead. Or it may have traveled slowly through fractured rock to the roots of a wildflower on a shady cliff. Perhaps it joined with others in a torrent to sweep plants and soil into the river or carry a seed to a distant shore. Or maybe it quietly seeped through the skin of a toad buried in a sandy bench above the river. Its near future might send it up the trunk of a cottonwood tree, down the gullet of a thirsty deer, or into the vast ocean. Its residence in the river might last minutes or years. Rivers are thus part of a great, powerful water cycle that rearranges the landscape, modifies the weather, enables life, and binds ecosystems together.

Connecticut is abundantly endowed with streams that provide pleasant and biologically unique contrasts with the upland landscape (Fig. 4.1). The flowing waters of the state's major rivers and smaller streams (creeks or brooks) amount to an impressive 8,400 miles statewide (Figs. 4.2 and 4.3). According to the National Wetlands Inventory, these encompass more than 7,100 acres of riverine deepwater tidal waters and 8,200 acres of nontidal deepwater streams (excluding streams so small as to appear linear on aerial photographs).

Most of the state is drained by the Housatonic, Connecticut, and Thames rivers (Fig. 4.4). Feeding these rivers are a few medium-sized tributary rivers (Fig. 4.5) and myriad smaller streams originating primarily as steep rocky creeks in the uplands (Fig. 4.6). Numerous small rivers drain the remainder of the coastal region.

Scientists refer to the smallest perennial streams as first-order streams. These headwater streams are generally not more than 5 feet wide and less than a mile long, and drain small watersheds. First-order streams may be small but they are many—about 40 percent of the total length of all perennial streams is made up of first-order streams. Below the junction of two first-order streams, the combined flow is a second-order stream (often 5 to 10 feet wide), which have only first-order streams as their tributaries. A third-order stream (10 to 25 feet wide) results from the union of two second-order streams, and so on. Sixth-order streams generally are major, navigable rivers. At their mouths, the Connecticut and Thames rivers are sixth-order streams, the Housatonic, Quinebaug, and Farmington are fifth-order, and the Naugatuck, Quinnipiac, Hockanum, Salmon, and Shepaug rivers are fourth-order.

The volume of water in a stream varies widely from place to place and season to season. The Salmon River in central Connecticut illustrates the seasonal pattern of streamflow in a medium-sized river. Median streamflow peaks in March at about 300 cubic feet per second (cfs) and is minimal in August and September at 30 to 40 cfs. Not far from the Salmon River, a smaller stream that I monitor stopped flowing in the dry summer of 1999, renewed a good steady flow after rains in September, and continued to flow through fall, winter, and spring. In 2000 during a cool, wet summer, the creek flowed all summer long. In cold winters, it freezes completely at the surface.

The concentration of ocean-derived salts in riverine ecosystems is, by definition, less than 0.5 parts per thousand (ppt) during average low flow conditions, so the lowest, brackish parts of rivers that join Long Island Sound are not

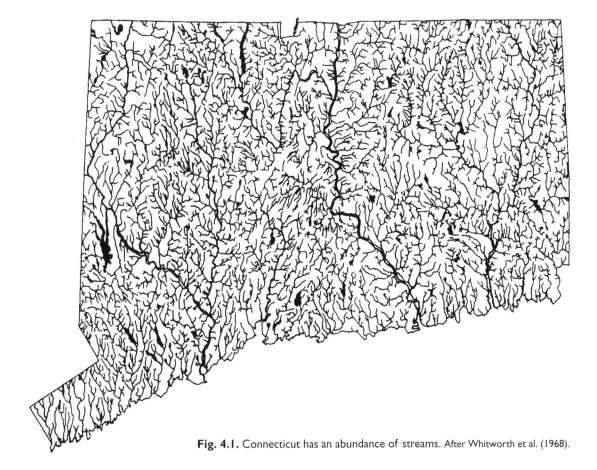

Fig. 4.1. Connecticut has an abundance of streams. After Whitworth et al. (1968).

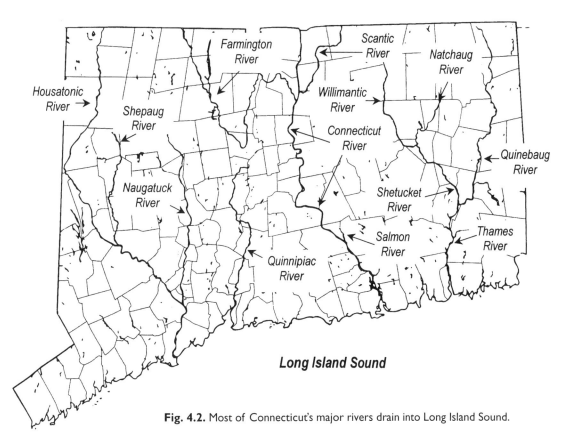

Fig. 4.2. Most of Connecticut's major rivers drain into Long Island Sound.

Fig. 4.3. The Connecticut River is Connecticut's largest river (shown above in Haddam). The sedge in the foreground is three-square.

Fig. 4.5. The upper Housatonic River (right) is an example of a medium-sized river. Here it cuts through marble bedrock in Kent.

Fig. 4.4. Drainage basins of Connecticut (below). Most of the state is drained by the Housatonic, Connecticut, and Thames rivers.

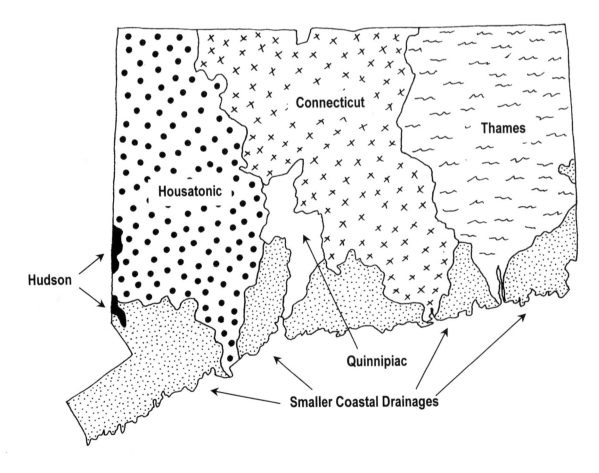

Fig. 4.6. Numerous small rocky streams drain Connecticut's uplands. Here, in northwestern Connecticut, spring salamanders inhabit the cold-water seepage pool to the left of the small stream.

included in riverine ecosystems. Riverine ecosystems include not only open flowing waters but also areas of submerged aquatic vegetation and stands of emergent plants that do not remain standing through winter. In this chapter I also discuss stream-influenced communities such as stream-edge and floodplain wetlands and stream shallows dominated by trees, shrubs, and/or persistent emergent herbaceous plants. Technically such wetlands are part of the palustrine system, but I find it more convenient to treat them here.

Open Waters and Aquatic Beds

Vegetation

Deep river channels often are sparsely vegetated, but shallows and slow-flowing areas can be rich in submerged and floating vegetation. Beds of aquatic vegetation in major, low-gradient rivers, including freshwater tidal areas, often contain water-celery, various pondweeds, hornwort, waterweed, naiad, water-milfoil(*), water stargrass, water-starwort, and, particularly in coves, yellow water-lily and white water-lily. Water shamrock* is distinctive but localized (e.g., Bantam River). Riverweed, a relatively rare aquatic perennial plant, may be attached to submerged or seasonally submerged rocks in fast-flowing water. Horned pondweed occurs in tidal waters. In the Connecticut River system, sites with relatively high diversity of submerged or floating vegetation include Hamburg Cove, Selden Cove, and the lower Salmon River.

These plants are important food producers for riverine ecosystems. On a smaller scale, diatoms and algae growing among stream-bottom sediments and attached to rocks and plants also play an important role in converting the Sun's energy into organic form. Some of this aquatic vegetation, plus nutrients represented by leaves and other plant material that have fallen into the water, is transported downstream by water flowing from tributaries to larger mainstem rivers and estuaries, where it is used by organisms far from the production sources.

Fauna

Invertebrates. Small crustaceans known as zooplankton (Fig. 13.4) are barely visible but ecologically important in large rivers. Zooplankton (and phytoplanton) populations are much reduced or absent in smaller, rushing, headwater streams.

Representative river zooplankton include water "fleas" (cladocerans) and calanoid, cyclopoid, and harpactocoid copepods. These are numerous in late spring and early summer. They move into the upper parts of the water column during periods of darkness and occur mainly near the bottom during daylight hours. Zooplankton feed on phytoplankton (photosynthetic organisms that are free-floating in the water column) and in turn serve as an important food source for young fishes.

In addition to zooplankton, rivers and creeks harbor a diverse array of larger invertebrates, including freshwater sponges, various flatworms, the sometimes massive colonies of the ectoproct *Pectinatella magnifica*, nematodes, various snails, fingernail clams, unionid mussels, oligochaete worms, leeches, isopods, various amphipods, ostracods, crayfish, nematodes, water striders, adult beetles, and the immature stages of mayflies, dragonflies, damselflies, stoneflies, fishflies, dobsonflies, beetles (diving, whirligig, water-penny, riffle, crawling water, and water scavenger), spongillaflies, various caddisflies, crane flies, mosquitoes, biting midges, sand flies, black flies, and chironomid midges. These small animals inhabit bottom sediments, cling to rocks, wood, or plants, or live among submerged masses of decaying leaves.

Immature stages of many kinds of stream insects periodically emerge from hiding places and drift downstream during periods of darkness. These drifting insects, augmented by insects that may fall into the water from above, often are a significant element in the diet of stream fishes. Adults of some drifting insects may fly upstream to lay eggs. Thus populations do not become depleted in upstream areas.

These invertebrates vary from rare to very abundant, but many are small, secretive, or seldom noticed except by specialists. For example, in the freshwater section of the Connecticut River in Haddam, the small, bottom-dwelling oligochaete worm *Limnodrilus hoffmeisteri* is tremendously abundant, but few people are aware of its existence.

Of these river invertebrates, mollusks are among those most likely to attract attention. Careful search of rocks, plants, shells, detritus, wood, or sand or silt bottoms in the Connecticut River might yield a dozen species of native snails. Most of these exist primarily in quiet waters, but the one-inch-long piedmont elimia (or Virginia river snail) lives on rocks in flowing water in a small section of the river at the northern edge of the tidal zone north of Hartford. On a smaller scale, you might find tiny freshwater limpets (Ancylidae: creeping ancylid, a stream specialist) on shells, rocks, or wood. Asian clams* are numerous in the Connecticut River and tributaries. Exotic mysterysnails of Asian origin are also established in the river's tidal freshwaters. They grow up to two inches long and live on shallow, sandy, or muddy bottoms.

Many species of tiny fingernail clams inhibit the river bottom in the freshwater section of the river, especially in sandy substrates, but these are dwarfed by the larger freshwater mussels, including the eastern pondmussel, tidewater mucket, eastern lampmussel, eastern elliptio, triangle floater, and alewife floater. These filter feeders include species that may reach several inches in length. All of them depend on fishes as the host for the mussel's minute larval stage. Two mussels, the dwarf wedgemussel and yellow lampmussel, are now very rare or extirpated in the lower Connecticut River, evidently victims of water pollution. In fact, freshwater mussels are one of the most endangered groups of animals worldwide, and the North American freshwater fauna as a whole, including mollusks, crayfishes, fishes, and amphibians, is of grave conservation concern.

Among crustaceans, crayfish generally are the largest river inhabitants, but even bigger juvenile and adult blue crabs move far up freshwater tidal sections rivers in summer. In the Connecticut River, I found blue crabs as far upstream as Haddam.

One of the most surprising river invertebrates is the ectoproct *Pectinatella magnifica*, a colonial animal consisting of rosettes of reddish zooids embedded in gelatinous material that may grow to several inches or several feet across. These harmless tentacled animals feed on minute organisms and organic material that they extract from the water.

At various times of the year the flying adult stages of insects with aquatic larvae may fill the air. Particularly abundant are the many species of harmless mosquitolike flies known as midges, massive numbers of which sometimes emerge and fly in early summer. Less abundant but individually more impressive are lively dragonflies and damselflies,

Fig. 4.7. Beds of floating and emergent plants in quiet rivers and coves support an abundance of life.

regular river companions on any summer boating excursion. Dragonflies and damselflies are predators as both aquatic larvae and flying adults. Larvae generally live in the water for one to three years before transforming in spring or summer into flying adults, which in New England do not survive the winter.

Beds of water-lilies or pickerelweed in quiet rivers and coves support a rich assemblage of insects (Fig. 4.7). Flowering water-lilies and pickerelweeds often attract nectar and pollen feeders such as bumblebees, honey bees, halictid bees, and syrphid flies. These insects also pollinate the flowers of river-edge plants such as buttonbush and basswood.

The often ragged appearance of water-lily leaves in summer reflects the ecological importance of these plants as animal food (Fig. 5.6). Water-lily leaf beetles are closely linked with water lilies and other aquatic plants, which furnish the diet of both larvae and adults. You may find the beetle's eggs bordering small holes on the underside of the floating leaves. The larvae of several moth species (Pyralidae: Nymphulinae), in a radical departure from the usual land-dwelling lifestyle of butterfly and moth caterpillars, are highly aquatic and feed on the leaves of water-lilies, pickerelweed, and cattails. Look for oval holes cut in the leaves of water-lilies; larvae use the cut-out pieces to build a case that may be attached to the leaf underside. After a period of feeding, the caterpillars swim to shore, overwinter in leaf litter, pupate in spring, and eventually emerge as moths. The leaves and spongy leaf stems of water-lilies and pickerelweed also provide egg laying sites for waterscorpions, backswimmers, and certain dragonflies such as the common green darner. Attached to the undersurface of water-lily leaves you might also find grazing snails and the egg clumps of various beetles, damselflies, caddisflies, snails, and water mites. These abundant invertebrate populations and the thick growth of water-lily beds feed and shelter many small fishes.

Streambeds of gravel, cobble, or boulders, and sandy river bottoms with dense growths of water-celery, are particularly rich in invertebrate diversity and abundance. Soft sediments may harbor vast numbers of organisms such as oligochaete worms, but invertebrate diversity may be much less than in coarser substrates where oxygen availability is higher.

Fallen trees in streams shelter various invertebrates and fishes and may trap organic material that provides an important basis for the aquatic food web. Fallen leaves from streamside plants often are an important nutrient source in creeks and small streams. Such organic debris supports populations of decomposers, including bacteria, fungi, and debris-feeding insects and crustaceans such as isopods, larvae of some flies, caddisflies, and stoneflies. Larvae of some caddisflies and stoneflies graze on diatoms and algae that grow as a film on submerged surfaces. From a biodiversity conservation perspective, it is a bad idea to remove natural plant debris from streams.

The populations of shredders, collectors, scapers, grazers, and other small animals in streams serve as important food sources for fishes and predatory invertebrates, including beetles, dragonflies, damselflies, dobsonflies, and others. Among the notable riverine dragonflies in Connecticut are three rare, localized species—midland clubtail, riverine clubtail, and arrow clubtail—that recently were discovered living along sandy beaches of the Connecticut River. Under cover on shore you might find a hellgrammite, the strong-jawed larval stage of the dobsonfly, or the similar larva of a fishfly. After spending a number of years living among stones on the river bottom, these impressive (more than 3 inches long) aquatic predators crawl ashore to pupate on land.

Permanent rivers and creeks have the highest invertebrate diversity, but even intermittent streams that shrink to isolated pools or dry up completely support an abundance of invertebrates, even in the summer dry phase. In August, during a casual five-minute survey of a dry/damp section of a small rocky streambed, I found many adult ground beetles of several species, adult diving beetles, abundant beetle larvae, a few 1.5-inch fishfly larvae, a dragonfly nymph, a giant water bug, water striders, a ground cricket, the cases of a few species of caddisflies, water mites, numerous millipedes, and several earthworms. Some of these are nonaquatic species that seasonally colonize dry streambeds.

Among the nonaquatic insects that might catch your eye while you explore the banks of larger rivers are the tiger beetles. These half-inch-long predators stalk sandy river beaches. Both the adult and burrowing larval stages prey on flies, ants, and other insects. One species, the puritan tiger beetle, is a rarity, restricted to just scattered areas along the Connecticut River in Connecticut and Massachusetts and a few areas in the Chesapeake Bay area. The adults emerge in early summer after overwintering twice in the larval stage. Predatory robber flies often hunt near active groups of adult tiger beetles, seeking similar prey.

Fishes. Fishes, of course, are among the better known and appreciated inhabitants of riverine ecosystems. A stream's fish fauna is affected by a wide array of historical, physical, chemical, and biological factors, so not all streams contain the same assemblage of fish species.

Fishes in smaller, high- to moderate-gradient streams a few to several yards wide, which tend to have cool, clear water and rocky-gravelly substrates, include brook trout, brown trout*, chain pickerel, golden shiner, common shiner, blacknose dace, longnose dace, creek chub, fallfish, white sucker, creek chubsucker, slimy sculpin, pumpkinseed, tessellated darter, American eel, and others, sometimes even fairly large largemouth bass* and fishes more typical of larger streams. Fishes in fast-flowing streams tend to be strong swimmers (trout, sucker), streamlined bottom-dwellers that are able to brace themselves against the current (darter, sculpin), or species that can effectively avoid strong flows by using pools and backwaters (dace and other minnows).

Headwater streams often go dry in summer, leaving fishes isolated in pools or eliminating them entirely if the pools dry up or get too warm or depleted of oxygen. Blacknose dace, brook trout, and other fishes easily recolonize after flows return, if not blocked by dams. In these smallest streams, insects that drop into the water from streamside vegetation may be important fish foods.

Compared to smaller streams, large rivers have a greater diversity of habitats and resources, and usually warmer water and less environmental fluctuation. Large rivers support large, piscivorous fishes that depend on higher riverine productivity, plus many of the fishes present in smaller streams (except cold water fishes such as brook trout).

Notable among the fishes of the lower freshwater portion of the Connecticut River and its coves are several large migratory species such as the sea lamprey, shortnose sturgeon, Atlantic sturgeon, Atlantic salmon, American shad, alewife, blueback herring, hickory shad, and striped bass. In the 1960s, blueback herring, Atlantic menhaden, white perch, spottail shiner, brown bullhead, and white catfish* were the most abundant species in the Connecticut River between Higganum and Essex. Other common species included the alewife, golden shiner, American shad, banded killifish, bluegill, mummichog, and yellow perch. Summer surveys by my classes and others from 1985 to 2000 often yielded spottail shiner, juvenile clupeids (American shad, blueback herring, alewife), banded killifish, tessellated darter, white perch, yellow perch, common carp*, golden shiner, largemouth bass*, smallmouth bass*, American eel, rock bass*, bluegill*, northern pike*, brown bullhead, channel catfish*, and other species in smaller numbers. Sheltering, food-rich beds of submerged plants such as water-celery harbor more fishes than do barren areas. Connecticut's other large rivers have a similar if somewhat less diverse fish fauna.

Stream fishes segregate ecologically, based on differences in body structure and physiological tolerance. Diet and feeding behavior often reflect mouth and body shape. Thus suckers are capable bottom-oriented suction feeders and streamlined, toothy trout are fast-moving predators. And juvenile fishes, with their different foods and vulnerability to predators, generally are ecologically distinct from adults of their species.

Different stream-bottom substrates serve fishes in different ways. Gravel, cobble, and boulder bottoms provide both feeding and spawning sites for stream fishes. Sand and silt substrates tend to be used only for feeding.

In addition to substrate conditions, the stream fish fauna is importantly affected by water temperature, with cooler water supporting trout, dace, suckers, sculpins, herrings, and shad, and warm waters more likely to harbor species such as catfishes*, bullheads, bass*, sunfish(*), and carp*. And of course turbidity, flow rate, vegetation, and other factors also influence the fish fauna. Most fishes require access to a diversity of habitats in order to complete their life cycles.

Stream fish faunas often are notably dynamic. They vary from season to season and from year to year as conditions change gradually or catastrophically and as a result of seasonal migrations.

Predation is an important factor in the lives of stream fishes. As a general rule, the life of a stream fish ends in the jaws or stomach of a predatory insect, another fish, bird (cormorant, kingfisher, merganser, heron), or mammal (mink, otter).

Amphibians and Reptiles. The mudpuppy is the only Connecticut amphibian preferentially occurring in large rivers. Its status as native or introduced in the state is debatable. This gilled amphibian occurs in the Connecticut River in northern Connecticut and feeds on various aquatic invertebrates. Several frog species occur in streamside marshes and floodpools but generally not in the main channel of rivers.

Characteristic amphibians of small streams are the two-lined salamander, dusky salamander, and ubiquitous green frog. Adult and newly metamorphosed dusky and two-lined salamanders, and juvenile green frogs can be abundant in summer under rocks in intermittent streamcourses long after streams have stopped flowing. The spring salamander is a habitat specialist that favors small cold streams in the northern part of state. All of these feed largely on invertebrates.

Large streams harbor a few reptiles such as snapping turtles, painted turtles, wood turtles, and northern water snakes (which also prowl small streams). These rely on the high productivity and shelter of marshes and slower flowing waters and so tend to be scarce or absent in open river waters and smaller headwater streams.

Birds. Birds are perhaps the most visible of river fauna. Waterfowl in particular make frequent use of the lower portions and coves of the state's major rivers. Get out your binoculars and you might see mallard, American black duck, gadwall (fall to early spring), American wigeon (mainly fall to early spring), common merganser (especially in winter), hooded merganser (fall to early spring), common goldeneye (mainly late fall and winter), ring-necked duck (especially winter and during spring migration), canvasback (fall to early spring), redhead (mainly fall and winter), greater scaup and fewer lesser scaup (fall to early spring), mute swan*, and Canada goose. Waterfowl present in the largest numbers in winter in coves in lower tidal rivers include canvasback, mute swan*, greater scaup, hooded merganser, American black duck, and mallard.

These birds have diverse diets. For example, common goldeneyes dive to the bottom to feed on various crustaceans and mollusks, whereas black ducks and canvasbacks mainly eat wetland vegetation. Canada geese often graze on river-bank plants, and mute swans* gather in large flocks in river coves and feed heavily on submerged vegetation. Introduced resident Canada geese and mute swans now nest along Connecticut rivers and have a year-round impact on aquatic and wetland plants (see Chapter 20). The fish-dominated diets of common and hooded mergansers differ dramatically from other waterfowl.

In addition to mergansers, many other river birds depend on fishes, especially the multitudes of small fishes that cruise river shallows. Great blue herons, great egrets, snowy egrets, and green herons stalk fishes in river edges and coves in the warmer months. Throughout the year, double-crested cormorants dive and pursue their fish prey beneath the surface, joined in winter by great cormorants and in fall and winter by small numbers of pied-billed grebes. You can find plunge-diving belted kingfishers feeding on small fishes throughout much of the year. Other birds, such as ospreys, seek larger river fishes. Ubiquitous herring, ring-billed, and great black-backed gulls feed mostly on dead fishes. Gulls also often take advantage of foods discarded by humans.

Invertebrates in river shallows and on shores and exposed sand- and mudflats attract spotted sandpipers (mainly mid-spring to early fall), migrating least sandpipers (mainly late summer and early fall), and killdeer (mainly late winter to fall). Look for tree swallows, bank swallows, barn swallows, and rough-winged swallows as they feed on insects over the water from spring through early fall.

In contrast to the diverse bird fauna of large rivers, far fewer birds use pools of small headwater streams. I occasionally found great blue herons, green herons, and kingfishers at pools where small fishes and frogs were available.

The bald eagle is one of the most impressive of Connecticut's river birds. Dozens of eagles range along major rivers in winter and early spring, searching for dead or crippled fishes and waterfowl. In recent years, eagles have begun nesting along the Connecticut River and elsewhere. This reflects the resurgence in eagle numbers and expansion of breeding range that have occurred throughout North America since the banning of DDT use in the United States in 1972. However, frequent human disturbance may be limiting their use of the Connecticut River.

Mammals. Beavers, river otters, and muskrats are the mammals you are most likely to encounter out in the open waters of rivers. Otters rest on shore and leave behind droppings full of fish scales and crayfish parts. Hungry harbor seals occasionally make brief sojourns up large rivers during their winter residency in Long Island Sound. Smaller streams have smaller mammals, including water shrews that freely enter the open water of small streams as they search for insect prey. At dusk in the warmer months, bats of several species visit rivers to drink and often forage on moths, beetles, flies, and other small flying insects over the water. See the following sections on marshes and beaver-influenced wetlands for further comments on the ecology of other river-associated mammals.

Riverine and Floodplain Marshes

Generally only people paddling small boats get a good close look at riverine marshes. From land, thick vegetation and wet, muddy ground may inhibit access, and large boats approaching from river channels risk grounding in the shallow water.

As defined here, these wetlands include marshes dominated by persistent or nonpersistent emergent herbaceous plants occurring along freshwater sections of tidal or nontidal streams. Areas of persistent vegetation (that is, where plants remain standing in winter) are, in many classifications, somewhat arbitrarily included in the palustrine category of wetlands, but for convenience, I discuss these areas here under riverine systems—after all, the existence of both the nonpersistent and persistent wetland vegetation is dependent on the stream. River-margin marshes are most extensive along the Connecticut River, which supports about 1,000 acres of freshwater marsh from Selden Creek northward to Hartford.

Research by Nels Barrett, Ken Metzler, and Ron Rozsa has produced a great deal of information about vegetation patterns in Connecticut's tidal freshwater marshes, especially along the lower Connecticut River.

Vegetation

Freshwater Tidal Marshes. Freshwater river marshes, especially tidal areas, harbor a high diversity of plant species. For example, salt marshes at the mouth of the Connecticut River contain fewer than 20 species of plants and brackish marshes just a few dozen, whereas tidal freshwater marshes of the lower Connecticut River harbor more than 150 plant species. Factors probably contributing to the high plant diversity in freshwater tidal marshes include the very low salinity (relatively few plants can cope successfully with salt water) and the greater availability of oxygen and nutrients due to the "refreshing" action of the tides, which may allow more plant species to coexist (some of these plants may not occur in nontidal marshes with less favorable oxygen and nutrient conditions).

The distribution of riverine tidal marsh plants is related to elevation with respect to water level (Figs. 4.8 and 4.9). The emergent plants in *regularly submerged marshes* include arrowheads, soft-stem bulrush, common three-square, wild rice, pickerelweed, golden club (Fig. 4.10), arrow-arum (Fig. 4.11), yellow water-lily, soft rush, river bulrush, cattails, sweetflag, common reed, three-way sedge, freshwater cordgrass, false pimpernel, water purslane, bur-reed, salt-marsh

Fig. 4.8. Tidal freshwater marshes may exhibit distinct vegetation zonation that reflects elevation and consequent frequency and duration of tidal flooding.

Fig. 4.10. Golden club is a unique element of the freshwater tidal marsh flora.

hemp, water parsnip, bur-marigolds, sneezeweed, water smartweed, rice cutgrass, spikerush, and water-plantain.

At higher elevations, the vegetation of *irregularly flooded tidal marshes* generally includes broadleaf cattail, narrowleaf cattail, halbeard-leaved tearthumb, arrow-leaved tearthumb, bedstraw, water smartweed, water horsetail, sensitive fern, marsh fern, royal fern, river bulrush, sweetflag (Fig. 4.11), common reed (especially in disturbed areas), purple loosestrife* (an invasive species), bur-reed, water millet, reed canary grass, bluejoint grass, groundnut, bur-marigold, false nettle, tussock sedge, and other *Carex* sedges. Dominant species vary in different marshes, and not all species are present in every marsh. Note that some plants occur in both low marsh and high marsh.

A diversity of woody plants often borders the upper edges of freshwater tidal marshes of the Connecticut River and its major tributaries. Common species include buttonbush, wil-

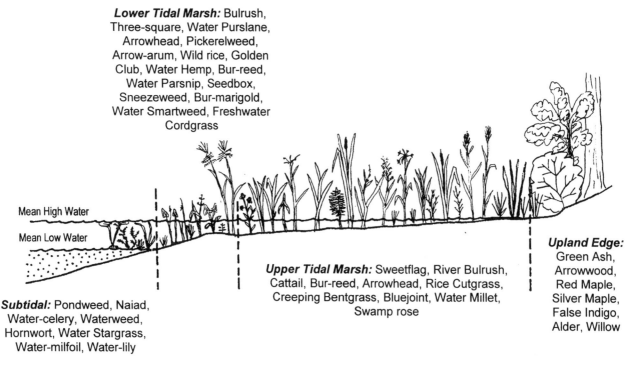

Fig. 4.9. Examples of plants in tidal freshwater marshes. See text for further information.

Fig. 4.11. Arrow-arum (foreground) and sweetflag (background) can be dominant components of freshwater tidal marshes.

Fig. 4.12. Many floodplain marshes exhibit dramatic seasonal changes between early spring (upper) and late summer (lower).

lows, speckled alder, northern arrowwood, nannyberry, swamp rose, red maple, silver maple, false indigo*, silky dogwood, grapes, pin oak, American elm, and green ash. These plants provide good nesting and feeding sites for small birds.

Tidal marshes and other floodplain marshes often exhibit striking seasonal changes in appearance. Death and decomposition of many tender wetland plants such as pickerelweed in fall means that marshes that were thick with vegetation in summer often look like open ponds, lakes, or mudflats in winter and early spring (Fig. 4.12). Wild rice, an annual, is hardly evident until late spring, then grows fast in early summer. When flowering and fruiting in late summer it reaches heights of 10 feet or more and is among the most impressive of tidal marsh plants, especially when enlivened with large noisy flocks of grain-eating blackbirds. Seasonal change in tidal marshes is less dramatic where cattails, river bulrushes, common reed, or other persistent emergent plants are present.

Nontidal River Marshes. Nontidal marshes and their margins also support a diverse flora (Fig. 4.7). If you paddle a canoe or kayak along a slow river in northwestern Connecticut, you may see, in addition to the submerged and floating plants mentioned in the section on the open waters of rivers, many of the following emergent and riparian plants: pickerelweed, bur-reed, arrow-arum, arrowhead, soft-stem bulrush, smartweeds, reed canary grass, common reed, sedges, broadleaf cattail, bindweed, blue flag, cardinal-flower, fringed loosestrife, purple loosestrife*, willows, smooth alder, speckled alder, arrowwood, red osier, buttonbush, winterberry, honeysuckle, red maple, and American elm.

Energy Flow. Energy and nutrients in marsh vegetation reach consumers such as predatory fishes, birds, and mammals primarily through a detritus-dependent array of decomposers (e.g., herons eat fishes that eat amphipods that eat marsh-plant detritus that is also broken down by fungi and bacteria). Overall, relatively little of the living marsh plant biomass is consumed by plant-eating animals. Much more is cycled into animal populations via decomposers and detritus feeders that consume dead plant material. This is fundamentally different than the situation in, say, a grassland ecosystem in which the predators feed primarily on small animals that eat live plants (e.g., birds eat grasshoppers that eat live grass). Among common marsh plants, soft species such as yellow water-lily, arrow-arum, pickerelweed, arrowhead, and wild rice break down quickly, whereas coarse sedges (*Scirpus* spp.), cattails, and common reed slowly decompose.

Fauna

River marshes dominated by emergent plants provide important wildlife habitat. Some members of our river fauna are staggeringly abundant, whereas others are rare and seldom observed. The animals you might see from your boat or from shore vary considerably, depending on such factors as water depth and temperature, the type of vegetation that is present, substrate conditions (for example, muddy, sandy, or rocky), and the time of year.

Invertebrates. Snails graze on microscopic life on the submerged parts of water plants and other objects. Crustaceans in river marshes include numerous scavenging amphipods, isopods, and ostracods ("seed shrimp"), and omnivorous crayfish also are present. In summer, blue crabs move into tidal freshwater streams and marshes, though almost all the ones I saw in fresh water were dead. Insects are represented by a wide variety of predatory, scavenging, and herbivorous insect larvae or adults, including mayflies, dragonflies, damselflies (big bluet occurs in Connecticut mainly in marshes along the lower Connecticut River), stoneflies, water striders, water boatmen, backswimmers, giant water bugs, water measurers, waterscorpions, dobsonflies, caddisflies, diving beetles, crawling water beetles, water scavenger beetles, water-lily leaf beetles, whirligig beetles, midges, punkies, crane flies, and sand flies. Shallow waters often support enormous numbers of water boatmen, but in summer some river-cove shallows become remarkably warm (over 100°F) and may be devoid of any insect life.

A tramp through a high river marsh during its unflooded phase might yield various plant-eating insects, including grasshoppers, ground crickets, various plant-juice-sucking homopteran insects, and beetles. In summer and early fall, the "tick-tick-tick-tick-buzzzzzzzzes" of meadow katydids are a conspicuous sound in streamside vegetation. Two species of introduced mantids (better known as "praying mantis") prey on other marsh insects.

Fishing spiders are common in many freshwater marshes, including those along rivers. These large spiders use their legs to row or gallop over the water surface, dive easily under water, and sometimes capture and eat prey items as large as small fishes and amphibian larvae.

Where great water dock and curly dock* are present in marshy floodplains, you might look for the "floppy flying" bronze copper, a rarity in June and July. Flood control projects may have reduced available habitat for this butterfly.

In addition to these invertebrates, many of the species or groups discussed in the section on open waters and aquatic beds also occur in adjacent emergent marshes. See Chapters 5 and 6 for information on invertebrates in nonriverine marshes.

Fishes. Many open-water fishes, discussed previously, make use of the less densely vegetated sections of flooded river marshes. These marshes are especially important as nursery areas. Small to large schools of banded killifish can be particularly numerous in shallow tidal waters. With a rising tide, these fishes move over flooded mudflats and among emergent plants in water less than an inch deep, where they may avoid predatory fishes (sometimes by leaping briefly onto shore) but often fall prey to great blue herons, green herons, and other wading birds.

Amphibians. River marshes and floodplain pools provide breeding habitat for several kinds of frogs and toads. American toads, spring peepers, gray treefrogs, green frogs, and pickerel frogs can be heard singing in spring, followed by summer-breeding bullfrogs. Northern leopard frogs are restricted in central Connecticut to floodplain wetlands and meadows, especially along the Connecticut River. Bullfrogs and pickerel frogs can breed successfully in water containing predatory fishes, but the other species rely on fish-free water for larval growth and survival. Most of these amphibians are more common in nontidal marshes than in tidal waters. Amphibians are active primarily in the spring and summer and are dormant in winter.

Reptiles. Where vegetation isn't too thick, you might see snapping turtles, painted turtles, and northern water snakes. There are no venomous snakes in river marshes, though the fish- and frog-eating northern water snake often is mistaken for one. As is true of amphibians, reptiles are active primarily in spring and summer and spend the winter underwater (turtles) or underground in upland sites near water (water snake).

Birds. River marshes support a rich and abundant avifauna. Marshes along the lower Connecticut River generally support about 20 species of breeding birds, a roughly equivalent number of nonbreeding summer visitors, and many additional species that are present primarily during migration.

Birds that actually nest in river marshes range from waterfowl to songbirds. Common nesters include mute swans*, Canada geese, mallards, rails, marsh wrens, common yellowthroats, song sparrows, swamp sparrows, and red-winged blackbirds. Marsh wrens, more easily heard than seen, enliven most stands of cattails, bulrushes, and other tall thick marsh plants. The songbirds take advantage of the marsh's excellent nesting cover and bountiful supply of seeds, insects, and other invertebrates.

Marsh-dwelling dabbling ducks, such as American black ducks, mallards, and wood ducks, have a varied diet of shallow water insects, mollusks, worms, crustaceans, algae, and the leaves and seeds of wetland plants. Wood ducks also eat large numbers of acorns in bottomland forests, and mallards may supplement their diet with waste corn in fields. These three ducks all nest within the river/cove ecosystem, though wood ducks require tree cavities or nest boxes.

The rich crustacean, mollusk, and worm fauna of tidal marshes and channel banks support nesting populations of rails (king rail, Virginia rail, sora). Migrating soras are common late summer-early fall visitors to marshes with wild rice, a favored seasonal food.

In summer, the rich food resources of river marshes also often attract large numbers of birds that nest in other habitats. Typical summer marsh nonbreeders include green herons, great egrets, snowy egrets, great blue herons, and common grackles, and living nearby or flying over these marshes you are likely to see herring gulls, great black-backed gulls, ospreys, tree swallows, bank swallows, and barn swallows. American black ducks, mallards, blue-winged teals, and green-winged teals are the most numerous migratory waterfowl in the lower Connecticut River marshes.

Spotted sandpipers forage on the banks of tidal marshes. Migrating least sandpipers often make use of mud flats exposed at low tide in late summer and early fall. In early fall, tall grasses of the marshy islands in the lower Connecticut River are important nighttime roost sites for thousands of migrating swallows that have spent the day feeding on flying insects over a wide area surrounding the river.

A newcomer now living throughout the state's riverine marsh ecosystems is the introduced mute swan. This exotic species nests commonly along the river and has undergone a population explosion in Connecticut. Biologists suspect that this aquatic plant-eating bird may be harming the food resources needed by native waterfowl. Introduced swans and populations of introduced Canada geese (the latter is native as a migrant but is not a native breeder or summer resident) can greatly affect wetland vegetation through intense grazing. For example, in summer I saw large areas where new growth of wild rice had been leveled by grazing geese. Wild rice marshes are valuable as food resources for native waterfowl, shorebirds, and rails, so the impact of the introduced waterfowl may be considerable.

Considering all the marshes (freshwater and brackish) occurring along the Connecticut River in central and southern Connecticut, Robert Craig found that the number of breeding bird species tends to increase as marsh size increases, whereas the number of species using the marshes primarily for feeding tends to be highest in marshes not isolated from other marshes. Big marshes provide abundant resources that can support larger, more viable populations of each species and are more likely to meet the minimum requirements for species that require large areas. Marshes smaller than about 50 acres tend to have noticeably depauperate breeding bird faunas; marshes on the order of 100 to 250 acres or more have the largest number of breeding bird species. Marshes that are not isolated allow feeding birds to go easily from marsh to marsh as they search for food, increasing feeding opportunities with minimal effort and time. Most rare marsh dwellers occur in the larger marshes, which thus are important contributors to biodiversity.

Mammals. Wild mammals tend to be secretive or nocturnal and so are not often seen, but quiet boaters may get a glimpse now and then in early morning or evening. Grass-eating meadow voles are well hidden in their runways in thick grass of high marshes. Star-nosed moles, resembling something out of Dr. Seuss, are common though seldom-observed burrowing and semiaquatic mammals of riverside freshwater wetlands, where they prey on worms and various aquatic invertebrates. White-tailed deer quietly visit rivers and cove edges to drink and forage on wetland plants that provide important sources of sodium. Fruiting plants and rich populations of invertebrates, frogs, and fishes attract ubiquitous raccoons. Stealthy minks prey on muskrats and other aquatic animals, whereas muskrats feed primarily on the succulent parts of cattails and other marsh plants and sometimes on river mussels. Beavers often feed on soft river-marsh plants in addition to their better-known diet of woody streamside vegetation.

Floodplain Forests

Floodplains are low-lying areas that become inundated when rivers overflow their banks. Connecticut's floodplains are mostly adjacent to the Connecticut River and its major tributaries from Middletown north. Under natural conditions, they are dominated by forests of deciduous trees (Fig. 4.13). These forested wetlands are technically regarded as palustrine wetlands, but I prefer to discuss them here because they are so closely tied to the riverine system.

Floodplain forests exist under an exceptionally dynamic hydrological regime. During peak flows in spring, and sometimes in other seasons, when rivers may overtop their banks, it is possible to explore these forests by canoe or

Fig. 4.13. Large silver maples dominate this floodplain forest. Stephen P. Broker.

Fig. 4.15. Horizontal marks on the trunks of floodplain trees indicate the water level during a recent flood stage. Photographed in winter before spring flooding.

kayak (Fig. 4.14). In late summer, you have to explore on foot, and it is impressive to see high-water marks on tree trunks, sometimes above your head (Figure 4.15).

When a river rises above its banks, it spreads out over a wide area, reducing the velocity of the water, particularly in areas away from the river channel. These slow-moving waters carry and deposit fine sediments in the river's floodplain, sometimes far from the river channel. Water moves faster closer to the river channel, and this faster flow carries not only silt but also sand and other coarse materials. These heavier materials are deposited close to the river because the water velocity farther away is too slow to move them. This differential deposition results in the formation of natural sandy levees bordering the river. In this zone, plants must cope with the mechanical damage (e.g., from floating logs, ice, and debris), erosion, and sedimentation that accompany flooding.

Vegetation

As in marshes, floodplain vegetation largely reflects the frequency, duration, and timing of flooding (Metzler and Damman 1985). Inevitably, human influences are also evident (see following Conservation section).

Fig. 4.14. A small boat is essential for exploring floodplain forests during spring flooding.

Sandy levees that border the Connecticut River have a diverse woody flora that may include eastern cottonwood, swamp cottonwood, black willow, elms, American sycamore, box elder, silver maple, red maple, green ash, swamp white oak, pin oak, bitternut hickory, basswood, butternut (locally), riverbank grape, dewberry, buttonbush, and poison ivy. Catalpa, white mulberry, black locust, false indigo, and Asian bittersweet are frequently occurring exotics on these levees. Less common along tidal rivers of the Connecticut River system is bladdernut, a large shrub that becomes adorned with large, three-lobed, bladdery fruits by mid-summer.

In contrast, a single species—silver maple, the quintessential floodplain tree—dominates the muddy, fine-grained soils of the lower parts of the floodplain behind the levees, where flooding occurs most often and lasts the longest. Silver maple seedlings tolerate flooding and can be abundant in low areas subject to ponding, but few of these ever attain mature size. Larger maple saplings grow mainly along the river edges of the forest.

Green ash, pin oak, and American elm join silver maple as the trees you are most likely to see while boating in tidal rivers through floodplain forests, such as the Mattabesset River, a major Connecticut River tributary. Red maple occurs in some areas. Shrubs are few in low floodplains, but in favorable sites you might encounter scattered buttonbush and sometimes spicebush, silky dogwood, alder, northern arrowwood, false indigo*, swamp rose, or various other wetland species. Common river-margin vines include poison ivy, riverbank grape, fox grape, bur-cucumber, wild cucumber, hog-peanut, and groundnut.

The herbaceous understory of a floodplain forest also varies with the duration of flooding and availability of sunlight. It's often easy to find false nettle, wood-nettle, and sensitive fern. Among the rarities, you might find green dragon—a plant that in late spring or early summer has a long "tongue" projecting several inches beyond a hoodlike structure. Long-lasting high water sometimes supresses the growth of understory plants in floodplain forests. Late spring flooding and the development of heavy shade beneath floodplain trees well before the end of May afford little time for spring wildflowers to develop.

In summer, after water level has dropped, a stroll along a sandy or silty floodplain riverbank generally yields an unpredictable array of warm-season grasses, wetland plants, weedy species, and seedlings and saplings of cottonwood, black willow, and silver maple. The specific plants you find depends on how long the shore was flooded that year and which plant seeds happened to be present.

Overall, the flora of floodplain forests is modestly diverse. Surveys of tracts in southern New England have yielded about 130 to 180 species of vascular plants.

Fauna

The invertebrate fauna of floodplain forests is too diverse to review adequately here. Many of the species are the same as those in upland deciduous forests. Among the more unique species are horsehair worms, which sometimes appear in forest pools.

Many riverine fishes are seasonal visitors in floodplain forests. They move in with flood waters and depart when the water recedes.

Reptiles and amphibians are relatively scarce in dense floodplain forests. Green frogs and painted turtles are perhaps most common; spring flooding and heavy shade in summer discourage most potential inhabitants. Sunny floodplain shrub swamps and marshes harbor many more amphibians and reptiles.

The summer avifauna consists primarily of migratory, canopy-dwelling songbirds that winter in the tropics, such as essentially all of those that occur in other deciduous forests (see Chapter 7). Notable nesting birds of floodplain forests include wood duck, red-bellied woodpecker, Acadian flycatcher, Carolina wren, warbling vireo, cerulean warbler, Kentucky warbler, and hooded warbler. Belted kingfishers and bank swallows frequently dig their nesting burrows in steep sandy banks along rivers, and rough-winged swallows later may use these for nesting. Fish-eating double-crested cormorants, great blue herons, green herons, ospreys, bald eagles (mostly in winter), and belted kingfishers are conspicuous among the birds that perch in river-edge trees. Spotted sandpipers forage and sometimes nest along river shores of floodplain forests.

The margins of small, rushing, forested streams with steep embankments, which might be considered a very narrow floodplain forest but more often are referred to as riparian areas, often harbor nesting Louisiana waterthrushes. Acadian flycatchers nest on hemlock branches overhanging streams in ravines. Winter wrens seek prey along such areas in colder months.

Characteristic mammals of floodplain forests include the aquatic and wetland species discussed in previous sections of this chapter, plus upland mammals such as opossums, shrews, various bats, gray squirrels, and white-footed mice.

Beaver-Influenced Wetlands

When a beaver moves in and establishes its home in a stream system, dramatic physical and biological changes usually result (Fig. 4.16). A beaver dam changes a flowing stream into a small impoundment with increased water surface area and decreased water velocity. The dam holds back upstream runoff and reduces flooding in areas downstream. Flooding

Fig. 4.16. When beavers dam flowing water, they change streams and forested wetlands into ponds, dead-wood swamps, or marshy wetlands. Here, a beaver dam flooded a white pine and red maple swamp and killed the trees, which now support nesting great blue herons and a rich assemblage of cavity-nesting waterfowl, woodpeckers, and songbirds.

of former wetlands and uplands kills trees and other wetland and upland plants but creates conditions suitable for growth of aquatic plants and expansion of wetland vegetation in areas where drier uplands had been. Beavers actively redistribute wood in the system as they harvest tree branches and shrubs on land and drag them into the water for use as winter food or to make lodges. Sometimes beavers excavate small canals in pond edges to facilitate travel to and from their logging operations. Summer water temperatures increase—a wide beaver pond with deforested margins heats up much more readily that does a narrower, flowing stream shaded by riparian trees.

Beaver impacts may extend far beyond their period of residency. Hydrological and successional vegetation changes that occur when beavers deplete local food resources and eventually abandon a site result in a series of habitats that are vastly different from the forested stream or swamp that would exist in a beaverless landscape. When the unmaintained beaver dam ultimately fails, water level drops, mudflats may be exposed, and the site may develop into a wet meadow, which with enough time might develop into a red maple swamp. After food resources recover, beavers may re-colonize the site, re-establish a dam, and recreate the pond, allowing the expansion of marshy vegetation. Thus beavers generate diverse habitats that vary over space and time. These ponds, marshes, mudflats, and wet meadows support many wildlife species and contribute importantly to regional biodiversity.

In western Massachusetts, where conditions are similar to northwestern Connecticut, McMaster and McMaster determined that 15 beaver-influenced wetlands supported a rich vascular flora of 231 species. Plants varied quite a bit from site to site (there were about 30 to 80 species per site). Present at all sites were red maple, the sedge *Carex stipata*, spotted jewelweed, rice cutgrass, marsh purslane, sensitive fern, silky willow, woolly sedge, meadow-sweet, and hardhack. Open-water areas had lesser duckweed, common bladderwort, water-shield, naiad, yellow water-lily, water smartweed, pondweeds, bur-reed, bulrushes, and *Carex* sedges. Mud flats generally were dominated by rice cutgrass, yellow loosestrife, or spikerush. *Carex* sedges, bulrushes, *Juncus* rushes, goldenrod, Joe-Pye-weed, and jewelweed were typical dominants in wet meadows upstream of open water. Bluejoint grass, tussock sedge, reed canary grass, or woolly sedge sometimes created hummocky microtopography in these areas. Drier beaver meadows that rarely flooded during the growing season supported *Carex crinita*, spikerush, rattlesnake grass, soft rush, and broadleaf cattail in association with swamp aster, spotted Joe-Pye-weed, bedstraw, jewelweed, meadow-sweet, winterberry, speckled alder, and occasionally saplings of yellow birch, black birch, paper birch, gray birch, and black cherry.

The fauna of beaver-influenced wetlands is much like that of riverine marshes (see previous section) and wet meadows (Chapter 6), augmented by birds and other species that use standing trees killed by beaver-caused flooding (e.g., cavity-nesting wood ducks, hooded mergansers, tree swallows, bluebirds, woodpeckers; nesting colonies of great blue herons). Brown creepers sometimes nest behind the loose bark of trees killed by flooding. Upstream from a beaver dam, conditions deteriorate for fast-water stream dwellers or those that need cool water or clean rocky substrates (e.g., certain stoneflies, spawning trout), but circumstances improve for still-water wildlife and those that favor soft substrates (e.g., breeding frogs, turtles, ducks).

Conservation

Many streams in Connecticut have been degraded by human activities. Some of the impacts are direct, as when a dam is built, but many of the stresses on streams originate on land.

Dams. All of the state's largest rivers have been dammed, though the small dam on the Connecticut River is now breached. Dams and impoundments interfere with fish movements and drastically alter flow, temperature, and other physical conditions, making it impossible for some native fishes, freshwater mussels, and other invertebrates to complete their life cycles.

Land Development. Agriculture and land development often result in erosion that causes excessive deposition of sediments in stream bottoms, sometimes eliminating fish spawning sites and habitat for gravel- and cobble-dwelling invertebrates. Development also leads to increased road sanding in winter, which in turn results in sedimentation, as well as chemical pollution, when rains wash salty sands into streams. Eroded sediments that enter bodies of water decrease water clarity and may interfere with fish feeding behavior.

Clearing of vegetation not only results in erosion and sedimentation and eliminates an important source of nutrient input but also removes shade and may cause increased water temperature, reducing populations of cool-water salamanders, fishes, and invertebrates. Removal of vegetation from riparian corridors also degrades the cover, shade, moisture, and cool temperatures needed by amphibious salamanders that use both aquatic and streamside habitats.

Most floodplain forests in Connecticut have been cleared for agriculture and other human uses. Whole cities (for example, Hartford) have been built on the higher parts of floodplains that formerly supported rich forests of sugar maple, white ash, basswood, American elm, bitternut hickory, and other species. Levees, stream channelization, and vast areas of pavement prevent floodplains from conducting their free flood control services for downstream communities. Floodplain forests are now scarce and unfortunately still threatened by clear-cutting for low-value agricultural purposes, airstrip management, or golf course development. Even protected sites have been cut and cleared in recent years. Floodplain forests are in great need of better protection.

Pollution. Water pollution has degraded Connecticut rivers for centuries. In many areas, waterways served as receptacles and conduits for raw sewage until well into the twentieth century. In Willimantic, dye-containing effluent from thread-manufacturing mills imparted startlingly bright colors to local rivers. Chemical and copper wastes from early brass manufacturing and other industries flowed freely into the Naugatuck and Housatonic rivers and Long Island Sound. The Naugatuck River reportedly is devoid of stoneflies (indicators of good water quality), evidently due to pollution from industrial sources. Sediments and fishes in several rivers and reservoirs, especially in western Connecticut, are contaminated with PCBs (polychlorinated biphenyls) and mercury deriving from past human industry. Today, marinas are perpetual sources of various pollutants.

In 1991, the U.S. Geological Survey (USGS) began an investigation of the Connecticut, the Housatonic, and the Thames river basins under the National Water-Quality Assessment Program. Water quality in these areas has been adversely affected by human activities. Waste materials from cities and industries have created serious water-quality problems on major rivers. Nonpoint pollution sources, such as residential septic systems, storm runoff, and agricultural areas, have affected small streams and principal aquifers. Water and sediments in parts of the study area have been contaminated by nutrients, pesticides, trace metals, or synthetic organic chemicals. The presence of contaminants affects the suitability of water for drinking, industrial use, recreation, or aquatic life. In addition, the quality of fresh water in this study area also affects Long Island Sound, which receives water from all major streams in Connecticut. The movement of nutrients, especially nitrogen, into the Sound is a major regional concern. The USGS is currently compiling available information on water quality and is conducting an intensive sampling program to characterize water-quality conditions in the study area.

Non-native Species. As in many ecosystems, exotic species are a problem. Non-native species now dominate the fish faunas of many rivers. Introduced fishes, clams, and other invertebrates may negatively affect native species through competition, predation, and disease transmission. Exotic plants change the fundamental nature of the ecosystem and tend to drive it toward biological impoverishment (see the following section on the Connecticut River for further discussion of exotic plants).

Among the exotic threats to the river ecosystem is the Asian clam*, which became exceedingly abundant in the Connecticut River and some tributaries during the 1990s. Where abundant, Asian clams can alter ecosystems by filtering enormous quantities of phytoplankton, and in some circumstances they can displace native mollusks.

River ecologists have been on the lookout for the appearance of the zebra mussel, a European native that has become established elsewhere in the United States (and in Twin Lakes in Connecticut). This dark and light banded clam, usually an inch long as an adult, has the capacity to form dense aggregations that can overwhelm and kill native mussels and alter aquatic ecosystems through their enormous water-filtering capacity. Fortunately, the Connecticut River has marginal conditions for the establishment of zebra mussels.

Sometimes native wildlife may pose a conservation problem. For example, currently high beaver densities result in reflooding of abandoned beaver ponds and meadows after a short time interval. This reduces availability of late successional beaver wetlands (meadows). Species that favor such sites, such as adder's-tongue ferns, may thus be threatened by burgeoning beaver populations. However, bear in mind that over the long term beavers play an important role in generating habitat for many wet-meadow species.

Protection. Adequate protection of stream habitats involves many measures, such as maintaining vegetated riparian zones within at least 100 feet of a perennial watercourse; placing septic systems at least 100 feet away from a stream; controlling stormwater runoff from urban areas such that erosive torrents and pollutants do not reach streamcourses; using erosion control measures during construction, excavation, and filling activites; maintaining natural flow regimes below dams; providing a way for fishes and other aquatic animals to move freely past dams and culverts; proper treatment of water flowing through waste-water treatment facilities and industrial sites; and prohibiting the introduction of (and/or removing) non-native organisms. Additionally, timber harvest should not occur within about 50 feet of a stream, with wider buffers used on steeper slopes, and cuts should be restricted to a light partial harvest in areas 50 to 100 feet from a stream. The uncut and lightly harvested areas will protect habitat for aquatic and riparian organisms and serve as corridors connecting uncut patches of forest. Additionally, stream-side vegetation is important because fallen woody debris is a crucial habitat component for trout and other fishes. For the most effective conservation of stream-associated wildlife, intensive clearing and development should not occur within at least several hundred feet of a stream edge.

River conservation in Connecticut recently was given a boost. On 25 October 2001, the National Oceanic and Atmospheric Administration (NOAA) and Trout Unlimited announced a new partnership to restore habitat vital to the conservation of America's coastal fisheries. Under the partnership, NOAA will provide up to $1 million over a three-year period in support of Trout Unlimited habitat restoration projects. The NOAA/Trout Unlimited partnership, managed through NOAA's Community-Based Restoration Program, already has provided $210,000 in first-year funding to support Trout Unlimited fish habitat projects selected in Embrace-A-Stream and other Trout Unlimited coastal fishery programs. Projects initially selected for funding included the Norwalk River in Connecticut.

Additional river conservation topics are discussed in the following section on the Connecticut River. Local river-oriented conservation activities in Connecticut are too diverse and extensive to review here. To learn more about river conservation and to get involved, contact your local watershed association.

The Connecticut River

Physical and Hydrological Aspects

After Long Island Sound, the Connecticut River is the state's most significant body of water. The river took its present form several thousand years ago at the end of the Pleistocene ice age. The river begins just above the Fourth Connecticut Lake (actually a small pond) near the Canadian border in northern New Hampshire, forms the border between New Hampshire and Vermont, and flows through Massachusetts and Connecticut to its mouth at Long Island Sound between Old Lyme and Old Saybrook, approximately 410 miles from the headwaters (Fig. 4.17). The Connecticut is the longest river in New England, but not quite as long as Lake Michigan or the Mississippi River from its mouth to the southern border of Arkansas. Thirty-six major tributaries join the river, the largest being the Chicopee, White, Deerfield, West, and Farmington rivers; only the last one named is in Connecticut. The drainage basin has an area of 11,760 square miles.

With an average flow of 20,000 cubic feet per second, the Connecticut's discharge is comparable to that of the Hudson and Delaware rivers. On the United States east coast, only the Susquehanna River carries a larger volume of water. The river provides nearly 70 percent of the freshwater input to Long Island Sound.

The highest point along the river valley is Mt. Ascutney (3,144 feet) in southern Vermont. The elevation at the headwaters is 2,650 feet, but the river drops two-thirds of its elevation in the first 60 miles. River elevation is only 700 feet as it passes the White Mountains in northern New Hampshire. In Connecticut, the river is only 40 feet above sea level at the Connecticut-Massachusetts state line, and the river is predominantly placid as it drops to sea level at Long Island Sound, except at Enfield where there is a small rapid.

The mouth of the river is about 4,350 feet wide between Lynde Point and Griswold Point (Figs. 4.18 and 4.19). North of the mouth region, maximum river width is 2,100 feet at Longmeadow, Massachusetts.

The river is fairly straight north of Hartford, quite curvy between Hartford and Middletown as it meanders over its floodplain, and less curvy south of Middletown, where the river leaves a broad valley and cuts southeastward to Long Island Sound through a hilly region.

Tidal influence extends as far north as Windsor Locks, about 56 miles north of the river mouth. Mean tidal range is

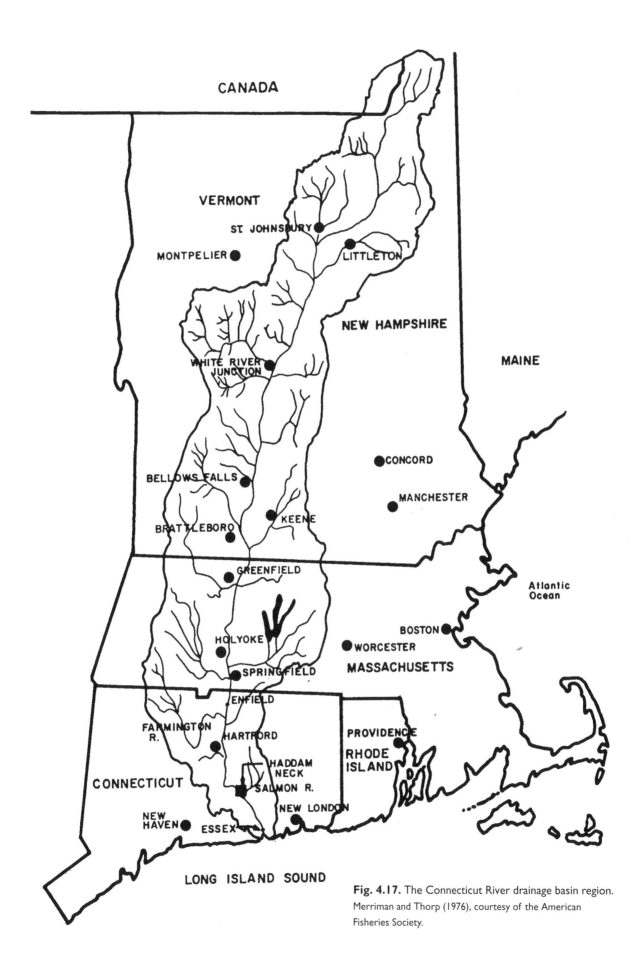

Fig. 4.17. The Connecticut River drainage basin region. Merriman and Thorp (1976), courtesy of the American Fisheries Society.

Fig. 4.18. Aerial photograph of the mouth of the Connecticut River in 1991.

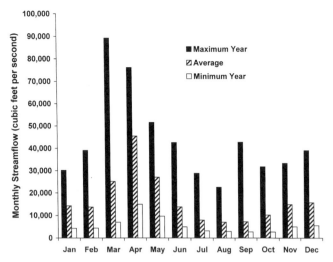

Fig. 4.20. Monthly streamflow of the Connecticut River at Thompsonville, in north-central Connecticut. Data from U.S. Geological Survey.

Fig. 4.19. Mouth of the Connecticut River and Long Island Sound, viewed from the north. Paul J. Fusco/Connecticut DEP Wildlife Division.

about 2 feet near the river mouth and 1 foot at Hartford; during periods of low flow, tidal range may be as much as 3.5 feet at the mouth. Slack tide occurs about 3.5 hours later at Hartford than at the mouth.

In most years, maximum flow occurs between late March and early May, and minimum flow is in late summer or early fall (Fig. 4.20). Above the tidal section, at Thompsonville, Connecticut, the recorded daily flow has varied between the extremes of about 282,000 cubic feet per second (cfs) (March 1936 flood) and 950 cfs (October 1963). Very high flows sometimes occur after heavy rains in late summer and early fall, or during winter thaws when melting snow may be accompanied by abundant rainfall. Tidal changes affect the flow in the lower river. The influence of the tide is greatest during periods of low flow.

Salinity incursions at low flow occur north to around the Route 82 bridge at East Haddam. During high flows, salinity may drop to zero parts per thousand at all river depths as far south as the I-95 bridge near the river mouth, even at high tide.

Water surface temperature reaches a peak of about 80 to 84°F in summer and drops to just above freezing in winter. In recent years in Haddam, the surface froze all the way across the river during the coldest periods of the coldest winters.

Dams and Flood Control

The Connecticut River, like most rivers, periodically spills over its banks and floods adjacent lands. Floods often are viewed as natural "disasters," but in fact they are a normal, natural phenomenon. Among other functions, flooding results in the development of unique ecological systems with significant economic and recreational value, affects local hydrology, facilitates the movements of fishes and other organisms, creates breeding habitat for various animals, and provides nutrients to soils used for agriculture.

Alteration of the river's flow began in the 1790s when the first dam was built at Turner's Falls, Massachusetts. The first (and only) dam on the river in Connecticut was built at Enfield in 1829. Today there are 16 functional obstructions on the river between its mouth and origin, impounding nearly 200 miles of its length, and many others on tributaries, for a total of nearly 1,000 dams in the watershed. Many former rapids are now dam sites or drowned beneath reservoir waters.

These dams blocked fish access to upstream spawning areas and caused a severe reduction in populations of Atlantic salmon, American shad, and other migratory species. The installation of fish ladders at several dams in the 1970s and 1980s has made it possible for these fishes once again to ascend the river and some major tributaries.

Flooding along the Connecticut River, especially in 1927, 1936, 1938, and 1955, following simultaneous spring snow melt and heavy rain, or associated with hurricanes, led to the installation of extensive flood control structures beginning in the late 1930s, with further measures added in the early 1960s. Currently, flood control systems along the Connecticut River drainage include 16 major dams on tributary streams and seven major projects in river valley cities. The reservoirs are controlled and coordinated by the U.S. Army Corps of Engineers in Waltham, Massachusetts. Flood control reservoirs reduce flooding and may also provide public water supplies and hydropower. They replace river-oriented recreation and fisheries with lake-associated activities. Dams may allow development of downstream cold-water fisheries.

Flood control reservoirs generally function by storing water during periods of heavy runoff and releasing it more gradually at a later time when flow normally would be low. In other words, the dams dampen the fluctuations in stream flow, lowering the peaks and increasing the basal flow rate. The Connecticut River still rises onto its floodplain in Connecticut in some years (Fig. 4.14), but the frequency, height, and duration of flooding have been reduced.

Flood control dams are severely destructive to natural riverine ecosystems, with impacts extending far above and below the dam. Riverine organisms are impacted by the dam as a physical barrier to movement and by altered stream flow, temperature, and turbidity. These changes may interfere with the normal life cycles of stream organisms. For example, the water coming out of a reservoir may be too cold for fish or mussel reproduction downstream. Organisms that depend on periodic flooding are detrimentally affected by typical management protocols. Fishes, frogs, and toads that breed in streamside flood pools may have their breeding and nursery habitat eliminated. The habitat of the endangered puritan tiger beetle has been degraded or destroyed by reservoirs and shoreline alterations. Availability of sandy riverbank habitat has been reduced by flood control dams that reduce spring flooding and allow vegetation to develop on sandy river banks and beaches. Eastern cottonwood regeneration is now limited because dams have reduced the natural flooding that creates the sediment bars on which cottonwood seedlings often become established. Because of fluctuating water levels, flood control reservoirs often lack a marshy littoral zone and biologically are relatively depauperate compared to natural lakes.

Local flood control projects in cities along the Connecticut River include a system of levees and dikes, which simply prevent rising water from flowing into areas on their landward side, as long as the water level does not overtop the structures. The use of levees in populated areas is favorable in that fewer dams are needed upstream and, because cities are already heavily developed, local environmental damage and alteration are minimized. Problems with levees include: possible buildup of water on the landward side of the levee during local storms (pumps are used to move water past the levee and into the river); increased upstream flooding due to a bottleneck effect when levees are built on both sides of the river; increased accumulation of sediments (under certain conditions) due to channel narrowing (sediments no longer can spread out over a wide floodplain, which raises the river bed and makes the levees less effective); and potential for locally severe flood effects should a dike or levee breach and allow large volumes of high-velocity water to flow into a nominally flood-protected area.

Hartford has an extensive system of earthfill dikes, concrete floodwalls, and several pumping stations. In addition, various conduits and diversion channels move flood waters under or around the city. Several cities north of Hartford also have installed similar local flood control structures. These devices have been effective in greatly reducing damage during major floods. Deepening, widening, and/or straightening of the river channel fortunately are not important flood control measures along the Connecticut River.

In the 1960s, additions to the Connecticut River flood control system, consisting of seven new dams, were proposed. These dams would have allowed river cities to avoid inundation even in the event of a worst-case scenario flood (one in a thousand years). The dams were never built, due largely to the great environmental damage that would result. Instead, raising the elevation of the dikes has been viewed as a more viable alternative if structural protection is desired. Actually, improved flood warning systems, restriction of development in the lower floodplain, and the preservation of natural flood-water storage areas (wetlands) probably are the best ways to cope with floods, and these methods have been promoted in recent years.

Many dams south along the Connecticut River to Holyoke, Massachusetts, contain turbines that use flowing water to generate electricity and, at least historically, to power mills and other industry.

Quabbin Reservoir, on the Swift River of the Chicopee River drainage, stores and diverts water from 86 square miles of the Connecticut River watershed for use in the greater Boston area.

Power Plant Impacts

A nuclear power plant at Vernon in southern Vermont and a few fossil fuel steam-electricity power plants farther south in Massachusetts and Connecticut affect the river primarily by drawing water from the river for cooling and by releasing warm water. Up to 6 percent of the river water flowed through the Haddam Neck, Connecticut, nuclear power plant that was shut down in July 1996. The heated effluent from a power plant can be much warmer than river temperature. It was a strange sensation to paddle through the steamy warm effluent at Haddam Neck on a cold day.

Power plant cooling systems suck in and kill (through heat and mechanical damage) many tiny fishes (e.g., alewife and blueback herring), fish eggs, and zooplankton. However, ecological studies conducted from the mid-1960s to the early 1970s concluded that the power plant at Haddam Neck had no major impact on the river biota. Migratory fishes, such as American shad, appeared to avoid the warm-water plume coming from the power plant and into the river. Several species of native and introduced fishes may be attracted to and abundant in the heated water leaving power plants.

A certain amount of radioactive material was released into the river via the warm-water discharge canal at Haddam Neck, but this was reported to be of sufficiently low level as to be harmless to humans.

Bridges and Canals

Considering its entire length in New England, about 80 bridges cross the Connecticut River. The first one was built in 1784 at Bellows Falls, Vermont, across to New Hampshire. Twenty of the bridges carry only railroad traffic. Picturesque covered bridges (often built of eastern white pine) span the river in a few places north of Massachusetts.

In some areas, canals parallel the river. Large canals that formerly serviced mills are still conspicuous in Holyoke, Massachusetts. In Connecticut, the Windsor Locks canal, completed in 1829 and now existing as a nonfunctional historical structure, was built to carry boat traffic around the Enfield rapids.

Boat Traffic

Until Adriaen Block sailed up the river in 1614, only small Native American boats graced the river. Sailing vessels became numerous in the river during the colonial period. The quiet was broken by gunfire during the American Revolution and vanished as regular steamboat service on the Connecticut River began in 1823 and continued until the early 1930s.

Today, boat traffic consists of many pleasure vessels in summer and oil barges year-round. And many canoeists and kayakers enjoy the many coves and side channels along the lower river. Good paddling sometimes exists on the river itself, but in the warmer months, large numbers of power boaters make river paddling hazardous, and summer boat traffic is a continual source of disturbance to water birds. Some high-powered watercraft emit an ear-splitting roar that disturbs both people and wildlife. Large waves generated by power boats crash onto shore and can kill rare, delicate dragonflies emerging from the water onto sandy shores in summer (see Chapter 14). Heavy recreational use of sandy beaches and sand bars by boaters disrupts or eliminates bird nesting and resting areas, and tramples populations of some of the river's rare insects (e.g., puritan tiger beetle, various clubtail dragonflies). Oil barges and Coast Guard ice-clearing activities periodically disturb the bald eagles and ducks that feed and rest along the river in winter.

The oil barges carry home- and factory-heating oil, diesel and jet fuel, gasoline, coal, and liquid asphalt to several ports of delivery between Middletown and East Hartford. Barge traffic is heaviest in fall and winter. A tug pushes a loaded barge upriver and pulls it empty downriver. A qualified captain, tested and certified specifically for the Connecticut River, is required to pilot the tug-barge tandem on the river, where rock outcroppings, submerged shoals, an often-narrow channel, and tricky currents make navigation challenging. The Coast Guard monitors all oil barge traffic, and the state Department of Environmental Protection (DEP) also regulates the transport and delivery of petroleum products on the river. Booms are deployed around the cargo at the delivery point, as required by the Oil Pollution Act of 1990. The DEP, together with voluntary cooperatives, have containment and cleanup materials in readiness should a spill occur.

In addition to petroleum products, the river has served as a transportation corridor for brownstone, cut from the quarry at Portland beginning in the early 1800s, and logs, cut in northern New England and rafted down the river in the 1800s and early 1900s. Big-log drives ended by early 1900s with the construction of hydroelectric dams. Smaller pulpwood logs were driven on the upper river until the 1940s. By the mid-1800s, railroads reduced the importance of the river for commercial transportation.

Sewage Treatment

An increasing number of sewage treatment facilities have appeared along the river in recent decades. As an example of the functioning of these facilities, the Middletown Water Pollution Control Plant processes millions of gallons of sewage each day, using an activated sludge process that incorporates settling tanks for removal of solids, decomposition of organic matter by microorganisms in aerated tanks, and chlorination of the effluent before it is released into the river. Removed solids are moved by rail to the Mattabassett District Treatment Plant for incineration. Sewage treatment plants have functioned importantly in transforming the river from a smelly conduit for raw sewage to a stream with swimable water, though at times I have detected a distinct sewer odor as I paddled south of Middletown. Recent efforts to upgrade sewer systems such that storm water runoff does not overwhelm sewage treatment plants should alleviate the problem.

Dredging

The U.S. Army Corps of Engineers periodically arranges for dredging of the navigation channel in the river to maintain a minimum depth of 15 feet at low tide (some of the oil barges, when fully loaded, draw 12.5 feet). Historically, some of the coves and anchorages in the lower river were dredged. The impacts of today's dredging in the Connecticut River are not well documented but appear to be negligible (Fig. 4.21). Formerly, dredged sediments were deposited on wetlands. Fortunately, that destructive practice is no longer permitted. Sediments dredged from the river and from marinas are instead deposited in designated sites in Long Island Sound. In 2001, the U.S. Environmental Protection Agency and U.S. Army Corps of Engineers were developing an environmental impact statement concerning the designation of one or more disposal sites in Long Island Sound (see www.epa.gov/region01/lisdreg).

Other Connecticut rivers also have been dredged. For example, Navy contractors dredged the Thames River in 1995 and 1996 to accommodate huge Seawolf-class submarines. About 900 million cubic yards of dredged material was deposited at a site in Long Island Sound.

Fig. 4.21. Dredgers periodically deepen the Connecticut River's navigation channel.

Invasive Plants

The Sylvio Conte National Fish and Wildlife Refuge recently obtained a grant from the National Fish and Wildlife Foundation to work with partners to develop and coordinate an ecosystem-wide strategy for monitoring and controlling invasive plants. The strategy is set forth in "The Connecticut River Watershed/Long Island Sound Invasive Plant Control Initiative Strategic Plan."

Non-native aquatic plants that are widely recognized as being invasive throughout the watershed include fanwort, Brazilian elodea, Eurasian water-milfoil, curly pondweed, and water chestnut (Fig. 4.22). In the 1990s, water chestnut became established in the Connecticut River system at Holyoke (Massachusetts) and in the vicinity of Hartford and Glastonbury. In August 2001, I found (and removed) a single fruiting plant in the Connecticut River in Haddam. Water chestnut has also been found in northeastern Connecticut. This Old World native lives in shallow water of lakes, ponds, and slow streams. It is an annual that spreads easily by water- or animal-transported seeds or when dislodged plants move with water currents. The seeds are viable for up to a year. Water chestnut, not to be confused with a different plant of the same name used in cooking, already has demonstrated that it can dominate aquatic ecosystems and choke out native species.

The Connecticut Department of Environmental Protection has initiated an aggressive eradication campaign against water chestnut, involving removal by machine and by hand, but it may take several years to eliminate the known populations of this prolific plant. Any sightings of water chestnut in Connecticut should be reported to the DEP's Office of Long Island Sound Programs in Hartford

Fig. 4.22. Water chestnut is an invasive exotic plant in the Connecticut River system. A spiny fruit is shown at bottom center.

Fig. 4.23. Recent marsh restoration activities involving removal of phragmites are evident as light areas in this aerial photograph of Lord Cove. Same area as shown in figure 3.30. Paul J. Fusco/Connecticut DEP Wildlife Division.

(http://dep.state.ct.us/olisp1). It is hoped that any new populations can be eradicated before they become too large and difficult to control.

Wetland plants widely recognized as invasives in the Connecticut River drainage include purple loosestrife and common reed (Fig. 3.30). The Connecticut Botanical Society has conducted a survey of purple loosestrife along the Connecticut River and is assisting the Connecticut Department of Environmental Protection in collecting information on invasive species and rare plants. Research on biological control of purple loosestrife is being carried out through cooperative efforts between the University of Connecticut Department of Plant Science and USDA Animal and Plant Health Inspection Service Cooperative (APHIS) (Agricultural Pest Survey).

Partnerships among the Connecticut Department of Environmental Protection Office of Long Island Sound Programs, Connecticut Department of Environmental Protection Wetlands Habitat and Mosquito Management, U.S. Fish and Wildlife Service Partners for Wildlife Program, the Connecticut Chapter of The Nature Conservancy, local towns, and private landowners have been formed to share the on-going work to control the spread of common reed in the tidal wetlands of the Connecticut River (Fig. 4.23).

International and National Importance

The international importance of the wetlands of the lower Connecticut River (Fig. 4.24) was formally recognized

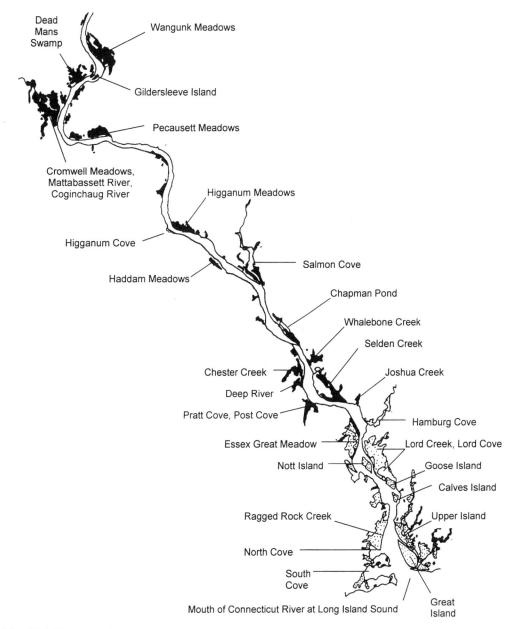

Fig. 4.24. The lower Connecticut River and its tidal wetlands contribute uniquely to the state's biological diversity. Stipple: estuarine marsh. Black: freshwater tidal wetlands. Modified from Metzler and Tiner (1992).

in 1994 under a designation of the "Convention on Wetlands of International Importance especially as Waterfowl Habitat," a treaty involving 81 nations. The designation does not impose any specific restrictions or regulations. It serves mainly to emphasize the importance of the ecosystem and directs that the ecological character of the system should not be altered or destroyed.

On 30 July 1998, President Clinton designated the Connecticut River as an American Heritage River, one of just 14 rivers honored nationwide. The designation is the result of a valley-wide nominating effort led by the Connecticut River Watershed Council and supported by over 250 communities and institutions and by the Congressional delegation and governors of the four watershed states.

The Sylvio Conte National Fish and Wildlife Refuge was recently established along the Connecticut River. Administered by the U.S. Fish and Wildlife Service, the refuge includes an array of small areas of critical habitat along the whole length of the river and operates in conjunction with public and private efforts to protect and enhance the quality and natural functioning of the river ecosystem.

The Connecticut Chapter of The Nature Conservancy (TNC) has embarked on a major program ("Tidelands of the Connecticut River") involving land acquisition, habitat restoration, environmental monitoring, and research along the lower Connecticut River. TNC already has protected several important parcels along the tidal section of the river.

A "Gateway" conservation zone along the lower Connecticut River encompasses riverside portions of Chester, Deep River, East Haddam, Essex, Haddam, Lyme, Old Lyme, and Old Saybrook. Public Act 73–349 provided for the protection of this area through a combination of state and local action. This measure was developed as an alternative to a proposal to establish a "Connecticut River National Recreation Area" under the U.S. Department of the Interior. Many residents of the lower valley feared that a federally sponsored recreation area would result in too many tourists, consequent degradation of river beauty and natural resources, and loss of local autonomy. In response to these objections, the national recreation area idea was abandoned and the bill passed by the state legislature provided for the protection of the area through a regional conservation compact among the towns and through state purchase of scenic easements and development rights. Most of the responsibility for planning, establishing, and administering the program was left to the eight towns named above. Detailed plans were worked out by a Gateway Committee, which consisted of one representative from each town, one from each of the two regional planning agencies, and one representative of the state Department of Environmental Protection.

The Connecticut River Estuary Regional Planning Agency (CRERPA) was established almost 30 years ago to serve the nine towns in the lower Connecticut River region. Under state law, towns in the state's fifteen planning regions have organized regional planning agencies to act as a forum for addressing issues of regional concern and to prepare a Regional Plan of Development.

Epilogue

Native Americans lived along the Connecticut River for several thousand years prior to the arrival of the first Europeans. They used the river as a source of food, especially during fish spawning migrations, and for transportation, but their impact on the river ecosystem probably was minimal. When Adriaen Block sailed upriver as far as the rapids at Enfield in 1614, several years before the arrival of the *Mayflower* at Plymouth Rock, he experienced a clean, free-flowing river. Block reported that few people lived near the river's mouth, but an agricultural community and a fortlike village occupied the area near present-day Hartford. Several thousand Native Americans lived along the Connecticut and Massachusetts sections of the river.

Following Block's voyage, the river ecosystem took a severe battering that lasted three and a half centuries. The river was dammed, polluted, and degraded by non-native invasive plants, and the river's floodplain forests and meadows were largely supplanted by a flood of culture and agriculture. A fortunate result of the deplorable condition of the river by the mid-1900s was increasing recognition and appreciation of its remaining natural values and a realization that something could be done to restore them. Thanks to efforts by the Connecticut River Watershed Council, The Nature Conservancy, U.S. Fish and Wildlife Service, Connecticut Department of Environmental Protection, local towns, and other groups and individuals too numerous to mention, vast improvements have been made in water quality, and further protection and active restoration of the river and its watershed are now in full swing. Instead of inheriting a squandered resource, our descendents can look forward to many fine days spent boating, bird-watching, fishing, hunting, crabbing, or otherwise enjoying the riches of the Connecticut River.

Chapter 5

Lakes and Ponds: Lacustrine Ecosystems

From land, an open expanse of water is a compelling view, and the appeal of the scene is only enhanced when green forested hills surround the blue liquid. Thus lakes and ponds, or lacustrine ecosystems, are not only ecologically unique (as we shall see) but also sublimely attractive.

Physical Characteristics

As defined by ecologists, lacustrine systems are large bodies of nonflowing water, including not only lakes but also reservoirs, large ponds, and similar nonsaline bodies of water with a surface area of more than 20 acres and a maximum depth of more than 6.6 feet (2 m) at low water (Fig. 5.1). Ponds and wetlands that are smaller than 20 acres are included in the lacustrine system if the shoreline is dominated by bedrock or if the pond or wetland has an active wave-formed shoreline, or if maximum water depth is more than 6.6 feet at low water. Deep water lakes and ponds are ecologically very different from shallow, thickly vegetated marshes, but these ecosystems clearly grade into one another in the form of marshy ponds and shallow lakes. The quantitative thresholds for defining lacustrine systems are somewhat arbitrary but provide a convenient, objective means of subdividing the continuum into practical units for discussion, conservation, and other uses.

Lacustrine ecosystems encompass open water areas and beds of submerged, floating, or nonpersistent emergent vegetation. Shallower bodies of water, or lakeside or pond-like areas dominated by emergent plants that remain standing through winter, are ecologically similar to marshes and are included in the palustrine ecosystem (see next section), but for convenience I mention some of the representative emergent-zone biota here as well.

Most of Connecticut's natural lakes consist of preglacial bedrock basins that were dammed by glacial drift, but some ponds are glacial kettles—areas where a massive chunk of ice was buried in sediment, forming a water-filled, kettle-shaped depression after the ice melted.

Lakes remain lakes because they receive water from some outside source, such as a stream or groundwater discharge (or, to a much lesser extent, overland water flow or rain). Lakes maintain a relatively stable water level when these inputs match water losses from stream outflow, direct evaporation, evapotranspiration (loss of water from aquatic plants), and groundwater recharge.

Connecticut's relatively small number of natural lakes and large ponds (about 70) are now outnumbered by at least 85 large reservoirs created by the damming of streams. There are over 200 water supply reservoirs in the state. Counting small ponds and reservoirs, Connecticut has nearly 20,000 natural and human-made lakes, ponds, and reservoirs. Litchfield County's Bantam Lake, with a surface area of 916 acres and a maximum depth of less than 25 feet, is the largest natural lake. Candlewood Lake in Fairfield and Litchfield counties, with a surface area of 5,420 acres, is the largest reservoir.

Reservoirs with stable water level generally are similar ecologically to natural ponds and lakes. Large reservoirs with

Fig. 5.1. East Twin Lake in Salisbury is a large, hard-water lake with a maximum depth of about 75 feet.

major seasonal or annual changes in depth, such as those that serve as a source of drinking water, tend to be biologically depauperate. They do not develop a rich, marshy, shallow-water zone and, with their steep barren shores, can be downright ugly at low water.

Most ponds and lakes are ice-covered for at least part of the winter. The ice may last most of winter in the coldest years and at higher elevations, but in recent years winter ice has been short-lived in much of the state.

Water temperature generally is just a few degrees above freezing from top to bottom in winter, when the low angle of the Sun and low air temperatures do not produce any significant warming. During the coldest periods, surface waters are at or very near freezing, and water temperature may increase gradually to about 39°F near the bottom (water is densest or heaviest at this temperature).

In early spring after ice-out, as air temperatures rise and the Sun begins to exert more influence due to its higher position in the sky, surface waters gradually warm several degrees and spring winds mix water thoroughly such that temperature remains fairly uniform at all depths. By late spring, the temperature in the first several feet near the surface rises significantly above that at deeper levels, due to greater effect of the Sun and warmer air on surface waters, acting in conjunction with reduced winds and thus less mixing. Also, warmer surface water is less dense than deeper colder waters, so mixing is inhibited. Temperature stratification, with a distinct thermocline (zone of rapid temperature change with depth), develops fully and is maintained throughout summer as warm air and strong sun keep surface waters warm (Figs. 5.2 and 5.3).

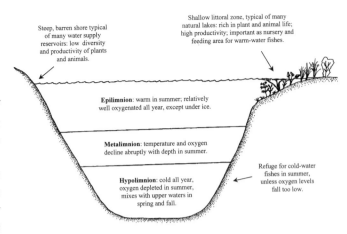

Fig. 5.3. In summer, lakes become thermally stratified. Not drawn to scale; vertical dimension is exaggerated.

In autumn, reduced air temperatures and a weaker (low angle) Sun result in cooling of surface waters. Cooled water has increased density, so it sinks and mixes deeper into the lake, aided by autumn winds. Throughout autumn, the water continues to mix and cool, with temperatures roughly the same from top to bottom. By winter, the water is again just a few degrees above freezing at all depths.

Despite the fact that warm water holds less dissolved oxygen than cold water, the colder, deeper parts of lakes may become depleted of oxygen for much of the summer. Microbial decomposition of organic material that sinks to the bottom consumes available oxygen, and lack of mixing in summer means that oxygen diffusing into surface water from the air is not available to replenish oxygen supplies in deeper water. The dim light in deeper waters prevents significant oxygen production by any photosynthetic phytoplankton that may be present.

Autumn mixing (described above) and reduced decomposition resulting from cold temperatures may eliminate the anoxic conditions that develop in summer, but low oxygen levels may develop again in winter in an ice- and snow-covered lake as a result of the "sealed" conditions that restricted oxygen replenishment from the atmosphere and the low rate of photosynthesis (due to dark conditions), combined with oxygen use by organisms. Such oxygen depletion ends with ice melt and spring mixing. These patterns of seasonal change in temperature and dissolved oxygen are typical of many relatively deep productive ponds and lakes, but not all lakes exhibit these exact characteristics.

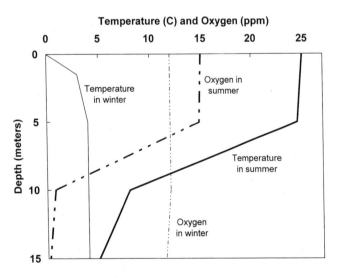

Fig. 5.2. Temperature and oxygen conditions in deep lakes and ponds vary greatly with season and depth. Not all lakes conform precisely to these values, but most lakes fit the general pattern. After Deevey (1951).

Lakes vary in their chemical characteristics, depending on local geology, atmospheric inputs (e.g., acidity of rainfall), and the quality of inflowing streams and groundwater, and this variation affects a lake's plant and animal life. At one extreme are soft-water lakes, which are acidic (pH usually less than 7) and contain low amounts of dissolved calcium (less than 10 mg of calcium per liter) and other dissolved salts.

At the other end of the spectrum are hard-water lakes, with basic water (pH above 7) that is rich in salts (more than 25 mg of calcium per liter). In Connecticut, most lakes and ponds have a pH of 6.0 to 8.0, with a minority more acidic or basic. Calcium concentration ranges from less than 1.0 mg/l to more than 40.0 mg/l. Hard-water lakes are typical of the limestone region of western Connecticut.

All else being equal, hard-water lakes tend to have greater plant and animal diversity and biomass than do soft-water lakes. However, other nutrients, particularly nitrogen and phosphorus, can have dramatic effects on the quantity and diversity of lake biota. Low availability of these nutrients can limit lake productivity, whereas abnormally high inputs, such as may result from runoff or groundwater discharge containing fertilizers or sewage, often yield undesirable algal (cyanobacteria) blooms or excessive growth of submerged and floating water plants.

Vegetation

Canoeists paddling through beds of floating and submerged plants (Fig. 5.4) are likely to find duckweeds (Lemnaceae), yellow water-lily, white water-lily, water-shield, floating-heart, pondweeds, bladderworts, hornwort, fanwort*, naiads (annuals that appear in late summer), water-celery, waterweed, water-milfoils(*), manna grass, and algae. In calcareous lakes and ponds in western Connecticut, astute botanists might recognize Hill's pondweed and slender pondweed.

Fig. 5.5. The margins of lakes and ponds with stable water levels support diverse plant life, which in turn provides important feeding, breeding, and resting areas for numerous animals. In contrast, water supply reservoirs are subject to seasonal draw-downs and generally have barren margins that support fewer wildlife species.

Emergent and shoreline plants often present a boater with a visual feast of flower color and leaf shape (Fig. 5.5). Characteristic species include arrowheads, three-way sedge, spikerush, three-square, bur-reeds, woolly sedge and other bulrushes, twig-rush, smartweeds, pickerelweed, arrow-arum, meadow-beauty, yellow-eyed grass, pipewort, New York ironweed, water lobelia, cattail, marsh fern, royal fern, and water-willow. Most of the herbaceous plants wither in autumn and regrow in spring and summer. These plants define and are most numerous in the littoral zone—the shallow area with good light penetration around the margin of a lake or deep pond. As indicated previously, "ponds" or "lakes" that have these plants extending virtually all the way across them, due to shallow water, are regarded by ecologists as palustrine systems or marshes. Woody plants that grow along pond and lake shores are mentioned in the shrub swamp section of Chapter 6.

Though detailed treatment of microscopic life is beyond the scope of this book, it is important to realize that diatoms, algae (especially green algae, Chlorophyta), chlorophyll-containing flagellates, cyanobacteria, and other microorganisms that coat rocks, bottom sediments, plants, and other submerged objects, and phytoplankton free-floating in the water, are important producers and food sources for small animals in lacustrine ecosystems. Planktonic algae change seasonally; dominant kinds may include golden-brown algae in the colder months, green algae in early summer, and cyanobacteria (blue-green algae) in late summer.

Fig. 5.4. The shallow portions of lakes and ponds often contain biologically rich beds of emergent, floating, and submerged vegetation. Here soft-stem bulrush emerges above white and yellow water-lilies.

Fauna

Invertebrates. Careful searching of the surface of a lake or pond, especially near beds of aquatic plants, may yield tiny scavenging springtails and more obvious water striders and whirligig beetles. In summer (by late June) a canoe may stir up water-lily leaf beetles and numerous small white moths from atop lily pads, along with many predatory adult damselflies and dragonflies. The water-lilies are essential as food for the beetles and moth larvae.

In the water column are minute, phytoplankton-eating copepods and cladocerans. Water boatmen, backswimmers, giant water bugs, and diving beetles are common predatory swimmers.

If you study samples collected from lake bottoms, or carefully examine the bottom while snorkeling, you likely will find filter-feeding freshwater mussels (*Anodonta*) and smaller fingernail clams; grazing snails (abundant in hard-water lakes in western Connecticut); the predatory immature stages of dragonflies, damselflies, alderflies, and beetles; parasitic or predatory leeches; larval mayflies of various food habits; herbivorous or detritus-feeding midge larvae; grazing, herbivorous, or filtering caddisfly larvae; scavenging amphipods and water mites; omnivorous crayfishes; detritus-feeding oligochaete worms; and scavenging or micropredatory flatworms.

Invertebrates in beds of submerged and floating aquatic vegetation (Fig. 5.6) in lakes and ponds are basically the same as those in tidal freshwater marshes (see Chapter 4).

Samples from lakes and ponds viewed under a microscope reveal a biologically diverse world of tiny organisms, but these—protozoa, rotifers, and such—are beyond the scope of this book.

Fishes. The fish fauna of Connecticut's lakes and large ponds has been highly altered by the introduction and establishment of non-native species. Today, the fishes most likely to be found in lakes include largemouth bass*, smallmouth bass*, bluegill*, pumpkinseed, black crappie*, golden shiner, yellow perch, white perch, brown bullhead, white catfish*, northern pike*, chain pickerel, common carp*, brook trout, rainbow trout*, brown trout*, and kokanee*. Some of these, including largemouth bass and yellow perch, are ubiquitous, whereas others, such as pike and kokanee, inhabit just a few stocked lakes and reservoirs. Most of the introduced species are predatory sport fishes that feed on other smaller fishes and invertebrates.

The majority of these fishes are warm-water species that do well in Connecticut's many warm lakes and ponds. Trouts and other salmonid fishes tend to be in cold-water lakes, typical of higher elevations in northwestern Connec-

Fig. 5.6. These water-lily leaves exhibit heavy use by pyralid moth caterpillars (large oval holes), water-lily leaf beetles (small holes), and other animals.

ticut, and in large, deep "two-story" lakes that in summer have adequate deep, cold-water habitat. However, these cold-water fishes may die off if oxygen levels are depleted by decomposition of excessive algal growth (see "Conservation"). Warm-water fishes reside at shallower depths in temperature-stratified lakes.

The bottom-dwelling ichthyofauna of warm lakes, which tend to be turbid, may be dominated by catfish(*), carp*, and suckers that do not rely on sight for feeding, whereas vision-feeders such as salmonids often live at great depths in deeper lakes with clear, cool water.

Compared to natural lakes, reservoirs usually are subject to large water level fluctuations, and these tend to have depauperate fish populations. Changes in water level may expose shallow-water fish nests or submerge them under unsuitably deep water, interfering with reproduction. Unstable water level also inhibits the establishment of a vegetated littoral zone, which provides important spawning sites, nursery habitat, and food resources. Large drops in resevoir level can block access to tributary streams that may be essential for spawning of lake-dwelling fishes.

Fish populations often have major impacts on zooplankton and coexisting fish populations. For example, fish predation generally eliminates large, conspicuous cladoceran (water-flea) species and favors smaller species with protective spines. And competition among similar species often imposes restrictions on the habitat use and prey of the coexisting fish species. Thus introductions of exotic fishes can have large and sometimes unpredictable consequences for lake ecosystems.

Amphibians. Because many amphibians cannot withstand fish predation and require higher food densities than are available in open lake waters, salamanders and frogs tend not to be dominant members of the open-water lake fauna, but they may thrive where marshy edges provide adequate cover and sustenance. The amphibians most likely to be seen include eastern newts, pickerel frogs, green frogs, and bullfrogs—species whose skin secretions make them at least somewhat unpalatable to fishes. As adults, all of these depend primarily on invertebrates for food. Larval frogs feed on algae, detritus, and other minute organisms.

Reptiles. Many lakes and large ponds are inhabited by omnivorous snapping turtles, abundant painted turtles (often basking on logs), spotted turtles, and stinkpots. The snake most likely to be seen is the harmless, fish- and frog-eating northern water snake, which basks on sunny shores, rocks, and logs. All of these are most common in marshy lake edges where food is plentiful.

Birds. Birds of lakes and ponds include an array of fish eaters such as common loons, pied-billed grebes, double-crested cormorants, great blue herons, great egrets (late summer), green herons, hooded mergansers, common mergansers, ring-billed gulls, ospreys, and belted kingfishers. Some birds, such as mute swans*, Canada geese, wood ducks, American black ducks, and mallards, feed on plant material, including seeds of wetland plants, whereas buffleheads and ring-necked ducks eat mainly invertebrates. Waterfowl often seek shelter under plants overhanging lake and pond edges. Open shallows and shores host foraging migrant shorebirds such as greater and lesser yellowlegs and solitary and least sandpipers, as well as migrant or locally nesting spotted sandpipers, all of which feed mostly on invertebrates. Scavenging crows, and sometimes ravens, also visit lacustrine shores. From spring through fall, swallows of several species forage on flying insects over water and sometimes dip in for a drink or quick bath. Of course, numerous additional birds frequent lakeside marshes and swamps. These are discussed in Chapter 6.

Mammals. The usual large mammals of major water bodies—beavers, muskrats, minks, and river otters—swim the open waters of lakes. Beavers and muskrats depend primarily on plant resources whereas the minks and otters are top carnivores that require plentiful animal prey. Muskrats also harvest and eat many freshwater mussels, evidenced by piles or concentrations of shells in shoreline middens. Omnivorous raccoons prowl shorelines. Star-nosed moles sometimes dive in shallows as they forage on aquatic invertebrates. Several bat species feed on insects above water in summer and dip down for a drink, especially just after leaving their daytime perch as they begin their nocturnal foraging.

Conservation

A natural lake with clean water and native biota is a rare thing. Globally and locally, most lakes have been degraded by a wide range of intentional and inadvertent human actions. Our remaining high-quality lacustrine ecosystems deserve high priority for protection.

The effects of dams on rivers are relatively well known, but dams also impact lakes and ponds. For example, the water level of more than 70 percent of Connecticut's natural lakes has been raised by damming, apparently under the presumption that if some water is good, more must be better. Fewer than 20 natural lakes and large ponds are unaltered by dams. Deepened lakes tend to have fluctuating water levels, barren shores, decreased native biodiversity, and/or high levels of disturbance from motorized watercraft and other human activity.

Lakes are affected by human activites in surrounding uplands, and these events may be recorded for posterity in lake-bottom sediments. Cores taken from bottom sediments of Linsley Pond in North Branford show that the initiation of farming in 1700 resulted in only minor changes in the lake biota. Changes in agricultural activity at a nearby farm in the early 1900s caused a rapid shift to eutrophic diatom and midge assemblages, likely related to increased nutrient runoff into the pond. In 1960, construction of suburban housing caused major changes in the zooplankton community as the pond became hypereutrophic (excessively fertilized). Subsequent studies have revealed increased urbanization, residential development, and consequent chemical and biological degradation in lakes in many watersheds in Connecticut over the last decade.

Excessive inputs of nutrients from agricultural, urban, or residential sources (such as faulty septic systems) into lakes and ponds (particularly small, shallow impoundments) commonly result in excessive growth of algae and aquatic plants such as Eurasian water-milfoil. These in turn often result in applications of chemical toxins by humans wanting to rid the water of the thick algae and plants. Use of chemicals to control organisms, including not only water weeds or algae but also mosquitoes or unwanted fishes) in the public or private waters of the state requires a permit from the Pesticides Management Division (Bureau of Waste Management, Department of Environmental Protection). Applications are evaluated on the potential for causing unreasonable adverse effects on humans or the environment. Only chemicals registered for aquatic sites may be used. (Researchers are assessing the possibility of using biological controls such as a moth *[Acentria ephemerella]* whose larvae feed on the growing tips of Eurasian water-milfoil, but so far this is only in the experimental stage).

Erosion associated with land development often affects lakes and ponds by increasing turbidity and sedimentation. Murky water reduces light penetration and can kill aquatic plants and reduce their productivity, as well as the productivity of the entire ecosystem. Turbidity also can interfere with the feeding of vision-dependent fishes, and sediments can smother fish eggs.

Lakeshore development not only may cause excessive inputs of nutrients, pesticides, and sediment but also often eliminates natural wetland vegetation and usurps habitat that would otherwise be used by many kinds of wildlife. Ecologically sound development leaves lakeside vegetation intact. Lake access for nearby homes is best limited to narrow paths.

Lake problems come not only from the land but also from the air. For example, most mercury contamination of the environment comes from regional airborne sources. Elevated and potentially hazardous levels of mercury in the tissues of fishes is a widespread problem in New England and New York. Acid deposition from atmospheric pollutants can harm fishes and their food resources in lakes and streams. Acidification also leaches from soils metallic toxins (e.g., aluminum, mercury) that are detrimental to fishes and other aquatic animals. Fortunately, lake acidification is not currently a significant problem in Connecticut.

Massive fish kills likely related to toxins sometimes occur in Connecticut lakes. A large-scale fish kill, primarily affecting yellow perch, occurred shortly after Lake Pocotopaug in East Hampton was treated for algae in June of 2000. Apparently an aluminum-containing product being used to treat the lake caused the kill. In late December 1999, large numbers of several fish species were found dead and dying in Lake Pocotopaug. No obvious cause of the kill was identified, but the affected fishes had mucus-covered gills. Perhaps the mucus suffocated the fishes, but the cause of the excessive mucus secretion was not determined.

Some fish kills in lakes result from oxygen depletion rather than chemical toxins. For example, a fish kill at Upper Bolton Lake in Coventry and Vernon, discovered in March 2000 when the lake's ice began to melt, affected 1,200 to 1,500 fishes, including bluegill*, pumpkinseed, largemouth bass*, and chain pickerel. The fish mortality was attributed to winterkill. Winterkill generally occurs in very shallow, nutrient-enriched ponds with an abundance of aquatic vegetation. When snow blankets an ice-covered lake of this type, sunlight penetration and plant photosynthesis are reduced while respiration by living organisms and decomposers keeps consuming oxygen. Dissolved oxygen levels in the water may fall below levels necessary for fish survival.

Water quality in lakes and ponds in Connecticut state parks is an ongoing problem. In recent years, several of the state's swimming areas were closed due to unacceptably high bacterial levels.

Exotic invasive organisms sometimes cause problems in lakes. As previously mentioned, prolific growths of Eurasian water-milfoil commonly choke shallow lake waters, interfering with recreational use and sometimes leading to oxygen depletion when the plants die and decompose. Recently, zebra mussels* were found in East and West Twin Lakes in northwestern Connecticut. Based on experience elsewhere in North America, this highly invasive, quick-spreading species has the potential to quickly increase its range and abundance and dramatically change the ecology of lakes and other inland waters where calcium (Ca^{2+}) concentrations exceed 12 mg/L. Every effort should be made to remove and kill any zebra mussels you find and to prevent their transportation on boats and trailers.

On the positive side, Connecticut now has its first "Heritage Lake." The 680-acre Lake Waramaug, in the towns of Kent, Warren, and Washington in Litchfield County, was designated a Heritage Lake by the Department of Environmental Protection in September 2000. The Heritage Lake program, created by 1999 legislation (Public Act 99-135) does several things. It makes designated lakes (which must be located in two or more municipalities) available for a program of preservation and enhancement of the historic, cultural, recreational, economic, scenic, public health, and environmental values; encourages partnerships and agreements with municipalities that are contiguous to the lake; directs programs, grants, and technical assistance to support the lake; coordinates state and municipal activities and resources to preserve, protect, and restore the lake and its shoreline; provides municipalities with access to existing scientific data and information relating to the lake; and cooperates with the municipalities to promote and encourage public use and enjoyment of the resource.

The Connecticut Department of Environmental Protection's Bureau of Water Management has published a useful book *Caring for Our Lakes,* and DEP also provides technical assistance to towns and lake associations requesting information on management of aquatic nuisance species.

The Department of Environmental Protection collects and makes available information about the physical and biological characteristics of many lakes and large ponds in Connecticut. You can access detailed information about these lakes and ponds through the DEP's website (http://dep.state.ct.us/cgnhs/lakes/lakepond.htm). The DEP bookstore in Hartford has publications containing similar information.

The Connecticut Federation of Lakes is a membership organization concerned with lake conservation and management.

Chapter 6

Inland Wetlands: Palustrine Ecosystems

Palustrine ecosystems are better known as marshes, swamps, and bogs. They include all nontidal wetlands dominated by trees, shrubs, or emergent herbaceous vegetation, as well as all such tidal wetlands in which the concentration of ocean-derived salts is less than 0.5 ppt. Also included are nonsaline open-water areas and pondlike waters that are smaller than 20 acres and less than 6.6 feet (2 m) deep in the deepest part at low water. (See Chapters 4 and 5 for further discussion of wetlands associated with rivers, lakes, and ponds.) Based on the National Wetlands Inventory mapping, palustrine wetlands encompass 152,000 acres statewide (nearly 90 percent of Connecticut's wetlands), though the actual acreage is much larger, perhaps as much as three times larger.

The term "wetland" encompasses a wide variety of areas that are subject to regular or periodic flooding, collect surface water, or are subject to groundwater discharge. In Connecticut, wetlands include land (including submerged land) with soil types designated as poorly drained, very poorly drained, alluvial, or floodplain by the National Cooperative Soil Survey. These soils (except infrequently flooded floodplain soils) are referred to as "hydric" soils. Such areas often are dominated by hydrophytic plants but may or may not be vegetated. Hydrophytic plants are those that thrive in water or where excessive water content of the substrate causes at least periodic oxygen deficiency (see "Wetland Indicators" and "Plants and the Wetland Environment" later in this chapter).

Palustrine wetlands dominated by woody plants often are called swamps whereas most of those with mainly herbaceous vegetation are termed marshes, or wet meadows. In seasonally dynamic wetlands, such as inland marshes, meadows, and red maple swamps, water level generally is highest in spring and lowest in mid- to late summer. The deep portions of ponds, lakes, and rivers are generally regarded as aquatic systems or watercourses rather than wetlands.

Freshwater Marshes and Meadows

Freshwater herbaceous wetlands, often associated with ponds and lakes, occur throughout the state (see Chapter 4 for information on tidal freshwater marshes). Standing water may be permanent, semipermanent, or seasonal, or the soil may simply be saturated and without standing water. Perhaps the richest of all marshes are those that have a roughly equal mixture of permanent open water and thick marsh vegetation.

Vegetation

By definition, these wetlands are dominated by herbaceous (nonwoody) plants (Figs. 6.1 to 6.3). Plantlife varies greatly in species composition, depending on duration and frequency of flooding and soil saturation. In wetter, marshy areas, the long list of emergent plants extending above the water includes tussock sedge, woolly sedge, cattails, water-willow, common reed, pickerelweed, arrowheads, arrow-arum, bur-reeds, soft-stem bulrush, other bulrushes, soft rush, bluejoint grass, reed canary grass, smartweeds/tearthumbs, spotted jewelweed, blue flag, yellow iris*, skunk cabbage, water-horehound, marsh fern, sensitive fern, crowfoot, swamp milkweed, boneset, asters, blue vervain, New York ironweed, mermaid-weed, purple loosestrife*, false nettle, and many others. Submerged and floating plants that border and intermingle with these taller emergent plants along the edges of streams, ponds, and lakes were discussed in Chapters 4 and 5.

Wet meadows, which are not so wet as marshes, can include many of the plants mentioned above and may be dominated by sedges, grasses, or rushes, or a mixed assemblage of wetland plants. In late summer, some mixed-species wet meadows are spectacular, with tall, colorful

Fig. 6.1. Freshwater marshes with a balanced mixture of vegetation and open water support the highest levels of biodiversity.

Fig. 6.3. This thick cattail marsh developed from a red maple swamp as a result of a long-term rise in water level, which killed the maples.

Fig. 6.2. This large wet meadow includes grasses, sedges, and a mixture of other herbaceous plants.

stands of pink, purple, yellow, and white flowering Joe-Pye-weed, New York ironweed, New England aster, boneset, and goldenrods.

Persistent flooding may exclude trees and shrubs from marshes and meadows, but drier areas eventually may be invaded by woody plants, which also invade when drought reduces water levels. Hummocks formed by tussock sedges that have colonized shallow water often become sites on which swampland trees and shrubs become established (Fig. 6.5). Woody plants associated with the marshes and wet meadows include highbush blackberry, poison ivy, swamp rose, buttonbush, winterberry, poison sumac, highbush blueberry, alders, willows, red maple, spiraea, silky dogwood, northern arrowwood, and sweet pepperbush. Morrow's honeysuckle* has extensively invaded some areas transitional between wet meadows and adjacent moist forests or forested wetlands.

Metzler and Tiner reported that the distinctive water chemistry of the limestone areas of western Connecticut supports a number of wetland plants of relatively restricted occurrence in the state, including Muhly grass, the bulrushes *Scirpus pendulus* and *Scirpus acutus,* capillary beakrush, various localized *Carex* sedges, shining ladies-tresses, purple avens, spreading globeflower, fringed gentian, cotton-grass, grass-of-Parnassus, large yellow lady's-slipper, showy lady's-slipper, buckbean, pitcher plant, and twigrush. Shrubs in these calcareous wetlands may include red osier, swamp birch, swamp buckthorn, hoary willow, and autumn willow, as well as various more widely distributed species.

Fauna

Invertebrate diversity and abundance in freshwater marshes can be very high. Careful search might yield freshwater sponges, fingernail clams, various snails, water mites, isopods, amphipods, midges, phantom midges, caddisflies, mayflies, dragonflies, damselflies, water boatmen, backswimmers, giant water bugs, waterscorpions, diving beetles, water scavenger beetles, and others. Through feeding, these animals convert detritus, microorganisms, plant material, phytoplankton, zooplankton, and smaller animals into morsels that in turn attract hungry frogs, turtles, birds, mammals, and other charismatic animals that we may more readily appreciate.

Some Lepidoptera specialize on marsh and wet meadow plants. Caterpillars of the broad-winged skipper feed on phragmites, and tussock sedge and other sedges are the major foods of black dash and mulberry wing caterpillars. These skippers can be seen nectaring on swamp milkweed in July and early August.

The high productivity, sunny conditions, and thick, low vegetative cover of marshes support a rich and abundant herpetofauna, including snapping turtles, painted turtles, spotted turtles, stinkpots, northern water snakes, ribbon snakes, common garter snakes, spring peepers, gray treefrogs, bullfrogs, green frogs, and pickerel frogs. Several fish species typical of ponds and small lakes make use of adjacent marsh habitats during flood periods.

Marshes are important bird habitats. These birds include many of those associated also with shrub swamps, but they tend to be species that are absent from forested habitats. Relatively small marshes might yield a list of about 30 species of wetland birds, plus many others from adjacent habitats. Mallards are by far the most common waterfowl nesting in marshes and other inland wetlands. Canada geese also nest commonly, while black ducks and pied-billed grebes are much less common as nesters. Waterfowl generally associate with marshes that have persistent open water.

Marsh plants support the nests of various songbirds, such as marsh wrens, red-winged blackbirds, common grackles, swamp sparrows, and cedar waxwings. Song sparrows, American goldfinches, and common yellowthroats often nest in thickets adjacent to marshes. Marshes with standing snags, such as result when beaver dams flood swamps and kill trees, may attract nesting wood ducks, hooded mergansers, eastern screech-owls, woodpeckers, tree swallows, and bluebirds (see Chapter 4 for information on beaver-influenced wetlands). Wilson's snipe is a regular but secretive user of marshes and wet meadows during spring and fall migration. Sedge wrens nest primarily in wet meadows with scattered shrubs, but this species is very rare in Connecticut.

Most bird species are present mainly or only from spring to fall, when food resources (insects and other invertebrates; foliage, fruits, and seeds of wetland plants) are most available. A few hardy birds, mainly song sparrows and swamp sparrows, are present in winter when cold and ice severely reduce food availability.

The usual wetland mammals—burrowing, invertebrate-eating star-nosed moles, herbivorous muskrats, omnivorous raccoons, and fish- and frog-eating minks—live in marshes and wet meadows. Beavers inhabit (and may create) deep marshes associated with streams or lakes. White-tailed deer and, increasingly, moose visit to forage on sodium-rich wetland plants (see Chapter 21).

The fauna of the drier representations of wet meadows is similar to that of grasslands (see Chapter 7).

Shrub Swamps

Shrub swamps frequently occur as transitional areas between upland areas and ponds, streams, or marshes, or they may be isolated in upland depessions, even on hilltops (Fig. 6.4). By definition, these wetlands are dominated by woody vegetation less than 20 feet tall (or at least they have significant amounts of such vegetation). Not uncommonly, the shrubs around the swamp margins are densely spaced, tall, and very difficult for a person to penetrate, as I discovered while radio-tracking timber rattlesnakes that sometimes moved into these habitats in summer.

Vegetation

If you venture into Connecticut's shrub swamps, you may find just a few shrub species or many, depending on the location. Likely you'll see some of the following: buttonbush, leatherleaf, speckled alder, smooth alder, pussy willow, spiraeas, highbush blueberry, maleberry, fetter-bush,

Fig. 6.4. Upper: Shrub swamps may exhibit large seasonal changes. This buttonbush swamp dried up completely during a summer drought. Lower: The boggy shrub swamp in the foreground is dominated by leatherleaf, sweet gale, and sheep laurel.

chokeberries, northern wild-raisin, arrowwood, nannyberry, swamp rose, sweet pepperbush, swamp azalea, sheep laurel, common winterberry, smooth winterberry, mountain-holly, shadbush, silky dogwood, gray dogwood, red osier, common elder, sweet gale, and/or poison sumac. Frequent associates of the shrubs include marsh fern, sensitive fern, royal fern, cinnamon fern, skunk cabbage, tussock sedge, three-way sedge, cattail, sphagnum moss, many other herbaceous plants, and semi-shrubby water-willow (swamp loosestrife). As with the shrubs, the herbaceous species vary widely among different swamps. Most of the plants grow on hummocks that extend above the water. Swamps in the limestone valleys of western Connecticut sometimes contain swamp birch, swamp buckthorn, and various willows.

Red maples often occur in shrub swamps, but flooding by high water in spring or summer in wet years, or flooding behind beaver dams, together with feeding by deer and rabbits, may kill or stunt maples that germinate and become established during dry periods. However, under certain conditions, especially where clumps of tussock sedge create favorable sites for colonization by woody plants, red maple can survive, grow to large size, and eventually dominate, coverting the shrub swamp to a forested swamp (or a wet meadow may become a shrub swamp with small maples, then develop into a forested swamp as the maples mature).

Fauna

Invertebrates in shrub swamps include representatives of virtually all of the groups found in lakes and ponds (see Chapter 5), freshwater marshes (see preceding section), and vernal pools (see following). The species makeup varies among these habitats, but in most cases it takes some expertise to detect these differences.

Fishes generally are absent from shrub swamps, except in flooded swamps bordering a lake or river. Gray treefrogs and green frogs often inhabit shrub swamps, and vernal pool species also may be common. Spotted turtles, painted turtles, northern water snakes, and eastern ribbon snakes are the most frequently observed reptiles. Locally, some timber rattlesnakes make extensive use of these habitats in spring and summer.

Bird life in shrub swamps is quite different from that in nearby forested habitats. A 1995 census of breeding birds in a 20-acre tract of shrub swamp/sedge hummock habitat in northwestern Connecticut by Andrew Magee yielded 35 species and 130 territories as follows: swamp sparrow (32 territories), red-winged blackbird (27), yellow warbler (23), common yellowthroat (18), American goldfinch (10), willow flycatcher (9), gray catbird (9), song sparrow (9), eastern kingbird (6), cedar waxwing (6), tree swallow (3), mallard (2), marsh wren (2), veery (2), American robin (2), and 13 additional species represented by at least 0.5 territories. Swamps with standing snags may attract nesting wood ducks, hooded mergansers, eastern screech-owls, woodpeckers, tree swallows, and bluebirds. Most shrub swamp birds are migrants that use these habitat primarily in spring and summer when nesting cover and food (mainly insects) are maximally available.

Shrub swamp mammals include the characteristic wetland species (star-nosed moles, muskrats, minks, and raccoons), and other more terrestrial species visit these wetlands when standing water is not present.

Forested Wetlands

Forested wetlands are the state's most common type of palustrine wetland. More than 60 percent of the total palustrine acreage is forested or forest/shrub. Some forested wetlands occur along permanent streams or adjacent to coastal marshes, but the ones discussed here are typical of wet depressions in uplands (see "Floodplain Forests" in Chapter 4 for information on swampy forests occurring along rivers). These often hummocky habitats are defined by a predominance of woody vegetation over 20 feet tall.

Vegetation

Red Maple Swamps. The overwhelmingly dominant tree in Connecticut's forested wetlands, aside from floodplain forests (treated separately here), is red maple, a plant with a wide distribution and broad ecological tolerance. It occurs in both wetlands and uplands throughout the eastern United States and southernmost Canada west to Texas and Minnesota. Red maples dominate swamps ecologically, despite being relatively small trees. In most swamps, red maples attain heights of only 40 to 50 feet even after a hundred years.

Red maple swamps occur in diverse hydrogeologic conditions. Soils may be seasonally or temporarily flooded or seasonally saturated. Some swamps are small, such as those in isolated basins over bedrock or in glacial till, or in seepage areas, whereas others may cover large areas on former glacial lake beds or along streams or lake edges in valley bottoms. In a particular swamp, water level varies greatly from month to month and from year to year, and some swamps are much wetter than others. Generally, water level is highest in spring and lowest in late summer or early fall, so plan your explorations accordingly. Often surface flooding lasts from October or November through June, but in some swamps or during dry years, free-standing water may not be present in any month.

An easily seen swamp type along upland drainages and seepages has red maple, spicebush, skunk cabbage, and marsh-marigold as the common species. The yellow flowers of spicebush and marsh-marigold, and the large, growing leaves of skunk cabbage, are particularly conspicuous in April.

If you visit other deciduous forested wetlands, such as those on former glacial lake beds or along streams or lake edges in valley bottoms, you'll find a more diverse assemblage of plants that might include, in addition to red maple, trees and shrubs such as yellow birch, black birch, American elm, black gum, black ash, swamp white oak, pin oak, basswood, sweet gum (southwestern Connecticut), swamp cottonwood, eastern hemlock, eastern white pine, spicebush, highbush blueberry, fetter-bush, maleberry, swamp azalea, sweet pepperbush, winterberry, chokeberry, speckled alder, arrowwood, nannyberry, northern wild-raisin, or poison sumac. Great laurel is a large, locally dominant shrub in several swamps in eastern Connecticut. Some swamps have abundant viney growth of greenbriar, riverbank grape, fox grape, Virginia creeper, or Asian bittersweet*.

The herbaceous flora in red maple wetlands varies widely from swamp to swamp. Common species include skunk cabbage, false hellebore, marsh-marigold, swamp dewberry, shining firmoss, cinnamon fern, sensitive fern, spinulose wood fern, royal fern, marsh fern, jack-in-the-pulpit, spotted jewelweed, northern white violet, goldthread, turtlehead, water-horehound, and various mosses and sedges. Many additional species are locally common.

In western Connecticut in areas dominated by a limestone-based geology, Metzler and Tiner found that the characteristic assemblage of plants is different than that just described and includes not only widespread species such as red maple, American elm, poison sumac, winterberry, silky dogwood, red osier, spicebush, American hornbeam, skunk cabbage, false hellebore, and sensitive fern, but also black ash in greater abundance, plus plants indicative of the alkaline conditions, including purple avens, swamp saxifrage, northern swamp-buttercup, miterwort, tufted loosestrife, swamp thistle, and certain specialized *Carex* sedges. Except for highbush blueberry, shrubs of the heath family (Ericaeae) tend to be scarce. Swamps with calcium-rich soils or nutrient-rich groundwater discharge have the richest herbaceous flora.

Red maples and other woody plants in mature or maturing swamps often grow on hummocks formed by tussock sedges (Fig. 6.5) that were present in a pre-swamp wet meadow. In young red maple swamps that have developed directly from wet meadows, shrubs may be virtually absent.

Fig. 6.6. This Atlantic white cedar swamp has a dense shrub layer (great laurel, mountain-holly), but other cedar swamps may have a more open understory of scattered sweet pepperbush, or sphagnum and ferns.

Fig. 6.5. Upper: The red maple swamp in the background will expand if trees continue to become established in the tussock sedge wet meadow in the foreground. Stephen P. Broker. Lower: Red maples in this forested wetland become established on sedge tussocks of a marshy wetland.

Some red maple swamps go through a shrub swamp stage before the red maples attain dominance. Woody and herbaceous swamp plants often colonize mounds that form as a result of tree blow-downs, which cause soil to be pulled up by the spreading root system and deposited as a mound at the base of the fallen tree. A deeper pool generally forms adjacent to the mound where the soil was pulled out.

Evergreen Swamps. Conifers are present in and may dominate some forested wetlands. Eastern hemlock often dominates evergreen and mixed swamps in the Northwest Highlands region, where black spruce and red spruce are localized associates. Among other swamp conifers, eastern white pine is widely distributed. Deciduous red maple and yellow birch often are mixed with the conifers. Understory shrubbery may be poorly developed under hemlock or pines. Goldthread, skunk cabbage, cinnamon fern, spicebush, mountain laurel, star-flower, peat mosses, and liverworts may be common in evergreen and mixed swamps.

Northern white cedar occurs in some calcareous swamps of the Northwest Highlands. Metzler and Tiner found that American hornbeam, foamflower, miterwort, star-flowered false Solomon's-seal, and tufted loosestrife are typical associates in these swamps.

Dense stands of Atlantic white cedar are patchily distributed in southern and especially eastern Connecticut in glacial kettles, moraine hollows, and old lake beds (Fig. 6.6). Thirty-three of the thirty-nine recently extant stands are east of the Connecticut River. These acidic swamps are semi-permanently to seasonally flooded or saturated and are generally wetter than adjacent red maple swamps. Great laurel, highly localized in Connecticut, occurs with the cedars in some of these swamps. Most of the other plants you're likely to find are widespread in many wetland types: red maple, eastern hemlock, black gum, white pine, gray birch, yellow birch, mountain laurel, shadbush, highbush blueberry, chokeberry, dangleberry, maleberry, fetter-bush, swamp azalea, inkberry, smooth winterberry, mountain-holly, common winterberry, sweet pepperbush, spicebush, poison sumac, *Carex* sedges, sarsaparilla, sundew, pitcher plant, star-flower, partridge-berry, goldthread, marsh St. John's-wort, skunk cabbage, cinnamon fern, royal fern, Massachusetts fern, Virginia chain-fern, netted chain-fern, marsh fern, wild lily-of-the-valley, sphagnum, and liverworts. Inkberry, creeping snowberry, and black spruce are rare associates. You can examine an Atlantic white cedar swamp, with an understory of great laurel, along a boardwalk in Pachaug State Forest in Voluntown. For further information on cedar swamps, see the "Conservation" section at the end of this chapter.

See "Bogs and Fens" for further information on wetlands that include spruce.

A Delicate Balance. Forested swamps often reflect a delicate balance in local water level. Small increases in water level, such as may result from dam building by beavers, may kill trees and cause a shift to a more open wetland community. A minor drop in average or seasonal water level may allow invasion by plants of moist uplands and a loss of wetland species.

Fauna

Forested swamps have a diverse invertebrate fauna and a few unique species. Hessel's hairstreak, a rare, emerald-green butterfly, occurs solely in or near swamps with Atlantic white cedar, the only larval food source. The ringed boghaunter, a rare dragonfly, also may occur in cedar swamps with abundant sphagnum. Many of the species discussed in the sections on vernal pools and deciduous forests also inhabit forested swamps. Riverine species typical of small streams usually are present in streams through swamps.

Fishes are generally absent unless the swamp has a stream in it or is connected to a permanent body of water such as a pond or lake. Fishes may move from these aquatic habitats into swampy forests during periods of high water. Swamp darters inhabit swampy bodies of water in eastern Connecticut.

Amphibians and reptiles of forested wetlands generally are those associated with vernal pools, which frequently occur within forested wetlands, and those common in deciduous forests (e.g., redback salamander in moist but not wet sites). Additionally, four-toed salamanders and (locally) blue-spotted salamanders inhabit forested wetlands, as do gray treefrogs. Four-toed salamanders lay eggs in mossy hummocks. Flooded areas are sure to harbor ubiquitous green frogs. In mid- to late summer, when these wetlands have dried out, American toads and, especially when conditions are moist, wood frogs and spring peepers often forage in forested swamps.

Forested wetlands are rich in bird life. Birds that nest in deciduous forested swamps include wood ducks, common mergansers, hooded mergansers, red-shouldered hawks, broad-winged hawks, eastern screech-owls, great horned owls, barred owls, ruby-throated hummingbirds, red-bellied woodpeckers, downy woodpeckers, hairy woodpeckers, pileated woodpeckers, blue jays, Carolina wrens, black-capped chickadees, tufted titmice, blue-gray gnat-catchers, veerys, gray catbirds, yellow-throated vireos, warbling vireos, red-eyed vireos, northern waterthrushes, Louisiana waterthrushes, common yellowthroats, Canada warblers, ovenbirds, black-and-white warblers, hooded warblers, and others. Swamp trees and understory plants offer cavities and other shelter for nesting as well as abundant food (insects, fruits, mice).

The number and precise combination of bird species in a swamp varies with the age of the trees, size and wetness of the swamp, and thickness of shrubby vegetation. Generally, the bigger the swamp the greater the number of bird species, and large, vegetationally heterogeneous swamps tend to have the largest number of bird species. Researchers in southern New England have determined that extensive forest cover in a region enhances the number of forest interior bird species that use a particular swamp, and even small swamps in a forested landscape may support a diverse avifauna. Wetter areas tend to have a larger number of bird species and greater bird abundance than do drier areas with similar vegetation structure. During the breeding season, you're likely to find at least 15 to 20 bird species in any sizeable tract of forested swamp, and 40 to 50 species regionally. Some of these birds are resident year-round, while others are long-distance migrants present only in spring and summer.

Most of the breeding birds in red maple swamps also nest in evergreen swamps. In New Hampshire, a 14.6-acre tract of Atlantic white cedar swamp had 21 to 23 nesting pairs of 13 to 16 species. Canada warbler was the most common nesting species. Diversity in Connecticut cedar swamps is similar.

No mammals live only in forested wetlands. The typical species are those widely distributed in various wetland communities. The most common large mammal is the white-tailed deer, which feeds in swamps and uses them for refuge. Beavers sometimes colonize, dam, and flood forested wetlands, changing them to open water and dead-wood wetlands (Fig. 4.16). The small mammal fauna includes water shrews, short-tailed shrews, star-nosed moles, southern flying squirrels, gray squirrels, red-backed voles, woodland jumping mice, and white-footed mice.

Peatlands: Bogs and Fens

Bogs and fens are wetlands with highly organic soils. These wetlands, also known as peatlands, are rare and scattered in Connecticut, occuring mainly in the northwestern part of the state. Peatlands developed in Connecticut during early post-Pleistocene deglaciation. After periods of probable fluctuation, they likely reached their current extent perhaps a couple thousand years ago. Today the state has marginal conditions for peatland development.

Peatland soils are rich in organic material because the accumulation of dead plant material exceeds the rate of decomposition. This low rate of decomposition results from high acidity, which inhibits decomposers such as bacteria. The high acidity of bogs results in part from the physiology of sphagnum moss, the quintessential bog plant.

Sphagnum increases acidity because it releases hydrogen ions as it extracts minerals from the water. Sphagnum contributes to the accumulation of dead plant material (peat) because it contains a large amount of complex lignin-like compounds that are difficult for decomposers to utilize.

Other notable sphagnum characteristics include its exceptional ability to hold water (15 to 20 times its dry weight), good growth during cool seasons, and high nutrient content. Sphagnum increases mainly through vegetative reproduction and much less by spore production and germination. There are many species of sphagnum, and they differ in their moisture requirements.

Bogs

Bogs generally occur in areas where summer climate is cool and moist, and nutrient and dissolved oxygen input are low, as in gravelly or sandy areas where nutrients come mainly from rain and other precipitation (Fig. 6.7). Some bogs are in depressions ("kettles") left as glacial features (for example, a place where a block of ice was left imbedded in the ground). Others are between drumlins (long, streamlined, glacially formed hills) or in other poorly drained areas. Bogs are more extensive in northern New England than in Connecticut, where bogs are primarily in the Northwest Highlands.

Bogs sometimes develop from open-water ponds that gradually fill with muck on the bottom as a floating mat of vegetation grows inward from the pond margin. Eventually open water may disappear and the wetland may become a

Fig. 6.7. Bogs in Connecticut generally are small, have thick vegetation including woody species, and contain plants that are rare or absent in other habitats. Here small black spruce trees protrude above the shrubs, and several tufts of cotton-grass are visible in the foreground.

shrubby or forested bog, a process than can take 10,000 years. Boggy areas can form in small or large depressions or adjacent to large bodies of water. For example, boggy mats of vegetation that undulate when walked upon sometimes develop along the south side of a small pond, in the cool shade of adjacent trees, or they may occur adjacent to highland lakes and larger ponds.

Species diversity in bogs is low because of the harsh conditions—not many species can withstand the acidity and/or low nutrient availability. Sphagnum, the main peat maker, is the standard element of bog vegetation. Woody plants in Connecticut's bogs include black spruce; American larch; white pine; pitch pine; eastern hemlock; Atlantic white cedar; a rich assemblage of heaths, such as leatherleaf, sheep laurel, bog laurel, bog-rosemary (rare), Labrador-tea (rare), large and small cranberry, creeping snowberry (rare), dwarf huckleberry (rare), blueberry, maleberry, swamp azalea, rhodora, and other Ericaceae; sweet gale; poison sumac; black chokeberry; paper birch, yellow birch, gray birch, bog birch (rare); red maple; winterberry; and mountain-holly. Trees are often stunted (old but small) in the harsh bog environment. Among the herbaceous plants, look for water-willow, pitcher plant, round-leaved sundew, bladderworts, various orchids, goldthread, wild calla, various *Carex* sedges, cinnamon fern, Virginia chain-fern, and cotton-grass. Not all of these plants occur in every bog, but most of them occur in kettle bogs. Leatherleaf, sweet gale, and sheep laurel often dominate boggy mats in and bordering ponds in the Northwest Highlands. You can get a little taste of the north country by visiting a late-successional black spruce bog along a boardwalk in Mohawk State Forest (Fig. 6.8).

Fig. 6.8. The trees in this view of a late successional bog include black spruce and American larch.

Fig. 6.9. Fens (foreground) are rich in herbaceous plants and shrubs.

Fens

Fens are peatlands whose water input comes primarily from groundwater and surface water (Fig. 6.9). These waters often have higher nutrient content than does the rainwater that supplies bogs, so many fens have more favorable growing conditions for plants. However, fens vary in their characteristics. Rich fens have a pH above 6.0, many calcium-loving species, and little or no peat moss. Poor fens are more acidic and have more peat mosses and other bog plants.

Fens are wet but not submerged with water throughout the year. They may be level or sloping, and a thick peat layer may or may not be present. Fens are a rare community type in Connecticut, represented mainly by calcareous fens in western Connecticut.

Typical plants in rich calcareous fens in western Connecticut include various *Carex* sedges (*C. lasiocarpa, C. aquatilis, C. lacustris, C. interior, C. leptalea, C. flava*), twig-rush, sphagnum peat mosses, narrowleaf cattail, pitcher plant, sundew, bog birch, swamp azalea, leatherleaf, large cranberry, highbush blueberry, and buttonbush (Metzler and Tiner 1992). In western Connecticut, fens on spring-fed slopes with nutrient-poor, alkaline water support a number of rare plant species.

Open fens are ephemeral habitats that depend upon natural disturbances such as wind storms, periodic inundation by beaver ponds, grazing by moose and other large herbivores, and especially fire to keep them from turning into wooded swamps dominated by shrubs or red maples. Through intentional burning, humans have long played a role in keeping fens from being taken over by woody plants. Recently, with less burning, many open fens have developed into shrub fens and red maple swamps, which do not provide suitable conditions for some of the rare plants found in open fens. In northern New England, the Nature Conservancy is using prescribed burns to conserve these habitats and the rare species they support.

Carnivorous Peatland Plants

Generally we don't think of plants as carnivores, but some are passive predators as well as photosynthesizers, and peatlands are good places to find them. Among these oddities are pitcher plants, sundews, and bladderworts, some of the last of which occur not only in peatlands but in many lakes and ponds as well.

Pitcher plant (*Sarracenia purpurea*) is an herbaceous, rosette-forming perennial that can live up to half a century (Fig. 6.10). In spring, large plants produce a single-flower infloresence pollinated by bees and sarcophagid flies. The seeds disperse in the wind and water in winter. Like other carnivorous plants, pitcher plant has a weakly developed root system.

The most unusual aspect of this plant is its vaselike leaves. New leaves open in late spring and begin accumulating

Fig. 6.10. Pitcher plants capture insects in their vaselike leaves.

rainwater. Old leaves overwinter as fluid- or ice-filled pitchers, then begin to rot by the end of their second summer. Small animals, such as ants (often the bog specialist *Myrmica lobifrons*), beetles, snails, and other invertebrates enter the pitchers, attracted by the nectar glands around the rim of the pitcher. The red veins in the leaves and ultraviolet reflectance patterns may advertise the location of the nectar glands. Downward-pointing hairs and a sticky, sluffing lining inside the leaf interfere with the efforts of animals to leave the pitcher, and some of the visitors become permanently trapped and die in the pitcher water.

These dead organisms are the basis for a miniature ecosystem in the pitcher water. The larval state of a mosquito (*Wyeomia smithii*) develops in the fluid and filter-feeds and browses on rotifers (usually *Habrotrocha rosa*), protozoa, bacteria and detritus in the water. This mosquito does not need to consume a blood meal in order to produce a batch of eggs. Midge and sarcophagid fly larvae (*Metriocnemus knabi*, Chironomidae; *Fletcherimyia fletcheri*, Sarcophagidae) also thrive in the pitcher water, feeding on the remains of insect carcasses and detritus at the top (sarcophagid) or bottom (midge) of the pitcher water. Sarcophagids may cannibalize each other until only one remains, and they may prey on mosquito larvae and smaller organisms. The mosquitoes and midges overwinter as larvae that may be encased in ice. Sarcophagid larvae leave the pitchers and apparently overwinter as pupae in sphagnum moss. Slime mites (*Sarraceniopus gibsoni*) join the dipterans as carcass feeders in the pitcher water. Many of these animals reproduce only in pitcher plant water.

The waste products of the pitcher-water organisms become nutrients that can become absorbed by the plant (though digestive enzymes released by *S. purpurea* are weak or absent). In turn, the plant creates a favorable environment for these animals by removing potentially toxic metabolic wastes and supplying oxygen.

Despite the adaptations for and benefits of "carnivory," many pitchers fail to capture any insects over periods of several weeks, even in summer. Young pitchers secrete more nectar and tend to capture more prey than do older pitchers. Pitchers in their second season continue to capture prey. Some potential prey are intercepted by spiders that build their webs across the pitcher opening.

Probably animal consumption helps compensate for the relatively low nutrient (nitrogen) availability in bog soils; it seems to be important for the production of flowers and fruits. Although pitcher plants get most of their nitrogen from the pitcher fluid, relatively little of it comes from animal prey. Nitrogen deposition from the atmosphere is more important. However, pitcher plants thrive on low-nitrogen conditions, and recent studies suggest that excessive nitrogen deposition, which can result from air pollution, may result in altered pitcher leaf morphology and excessive flower production and might lead to declines in pitcher plant populations.

Additional pitcher plant–associated insects include two rare moths—**pitcher plant moth** (*Exyra rolandiana*) and **pitcher plant borer** (*Papaipema appassionata*)—and a tortricid moth (*Endothenia hebesena*). Larvae of the first two species feed on the leaves and rhizomes, respectively, while caterpillars of the tortricid feed on the seedheads.

Sundews (*Drosera spp.*), like pitcher plants, do not require an animal diet, but carnivory does seem to enhance their growth and reproduction (Fig. 6.11). The animal traps of sundews consist of sticky filaments on the leaves. When an insect becomes stuck, the filaments slowly envelop it. It is broken down gradually by enzymes secreted by the plant and by microorganisms, and the nutrients are absorbed. Sundews die back with the first frost. They overwinter as live roots and leaf buds. I found sundews on mossy, partially submerged logs or on the banks of, or hummocks in, ponds and lakes.

Bladderworts (*Utricularia spp.*) occur in peatlands, wet soil, ponds, and lakes (Fig. 6.12). Most bladderworts are rootless aquatic plants that consist of branching stems and fine, much-divided leaves having small bladders that act as simple mechanical traps for small aquatic animals such as zoo-

Fig. 6.11. Sticky droplets on round-leaved sundew leaves capture insects that serve as food for the plant.

plankton. An animal brushing against the hairlike trigger on a bladder causes it to open a small "door" and suck in water and anything in it. This suction results from the partial vacuum (low hydrostatic pressure) maintained inside the bladder by active cellular processes. Immediately after opening, the door closes as pressure inside and outside the bladder equalizes. Any animals caught in the bladders are digested by enzymes released by the plant and by bacterial action. Bladders may contain up to hundreds of tiny organisms.

Fig. 6.12. Aquatic bladderworts have a simple stem bearing one or more flowers. The submerged, finely dissected leaves bear tiny submerged bladders that trap small aquatic animals.

Peatland Orchids

Several kinds of orchids occur mainly in peatlands and similar habitats. Fringed orchids, grass-pink (Fig. 6.13), rose pogonia, dragon's mouth, and ladies'-tresses may be locally common in bogs or wet meadows. See "Understanding Orchids" at the end of Chapter 9 for further information.

Peatland Fauna

Other than pitcher plant specialists, most of the animals of bogs and fens are species that occur in a wide range of habitats. However, many aquatic species are excluded from bogs by excessive acidity.

The peatland herpetofauna in Connecticut includes spotted salamanders, four-toed salamanders, eastern newts, green frogs, northern leopard frogs, pickerel frogs, spring peepers, painted turtles, snapping turtles, spotted turtles, common garter snakes, northern water snakes, and redbelly snakes. A visit to a peatland in spring or summer might

Fig. 6.13. Grass-pink, a peatland orchid, has yellow-tipped filaments (pseudopollen) that deceive pollen-seeking bees into landing on the flower and pollinating it.

Fig. 6.14. Bog turtles in Connecticut are restricted to calcareous fens.

Fig. 6.15. Vernal pools support a unique assemblage of wildlife.

yield birds such as those associated with wet meadows, shrub swamps, and forested wetlands. You may also find migrants and locally nesting species, including olive-sided flycatchers, blue-headed vireos, black-throated green warblers, yellow-rumped warblers, golden-crowned kinglets, dark-eyed juncos, and white-throated sparrows. Shrews, star-nosed moles, voles, and southern bog lemmings live in tunnels and runways among mosses and other low plant cover. Many additional birds and mammals live in peatlands that have an abundance of trees and shrubs.

With the pitcher plant fauna already mentioned, a few species are peatland specialists. Bog turtles (Fig. 6.14) in Connecticut occur only in calcareous fens, and Dion skippers are associated with *Carex* sedges (larval food source) in these fens and other alkaline to neutral wetlands in western Connecticut. Another peatland species is the bog copper, whose caterpillars eat only cranberry leaves. The ringed boghaunter is a rare dragonfly that breeds only in fishless, permanent sphagnum pools in bogs and fens. Certain damselflies and dragonflies, such as amber-winged spreadwing, petite emerald, brush-tipped emerald, white corporal, frosted whiteface, crimson-ringed whiteface, and Hudsonian whiteface, are most often associated with bogs, fens, or peaty margins of ponds or lakes, generally where pitcher plants grow. The sphagnum ground cricket appears in boggy wetlands with sphagnum moss.

Vernal Pools

Vernal pools are one of our smallest yet most interesting and important wetland types (Figs. 6.15 and 6.16). They exist in scattered sites in highlands and lowlands statewide. Usually vernal pools are relatively small, temporary, relatively isolated bodies of water occupying depressions. Some do not dry up every year, and some do not hold water every year. Traditionally, in New England, the term "vernal pool" has been applied to only those pools that support breeding populations of certain amphibians (see following paragraphs) and/or fairy shrimp. However, vernal pools are not absolutely discrete entities but rather represent a part of a wetland continuum that ranges from small, shallow, short-lived pools that contain few if any typical vernal pool animals to semi-permanent to permanent ponds that may be inhabited not only by typical vernal pool species but also by plants and animals characteristic of permanent waters. This discussion focuses on vernal pools in the traditional sense and includes both vernal pools in the strict sense (i.e., those that fill in spring) and autumnal pools (those that fill in fall). The end of this section includes a brief discussion of other types of temporary pools and ponds. *A Field Guide to the Animals of Vernal Pools* (Kenney and Burne 2000) is useful for identifying most of the amphibians, reptiles, and invertebrates you're likely to find in vernal pools.

Sources of Water and Physical Conditions

The water in a vernal pool may come from overland flow (runoff), interflow through the soil (or other substrate above groundwater), groundwater, or direct precipitation such as rain or snow. In many basins, standing water is present from about October or November through early or mid-summer. These pools often partially fill in summer (then may dry again) and in fall and attain maximum size and depth in spring. Small basins and some large ones sometimes are dry in fall and winter and fill in spring.

When fully filled, most vernal pools are less than 100 feet in diameter and not more than a few feet deep. Water temperature ranges from just above freezing in winter to the mid-80s°F in summer in some pools.

Fig. 6.16. A vernal pool in early spring of 1996 (upper) and the same basin in early summer of 1997 (lower).

Recognizing Dry Vernal Pool Basins

Vernal pool basins may be recognized as such during the dry period (Fig. 6.16) by various clues, such as a low area lacking vegetation (in contrast to trees and shrubs of the surrounding forest), sometimes marked by water-stained leaf litter in the lowest part of the basin; clams or other aquatic invertebrates among leaves or debris on the basin bottom; strands of filamentous algae draped over twigs or other plant material; water marks on tree trunks; or the presence of obligate wetland plants such as buttonbush, bur-reed, or *Carex* sedges in a low area surrounded by upland plants. I have heard that false hop sedge *(Carex lupuliformis)* is a good indicator of a vernal pool. In late summer, some vernal pool basins have significant growth of marsh fern, rice cutgrass, and beggar-ticks. In general, emergent plants and submerged aquatic vegetation are scarce in vernal pools, but sometimes bur-reed or *Carex* forms significant patches. Algal production is less than in permanent ponds and is greatest in spring before development of the forest canopy shades the pool. Some vernal pools exist within shrub swamps or forested swamps and have more diverse vegetation, such as red maples and shrubs (highbush blueberry and others), the latter usually growing on soil mounds such as form at the base of fallen red maples (see preceding sections).

Invertebrate Fauna

Vernal pools support a rich and abundant invertebrate fauna. Species richness generally is highest in pools that have a relatively long hydroperiod (duration of surface water). Pools that last several months have the greatest wildlife diversity. Short-lived pools support fewer species but are nevertheless ecologically important (see "Conservation" at end of chapter).

Careful surveys may yield more than 100 species in a single pool, including microcrustaceans (cladocerans, copepods), amphipods, isopods, ostracods, fairy shrimp, water mites, oligochaete worms, leeches, springtails, libellulid and aeshnid dragonfly nymphs, damselfly nymphs (e.g., familiar bluet, emerald spreadwing, slender spreadwing), caddisflies, mayflies, adult and larval beetles (diving, whirligig, water scavenger, minute moss, marsh, and crawling water beetles), water boatmen, backswimmers, giant water bugs, waterscorpions, water striders, midge larvae, phantom midge (gnat) larvae, mosquito larvae, fishfly larvae (rare), various snails (e.g., thicklip rams-horn, lance aplexa, vernal physa, and pygmy fossaria), and fingernail clams. My vernal pool surveys revealed that adjacent pools only a few steps apart are sometimes inhabited by significantly different assemblages of species in any particular year.

Probably many invertebrate species that live in vernal pools are restricted to these habitats. I suspect that further study will reveal that the vernal pool invertebrate fauna is more unique than is now known. However, some vernal pool animals, especially predatory insects and certain isopods and amphipods, also are among the typical residents of permanent ponds.

Freshwater invertebrates such as sponges, alderflies, most stoneflies, and caseless caddisflies seldom if ever occur in vernal pools.

Amphibians and Other Vertebrates

Vernal pools are particularly notable for their unique assemblage of amphibians. Spotted salamanders, marbled

Fig. 6.17. Wood frogs communally deposit their egg masses in vernal pools typically in March.

salamanders, Jefferson salamanders, and wood frogs are the stereotypical breeding species that produce significant numbers of offspring in no other habitat. Single pools support breeding populations of up to several hundred wood frogs (Fig. 6.17) and often 100 to 200 spotted salamanders, with the other species generally in lesser numbers. These vernal pool amphibians depend on pools that contain water from at least March through August (marbled salamanders require water from fall through at least June). Spring peepers, gray treefrogs, eastern newts, and occasional other frogs also use these pools for feeding or maintaining water balance. I sometimes found redback salamanders (an upland species) in vernal pool basins during the dry phase.

Spotted turtles make extensive use of vernal pools for foraging, basking, and mating. Box turtles occasionally visit these pools to drink or forage. Painted turtles commonly use semi-permanent pools close to larger, permanent waters. Wood ducks, black ducks, and mallards regularly visit vernal pools when they contain water. Spotted turtles and ducks seem to be attracted to the seasonally rich invertebrate and larval amphibian food resources in the pools, as are various snakes that sometimes forage in vernal pool basins. Mammals are not prominent users of vernal pools, but various generalist and wetland species such as raccoons sometimes visit these wetlands.

Vernal pools, being relatively small and ephemeral, do not provide suitable habitat for fishes. Most of the amphibians and invertebrates that are restricted to vernal pools for successful breeding cannot survive and reproduce in the presence of predatory fishes, so the mutually exclusive distributions of the typical vernal pool animals and fishes is not surprising.

Seasonal Changes in Fauna

Longer-lasting vernal pools exhibit a somewhat predictable sequence of inhabitants, although many pools do deviate from the generalized pattern described here. Some vernal pool invertebrates are present soon after the basin fills whereas others may not appear until months later. Herbivores and detritivores that feed on leaf-litter microbes (especially fungi, bacteria, and protozoa) or on algae in the water column may dominate at first. In fact, leaf litter from surrounding trees and shrubs is a critically important basis for vernal pool food webs. Leaf litter is colonized by microbes not only when submerged but also when exposed to air after late spring to early summer shrinkage of the pool or when leaves fall into dry pool basins in autumn. Such material may be colonized by other microbes when the pools fill and the litter is flooded. Dissolved nutrients from this material may spark rapid algal growth before surrounding trees grow new leaves and shade the pool. The microbe-rich litter supports invertebrate detritivores such as midge and caddisfly larvae, as well as isopods, amphipods, grazing snails, and other bottom dwellers.

Aedes mosquito larvae hatch early in spring when rising water levels flood pool margins where females deposit most of their eggs. Partially grown fairy shrimp may be present as early as February. Early-season vernal pool samples in March often contain fairy shrimp, copepods, cladocerans, ostracods, mature isopods, caddisfly larvae, dytiscid beetle larvae, chironomid midge larvae, phantom midge larvae, snails, and marbled salamander larvae that hatched the previous fall. Wood frogs (Fig. 6.17), spotted salamanders, and Jefferson salamanders deposit many egg masses in March.

By the end of April, larval damselflies, dragonflies, and caddisflies emerge from eggs and begin to feed on small crustaceans, and *Caecidotea* isopods release their young. Some larval caddisflies prey on amphibian eggs. Adult water striders, backswimmers, diving beetles, water scavenger beetles, and whirligig beetles may also arrive in April. Bottom samples may yield flatworms and oligochaete worms. And April is the month when many amphibian larvae emerge from eggs that were laid in March.

By May, fairy shrimp have produced their eggs and died, some caddisflies complete development and pupate, and adult mosquitoes may emerge. Pools may contain large numbers of ostracods, copepods, water fleas (cladocerans), midge larvae, marsh (scirtid) beetle larvae, and water mites. Predatory insects such as additional beetle species, giant water bugs, broad-shouldered water striders, water boatmen, and water striders, which earlier were scarce, may increase rapidly through immigration of adults that wintered in permanent water. These migrant predators have no specializations for vernal pools, but they feed on the pool's high-quality food and often complete their larval development before the pools dry. Migrant green darner dragonflies may deposit eggs in large pools in late April and May, and their larvae may be numerous by the end of May.

In June, vernal pools often begin to shrink as warm, dry weather increases evaporation and water tables fall. *Anopheles* mosquito larvae may be present throughout the summer, as long as there is standing water. Adult dragonflies and damselflies may emerge from some pools in June and July. Marbled salamanders and wood frogs metamorphose and leave the pools in June and July. By mid-July, smaller pools have dried and, in larger pools, invertebrates have completed development and emerged or have laid drought-resistant eggs.

By early August, even large vernal pools may dry, killing some of the insect larvae that remain. These may include larvae of typical permanent-pond dragonflies that sometimes breed successfully in large vernal pools in wet years when water remains throughout summer. Some insect larvae survive pool drying (see following discussion). Most newly metamorphosed spotted salamanders and Jefferson salamanders leave the pools in July and August, but departure may be delayed into early fall if the water lasts.

Summer rains may temporarily refill dry pools, and certain mosquitoes take advantage of these conditions and produce a new brood. Drought-resistant arthropod eggs present in pool basins do not hatch in summer because they require a period of cold before breaking dormancy and because their position near pool margins means that they are not immersed by partial pool refilling in summer.

In August and September, marbled salamanders move to dry pool basins and deposit eggs under rocks, logs, and leaf litter. Females attend their eggs until fall rains fill the basins and flood the nest sites. Then the females depart and move into the surrounding woods. Soon after flooding, the eggs hatch, and larvae feed and grow until winter's cold forces them into dormancy at the pool bottom. With warming temperatures in late winter, they begin feeding again.

Colonization

The methods by which animals colonize vernal pools are wonderfully diverse. Adult stages of beetles, bugs, dragonflies, damselflies, and flies arrive simply by flying. Amphibians, and sometimes crayfish, leeches, and amphipods can reach some pools by overland movements of adults. Algal spores, resting eggs of rotifers, fairy shrimp, cladocerans, and flatworms may arrive in mud attached to more mobile animals. Eggs of fairy shrimp, cladocerans, and ostracods may arrive with waterfowl (in the gut, on feathers, or in attached mud), or perhaps on the muddy limbs of raccoons or deer. Amphipods and clams may cling to feathers or fur. Clams, water mites, and ostracods sometimes hitch rides on backswimmers and water boatmen, and clams may arrive clamped onto the toes of amphibians that have visited other pools. Dragonflies serve as transportation for *Arrenurus* water mites. Winds sometimes can carry and disperse small organisms. Temporary stream connections may carry in insect larvae, snails, aquatic worms, and other species.

Dealing with Drought

Dewatering is a critical ecological factor, and vernal pool organisms deal with it in various ways. Fairy shrimp, mayflies, cladocerans, and many midges survive the dry phase as dessication-resistant eggs. In fact, vernal pool fairy shrimp require habitat drying before the eggs can hatch. Other microcrustaceans, amphipods, isopods, and some midges may survive as immatures hidden under leaf litter, wood, or rocks, or burrowed in the basin bottom during its dry phase, and adult snails, worms, beetles, and bugs may survive there as well. Shade from the surrounding forest is important in preventing excessive drying that would kill these animals. Adult beetles and bugs additionally have the option of simply flying away from drying pools and spending the dry period in nearby permanent waters.

Adult amphibians are not affected by pool drying because they live on land most of the year, regardless of the presence of water in the pool basin. Algae survive prolonged exposure to drying as modified vegetative cells that develop thickened walls and mucilage sheaths and accumulate oils in cells. Spores of bacteria and fungi, plant seeds, eggs and early embryos of some crustaceans, and larvae of certain insects also tolerate partial drying. Certain tiny organisms, such as protozoans, rotifers, nematodes, and tardigrades, tolerate substantial dessication at any stage of their life cycles.

See "Conservation" at the end of the chapter for further discussion of vernal pools.

Other Types of Temporary Pools

Floodplain Pools. Springtime flooding along river floodplains produces temporary pools that are critical for the reproduction of certain amphibians. These pools serve as high-quality breeding habitat for northern leopard frogs and American toads.

Temporary pools in floodplains and other open areas also are a primary breeding habitat for the larval stages of two dragonflies, the wandering glider and spot-winged glider. Most dragonflies are associated with permanent bodies of water.

Sand Plain Kettles. Some sand plain areas include bowl-like or shallow, saucerlike basins that are occupied by various kinds of wetlands, ranging from short-lived pools to bogs to permanent ponds. The basins generally are "kettles," which formed as a result of sand being deposited around and over a huge block of glacial ice at the end of the Pleistocene.

Pools that form in sandy kettles only after torrential rains provide breeding habitat for eastern spadefoots. These toads have been eliminated from much of their range in Connecticut by habitat destruction. The relatively fast-drying pools used by spadefoots may form in spring or summer. The pools' existence is too short and erratic to support populations of typical vernal pool amphibians.

Some of these basins contain semi-permanent ponds that dry up only during drought. Yellow water-lilies may grow in the lowest parts of the basin. Exposed shores may have a short turf of spikerush (*Eleocharis acicularis*), plus bluejoint and *Panicum* grasses, *Carex* sedges, rice cutgrass, dwarf St. John's-wort, smartweed, marsh cress, and nodding bur-marigold. The high-water line of one pond I often visit is marked by tall eastern cottonwoods. Periodic drying eliminates fishes that may have been introduced or that colonized via intermittent outlet streams that flow into permanent rivers during wet periods. Drying makes these wetlands important breeding habitats for amphibians and invertebrates that do poorly in permanent ponds with predatory fishes. In their fishless condition, these ponds are rich in aquatic life and highly productive, supporting an array of species such as fairy shrimp and clam shrimp that thrive in temporary pools and painted turtles and bullfrogs, including large larvae, usually found in permanent ponds.

In fall, where semi-permanent kettle ponds had dried, I found live adult diving beetles, dragonfly nymphs, planorbid snails, and eastern newts under wood or rocks on the basin bottom. Others animals, such as painted turtles, walk overland to permanent water when the pond dries.

The edges of some of these ponds host rare plants typical of coastal plain pond shores.

Tree-hole Pools. Temporary pools also exist in tree holes. Common inhabitants include larvae dipterans (e.g., tree-hole mosquito, chironomids, biting midges, craneflies, moth flies, wood gnats, syrphid flies) and beetles (helodids, scirtids, and pselaphids). Amphibians in Connecticut do not use tree-hole pools for breeding, but gray treefrogs use them for maintaining water balance in summer.

Wetland Indicators

In Connecticut, all wetlands are subject to protective regulations, and if you are planning to develop your property you will need to know if the area includes wetlands. Of course, this is often easy enough if the site has standing water for most or all of the year. But many wetlands, such as vernal pools, shrub swamps, and red maple swamps, may be seasonally dry and not obviously a wetland, yet they are still subject to regulation. So how can these be identified during the dry period?

Connecticut regulations state that wetlands are defined by soil characteristics and are to be delineated on a site development map by a certified soil scientist. Mucky or peaty organic soils and poorly drained and very poorly drained mineral soils are characteristic of wetlands, and mineral soils of these kinds may develop certain features (such as gray, greenish, or bluish-gray color or gray-yellow-brown mottling) that identify them as wetland or hydric soils. However, anyone who takes the time to learn some botany can make good predictions about whether an area is a wetland based on the kinds of plants that are present. In fact, indicator plants often are the primary evidence used to map wetlands (see also the preceding discussion of vernal pools for information on other types of wetland indicators).

Some plants, known as obligate hydrophytes (OBL), virtually always occur in aquatic habitats or wetlands and are reliable indicators of such habitats. Examples of the approximately 430 species of obligate hydrophytes in Connecticut include the trees and shrubs are indicated as OBL in Chapter 9, plus many herbaceous plants, such as water-lilies, hornwort, water-celery, pondweeds, water-plantain, grass-pink, blue flag, bur-reed, skunk cabbage, cattail, and wild rice.

In addition, the state hosts some two dozen species of exotic obligate hydrophtes, such as fanwort, yellow iris, water shamrock, Eurasian water-milfoil, watercress, and creeping yellow cress. Though not part of the native flora, these too indicate aquatic or wetland habitats subject to regulation.

There are numerous additional herbaceous and woody plants that may occur in wetlands as well as in nonwetland situations. For example, red maple and black gum are typical swamp trees, but they grow in uplands as well. Ecologically flexible plants that regularly live in uplands but more often occur in wetlands are known as facultative wetland plants (FACW). Black gum is a FACW species, but red maple, despite its dominance in swamps, more often occurs in uplands, so it is designated as FAC, which refers to plants that occur in wetlands 34 to 66 percent of the time.

Additional examples of Connecticut's approximately 275 native species of facultative wetland (FACW) plants, in addition to those woody plants indicated as FACW in Chapter 9, include bluejoint grass, jack-in-the-pulpit, jewelweed, nodding ladies'-tresses, tall meadow-rue, New England aster, and New York ironweed. The Connecticut flora includes around 30 additional exotic species of facultative wetland plants, such as lady's thumb, yellow cress, marshmallow, and purple loosestrife.

Areas dominated by multi-species assemblages of OBL or FACW plants likely are wetlands that fall under the scope of state and local wetlands regulations. Additionally, lands within 50 to 100 feet of such wetlands (upland review areas) usually are regulated as well. You must contact your town's wetlands agency and apply for a permit before altering or developing any regulated wetland or upland review area.

Plants and the Wetland Environment

The vast majority of plants cannot survive flooding or saturated soils for very long, especially during the growing season. Herbaceous plants may succumb to flooding in just a few days. Young woody plants may survive a few weeks, and mature upland trees may live months or years, but eventually they die. Yet certain plants thrive in and may actually require flooded or saturated soils. How do they do it?

Before we consider how aquatic and wetland plants cope with the saturated soil conditions that kill most plants, let's review what happens physically and chemically in waterlogged or flooded soils. We need consider only the upper soil levels, since the vast majority of the roots of woody wetland plants are in the upper 18 inches of the soil, and herbaceous plant roots are mainly in the upper 6 inches.

Wetland Soils and Plant Stress

Movement of oxygen and other gases in soils occurs chiefly by the slow process of diffusion. Diffusion of oxygen through well-drained upland soils occurs relatively quickly due to the extensive network of tiny air spaces. As oxygen is depleted by the respiration of soil organisms, it is continuously replenished by diffusion from above. In waterlogged soils of wetlands, the air space is occupied by water and diffusion of gases occurs 10,000 times more slowly than in porous soils. In freshly flooded soils, respiring aerobic microorganisms may reduce the oxygen concentration to zero within a few hours (or it may take a few days). Once the soil is depleted of oxygen, the diffusion rate from the external atmosphere is insufficient to maintain the supply for oxygen-dependent microorganisms. These die off and are replaced by a new population of anaerobic microorganisms.

In completely flooded soil, anaerobic conditions prevail to within a few centimeters of the soil surface. This deep, chemically reduced layer often is bluish-gray to greenish-gray. The lack of oxygen in the deep layer prevents normal aerobic root respiration and affects the availability of plant nutrients and the presence of toxic substances in the soil (see following discussion). A thin layer of oxidized soil (often brown or brownish-red) may exist at the surface, just below the water-soil interface. This layer often is important in nutrient cycling in the wetland ecosystem. High carbon dioxide concentrations may occur in waterlogged soils, but rarely if ever does soil carbon dioxide reach lethal levels.

When the oxygen supply is limited, some soil microorganisms make use of electron acceptors other than oxygen for their respiratory oxidations. This results in the conversion of numerous compounds into a state of chemical reduction and is reflected in a lowering of the oxidation-reduction (redox) potential. In newly flooded soils, typical events include oxygen depletion, nitrate reduction, and sulfate reduction, with increases in reduced manganese, reduced iron, hydrogen sulfide, methane, and available ammonium ion. Phosphorus, which tends to be tied up in organic matter or in inorganic sediments, generally increases in soluble form in flooded, anoxic soils.

Nitrogen often is the nutrient in shortest supply in flooded soils. During decomposition of organic matter, organically combined nitrogen is converted to ammonium ion. Ammonium ion may be absorbed by plants or by anaerobic microorganisms, immobilized on soil particles, or oxidized in the surface soil layer by bacterial nitrification (or around plant roots where oxygen may be available; see following). Waterlogged soils may lose substantial nitrogen through the process of denitrification, a process in which certain facultative anaerobic microorganisms use nitrate as an oxygen source in respiration, releasing gaseous nitrogen or nitrous oxide. Counteracting this nitrogen loss, other free-living microorganisms may capture and fix gaseous nitrogen. Nitrogen may become available in the soil through the internal atmosphere of aquatic plants, with loss of nitrogen to the soil around the roots.

Reduction of sulfate to sulfide occurs through the action of obligate anaerobic bacteria. In iron-containing soils, the presence of ferrous iron, and some other abiotic mechanisms, limits the level of sulfide. Methane production results from the action of bacteria on carbon dioxide under highly reduced (anaerobic) conditions.

Many of these characteristics of wet soils create problems for growth and survival of plants. Soil oxygen deficiency per se apparently is not a problem for wetland plants (see next paragraph), though it is for nonwetland species, which may suffer from the buildup of the toxic end-products of anaerobic metabolism. Some plants are excluded from wetland soils by the toxic effects of high iron, manganese, and/or sulfide concentrations.

Waterlogging syndrome in nonwetland plants may involve leaf discoloration and, surprisingly, massive leaf water loss (plants reduce water uptake, probably due to a reduction in root metabolism caused by anoxia). Anoxia also prevents normal nutrient uptake. High levels of carbon dioxide in the soil result in reduced root permeability and wilting in nonwetland plants.

Flooding is most harmful in summer but may be tolerated in winter or early in the growing season when temperatures are colder and plant metabolism (and consumption of oxygen by soil organisms) is much lower.

Coping with Flooding and Waterlogged Soils

Herbaceous Plants. Many herbaceous wetland plants thrive in waterlogged soils and can produce extensive and healthy root systems in all but the most anaerobic of soils. These plants depend on one or more of the following characteristics: the ability to exclude or tolerate soil-borne toxins; the provision of air-space tissue to allow diffusion of oxygen from the atmosphere to the roots; the ability to metabolize anaerobically and tolerate an accumulation of anaerobic metabolites; and the ability to respond successfully to periodic soil flooding.

Plants can exclude some toxins (e.g., iron) by oxidizing them outside the roots. This may be accomplished by the loss of oxygen from the roots (oxygen is obtained from above-ground air and diffuses through the plant's air spaces). Wetland plants have a greater ability to release oxygen from the roots than do nonwetland plants. In essence, the wetland plants may create a zone of aerobic conditions around the roots. This greatly improves the root environment compared to the hostile conditions of anaerobic soils. Some or most oxidation in the vicinity of the roots may be due to enzymes in the root cells. In some cases, symbiotic bacteria (*Beggiatoa*) may oxidize hydrogen sulfide to sulfur and thus protect plant roots. In turn, the plant roots may release certain enzymes required by the bacteria.

In fully waterlogged soils, the only significant path of oxygen to the roots is through the intercellular space in the plant itself. In herbaceous wetland plants, over half the volume of the plant may consist of air space. The internal air spaces are most extensive in plants that have developed in low-oxygen soils. Plants without extensive internal air spaces generally are intolerant of wet soils and have poor ability to oxidize inorganic toxins around the roots. The spongy structure of many herbaceous wetland plants also may be beneficial in reducing the plant's tissue mass and thus reducing metabolic demands, while at the same time affording sufficient mechanical strength. The amount of oxygen within the plant's internal air spaces apparently cannot be viewed as a "reservoir," since the oxygen in this space is quickly depleted under conditions when replacement is not possible. The oxygen is replenished through photosynthesis in the foliage, and oxygen also may enter through openings in the leaves and/or stem.

Under certain conditions, the roots and underground tissues of wetland plants are subject to anoxic conditions (e.g., when submerged after leaf die-back in fall). Under these conditions, anaerobic metabolism predominates and byproducts such as ethanol or lactic acid may accumulate. Some wetland plants produce anaerobic end-products that are not as toxic as is ethanol (e.g., shikimic acid in pond lilies and irises, glycerol in alder). Under anaerobic conditions, wetland plants also may accumulate malic acid instead of ethanol; the former is nontoxic and remains in the plant until the return of aerobic conditions allows its removal. Some plants are so well adapted for wetland conditions that they actually show enhanced growth when oxygen levels around the roots are decreased.

Additional adaptations in herbaceous wetland plants include the development of adventitious roots in zones where the redox potential is less severe. Adventitious roots may form in common reed, purple loosestrife*, and docks (*Rumex* spp.). Tussock sedge has both deep and shallow roots that may shift function depending on the extent and duration of flooding. Cattails, arrowheads, and bulrushes may produce deep roots when dry conditions permit, and metabolic adjustments and good aeration of these roots allow them to survive when flooding returns. Other plants show stem growth adjustments. For example, floating-heart exhibits rapid stem elongation in response to rising water level, keeping the leaves at the surface, a more aerobically favorable environment.

How do constantly submerged aquatic plants deal with lack of exposure to the air? Generally they have thin, finely divided leaves that lack the waxy coatings typical of land plants. This makes it possible for plants such as hornwort and water-milfoil to absorb water and mineral nutrients directly from the water, so they are not completely dependent on the roots.

Fig. 6.18. Wetland plants, such as this toppled arrowwood, typically have shallow root systems.

Woody Plants. Relatively few woody plants grow in permanently flooded soil. As with herbaceous species, an adequate ventilating system appears to be essential for growth and survival in waterlogged soils. Roots do not grow in anaerobic soils but, once they are established, they can survive sometimes for months, depending on the species.

Woody wetland plants generally develop wide-spreading root systems near the surface of the substrate where oxygen is least limiting (Fig. 6.18). Versatile plants like red maple may have shallow roots in wetlands and a deep taproot in uplands.

Some woody plants have lenticels that enlarge in response to flooding or saturated soils. Lenticels are perforated, wartlike structures in the bark that allow gas exchange between the air or water and the interior of the plant. Lenticel enlargement may be accompanied by an increase in diameter and/or fluting or buttressing of the lower stem, which may provide more area for lenticels and may enhance stability of shallow-rooted trees. These responses have been observed in pitch pine, white pine, eastern hemlock, green ash, white ash, black willow, cottonwood, American elm, red maple, black gum, sycamore, and others. (Herbaceous plants also may show increased stem diameter in flooded soils.) Another response to flooding is the development of multiple trunks. Stem buds break dormancy and give rise to new branches, a multitude of which increases the area available for gas exchange with the atmosphere.

The path for oxygen movement through the stems of most woody species appears to be empty conducting elements. In green ash seedlings gases move freely through the inner bark (cambium), which contains intercellular spaces. This aeration mechanism may be essential for survival of prolonged inundation. In sycamores, tulip-trees, and sweet gums, intercellular spaces are either absent or so small that their continuity across the cambium is uncertain.

Root oxidizing activity has been demonstrated in woody wetland species such as black gum, green ash, birch, and willow. Similar experiments failed to demonstrate such activity in sweet gum, tulip-tree, and sycamore; these are generally floodplain species that can withstand only short periods of inundation, even in slowly moving aerated water. In swamp tupelo, tolerance of flooding is attained through a combination of oxidation of the soil around the roots (oxygen entering through lenticels), anaerobic respiration at low oxygen concentrations, and tolerance of high carbon dioxide concentrations in the new roots after flooding. Willows in which the lenticels were progressively blocked showed a gradual reduction in oxygen loss through the roots (another study found that the leaves may be as important as the lenticels as an oxygen entry point for root aeration).

Certain species (e.g., willows, alders, buttonbush), upon flooding, may rapidly produce adventitious roots that arise from hypertrophied lenticels just below the water surface. These roots may differ biochemically from the tree's initial roots, the former being more tolerant of anaerobic conditions. Other woody plants that produce adventitious roots in response to prolonged increase in water level include buttonbush, false indigo*, green ash, black willow, cottonwoods, tulip-tree, American elm, and black gum. The roots may functionally replace submerged soil roots that die from anoxia, serving to absorb water and nutrients, and they may have enlarged lenticels and internal air spaces that facilitate gas exchange. Some flooded plants grow new roots near the surface of the soil where more oxygen is available.

Many plants that occur in both wetlands and uplands, such as black gum and red maple, have genetically differentiated ecotypes that differ in their tolerance of flooding and water-saturated soils.

Reproduction in the Water

Most wetland and aquatic plants have flowers, but much of their reproduction is vegetative (asexual). New shoots and new plants form from sprouts and detached parts of existing plants, with no pollination or genetic recombination. For example, many water-willow plants arise through rooting and growth of the tips of arching stems. Sweetflag proliferates from buds on the rhizome. Arrowhead produces root tubers that sprout into plants the following year (if not eaten by ducks). Some plants, such as bladderworts and hornwort, form specialized budlike structures (turions) that detach from decaying plants, rest on the bottom through winter, and develop into new plants the next spring. Detached pieces of water-weed and some other submerged plants can grow into new plants.

Despite the dominance of vegetative reproduction in aquatic and wetland plants, sexual reproduction through cross pollination does occur. On land, most plants are pollinated by the wind or by insects. They produce seed-containing fruits that fall to the ground or are dispersed by animals, and seeds germinate at the soil surface. But what about aquatic and wetland plants that may be wholly or mostly submerged in water and whose seeds may not reach land? How is pollination accomplished? Where does germination occur?

Some aquatic and wetland plants, such as pickerelweed, white and yellow water-lilies, water-plantain, bladderworts, and water-willow, are essentially like many upland plants in that they have showy nectar- and pollen-containing flowers that extend into the air and are pollinated by flying insects. Some, such as marsh-marigold, have patterns visible under ultraviolet light (UV), which direct UV-sensitive pollinating bees to the nectar in such a way as to ensure pollination. The sweetly fragrant flowers of swamp rose and white water-lily attract beetles, while the foul odor of skunk cabbage entices pollinating flies. The red tubular flowers of cardinal flower lure pollinating hummingbirds and swallowtail butterflies. Other aquatic and wetland plants, such as wild rice, bur-reed, and water-milfoil, extend flowers far above the water and use wind instead of insects to transport the pollen.

Wetland plants may have separate pollen flowers and seed flowers on the same plant (e.g., bur-reed, cattails) or on different plants (e.g., willow). In some plants with the male and female parts in each flower (e.g., white water-lily, buttonbush), the different parts mature at different times. In others, including wild rice, pollen flowers are below the seed flowers. Both arrangements reduce the probability of self-fertilization. The nonfloral parts of many of these plants are highly modified for the aquatic environment, but the flowers are much the same as those of upland plants.

Other water plants, such as water-celery (tapegrass), exhibit a more amphibious form of reproduction. Female flowers are at the end of a long stalk and open at the surface of the water (Fig. 6.19). The male flowers are produced underwater and eventually break off and float to the surface, where they release pollen. Pollen floats on the surface and may contact a female flower just above the water surface. Later, the stalk of the fertilized female flower retracts and pulls the developing fruits beneath the water surface.

Similarly, pollen from male flowers of *Elodea* waterweeds reaches female flowers by floating on the water surface. However, male flowers generally are very rare in these

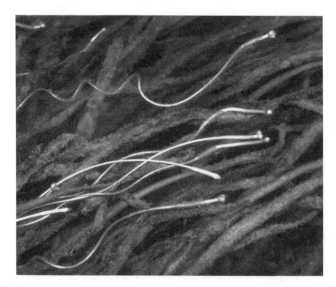

Fig. 6.19. Water-celery flowers are pollinated at the water surface. Here several cordlike flowering stalks extend to the surface above the straplike leaves. A tidal current flowed from left to right.

plants, and vegetative reproduction by fragmentation is the norm.

Pondweeds (*Potamogeton* spp.) are submergent or floating water plants that exhibit a variety of reproductive mechanisms. The petal-less flowers of some pondweeds are at or just below the surface of the water, the movement of which transports pollen from one flower to another. The whitish dustlike pollen may coat the surface of ponds where pondweed is abundant. Other pondweeds extend their flowers above the water suface and depend on wind pollination. These avoid self-pollination by ripening the male and female parts of the flowers at different times. After flowering, the spikes may lean down into the water as the fruits mature (the nutlets and rhizome tubers of these pondweeds are a frequent food of waterfowl and rails). Some pondweeds do not produce much seed and reproduce primarily through vegetative means, such as by sprouting from stem fragments.

Hornwort or coontail (*Ceratophyllum*) is one of the few aquatic plants in which cross pollination occurs completely underwater (Fig. 6.20). Anthers from male flowers break off and float to the surface. Released pollen sinks and contacts the stigmas of the separate, submerged female flowers.

Seed dispersal in aquatic and wetland plants may be accomplished by wind, water currents carrying floating seeds, or animals. For example, cattails rely on the wind to carry their tufted seeds to new locations, arrow-arum seeds (Fig. 6.21) often float away from the parent plant, and some of the wetland plant seeds eaten by migratory ducks pass through the gut unharmed and end up far from their source. Passage through an animal's gut may actually facilitate germination.

Fig. 6.20. Hornwort is a common submerged plant. Its tiny flowers (not visible here) are pollinated underwater.

The barbed seeds of bur-marigolds disperse after snagging onto mammal fur or bird feathers. Tiny seeds such as those of invasive purple loosestrife may get moved long distances in the mud that clings to the feet of birds or mammals, while the dustlike seeds of orchids readily disperse in the air. Jewelweed fruits open explosively and fling the seeds as far as several feet from the parent plant.

Many species of wetland plants, including arrow-arum, cattails, wild rice, buttonbush, American elm, and black willow, can germinate and begin seedling growth under water.

Fig. 6.21. These arrow-arum fruits will disperse by floating with a rise in water level.

Others, such as black gum, green ash, and sycamore, germinate more successfully after water levels drop, allowing seedlings to get their leaves into the air. Submergence of young growing plants may increase ethylene production, which results in more rapid stem growth and enhanced ability to project stem tips and leaves above the water.

Conservation

Like tidal wetlands, inland wetlands in Connecticut have been extensively altered by humans. About 40 to 50 percent of the state's freshwater wetlands have been lost as a result of filling, draining, dredging, or flooding by reservoirs. Other wetlands have suffered from pollution, tree cutting, and invasion of non-native plant species. Increased recognition of wetland values, and better protective regulations (and enforcement), are vital to preventing further losses and degradation. Restoration of damaged wetlands, and creation of new ones as mitigation for unavoidable losses and to offset historical losses, should be high on the conservation agenda.

Values

Natural wetlands provide many benefits to wildlife and humans. As the previous sections have shown, wetlands support unique, diverse, and abundant plant life, and wetland productivity supports vast numbers of animals, including many species of mammals, birds, reptiles, invertebrates, and plants, and most fishes and breeding amphibians. Associated with these wildlife values are many recreational and commercial opportunities, including shellfishing, fishing, hunting, and nature study. Wetlands help maintain water quality by removing pollutants and sediments. Wetlands may act as sources of nutrients, nutrient sinks, or nutrient transformers, depending on the wetland type, season, ecosystem condition, and type of nutrient. Other notable values include water retention aiding in flood control, protection from hurricane-associated wave damage, erosion control, groundwater recharge, water supply, resources for education and research, and scenic beauty.

Regulations

Passage and implementation of Connecticut's Inland Wetlands and Watercourses Act in 1972 reduced the amount of degradation and outright destruction of wetlands that formerly occurred with little or no review or consideration of the impact on wetland values or other property. Under the Act, most alterations and potentially harmful activities in inland wetlands and watercourses are now illegal without a permit. Anyone considering an alteration in or adjacent to an inland wetland or watercourse must contact their town wetlands commission to find out if a permit is required.

Construction, alteration, repair, or removal of dams, dikes, reservoirs, and similar structures is regulated by the Inland Water Resources Division under CGS Sections 22a-401 through 22a-411. Activities that cause, allow, or result in the withdrawal from, or the alteration, modification, or diminution of, the instantaneous flow of the waters of the state also are regulated. In general, a permit is required to conduct activities that result in the alteration of surface water flows, and for withdrawals of surface and groundwater exceeding 50,000 gallons in any 24-hour period. These activities are regulated by the Inland Water Resources Division under CGS Sections 22a-365 through 22a-379.

Further information on these and other regulations pertaining to wetlands and watercourses can be obtained from the Connecticut Department of Environmental Protection.

Non-native Invasive Plants

Despite these regulations, major threats to inland wetlands still exist. Important among these are non-native invasive species that threaten the ecological integrity of wetlands. Common reed (Fig. 6.22) and purple loosestrife* in particular have invaded and become dominant in many wetlands. Common reed is native to Connecticut, but the native form has been replaced over the past century by an aggressively invasive, dense-growing form that was inadvertently introduced from Eurasia and now is common is areas not previously known to have common reed. This tall grass spreads primarily by underground rhizomes. Purple loosestrife, another invader from Eurasia, is a large, robust plant that requires cross-pollination by insects. It spreads mainly as a result of its prodigious seed production. Seeds remain viable for years.

Common reed and purple loosestrife often invade disturbed wetlands and their margins. Expansion of these plants has been accompanied by biological impoverishment: monocultures of common reed replace relatively rich native communities, and productivity of native plants is reduced as loosestrife increases in dominance. Profusely flowering purple loosestrife might also negatively affect native plants by outcompeting them for pollinators.

The makeup of the plant community is not the only thing affected by invasive exotics. A myriad of animals that are tied closely to the native plants may decline in abundance or disappear altogether as their native plant hosts are

Fig. 6.22. Common reed (left) encroaches on a cattail marsh.

replaced by exotic species. Most native wildlife coevolved with native plants and are less able to make use of exotic plants, or their movements may be hindered by the physical structure of the exotics. Altered ecological processes may extend the impact of the exotic plants beyond the area in which they are present (for example, via changes in decomposition and the export of nutrients from the system, or through impacts on migratory birds that depend on native plants as food sources).

Exotic-dominated communities are not without value. Some may help retard erosion and benefit water quality. And they do fulfill some of the ecological functions of the native plants they replace (see Chapter 3). But often they are but a cheap imitation of the biological richness and intricacy of the native ecosystem.

Because of this negative impact of exotic plants, efforts are being made to find ways to manage them such that their effects are minimized. Some success in controlling common reed in freshwater wetlands has been attained by cutting the plants and carefully applying herbicides (see Chapter 3). Another approach being used against invasive plants involves studying the plants in their native habitats (usually Europe or Asia), identifying the pathogens, parasites, predators, or herbivores that limit them there, and introducing these organisms into areas where the plants need to be controlled. Research thus far indicates that leaf-eating beetles (*Galerucella* spp.), root-mining weevils (*Hylobius transversovittatus*), and flower-feeding weevils (*Nanophyes marmoratus*)—native to Europe—may be helpful control agents for purple loosestrife. However, as we have often learned the hard way, introductions must be done cautiously because the pathogens and herbivores introduced to control plant pests may instead or additionally attack native species and become yet another pest.

Habitat Protection

In general, plant species richness of marshes, swamps, and other wetlands increases as the size of the wetland increases. Animal species richness also is higher in larger wetlands. Because of their richness (due in part to the presence of certain rare species that do not use small wetlands) and their favorable prospects for long-term viability (small tracts support small populations that are prone to extirpation), large wetlands and large wetland/upland mosaics warrant high conservation priority.

Vernal Pools. The foregoing does not mean that small, isolated wetlands such as vernal pools are expendable or less valuable. As we have seen, vernal pools are unique and support an array of species that may be virtually absent from larger wetlands.

Vernal pools also warrant special attention because of their high vulnerability to damage. Many vernal pools have been destroyed or degraded by filling and draining done in conjunction with land development projects. Although wetland protection laws now have reduced these impacts, many vernal pools are nevertheless filled or disturbed by people who do not recognize them as regulated wetlands.

Vernal pool ecosystems include more than just the water. Amphibians that breed in vernal pools absolutely require surrounding upland wooded areas for summer feeding and winter hibernation sites. Required habitat features include fallen wood, leaf litter, and uncompacted soil perforated with holes formed by decayed root systems or dug by small burrowing mammals. Adult salamanders often move far from vernal pool breeding sites to their summer and winter habitats (Fig. 6.23). Similarly, wood frogs commonly move to upland feeding and wintering areas that may be several hundred yards from the vernal pool breeding site. These amphibians generally (but not always) return to breed in their natal pool. Juvenile salamanders and frogs may disperse a half mile or more from their natal pool.

The narrow buffer areas usually established around vernal pools and other wetlands are too small to adequately protect vernal pool amphibians. Clearing of wooded areas surrounding vernal pools may eliminate several required habitat features, such as sources of shade important in preventing premature drying of the pool before amphibian larvae have completed their development; upland feeding habitat of adults; summer cover and winter hibernation sites; sources of the leaf litter that provides important nutrient inputs to the vernal pool ecosystem as well as nesting cover for marbled salamanders; sources of shade important in maintaining temperature and moisture conditions needed by vernal pool amphibians when they are on land;

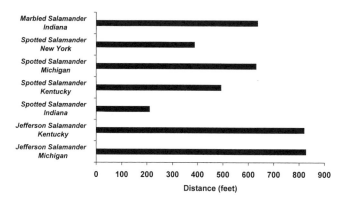

Fig. 6.23. Average terrestrial migration distance from breeding site for vernal pool salamanders, based on intensive studies of radio-tagged or isotope-tagged individuals. Vernal pool ecosystems include not only the pool but also a large area of land around the pool.

and sources of woody debris used as amphibian egg attachment sites.

Clearing of wooded upland habitat can cause significant declines in or extirpation of the vernal pool amphibian populations. In fact, available information indicates that decreases in forest cover and increases in road density within a mile or two of a wetland result in declines in species richness not only of amphibians but also other vertebrates and wetland plants.

Golf course development adjacent to vernal pools may result in high levels of nutrient inputs from fertilizer applications or fecal droppings of flocks of golf course–associated geese. These impacts are not compatible with the maintenance of viable populations of typical vernal pool species. Pesticides applied to golf course lawns could contaminate nearby waters and might pose toxicity problems for vernal pool amphibians and their food resources. And of course, the thick turf of grassy fairways does not meet the upland habitat needs of vernal pool amphibians.

Replacement of the vernal pools with deeper permanent ponds, such as found on most golf courses, is not suitable mitigation—many vernal pool amphibians and invertebrates cannot maintain viable populations in permanent-water habitats that ultimately are invaded by or are stocked with predatory fishes and/or other incompatible aquatic species.

Selective timber harvest near vernal pools, if done carefully, can be accomplished with minimal harm, but the vernal pool basin and the adjacent habitat extending 100 feet from the edge of the basin should be left completely undisturbed. For ideal protection, buildings, roads, driveways, orchards, gardens, croplands, lawns, parking areas, and other permanent clearings or structures should not be located within approximately 700 feet (215 m) of a vernal pool.

Road salt that washes off pavement and into vernal pools and other wetlands is a concern in many areas. In some situations, greatly elevated salt concentrations may extend hundreds of yards from the roadway. High salt concentrations can reduce the species richness and abundance of native wetland plants and may render the wetlands unsuitable for certain invertebrates and amphibians.

Strict conservation measures are appropriate for high-quality vernal pool ecosystems (that is, the pool basin and surrounding uplands) supporting relatively large numbers of vernal pool amphibians. Areas with small pools that may contain few amphibian egg masses may seem to warrant lesser protection, but caution is in order—such pools often are not very far from a major, productive vernal pool. Additionally, even a small pool may serve as a successful amphibian breeding site in wet years, when it may serve as an important periodic link between populations in widely separated pools. Although vernal pool amphibians tend to breed in their natal pool, they sometimes do not, and this dynamism establishes population and genetic links among an array of pools. In fact, persistence of amphibian populations in a region often depends on the existence of a network of small wetlands that allow inter-wetland movement and recolonization of sites in which drought has extinguished a local population. Elimination of too many small wetlands may prevent recolonization and can result in permanent loss of amphibian species in an area. What happens to one pool can impact populations in other pools. One consequence of the recolonization dynamics of vernal pool amphibians is that the conservation value of a small, somewhat isolated wetland may be greater than that of a wetland of equivalent size located adjacent to another wetland. Also, there are important mitigation implications. For example, creation of one 1-acre pool likely would be inadequate mitigation for the loss of four scattered quarter-acre vernal pools.

Atlantic White Cedar Swamps. Cedar swamps are of particular conservation concern. During the 1900s, the number of stands declined from eighty-six to thirty-nine. The decline was due primarily to conversion to agricultural or commerical uses, including selective logging. These swamps may require management to ensure their perpetuation. Some biologists believe that cedars may be replaced by red maple if lowered water levels and/or openings among the cedars allow the maples to colonize the swamp. Others point to fire as a necessary element in stand regeneration. In a swamp in Massachusetts, fine-resolution pollen analyses covering a 1,000-year period indicate that Atlantic white cedar did not persist for more than a century or two without a fire. In the absence of fire, or lacking canopy-removing disturbances such as logging, which create favorable conditions for regeneration, cedar stands in Connecticut may slowly decline further, even with protection. Atlantic white cedar swamps also can be damaged by permanent flooding such as might result from new beaver dams or road construction that backs up water behind inadequate culverts.

Chapter 7

Uplands: Terrestrial Ecosystems

As we saw in Chapter 6, Connecticut has many wetlands, but upland ecosystems are far more extensive (Fig. 7.1). Uplands have well-drained soils, and the ground is rarely saturated with water. Some upland areas occasionally may be flooded or saturated but, if so, this happens so infrequently that characteristic wetland plants are absent.

Deciduous Forests

Connecticut is a land of trees, and deciduous trees—leafless for several months each year—dominate most of the landscape (Figs. 7.2 and 7.3). Aside from Long Island Sound, there is no spot in the state that is far from an oak, maple, birch, or other deciduous tree.

Fig. 7.1. Forested uplands dominate the Connecticut landscape. Paul J. Fusco/Connecticut DEP Wildlife Division.

Fig. 7.2. Deciduous forest is the dominant ecosystem in Connecticut.

Factors Affecting Forest Composition and Development

Soil Relationships. The plant species in a deciduous forest vary with soil characteristics, especially moisture, temperature, and chemistry (e.g., acidity, fertility), which in turn are affected by vegetation. Accordingly, the oaks, hickories, and associated vegetation of warm, dry soils of sandy or rocky south-facing slopes contrast with the sugar maples and beeches growing in rich humus soils or on cooler north-facing slopes. (In the Northern Hemisphere, south-facing slopes face toward the Sun more so than north-facing slopes, and the heating and drying effect of the Sun is greater on south-facing slopes.) Soils of deciduous forests on south-facing slopes are relatively hot and dry in summer, and they lose their snow cover quickly after winter storms, even when adjacent wooded flatlands retain a thick blanket of snow.

Fig. 7.3. Deciduous forests are leafless for half the year.

Moist deciduous forests often have well-developed plant layers or stratification and a large diversity of woody plants (Table 7.1, fourth column). Trees often grow to large size under the favorable growing conditions. Tall trees form an umbrella-like canopy over smaller shade-tolerant trees, a still shorter shrub layer, and a low herbaceous layer of wildflowers and ferns. A distinct layer of dead leaves and scattered rotting logs covers the forest floor. Soils are thick, humus-rich, and often rocky.

Sites with intermediate moisture availability may support similar plant species in lesser abundance plus others that tolerate drier soils (Table 7.1, third column). American chestnut, formerly a dominant species and now occurring mainly as sprout saplings, was reduced to a vestige by an introduced fungal blight disease in the early 1900s. In some areas, mountain laurel forms dense thickets that contrast with nearby relatively open areas that may have a cover of lowbush blueberry and huckleberry shrubs through which you can easily walk. Laurel-dominated areas are rather monotonous, tending to have few or no other plants other than large canopy trees. In central Connecticut, red maple, red oak, and black birch often are the most common trees in mid-slope forests, but many other species often are as numerous as one or more of these. The understory of some oak forests is occupied by numerous young white pines, growing slowly but steadily toward eventual dominance of the site.

Drier forests on south-facing hillsides and ridge tops commonly have simpler structure and consist primarily of openly spaced trees, sometimes with a layer of low shrubs such as huckleberries and blueberries, or a grasslike carpet of sedges between the trees (Fig. 7.4; Fig. 7.12). Soils may be thin. Hilltop trees are generally rather small due to slow growth in dry soils. Relatively dry hilltop forests are less diverse in woody flora than are moist areas, but they have a relatively unique assemblage of plants, many of which are shared with sand plain forests (Table 7.1, column 2). Oaks are more tolerant of low-moisture conditions than are most other deciduous forest trees and so tend to dominate drier sites and increase in relative abundance during long droughts.

You can often see different stages of vegetation development on hilltops and dry rocky slopes. Bare rock is initially colonized by mosses and lichens, which trap wind-blown dust and begin soil formation (Fig. 7.5). Soon sedges, grasses, lowbush blueberries, huckleberries, chestnut oaks, and others colonize. These trap more dust and deposit leaves and other debris, further enhancing soil and vegetation development. The rock exposure becomes progressively smaller and may disappear entirely as the thin soil layer and vegetation continue to expand.

Table 7.1. Ecological Distribution of Some Trees and Shrubs in Connecticut

Species	Dry sites (ridgetops, sandplains)	Intermediate	Moist sites (bottomlands, near wetlands)	Wetlands
Scrub oak	xxxxx			
Dwarf chestnut oak	xxxxx			
Bearberry	xxxxx			
Pitch pine	xxxxx	—	—	—
Scarlet oak	xxxxx	xxxxx		
Red-cedar	xxxxx	xxxxx		
Common juniper	xxxxx	xxxxx		
Pin cherry	xxxxx	xxxxx		
Black huckleberry	xxxxx	xxxxx		
Lowbush blueberry	xxxxx	xxxxx		
Sweet fern	xxxxx	xxxxx		
Bayberry	xxxxx	xxxxx		
Pasture rose	xxxxx	xxxxx		
Shining sumac	xxxxx	xxxxx		
Black oak	xxxxx	xxxxx	—	
White oak	xxxxx	xxxxx	—	
Pignut hickory	xxxxx	xxxxx	—	
American chestnut	xxxxx	xxxxx	—	
White pine	xxxxx	xxxxx	—	—
Chestnut oak	xxxxx	xxxxx	—	
Shadbush	xxxxx	xxxxx	xxxxx	
Black cherry	xxxxx	xxxxx	xxxxx	
Black birch	xxxxx	xxxxx	xxxxx	
Aspens	xxxxx	xxxxx	xxxxx	
Gray birch	xxxxx	xxxxx	xxxxx	xxxxx
Red maple	xxxxx	xxxxx	xxxxx	xxxxx
Sheep laurel	xxxxx	xxxxx	xxxxx	xxxxx
Red oak	xxxxx	xxxxx	xxxxx	
White ash	(trap-rock)	xxxxx	xxxxx	
Shagbark hickory	(trap-rock)	xxxxx	xxxxx	
Sugar maple	(trap-rock)	xxxxx	xxxxx	
Hop-hornbeam	(trap-rock)	xxxxx	—	
Eastern hemlock	—	xxxxx	xxxxx	—
Paper birch	—	xxxxx	xxxxx	
Striped maple	—	xxxxx	xxxxx	
Sassafras		xxxxx	xxxxx	
Hazelnuts		xxxxx	xxxxx	
Mountain maple		xxxxx	xxxxx	
Basswood		xxxxx	xxxxx	
Mountain laurel		xxxxx	xxxxx	
Maple-leaved viburnum		xxxxx	xxxxx	
Mockernut hickory		xxxxx	xxxxx	
Highbush blueberry		xxxxx	xxxxx	xxxxx
Grapes		xxxxx	xxxxx	xxxxx
American beech		—	xxxxx	
Yellow birch		—	xxxxx	
Flowering dogwood		—	xxxxx	
Tulip-tree		—	xxxxx	
Bitternut hickory		—	xxxxx	
American hornbeam		—	xxxxx	
Pink azalea		—	xxxxx	
Witch-hazel		—	xxxxx	
Black gum		—	xxxxx	xxxxx
Sweet pepperbush		—	xxxxx	xxxxx
Butternut			xxxxx	
Spicebush			xxxxx	xxxxx
Silky dogwood			xxxxx	xxxxx
Gray dogwood			xxxxx	xxxxx
Red osier			xxxxx	xxxxx
Fetter-bush			xxxxx	xxxxx
Maleberry			xxxxx	xxxxx
Swamp azalea			xxxxx	xxxxx
Common elder			xxxxx	xxxxx
Northern arrowwood			xxxxx	xxxxx
Common winterberry			xxxxx	xxxxx
Pin oak			xxxxx	xxxxx
Swamp white oak			xxxxx	xxxxx
American elm			xxxxx	xxxxx
Wild black currant			xxxxx	xxxxx
Box elder			xxxxx	xxxxx
Green ash			xxxxx	xxxxx
Black ash			xxxxx	xxxxx
American sycamore			xxxxx	xxxxx
Eastern cottonwood			xxxxx	xxxxx
Alders			xxxxx	xxxxx
Willows			xxxxx	xxxxx
Chokeberry			xxxxx	xxxxx
Spiraea			xxxxx	xxxxx
Marsh-elder			xxxxx	xxxxx
Sweet gale			xxxxx	xxxxx
American larch			xxxxx	xxxxx
Black spruce			xxxxx	xxxxx
Silver maple			xxxxx	xxxxx
Poison sumac				xxxxx
Swamp rose				xxxxx
Smooth winterberry				xxxxx
Mountain-holly				xxxxx
Swamp buckthorn				xxxxx
Atlantic white cedar				xxxxx
Cranberry				xxxxx
Bog-rosemary				xxxxx
Labrador-tea				xxxxx
Bog laurel				xxxxx
Leatherleaf				xxxxx
Buttonbush				xxxxx

xxxxx = primary distrubution. ——— = secondary distribution.

Fig. 7.4. Dry hillside or ridgetop forests often have small, widely spaced oak and hickory trees and an understory of huckleberry, blueberry, or woodland sedge.

Some deciduous forest plants are fairly good indicators of soil acidity. Indicators of strongly acidic (pH 4 to 5) soils include areas with abundant mountain laurel, azaleas, blueberries, trailing arbutus, wintergreen, partridge-berry, and pink lady's-slipper. In the southwestern quarter of the state, deciduous forests often are dominated by basswood, sugar maple, tulip-tree, yellow birch, white ash, butternut, and black cherry, and the blueberries and huckleberries so common elsewhere are replaced by flowering dogwood, witch-hazel, spicebush, American hornbeam, and hop-hornbeam, with a rich herbaceous flora, that together reflect the less-acidic soil conditions in this region.

Vegetation generally reflects soil chemistry differences deriving from local geological variations. For example, mature upland forests in northwestern Connecticut, where soils may be quite acidic, often are dominated by eastern hemlock and American beech, with additional species including sugar maple, white ash, yellow birch, and white pine. In contrast, white pine and sugar maple often are the dominants in adjacent lowland forests on calcareous soils.

Shade Tolerance. Availability of sunlight—independent of soil moisture—is another important influence on plant distribution. Understory tree saplings and other plants in deeply shaded areas differ from those of sunnier forest edges, "thinned" forests, and forest gaps. Very shade-tolerant seedlings, saplings, or sprouts of sugar maple or American beech may be common under a dense canopy of maples, beeches, or other trees, but sun-loving pitch pine and tulip-tree are scarce or absent on shady forest floors that do not meet their germination requirements. Also, differences among species in the survivorship and growth of tree saplings, which in part reflect shade tolerance, strongly influence the species composition of trees that grow up into mature trees. Sunny gaps in a forest, resulting from death of a large overstory tree, provide favorable growth conditions for black cherry, white oak, red oak, white pine, and other species that grow slowly in shade. Shading is a critical factor in forest community dynamics.

The vegetation of new forest openings may include a variety of sun-loving plants that were not present in the area before the opening was created. Where did these plants come from? Some may have originated from seeds dispersed by the wind over long distances, or transported to the site by squirrels, jays, or other animals, whereas others germinated from seeds that had been dormant in the soil for many years.

Mature trees casting heavy shade, and sunny forest gaps are both natural expressions of deciduous forest ecosystems and contribute importantly to their biological diversity.

Fig. 7.5. Lichens and mosses colonized this rock, setting the stage for soil formation and allowing colonization by grasses, forbs, and woody plants.

Microrelief. On a smaller scale, plants growing on a mound of bare soil exposed at the base of a fallen tree tend to be different than those occuring in the treefall pits next to these mounds or on nearby undisturbed, leaf-covered forest floor. The different microsites develop different soil characteristics that affect plant distributions. For example, relative to treefall pits, adjacent mounds tend to be drier, covered by less leaf cover and winter snow, lower in nutrients and organic matter, warmer in summer, colder in winter, and more subject to frost heaving.

Fig. 7.6. This 25-foot-tall black birch, photographed in 2002, became established on the bare soil mound at the base of a tree toppled by Hurricane Gloria in 1985. Roots eventually grow downward into the forest soil.

Fig. 7.7. Erosion of the soil mound (or decomposition of the log), on which this black birch became established, has left the tree on "stilts."

An easily observed example of this microsite phenomenon is provided by black birch, which readily colonizes the bare soil mounded at the base of wind-toppled trees (Figs. 7.6 and 7.7). Many mature black birches in Connecticut forests grow on tree-fall mounds formed as a result of the 1938 hurricane, and a cohort of younger birches stand on mounds caused by Hurricane Gloria in 1985. Adjacent undisturbed areas generally lack black birch seedlings because the seedlings do not produce a root long enough to penetrate a thick mat of leaves. Instead, tree seedlings that successfully establish in leafy forest floors often are large-seeded plants that have sufficient resources to put down a long tap root through the leaf mat or that have been "planted" (stored but not eaten) by squirrels or jays. Thus black birch may be scarce in undisturbed forests but may increase at the expense of oaks in forests where fallen trees are common. In addition to black birch, brambles (*Rubus* spp.) sometimes are common on soil mounds in forests but scarce in nearby undisturbed sites.

Logging and Fire. Historical logging and fire have affected forest composition throughout the state. Past tree-cutting associated with agricultural expansion in the 1700s and 1800s and, later, charcoal production that helped fuel the industrial revolution affected virtually every forest stand such that most forests today are dominated by trees not much more than a hundred years old.

Many deciduous forest trees, including oaks, hickories, maples, birches, basswood, chestnut, and others, readily send up new sprouts from the stumps or lower stems of trees that have been cut or injured by insects or disease. Most of the small clusters of young but mature trees common in today's forests originated from the stumps of trees cut several decades ago (Fig. 7.8). Repeated cutting in the past, together with fires (see following), probably eliminated from some areas nonsprouting trees such as hemlock and contributed to the dominance of oaks and other trees that sprout from stumps.

Despite historical logging, today's forest do include some

Fig. 7.8. Many deciduous forests include clusters of trees that originated as stump sprouts following logging.

Fig. 7.9. This large white oak developed its massive spreading limbs in the absence of competing trees. Now it is surrounded by slender trees of a young forest that replaced an abandoned field.

large trees. Old, thick-but-short-trunked "wolf trees" with wide spreading branches, indicating their former presence and development in open areas without nearby competing trees, are not uncommon in young forests (Fig. 7.9). These wolf trees were spared during the last tree-cutting episode. The younger trees developed long trunks and few major side branches as they sought light above adjacent competing trees.

Lands that were cleared for pastures and hayfields and then became reforested following the decline in agriculture sometimes can be recognized by dead and dying, shade-intolerant red-cedars (typical colonizers of turf) growing beneath a canopy of deciduous trees (Fig. 7.10). Old stone walls often transect such areas. Several other trees and shrubs invade and flourish in former pastures. Sassafras spreads to openings through sprouts that arise from the expanding root system of trees along forest edges. Common juniper and white pine are typical colonizers of pastures in the northern hills of Connecticut; they persist because cattle avoid eating them. Gray birch often colonizes and thrives in bare spots in old fields. Even tree species characteristic of old, mature forests can be among the initial colonists of abandoned pastureland (for example, as a result of jays or squirrels "planting" nuts or seed dispersal by the wind). Today, multiflora rose, autumn-olive, and various other exotics run rampant in some old fields. Most of these exotics have bird-dispersed seeds (see Chapter 20).

Logged or burned areas often are colonized by gray birch, black birch, black cherry, or pin cherry, species that can establish themselves in mature forests only where disturbance has removed the canopy or exposed bare ground. Paper birch, most common in northern Connecticut, also indicates former disturbance. These plants may appear quickly when long-dormant seeds, newly exposed to the Sun, finally germinate, or through effective dispersal of seeds by the wind (birches) or birds (e.g., pin cherry). Pollen data from sediment cores taken from pond bottoms indicate that birch abundance increased greatly after European colonization, a response to increased forest disturbance. Higher incidence of fires during the colonization period probably caused a decrease in the abundance of eastern hemlock and American beech. Cutting for the tanning industry also contributed to the decrease in hemlock.

Today, fire is not a very significant factor influencing Connecticut's deciduous forests, due to effective suppression by humans and the often wet conditions. But Native Americans used periodic if not frequent burning to manage some forests for a favored open understory, wildlife habitat, and food plants. Past burning accounts to some degree for today's widespread vegetation of oaks, hickories, and red maple, all species that sprout after being burned, and large

Fig. 7.10. Dead red-cedars (right) in this deciduous forest indicate that the site formerly was a grassy field or clearing.

areas of lowbush blueberry, black huckleberry, and other fire-thriving plants. Burning by Native Americans also created and maintained largely treeless areas along the coast and in river valleys. Among deciduous trees, American beech is one of the least fire-tolerant species, and it, along with its frequent moisture-loving and fire-intolerant associate, eastern hemlock, is largely restricted to moist sites that do not readily burn.

Severe Weather Events. Hurricanes periodically have major impacts on Connecticut's deciduous forests, though these forests are much less susceptible to wind damage than are mature conifer stands. However, a stand of old white oak, black oak, shagbark hickory, and red maple in Stonington was destroyed by the hurricane of 1938. Trees downed by Hurricane Gloria in September 1985 are conspicuous in many forests today (Fig. 7.6).

Dominant canopy trees may have their leaves stripped and limbs broken by hurricane-force winds, or they may be toppled and killed by excessive root breakage. Understory plants respond to the resulting increase in light availablility with increased growth, and they may change status from a stunted minor element in the community to a dominant. Dormant seeds of light-loving plants get a chance to germinate, grow, and reproduce. Winter storms that coat woody plants with heavy burdens of ice often result in broken limbs that similarly prune the vegetation, let more light into the understory, and improve conditions for certain understory vegetation. On thin soils, strong winds may uproot trees and expose bare rock, setting the stage for recolonization by lichen species that begin the process of vegetation development on such inhospitable sites, as discussed earlier.

Fig. 7.11. A spring drought killed the leaves of this chestnut oak on a dry hilltop.

Drought can stress or kill plants and may open the way for competing xerophilic species or render the plants susceptible to pathogens. A pronounced drought in the spring of 2001 killed the newly grown leaves of many trees (Fig. 7.11).

Biotic and Other Influences. The current distribution and relative abundance of deciduous forest trees and shrubs reflects not only their interactions with the physical environment but also biotic factors such as the impacts of herbivorous mammals including white-tailed deer (Chapter 21) and insects, non-native invasive species (see "Conservation" at the end of this chapter; end of Chapter 9), and the effects of seed-dispersing birds and mammals (see "Acorn Ecology," this chapter; "Fruits and Fruit-eating Birds," Chapter 20). Pathogens can be particularly influential. For example, increased abundance of oaks in the 1900s resulted from the massive die-off of competing American chestnuts caused by the chestnut blight fungus. Additionally, there are new influences at work, including acid deposition caused by air pollution, an increased level of atmospheric carbon dioxide resulting from the burning of fossil fuels and worldwide deforestation, and global warming, Forest patterns that we see today reflect complex interactions of physical, biological, and anthropogenic influences and often defy simple explanations.

Deciduous Forests of Trap-rock Ridges

Trap-rock (basalt) ridges in central Connecticut warrant special mention because they support a particularly diverse

Fig. 7.12. This woodland dominated by oaks, hickories, and white ash is on the summit of a trap-rock ridge. The ground is carpeted by a turf of sedges. Shrubs are almost totally absent.

forest flora and fauna on nutrient-rich soils (Figs. 1.6 and 7.12). These forests are mostly deciduous but may also include significant stands of hemlock.

Characteristic ridge-top species include red-cedars near cliffs, various oaks and hickories, white ash, and additional trees and shrubs listed in Table 7.1. Species generally associated with moist soils (for example, striped maple) occur sporadically on the summits of some trap-rock ridges. Sugar maples may live for decades on ridge-tops before being killed by severe drought.

On the west side of the trap-rock ridges, soils are relatively shallow, sun-exposed, warm, and dry, as are conditions at the upper edge of the talus. Trees on talus slopes of trap-rock ridges include black birch, gray birch, hop-hornbeam, white ash, chestnut oak, red oak, sugar maple, and basswood, sometimes with pockets of eastern hemlock and paper birch on benches at the base of steep talus. At the base of the talus, weathered basalt from the talus slope above forms a deep, rich soil, often heavily wooded by deciduous trees and hemlocks. The feldspar minerals in the basalt include a high level of calcium and contribute to the development of soils rich in plant nutrients. The area at the base of the talus slope is kept from getting too warm not only by tree shading but also by the cool heavy air that flows downslope deep in the talus. The cool air discharging from the bottom of the talus on warm summer days is impressive (and refreshing if you know where to find it). On a balmy April 24, when air temperature was 76°F and the snow was long gone, I found that the air flowing out from the base of a sunny talus slope was 33°F—and on a warm July 27 at 88°F the air discharging from the talus was a chilly 47°F! Presumably the air is so cold because it flows over ice or recently thawed ground beneath the talus. Ice commonly persists low on the talus slope long after it has melted elsewhere.

These diverse temperature and moisture conditions along trap-rock ridges, together with the rich soils, support a strikingly diverse wildflower flora, often including wild columbine, rock-cresses, bloodroot, Dutchman's breeches, blue cohosh, wild ginger, red trillium, hepatica, wood anemone, rue anemone, early saxifrage, spring beauty, violets, and others that are much less common on nearby hills with a different geological origin. The ridges and their flora are also focal points of butterfly activity. Spring-flying butterflies on trap-rock ridges include falcate orangetip, brown elfin, mourning cloak, Compton tortoiseshell, Edwards' hairstreak, juniper hairstreak, American lady, eastern tiger swallowtail, Juvenal's duskywing, and others. Trap rock ridges additionally include rich wetland ecosystems (such as vernal pools) and provide unique habitats for cliff-associated species.

Forest Floor Flora and Fungi

Fallen leaves often dominate the floors of deciduous forests, but low-growing plants are also always present, though sometimes only sparsely. Some of the more characteristic semi-woody and nonwoody plants on the shady floor of typical deciduous forests in Connecticut include Christmas fern, wood ferns, and many other ferns, woodland sedge, other *Carex* sedges, shining firmoss, running clubmoss, common running clubmoss, tree clubmoss, wintergreen, trailing arbutus (now somewhat rare and localized), pipsissewa, spotted wintergreen, partridge-berry, wild sarsaparilla, white wood aster, Indian pipe, wood anemone, violets, dwarf ginseng, and a wide array of other herbaceous wildflowers. Dense patches of moss and lichens often are present on rocks and at the bases of tree trunks. A rich array of fungi and bacteria do their mostly hidden work of breaking down fallen leaves, wood, and other organic matter on the forest floor.

Above, on the surfaces of living leaves of forest trees, a diverse community of fungal species live and grow. Some attack living tissue while other more benign species are nourished by wind-deposited pollen, spores, and dust, as well as by exudates from the leaves. Some of the fungi are somewhat insecticidal or physiologically disruptive to insect herbivores, protecting the leaves and fungi from damage and ingestion.

Autumn Coloration and Leaf Fall

Deciduous forests of New England are renowned worldwide for their brilliant autumn foliage. The leafy palette includes stunning arrays of red, orange, gold, yellow, maroon, purple, brown, and green. What accounts for this visual feast?

In September and October, as nights grow longer and temperatures decrease, several changes occur within the leaves of deciduous trees and shrubs. Production of chlorophyll, the compound that is involved in capturing sunlight for photosynthesis (the production of sugars and oxygen from carbon dioxide and water), declines and existing chlorophyll breaks down. As a result, leaves lose their green color. At the same time, leaf stomates ("air valves") close, stopping or at least greatly reducing the movement of oxygen and carbon dioxide between the leaf and the surrounding air. Additionally, specific biosynthetic processes are at work, and hormonal changes cause leaf attachments to weaken. All of these processes lead to the changes we see in leaves in autumn.

Color Change. The predictable timing of leaf color change suggests that under natural conditions seasonal changes in light must be more influential than temperature in causing color change. Temperature variations seem to affect color intensity (see following discussion) more than the timing of color change. So how do plants, which lack eyes or other obvious light receptors, register light change and respond appropriately? A light-absorbing pigment known as phytochrome, which is widespread in plant cells, is the critical element. Phytochrome changes form and function with changes in illumination. In response to the increasingly longer nights of late summer and early fall, phytochrome exerts hormonelike effects that turn on and off the various plant processes that affect color change.

The color that leaves become after the breakdown of chlorophyll depends on what other pigments are present. Dark red, scarlet, and purple tones, such as those found in the leaves of shining sumac, black gum, some oaks, and maple-leaved viburnum, result from anthocyanin pigments that usually are dissolved in the cell sap. Anthocyanins are a byproduct of highly active metabolic processes that are involved in the retrieval of nutrients from the leaf to the stem. They form when flavonols combine with sugars that have accumulated in the leaves as a result of a reduction in the flow of sugars from the leaf to the tree as the breakage or abscission zone forms at the base of the leaf stalk (see following). In many trees, these pigments are present only in fall. Anthocyanins yield red colors in acidic conditions, violet at neutral pH, and blue in alkaline solutions. Concrete sidewalks plastered with wet leaves often become temporarily stained by water-soluble anthocyanins.

Fig. 7.13. Light strongly affects the development of red autumn color. Here a wet sweet gum leaf had another leaf stuck to its upper side. Sun-exposed parts of the leaf turned vivid red while the covered parts turned yellow.

Sunlight clearly affects the development of anthocyanins (Fig. 7.13). You can see this by examining the leaves of flowering dogwood and highbush blueberry. Portions of the leaves exposed to direct sunlight turn deep red, whereas the shaded parts of the same leaves (such as where one leaf is just above another) remain green.

Yellow and orange colors, such as those characteristic of maples, hickories, and birches, are due to the presence of carotenoids, pigments in the chloroplasts that are involved in the absorption of light energy. Carotenoids also help protect chlorophyll from being sun damaged. Carotenoids are relatively nutrient poor, are incompletely broken down (if at all), and do not yield much that the plant can store.

Some leaves (such as those of some oaks) turn brown as the green chlorophyll disappears. The brown color is due largely to the presence of tannins and the absence or scarcity of other pigments. Other colors, such as those of beeches and various oaks, result from an interaction of tannins, anthocyanins, and/or carotenoids.

In central Connecticut, most species are in peak fall color in mid-October. Bright, dry, sunny days and cool but not freezing nights yield the brightest fall colors, especially the reds. Severe early frosts reduce the reds. Cloudy days and warm nights result in slow breakdown of chlorophyll and slow production of anthocyanins—and relatively dull fall colors.

Leaf Fall. Environmental changes in autumn also induce chemical changes that result in the formation of a zone of breakage (abscission) at the base of the leaf stalk (petiole). Compound leaves such as those of ashes and sumacs may also form a breakage zone at the base of each leaflet. In the breakage zone, cells lose their adhesion to one another, and gradually a layer of protective fatty or corky material forms over the detachment site on the stem. Eventually the pressure of wind or rain breaks the ever-loosening connection and the leaf falls to the ground. Even in the absence of strong winds or rain, cold snaps appear to result in pulses of leaf fall of leaves that have a weakened abscission zone. The processes that result in leaf abscission are under the control of the plant hormones (ethylene and, in some cases, abscissic acid) that cause senescence and cessation of normal growth.

Though reduced daylengths and temperatures are the primary controls on leaf fall, other factors, such as water deficit, low light intensity, mineral deficiency, or injury, may influence hormone levels and can cause premature leaf detachment. For example, severe drought and heat caused many new, partly grown, green leaves to fall from tulip-trees and red oaks in May 2001.

Before the leaves fall, much of their nutrient content is withdrawn and stored within the perennial tissues of the plant. Transported nutrients include certain carbohydrates, nitorgen, sulphur, iron, phosphorus, potassium, magnesium, and manganese, derived from the breakdown of proteins, nucleic acids, and porphyrins. In deciduous species, this withdrawal of nutrients occurs shortly before abscission. In evergreens, nutrient recovery occurs over a longer period. Generally about half of the nitrogen and about one-third of the phosphorus is withdrawn before the leaf is shed.

Bud formation and development of cold resistance are not part of autumn color phenomena. Bud formation occurs in summer, well before the leaves are shed. Cold resistance develops after leaf fall, as the plant becomes exposed to increasingly cold temperatures.

In central Connecticut, leaf fall is heaviest in mid- to late October. By November, nearly all forest trees are leafless except for the oaks and beeches, especially small trees and the lower branches of larger trees, to which dead leaves may remain attached through most of the winter. Black willows may still hold some greenish leaves in mid-November. Bright pink-red leaves of highbush blueberry may still cling to these wetland shrubs in December and January. Planted trees in residential areas, trees along roads, and exotic plants such as Asian bittersweet often hold onto their leaves longer than do native forest trees.

Why Drop Leaves? Deciduous forest trees can be leafless for half the year or more. The long leafless period presumably is adaptive in allowing the plants to better survive winter conditions. Big, broad leaves are wonderful photosynthesizers in summer when water is generally plentiful, but in winter, retained leaves with a large surface area can cause problems by allowing excessive evaporation at a time when cold or frozen soils inhibit water uptake by roots. Probably the deciduous habit evolved as drought-avoidance behavior, effective not only in cold winters but also in warm climates that have a distinct dry season. However, the deciduous habit works best in environments that have the abundant available resources necessary for regrowing a whole new crop of leaves every year. Evergreens tend to predominate in sites where available resources are limited, and they combat drought periods in part by having needlelike leaves that lose relatively little water. Yet some broadleaf trees and shrubs can also resist drought without losing their leaves. For example, broadleaf evergreens are conspicuous in summer-drought climates that have mild winters (e.g., coastal California). These plants have tough, thick, "varnished" leaves that can greatly reduce water loss.

Deciduous species also have larger-diameter vascular vessels than do evergreens and, for physical reasons I won't go into here, are thus more vulnerable to disruption of water supply to the leaves as a result of freezing, which would limit the usefulness of leaves in winter. Additionally, dropping leaves and growing new ones the next year gives plants a chance to replace leaves damaged by herbivores, fungi, and microbes. This maximizes the plant's capacity for photosynthesis, growth, and reproduction. Finally, the deciduous habit minimizes the surface area on which ice and snow may accumulate and thus reduces winter damage.

Leaves have to be replaced sooner or later, even by evergreens, which drop some of their leaves every year. White and pitch pine leaves drop after two or three years, hemlock leaves live three to six years, and spruces and firs retain needles seven to ten years.

Native Fauna

Invertebrates. Innumerable invertebrates inhabit deciduous forests. Many of these are insects that feed on foliage or plant juices or bore into wood. To appreciate the abundance and impact of arboreal plant eaters all you have to do is place a white pan or sheet on the forest floor and come back a few days later to see a sample of the insect frass (excrement) that has rained down from the vegetation. Or simply examine a sample of leaves, a high percentage of which will show evidence of insect herbivory.

Forest canopy insects are not only numerous but diverse. For example, researchers sampling the canopy of mixed forest in eastern Connecticut found 36 species of long-horned beetles; all but a few of these were common species in the region.

In late summer, singing katydids and crickets become a conspicuous part of the forest insect fauna—true katydids in the trees, bush katydids and tree crickets in shrubs or low in trees, and ground crickets among the leaf litter. Joining these are ground-dwelling camel crickets.

Other invertebrates, such as isopod crustaceans, millipedes, mites, certain beetles, springtails, slugs, terrestrial snails (some with shells up to an inch in diameter), and other smaller organisms, including especially fungi and bacteria, subsist primarily or at least in part on the abundant decaying leaf litter or decomposing wood in deciduous forests. Some of these organisms feed heavily on fungi and bacteria. In fact, the vast majority of the biomass in these forests ends up being consumed by decomposers rather than by herbivores that eat living plant material. Earthworms process organic material in or at the surface of the soil. These abundant small decomposer animals, together with the herbivores, make the energy and nutrients accumulated in plant biomass available to higher-level consumers, such as shrews, moles, bats, squirrels, and most forest birds, reptiles, amphibians, spiders, centipedes, and predatory insects, that cannot subsist on living foliage or rotting plants.

Ants play many roles in deciduous forest ecosystems, ranging from predator and scavenger to seed eater and seed disperser. Some ants tend herds of honeydew-secreting, plant-feeding aphids, favoring the insects at the expense of the plants. The honeydew and aphids are to the ants as milk and Holsteins are to the dairy farmer. Some ants attack caterpillars and can reduce herbivory on plant leaves, thus affecting the vigor and survival of individual plants near ant nests.

Amphibians and Reptiles. Deciduous forests are home to many salamanders, such as eastern newts and Jefferson, blue-spotted, spotted, marbled, and redback salamanders, the last generally being the most abundant of any vertebrate species occurring in Connecticut's forested habitats (Fig. 7.14). American toads, gray treefrogs, spring peepers, and wood frogs are the characteristic anurans. All of the amphibians of deciduous forests, except redback and slimy salamanders, breed in aquatic and wetland habitats, especially vernal pools. Forest-floor invertebrates are important foods for most amphibians.

Fig. 7.14. Redback salamanders are the most abundant vertebrate in most deciduous forests. This one lacks the red stripe on the back.

Relatively few reptiles live primarily in mature deciduous forests. Species most likely to be encountered are the eastern box turtle, racer, ringneck snake, eastern rat snake, milk snake, common garter snake, and, locally, the timber rattlesnake. All of these so-called "cold-blooded" animals need Sun-derived warmth and tend to be most common where openings or wide spacing of trees allow sunlight to penetrate to the ground. Except for the omnivorous box turtle, these reptiles are predators that subsist mainly on smaller animals.

Birds. Deciduous forests are rich in bird life. In summer, a typical large tract of mature forest harbors such birds as ruffed grouse, wild turkeys, barred owls, eastern screech-owls, downy woodpeckers, red-bellied woodpeckers, eastern wood-pewees, great crested flycatchers, blue jays, black-capped chickadees, tufted titmice, white-breasted nuthatches, veerys, wood thrushes, red-eyed vireos, black-and-white warblers, ovenbirds, scarlet tanagers, rose-breasted grosbeaks, and often many others.

Some of the less common avian inhabitants of mature forests occur only in extensive forest, and they become scarce or are absent when forests are interrupted by openings or occur in small patches. Examples of these include hermit thrushes, yellow-throated vireos, cerulean warblers, worm-eating warblers, brown creepers, and blue-gray gnatcatchers. Some typical deciduous forest species, such as ovenbirds, scarlet tanagers, and wood thrushes, populate large forests more densely than smaller tracts. See Chapter 20 for further information.

A 1995 census of birds in a 25-acre tract of second-growth hardwood forest in northwestern Connecticut by Andrew Magee yielded 93 breeding pairs of 40 species, as follows: red-eyed vireo (17 territories), ovenbird (16), veery (8), American redstart (6), scarlet tanager (3.5), eastern wood-pewee (3), American robin (3), wood thrush (2.5), gray catbird (2.5), yellow-bellied sapsucker (2), downy woodpecker (2), black-capped chickadee (2), tufted titmouse (2), blue-winged warbler (2), chestnut-sided warbler (2), red-bellied woodpecker (1.5), American crow (1.5), common yellowthroat (1.5), eastern towhee (1.5), and 16 additional species represented by at least 0.5 territories on the study plot.

Resident grouse and turkeys eat much plant material (mainly buds, nuts, and seeds), but the vast majority of deciduous forest birds are migratory species that feed on insects. In fact, birds may play an important role in reducing populations of leaf-eating insects. Some nonmigratory birds switch to seeds and fruits when insects are scarce.

Mammals. Shrews, bats, eastern chipmunks, gray squirrels, southern flying squirrels, white-footed mice, red-backed voles, woodland voles, weasels, and gray foxes live in forest interior habitats throughout Connecticut. Deer take shelter in forests but forage primarily along forest edges and openings. Fishers have become relatively common in large forest tracts in the northern and central parts of the state. Many other mammals travel through forests on their way to other more preferred habitats, and some others use forests on an occasional basis or in sparse numbers. Most deciduous forest mammals are carnivores or insect or fruit/nut eaters; relatively few feed on foliage.

Native mammals can affect the composition and regeneration of their forest habitat. Squirrels and other small mammals, as well as blue jays, plant many tree seeds, some of which are not retrieved and grow into trees. Heavy browsing by deer kills or significantly reduces survival rate of seedlings and inhibits sprout growth. Intense grazing by large deer populations can also lead to local extirpations of favored herbaceous food plants in the forest understory. Rodents feed on and spread the spores of mycorrhizal fungi that form critical, mutualistic associations with plant roots, without which the trees fare poorly (see Chapter 8).

Acorn Ecology

Oaks are a predominant component of Connecticut's deciduous forests, and acorns are among the most distinc-

Fig. 7.15. Acorns of chestnut oak (upper left), red oak (upper center), black oak (lower left), scarlet oak (lower center), and pin oak (right).

tive of all plant propagules (Fig. 7.15). These large fruits consist of a cap and a shell enclosing two seedling leaves (cotyledons) attached to a tiny seedling ("embryo") that is located at the apex of the acorn between the cotyledons. The cotyledons—the "meat" of the acorn—contain the resources that are used by the seedling to establish a large root after the acorn germinates.

Acorns mature and fall from oaks in late summer and early fall. In mid-September, the ground may be strewn with acorns while trees still retain many on the branches. By mid-October, the vast majority of mature acorns often have fallen to the ground, but red oaks, at least, sometimes hold many acorns through mid-November. Being on the ground before most forest leaves have fallen probably increases the acorn's chance of survival; successful rooting and seedling establishment likely are more difficult for acorns lying on top of a thick layer of leaves, where the acorns also are more prone to desiccation. Also, visually oriented acorn predators are more likely to overlook acorns hidden beneath fallen leaves, although gray squirrels—important acorn dispersal agents (see following section) with an acute sense of smell—have little trouble finding leaf-covered acorns. Acorns of the white oak group, such as white oak and chestnut oak, generally take one season to develop to maturity whereas those of the red/black oak group, including black oak, red oak, scarlet oak, pin oak, scrub oak, and others, mature after two seasons on the plant.

Acorns contain variable amounts of tannin, a widespread plant compound that if ingested may intereferre with normal digestion by binding proteins and rendering them less available, inhibiting digestive enzyme activity, causing damage to the gut lining, or having toxic effects. The amount of tannin varies among oak species and among individuals of each species, making some acorns more or less palatable than others. In general, acorns of the black/red oak group have higher tannin content than white oak acorns.

Despite the tannins, acorns can be high in energy content; the lipid content of red oak species is 18 to 25 percent of dry mass; and 5 to 10 percent in white oaks. Acorns are relatively low in protein but when mixed with other foods are a critical energy resource for numerous vertebrates and invertebrates, including ruffed grouse, wild turkey, blue jay, woodpeckers, eastern chipmunk, flying squirrel, gray squirrel, white-footed mouse, deer mouse, black bear, white-tailed deer, weevils (*Curculio* spp., *Conotrachelus*), sap beetles, and the larvae of acorn moths and filbert worms. These animals have evolved digestive mechanisms for overcoming, at least to some degree, the detrimental effects of tannin (see Chapter 21).

Because of this dietary popularity among so many animals, there is a low probability that any given acorn will survive to produce an established oak seedling. However, the long reproductive lifetime of an oak, which may last centuries, helps ensure that it will leave at least some offspring.

Variability in Acorn Production. Oaks erratically produce large acorn crops. A single oak tree may produce several thousand acorns in one year, and few or none in other years. White oak acorn production is relatively low every other year, with a large seed crop about every fourth year. Regionally, oaks of a particular species often are synchronous in their volume of acorn production. And oaks of different species may synchronously produce large or small numbers of acorns. Favorable or unfavorable conditions during the flowering period may help (but do not fully) explain why oaks of different species are synchronized over large areas. For example, a late frost could damage flowers and cause low acorn production for multiple oak species over a large area. However, different oak species in a community also frequently are not synchronized in their levels of acorn production, in part because black/red oaks require two years for acorn development and white oak acorns develop in one year.

Unpredictability in acorn crop size may facilitate successful establishment of oak seedlings. In years with few acorns, acorn eaters may experience poor nutrition or even starvation, and their populations may decline. A bumper crop the next year may result in more acorns than the reduced populations of acorn eaters can consume, so acorns have a better chance of becoming oak seedlings. Also, huge acorn crops may stimulate certain acorn eaters such as squirrels and blue jays to store excess acorns, and storage sites often are suitable for germination and seedling establishment, so oak reproductive success is enhanced even further. Profuse flowering by oaks in a single year possibly may also be beneficial by increasing the efficiency of wind pollination.

In central Connecticut, moderate to large crops of acorns were produced in 1989, 1993, 1994, and 1997, but very few were present in 1995 and 1996. Red oak, scarlet oak, black oak, white oak, and chestnut oak produced a huge crop of acorns in 1998. Scarlet and black oaks produced a moderate crop of acorns in 1999, but chestnut oak, white oak, and red oak dropped very few. A moderately large number of acorns (but few of white oak) appeared in 2000. Oaks of all species produced a notably large crop of acorns in 2001.

Acorn predators and dispersers. Acorn weevils and short-snouted weevils are among the most important acorn predators and may infest a high percentage of available acorns. My students and I found that often up to 80 to 90 percent of red oak acorns are infested, but sometimes a very low percentage of chestnut oak acorns contain weevil larvae.

Adult acorn weevils emerge from the ground and attack acorns while the acorn is still on the tree, beginning several weeks before the acorns ripen (Fig. 7.16). They use tiny teeth at the end of their long snout to bore into the acorn. Feeding on the nutmeat occurs as the weevil bores. Weevil feet have excellent traction that allows the insects to climb trees and maneuver on smooth acorns without slipping.

The female acorn weevil lays an egg inside each channel she makes. In about a week or two, the eggs hatch into larvae that feed on the acorn meat (Fig. 7.16). An infested acorn may contain one to several short, plump weevil larvae. In late summer or early fall, after the acorn falls, each full-grown larva exits the acorn through a tiny hole that it chews, then it burrows down into the leaf litter and pupates in the soil. The following year, or up to five years later, the adult weevils emerge and attack the acorn crop, though many pupae may be found and eaten by white-footed mice and short-tailed shrews.

Like adult acorn weevils, filbert moth larvae gnaw their way into acorns (Fig. 7.17). Other acorn-eating invertebrates such as short-snouted weevils, tiny sap beetles (Fig. 7.18), and acorn moths, gain access to the acorn meat through cracks and openings that are already present.

Blue jays harvest, store, and eat large numbers of acorns. When a choice is available, they often show a preference for small acorns, such as those of pin oak, from which they may harvest over half the crop. In late summer and early fall, they pluck mature acorns from the tree and carry them in their bill and expandable throat and esophagus. Small acorns can be carried in batches of up to about five at a time. Upon arrival at a storage area, which occasionally may

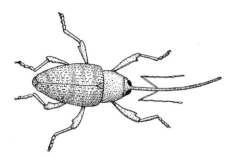

Fig. 7.16. Upper: Adult acorn weevils use their long snout to bore into acorns. Lower: Weevil larvae are major consumers of acorns.

Fig. 7.17. Filbert moth larvae chew their way into acorns.

be up to 1 or 2 miles away, the jay lands on the ground and disgorges the acorns. Then it takes them in the bill one by one and shoves them into the ground or among leaf litter or ground vegetation. Each cached acorn is covered with nearby plant debris. Jays retrieve and eat cached acorns in winter and early spring when other foods are scarce. Some of the acorns cached by jays are discovered and eaten by mammals. How do the jays find the hidden acorns? Probably they simply remember where they put them. Jays peck open acorns while holding them with their toes.

In late summer and fall, gray squirrels also harvest and store large numbers of acorns. The squirrels dig a small hole, place the nut in it, ram it firmly into the ground with their front teeth, then fill the hole and arrange the ground cover in an apparent effort to conceal their digging. Only one nut is placed in each hole. Later the squirrels use their memory and acute sense of smell to retrieve the buried nuts, which are an important source of food from fall through spring (Fig. 7.19).

Gray squirrels handle acorns of the white oak group differently from those of the red/black oak group. They often bite off the apical end of white oak acorns before burying them; acorns of the red/black oak group are not damaged in this way before burial. This difference in treatment of the acorns evidently is related to differences in germination characteristics. White oak group acorns germinate in fall and send down a long taproot, transferring the materials of the acorn meat deep into the soil (Fig. 7.20). Red/black oak acorns germinate in spring of the following year. Damage to the apical end of the acorn kills the oak "embryo" and prevents growth of the taproot. Squirrels that inflict this damage before burying the nut of white oak group acorns ensure that the acorn meat nutrients are not lost to root growth during storage. Interestingly, squirrels may revisit their caches in spring and excise the embryos of germinating acorns. Despite their special treatment of some white oak acorns, squirrels are more likely to immediately eat white oak acorns than acorns of the red oak group. Similarly, white-footed mice and flying squirrels tend to consume perishable

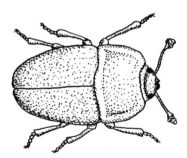

Fig. 7.18. Sap beetles (Nitidulidae: *Stelidota*) enter damaged or sprouting acorns. Adults (only a few millimeters long) and larvae feed on the acorn meat.

108 | 7. Uplands: Terrestrial Ecosystems

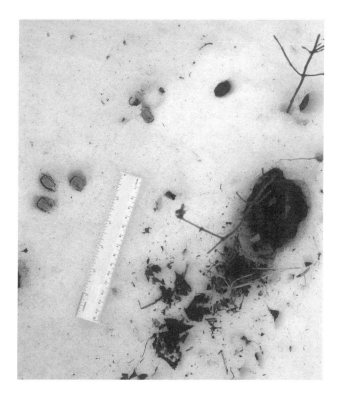

Fig. 7.19. In winter, gray squirrels retrieve and eat acorns that they buried in autumn. Note discarded strips of acorn shell at left.

Fig. 7.20. Acorns of the white oak group, such as these shiny chestnut oak acorns, germinate in fall whereas red oak, scarlet oak, and black oak acorns germinate in spring. An acorn weevil exit hole is visible in one of the red oak acorns. These acorns were collected in October.

white oak acorns immediately and cache less-perishable seeds. As a result, mammals generally disperse acorns of red/black oak species much farther from the parent tree than they move white oak acorns.

When feeding on acorns, squirrels and jays sometimes eat only a portion of the acorn meat (cotyledons) at the cap end, leaving behind the apical part that has a higher concentration of tannins and which contains the embryo. Acorn weevils also tend to feed mainly in the cap end. Many of these partially consumed acorns do successfully germinate.

One might wonder whether squirrels and other acorn eaters that store acorns can discriminate between acorns that are infested with weevil larvae and those that are not. A squirrel might not necessarily discard and might even favor a weevil-infested acorn if it was selected to be eaten and not buried. After all, the weevil larva itself could be a good source of nutrients. But it would not be advantageous for a squirrel to invest the energy in burying an acorn only to have much of it consumed from within by a weevil grub (which would likely vacate the acorn before it was retrieved by the squirrel). Recent experimental evidence indicates that gray squirrels can distinguish between infested and noninfested acorns. They often selectively cache sound acorns (benefiting the oaks) and eat the infested ones, and they usually eat the weevil larvae. In contrast to squirrels, white-footed mice apparently exhibit no preference between infested and noninfested whole acorns. In experiments in captivity, blue jays handled, opened, and consumed uninfested acorns significantly more often than infested ones, indicating that they can tell the difference and suggesting that weevil larvae may not enhance the nut's attractiveness as food.

Acorns are important in the late summer and fall diet of white-tailed deer. In fact, deer can remove a large proportion of fallen acorns, which are a critical high-energy food source for attaining good body condition and fat deposition prior to winter. For deer, winter is a time of food stress and declining body condition. Individuals that do not put on adequate fat in fall are unlikely to survive the winter. Recent research suggests that in years with poor acorn crops, deer tend to travel more in search of food and are more vulnerable to hunting mortality.

Wild turkeys similarly do better when acorns are plentiful, and research in Massachusetts has shown that the reproductive success of black bears is enhanced in years when acorns and other mast crops are abundant.

Oaks and acorns also are involved in the ecology of other forest species. White-footed mice, for example, sometimes eat or store up to two-thirds of the available acorn crop. When acorns are scarce, populations of white-footed mice decline. Reduced mouse populations result in less predation on gypsy moth pupae, allowing moth populations to increase. Heavy defoliation of oaks by gypsy moth caterpillars results in greatly reduced acorn production for the following 1 or 2 years, which in turn decimates mouse populations.

Populations of songbirds that do not eat acorns nevertheless may be affected indirectly (and in unexpected ways) by the size of the acorn crop. For example, small mammals such as white-footed mice and chipmunks eat songbird eggs, and predation on ground and shrub nests may increase when large acorn crops result in increased small mammal populations. Blue jays also consume acorns and small bird eggs and nestlings, and jay abundance in an area at least partially reflects acorn abundance. However, the effects of jays on other birds might be reduced if jay populations are limited by egg-eating small mammals. Additionally, certain hawks that prey both on small mammals and songbirds appear to increase in abundance when rodents are numerous, so they also may influence the relationship between acorn abundance and songbird populations. Theoretically, deer may figure in this interaction as well, since the abundance of acorns may affect their feeding on understory vegetation, which in turn may affect the availability of cover for ground-nesting birds. All this complexity (and it goes further than these examples) serves to illustrate why ecological studies of cause and effect often are inconclusive, but likely it is safe to conclude that variation in acorn and nut production by oaks and other mast producers has profound and pervasive impacts throughout the entire forest ecosystem.

Acorn Ants. Ants of the genus *Leptothorax* commonly nest inside old acorns that have been hollowed out by other insects. These ants are predators and scavengers that may eat springtails, pseudoscorpions, mites, and dead invertebrates found in forest soil and leaf litter. Other cavity-nesting ants (e.g., *Tapinoma sessile*) may use acorns as temporary nest sites for young colonies.

Coniferous and Mixed Forests and Woodlands

Vegetation

Forests dominated by conifers exist in patches throughout the state. Eastern hemlock and eastern white pine are the most common forest conifers. Balsam fir and red spruce—species characteristic of northern New England—are rarities in Connecticut's Northwest Highlands.

Eastern Hemlock Forests. Eastern hemlock is a primary component of many coniferous forests in Connecticut (Fig. 7.21). This species certainly declined with the extensive forest cutting and burning that accompanied historical phases of human settlement of the region, and hemlocks remained scarce even in areas that soon reverted to forest (cut or burned hemlock trees do not resprout but many

Fig. 7.21. This fine eastern hemlock forest has not been affected by hemlock woolly adelgids.

associated deciduous trees do). Then hemlocks gradually increased in abundance with continued reforestation and fire protection.

Hemlocks are very shade tolerant and can live several hundred years. Thus they can perpetuate themselves indefinitely if conditions remain stable. Hemlock stands are widespread in the cooler and moister conditions in the northern part of the state and occur commonly along stream gorges (but sometimes even on rocky south-facing slopes) in southern Connecticut. Thick leaf litter inhibits the establishment of hemlock seedlings, so hemlocks are most likely to colonize ledges, slopes, and other litter-free sites. Sometimes hemlocks become established on mossy, slow-decaying hemlock logs, but "nurse logs," on which a line of trees may become established, are rare in Connecticut.

Deciduous woody plants in hemlock stands often include American beech, yellow birch, black birch, red oak, white oak, witch-hobble (Northwest Highlands) plus American yew and many other species occurring in moist deciduous forests. Few herbaceous plants (e.g., star-flower,

Fig. 7.22. Eastern white pine can reach majestic heights in Connecticut's coniferous and mixed forests.

wild lily-of-the-valley) live in the deep shade and thick layer of needles under large hemlocks.

Unfortunately, non-native organisms have cast a dark shadow over the future of this magnificent tree. Hemlocks in Connecticut are being decimated, especially in southern Connecticut, by the hemlock woolly adelgid*, an exotic insect pest that arrived in Connecticut around 1985 (see Chapter 14), and these dying stands are being replaced by deciduous trees. Seedlings of black birch, red oak, and various other oaks are now common in hemlock stands killed by adelgids. There is evidence that loss of hemlocks increases nitrogen leaching from the soil and may lead to nutrient losses that could affect future forest productivity. Nutrient leaching and increased light resulting from adelgid impacts may cause significant changes in ecosystem processes. Populations of hemlock-associated species, such as the black-throated green warbler, blackburian warbler, and Acadian flycatcher, decline as deciduous trees replace adelgid-killed hemlocks. Hemlocks suffer not only from adelgids but also are intolerant of defoliation and suffer high mortality during severe gypsy moth* outbreaks.

Eastern White Pine Forests. Eastern white pine forms significant stands in some areas of Connecticut, especially in the northern and eastern parts of the state (Fig. 7.22). White pine reproduces best on bare sunny ground, so today's stands generally indicate past disturbance (and planting). In northern Connecticut, some white pine forests exist on former pasturelands that were abandoned in the mid-1800s. Pines can become established and grow slowly in small openings under heavy shade of mature pines or deciduous trees, sometimes reaching heights of only 6 feet after 20 to 30 years. In average sites in central Connecticut, white pines growing among mature deciduous trees are roughly 20 to 25 feet tall after 30 years. Eventually such pines may penetrate the deciduous canopy, grow more quickly, and attain dominant stature in the forest. However, today's large pines are "babes in the woods" compared to the behemoth pines that existed in pre-colonial Connecticut.

Sand Plain Communities. A notable type of mixed conifer–woodland is the sand plain community, characterized by sandy soils dominated by pitch pine, white pine, and oaks (scrub, black, scarlet, chestnut, white, dwarf chestnut, and post), with sunny spaces between the pines and oaks partially occupied by huckleberry, blueberry, sheep laurel, bearberry, sweet-fern, or young white pines (Fig. 7.23). Other locally occurring plants include American holly and prickly-pear cactus. The sandy soils of these woodlands do not retain water for very long and thus can be very dry in summer.

Dry conditions make sand plain communities susceptible to fire. The sand plain community indeed depends on fire for its perpetuation. Pitch pines require sunny open ground for successful germination and seedling establishment, and the shrubs sprout vigorously after fire and grow well in the sunny

Fig. 7.23. This sand plain community includes pitch pines, white pines, and several oak species. Fire is needed to maintain the distinctive qualities of this community.

Coniferous and Mixed Forests and Woodlands

areas between the pines. In the absence of fire or similar disturbances, pitch pines eventually are replaced by white pines or deciduous trees that germinate and grow in shaded areas beneath other trees and eventually shade out the pitch pines and other characteristic sand plain plants. Infrequent fire is a problem affecting Connecticut's few remaining sand plain communities, where prescribed burning is difficult due to the close proximity of residental and commercial areas. As a result, pitch pine is dying out in most of the state's sand plains. You can visit examples of these communities at Hopeville Pond State Park and Wharton Brook State Park.

Sand plain communities in Connecticut have been greatly reduced as a result of historical tree harvest for wood products, subsequent conversion of these areas to agricultural use through the early 1900s, and residential and commercial development. Studies of sand plain communities in the Connecticut River valley of Massachusetts indicate that pitch pines now occur almost exclusively on formerly plowed lands whereas the vast majority of scrub oak stands are in areas that were not plowed.

Pitch pine and scrub oak occur not only in sand plains but often also the thin soils of rocky hilltops and outcrops (Fig. 7.24). Both habitat types are subject to exceptionally dry conditions in which few other large woody plants can grow and successfully compete.

Influences of Humans and Hurricanes. Humans have increased the abundance of conifers in many places. Planted groves of white pine, Norway spruce, and European larch are frequent, especially bordering water supply reservoirs. These conifer stands serve as "leaf filters" that keep deciduous leaves from blowing into and accumulating in the water, where their decomposition results in undesirable contaminants. In contrast, conifer needles tend to drop straight down and do not end up in the water. Formerly, planted red pines formed significant stands in state forests and scattered other locations throughout the state, but the pines are now largely dead, decadent, or dying as a result of scale insect infestations (see Chapter 9), and in most areas natural processes are replacing them with deciduous trees.

Fig. 7.24. Pitch pine (rear), scrub oak (middle), and huckleberry and blueberry (front) dominate many dry hilltops.

Fig. 7.25. Cathedral Pines in Cornwall (photographed in 1991) was devastated by tornadolike winds in 1989.

Among the various forest types in New England, stands of tall conifers on exposed sites are the most vulnerable to wind damage. Awesome examples of this occurred in 1938 when, at the Bolleswood Natural Area at Connecticut College, a hurricane blew down over a hundred hemlocks 2 to 3 feet in diameter (similar events occurred in other areas). Similarly, on the afternoon of 10 July 1989, localized tornado-like winds devastated a magnificent stand of old white pines and hemlocks at the Cathedral Pines site in Cornwall (Fig. 7.25).

Characteristics of Conifers. Except for larches, all native conifers in Connecticut are evergreens. Larches grow a new crop of needles each year and drop all of them before winter. The other conifers also grow a new crop of needles each year, and they may drop numerous needles each fall, but the new ones are kept for several years, so the plants are always green.

Conifer reproduction in Connecticut is primarily sexual. Pollen-producing cones and seed cones are produced in greatest number in the upper part of the tree. The seed cones of all of our conifers are fertilized through wind-dispersed pollen. Abundant cone crops generally occur at intervals of 2 to 4 years.

Fauna

Stands of white pine and hemlock support a distinctive fauna. Many animals that are common in pure conifer stands are relatively scarce in hardwoods, but mixed forests of conifers and hardwoods tend to have most of the hardwood species plus many of the conifer specialists. Among the usual conifer-associated species are the red squirrel, red-tailed hawk, great horned owl, red-breasted nuthatch, blue-headed vireo, black-throated green warbler (most common in large tracts), blackburnian warbler, yellow-rumped warbler, magnolia warbler, pine warbler, and eastern pine elfin (a butterfly).

A 1995 census of breeding birds in a 26-acre section of old-growth hemlock and white pine in northwestern Connecticut by Andrew Magee yielded 43 species and 123 territories as follows: blackburnian warbler (19.5 territories), black-throated green warbler (13.5), ovenbird (13), veery (9), red-eyed vireo (8), yellow-rumped warbler (6), black-capped chickadee (5), wood thrush (5), black-and-white warbler (4), red-breasted nuthatch (3), hermit thrush (3), American robin (2.5), pine warbler (2.5), wild turkey (2), downy woodpecker (2), blue jay (2), American crow (2), brown creeper (2), blue-headed vireo (2), Canada warbler (2), scarlet tanager (2), purple finch (2), hairy woodpecker (1.5), and 12 additional species represented by at least 0.5 territories within the study plot.

Some of the typical or characteristic animals in sand plain communities include nesting pine warblers and eastern towhees, winter-visiting red-breasted nuthatches, crossbills, and pine siskins, buck moth, plus a wide assortment of other animals found in most shrubby habitats, pinelands, and oak forests.

Many planted stands of conifers, such as those around reservoirs, have a depauperate fauna because they tend to consist of trees that are all the same age, and they generally lack well-developed shrub and herb layers. Such monotonous conditions support only a small portion of the wildlife that would exist in a more diverse environment. These areas do provide significant feeding areas and shelter for a variety of birds and mammals.

Regionally in the northeastern United States nearly 100 bird species and almost 50 species of mammals are associated with hemlocks, though fewer than a dozen of each of these groups are strongly associated with hemlock, and none is restricted to stands of this species. Cool-water fishes such as brook trout and associated stream invertebrates upon which they feed benefit from the cool, stable physical conditions provided by riparian hemlock stands.

Forest Edges and Shrubby Old Fields

Forest edges, old pastures, and abandoned cultivated fields that have been invaded by shrubs or small trees, and similar disturbed areas are frequent components of the Connecticut landscape. Forest edges are extensive (e.g., along roads) and are increasing as house lots and roadways are cleared in deciduous forests. In contrast, shrubby old fields are ephemeral habitats that are declining in extent in Connecticut, having been reforested or converted to residential housing. Both edges and old fields tend to be rich in shrubbery and vines that provide abundant food (insects, small fruits) and thick cover for many birds and mammals.

Vegetation

Forest edge vegetation includes various forest interior trees that were present before the edge was cleared, plus others such as black cherry, sassafras, and flowering dogwood that are more typical of edges (Fig. 7.26). Edges often are much overgrown with vines such as greenbriar and poison ivy and the exotic Japanese honeysuckle and Asian bittersweet. Multiflora rose, European buckthorn, autumn-olive, Japanese barberry, Morrow's honeysuckle, and tartarian

Fig. 7.26. Forest edge vegetation often includes a thick growth of shrubs and vines, usually containing exotic species that thrive on disturbance and abundant sunlight.

Fig. 7.27. Old fields with a mixture of small trees, shrubs, and tall herbaceous plants attract a unique array of wildlife species, many of which are uncommon or absent in mature forests and grassy fields. Stephen P. Broker

honeysuckle are frequent exotic shrubs. Invasive exotic herbaceous plants such as garlic mustard, cypress sprurge, and Japanese knotweed have become common at forest edges, especially along untended roadsides.

Woody plants occurring as initial colonists of pastures (Fig. 7.27) and other grassy fields, and in cleared powerline corridors (Figs. 7.28 and 7.29) often include native species such as red-cedar (by far the most common species), common juniper, gray birch (on patches of bare soil), white pine (mainly in northern Connecticut), smooth sumac, staghorn sumac, shining sumac, bayberry, sweet-fern, flowering dogwood, gray dogwood, silky dogwood, grape, brambles, choke cherry, black chokeberry, and hawthorns, as well as non-natives including multiflora rose and autumn-olive. Many other woody species may colonize pastures and hayfields. Generally these colonists are sun-loving species that die out when taller trees shade them. Dense stands of shrubs that develop in deforested areas can resist effective invasion by tree species and sometimes are quite long lasting, as are some grassy areas on poor soils. Multiflora rose* and Japanese honeysuckle* may be especially effective at competitively inhibiting the establishment of native trees. Colonization of old fields by woody plants also can be inhibited by voles *(Microtus)* and other rodents feeding on seeds and seedlings.

The usefulness of native shrub stands in retarding reforestation in utility corridors often seems to be underappreciated by utility companies, which of course wish to prevent tall trees from growing up under power lines. All too often I found these corridors sprayed with herbicides that killed native shrubs that help inhibit reforestation. At other times, I found many flowering dogwoods and red-cedars that had been cut down, irregardless of the fact that these plants never grow tall enough to interfere with the high power lines in these areas. In one utility corridor, over a period of 15 years, I witnessed how such cutting and herbicide use converted an area that was exceptionally rich in native upland and wetland shrubs and animal wildlife—a real showcase—to a depauperate expanse dominated by exotic multiflora rose, common reed, and autumn-olive, with just a few remaining native woody plants.

Nonwoody plants also have significant impacts on vegetation development on old fields. For example, stands of little bluestem may be long lasting on poor soils, and exotic orchard grass can inhibit other herbaceous plants for several years.

Fig. 7.28. Managed utility corridors fragment forests but support shrubby, early successional vegetation that adds diversity to the landscape.

Fig. 7.29. Flowering dogwood (white flowers evident) and red-cedar (dark foliage), shown here in early May, can be abundant along utility corridors but are scarce or absent in adjoining deciduous forest.

Fauna

Amphibians generally are scarce in these shrubby, viney habitats, but you might find American toads, spring peepers, or pickerel frogs if a body of water or wetland is nearby. Reptilian inhabitants of old fields generally include eastern box turtles, racers, milk snakes, and common garter snakes, which are attracted to the favorable sun/shade mosaic and abundance of food.

The birds of shrubby old fields and forest-edge thickets make a long list: mourning dove, downy woodpecker, northern flicker, eastern phoebe, American crow, blue jay, Carolina wren, house wren, gray catbird, northern mockingbird, brown thrasher, eastern bluebird, American robin, cedar waxwing, European starling, blue-winged warbler, golden-winged warbler, yellow warbler, yellow-rumped warbler, chestnut-sided warbler, prairie warbler, American redstart, Nashville warbler (northern hills), yellow-breasted chat (coastal), white-eyed vireo (southern areas), northern cardinal, rose-breasted grosbeak, indigo bunting, eastern towhee, American tree sparrow, chipping sparrow, field sparrow, song sparrow, white-throated sparrow, dark-eyed junco, brown-headed cowbird, northern oriole (groves), house finch*, American goldfinch, and others. Prairie warblers and field sparrows are especially characteristic breeding species of shrubby or weedy utility corridors. Red-tailed hawks, great horned owls, black-capped chickadees, tufted titmice, white-breasted nuthatches, black-and-white warblers, scarlet tanagers, and additional species whose presence depends on the adjacent forest trees also commonly use forest edges. Mixed-species flocks of migrating songbirds often use structurally complex forest-edge habitats, which provide diverse feeding opportunities.

In the absence of disturbance, the bird life and other fauna and flora of these habitats change over time as vegetation succession proceeds. Studies by Robert Askins at the Connecticut College Arboretum documented changes in bird populations as an old field became forested over a period of thirty-seven years. Old field species such as brown thrasher, yellow-breasted chat, prairie warbler, American goldfinch, song sparrow, and field sparrow disappeared and were replaced by woodland species including eastern wood-pewee, blue jay, black-capped chickadee, tufted titmouse, wood thrush, veery, red-eyed vireo, ovenbird, and hooded warbler. Old fields with abundant fruiting shrubs attract greater use by migratory songbirds than do young forests that eventually replace the shrubby areas.

Most of the mammals and birds of forest edges and shrubby old fields are common, wide-ranging species that adapt well to disturbed situations. The usual species include opossums, various shrews, eastern moles, red bats, silver-haired bats, eastern cottontails, snowshoe hares, eastern chipmunks, woodchucks, gray squirrels, white-footed mice, woodland voles, coyotes, red foxes, gray foxes, weasels, striped skunks, and white-tailed deer.

Fields and Grasslands

In Connecticut, fields and upland grasslands exist almost exclusively in areas with past and present human disturbance (Fig. 7.30). These open areas often reflect past agricultural activity and generally require continued disturbance (e.g., mowing, grazing, fire) to counteract invasion by woody plants. Some of today's largely treeless landscapes originated thousands of years ago as a result of regular burning by Native Americans.

Vegetation

A plethora of annual herbaceous plants can be found in abandoned cultivated fields, along roadsides, and in hayfields and pastures throughout the state. Many of the plants characteristic of these habitats are not native to North America but were introduced, accidently or intentionally, from Europe or Asia. In open, sunny areas, these introduced species may exclude native species through superior competitive abilities, especially in the absence of most of the predatory, pathogenic, or competing organisms that control their numbers or growth in their native range.

Fig. 7.30. Most Connecticut grasslands are maintained by regular warm-season mowing that prevents successful nesting by grassland birds.

Most weedy annuals, such as pigweeds*, crabgrass*, spotted spurge, common beggar-ticks, daisy-fleabane, common cocklebur*, lamb's-quarters*, and pinkweed, are summer annuals (see Chapter 9). Common and giant ragweeds, the wind-dispersed pollen of which triggers late-summer hay fever in humans, may be all too common summer annuals in disturbed roadsides and fields (but ragweed seeds are valuable foods for many ground-feeding, seed-eating birds). Other common plants are winter annuals, including horseweed (also a summer annual), field pepper-grass*, shepherd's-purse*, common chickweed*, and charlock* (sometimes a summer annual).

Plants of weedy fields have well-developed dispersal abilities—important for organisms associated with ephemeral habitats. The seeds generally are extremely numerous, small, and may be attached to hairlike tufts that catch the wind, or the seed-containing fruits may have hooks or other velcro-like structures that allow dispersal in the fur of mammals (or the clothing of humans). These often drought- and heat-tolerant plants thrive in open sunny places and languish after larger, taller plants become established.

Several species of biennials grow in abandoned fields. Common evening-primrose and wild lettuce are native biennials. Exotics include wild carrot (Queen Anne's lace), poison-hemlock, common mullein, common burdock, and bull thistle.

Eventually, in the absence of cultivation or other churning of the soil, fields are invaded by herbaceous perennials such as goldenrods, milkweeds, birds-foot trefoil*, cinquefoil, horse-nettle, dogbane, asters, black-eyed Susans, chicory*, ox-eye daisy*, dandelions*, hawkweeds(*), Canada thistle*, cypress spurge*, common yarrow*, common tall buttercup*, ground-ivy*, heal-all*, pokeweed, curly dock*, wild strawberry, stinging nettle*, English plantain*, butter-and-eggs*, certain violets, white clover*, and red clover*. Native goldenrods and asters are conspicuous, as are many non-natives that thrive in sunny disturbed areas. Braken, a native fern, may be common in fields, abandoned pastures, and burned or logged land.

Soon, brambles and other shrubby or viney species, especially non-natives such as Asian bittersweet, autumn-olive, and multiflora rose, become established. Quaking aspen, bigtooth aspen, birches, black locust, and pin cherry often invade roadsides, and these same trees often are among the early woody growth in clearings that expose bare soil or that result from fire. Some trees, such as sassafras, invade clearings from forested edges by sprouts from the spreading root system. Even typical forest trees may quickly colonize, especially if seeds are transported by birds or the wind. Sandy soils in the partially shrubby stage of development may contain autumn-olive*, sumacs, sweet-fern, gray birch, black cherry, goldenrods, wild carrot*, common evening-primrose, bush-clover, and soapwort*.

If regularly mowed or burned, weedy fields eventually become predominantly grassland. Grasses are wind-pollinated flowering plants that have the growing part of the stem near the ground rather than at the tip. This allows grasses to withstand and quickly rebound from grazing or mowing.

Most large grassland habitats in Connecticut are associated with farmland and airports and depend on mowing for their continued existence. The flowering stems of often-mowed grasses may seldom if ever reach maturity or produce seed, but such grasslands nevertheless may persist and expand through extensive vegetative reproduction via horizontally spreading rhizomes.

A special form of photosynthesis adapts grasses to warm dry conditions. Termed C-4 photosynthesis, this process operates efficiently under bright light and warm conditions and uses relatively little water.

Fauna

Grasslands support a unique assemblage of wildlife. Field crickets, ground crickets, meadow katydids, and conehead katydids are among the conspicuous insects of grassy fields. Butterflies can be seasonally numerous where lack of recent mowing has allowed nectar-producing plants to grow up and flower. Grassy field edges may harbor such reptiles as eastern box turtles, and smooth green snakes occur locally in lush meadows, which may also host various frogs if there is water nearby. Grassland-specialized birds such as upland sandpipers, bobolinks, eastern meadowlarks, horned larks, and grasshopper sparrows are very localized nesters. Barn owls (rare) and red-tailed hawks hunt for rodents in grasslands. Ring-billed gulls, killdeer, American robins, red-winged blackbirds, common grackles, and European starlings* feed in mowed grassy fields throughout much of the year. Large areas of short grass, especially near the coast, attract migrant or wintering black-bellied and American golden-plovers, American pipits, horned larks, snow buntings, and lapland longspurs. Wild turkeys feed in fields adjacent to forest. Eastern moles, woodchucks, meadow voles, red foxes, coyotes, and white-tailed deer frequent large pastures.

Tilled fields support ephemerel feeding use by killdeer (which may nest), ring-billed gulls, and several other species (e.g., Canada geese, mallards, sparrows, snow buntings, eastern meadowlarks, red-winged blackbirds, horned larks) that come to feed on weed seeds and seeds in spread manure, especially in fall and winter. Seasonal pools in these fields attract mallards, Canada geese, and sometimes least and pectoral sandpipers.

Suburban Yards

Suburban yards, cut into forests or replacing abandoned fields, are an increasing element of the Connecticut landscape. Dense to scattered houses, planted non-native trees and shrubs, and all-too-extensive lawns typify most suburban areas. Common and conspicuous vertebrate animals of suburban yards, particularly those with lawns and some mature trees and shrubs, include eastern moles (aerating and debugging the lawn), little brown bats (debugging the air), gray squirrels, mourning doves (eating weed seeds), blue jays, American crows, black-capped chickadees and tufted titmice (visiting from nearby woods), northern mockingbirds, American robins, European starlings*, house finches*, and house sparrows*. Domestic cats and dogs can have a significant impact on the fauna of suburban yards through harassment of and predation on wildlife (see "Conservation" in Chapter 21). Yards with bird feeders are visited by additional species such as those discussed in Chapter 20. Suburban yards with nearby woods, fields, and streams may be visited by hundreds of other animal species that are more typical of natural areas. Highly manicured yards subject to pesticide use have an impoverished biota and deprive their owners of many opportunities for enjoyment of the natural world.

Cities

Cities may offer fine cultural amenities, but these assemblages of buildings, vehicles, and pavement are biologically depauperate and dominated by just one species (humans). Vacant lots and other spaces among pavement and buildings tend to be occupied by weedy plants native to Eurasia, the largest of which is ailanthus. Some old neighborhoods have large numbers of mature planted trees and even some trees that are relicts of the former native forest. I enjoy looking for these old trees but find it disturbing to contemplate the changes they have experienced.

Of course, cities do have some wildlife. Within urban areas, parks with remnants of natural communities and/or plantings may harbor gray squirrels and various species of common, disturbance-tolerant birds. House mice*, Norway rats*, rock pigeons*, European starlings*, and house sparrows* are the most observable vertebrate animals in the inner urban sections of most cities. Most of these species have been introduced to Connecticut from Eurasia. Near large bodies of water, ring-billed gulls and American crows may loiter around parking lots where human garbage is available. Big brown bats sometimes are fairly common in areas with large older buildings. Chimney swifts nest in chimneys of older city houses. Cockroaches*, silverfish*, and firebrats* are among the typical city insects occurring in buildings.

Pigeons and other city birds serve as prey for peregrine falcons, which now sometimes nest on ledges of tall buildings (e.g., the Travelers Tower in Hartford), dignifying the city but perhaps degrading the experience of seeing a peregrine falcon.

Conservation

A Basic Approach for Forest Conservation

Tracts of forest that have the greatest variation in slope direction, slope steepness, and soil moisture have the greatest amount of environmental diversity. Because species differ in their environmental requirements, it follows that a wider array of environmental conditions should support a greater diversity of plants and animals. Conservation biologists have therefore concluded that conservation of areas with a large amount of geomorphological heterogeneity should be an effective approach to protecting biodiversity, not only now but also in the future as ecological communities change with variations in climate and disturbance patterns.

Deforestation

A myriad of conservation issues involve upland ecosystems. Foremost among these is deforestation and other conversions of natural areas to commercial, residential, and agricultural uses.

Deforestation was rampant from the mid-1700s to mid-1800s and was associated with broad-scale agricultural development and charcoal production. By the early to mid-1800s, 70 to 80 percent of the state was deforested. At that time forest-associated wildlife was severely reduced.

In the late 1800s and early 1900s, reforestation began as many tracts of land devoted to agriculture were abandoned as people moved to cities to take manufacturing jobs, though logging continued to supply cordwood and charcoal for home hearths, locomotives, steamboats, and industry. Many deforested areas are now forest again (about 60 percent of the state is forested). These reforested areas currently are being reduced piece by piece, by extensive residential development. Also, many trees in these forests have attained commercially valuable size, and a new cycle of logging is underway.

Residential development is permanently removing forest habitat and negatively impacting remaining forest wildlife through fragmentation of extensive tracts of forest into smaller patches. A discussion of some specific effects of ongoing forest loss and fragmentation on birds is included in the conservation section of Chapter 20. In addition to these impacts are changes in water availability, sunlight, wind, pollinators, seed predators, herbivores, competitors, and other factors that may detrimentally affect remaining forest plants when they are exposed at forest edges.

Although reforestation renewed deciduous forest communities in many areas in the 1900s, most deforested sand plain communities were permanently lost to development. Protection and proper management, probably including periodic burning, are critical to the conservation of sand plain ecosystems

Forestry Practices and Missing Resources

A hole in a tree is a precious commodity for many animals (Fig. 7.31). In Connecticut, at least fifteen species of mammals, ranging from mice to fishers, regularly nest or take shelter in tree cavities, and several additional mammals use hollows in stumps or logs for shelter (see "Habitat" in Chapter 21). Woodpeckers, of course, make their own cavities in trees, but more than twenty species of native birds, ranging from wood ducks to bluebirds, depend on already-formed tree cavities as nesting sites (see "Reproduction" in Chapter 20). Additionally, gray treefrogs take refuge in tree hollows in summer, and several kinds of insects make important use of water-filled tree cavities (e.g., see "Vernal Pools" in Chapter 6).

It follows that forestry practices that remove dead, damaged, decayed, or diseased trees—the usual bearers of cavities—have a detrimental affect on numerous kinds of wild-

Fig. 7.31. Tree cavities, formed by microbial rot after a branch breaks off, or excavated by a bird, such as a pileated woodpecker, are essential for reproduction and shelter for many wildlife species. Paul J. Fusco/Connecticut DEP Wildlife Division.

life. Wildlife-friendly forest management involves maintaining at least one live cavity-bearing tree, 15 inches or more in diameter, per acre, plus at least a few large snags (standing dead trees that can provide cavities and other wildlife uses) per acre. Den trees and snags within 100 feet of a stream or wetland should not be removed. Additionally, provision should be made for the long-term availability of such trees. Beeches, birches, and maples readily form suitable cavities and are good candidates for wildlife use.

It is best to leave cavity trees as elements of scattered patches of uncut mature trees because such patches not only provide cavities and continued nut/seed/fruit production but also (through shading) provide cool, moist refuges that can shelter ground-dwelling amphibians and other small moisture-dependent animals, which can then repopulate the cut area as forest regrowth occurs. Such patches should be left even in otherwise clear-cut areas.

Ideally, in addition to these uncut patches, loggers should leave adequate uncut trees (living and dead) to ensure continuous availability of coarse woody debris (logs) on the floor of the future forest (Fig. 7.32). Tree-fall gaps—sunny forest openings with fallen logs and dense low vegetation, resulting from the toppling of dead or old, weakened trees—provide important resources (shelter, concentrations of insects and fruits) and ecological niches that otherwise may be scarce.

Logging operations often focus on removal of the largest trees, which of course have the greatest economic value. Removal of large trees has a negative impact on forest wildlife because mature trees generally are the ones that provide

Fig. 7.32. Logs on the forest floor provide shelter and food for a diverse array of forest animals.

suitable cavities, and they are the most prolific producers of nuts that are critical foods for many birds and mammals. Some of these valuable trees should be left for wildlife.

Grassland and Old Field Biota

Many grassland areas that formerly existed as a result of forest clearing for agriculture (or through burning by Native Americans) have been lost to reforestation and residential development. For example, a field of thick grass where I often found large numbers of meadow voles, plus red foxes and red-tailed hawks hunting them, as well as nesting eastern meadowlarks and bobolinks, now has houses, driveways, manicured lawns, and a distinct paucity of this disappearing wildlife community.

Most grasslands that have escaped development unfortunately are mowed during the nesting season, preventing successful nesting by grassland birds. With a compatible mowing schedule, such as that established in some areas at Bradley International Airport, populations of these birds can be maintained, adding unique elements to the state's biodiversity. Additional detrimental impacts of mowing include elimination of seed crops that would otherwise be used by migrating sparrows in fall and loss of food supplies (foliage, nectar) for butterflies and other insects. See the conservation section of Chapter 20 for further information.

Native sand plain grasslands, perpetuated by fire, are now gone from more than 90 percent of the historical range in New England and New York as a result of land development and changes in land use practices. These grasslands supported a now-rare assemblage of plants, including bushy rockrose, sand-plain agalinus, northern blazing star, and sickle-leaved golden aster, as well as a distinctive butterfly fauna.

Similarly, shrubby old fields that contribute uniquely to Connecticut's biodiversity are a declining commodity. Field sparrows, prairie warblers, and sun-loving reptiles—examples of some of many the species that are restricted to or thrive best in these habitats—have disappeared from many sites as a result of land development or natural habitat changes. If we wish to continue to enjoy a wide diversity of wildlife, efforts should be made to ensure the continuous availability of old field vegetation (though not at the expense of increased forest fragmentation). Large mosaics of contiguous grassland and old field habitats provide the best conditions for nonforest wildlife.

Non-native Invasive Species

Exotic plants, insects, and pathogens exert major impacts on Connecticut's deciduous forests, as throughout much of North America. Over the the past 500 years, some 2,000 insect species and 2,000 weedy plants of foreign origin have invaded the continent, mainly through inadvertent introductions. In Connecticut, these exotics have altered native forests through genocide (Dutch elm disease, chestnut blight [Figs. 7.33 and 7.34]), defoliation and killing of dominant canopy trees (gypsy moth, hemlock woolly adelgid), and competition with native species (many exotic plants).

Disturbed forests and forest edges often have a significant, and sometimes dominant, component of non-native vegetation. Common established exotic trees, shrubs, and vines of Connecticut woodlands include Japanese barberry (Fig. 7.35), multiflora rose, black locust, European buckthorn, European privet, Morrow's and tartarian honeysuckle, Japanese honeysuckle, Asian bittersweet, and winged spindle-tree. These may aggressively compete with and displace native plants. Except black locust, the seeds of all of these are spread readily to new areas by fruit-eating birds. Fruit trees such as apple and pear, lilacs, and other cultivated plants may be present in semi-open forests in the vicinity of former farms and settlements, but these are not invasive and do not pose a conservation problem.

The gypsy moth is periodically a highly influential non-native species in Connecticut forests. Gypsy moth caterpillars defoliate and sometimes kill trees, especially oaks, creating sunny openings in which previously inhibited or absent shade-intolerant species may thrive. Understory plants under defoliated canopies may exhibit increased growth and

Fig. 7.33. Chestnut blight fungus is one of the most influential invasive exotic organisms in eastern North America, responsible for the virtual elimination of American chestnut as a mature, productive, forest tree. This tree produced viable nuts shortly before succumbing to the blight evident in the swollen, cracked section of the stem near the top of the photograph.

Fig. 7.34. An old American chestnut stump: testimony for the often-devastating effects of non-native organisms.

Fig. 7.35. Japanese barberry, a non-native invasive plant, dominates the understory of this deciduous forest in eastern Connecticut.

survival as a result of increased light availability. It seems possible that the rain of caterpillar excrement (frass) that accompanies defoliation may enhance nutrient availability and contribute to better survival of understory plants that have been suppressed by the canopy trees, but recent research indicates that although frass nitrogen becomes incorporated into soils, most of it may be unavailable to plants and microorganisms.

Defoliation of overstory trees may reduce the suitability of the forest for certain animals such as birds that forage and nest in the forest canopy, whereas increased growth of shrubs and saplings in defoliated forests may favor thicket-dwelling birds such as the eastern towhee. In gypsy moth–defoliated forests in Pennsylvania, the abundance of black-capped chickadees, wood thrushes, blue jays, and Baltimore orioles decreased, whereas house wrens increased. Other studies in New England found that gypsy moth defoliation (but not excessive tree mortality) resulted in only minor, short-term changes in bird fauna. See Chapter 14 for further information.

Chapter 8

Algae, Fungi, and Lichens

Here begins a series of chapters dealing with many of Connecticut's major groups of wildlife species. This chapter covers two groups—algae and fungi/lichens—that are not closely related but are treated together simply for convenience. They do share the quality of being poorly understood by most people (myself included!).

Algae

The best known algae are the seaweeds of Long Island Sound. Algae also occur in fresh water, but these mostly microscopic algae are not treated in detail in this book.

Seaweeds are an important converter of solar energy into living biomass. Smaller algae known as phytoplankton also capture the Sun's energy and convert it into an organic form that can be used by zooplankton and other small animals. See Chapter 3 for a brief discussion of estuarine phytoplankton.

Most estuarine algae attach to hard surfaces such as rocks or shells in both subtidal and intertidal areas. Many seaweeds are long-lived and easily seen year round, whereas others grow anew each year and are evident only during particular seasons. For example, the lobed, pink, often dotted blades of *Grinnellia americana* occur only in August and September, and the thin, light green sheets of *Monostoma pulchrum*, resembling a delicate sea lettuce, appear only in spring and early summer.

Seaweeds feed and shelter a rich fauna of marine mollusks, crustaceans, and fishes, and they provide attachment sites for other algae and sessile animals. Some seaweeds have phenolic, terpenoid, or halogenated chemicals in their surface tissues that combat attacks by some herbivores. Swans, geese, various ducks, and coots are among Connecticut's largest consumers of seaweeds.

Seaweeds are important even when detached and decomposing. For example, decaying beach-cast seaweeds feed multitudes of amphipod beach hoppers and beach fleas, both of which are important food resources for many shorebirds.

Over 100 species of seaweeds or macroalgae have been observed along the coast of Connecticut. These saltwater algae, especially brown algae (Phylum Phaeophyta) and, to a lesser degree, red algae (Phylum Rhodophyta) include the largest representatives of the Protoctista kingdom. In fact, certain brown algae (kelp) are among the world's longest organisms, sometimes reaching more than 300 feet in length and thus nearly rivaling towering redwood trees (although Connecticut kelps do not grow this large).

Reproductive patterns in algae are too diverse to summarize comprehensively in this book, but here are a few examples for common Connecticut species, though these descriptions may be comprehensible only to readers with biological training. In rockwed and green fleece*, mature diploid individuals give rise to haploid gametes that unite in the water to form embryos that develop directly into adults. Sea lettuce and hollow green weeds represented by diploid sporophytes that produce haploid spores through meiosis. These germinate and grow into haploid gametophytes that produce haploid gametes through mitosis. In the water, the gametes unite to form a diploid sporophyte, and the cycle begins again. For those of you confused by these descriptions, the bottom line is that new seaweeds of these species arise from female gametes that have been fertilized in the water by male gametes.

The following section includes examples of some of Connecticut's most common seaweeds. Additional information on seaweeds is included in Chapter 3.

Green Algae

Green algae (Phylum Chlorophyta) range in size from tiny freshwater phytoplankton to larger, "leafy" marine algae such as sea lettuce. Common intertidal species tend to be fast-growing colonizers of newly exposed rocks.

Green fleece* *(Codium fragile).* Spongy branched alga native to western Pacific; inadvertently introduced in 1950s (probably via Europe, where it became established in the early 1900s); now locally abundant, attached to rocks or shells; individuals may live more than a year. *Note:* Heavy growths of green fleece can smother oysterbeds. In waters off Maine, green fleece is replacing kelp, especially where heavy infestations of encrusting lacey ectoprocts cause kelp defoliation during storms and thereby create openings in kelp beds.

Hollow green algae (*Enteromorpha* spp.). Several species in wide range of shallow-water habitats. Bright green, generally tubular; often abundant on shallow subtidal and intertidal rocks, shells, and other algae, even in waters with low salinity and moderate pollution.

Sea lettuce (*Ulva lactua*) (Fig. 8.1). Leafy; tolerant of diverse conditions, including polluted or low salinity water; attaches to rocks, shells, algae, or smooth cordgrass in intertidal and shallow subtidal zones, or floats in quiet water; major food source for mute swans* and migrating brant.

Mermaid's hair (*Cladophora* spp.). Fine branched filaments attached to objects or floating in intertidal and subtidal areas; with sea lettuce, thrives in waters with unnaturally high nitrogen inputs.

Brown Algae

Brown algae reach their pinnacle of development in cold and temperate oceans. They tend to dominate subtidal areas not subject to severe disturbances. These algae are relatively "tough-skinned," and many contain defensive chemicals known as phlorotannins that enhance their resistance to grazing snails and other herbivores. Large, tough species often grow slowly, live several years, and may outcompete other algae. Smaller and more delicate ones tend to be annuals that grow anew each year.

Kelp (*Laminaria saccharina*) (Fig. 8.2). Long blades often up to several feet long, "anchored" by cylindrical stalks to bottom rocks; exceptionally long blades develop in deep water, sometimes wash up on beaches; kelp beds, with *Laminaria saccharina* as most common species, occur along the Connecticut coast east of New Haven Harbor;

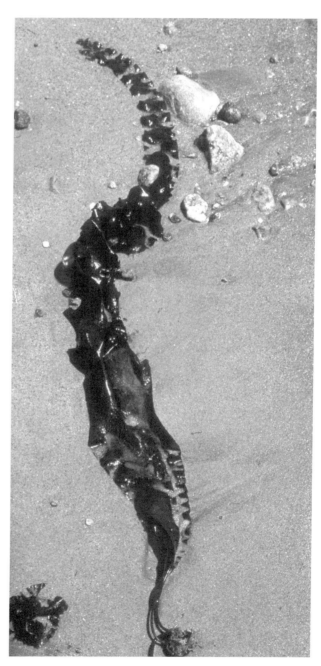

Fig. 8.1. Sea lettuce is very common in the shallows of Long Island Sound.

Fig. 8.2. Kelp is Long Island Sound's largest alga.

Fig. 8.3. Rockweed (left and right) and Irish moss (center) are tough intertidal algae and common in Long Island Sound. Here Irish moss has bleached to nearly white, but usually it is deep purplish red.

may be eliminated by dense populations of grazing sea urchins; sometimes covered by encrusting lacy ectoproct, introduced from the Pacific Ocean.

Rockweeds (e.g., *Fucus vesiculosus, F. distichus, F. spiralis*) (Fig. 8.3). Abundant on intertidal rocks, shells, or wood, or at base of smooth cordgrass along tidal creeks; thick masses provide cool, moist shelter for small animals (e.g., yellow periwinkle) during exposure to sun at low tide; perennial; like knotted wrack and Irish moss, may be inhibited from colonizing certain areas by existing growths of sea

Fig. 8.4. Knotted wrack (center) is a grazer-resistant brown alga. The smaller, attached clumps are the red algae *Polysiphonia lanosa*. Rockweed is barely visible at left.

lettuce, hollow green weeds, and laver, but sometimes colonizes new areas by producing new individuals through sexual reproduction.

Knotted wrack (*Ascophyllum nodosum*) (Fig. 8.4). Often attached to intertidal and subtidal rocks below highest rockweeds on exposed rocky shores subject to moderate wave action; also commonly attached to rocks and other objects in quiet, shallow subtidal cove waters; does poorly in heavy surf; other more palatable algae, such as sea lettuce, hollow green weeds, and *Polysiphonia lanosa*, may take advantage of knotted wrack's better herbivore defenses and live attached to this species, benefiting from the relative scarcity of grazers.

Crustose brown algae (*Ralfsia verrucosa*). Brown crust on rocks or shells in tidal pools or on objects in lower intertidal zone; tough, slow-growing, very resistant to damage from storms or herbivores.

Filamentous brown algae (*Ectocarpus* spp.). Small, branched, brown, hairlike; attaches to larger algae such as rockweed and knotted wrack or other solid objects.

Red Algae

Red algae contain chlorophyll *a* and phycobiliproteins, which produce the characteristic red color of these algae and facilitate the absorption of light energy that penetrates most deeply into the water, so it is not surprising that many of the algae living in the deepest water are reds. However, patterns of algal distribution reflect various factors, including tolerance of physical stress, impacts of herbivores, and competition, as well as the type and concentration of photosynthetic pigments. In bright intertidal waters, red algae may have lots of chlorophyll and appear brown, black, olive, or dark purple. In deep water or other dark sites, chlorophyll is scarce and red algae are pink, red, or purple.

Crustose red algae (*Hildenbrandia rubra*). Red crust on rocks or shells. *Phymatolithon laevigatum*, a coralline alga, forms pink crusts that produce a thick, chalky or stony layer on rocks (Fig. 8.5).

Dulse (*Palmaria palmata*). Tough, wide blades; attaches to rocks in water still deep at low tide; harvested commercially in Canada for human consumption.

Coral weed (*Corallina officinalis*) (Fig. 8.5). Erect, branched alga; pink (bleaching to white) limey crust of calcium and magnesium carbonates on surface; uncommon but generally on rocks in lower intertidal and subtidal zones; tough, slow-growing, very resistant to damage from storms or herbivores.

Irish moss (*Chondrus crispus*) (Fig. 8.3). Usually in dense masses in zone between rockweeds above and coral weed and kelp below; often in backwash areas, but versatile and

Fig. 8.5. Here are remnants of two of the toughest, slowest-growing algae: a crust-forming coralline algae (left) and the jointed branches of coral weed (right).

Fig. 8.6. Mushrooms of various forms are the spore-producing bodies of fungi. These were collected in early August.

often abundant in other sites; freeze sensitive, so mainly subtidal; grows to about 10 inches long in protected sites, shorter where wave action is strong; varies from purplish red to whitish; similar to brown macroalgae in being long-lived and slow growing, a good competitor, and resistant to herbivores; can spread through vegetative or sexual reproduction; used for human food and in manufacture of gels and thickeners.

Banded red algae (*Ceramium* spp.). Soft, fine, branched, common; appear beaded or banded; have pincerlike branch tips; attach to larger algae or other objects in diverse sites in intertidal to subtidal zones in shallow water, or float free.

Polysiphonia or tubed weeds (*Polysiphonia* spp.). Finely bushy with fine crosswise and lengthwise lines on filaments (most visible with hand lens); attach to rocks, shells, wood, algae, or eelgrass submerged at low tide; common in bays and quiet water. *Polysiphonia lanosa* lives only on knotted wrack (Fig. 8.4).

Fungi and Lichens

Fungi are neither plants nor animals but rather representatives of a distinct division of life that includes mushrooms, molds, yeasts, and lichens. Worldwide there are approximately 56,200 recognized species of nonlichenized fungi (e.g., mushrooms) and an additional 13,500 species of lichens. However, the full diversity of fungi is poorly documented, and some estimates place the actual number of fungal species as high as 1.5 million.

Fungi

Fungi have enormous economic and ecological importance. They are sources of foods (e.g., for humans, squirrels, and mice) and antibiotics, active agents in the fermentation of alcohol-containing beverages and the rising of bread, and, with bacteria, earthworms, and tiny arthropods, perform the critical functions of decomposition and nutrient transfer in natural ecosystems. Important things do indeed come in small packages.

Nearly all fungi require free access to oxygen, and all lack chlorophyll and cannot produce their own food. They nourish themselves by releasing enzymes and externally digesting organic material and absorbing the released nutrients. Saprophytic fungi digest dead and decaying material, such as downed logs and leaves on the forest floor (Fig. 8.6). Parasitic fungi attack living tissue of animals, plants, or other fungi. Dutch elm disease, chestnut blight, and athlete's foot are caused by parasitic fungi.

Mycorrhizal (mycotrophic) fungi live in a mutually beneficial association with the roots of living plants, with each partner in the relationship providing the other with some essential nutrients (Fig 8.7). For example, the fungus makes available to the plant nitrogen, phosphorus, potassium, calcium, and other substances that otherwise may not be readily available, and the plant is the source of carbohydrates, amino acids, and vitamins required by the fungus. Some mycorrhizal fungi contain bacteria that take in nitrogen from the air and convert it to a form that can be used by the fungus and plant. Mycorrhizae also increase water retention and may protect plant roots from disease microbes. Without the fungus, the plants grow poorly if at all, and the fungus may not be able to live at all unless associated with the plant root. Amazingly, mycorrhizae associated with different plants may interconnect and enable water and nutrients to pass from one plant to another. Many familiar mushrooms, such as boletes and chanterelles, form mycorrhizae.

Some mycorrhizae exist entirely or primarily within the roots of the plants, although in woody plants only tulip-tree,

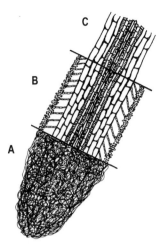

Fig. 8.7. Simplified view of a mycorrhizal fungus on a pine root: A = tip of plant root surrounded by a mass of fungal threads. B = Lengthwise cross-section of the root, its fungal sleeve, and its connections to the root. C = cross-section of the root as it would appear without the mycorrhizae.

maples, various members of the heath family, and a few others are known to have this arrangement. Mycorrhizae in which the fungal tissue is mainly outside the root occur in pines, beeches, oaks, birchs, willows, walnuts, and a few other families (Fig. 8.7).

Each fungus species tends to be associated with a specific substrate (e.g., particular kinds of decaying wood, organic soil, plant roots, skin, etc.) (Fig. 8.8). Thus fungi that attack living trees or ferment grapes do not assail people, and so on. Some fungi commonly grow on other fungi, such as decomposing mushrooms, which also may attract and serve as a larval food for small flies.

A mushroom is the fruiting (spore-producing) body of a fungus, which also comprises an often large network of fine filaments (hyphae; singular, hypha) that do the digesting and absorbing of nutrients. The network of filaments or mycelium is the initial stage in the life of a fungus.

The fruiting body of most fungi develops aboveground, but the fruiting bodies of hypogeous fungi (truffles) form underground. These fungi, when "ripe," release distinctive odors that attact squirrels and mice, which dig them out and eat them. The spores pass through the digestive system unharmed and are spread through the habitat in the mammals' feces.

A mushroom is produced only after sufficient nutrients have been absorbed and when environmental conditions such as temperature, moisture, pH, and daylength are appropriate. In Connecticut, mushrooms appear primarily from April through October, with each species exhibiting its own fruiting schedule. Many mushroom species are most evident a few days after soaking rains in late summer and

Fig. 8.8. Prolific fruiting bodies of fungi indicate that the fungus has accomplished extensive decomposition within this dead tree trunk.

early fall—the fruition of the seasonal digestion of the fallen leaves and wood on the forest floor. Other mushrooms (e.g., morels) appear only in spring.

Fungal spores are dustlike particles that generally survive for years under dry condtions but are killed by boiling. In a typical mushroom species, spores released from the fruiting body germinate and grow after they land on a suitable substrate. The spore grows into a filament or network of filaments. If a filament (hypha) comes into contact with a compatible filament of another individual, the two may fuse and form a two-nucleus cell that grows and forms a new mycelium. Eventually this may produce a mushroom. In the mushroom, the nuclei from the two different hyphae fuse, then meiosis occurs, reducing the diploid nuclei to haploid nuclei as the spores are formed. The mushroom decomposes after the spores are released. Some fungi produce spores without the need for fusion of different hyphae.

Not many mushrooms are poisonous, but these few can be deadly or cause serious liver or kidney damage. Among the deadly ones in Connecticut is the destroying angel (*Am-*

Fig. 8.9. Destroying angel *(Amanita virosa)* is one of relatively few fungi that are deadly to humans. See text for identification characteristics.

Fig. 8.10. Lichens thrive on physically harsh, barren substrates that exclude most organisms.

anita virosa), a common species in summer and early fall (Fig. 8.9). It's best to assume any pure white mushroom with a smooth cap, conspicuous collar of thin tissue on the stem (this may drop off), and a cuplike or saclike structure at the base, or anything similar, is this one. Immature amanitas resemble edible puffballs, but when sliced in half reveal developing cap, gills, and basal cup (volva), whereas puffballs are solid inside. If you believe that someone has eaten a poisonous mushroom, induce vomiting and call a physician, hospital, or poison control center.

Lichens

Lichens are associations of a fungus and a photobiont such as green algae or cyanobacteria (or both) (Fig. 8.10). A large diversity of fungi are components of lichens, as are a variety of algae and cyanobacteria. The photobiont provides the fungus with carbohydrates produced by photosynthesis whereas the fungus ameliorates the physical environment and allows the partnership to withstand environmental extremes. In some cases the fungus is mildly parasitic on the photobiont but other lichen fungi are aggressive parasitic exploiters of their photobionts. Lichens that contain cyanobacteria can "fix" gaseous nitrogen and convert it to nitrate that is used by the lichen and ultimately may fertilize the soil. Lichens obtain minerals by simple absorption through any part of their body. They do poorly in smog-laden air and so have been used as pollution indicators.

Lichens attach to rocks, bark, wood, soil, or other objects. Most are quite specific about the substrate on which they grow. With mosses, lichens are among the major colonizers of bare rock. They begin soil formation and create conditions that eventually allow the establishment of flowering plants. Lichens are important soil stabilizers, but many are delicate and easily damaged by vehicular and foot traffic.

Lichens may be crustlike (crustose), leafy (foliose), erect, mosslike, or hairlike (fruticose), or may comprise small scalelike lobes that are intermediate between crustlike and leafy (squamulose). They become dry, brittle, and dormant (but quite alive) during droughts. When moisture returns, they quickly absorb water (from rain or humid air), become soft or supple, change color, and resume growth. Color change in a hydrating lichen often tends toward green because the outer layer of the lichen becomes transparent and allows greater exposure of the photobiont layer.

Lichens grow slowly. For example, crustose forms expand outward less than 0.1 inches (0.5 to 2.0 mm) each year. Some lichens produce substances that inhibit growth of soil fungi, interfere with seed germination, deter browsing invertebrates, or erode rock.

Deer, various rodents, snails, slugs, and various other invertebrates feed on lichens to some degree. Hummingbirds, flycatchers, gnatcatchers, and other songbirds adorn their nests with lichens, with a camouflaging result. Lichenlike skin coloration allows gray treefrogs to "disappear" when perched on lichen-covered trees.

The fruiting bodies of lichens, which produce spores, are those of the fungus. The spore-producing structures generally release the spores into the air, and the spores disperse in the wind. Unless the spore germinates very close to the proper photobiont, a new lichen will not form. Lichens more often reproduce vegetatively through specialized fragments that can be dispersed by the wind or other agents.

Chapter 9

Plants

What is a plant? This may seem to be a trivial question, considering how easily most persons recognize various wildlfowers, shrubs, and trees as members of the plant kingdom. However, certain groups of organisms—mushrooms and other fungi, for example—commonly thought to be plants, are not. And there are large numbers of chlorophyll-containing organisms (e.g., various kinds of algae) that are regarded as plants in some classifications but not in others. Some recent classifications of living things include in the plant kingdom only multicellular organisms composed of nucleated cells that are enclosed in walls of cellulose and that usually contain chloroplasts, where photosynthesis occurs. Plants are further characterized by having life cycles that include alternating haploid and diploid generations and by retaining their embryos in maternal tissue. In simpler terms, if it's not an animal, fungus, bacterium, alga, or protozoan, it's a plant.

In this chapter, hydrophytes—plants that always or usually occur in aquatic habitats or wetlands—are indicated by standard acronyms as follows: OBL indicates plants that virtually always occur (more than 99 percent frequency of occurrence) in aquatic or wetland habitats; FACW indicates plants that usually exist in these habitats (67 to 99 percent frequency of occurrence).

Importance

To a large degree, plants make Connecticut's terrestrial and wetland ecosystems look and function as they do. Through the process of photosynthesis, plants capture energy from sunlight and convert it into an organic form that can be used as a nutrient source not only by the plants but also by animals and other forms of life. And plants decrease atmospheric carbon dioxide and generate gaseous oxygen, which the vast majority of life absolutely requires. (Algae, diatoms, and dinoflagellates play these roles in large bodies of water.) Plants also affect the environment by taking up soil moisture, releasing into the air large amounts of water vapor (a single large tree may transpire roughly 50 to 100 gallons of water each day).

Plants also are an important aspect of habitat. They provide shelter (insulation, shade, wind reduction), building materials, a moisture source, a place to lay eggs, bear young, or store food, and perches on which to hunt, avoid predators, or cool off. Plants additionally provide humans with fuel, fertilizer, drugs, pesticides, ornamentation, fragrance, and research material. Suffice it to say that plants allow most animals, including humans, to exist.

Biogeography

Hardly any of Connecticut's native plants occur west of the North America's Rocky Mountains, yet a surprisingly large number of Connecticut natives are indigenous to Europe and Asia, and there is a notable array of species and genera that occur in both Connecticut (and much of eastern North America) and eastern Asia. Approximately 65 genera of seed plants exhibit an eastern Asian–eastern North American distributional disjunction.

This seemingly improbable affinity dates back millions of years ago to the Tertiary period, when a warmer climate and a continuous land connection resulted in a broad expanse of similar environmental conditions extending from northeastern Asia across Alaska and into eastern North America. With subsequent cooling and drying of the climate and submergence of the connecting land, these plants became restricted to the warm, moist refuges in the eastern portions of the two continents, while drought-tolerant species evolved in western North America. Connecticut examples of this pattern include interrupted fern *(Osmunda claytoniana)*, maidenhair-fern *(Adiantum pedatum)*, sensitive fern *(Onoclea sensibilis)*, dogwoods *(Cornus* spp.), partridge-berry *(Mitchella repens)*, tulip-tree *(Liriodendron tulipifera)*, tupelo *(Nyssa sylvatica)*, blue cohosh *(Caulophyllum thalichtroides)*, lopseed *(Phryma leptostachya)*, sarsaparilla *(Aralia* spp.*)*, ginseng *(Panax* spp.*)*, skunk cabbage *(Symplocarpus foetidus)*, trillium *(Trillium* spp.*)*, and false miterwort *(Tiarella cordifolia)*. In a few cases (e.g., moonseed, *Menispermum canadense*), it is possible that the disjunction may derive from long-distance dispersal rather than range restriction. Some other plants, such as hay-scented fern *(Dennstaedtia punctilobula)* and walking fern *(Asplenium rhizophyllum)*, have their closest relatives in eastern Asia.

Nonvascular Plants

The plant kingdom includes two major divisions: nonvascular plants and vascular plants. Living nonvascular plants include only the mosses (Phylum Bryophyta, 10,000 species worldwide), liverworts (Phylum Hepatophyta, 6,000 species), and hornworts (Phylum Anthocerophyta, 100 species). These lack the roots, true stems, leaf veins, and tubular fluid-conducting vessels of vascular plants.

Mosses are common constituents of swamps and bogs. In upland habitats, mosses readily colonize sunny or shady bare ground or rock crevices and are important in preventing erosion, retaining water, and accelerating soil formation (e.g., by trapping dust). This facilitates colonization of these sites by other plants.

Familiar moss plants are haploid individuals (gametophytes) that produce sperm and/or eggs in separate structures among the upper leaves of the moss (Fig. 9.1). Rainwater or heavy dew is required for transport of the sperm to the egg for sexual reproduction. Fertilization of the egg by the sperm results in a diploid zygote that grows into a sporophyte that has at its top a spore capsule in which meiosis (spore formation) occurs. The filament and capsule of the diploid sporophyte remain attached to the haploid gametophyte. Spores are released from the capsule and germinate to form new haploid individuals.

Liverworts and **hornworts** have similar life cycles that are basically like that of mosses. Liverworts live in shady moist sites, with most species in the tropics, and have a flattened, lobed shape (gametophyte) and a less complex sporophyte than do mosses (sporophyte stays attached to a flowerlike structure of the female gametophyte) (Fig. 9.2). Haploid liverworts can also produce tiny spheres that germinate and grow into new haploid individuals. Hornworts, some of which have hornlike growths (sporophytes) extending from a flattened thallus (gametophyte), also inhabit wet or moist sites, as well as bare rock. Sometimes they are aided in nutrient acquisition by cyanobacteria that convert nitrogen in the air into organic form.

Fig. 9.1. Moss spores form in capsules at the end of long filaments. This view shows both old and newly growing spore capsules.

Fig. 9.2. Liverworts are simple lobed, flattened plants of moist, shaded habitats.

Vascular Plants

Most plants are vascular plants (about 250,000 species worldwide). These have true roots and stems, veined leaves, and conducting tissues called xylem and phloem. The xylem transports water and nutrients from the roots to the rest of the plant, whereas the phloem is the conduit through which sugar and other leaf products move to other parts of the plant. The vascular plant flora of Connecticut includes about 1,700 native species, plus an additional 925 established non-native species. Vascular plants comprise two major divisions, nonseed-bearing plants (true ferns and fern allies, such as moonworts and grapeferns, adder's-tongue ferns, firmosses, clubmosses, spikemosses, quillworts, and horsetails) and seed-bearing plants (e.g., conifers and flowering plants).

Fern Allies

Horsetails (Phylum Sphenophyta, with up to a few dozen species worldwide, 8 in Connecticut) are the sole surviving lineage of a plant group that flourished in Carboniferous coal-forming swamps some 350 million years ago. Indeed, some members of this group grew into large trees. Today the group includes only small herbaceous horsetails. Horsetails inhabit both wetland and upland soils. All of the species in Connecticut are widely distributed in the Northern Hemisphere. The silica-laden stems can be used as a nail file or fine abrasive.

The familiar cylindrical form of horsetails is a diploid sporophyte (Fig. 9.3). Production of spores through meiosis occurs in a cone at the top of the hollow stem. Spores germinate in the soil and develop into inconspicuous gametophytes, which produce the eggs and sperm, the union of which results in the origin of the diploid sporophyte. The spore-bearing stalk often is separate from the bottlebrush-like stalk that is involved in photosynthesis and from which the name "horsetail" derives (despite poor resemblance).

The state's most common and widespread horsetails are field horsetail (*Equisetum arvense*), in open or semi-open upland habitats, including disturbed areas; water horsetail (*E. fluviatile*), in freshwater wetlands and aquatic margins (OBL); and scouring-rush (*E. hyemale*), found in shady stream and pond borders, fields, woods, wet areas, embankments (FACW).

Firmosses (Family Huperziaceae, 1 species in Connecticut), **clubmosses** (Family Lycopodiaceae, about 1,000 species worldwide, 7 in Connecticut), and **spikemosses** (Family Selaginellaceae, 3 species in Connecticut) are not true mosses but rather are nonflowering, evergreen, perennial

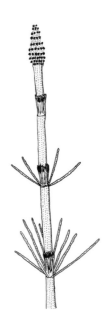

Fig. 9.3. Spores form in the top part of fertile stems of water horsetail.

(except one spikemoss) vascular plants. With horsetails, clubmosses were dominant tree-forming plants of the Carbonifeorus coal swamps. Today they are represented only by small nonwoody plants up to several inches tall. Often they grow in mixed forests near evergreens (mountain laurel, conifers) where fallen leaves of deciduous trees do not accumulate too deeply but where light is available when adjacent deciduous trees are leafless.

Most of these plants grow in upland areas, but some also occur in wetland soils. One Connecticut species—running clubmoss, *Lycopodium clavatum*—abundant in acidic soils of woods and clearings, is one of the world's most widely distributed plants, occurring not only in North America from Alaska to the east coast but also in South America, Eurasia, Africa, Indonesia, and New Guinea. Clubmosses are sometimes mistaken for small coniferous trees.

In these plants, the minute haploid gametophytes grow from dustlike spores and become soil-dwelling organisms often associated with mycorrhizal fungi. The familiar evergreen, moss- or pinelike sporophyte develops from the union of the gametophyte-produced eggs and sperm. Apparently, it takes several years for a spore to produce a gametophyte and several more years for the gametophyte to yield a new sporophyte.

The spore-producing area of clubmosses consists of conspicuous clublike structures projecting above the plants (Fig. 9.4). Firmoss spores are produced in inconspicuous structures interspersed among the leaves (Fig. 9.5). The spores, which mature in late summer or fall, are highly flammable (due to their fat content) and have been used in performance stunts for years. Clubmosses spread vegetatively

Fig. 9.4. Tree clubmoss is abundant in Connecticut forests. It has spore-forming cones at the ends of specialized stems.

Fig. 9.5. Shining firmoss lives in forests and swamps. Spore-forming structures are visible at mid-stem.

by producing new upright shoots from long horizontal rootstocks that grow out from existing plants. Because the old growth eventually withers, individual clubmosses can actually change their location over time.

Quillworts (Family Isoetaceae: *Isoëtes* spp.), like other fern allies, are nonflowering, perennial vascular plants (150 species worldwide; 4 to 5 species [plus hybrids] in Connecticut) (Fig. 9.6). These obligate wetland plants grow as tufts in lake, pond, or stream shallows or wet shores; all are obligate hydrophytes (OBL).

Moonworts and **grapeferns** (*Botrychium* spp.) and **adder's-tongue ferns,** formerly regarded as true ferns, are now viewed as fern allies. Most moonworts and grapeferns (7 species in Connecticut) have quite unfernlike fronds and somewhat grapelike clusters of spore-bearing structures on the fertile spike. Northern adder's-tongue (*Ophioglossum pulsillum*), recognized by its single, simple tonguelike blade adjoining the spore-bearing spike, occurs in wetlands or disturbed, early successional sites with moist soils, such as beaver meadows, acidic fens, pastures, or power-line corridors (Fig. 9.7).

Fig. 9.6. Quillwort often grows in shallow water. Abigail Rorer, from Tiner (1987).

Vascular Plants | 131

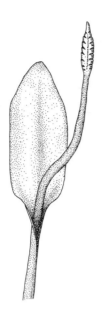

Fig. 9.7. Adder's-tongue is uncommon in wet meadows and similar sites.

Ferns

Ferns (Phylum Filicinophyta), a large group of several families of seedless vascular plants, are represented by 11,000 species worldwide and about 50 species in Connecticut. Most are recognizable as ferns even to the most urbanized people, but some, such as climbing fern (*Lygodium palmatum*), a vinelike fern with five-lobed "leaves," are quite unfernlike. The attractive climbing fern became rare as a result of being heavily collected for decorative use. In an effort to protect it, the 1896 state legislature passed a law banning such uses.

A large proportion of Connecticut's ferns, such as polypody (*Polypodium virginianum*) (Fig. 9.8) and Christmas fern (*Polystichum acrostichoides*), are widespread in but restricted to eastern North America. Among the most widely distributed ferns are royal fern (*Osmunda regalis*), a wetland species, and bracken (*Pteridium aquilinum*), a dryland fern, both of which range nearly worldwide.

The leafy parts of most fern species generally die back each year, and new fronds grow from a perennial rootstock the next year. Other ferns, such as Christmas fern, glandular wood fern (*Dryopteris intermedia*), marginal wood fern (*D. marginalis*), cliff-brakes (*Pellaea* spp.), and polypody, are evergreen. Some others are semi-evergreen.

Ferns vary greatly in their habitats. At one extreme, bracken and hay-scented fern (*Dennstaedtia punctilobula*) tolerate sunny, dry soils. When burned, bracken sprouts vigorously from the buried rhizome; it increases where the fire has removed fire-sensitive competitors. Others, such as

Fig. 9.8. Polypody is a ubiquitous, evergreen, rock-dwelling fern.

marsh fern (*Thelypteris palustris*) (OBL), sensitive fern (*Onoclea sensibilis*) (FACW), royal fern (OBL), ostrich fern (FACW), cinnamon fern (*Osmunda cinnamomea*) (FACW) (Fig. 9.9), and chain-ferns (*Woodwardia* spp.) (OBL or FACW), are restricted to wetlands or very moist soils.

Some ferns thrive on rocks of specific geochemistry. For example, purple cliff-brake (*Pellaea atropurpurea*), smooth cliff-brake (*Pellaea glabella*), slender cliff-brake (*Cryptogramma stelleri*), bulblet bladder fern (*Cystopteris bulbifera*), wall-rue spleenwort (*Asplenium ruta-muraria*), and walking fern (*Asplenium rhizophyllum*) are essentially limited to calcareous or circumneutral outcroppings, whereas hairy lip-fern (*Cheilanthes lanosa*), mountain spleenwort (*Asplenium montanum*), and rusty woodsia (*Woodsia ilvensis*) are mostly restricted to acidic/noncalcareous rock outcrops. Connecticut's rarest ferns are strongly associated with rock outcroppings. However, polypody, also a rock dweller, is very common.

Curiously little in the way of close ecological relationships between ferns and animals have been documented, though worldwide there are at least several hundred species of insects that eat ferns, and mites, millipedes, woodlice, and tardigrades have been collected on ferns and fern allies.

Fig. 9.9. Cinnamon fern, typically a wetland species, has distinctive fertile fronds.

Fig. 9.10. Fern spores are produced on distinct fertile fronds (see Fig. 9.9) or the underside of leafy fronds as shown here.

I've found walking ferns that had been heavily chewed by insects. Hairy trichomes on fern fronds and rhizomes may provide some protection from herbivores. Most uses of ferns are generalized and opportunistic; for example, deer feed on marginal wood ferns in winter, and veeries occasionally nest in clumps of cinnamon fern. The most specific interactions involve the larval stages of borer moths (*Papaipema* spp.) (Fig. 14.95), which bore into and feed solely on the stems, rhizomes, or roots of bracken, chain-ferns, sensitive fern, *Osmunda* ferns, and others. The attractive adult moths fly in late summer and early fall.

The familiar fern plant is a diploid spore producer (sporophyte) with leafy photosynthetic fronds. Haploid spores are formed by meiosis in distinct fertile fronds (Fig. 9.9) or in clusters of small sori (spore cases) on the underside of the leafy fronds (Fig. 9.10). The spores germinate and grow into small haploid gametophytes that produce sperm and eggs, the union of which results in the initiation of a new leafy sporophyte. Patient, sharp-eyed naturalists may want to look for the tiny heart-shaped gametophytes in moist ferny sites in summer.

Ferns vary in the percentage of the fronds that are fertile. For example, Tara Tracy found that among 30 clumps of Christmas fern sampled in fall, 20 to 67 percent (average 42) of the fronds were fertile.

Vegetative reproduction may be extensive in some ferns. Fronds often sprout from perennial horizontal rhizomes that extend from existing frond clumps. In this way, after many years, a single genetic individual may cover an extensive area. Or an individual can move across the landscape as new fronds sprout from new extensions of the rhizome and older parts of the rhizome and its fronds wither and die.

Fern fiddleheads are commonly harvested and eaten by humans, but caution is advised. Some ferns, such as bracken and some *Osmunda* ferns, contain carcinogens that may cause stomach cancers. Ostrich fern fiddleheads, marketed commercially in Connecticut, apparently are safe to eat.

The family Marsileaceae includes one non-native species, water shamrock (*Marsilea quadrifolia*), a perennial, aquatic, fern ally from Europe that is long-established in certain lakes and quiet streams in Connecticut (e.g., Bantam Lake). It resembles a "four-leaf clover."

Conifers (Gymnosperms)

Conifers (Phylum Coniferophyta, 550 species worldwide) include primarily needle-bearing trees and shrubs that produce naked, cone-enclosed seeds in which the embryo is embedded in the haploid tissue of the parent. In Connecticut, conifers are represented by yews, pines, larches, firs, hemlocks, spruces, cedars, and junipers. Most conifers are monoecious (male and female structures on the same plant), with the pollen-producing cones separate from the female cones where the seeds develop, and are wind-pollinated. Most are evergreen, with individual needles living several years, but larches are deciduous.

More so than most deciduous trees, conifers can thrive on nutrient-poor soils. Specific information on the natural history of Connecticut conifers is included in the following section, "Ecology of Trees, Shrubs, and Vines in Connecticut."

Flowering Plants

Flowering plants (Anthophyta or Angiosperms, 235,000 plus species worldwide) are the dominant plants alive today. In these plants, the gametophyte generation is microscopic and only the sporophyte generation is evident. With very few exceptions, the plant's ovules ("eggs") must be fertilized by pollen in order to produce a viable seed. The haploid pollen grains are transported to the female reproductive parts by the wind or more commonly by insects, sometimes by birds, bats, or other animals. The pollen grains develop extensions that convey the sperm nuclei to the ovules that are hidden deep inside the ovaries in the base of the flowers, where fertilization occurs. After fertilization, one or more ovaries (and sometimes associated structures) may increase tremendously in size and develop into a structure, such as a pod or fleshy fruit, that envelops the seed(s). The fruit may be large and colorful or quite inconspicuous. Production of fruits can be a major drain on the plant's resources, such that vegetative growth and storage of nutrients may be curtailed during fruiting years.

Flowering plants can be placed in a few major categories with respect to the life span and timing of growth of the individual plants. **Perennials** are plants that live and grow for multiple years. Some individuals may live thousands of years. It may take several years of growth before the plant flowers and produces seeds; thereafter, individuals of most perennials produce seed every year, though the amount of seed production often varies greatly among different years. Trees and shrubs, dominant in forest ecosystems, obviously are perennials, and so are a diverse assortment of herbaceous plants, including bloodroot *(Sanguinaria canadensis)*, goldenrods *(Solidago* spp.*)*, pokeweed *(Phytolacca americana)* (Fig. 9.11), and many others. The above-ground stems and foliage of most herbaceous perennials die back each year, but underground parts and roots survive the winter and give rise to the next year's growth.

Biennials germinate in spring and grow vegetatively the first year. Often they form a low-growing rosette of leaves that overwinters, and they flower and produce seeds the second year. Wild carrot* *(Daucus carota)*, mullein* *(Verbascum* spp.*)* (Fig. 9.12), bull thistle* *(Cirsium vulgare)*, and burdock* *(Arctium* spp.*)* are biennials.

Fig. 9.12. Mullein, a biennial, forms a rosette (right) the first year and produces a tall flowering stalk (left) the second year.

Fig. 9.11. Pokeweed is a native perennial that deteriorates with the first frosts of autumn.

Summer annuals germinate, grow to full size, flower, and produce seeds all in one year or growing season. Wild rice *(Zizania palustris)* (Fig. 9.13), though it may grow to a height of 10 feet, is an annual, as is the often tall-growing giant ragweed *(Ambrosia trifida)* and bur-cucumber *(Sicyos angulatus)*, a large vine.

Winter annuals germinate in late summer or early fall, winter as small ground-hugging leaf clusters, and produce flowers and set seed the next spring. Field pepper-grass* *(Lepidum campestre)* is a winter annual. Some winter annuals, such as common chickweed* *(Stellaria media)*, may germinate in early spring or late summer and may flower in late fall or winter if conditions are mild. Mature annuals that have flowered and fruited do not survive the winter.

Not all plants fit neatly into one of these categories. Some herbaceous plants, such as common winter-cress* *(Barbarea vulgaris)*, have flexible life histories and depending on conditions may be annuals, biennials, or short-lived perennials. Wild radish* *(Raphanus raphanistrum)* and horseweed *(Conyza canadensis)* may be either a summer annual or winter annual. White campion* *(Silene latifolia)* may be a summer annual, winter annual, biennial, or short-lived perennial.

Fig. 9.13. Flowering stalks of wild rice rise above pickerelweed in a freshwater tidal marsh.

In general, biennials and annuals are sun-loving plants associated with temporarily available habitats, such as newly cleared fields and other disturbed sites. They need good sun exposure in order to be able to grow, flower, and set seed in a short amount of time. Annuals and biennials produce large numbers of seeds that have excellent dispersal capabilities (often due to fuzzy appendages that catch the wind). This is important because these "fugitive" plants tend to die out when taller perennials invade a site and shade large areas of the ground.

Ecology of Trees, Shrubs, and Vines in Connecticut

Considering that all of Connecticut was glaciated and devoid of all traces of vascular plants as recently as about 15,000 years ago, the state has a surprisingly rich woody flora, including more than 250 native species. Virtually all of these plants arrived by their own natural dispersal mechanisms, though it is possible that a few may have been aided by humans who intentionally or inadvertently carried seeds with them as they moved into newly habitable areas following deglaciation. (Note: Wood makes up most of the volume of trees and shrubs, yet it is dead tissue, representing the cellulose walls of formerly living cells, bound and toughened with lignin.)

Connecticut's woody flora exhibits a diverse array of ecological patterns and interactions, as discussed in the following long section on the state's trees, shrubs, and vines. There is so much information here that you may have trouble "seeing the forest for the trees," so let's first take a brief look at some of the broad patterns of woody plant ecology in Connecticut.

Moist or moderately moist soils support the largest number of woody plant species in Connecticut, but many trees and shrubs also thrive in dry soil or wetlands (see Table 7.1). At the extremes, scrub oak and dwarf chestnut oak are essentially restricted to dry sites, whereas swamp rose, Atlantic white cedar, leatherleaf, cranberry, bog laurel, bog-rosemary, Labrador-tea, buttonbush, and few others exist only in wetlands (many other shrubs occur mainly in wetlands). The vast majority of trees and shrubs are intolerant or only moderately tolerant of dense shade. Eastern hemlock, American beech, sugar maple, American hornbeam, mountain laurel, and spicebush are among the relatively few highly tolerant species. See Chapter 7 for further discussion of habitat ecology and information on shade-tolerance-related vegetation dynamics.

The vast majority of woody plants in Connecticut are deciduous. Evergreens include all conifers except larch, plus bog-rosemary, bearberry, trailing arbutus, wintergreen, sheep laurel, mountain laurel, bog laurel, Labrador-tea, great laurel, cranberry, and just a few others. The leaves of deciduous plants unfurl and grow primarily from late April to mid-May. Most leaves fall from late September to early November, with a distinct peak in mid- to late October.

Pines, hemlocks, spruces, and other conifers are not flowering plants and thus lack showy blossoms. Among flowering plants, many have greenish or otherwise inconspicuous flowers. Some with noncolorful flowers have the reproductive parts in conspicuous catkins (e.g., willows, aspens, cottonwoods, sweet-fern, hickories, hazelnut, alders, birches, oaks, and others). The vast majority of other trees and shrubs have white flowers. Exceptions include sassafras, spicebush, witch-hazel, and northern bush-honeysuckle (yellowish flowers), and roses, steeplebush, sheep laurel, rhodora, pink azalea, lowbush blueberry, and cranberry (pink to purplish flowers). Tulip-tree has showy yellowish-green petals marked with orange-red.

All conifers are wind pollinated. Plants with dull-colored catkins are also wind-pollinated, as are most trees and shrubs with greenish, inconspicuous flowers. However, insects pollinate some plants with greenish flowers, such as greenbriar, poison ivy, sumacs, Virginia creeper, and grapes. Insects also pollinate essentially all native woody plants with white, yellow, or pink to purple flowers.

Wind-pollinated woody plants generally shed pollen in spring, often quite early, before the leaves develop in deciduous species (leaves might interfere with pollen movement). Some, such as alders, may flower before the vernal equinox. Most insect-pollinated trees and shrubs flower later in spring and sometimes into early summer. Standard summer bloomers include American chestnut, roses, brambles, spiraea, sumacs, basswood, prickly-pear, sweet pepperbush, spotted wintergreen, wintergreen, great laurel, swamp azalea, cranberry, buttonbush, northern bush-honeysuckle, and others. Latest of all are witch-hazel, groundsel-tree, and marsh-elder.

Woody plants produce fruits and seeds with various ecologies. Conifers produce seeds that are eaten (and usually destroyed) by birds, mammals, and insects. The seeds disperse in the wind, or sometimes get moved to suitable germination sites by food-storing mammals. Flowering plants exhibit a range of fruit characterisitcs. Oaks, beeches, chestnuts, hickories, and hazelnuts produce heavy, hard nuts that depend mainly on blue jays or squirrels for dispersal. Maples, ashes, tulip-trees, American hornbeam, birches, elms, and basswood have winged seeds that disperse in the wind. Some woody plants, such as spiraeas, sweet pepperbush, bog-rosemary, Labrador-tea, laurels, fetter-bush, maleberry, and rhododendrons, produce capsules that dry out and eventually split open to yield tiny seeds to the wind. Trailing arbutus uses ants to disperse the seeds, and witch-hazel shoots its fairly heavy seeds out of the pods. But the majority of trees and shrubs in Connecticut have small, fleshy fruits, usually red or blue-black, less often white or other colors, that are eaten and dispersed by many birds and mammals. See "Acorn Ecology" in Chapter 7 and "Fruits and Fruit-eating Birds," in Chapter 20 for further information on fruit characteristics and seed dispersal.

Overall, insects play an important role as tree and especially shrub pollinators, and birds and mammals are critical for seed dispersal. Fewer woody plants employ the wind for pollination and seed dispersal; most of these are trees, whose tall stature likely facilitates use of the wind.

Now let's go into specifics. The following pages include a brief ecological characterization of many of the trees, shrubs, and vines native to Connecticut, plus various non-native species that are widely established and maintaining populations without direct human assistance. Also included are some semi-woody plants and herbaceous vines. Not all species that you might find are covered here (I focused on the common species), but these species accounts do provide abundant examples of the impressive diversity and ecological relationships of Connecticut's woody flora.

The annotated list that follows is arranged taxonomically, following the sequence of families used by Magee and Ahles (1999), such that plants that are most closely related to one another (i.e., members of the same family) are grouped together.

Pines and Relatives (Pinaceae)

American larch, also called tamarack *(Larix laricina)*. Native in northern Connecticut, shade-intolerant tree, usually in boggy sites (FACW), where slow growing; lives up to a few hundred years. **European larch*** *(L. decidua)*, a widely planted tree, has spread from cultivation in some areas (Fig. 9.14). *Seed cones:* purplish red and green when pollinated by wind in spring, before foliage develops; often on same branch as yellow pollen cones, mature in one season, begin to open and release winged seeds to wind in mid-summer; seed dispersal may continue through fall and winter; open cones remain attached for up to several years; large crop every three to six years; individual prolific trees can produce as many as 20,000 cones. *Seeds:* eaten by crossbills and various other birds, red squirrels, mices, voles, and insects; germinate in spring/early summer; few seedlings survive. *Foliage:* eaten by woolly adelgids and larvae of larch sawfly and white-marked tussock moth; deciduous, brilliant gold in autumn (October–November). *Wood:* relatively hard, strong, and heavy, stands up to moisture without rotting. *Note:* American larch is near its southern limit in Connecticut. It

Fig. 9.14. European larch resembles American larch but has much larger cones. Both are deciduous conifers.

Fig. 9.15. Black spruce is restricted to isolated boggy wetlands.

grows as far north as trees survive, to arctic treeline in northern Canada. The bark is drilled by yellow-bellied sapsuckers and gnawed by porcupines. In the north country, Native Americans used the roots to sew birch bark together for canoes.

Black spruce *(Picea mariana)* (Fig. 9.15). A small, slow-growing, shallowly rooted tree, highly localized in boggy sites (FACW), primarily in northern Connecticut (widespread in boreal Canada); moderately tolerant of shade; lower branches, if in contact with soil, may take root and send up new vertical shoots. *Seed cones:* wind pollinated in spring (May); mature in one season, by end of summer; cones with seeds may stay attached to tree for years; plants only 6 feet tall may produce cones; large crop every two to several years; white-winged crossbills feed on the seeds of this and other small-seeded conifers, as do red squirrels. *Foliage:* evergreen. *Wood:* in Canada, used for rough construction, siding, paper pulp; a poor firewood. *Note:* This is the primary host of dwarf mistletoe (see "Parasitic Plants" later in this chapter). See the peatland section of Chapter 6 for further information on black spruce. **Red spruce** *(Picea rubens)* is an uncommon, shade-tolerant, shallowly rooted evergreen tree native to cool-moist sites in Connecticut's Northwest Highlands.

Red pine *(Pinus resinosa)*. A fast-growing, shade-intolerant tree; at southern edge of native range in Connecticut, where rare in sunny, sandy, or rocky areas of Northwest Highlands; in dense stands, grows straight and has live branches only near the top. *Seed cones* (Fig. 9.16): pollinated by wind in spring; usually on same tree as pollen cones, about 0.5 inches long after first season; close after pollination, fertilized about 13 months later, mature after two seasons of growth, remain on tree through third season or longer; open at maturity; winged seeds disperse up to several hundred feet in wind over several months; large seed crop every three to seven years; seed eaters include those listed for pitch pine. *Foliage:* evergreen, needles (in bundles

Fig. 9.16. Cones of Connecticut's native pines: white pine (bottom), pitch pine (upper right), and red pine (upper left).

Ecology of Trees, Shrubs, and Vines in Connecticut | 137

of two) live three to four years. *Wood:* light, relatively hard and strong (for a pine), often used for poles, pilings, pulp, and lumber. *Note:* The Civilian Conservation Corps planted many stands in Connecticut in the 1930s. These are now mostly dying or dead as a result of attacks by scale insects (*Matsucoccus resinosae*, Homoptera: Margarodidae) and adelgids of apparently Asian origin. Infestations may result in jumbles of dead, fallen timber. Many planted stands have been cut, others are being replaced by natural regeneration of hardwoods (planted red pines do not reproduce in most areas in Connecticut). Natural regeneration and seedling establishment are facilitated by fires, nearby seed-source trees, and adequate spring and summer rainfall.

Pitch pine *(Pinus rigida)* (Fig. 7.24). Widespread in sandy soils and thin soils of hilltops and rock outcrops; thrives in full sun, especially in areas subject to periodic fires; dies if excessively shaded by competing white pines or deciduous trees that grow taller; thick bark resists fire damage; new shoots sprout from stem buds after fire; cut trees may sprout from stumps; also occurs sparsely on bog hummocks. *Seed cones* (Fig. 9.16): pollinated by wind in spring; form on previous year's shoots; mature in two seasons, by end of summer; produced at young age; closed and open spiny cones remain attached to tree for many years; closed cones may open with heat of fire, releasing seeds onto barren soils ideal for germination and seedling establishment; large seed crop at intervals of several years; finches, nuthatches, woodpeckers, squirrels, mice, many other birds and mammals, and insects eat the seeds of this and other pines. *Foliage:* evergreen, needles in bundles of three; eaten by eastern pine elfin caterpillars. *Wood:* relatively hard for a pine, resists decay, suitable for rough construction and rustic flooring, not a good firewood. *Note:* Pine warblers and chipping sparrows often nest among pitch and white pines.

Eastern white pine *(Pinus strobus)* (Fig. 7.22). Common, tall-growing (rarely to 250 feet), pleasantly fragrant tree in dry to wet environments, often planted (as in groves around public water supply reservoirs); seedlings and saplings moderately tolerant of shade and may grow densely under mature trees. *Seed cones* (Fig. 9.16): pink-purple when pollinated by wind in May and early June; usually on same tree as yellow pollen cones; close after pollination; grow to about 1 inch at end of first summer, mature in late summer after two seasons of growth; large crop every three to five years; crops of 400 or more cones produced by healthy mature pines; release winged seeds (up to about 75/cone) mainly in late summer and early fall; seeds may be carried far by wind, germinate in spring; seeds eaten by white pine cone beetles and others mentioned under pitch pine; squirrels can bite off the cone scales to get at the seeds. *Foliage:* evergreen; numerous old needles (in clusters of five) drop in October after living two or three growing seasons; eaten by deer and eastern pine elfin larvae (deer avoid pines if other more preferred plants are available); usually survives defoliation by gypsy moth caterpillars. *Wood:* soft, light, but strong, prime as building material; the tall white pines of colonial New England gained fame (and infamy) as source of ship masts for Royal Navy. *Note:* Young pines grow slowly under taller trees for decades, but respond with vigorous growth if the overhead canopy opens up. Young (10 to 20 years old), sun-exposed pines may grow vertically 3 to 4 feet per year—in partial sun along forest paths I have found pines that had grown from 8 feet tall to more than 11 feet in one year; growth occurs during a short period in late spring and early summer. Pine weevils often attack and kill the leading shoot, resulting in a crooked trunk after a side branch takes over upward growth. Porcupines eat the bark, as do snowshoe hares. Sometimes I found numerous plant-juice-sucking spittlebugs on white pines in June. Trees may succumb to blister rust fungal disease, the alternate hosts of which are gooseberries and currants. (The fungus was inadvertently introduced with planting stock from Europe near the beginning of the twentieth century.) The roots of pines (and many other forest trees) often graft onto one another, allowing trees to share resources but also facilitating spread of pathogens.

Eastern hemlock *(Tsuga canadensis)* (Fig. 9.17). Very shade-tolerant, slow-growing, long-lived (to several hundred years) tree; common in cool-moist sites, streamside ravines, rocky slopes, and some swamps; intolerant of fire;

Fig. 9.17. Eastern hemlock produces small seed cones.

shallow rooted, vulnerable to drought and damage from road salt; drought-sensitive seedlings do best on moist substrates, may smother under the leaf litter of nearby hardwoods, so hemlock establishment is best on litter-free sites; seedlings and saplings grow very slowly in deep shade but may eventually become dominant. *Seed cones:* on tips of shoots, pollinated by wind in April–May; pollen cones and seed cones on same plant and often same branch; mature in one season, by end of summer, and persist until next year; open at maturity, release tiny, winged, wind-dispersed seeds during dry periods in fall and winter; large seed crop every two to three years; seeds germinate in spring, if not eaten by squirrels, chickadees, pine siskins, goldfinches, crossbills, juncos, and others. *Foliage:* evergreen, but each needle lives only a few years. *Wood:* relatively hard but brittle, suitable only for rough construction (e.g., barn siding); when burned produces little heat but many sparks. The tannin-rich bark resists decay; old stumps often have an intact sleeve of bark around the rotted wood. *Note:* Hemlock woolly adelgids are decimating hemlocks in Connecticut (see Chapter 14). Complete defoliation by gypsy moth caterpillars usually results in death of the tree, but hemlocks can survive partial defoliation. Groves provide the primary habitat in Connecticut for the black-throated green warbler; golden-crowned kinglets and Blackburnian warblers also often nest in hemlock trees. White-tailed deer use hemlock stands, which generally prevent deep snow accumulation on the ground, for winter shelter and food, but heavy deer browsing can result in poor regeneration.

White Cedars and Junipers (Cupressaceae)

Atlantic white cedar *(Chamaecyparis thyoides)* (Fig. 6.6). Long-lived, aromatic tree; in isolated acidic swamps (OBL) in a few dozen scattered locations, mainly in eastern and southern Connecticut; may grow in exceptionally dense stands; trees often on raised hummocks; seedlings grow best with ample light and moisture (killed by flooding). *Seed cones:* pollinated by wind in spring; mature from blue-purple to red-brown in late summer–early fall; seed and pollen cones usually on separate branches of same tree; winged seeds released mainly in fall and winter, may not germinate for two to three years. *Foliage:* evergreen; old foliage turns rusty or brown; eaten by deer and Hessel's hairstreak caterpillars. *Wood:* resists water decay and insects. See Chapter 6 for further ecological information. *Note:* **Northern white cedar** *(Thuja occidentalis)* is a rare, slow-growing, long-lived (to several hundred years), shade-tolerant, fire-intolerant, aromatic evergreen tree of moist or wet soil (FACW). Connecticut provides marginal conditions for natural reproduction of this northern tree, but planted trees do thrive.

Common juniper *(Juniperus communis)* (foreground, Fig. 7.28). Shrub of pastures, old fields, clearings (e.g., powerline corridors), and rocky hills. *Seed cones:* pollinated by wind in spring (small bees may collect pollen); pollen cones usually on separate plants; fruits berrylike, mature in two years by end of summer; high fat content, eaten by songbirds and rodents, though some remove flesh and eat only seed; uneaten cones may remain on the plant for a year or two; abnormally enlarged cones may contain (or formerly contained) feeding larvae of chalcid wasps. *Foliage:* evergreen; exposed winter foliage may turn reddish brown; often eaten by deer in winter.

Red-cedar *(Juniperus virginiana)* (Fig. 9.18). Common, small, narrow to widely spreading, shade-intolerant tree; in fields, old fields, forest openings, cleared roadsides, and developing forests that have not yet shaded out these old-field colonists. *Seed cones:* pollinated by wind in spring; pollen cones and seed cones usually on separate plants; fruits berrylike and blue-whitish, mature in one season, by end of summer, remain attached through winter; eaten and dispersed by cedar waxwings, finches, yellow-rumped warblers, mockingbirds, thrushes, and other birds, also eaten by raccoons, foxes, and various other mammals. Large

Fig. 9.18. Red-cedar, a common colonist of grassy fields, has berrylike seed cones.

numbers of seedlings often appear in fields far from any mature trees. *Foliage:* evergreen; new foliage spiny, mature foliage smooth-scaly; eaten by juniper (olive) hairstreak caterpillars and in winter by deer. *Wood:* long-lasting in contact with soil, used extensively for fence posts and formerly for pencil making. *Note:* Deer rub antlers on the main stem. Chipping sparrows, song sparrows, American robins, house finches*, and northern mockingbirds often nest hidden among foliage. **Cedar-apple rust** (*Gymnosporangium juniperi-virginianae*), a parasitic fungus, causes conspicuous galls on the foliage in spring or summer (Fig. 9.19). This rust also attacks apple foliage and fruits, so red-cedars usually are removed from areas near apple orchards. The initial fungal infection occurs in late summer and fall. The galls swell in spring and reach full size in fall. Small golden-brown "horns" (telia) emerge from the galls the next spring. They swell and become gelatinous orange with warm spring rains. In conjunction with rains, each gall may undergo multiple swelling and shrinking events before the horns finally disintegrate. Spores are released with each gelatinization of horns. The globular gall soon dies but remains attached to the branch for up to a year or more.

Greenbriars (Smilaceae)

Common greenbriar (*Smilax rotundifolia*) (Fig. 9.20) and **sawbriar** (*S. glauca*). Woody vines; common, often

Fig. 9.19. Cedar-apple rust is a fungal parasite of red-cedar.

Fig. 9.20. Common greenbriar is a native vine that thrives in forest edges and openings.

forming thickets at forest edges and openings; some thickets persist under tall forest trees, reminders of past disturbance. *Flowers:* malodorous (if odorous at all), greenish, pollinated by small anthomyiid and stratiomyiid flies, mining bees, halictid bees, and sometimes various beetles, mainly May to June; male and female flowers on separate plants. *Fruits:* ripen blue-black in summer, often remain attached through winter and well into spring; coastal greenbriar thickets produced huge fruit crop in 2000; eaten by ruffed grouse, thrushes, gray catbird, some other songbirds, and raccoons, especially in early or mid-spring after choice fruits are gone and before other food resources become available. *Foliage:* deciduous, but some leaves may remain attached through winter and well into next year; may turn yellow in fall. *Note:* Deer may browse the foliage and twigs but do not appear to make heavy use of greenbriars in Connecticut. Gray catbirds may nest in dense greenbriar thickets.

Willows and Poplars (Salicaceae)

Willows* (about 20 *Salix* spp.) are shade-intolerant, fast-growing shrubs and trees, which often grow adjacent to water. **Black willow** (*S. nigra*). Common scruffy tree on stream banks (FACW); readily sprouts from stumps and roots. **Pussy willow** (*S. discolor*). Frequent large shrub in swampy freshwater wetlands and shores (FACW). A few rare willows have state conservation designations. *Flowers:* pollinated by wind and insects (e.g., bumblebees, mining bees, syrphid flies), generally in late winter and spring; bees may do little actual pollinating, since seed catkins appear to attract few bees; nectar-secreting pollen catkins and seed catkins (Fig. 9.21) greenish to yellowish, usually on separate plants. *Seeds:* consistently good crops ripen mainly in late spring (by mid-May); short-lived, wind- and water-dispersed, with cottony tufts; germinate or die immediately after falling. *Foliage:* deciduous; develops early and falls late; eaten by caterpillars of Acadian hairstreak, mourning cloak, viceroy, and other Lepidoptera, and adults and larvae of the imported willow leaf beetle*; viceroy caterpillars overwinter along branches in silk-tied leaf fragments. *Wood:* flexible (good for baskets), weak, unsuitable for heavy construction, poor as firewood. *Note:* Willows can be excellent streambank stabilizers. The twigs break readily at the base and can take root if inserted into the ground. Deer, rabbits, and beaver eat the twigs and bark. Midges and sawflies cause galls (see "Galls and Gall Makers," Chapter 14). The bark contains salicin, the pain-relieving substance in aspirin.

Quaking aspen *(Populus tremuloides)* (Fig. 9.22) (the most widely distributed of North American trees) and **bigtooth aspen** *(P. grandidentata)*. Common, fast-growing, short-lived (individual stems often live 60 years or less), shade-intolerant trees; open areas such as road-

Fig. 9.22. Quaking aspen and other poplars flower conspicuously in early spring. Here the tiny fruits are just beginning to develop on the seed catkins in mid-April.

sides and forest edges. *Flowers:* pollen and seed catkins on separate plants; pollinated by wind in early spring (beginning by early April), before leaves appear (leaves of non-flowering aspens may be well-developed by late April); flowering may begin when a tree is less than twenty years old. *Fruits:* ripen in late spring or early summer, tufted, dispersed by wind; large crop of short-lived seeds at intervals of several years; catkins and buds are favored foods of ruffed grouse and porcupines. *Foliage:* deciduous; "trembles" in the wind on flattened leaf stalks; usually golden yellow in fall; begins to fall relatively early (bigtooth aspen); defoliated by various moth caterpillars; aphids cause globular galls on leaf stalks; fungal blight (*Apioplagiostoma populi*) results in death of all leaves on a shoot. *Wood:* soft, weak, brittle, nonsplintering, used for firewood, crates, particleboard, pulp for paper production, and sometimes veneer. *Note:* Aspens often grow in clones connected to a common root system. Clones may cover many acres and can persist for thousands of years, though they are relatively small in Connecticut. Aspens increase after fire through prolific sprouting from surviving roots. Seedling establishment requires bare ground; few seedlings become established in any given site (groups of small aspens are usually root suckers). Beavers (which relish aspen above all other plants), cottontails, snowshoe hares, and deer eat the leaves, twigs, and bark. Deer commonly rub their antlers on the trunk. Sapsuckers may riddle the

Fig. 9.21. Willow pollen catkins (smaller) and seed catkins (larger) grow on separate plants (here living side by side).

bark with holes. Woodpeckers readily excavate cavities in mature aspens and snags; many other birds and small mammals later use these holes. Various fungi commonly disfigure various parts of the tree. The smooth, greenish, chlorophyl-containing bark of aspen trees can photosynthesize to some extent even when trees are leafless.

Eastern cottonwood (*Populus deltoides*) (Fig. 9.23). Locally common, widespread, fast-growing, short-lived, shade-intolerant tree; grows mainly along rivers, also locally at pond margins and in moist swales; cut trees readily sprout from base. *Flowers:* male and female catkins on separate plants; wind-pollinated in early spring (early April), before leaves develop. *Fruits:* pea-sized by early May, dispersed on cottony tufts in May and June; short-lived seeds germinate on sand bars or other areas laid bare by high water. *Foliage:* deciduous; yellow in fall; leaf stalks flat, allow leaves to rustle pleasantly in the wind. *Wood:* light, weak, brittle, warps when drying. *Note:* Young trees may grow 4 to 5 feet per year. Old trees may have cavities used by woodpeckers, bluebirds, screech-owls, starlings*, and other birds. Other associates are similar to those of aspens.

Swamp cottonwood (*P. heterophylla*). Rare, straight- and clear-trunked tree; swampy areas (FACW) in central and southwestern Connecticut, where at northern extent of natural range.

Sweet-Fern, Sweet Gale, and Bayberry (Myricaeae)

Sweet-fern (*Comptonia peregrina*). Common shrub; mainly on dry, sandy soils; colonizes clearings, including power-line corridors; forms clones by sending up new shoots from spreading root stocks. *Flowers:* pollen and seed catkins on same or separate plants; wind-pollinated in early spring, before leaves develop. *Fruits:* seeds in clusters of brown nutlets in burlike covering; burs conspicuous by June; seeds mature by late summer to early fall; burs may stay attached through winter and the following spring. *Foliage:* deciduous; eaten by deer; moth caterpillar (*Acrobasis comptoniella*) joins leaves with silk, eats foliage, winters among leaves in silk-bound case; turns reddish-brown in fall; upper leaves stayed attached into winter. *Note:* Occasionally I have found grass nests of songbirds in the branches of this shrub.

Bayberry (*Myrica pensylvanica*) (Fig. 9.24). Common shrub in sandy areas along coast, also grows in inland clearings. *Flowers:* spring and early summer; pollen released to

Fig. 9.23. Cormorants perch in mostly dead eastern cottonwoods on the bank of the Connecticut River.

Fig. 9.24. Bayberry, a common component of coastal thickets, produces waxy fruits.

wind beginning in mid-May; male and female flowers mostly on separate plants. *Fruits:* waxy, mature in late summer, remain attached through winter and up to a few years, eaten by yellow-rumped warblers, tree swallows (sometimes thousands swarm into fruiting plants in September and October; see Chapter 20), and gray catbirds. *Foliage:* deciduous; reddish in fall, drops late; small plants often retain dense maroon foliage into winter. *Note:* Root nodules of bayberry, sweet gale, and sweet-fern contain nitrogen-fixing bacteria that facilitate growth in nutrient-poor soils that stress many other plants. **Sweet gale** *(M. gale)*. Deciduous shrub, most common in Northwest Highlands; acidic pond margins and wetlands (FACW).

Butternut, Walnut, and Hickories (Juglandaceae)

Butternut *(Juglans cinerea)*. Shade-intolerant, fast-growing, short-lived tree; sumac-like foliage; widely scattered but uncommon in moist lowland soils, along streams, and on talus slopes. *Flowers:* green, wind-pollinated in spring (e.g., latter half of May), as leaves develop; pollen flowers (on previous year's growth) and seed flowers (on new shoots) on same tree. *Fruits:* similar to walnuts but more elongate; reach full size by early to mid-July, mature in late summer or early fall; hairs on husk stain human skin brown; good crops develop every few years; squirrels open husk and eat nuts, or bury them for later use, though sweet oily kernel may soon go rancid; unretrieved nuts germinate after overwintering. *Foliage:* develops in late spring; deciduous; eaten by underwing moth caterpillars; yellowish in fall. *Wood:* scarce, beautiful, light, soft, does not crack or warp. *Note:* Adult and larval butternut curculios (beetles) feed on the nuts, shoots, and leaf stalks. Butternut secretes root toxins that inhibit the growth of many other plants. This tree is subject to fatal butternut canker disease, which has decimated populations throughout the range since the late 1960s. It is uncertain whether the *Sirococcus* fungus responsible for the disease is a non-native species or a native that has become a problem as a result of the trees being stressed by some other factor.

Mockernut hickory *(Carya alba)*. Mainly in uplands and coastal forests of central and southern Connecticut. **Bitternut hickory** *(C. cordiformis)*. Occurs in moist lowlands, coastal forests, and field edges. **Shagbark hickory** *(C. ovata)*. Usually in moist, fertile soils. **Pignut hickories** *(C. glabra* and *C. ovalis)* (Fig. 9.25). Common in uplands, not tolerant of shade. All are slow-growing trees (under forest conditions), sprout from stumps and roots, resist toppling by wind; moderate to low tolerance of shade; seedlings may survive only a few years under heavy shade. *Flowers:* greenish

Fig. 9.25. Pignut hickory is a common upland tree and a prolific nut producer.

catkins (Fig. 9.26), wind pollinated in spring (beginning in mid-May), as leaves develop; pollen flowers and seed flowers on same plant. *Nuts:* ripen in one season, in late summer; most fall in September and October, some may persist on trees through fall, even after dry husk splits open; produced in abundance every one to five years; large crop in Connecticut in 2001; eaten by hickorynut curculio larvae, gray squirrels, flying squirrels, chipmunks, and white-footed mice; animals seemingly ignore unpalatable bitternut hickory nuts. *Foliage:* aromatic, deciduous; yellow or gold, sometimes reddish, in fall; leaflets often fall before leaf stalks detach; leaves may turn brown and drop early in drought years; eaten by banded hairstreak, hickory hairstreak, and various

Fig. 9.26. Hickory flowers are wind pollinated in spring.

Ecology of Trees, Shrubs, and Vines in Connecticut | 143

underwing moth caterpillars; eriophyid gall mites cause numerous bumpy galls on the upper surface of the leaflets; aphidlike phylloxerids (*Phylloxera*) and cecidomyiid gall midges (*Caryomyia*) also induce gall formation on leaves. *Wood:* strong, tough, flexible, shock-resistant, heavy, splits with difficulty, decays quickly, produces tremendous heat when burned. *Note:* The nuts (except those of bitternut hickory) are strongly preferred over acorns by squirrels and chipmunks, which often gather virtually all viable nuts, leaving behind aborted ones and those infested by insects. Gray squirrels bury many nuts (some of which are never retrieved) and serve as an important dispersal agent. However, hickory seedlings and saplings can be rare, and few nuts may survive to germinate and grow. To see hickorynut curculio larvae, place several fallen hickory nuts in a dish and wait for them to squeeze out through circular exit holes in fall. Hickory bark beetles sometimes cause significant mortality, especially after trees are weakened by drought or fire; other beetles infest living and dead wood; many other insects feed on various parts of hickories. Sapsuckers drill holes in smooth bark and feed on oozing sap. Brown creepers may nest behind the peeling bark of shagbark hickory.

Birches, Hazelnuts, Hornbeams, and Alders (Betulaceae)

American hazelnut *(Corylus americana)* and **beaked hazelnut** *(C. cornuta)* (Fig. 9.27). Shrubs that tend to occur in clones formed through vegetative reproduction at forest edges and openings and under openly spaced trees. *Flowers:* wind pollinated in late March or April, before or as leaves appear; conspicuous male catkins release pollen to tiny female flowers (both present on same twig); eaten by ruffed grouse; midge induces gall formation in catkin scales. *Nuts:* fully formed but green in late July, ripe in late summer; disappear quickly, eaten by squirrels, chipmunks, white-footed mice, turkeys, grouse, woodpeckers, blue jays, and many other birds and mammals. *Foliage:* deciduous; eaten by rabbits, beavers, and deer. *Note:* Hazelnuts are widely distributed but easily overlooked, mistaken for young birches.

Eastern hop-hornbeam *(Ostrya virginiana)*. Small, shade-tolerant tree, exisitng mainly as scattered individuals in well-drained upland forest; intolerant of flooding; sprouts from stumps. *Flowers:* wind pollinated in spring (April and May), as leaves develop. *Nutlets:* in clusters of bladdery sacs, ripen in late summer, persist through fall and winter; germinate in spring; eaten by squirrels, ruffed grouse, turkeys, finches, and woodpeckers. *Foliage:* deciduous; pale yellow or dull gold in fall; dry leaves may stay attached into winter. *Wood:* exceptionally hard, tough, and heavy; makes excellent

Fig. 9.27. Beaked hazelnut nuts attract squirrels, turkeys, and other wildlife.

fuel, but small size of tree limits its utility. Note: Deer and beavers eat the foliage, twigs, and bark.

American hornbeam (ironwood, blue beech) *(Carpinus caroliniana)*. Small, smooth-barked, shade-tolerant tree of moist bottomlands; slow-growing but short-lived. *Flowers:* green, wind pollinated in spring (April and May), before or as leaves develop; pollen catkins and seed catkins in separate clusters on same plant; pollen catkins on previous year's twigs, seed catkins at tips of new shoots. *Nutlets:* in clusters attached to leaflike scales (Fig. 9.28); well developed by mid- to late July, ripen in late summer, and become conspicuous in fall after most leaves have fallen; some remain attached into winter; occur in large crops every three to five years (large crop in 2000); eaten by various birds (e.g., ruffed grouse, northern cardinal, evening grosbeak, American goldfinch) and mammals; gray squirrels, ruffed grouse, and wild turkeys eat buds and nutlets. *Foliage:* deciduous but some plants retain sparse dried leaves into winter; yellow, orange, red, or brown in fall; deer, rabbits, and beavers browse leaves and twigs. *Wood:* hard, tough, heavy, but rots quickly in soil. *Note:* Seedlings can become established on leaf litter under other trees.

Speckled alder *(Alnus incana)* (FACW) and **smooth alder** *(A. serrulata)* (OBL). Common, widely distributed, sun-loving shrubs; adjacent to water or in wet meadows; commonly in groups formed through vegetative reproduction and seedling establishment. *Flowers:* closed purple catkins conspicuous in winter; open and shed pollen to wind in late winter or early spring before leaves appear; pollen cat-

Fig. 9.28. American hornbeam, a small tree of moist bottomlands, produces clusters of nutlets attached to leaflike bracts (lower right).

Fig. 9.30. Black birch seed catkins attract goldfinches and other seed-eating birds in fall and winter.

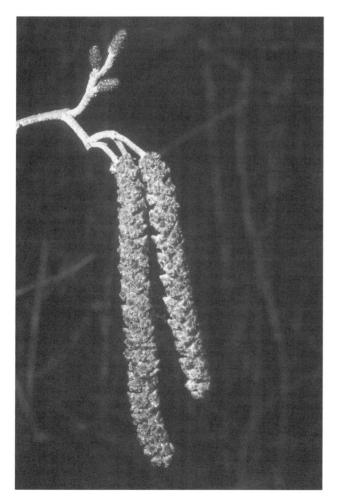

Fig. 9.29. Speckled alder pollen flowers (two at bottom) and seed flowers (three at top) open in late winter and early spring.

kins and seed catkins on same plant (Fig. 9.29). *Fruits:* cone-like, green and well developed by early summer; contain tiny winged seeds dispersed in fall and early winter; good seed crop every few years; fungus often causes elongated, malformed cone scales on long-lasting seed cones; redpolls, siskins, and goldfinches eat wind- and water-dispersed seeds. *Foliage:* deciduous; green, bronze, or brown in fall; adult and larval alder flea beetles chew holes in leaves (adults) or skeletonize them (larvae). *Note:* Woolly aphids, tended by ants attracted to honeydew secreted by aphids, feed on the sap (Fig. 14.33). Harvester (butterfly) caterpillars and lady beetles feed on woolly aphids. Beavers eat buds and bark and use wood for building material. Woodcocks forage for worms among alders, and alder flycatchers and other songbirds nest in alder thickets. Actinomycetes (*Frankia*), associated with root nodules (actinorhizae), convert nitrogen in the air to a form that can be used by plants; alders and their root microbes thereby enrich soil. Silver-spotted ghost moth (Hepialidae: *Sthenopis argenteomaculatus*) larvae feed in roots. At dusk in summer, adult males swarm in "dancing" flight near alders to attract females. The life cycle takes two years to complete.

Black birch *(Betula lenta)* (Fig. 9.30). One of Connecticut's most abundant trees; relatively short-lived; various situations, often in moist fertile soils but also on rocky or sandy ground; seedlings colonize soil freshly exposed at base of wind-thrown trees within forested areas (Fig. 7.6); increasingly numerous as oaks decline from reduced incidence of fire and gypsy moth defoliation. **Yellow birch** *(B. alleghaniensis)*. Moist but moderately well-drained, fertile soils; long-lived (for a birch), individuals may survive two hundred years; seeds germinate usually in spring, on various sites, including logs, which eventually rot and leave tree on root "stilts;" seedlings in hardwood leaf litter often die during summer drought. **Gray birch** (*B. populifolia*) (Fig. 9.31). Early-maturing, short-lived tree; usually less than 50 years old; may flower when less than 5 feet tall; colonizes clearings and old fields. **Paper birch** (white birch) (*B. papyrifera*). Well known for its peeling white bark; usually in rich fertile soils in cool sites, also cleared roadsides, mainly in northern and central Connecticut; does well on mineral soils in partial shade but establishes poorly on hardwood leaf litter; now blankets areas logged of pine and spruce over much of northern New England; in absence of disturbance, lasts only one generation before being replaced by other trees; few live longer than 150 years. **River birch** (*B. nigra*). Local and rare on streambanks. All except the last are common, widely distributed trees, and all except yellow birch are more or less shade intolerant. **Bog birch** or **swamp birch** (*B. pumila*). Shrubby, localized in boggy areas (OBL) in Northwest Highlands. Birch stumps, especially from young trees, readily send up shoots. *Flowers:* catkins, wind pollinated in spring (mainly between late April and mid-May) before leaves are fully formed; pollen catkins and seed catkins on same tree; some birches can produce mature fruits without seed development or ovule fertilization. *Seeds:* long-lasting female catkins of black birch, yellow birch, and paper birch shed winged nutlets in wind in fall and winter, with good crops every one to three years; seeds geminate in spring; river birch nutlets, released in late spring or early summer, may disperse on stream waters and germinate on muddy banks; eaten by nuthatches, American goldfinches, pine siskins, redpolls, and various other songbirds; ruffed grouse eat nutlets and buds. *Foliage:* deciduous; yellow in autumn, falls early, develops early in spring; foliage of young, nonflowering yellow birches develops before that of mature flowering trees; decays rapidly on the ground; eaten by tent caterpillars, gypsy moth larvae, and other moth caterpillars; beavers favor leaves, twigs, and bark. *Wood:* not especially remarkable, though yellow birch is valuable and serves well for interior finish and furniture; commonly used for firewood; paper birch wood is fairly hard and strong (less so than black and yellow birches) and can be used for lumber, generally put to less glamorous uses as such particleboard and pulpwood. *Note:* Birches generally require high light levels or bare ground for seedling establishment, so they thrive with disturbance (though established trees may be killed by fire). Tiny seedling roots cannot penetrate thick leaf litter. Removal of overstory hemlocks often results in germination of black and yellow birch seeds lying dormant in the soil. Trees on "stilts" usually result from erosion of the tree-fall soil mound on which the seedling became established (Fig. 7.7). Black birches grow slowly when shaded: In mature forest, my students and I determined that the terminal stem of sapling black birches grew an average of 3 to 6 inches per year over several consecutive years. Fungi may cause large cankers on black and yellow birch trunks, and other fungi cause minor damage to black, yellow, and paper birches. Eriophyid mites (*Cecidophytes betulae*) cause the formation of dense clusters of bud tissue on twigs. An aphid that makes galls on witch-hazel also uses birches (see "Galls and Gall Makers," Chapter 14). Deer browse the twigs in winter and nip off stump sprouts of yellow birch, but black birch is not a favorite food. Gnawing rodents, snowshoe hares, and cottontails may kill seedlings by girdling them. Tunneling by bronze birch borer (beetle) larvae may girdle, weaken, and kill birches. Ice accumulation often bends small paper and gray birches such that the tops become stuck to snow-covered ground. Yellow-bellied sapsuckers drill into birches and consume the oozing sap; hummingbirds also visit these sap wells. Black-throated green warblers may incorporate pieces of birch bark into their nests. Paper birch bark, sewn with larch root and sealed with pine, balsam, or spruce resin over a frame of northern

Fig. 9.31. Gray birch is a short-lived colonist of cleared areas. It thrives in newly developing forests, and its dead remains (shown here) may be common in mature forests.

white cedar, historically made a wonderful canoe and was also used for wigwam exteriors. Birch punkwood, taken from under decay-resistant bark of a rotting birch, and paper birch bark, make good fire tinder. Black birch is a natural source of wintergreen oil, and birch beer was traditionally brewed from sap (now replaced by synthetic chemicals).

Beeches, Chestnuts, and Oaks (Fagaceae)

American beech *(Fagus grandifolia)*. Common shade-tolerant, fire-intolerant tree; moist, usually rich soils; also currently growing under mature pitch pines in some sand plains that have not been recently burned; often forms clones through root sprouting; sprouts from stumps of young trees; growth is slow but annually variable in natural forest conditions; in a mature forest, my students and I determined that the terminal stem of saplings grew an average of 1.5 to 6 inches per year over several consecutive years; lifespan sometimes exceeds 350 years. *Flowers:* unisex, greenish-yellow, both sexes on same tree; wind-pollinated in spring (late April to early or mid-May) just before or after leaves emerge; may be killed by frost; sometimes visited by bees. *Nuts:* two to three per spiny bur (Fig. 9.32), ripen in one season, in late summer or early fall; drop in numbers after first hard frost; large crops occur every two to eight years (beeches in central Connecticut produced few nuts in 2001 and 2002); germinate from early spring to early summer; eaten by grouse, turkeys, woodpeckers, blue jays, American crows, squirrels, chipmunks, bears, and other birds and mammals; formerly fed millions of now-extinct passenger pigeons; some germinate after being transported and stored far from tree by blue jays; many meatless and seedless, discarded by foraging squirrels. *Foliage:* deciduous; tan, copper, or yellowish in fall; bleached-out leaves often still attached through winter and early spring; eaten by caterpillars of moths, including saddled prominent, gypsy moth, forest tent caterpillar, and Bruce's spanworm; rarely browsed by deer. *Wood:* strong, hard, heavy wood has been put to many uses; tends to warp when drying; good firewood but resists splitting; inferior to many other trees in almost every aspect. *Note:* Beech-drops (described later in this chapter) is a root parasite. Where limbs have broken off, beech often forms cavities used by wildlife; heart-rot fungi (e.g., *Stereum murrayi*) may speed interior decay. Beech blight aphids sometimes form white, woolly, honeydew-dripping colonies on the branches in late summer and early fall (see Chapter 14). Beech scale (*Cryptococcus fagisuga*) is an introduced pest. The small woolly females of this maleless species give rise to more females that feed by inserting their tubular mouthparts into the bark. This feeding may facilitate establishment of a fungus (*Nectria coccinea faginata*) that can cause bark cankers that may result in the death of some or all of the tree's branches. Additionally, cracks in the bark caused by *Nectria* may be invaded by other scale insects and by fungal species that decay the wood. Some beech trees are partially resistant to *Nectria* and survive.

American chestnut *(Castanea dentata)* (Fig. 9.33). Deep-rooted tree, a dominant forest species prior to early 1900s, growing to several feet in trunk diameter and several hundred years old; now occurs in upland woods only as saplings and small trees that eventually succumb to imported fungal blight that decimated formerly impressive stands; young cut or diseased trees readily sprout from base, keep coming back; small trees rarely live beyond 30 years or attain stem diameter of more than 6 inches; vigorous stands of abundant young chestnut sprouts sometimes develop after clear-cutting; decay-resistant, large stumps of pre-blight forest trees still exist in some areas (Fig. 7.34). *Flowers:* whitish, pollinated by wind, bumblebees, and other insects in early summer (late June or early July), after leaves have grown; pollen flowers and strong-scented seed flowers on same tree; cross pollination between different trees required for viable nut production. *Nuts:* usually three per bur, formerly in good crops every year and important as human food, now rare; ripen by late September or early October, eaten by deer, bears, squirrels, turkeys, and other wildlife. *Foliage:* deciduous; yellow in fall. *Wood:* valuable, many uses; straight-grained, decay-resistant, outer wood of stumps is very long-lasting; as blight spread out of control, most dead, dying, and still living chestnuts were cut and hauled away; high tannin content, valuable in the leather tanning industry. *Note:* Trees suffering from chestnut blight fungus (*Cryphonectria parasitica*) were first observed in New York City in 1904, and within a decade most chestnuts in Connecticut

Fig. 9.32. The nuts of American beech develop in spiny burs.

Fig. 9.33. Pollen flowers of American chestnut are now a scarce sight in early summer.

were dead or dying. By 1950, chestnut was blighted throughout its range. The fungus, whose spores are carried by wind, rain, or animals, enters the tree through cracks, wounds, or insect holes. Eventually it spreads through the inner bark and girdles the tree (Fig. 7.33). Evidence of attack includes yellowish or orange-brown cankers (scar tissue) in the bark. If necessary, the fungus can persist on other tree species until a new chestnut stem is available. Efforts have been made to combat the fungus by crossbreeding American chestnuts with other chestnut species, by introducing nonlethal strains of the fungus that might outcompete or, through "hybridization," reduce the lethality of the lethal form, and by locating and selectively breeding surviving chestnuts that may have some natural resistance to the fungus. Only time will tell whether the American chestnut can be restored to its former dominant position in eastern forests. I regret not getting to see this tree in its full glory.

White oak *(Quercus alba)*. Common, moderately shade-tolerant; widely distributed in various sites and soils; good competitor on soils with marginal nutrient content; individuals may live several hundred years, attain massive trunk and limb size; young trees grow slowly in shade of other trees, but can respond with quick growth if canopy opens. **Swamp white oak** *(Q. bicolor)*. Moist bottomlands (FACW), moderately shade tolerant. **Bur oak** or **mossy-cup oak** *(Q. macrocarpa)*. Rare, moderately shade-tolerant species in moist bottomlands of northwestern Connecticut. **Chinquapin oak** or **yellow oak** *(Q. muhlenbergii)*. Rare on limestone ridges and ledges in western Connecticut. **Chestnut oak** *(Q. prinus)* (Fig. 9.34). Common; often grows (slowly) on hilltops, also occurs locally in moist lowlands; somewhat shade tolerant, but sunny ground is best for seedling establishment and growth. **Post oak** *(Q. stellata)*. Uncommon; well-drained soils in some parts of southern Connecticut. All of these are trees, young stems of which readily sprout if cut. **Dwarf chestnut oak** *(Q. prinoides)*. Shrub of dry hills or sandy soils. These oaks are members of the white oak group, which in Connecticut also includes a few additional rare species. *Flowers:* greenish or yellowish seed flowers, in leaf axils of new shoots; wind pollinated in spring as leaves develop (early to mid-May); pollen flowers and seed flowers on same tree; dried-up pollen flowers litter ground in May; late spring frosts can kill flowers and reduce acorn production. *Acorns:* mature in one season, drop mostly in September and October, germinate in fall soon after falling; bumper crop (sometimes tens of thousands of acorns per tree) every two to six years or more; drought can reduce acorn production; contain less tannin than acorns of red-black oak group; eaten by blue jays, turkeys, grouse, squirrels, chipmunks, white-footed mice, bears, deer, weevils, moth larvae, and other wildlife; some acorns stored in the ground by squirrels and blue jays

Fig. 9.34. Chestnut oak is most prevalent on dry hilltops.

later germinate and become trees (see "Acorn Ecology," Chapter 7). *Foliage:* deciduous; brown, reddish-brown (white oak), or yellowish (chestnut oak) in fall; some dry leaves stay attached through fall and winter; contains much lignin and decomposes slowly, so oak leaves often constitute bulk of leaf litter in deciduous forests; eaten by caterpillars of Juvenal's duskywing, banded hairstreak, white M hairstreak, gypsy moth*, oakworm moth, and many other Lepidoptera. *Wood:* hard, strong wood of white oak is excellent for nearly all uses, from flooring and furniture to fuelwood and barrel staves; durable in contact with soil; bark is rich in tannin; outer wood of oak stumps is long-lasting; tunneled by wood-boring beetles. *Note:* Cynipid wasps lay eggs and induce gall formation on leaves and twigs (see Chapter 14). White oak group seedlings grow, mainly in the root system, until freezing weather ensues, but many seedlings die in winter due to desiccation from inadequate rooting. Oak stems often sprout after fire, and fires create favorable conditions for seedling establishment.

Scarlet oak *(Q. coccinea).* Shade intolerant; usually in dry soils. **Pin oak** *(Q. palustris).* Shade intolerant, relatively short-lived for an oak; mainly along streams or in other poorly drained soils (FACW) (also planted elsewhere). **Northern red oak** *(Q. rubra)* (Fig. 9.35). Common and widespread in moist soils but lives in diverse situations; grows best when only lightly shaded, more tolerant of shade than are black and scarlet oaks; can grow to nearly 20 feet in 10 years. **Black oak** *(Q. velutina).* Typical of dry, sandy soils; also in many other sites; shade intolerant. All of these are common, widely distributed trees, pin oak being the most limited in distribution. **Scrub oak** *(Q. ilicifolia)* (Fig. 9.36). Locally common shrub; hilltops, rock outcroppings, and sand plains. Characteristics of this red oak–black oak group are similar to those of the white oak group, except that the

Fig. 9.35. These northern red oak acorns are sprouting in mid-April. In warm years, they may sprout in March.

Fig. 9.36. Scrub oak, like other oaks, flowers in May.

acorns mature at the end of the second season (some falling as early as late August and others as late as the end of October or early November), and the acorns germinate in March or April rather than in fall. A long taproot allows seedlings to withstand drought. *Foliage:* deciduous; red, brown, or yellowish in fall (varies among species); green leaves may be present well into November; Edwards' hairstreak and buck moth caterpillars feed on scrub oak foliage. *Wood:* red oak is high quality but not so strong, hard, or as durable in contact with soil as that of white oak; scarlet oak wood is stronger than white oak but usually knotty; pin oak has knotty wood, inferior to red oak. *Note:* Red oak and black oak species commonly hybridize with each other, making identification difficult. Bumblebees may sip sweet secretions associated with cynipid wasp–induced galls on red oak acorns. Other associates are as in the white oak group.

Elms and Hackberries (Ulmaceae)

Slippery elm *(Ulmus rubra)* and **American elm** *(U. americana)* (FACW) (Fig. 9.37). Fast-growing, somewhat shade-tolerant, drought-intolerant trees; most common in deep, moist soils of bottomlands, readily sprout from stumps. *Flowers:* green, red-tinged, bisexual; wind pollinated in March or April, well before leaves develop; some elms can produce mature fruits without fertilization or seed development. *Fruits:* winged, wind-, or water-dispersed, produced every year; reach full size in April, mature in May, with large crops every two to four years; seeds germinate soon after falling or may stay dormant until next spring; eaten by wild turkeys, grouse, various songbirds, and squirrels; when better

Fig. 9.37. American elm fruits are well developed in early spring before the leaves emerge.

foods are unavailable, gray squirrels eat buds and may clip large numbers of twig tips in spring to obtain seeds. *Foliage:* deciduous; gold or yellow in fall; decomposes quickly, relatively high in potassium and calcium; eaten by question mark and eastern comma caterpillars and many other insects; deer and rabbits browse the leaves and twigs. *Wood:* American elm is heavy, hard, strong, difficult to split, warps after cutting, and rots quickly; mucilaginous inner bark of slippery elm historically used for poultices. *Note:* Baltimore orioles sometimes build pendulous nests on the drooping branches of large elms. Native elms, particularly American elm, have been decimated by Dutch elm disease, first diagnosed with certainty in the United States in 1930. The disease is caused by the introduced fungus *Ophiostoma novoulmi* (also referred to as *Ceratocystis ulmi*) and spread by native and introduced elm bark beetles, which tunnel in the bark. (The introduced fungus and bark beetle came to the United States in a shipment of logs from Europe in the early 1900s.) Sometimes the disease spreads when the roots of adjacent trees graft together. The disease damages the tree's water-conducting system. Despite the disease, small, reproductive trees are still common, especially near streams.

Hackberry *(Celtis occidentalis).* Uncommon, moderately shade-tolerant, deep-rooted tree that does best in rich soils of moist lowlands; stumps of young trees readily send up sprouts. *Flowers:* greenish, wind pollinated in mid-spring as leaves develop; clustered pollen flowers and solitary seed flowers (sometimes bisexual flowers) on same plant. *Fruits:* red-purple, dryish, ripen in late summer, some persist into winter; numerous in most years; eaten by various birds and mammals, such as woodpeckers, turkeys, thrushes, cedar waxwings, squirrels, opossums, foxes, raccoons, and skunks (the latter four eat fallen fruits). *Foliage:* deciduous; yellow in fall; eaten by caterpillars of American snout and emperor butterflies. *Wood:* tough, flexible. *Note:* Witches' brooms (dense twig clusters) are associated with presence of a powdery mildew fungus (*Sphaerotheca*) and an eriophyid mite (*Aceria celtis*). Jumping plant lice (*Pachypsylla* spp.) cause galls on the buds, leaf stalks, and leaves, including a nipple-like gall on leaf undersides. I found large nymphs with wing buds in nipple galls in mid-October. Gall flies (Cecidomyiidae) cause thorny galls on leaves.

Mulberries (Moraceae)

White mulberry* *(Morus alba).* Small tree or large shrub; naturalized on floodplains, roadsides, and other deep, moist soils. **Red mulberry** *(M. rubra).* Small, shade-tolerant tree; rich woods; a rarity in Connecticut. *Flowers:* drab, wind pollinated; in dense catkins, mainly mid- to late spring, before or as leaves develop; pollen flowers and seed flowers in separate catkins on same or different plants. *Fruits:* white, purple, or black in white mulberry, red or black in red mulberry; may form without pollination; ripen throughout summer; eaten by gray catbirds, American robins, goldfinches, grosbeaks, woodpeckers, raccoons, squirrels, foxes, and virtually all other fruit-eating birds or mammals, which disperse the seeds. *Foliage:* deciduous; brilliant yellow in late October (white mulberry). *Wood:* soft, weak; rot resistant, good for fence posts.

Barberries and Relatives (Berberidaceae)

Japanese barberry* *(Berberis thunbergii).* Common shrub; deciduous forests, forest edges, and other situations. *Flowers:* yellow, pollinated by bees and other insects in mid- to late spring, beginning by late April or early May. *Fruits* (Fig. 20.62): ripen red in summer; many last through winter and into early spring; eaten by various songbirds, which disperse the seeds, accounting for rapid spread into new areas. *Foliage:* deciduous; develops early in spring; red in fall, lasts well into autumn. *Note:* This plant is slightly to highly invasive in natural areas. It competes with native plants in conditions ranging from sunny and dry to shady and moist. In some forests, the understory is clogged with thickets of this nuisance plant (Fig. 7.35).

Magnolias and Relatives (Magnoliaceae)

Tulip-tree *(Liriodendron tulipifera)* (Fig. 9.38). Magnificently large, shade-intolerant tree; common in deep, moist,

Fig. 9.38. Tulip-tree fruit clusters are fully grown but tightly closed in September.

bottomland soils and moist rocky slopes. *Flowers:* tuliplike, bisexual, orange and green, not fragrant; pollinated by nectar-seeking bees, flies, and beetles in late spring (late May or June) after leaves have grown. *Fruits/seeds:* as might be expected of a good colonist of open areas, produces many seeds with good dispersal capability; large seed crop occurs almost every year; up to several thousand cones per tree; wind breaks apart conelike fruit clusters and widely disperses winged seeds from fall through winter to as late as May of next year; seeds remain viable for up to several years; eaten by squirrels, white-footed mouse, and strong-billed songbirds such as northern cardinal and purple finch; some fruits may form without fertilization or seed development. *Foliage:* deciduous; begins to develop as early as mid-April in warm years; yellow in fall; some trees retain many leaves into early November; accumulates large amounts of calcium; eaten by caterpillars of the eastern tiger swallowtail, tulip-tree silkmoth, and tulip-tree beauty, but generally not favored by gypsy moth* caterpillars; deer eat fallen flowers, browse succulent new growth of seedlings and saplings. *Wood:* soft, light, suitable for cabinets, crates, fuelwood, paper pulp, and such, but not for supporting beams; used historically to make huge single-log canoes. *Note:* Tulip-tree is near its northern limit in Connecticut. It is the tallest (up to about 200 feet) and most massive of hardwoods in eastern North America. The world's only other *Liriodendron* lives in China. Seedlings can be numerous in moist soils in sunny openings among trees. Stumps readily send up fast-growing sprouts after logging. Even large trees grow quickly: I found that the top branches of a 100-foot-tall tree grew at least 2 feet per year. Sapsuckers drill holes in bark and sip sap.

Sassafras, Spicebush, and Relatives (Lauraceae)

Sassafras (*Sassafras albidum*). Small, common tree; often grows in clones along forest edges, fast-growing sprouts from spreading root system readily invade openings. *Flowers:* greenish-yellow, pollinated by insects, especially flies, in spring (April and May), before or as leaves develop; pollen flowers and seed flowers, similar in appearance but with only one sex functional, usually on separate plants. *Fruits:* ripen dark blue in late summer and early fall, often in small quantities; supply limited but attractive food resource for wild turkeys, woodpeckers, flycatchers, gray catbirds, and other birds; seeds may remain viable in soil for several years, germinate in spring. *Foliage:* deciduous; yellow, orange, red, or pink in fall; may change directly from green to yellow or deep red, with both occurring on adjacent trees; eaten by caterpillars of promethea moth and spicebush swallowtail, which pupate on branches, where they may be readily found when trees are leafless in winter; also eaten by tulip-tree beauty caterpillars (in 1936, they severely defoliated sassafras trees in Connecticut); intense feeding by Japanese beetles skeletonizes leaves; contains difficult-to-digest phenolic compounds that may inhibit herbivory by most other insects and mammals; deer and rabbits sometimes browse tops of young sprouts but, mixed with other foods, sassafras apparently does not cause much digestive upset. *Wood:* soft, weak, but durable; produces many sparks when burned. *Note:* Sassafras's closest relatives (two species) are in eastern Asia. The leaves may be simple or lobed, with lobed leaves most common on young trees (Fig. 9.39). In Connecticut, Chris Pureka (unpublished) determined that simple, unlobed leaves, the most frequent leaf type, incurred less herbivory than did two-lobed leaves, and three-lobed leaves had highest levels of insect herbivory (averaging about 6 percent of the leaf area, measured in fall), as well as largest leaf size (area). Sassafras roots emit chemicals that may suppress growth of other plants. Oil of sassafras, obtained from root bark, has been used as a soothing agent, health-promoting tonic, and flavoring, but it could be a carcinogen in massive doses. Believing tales that sassafras possessed "cure-all" properties, the English harvested and shipped back home large quantities of sassafras from Connecticut coast in the early 1600s.

Spicebush (*Lindera benzoin*). Common large shrub; adjacent to water and in other moist soils, such as in the understory of swampy areas (FACW); groups may form through root sprouts. *Flowers:* unisexual, yellow; pollinated by bees, bee flies, small beetles, and other insects, highly conspicuous from mid- to late April until early May, before deciduous swamp trees develop leaves; pollen flowers and seed flowers on different plants. *Fruits:* nutritious, begin to ripen red in late August; both red and green fruits present in mid-September (Fig. 20.58); ripe fruits attract thrushes and other songbirds that disperse seeds; exceptionally large crop in 2002. *Foliage:* deciduous; yellow in fall; aromatic;

Fig. 9.39. Sassafras may have lobed and unlobed leaves. It forms spreading clones in forest edges and openings.

Fig. 9.40. Wild black currant *(Ribes americanum)* is widespread but uncommon.

phenolic compounds may inhibit herbivory to certain degree, but spicebush swallowtail caterpillars eat and actually prefer the foliage, having evolved the ability to tolerate or detoxify plant chemicals (look for caterpillars in folded leaf shelters in summer); sometimes heavily browsed by deer.

Gooseberries and Currants (Grossulariaceae)

Gooseberries and **currants**(*) *(Ribes* spp.) (Fig. 9.40). Historically represented in Connecticut by several species of shrubs, plus a few non-native species; widely distributed (but uncommon to rare) in woods, swamps, thickets, and roadsides, mainly in moist, sunny areas; may form small thickets as new stems arise from spreading rootstocks. *Flowers:* various colors, produce nectar and pollen in spring, attract pollinating bees and other insects. *Fruits:* varied colors, ripen in summer; fruits or seeds eaten by birds and mammals. *Foliage:* deciduous; gray comma, a butterfly not recently seen in Connecticut, depends on *Ribes* as larval food source. *Note:* Fungal blister rust disease of white pine requires *Ribes* to complete its life cycle, so gooseberries and currants were eradicated in some areas, though eradication often proved ineffective or unneccesary as a control method. The fungus reached the United States with diseased planting stock from Europe. As a result of eradication efforts, several of Connecticut's native *Ribes* species are now extirpated or very rare.

Witch-Hazel and Sweet Gum (Hamamelidaceae)

Witch-hazel *(Hamamelis virginiana)*. Large shrub; common in moist, shaded areas in forests and forest edges; groups of stems may occur together as a result of sprouting from underground runners. *Flowers:* yellow; appear in fall (late September to November), in stark contrast to all other woody plants in region; apparently cross pollinated by fungus gnats, other flies, or small wasps, or may self pollinate. *Fruits* (Fig. 9.41): begin to develop in spring, when actual fertilization occurs, well developed by early July, mature in September to October, open in October or early November; as they dry, pods exert pressure on black seeds, eventually propelling them up to 20 to 30 feet; empty pods stay attached for a year or more; vast majority of huge seed crop in 2000 was expelled by the end of October; flowering very sparse in 2000, fruits correspondingly scarce in 2001; prolific flowering in 2001 led to huge crop of fruits in 2002; gray squirrels bite open pods and eat seeds in late summer; seeds also eaten by bobwhite, grouse, turkeys, and probably ground-feeding sparrows and mice. *Foliage:* deciduous; yellow in fall; moth caterpillars roll or fold parts of leaves and feed inside shelter. *Wood:* hard but of no special use; deer, rabbits, and beavers may nibble twigs or gnaw bark. *Note:* Witch-hazel generally is not greatly harmed by insects or diseases. Aphids cause growth of conical galls on leaves and spiny galls on buds; these aphids generally migrate back and forth between birches and witch-hazel (see "Galls and Gall Makers," Chapter 14). Conical leaf galls develop in early to mid-May as the leaves near full size. An extract from bark, twigs, and leaves, combined with alcohol, is still sold today as an antiseptic and skin soother.

Fig. 9.41. Witch-hazel fruits were abundant in 2000 and 2002 but very scarce in 2001.

Sycamores and Plane-Trees (Platanaceae)

American sycamore (*Platanus occidentalis*). Common tall tree; streamsides, adjacent bottomlands, and other poorly drained soils (FACW); low to moderate shade tolerance, needs open, moist, sunny ground for germination and seedling establishment; sprouts emerge from stumps of young trees. *Flowers:* wind-pollinated in spring (May) as leaves develop; pollen flowers red, seed flowers greenish, both occur on same tree. *Fruit:* spherical fruit clusters (Fig. 9.42) stay on tree through fall and winter, eventually disintegrate in spring while still attached and disperse in wind; seeds sometimes eaten by finches and squirrels. *Foliage:* deciduous; grows relatively late in spring; dull yellow or tan in fall. *Wood:* hard, tough, resists splitting; decays readily, tends to warp, too weak for providing support to heavy structures. *Note:* In trunk girth, this is North America's largest deciduous tree. You can see spectacular sycamores, recognized by their gleaming "desert camouflage" bark, along roadsides near bottomland streams. One along the Farmington River in Simsbury is nearly 26 feet in circumference, and the branches spread 138 feet! Old trees often have cavities used by nesting wood ducks, raccoons, and other wildlife. Chimney swifts formerly roosted communally in hollow sycamores. The fuzzy, pale yellow caterpillars of the sycamore tussock moth (*Halysidota harrisii*) may be restricted to sycamores. In central Connecticut, many sycamore leaves withered and died in June 2000 and 2003, perhaps a result of anthracnose fungal attack. Both years were exceptionally wet in spring. A new crop of leaves grew, but the trees ended up with clumps of leaves and numerous bare twigs.

Fig. 9.42. These American sycamore pods survived winter and soon will disintegrate.

Roses and Relatives (Rosacae)

American mountain-ash (*Sorbus americana*). Uncommon, moderately shade tolerant, deciduous shrub or small tree of rocky ridges or moist woods; multiple stems may arise from common root system. *Flowers:* whitish, bisexual, pollinated by beetles and other insects in late spring after leaves develop. *Fruits:* orange-red; eaten by gray catbird, grosbeaks, cedar waxwings, grouse, and other birds and mammals, but some persist into winter. *Foliage:* gold in fall. *Wood:* soft, weak.

Pasture rose (*Rosa carolina*). Dry, rocky or sandy soils of clearings, pastures, and partly open areas. **Multiflora rose*** (*R. multiflora*) (Fig. 9.43). Native to eastern Asia; common shrub in old fields, deforested areas, fencelines, and forest edges. **Northeastern rose** or **shining rose** (*R. nitida*). Localized in boggy or swampy situations or other moist soils (FACW). **Swamp rose** (*R. palustris*). Common adjacent to water and in wetlands (OBL). **Salt-spray rose*** (*R. rugosa*) (Fig. 9.44). Native of eastern Asia; locally common shrub forming thickets on sandy dunes and other nonforested coastal areas. **Wild rose** (*R. virginiana*). Wet or dry soils of clearings, pastures, and various semi-open areas. A few additional native species are more localized, and several roses native to Europe also grow wild in Connecticut. All are tallish shrubs. *Flowers:* white (multiflora), pink or white (salt-spray), or usually pink (others); pollinated by nectar- and pollen-seeking bees, beetles, and other insects, June to

Fig. 9.43. Multiflora rose is an abundant non-native invader of clearings and old fields.

August (salt-spray rose: late May to early October; mutliflora rose: late spring to early summer). *Fruits:* ripen red in late summer, may persist on plants for several months; eaten by ruffed grouse, American robin and other thrushes, mockingbird, and others, mainly in winter and spring; mockingbirds defend territories containing fruit-laden multiflora rose thickets; mice gnaw into fruits and eat seeds. *Foliage:* deciduous, develops early in spring in multiflora rose, much later in swamp rose. *Note:* Multiflora rose effectively outcompetes native plants, monopolizes soil nutrients, and may form dense extensive thickets in abandoned fields. Voles and rabbits gnaw bark or nip twig tips of roses. Songbirds such as mockingbirds, gray catbirds, cardinals, and song sparrows nest in thickets, which also often shelter rabbits. Cynipid wasps of the genus *Diplolepis* induce galls of many kinds (see "Galls and Gall Makers," Chapter 14).

Fig. 9.44. Salt-spray rose, an exotic from Asia, is a common element of coastal thickets.

Blackberries, raspberries, and **dewberries** ("brambles," *Rubus* spp.) (Fig. 9.45). Include many native species of common shrubs and low-growing vines; in clearings and various semiopen situations (several FACW). **Wineberry*** *(R. phoenicolasias)*. Common along coast, increasing inland. *Flowers:* generally white (purple in *R. odoratus*), form on second-year stems; pollinated by bees and other insects from late spring to mid-summer (varies with species). *Fruits:* ripen red, purple, or black, varying with species, from early summer (some species) through early fall; some blackberries produce fruits and seeds without fertilization of ovules; eaten by many seed-dispersing birds and mammals. *Foliage:* deciduous; red in fall. *Note:* Deer and cottontails browse the stems. Plants produce new stems each year, and these live two seasons; stems that have flowered die back, to be replaced by new stems from the often perennial root system.

Red chokeberry *(Aronia arbutifolia)*. Freshwater wetlands, wet woods, and various thickets (FACW). **Black chokeberry** *(A. melanocarpa)*. Various thickets, swamp borders, tussock sedge clumps, forest clearings, and dry rocky hilltops (FACW). **Purple chokeberry** *(A. prunifolia,* or perhaps a hybrid of other *Aronia* species). Near water and in other wet soils (FACW). All are shrubs and may grow in clusters formed through vegetative reproduction (root sprouts). *Flowers:* white, sometimes pink-tinged; pollinated by insects, especially mining bees and bumblebees, in spring (or may self pollinate), generally beginning in May, though in warm sites black chokeberry flowers may open in late April. *Fruits* (Fig. 9.46): ripen red, purple, or black (as per common name) in summer (some beginning by mid-July); may persist, dry and shriveled, through winter and into early spring; eventually various frugivorous birds and mammals may eat them. *Foliage:* deciduous; red or orange in fall; tiny mites live in galls on leaves.

Fig. 9.45. Brambles, such as this black raspberry, shown fruiting in early summer, are common in disturbed areas.

Fig. 9.46. Chokeberries are widespread shrubs often occurring in wetlands.

Beach plum *(Prunus maritima)* (Fig. 9.47). Locally common shrub in sandy soils along coast. *Flowers:* white, highly conspicuous before and as leaves develop in spring (e.g., early May); pollinated by bumblebees and other insects. *Fruits:* ripen purple in mid- to late summer (August), provide food for various mammals and birds. *Foliage:* deciduous. *Note.* Other native plums, which are generally associated with streambanks or roadsides, are uncommon *(P. americana)* or extirpated *(P. alleghaniensis).*

Pin cherry *(P. pensylvanica).* Common, shade-intolerant, small tree or large shrub; thrives in clearings and burned areas; stumps readily sprout. *Flowers:* white, pollinated by bees, bee flies, and other insects in spring-early summer (beginning in late April or early May), as leaves develop. *Fruits:* sour, ripen red in July or August, may persist into winter; seeds remain viable for many years and germinate with exposure of ground to sun; eaten by wide range of birds and mammals. *Foliage:* deciduous; sometimes purplish red in fall. *Wood:* soft, weak, can be used for fuel or paper pulp. *Note:* Pin cherry is short-lived and lives only a few decades.

Black cherry *(P. serotina)* (Fig. 9.48). Common, shade-intolerant tree; clearings, field borders, rock outcroppings in forests, and ridge tops; stumps readily sprout; can be major component of secondary growth following logging, fires, or other disturbances. *Flowers:* white, pollinated by nectar-seeking bees, flies, beetles, and other insects in spring (peak in mid- to late May, later along coast than inland), just after leaves are full grown; some cherries produce mature fruits without fertilization or seed development; late spring frosts may damage flowers or young fruits, with no fruit maturation on those stalks. *Fruits:* ripen dark purple from late July through early September; abundant every few years; seeds may germinate up to two years after fruits fall; serve as food for a great many birds and mammals (ranging from mice and chipmunks to raccoons and bears); white-footed mice open pits and eat seeds, also store large numbers of pits (in my basement I found an old boot that mice had filled with cherry pits). *Foliage:* deciduous; varies from yellow to orange-red in fall; some plants hold many green leaves well into November; eaten by caterpillars of the eastern tiger swallowtail, red-spotted purple, coral hairstreak, striped hairstreak and many other Lepidoptera, but avoided by gypsy moth caterpillars; mites cause long, slender or lumpy galls to form on leaves in spring (Fig. 14.107); deer, rabbits,

Fig. 9.47. Beach plum flowers conspicuously in spring in coastal thickets.

Fig. 9.48. Black cherry often thrives among the vegetation in disturbed areas. Mrs. E. Rowan, from Lounsberry (1900).

Ecology of Trees, Shrubs, and Vines in Connecticut | 155

Fig. 9.49. Black knot is a fungal parasite commonly afflicting native cherries.

and voles often nip off seedlings and sprouts. Small trees are often defoliated by eastern tent caterpillars, which hatch from overwintered eggs and begin to feed on leaves right after leaf buds open in early spring. I found small black cherry trees that were completely defoliated by hundreds of caterpillars by the first of May. On some cherry plants, *Formica* ants may attack and kill the small caterpillars and thereby reduce the level of defoliation. Ants are attracted to nectaries present along edges of newly emerged leaves; cherry plants near ant nests attract ants and have fewer caterpillars and are less likely to be defoliated than plants without a nearby ant colony. Three to four weeks after leaf emergence, caterpillars are too big for ants to subdue, and by this time cherry leaves no longer have active nectaries. This is an example of a temporary, facultative, mutually beneficial relationship between ants and plants. *Wood:* moderately heavy but weak, beautifully colored, fine-textured, does not warp after seasoning; makes fine cabinets and paneling, but most trees today are too small, bent, and knotty to be of much value. *Note:* The seeds may germinate on bare soil or under leaf litter, but seedlings in forests often die within a few years unless competitors are removed or a tree-fall gap develops. A soil fungus specific to black cherries may kill young plants, especially those close to existing mature trees. Survivors may live more than 250 years. Black knot (Fig. 9.49), an apparently native fungal disease, causes rough, black enlargements on twigs of cherries and plums (primarily black cherry). Swollen, ruptured bark results in an approximately three-fold increase in twig thickness and may result in die-back of affected branches. Large cankerous swellings, with bark intact, may appear on trunk. The disease spreads via wind-blown spores, first noticeable in fall as swellings on (usually green) shoots that were infected in spring, when spores germinate, penetrate the bark, and begin to release growth-regulating chemicals. Knotty growth resumes enlargement the next spring and turns hard and black the next winter. Fruiting bodies develop and release spores the following spring, two years after initial infection (sometimes spore production occurs after one year). If the branch stays alive, the knot may enlarge annually, growing toward the base of the branch. A heavy infestation may stunt the tree or kill it. Sometimes a white or pink fungus of another species (*Trichothecium roseum*) covers black knot swellings, or the knot may be infested with mites or insects, such as larvae of the dogwood borer (a moth).

Choke cherry (*Prunus virginiana*). Large shrub or small tree; along roads and streams and in old fields and forest edges, habitats that reflect low tolerance of shade; may form clones from common root system; stumps readily sprout. *Flowers:* white, pollinated by insects in late spring-early summer. *Fruits:* ripen red to blackish in mid- to late summer and, though usually sour, sustain many birds and mammals. *Foliage:* deciduous; yellow to bronze in fall. *Wood:* heavy, hard.

Meadow-sweet (*Spiraea alba*). Common shrub; moist or wet soils in open areas (FACW). **Steeplebush** or **hardhack** (*S. tomentosa*). Fairly common shrub of fields and dry to wet meadows (FACW). A few additional spiraea species have escaped from cultivation to roadsides and old cellar holes. *Flowers:* white or pinkish, appearing in summer (beginning by early July) and into early fall in meadow-sweet; usually pink, arranged in spire, blooming in summer (mainly mid-July to late August) in steeplebush; pollinated by bees, beetles, and other insects; on flowering spike, older flowers at top tend to be functionally male while recently opened flowers lower down are female; bees tend to fly from top of one spike to bottom of another, then work upward, so pollen generally moves from one plant to female flowers of another, resulting in cross pollination. *Fruits:* dry capsules; long-lasting; winter winds shake plants and disperse seeds. *Foliage:* deciduous; cecidomyiid midges induce formation of spiny rounded galls at twig tips and podlike galls on leaves; deer sometimes browse spiraeas.

Hawthorns (*Crataegus* spp.) (Fig. 9.50). Thorny, moderately shade-tolerant shrubs or small trees; old fields, forest understory and edges, coastal forests, and various thickets. *Flowers:* white, rarely pink, often stinking; pollinated by insects in May and early June as leaves develop, or fruits and seeds may develop without fertilization. *Fruits:* applelike but small, ripen purple, red, orange, or yellow in late summer, may persist for several months; sometimes eaten by ruffed grouse or songbirds, but not particularly attractive to birds or mammals. *Foliage:* deciduous; orange or red in fall. *Note:*

Fig. 9.50. Hawthorns produce small applelike fruits.

Distinguishing among the many hawthorn species is difficult. Hawthorns are susceptible to fungal pathogens and insect damage. Thickets provide cover for birds and mammals.

Shadbush, serviceberry, or **juneberry** *(Amelanchier arborea, A. canadensis* (FACW), and *A. stolonifera)*. Fairly common shrubs or small trees; most often at forest and shrub swamp edges, rock outcroppings, and fencelines where not completely shaded by nearby trees. Roundleaf shadbush *(A. sanguinea)* is a rare species. *Flowers:* white, showy (Fig. 9.51), sweetly fragrant; pollinated by small bees and other insects in spring (between mid-April and mid-May)—when shad are migrating upstream—before or as leaves develop. *Fruits:* red-purple or black, ripen in June or July (hence one of the common names), nourish wildlife such as ruffed grouse, woodpeckers, thrushes, catbirds, waxwings, orioles, squirrels, bears, and many other animals that disperse the seeds in their droppings. *Foliage:* deciduous; yellow, orange, bronze, or red in fall; deer and rabbits browse foliage and twigs. *Wood:* heavy, hard, strong, but tends to crack and warp as it dries. *Note:* Shadbushes often hybridize with each other, confounding identification.

Beans, Peas, and Relatives (Fabaceae)

False indigo* *(Amorpha fruticosa)*. Shrub, native to eastern United States but not the Northeast; widely established in Connecticut, mainly along major rivers (FACW). *Flowers:* purple, insect-attracting spikes; appear in late spring or early summer. *Fruits:* small seed-bearing pods present in late summer, fall, and winter. *Foliage:* deciduous.

Black locust* *(Robinia pseudoacacia)*. Thorny, shade-intolerant tree; moist soils along roadsides and streams, in pastures, and other sites; grows fast, flowers young, and may form clones through root sprouts. *Flowers:* bisexual, white, fragrant; pollinated by nectar-seeking bees and other insects, in mid-May to mid-June, after leaves develop. *Fruits:* most seeds drop from attached pods before December, but some pods with seeds (four to eight per pod) persist into winter and following spring; large seed crops every one to three years. *Foliage:* deciduous; eaten by caterpillars of zarruco duskywing (not seen in Connecticut in recent years) and silver-spotted skipper; deer browse young growth. *Wood:* no less than the best there is—hard, heavy, incredibly strong and stiff, durable in contact with soil, and excellent for fuel; tunneling by locust borer beetle larvae weakens branches (easily broken by wind) and usually makes wood unsuitable for boards (look for big, black, yellow-banded adult beetles on goldenrods in late summer). *Note:* As in other members of the legume family, nitrogen-fixing bacteria in root nodules make nitrogen available to plants, allowing growth in poor

Fig. 9.51. Shadbush is most conspicuous when flowering in spring.

Ecology of Trees, Shrubs, and Vines in Connecticut | 157

soils and improving soil fertility as vegetation decomposes. Locusts are little used by wildlife other than those described.

Ailanthus and Quassias (Simaroubaceae)

Ailanthus* (tree-of-heaven) (*Ailanthus altissima*). Malodorous, fast-growing, shade-intolerant tree; locally common along roadsides, in vacant lots, locally on sandy coastal dunes, and in other open areas; quickly colonizes wide cracks and joints in concrete or asphalt; clones may form through sprouting of underground parts, and these clones may outcompete native plants. *Flowers:* greenish; seed flowers and foul-odored pollen flowers on separate trees, present in late spring and early summer, after leaves develop. *Fruits:* abundant; each seed enclosed in twisted, yellow-brown or reddish wing, readily dispersed by wind. *Foliage:* deciduous; resembles that of smooth sumac or walnut, but each leaflet is untoothed except for a pair of gland-tipped teeth at base; causes dermatitis in some people; eaten by caterpillars of huge ailanthus silkmoth* and colorful ailanthus webworm moth. *Note:* The bark and leaves contain compounds toxic to other plants, facilitating the formation of dense thickets devoid of other species. Ailanthus silkmoth, introduced in United States in 1861, is known in Connecticut from urbanized areas where ailanthus grows (e.g., New Haven), but it has declined drastically over the past couple of decades and now may be extirpated from Connecticut (and perhaps throughout the northeastern United States).

Sumacs, Poison Ivy, Cashews, and Relatives (Anacardiaceae)

Poison ivy (*Toxicodendron radicans*). Familiar to all by bad reputation; common vine or shrub in clearings, forest edges, and other disturbed, open, or semi-open areas; grows as low ground plant or may climb high into trees (adhering with hairlike aerial roots), often forming clones connected to common root system. A less common, bottomland species, *T. rydbergii*, lacks aerial roots and does not climb. *Flowers:* greenish, pollinated by small bees and flies in late spring and early summer. *Fruits:* whitish or tan, ripen in September, may persist into winter; eaten by flickers, ruffed grouse, and songbirds of many species (without ill effect); squirrels sometimes cut fruit clusters, remove fruit flesh, and eat and thereby destroy seeds, but sometimes they disperse seeds by inadvertently dropping fruits they are carrying. *Foliage:* deciduous; red when growing in spring, green in summer, and red or gold in fall; foliage or stems eaten by rabbits, deer, and moth caterpillars.

Note: Most people develop a rash in an allergic response to contact with the resinous oil in poison ivy and poison sumac. The rash usually appears a day or two after contact. Oil droplets in the smoke of the burning plants easily reach eyes, lungs, and other sensitive areas, with unpleasant and sometimes dangerous results, including systemic effects. In general, the response is proportional to the amount of oil that contacts the skin and the amount of skin affected. The rash may develop from contact with dormant plants in winter or even leaves or branches that have been dead for years. Direct or indirect (e.g., from a pet or tool) contact with the oil is necessary; one cannot develop a rash from just being near the plants.

The oils of poison ivy and poison sumac are chemically similar (the potent factor being urushiol), and if you are sensitive to one plant you'll also react to the other. Despite the name, these plants contain no poisons. The rash is an immune response and not an effect of toxicity. Upon contact with skin, substances in the oil undergo changes and bond with skin proteins. Our immune system perceives this as a threat and launches a defensive reaction aimed at not only the intruding molecules but also the affected skin, which becomes inflamed, itches, and blisters, then oozes when the blisters rupture (the harmless oozing fluid does not spread the rash). The response normally requires an initial contact that has no obvious effect but which evokes later sensitivity to urushiol.

To minimize the rash, rinse affected skin areas with lots of cold water, wash with cold soapy water, and/or rinse with alcohol—as soon as possible. Some people believe that the sap of jewelweed or plantain effectively combats the rash. In severe cases, contact a physician.

The feasibility of eliminating poison ivy plants in your yard depends on the extent of the patch. In small patches, poison ivy can be removed by cutting, chopping, and grubbing, if repeated diligently several times as certainly will be necessary (any roots that are left will send up new shoots). Planting competitive plants that will cast heavy shade or form dense root systems (e.g., grass) also may help. Of course, caution is in order. Remember that your skin, clothing, gloves, and tools will pick up the rash-causing oil if they touch any part of the plant, and the oil retains its potency for years. Though washing does help, the oil can withstand laundering to some degree. So you may want to wear old discardable clothing to work on thick patches and, later on, remember to avoid touching the working end of your tools until the oil has had ample opportunity to wear off.

What about chemical control? Effective herbicides, best used in spring and early summer, are available at garden centers. But these can cause health and environmental problems if improperly used.

Not everyone regards poison ivy as a plant to eradicate. Biologists, of course, recognize its significant role in ecosystems (e.g., as a food source for fruit-eating birds), though they may not want it in their own backyards. But British gardeners may be poison ivy's greatest advocates—they reportedly have imported and cultivated it as a garden plant, appreciated for its colorful foliage.

Poison sumac *(Toxicodendron vernix)* (Fig. 9.52). Large wetland or riparian shrub (OBL). *Flowers:* yellowish-green, pollinated by insects in late spring and early summer (June). *Fruits:* yellowish, greenish, or whitish; ripen in late summer, may persist through winter; attract various frugivorous songbirds, particularly in winter. *Foliage:* deciduous; yellow, orange, or red in fall. *Note:* See poison ivy account for information on allergic rash caused by this plant.

Staghorn sumac *(Rhus typhina)*. Shrub or small tree in old fields and other open areas. **Smooth sumac** *(R. glabra)*. Shrubby tree in old fields and clearings. **Shining sumac** *(R. copallinum)* (Fig. 9.53). Shrub in dry soils, often at forest edges, in sandy coastal thickets, or upland of salt marshes. All are common and may form spreading clones through sprouting of underground runners; growth is fast, facilitated by production of airy, pithy branches that economize on energy and nutrient use. *Flowers:* greenish red, greenish, or yellowish; pollinated by bumblebees, honey bees, other bees, beetles, and other insects after the leaves have grown; smooth

Fig. 9.53. In coastal thickets in mid-September, shining sumac fruits are red while the leaves are still green.

and staghorn sumacs flower mainly in June and July and may have ripe fruits when shining sumac is just beginning to flower (July to August); pollen and seed flowers often on separate plants; smooth sumac and staghorn sumac stems only knee high may flower and produce robust fruit clusters. *Fruits:* ripen red in late summer; many fruits of smooth and staghorn sumacs persist through winter and into spring or even the next summer, hence apparently not a top choice among fruit-eating birds, but a wide variety of birds, ranging from ruffed grouse to various songbirds and woodpeckers, eat them, as do gray squirrels; many birds readily eat shining sumac fruits. *Foliage:* deciduous; orange, red, or purple as early as late September; new spring foliage also may be red; red-banded hairstreak caterpillars eat shining sumac foliage; deer browse twigs and leaves. *Wood:* light, soft, brittle; bark contains much tannin, has been used for curing leather. *Note:* Nonsocial wasps hollow out the stems and use them as brood chambers. Cottontails eat the bark in winter.

Hollies (Aquifoliaceae)

Common winterberry *(Ilex verticillata)* (Fig. 9.54). Common shrub in freshwater wetlands and adjacent to ponds and streams (FACW). **Smooth winterberry** *(I. laevigata)*. Uncommon shrub; swampy areas (OBL). *Flowers:* whitish; appear in late spring and early summer; nectar attracts bumblebees, yellowjackets, paper wasps, and other small bees and wasps; pollen flowers and seed flowers usually on separate plants. *Fruits:* ripen red by September, long persisting; some stay attached through winter and into spring; individual plants do not produce fruits every year; eaten primarily in fall and winter by American robins, eastern bluebirds, white-throated sparrows, and most other fruit-eating birds; white-throated sparrows often eat pulp

Fig. 9.52. Poison sumac (here fruiting in early October) is restricted to wetlands.

Fig. 9.54. Winterberry flowers in early summer and holds abundant red fruits in fall and winter.

Fig. 9.56. Asian bittersweet is a non-native vine that overwhelms native plants. Birds disperse the seeds.

and seeds and discard skin. *Foliage:* deciduous; dull in common, shiny in smooth; yellow in smooth in fall; common may turn brown or blackish before falling in autumn.

Mountain-holly *(Nemopanthus mucronata)* (Fig. 9.55). Shrub of boggy wetlands (OBL), primarily in northwestern Connecticut; uncommon and localized in and near swamps elsewhere. *Flowers:* long-stalked, greenish white or yellowish, appear in late spring (e.g., early to mid-May), after leaves at least half grown; pollen flowers and seed flowers usually on separate plants. *Fruits:* ripen purplish-red in August; presumably eaten by birds. *Foliage:* deciduous; yellow in fall.

Bittersweets and Relatives (Celastraceae)

American bittersweet *(Celastrus scandens).* Uncommon but widely distributed vine in forest edges and streamsides;

Fig. 9.55. This mountain-holly was flowering in a bog in early May.

presumably more common before arrival of Asian bittersweet. **Asian bittersweet*** *(C. orbiculatus)* (Fig. 9.56). Common vine in forest edges, thickets, and other sunny habitats. *Flowers:* greenish or cream-colored, inconspicuous; pollinated by insects in mid- to late spring (May to June); pollen flowers and seed flowers mostly on separate plants. *Fruits:* yellow-orange pods; open to reveal scarlet arillate seeds; ripen in early fall, persist through winter, sometimes eaten by birds, which seem to ignore fruits in fall but eventually eat enough to disperse the plants. *Foliage:* deciduous; yellow-gold; that of Asian bittersweet especially noticeable in mid- to late October as bright foliage contrasts with often duller hues of plants upon which bittersweet grows. *Note:* Asian bittersweet often overgrows and stangles trees that support it. It may hybridize with American bittersweet, threatening the genetic integrity of the native species.

Bladdernuts (Staphyleaceae)

Bladdernut *(Staphylea trifolia)* (Fig. 9.57). Uncommon shrub or small tree of moist, fertile soils of forest edges, wooded stream banks, and rocky woods. *Flowers:* greenish-white, bisexual, open by early May. *Fruits:* in central Connecticut I found large, full-sized, inflated, papery seed capsules in late July and August. *Foliage:* deciduous.

Maples (Aceraceae)

Box elder *(Acer negundo).* Small deciduous tree of streamsides and other moist bottomlands; reported to be

Fig. 9.57. The distinctive pods of bladdernut develop by late July.

Fig. 9.58. Red maple seed flowers (lower) and pollen flowers (upper) add color to forests and swamps in early spring.

native in Connecticut only in Housatonic River valley, now also uncommon to rare in scattered other locations; cut trees readily send up sprouts; needs sunlight for germination and seedling establishment. *Flowers:* yellow-green, pollinated by wind and insects in early spring, just before or as leaves develop; pollen flowers and seed flowers on separate trees. *Fruits:* winged; ripen from green to brown in late summer and early fall, stay attached in winter, blow away in wind; large crop almost every year; seeds eaten by evening grosbeaks, other strong-billed birds, and small mammals. *Foliage:* yellow in fall. *Wood:* soft, light, weak, and brittle.

Striped maple *(A. pensylvanicum)*. Large shrub or small tree of shady and semi-shady areas with cool, moist soils, but occurs rarely on tops of trap-rock ridges in central Connecticut; most common in northwestern Connecticut. *Flowers:* yellowish, pollinated by wind and insects in spring (early May) as leaves reach full size; pollen flowers and seed flowers may be on same or separate plants. *Fruits:* winged; ripen in late summer or early fall and soon disperse; seeds eaten by ruffed grouse and presumably squirrels, chipmunks, and other granivores. *Foliage:* deciduous; yellow in fall. *Wood:* soft, rather light. *Note:* Strong-billed birds eat the buds. Deer, rabbits, beavers and porcupines feed on the bark. The plant withstands heavy browsing by deer and often is damaged by antler-rubbing deer.

Norway maple* *(A. platanoides)*. Commonly planted tree in yards, towns, and along roads. *Flowers:* yellow, first appear in April before foliage develops or when leaves are small, before sugar maple. *Fruits:* immature fruits may have fully formed "wings" by mid-May, ripen in late summer; large crop almost every year; seeds germinate in early spring, if not eaten by gray squirrels. *Foliage:* deciduous; green or yellow (sometimes red) in fall; consumed by various insects; leaves of urban trees often remain attached into November. *Note:* This tree is naturalized and reproducing strongly in some towns and natural areas, especially near disturbed areas, but it is not invasive in most places.

Red maple *(A. rubrum)*. Ubiquitous, moderately shade-tolerant tree in a wide array of wet and dry sites; grows fast initially, relatively short-lived by tree standards, but may live more than a century; many trees originate from stump sprouts and root suckers; in swamps, seedlings become established in dewatered depressions in late summer, but many are killed by flooding the following spring or when beaver dams raise water level for an extended period. *Flowers:* see note. *Fruits:* winged; ripen and disperse in wind in late spring; large crop almost every year; fruit fall begins early to mid-May; seeds germinate right away or a year later. *Foliage:* deciduous; yellow, gold, orange, pink, and/or red (strongly acidic soils) in fall, turning color as early as mid-September. *Wood:* somewhat soft, medium weight, not especially strong or durable. Animal associates are similar to those of sugar maple.

Note: Long before most trees display obvious signs of life, the red or orange-yellow haze of flowering red maples breaks the winter-long dominance of the brown and gray tones of the leafless forest. Red maple is indeed one of the earliest flowering trees (but not quite as early as silver maple). It flowers in March or April (and into early May at higher elevations), well before the leaves develop. Wind carries the pollen, and bees, which often visit pollen flowers in April, may pollinate if they switch to seed flowers while still carrying pollen. Pollen flowers fall from the trees in large numbers in mid- to late April. Red maple flowers almost always are either male (yellowish red or orange) or female (bright red) (Fig. 9.58). That is, each flower has only the male parts (pollen-producing stamens) or only the female parts (pistil) but rarely both. Each individual plant may have only male flowers, only female flowers, or flowers of both sexes (bisexual). Generally, most plants are males and the remainder are divided between females and bisexuals, with females usually outnumbering bisexuals. From year to year, individual plants tend to stay the same sex, but occasionally they change, most often between bisexual and male or female. Once a tree starts flowering, it tends to flower every year. The first year a tree flowers, it usually is either a male or female, with bisexuals rare.

Why do plants with only male flowers predominate? A study in Michigan evaluated and found no support for several possible explanations. Overall, the ratio of males to females in several populations was not consistently related to moisture or nutrient conditions. For moist sites, the sex ratio did not appear to be affected by age of the stand. Sex ratio also did not appear to be affected by the origin of the tree; trees that originated from seed were no more likely to be one sex or the other than were trees that started as sprouts. Males and females did not differ in the age at which they started flowering, and they did not exhibit differences in mortality rate. Additionally, the maples did not exhibit any biased pattern of sex change that would account for the male-biased sex ratio. However, one trend was evident. In dry sites, the male-biased sex ratio became stronger the older the stand was, suggesting that older trees in dry sites experience low nutrient availability and may be more likely to produce male flowers because it takes fewer resources to produce pollen than it does to produce fruits, and a male tree may still have good reproductive success if its pollen can reach female trees in the vicinity (for example, in a nearby wetland). Production of pollen rather than seeds in dry sites also may be advantageous because seedling establishment is difficult in dry sites.

Silver maple *(Acer saccharinum).* Relatively short-lived tree of floodplains and bottomlands near water (FACW); dominates floodplain forests along Connecticut River (Fig. 4.13); stumps and fallen trees along streams readily send up vertical shoots that become trees. *Flowers:* greenish-yellow or reddish, pollinated by wind and possibly insects in March and April, before red maple. *Fruits:* winged; mature and fall in spring (beginning by early May), around time leaves fully grown; good crop most years (Fig. 9.59). *Foliage:* deciduous; pale yellow, brown, or rarely orange or red in fall. *Wood:* hard, strong but brittle, medium weight. *Note:* Large trees often have cavities used by squirrels, raccoons, wood ducks, and other animals. Other associates are similar to those of sugar maple.

Sugar maple *(A. saccharum).* Long-lived, shade-tolerant tree that thrives in fertile, moist soils; distribution is best observed in latter half of April when flowers are conspicuous in largely leafless forest; seedlings sometimes densely abundant in shade of mature forests and forest gaps, but these grow slowly and nearly all die; in mature forest, my students and I determined that stem tips of sapling sugar maples grew an average of only about 3 or 4 inches per year over several consecutive years; sprouts from cut or damaged trees may grow faster than this; many large, old trees along country lanes now have extensive internal decay and break in

Fig. 9.59. Silver maple has well-developed fruits in early spring before the leaves are fully grown.

Fig. 9.60. Sugar maple flowers are conspicuous in late April and early May.

strong winds. *Flowers:* greenish-yellow, first appear in mid- to late April; pollinated by wind and insects (e.g., bees) in April and early May before or as leaves develop (Fig. 9.60); in each flower usually only stamens or pistil functional; both kinds of flowers generally present on same tree; some maples produce mature fruits without fertilization or seed development. *Fruits:* winged, ripen in late summer, may stay on tree through fall; disperse mainly in wind; produced in abundance every 2 to 5 years; seeds germinate in spring; ruffed grouse, evening grosbeaks, northern cardinals, and other similarly strong-billed birds, squirrels, chipmunks, and various other birds and small mammals eat the buds, flowers, and a minority of seeds. *Foliage:* deciduous; yellow, orange, or red in fall; fallen leaves rich in nutrients, enrich soil; pear thrips feeding interferes with leaf development, can reduce volume and sugar content of maple sap; deer browse twigs, foliage, and seedlings; porcupines, cottontails, and snowshoe hares feed on twigs, leaves, branches, and bark. *Wood:* tough, strong, shock resistant, with medium hardness and weight—fine for many uses, including fuel. *Note:* Old sugar maples and red maples often rot where dead limbs have broken off; chickadees and other birds may excavate these sites, which become shelter or nest locations for the excavators, screech-owls, squirrels, and various other animals. Sugar maples fare poorly when exposed to excessive de-icing salt runoff from roads, and they suffer from acid deposition resulting from air pollution, which has detrimental impacts on soil nutrients and soil fungi that help tree roots absorb nutrients and water.

The sugar maple is one of the few Connecticut trees that is exploited commercially for something other than its wood. Maple syrup production is a small industry in Connecticut, but it is valued by many who enjoy locally produced natural foods. Maple sap flows whenever freezing nights are followed by days above freezing. The cause of sap flow is not pecisely understood, but flow occurs when pressure inside the tree increases on warm sunny days, forcing sap out through the taphole or other wounds. On freezing nights, the tree cools and pressure drops; this results in water uptake by the roots, readying the tree for the next day's sap flow. Sap flow stops if temperatures stay continuously above or below freezing. Flows tend to be intermittent, depending on weather. The average sap yield per tree is about 10 to 20 gallons, but very productive sugar maples may yield up to 40 gallons, and vacuum application can increase the yield. Open-grown trees with a wide crown generally produce the most sap. The sugar in the sap is very dilute (usually 2 to 3 percent sugar, but up to 6 percent), which is why the sap must be boiled to concentrate the sugar. About 40 gallons of sap yield 1 gallon of syrup. The sugar is derived from starches accumulated the previous summer. Conversion to sugar occurs in late winter in preparation for the tree's nutrient requirements for the next growing season. The early European colonists of northeastern North America learned to harvest maple sap for sugar from Native Americans. Humans may have learned the technique from red squirrels, which often bite into the bark of sugar maples (much less often red maples), causing the sap to flow out in a streak onto the surface of the bark. Later, after most of the water has evaporated, the squirrels return and lick up an energy-rich snack of concentrated sugar residues. Many kinds of birds, plus chipmunks and various insects such as wasps, owlet moths, and flies, drink the sap that flows from wounds in maples, birches, and other trees in late winter and early spring. Songbirds and hummingbirds may feed on insects attracted to the sap.

Buckthorns (Rhamnaceae)

Alder-leaved buckthorn (*Rhamnus alnifolia*). Freshwater wetlands (OBL) in western and central Connecticut. **Common buckthorn*** (*R. cathartica*) (Fig. 9.61). Old fields, limestone wetlands, and forest edges (often in neutral to alkaline soils). **Glossy (European) buckthorn*** (*Frangula*

Fig. 9.61. Common buckthorn is a non-native invasive shrub.

alnus). Damp areas, field borders, coastal woods, and roadsides; has become more widespread and abundant in open and semi-open areas in eastern North America over past few decades. All are shrubs or small trees. Dense stands of non-native buckthorns may replace native vegetation by shading out native tree seedlings, shrubs, and wildlfowers. *Flowers:* greenish to yellowish, bisexual or unisexual; pollinated by bees, flies, and other insects mainly in late spring (May to June). *Foliage:* deciduous; that of common buckthorn develops early (late April to mid-May, before most other woody deciduous plants) and stays green and attached well into fall. *Fruits:* black; common buckthorn ripens by mid-September, may remain on plants until spring; glossy buckthorn ripens throughout summer, most have fallen by early November; eaten and seeds readily dispersed by woodpeckers, gray catbirds, American robins, thrushes, rose-breasted grosbeaks, mice, and other animals.

Grapes and Relatives (Vitaceae)

Virginia creeper *(Parthenocissus quinquefolia).* Common vine; attaches to trees and other objects in semi-wooded areas by means of adhesive disks. *Flowers:* greenish; bisexual or sometimes unisexual, pollinated by bees (e.g., bumblebees, metallic green halictids) and other insects in late spring and early summer. *Fruits:* ripen in late summer and early fall, attract various frugivorous birds. *Foliage:* deciduous; turns red in September or October, often conspicuous against foliage that is still green; eaten by caterpillars of Pandora sphinx moth and other lepidopterans.

Fox grape *(Vitis labrusca).* Forest edges and openings and along fences and streambanks. **Summer grape** *(V. aestivalis).* Common in dry woods and thickets. **Riverbank grape** *(V. riparia).* Localized along streams and roadsides (FACW). **New England grape** *(V. novae-angliae).* Uncommon, probably a hybrid between fox grape and riverbank grape; localized in thickets (FACW). All are vines that may climb high into trees. *Flowers:* small, fragrant, greenish; each plant generally has both bisexual and unisexual flowers; until shed, petals cover sexual parts of flower; pollinated by insects primarily in June. *Fruits:* ripen purple mainly in late summer to early fall, some persist into winter; eaten and seeds dispersed by wide assortment of frugivorous mammals and birds; woodpeckers and other birds pluck fruits from vines, whereas mammals often take advantage of fallen fruits; in fall, fecal droppings of raccoons and foxes often contain many grape seeds and skins. *Foliage:* deciduous; turns yellow in autumn; eaten by Pandora sphinx moth larvae, as well as many other insects; gall midges (Cecidomyiidae) cause galls on the leaves or tendrils. *Note:* Various thicket-dwelling songbirds (catbirds, mockingbirds, cardinals) incorporate grape bark strips into nests. Commercially valuable Concord grape was derived from fox grape, which provides a fabulous (but seedy) trail snack at summer's end.

Basswoods and Lindens (Tiliaceae)

Basswood *(Tilia americana).* Well-rooted, fast-growing, shade-tolerant tree of moist soils, including wetland edges and suitable sites on talus slopes; rebounds from cutting by sprouting prolifically from stumps. *Flowers:* creamy white, fragrant; pollinated by nectar-seeking bees, often in large numbers, in early summer (June to July), after leaves have grown. *Fruits* (Fig. 9.62): round, hard "nutlets," attached to leaflike bract; ripen in late summer, some persist on trees through winter; produced in abundance almost every year; often wind-dispersed; little information on use by birds, eaten occasionally (and sometimes dispersed) by squirrels and chipmunks; seeds may not germinate for years; germination is best in mineral soils (unlike older individuals, seedlings have deeply lobed leaves). *Foliage:* deciduous; light yellow or brown in fall; eaten by caterpillars of yellow-banded underwing, linden looper, and other moths, but insects rarely do serious damage; fallen leaves add nutrients to soil, more so than most other trees. *Wood:* soft, light, weak, brittle; fine for carving, boxes, furniture cores, paper pulp, and other uses; inner bark yields fibers that can be made into thread or rope. *Note:* This tree has good resistance to fungal infections and diseases, but it is sensitive to road salt. Rabbits and deer snip off and eat basswood sprouts.

Fig. 9.62. American basswood nutlets become conspicuous in November after leaves have fallen.

Cacti (Cactaceae)

Prickly-pear *(Opuntia humifusa)* (Fig. 9.63). Rare, low, spiny, succulent plant of coastal sandy soils; Connecticut's only native cactus; fleshy pads have clusters of short, barbed bristles but usually no large spines. *Flowers:* showy, yellow; pollinated by insects in early summer. *Fruits:* fleshy, sweet, many-seeded; persist through spring of next year. *Note:* Wildlife use in New England likely includes nibbling of fruits and pads by cottontails and seed-eating by rodents. See "Plants in Winter" later in this chapter for further information.

Oleasters (Elaeagnaceae)

Autumn-olive* *(Elaeagnus umbellata)* (Fig. 9.64). Shrub native to eastern Asia; formerly planted for ornamental uses, erosion control, soil improvement, and "wildlife enhancement"; now spreading rapidly without human assistance. *Flowers:* peak in mid- to late May. *Fruits:* ripen red in often enormous numbers in late summer and early to mid-fall; eaten by birds, which have dispersed this plant throughout

Fig. 9.63. Prickly-pear is Connecticut's only native cactus. Several fruits from the previous year's flowering are visible in the May photograph.

much of state. *Foliage:* deciduous; leaves develop early (in warm years, beginning in late March or early April); some leaves may remain attached through late fall. *Note:* This plant is becoming increasingly abundant along roadsides, in old fields, and other disturbed sites. It is a major competitive threat to native biodiversity in open sandy soils. Autumn-olive's microbe-mediated ability to extract gaseous nitrogen and efficiently convert it to organic form in soil can adversely affect native plant communities that depend on low soil fertility.

Dogwoods (Cornaceae)

Dogwoods *(Cornus* spp.). *Flowers:* generally white, pollinated by mining bees, halictid bees, bumblebees, syrphid

Fig. 9.64. Autumn-olive is an invasive non-native shrub that is increasing greatly in distribution and abundance. Birds eat the fruits and disperse the seeds.

flies, and beetles in late spring and early summer. *Fruits:* clusters provide food for various birds (wild turkeys, ruffed grouse, woodpeckers, gray catbirds, thrushes, cedar waxwings, northern cardinals, and various other songbirds) and mammals, many of which serve as good dispersal agents for the seeds. *Foliage:* deciduous; red in fall; fallen leaves quickly decompose. *Note:* Deer, cottontails, and beaver eat the twigs and bark. Branches of shrubby dogwoods sometimes cradle the nests of willow and alder flycatchers, yellow warblers, gray catbirds, American goldfinches, and other shrubland songbirds.

Silky dogwood *(Cornus amomum).* Fast-growing shrub of freshwater borders and wetland or moist thickets (FACW). *Fruits:* blue or blue-cream-colored; tremendously abundant in some years, scarce in others, ripen in late summer (by late August), may persist through October and into November. See "Dogwoods," above.

Flowering dogwood *(Cornus florida).* Small, slow-growing tree; common at forest edges and sometimes along powerline corridors (Fig. 7.29); does best in moist soils; leaves may droop during summer drought; sometimes forms small groups of stems attached to common root system, especially after a tree is cut or heavily damaged. *Flowers:* small, yellowish, surrounded by showy white bracts; bracts may begin to open in late April, reach peak form in early or mid-May; pollinated by insects. *Fruits:* ripen red in August, persist into early fall (October); white-throated sparrows sometimes bite off flesh and leave seed behind (Fig. 20.63); gray squirrels remove pulp and eat seed; seeds overwinter and generally germinate in spring. *Wood:* exceptionally heavy, hard, and shock resistant; on tool handle, takes a pounding without splitting; wears smooth with use. *Note:* Since the 1970s, dogwoods have been attacked and disfigured by anthracnose fungus (*Discula destructiva*), which may have been introduced on nursery stock from Asia. Purple-edged lesions appear on foliage and may spread into wood, where eventually cankers may girdle and kill tree. Fruit-eating birds that visit dogwoods possibly play a role in spreading the disease. Large die-offs have occurred. For example, in five long-term study plots in central Connecticut, the number of live dogwoods decreased from 661 in 1927 to 603 in 1977, then plummeted to 82 by 1987. Yet dogwoods remain common in certain open habitats such as powerline corridors, where perhaps relatively dry conditions inhibit fungal growth. Flowering dogwoods play an important role in concentrating and retaining calcium in forest ecosystems. Where dogwoods are numerous, a large population decline could lead to leaching of calcium from acidic soils (such as are typical of most of Connecticut). Acidic soils retain calcium poorly. There is evidence that calcium depletion in acidic soils may lead to declines in populations of snails and other calcium-loving biota, which in turn may affect other parts of the ecosystem. The acidic rain that is so prevalent in Connecticut only exacerbates the problem. See "Dogwoods," above.

Gray dogwood *(C. racemosa)* (Fig. 9.65). Common shrub of roadsides, old fields, and sunny wetlands; often forms dense stands connected to a common root system. *Fruits:* white, fat-rich, on bright red stalks; most abundant in September and October, though some may persist through November. *Foliage:* eaten by caterpillars of cecropia moth, white-marked tussock moth, and other Lepidoptera. *Note:* Swellings at the ends of upper branches are galls caused by gall midges. See "Dogwoods," above.

Red osier *(C. sericea).* Shrub of freshwater wetlands (FACW) and their borders. *Fruits:* ripen whitish or bluish by August. *Note:* See "Dogwoods," above.

Tupelos (Nyssaceae)

Black gum (tupelo) *(Nyssa sylvatica).* Moderately shade-tolerant tree in moist soils in coastal forests, along streams, ponds, and swamps, and sometimes in drier sites along field borders and on rocky slopes (FACW); stunted on dry sites. *Flowers:* small, greenish, appear in mid- to late spring as leaves are growing or shortly thereafter; pollen flowers and seed flowers (which often have aborted stamens) generally on different trees; pollinated by bees. *Fruits* (Fig. 9.66): still green in late July, most ripen blue-black by early to mid-September, when leaves are vividly colored; some remain attached through winter; bitter, but eaten by wood ducks, grouse, turkey, woodpeckers, various songbirds, squirrels, foxes, raccoons, bears, and other mammals; seeds over-

Fig. 9.65. Gray dogwood fruits are high in fat, and most disappear from the plants by the end of October.

Fig. 9.66. The leaves of this black gum were beginning to turn red in mid-September just as the fruits ripened.

winter and generally germinate in spring. *Foliage:* deciduous; develops relatively late, may remain leafless through mid-May in coastal forests; bright red, orange, or yellow, sometimes mixed with green, in September, sometimes as early as late July; deer browse young foliage and twigs. *Wood:* medium weight, softness, and strength; warps easily; decays quickly in contact with soil; impossible to split; toughness makes it good for heavy duty uses such as handles, chopping bowls, and factory flooring. *Note:* Despite the name, not a drop of gum can be squeezed out of this plant. Old trees often have hollows used by nesting wood ducks and hooded mergansers, squirrels, raccoons, owls, and other wildlife.

Sweet Pepperbush and White Alders (Clethraceae)

Sweet pepperbush *(Clethra alnifolia)* (Fig. 9.67). Common shrub of acidic wetlands and moist soils (FACW); often grows in large stands connected to a common root system. *Flowers:* white, strongly fragrant; pollinated by bumblebees, carpenter bees, honeybees, bald-faced hornets, yellowjackets, beetles, and other insects in summer, beginning by late July and extending through August and into September; many individual stems do not flower every year; small slender beetles may be abundant on flowers; bees most often visit flowers exposed to full sun; on each flowering spike, lower flowers open first and initially are functionally male, then female part matures, while flowers higher up on spike are still male; cross-pollination between different spikes and often different plants occurs when bees move up-

Fig. 9.67. The white flowering spikes of sweet pepperbush adorn shrub swamps and moist thickets in summer.

ward on each spike, ensuring that pollen moves from spike to spike rather than from male flower to female flower on same spike, as would happen if insect moved downward; however, I frequently saw bees and beetles move downward. *Fruits:* dry, many-seeded capsule; persist through winter and do not attract wildlife in any conspicuous way. *Foliage:* deciduous; yellow or brown in fall. *Note:* This plant's reliance on ample soil moisture is evident when the leaves droop during dry spells in summer.

Wintergreens (Pyrolaceae)

Spotted wintergreen *(Chimaphila maculata)* (Fig. 9.68). Common, broad-leaved, semi-woody plant of well-drained forest floors. *Flowers:* white or pink, sweet-smelling; pollinated by bees and flies, mainly in July. *Fruits:* capsules contain dustlike seeds, do not seem to attract fruit- and seed-eating animals. *Foliage:* evergreen.

Ecology of Trees, Shrubs, and Vines in Connecticut | 167

Fig. 9.68. Spotted wintergreen flowers in summer on forest floors.

Heaths (Ericaceae)

Bog-rosemary *(Andromeda polifolia).* Rare shrub in unshaded areas in a few boggy sites (OBL) in northwestern Connecticut; more common farther north in New England. *Flowers:* white-pink, bisexual; pollinated by bees and small flies in late spring; buds are conspicuous the previous summer. *Fruits:* dry, brown, globular; persist through winter, release seeds to wind. *Foliage:* evergreen; leaves have turned-under edges, white fuzz on underside. *Note:* Toxins (acetlyandromedol) protect these plants (also Labrador-tea and leatherleaf) from most herbivores.

Bearberry *(Arctostaphylos uva-ursi).* Prostrate shrub with a localized distribution in sunny, sandy/gravelly areas and on well-drained rock outcroppings and tops of cliffs (Fig. 9.69). *Flowers:* white; pollinated by insects, or self pollinated, in mid- to late spring. *Fruits:* ripen reddish in July, may persist through fall and into early winter; consumed by wild turkey, ruffed grouse, other birds, and bears; small mammals also likely consume the fruits or seeds. *Foliage:* evergreen; reddish brown in fall; eaten by hoary elfin larvae,

but this butterfly, which flies in April and May, is apparently extirpated from Connecticut; deer browse twigs and foliage.

Leatherleaf *(Chamaedaphne calyculata).* Small shrub of shallow freshwater and boggy wetlands; grows in sunny areas in water, in mats of mosses and sedges, or on stumps or logs in the water (OBL). *Flowers:* white, pollinated by bumblebees and other insects mainly from late April through May (see Fig. 9.76); seed set is much higher when

Fig. 9.69. Two uncommon, low-growing, woody plants: bearberry (upper), here growing among cliff-top rocks of a trap-rock ridge; and trailing arbutus (lower), flowering in early spring along a forest path. Both are caterpillar food for the rare or state-extirpated hoary elfin (butterfly).

flowers cross-pollinated than if self-fertilized. *Fruits:* flattened seed capsules mature in July. *Foliage:* almost evergreen; red in winter, then falls the next year after new leaves have grown; deer sometimes nibble foliage and twig tips. *Note:* See bog-rosemary.

Trailing arbutus *(Epigaea repens)* (Fig. 9.69). Uncommon prostrate shrub; grows locally in clones mostly in dry, acid soils of oak or pine woods and clearings, including forest paths and rural roads. *Flowers:* white, pollinated by bumblebees and other insects beginning in late March or early to mid-April and continuing into May in some areas. *Fruits:* ants eat tissue in opened seed capsules and may disperse seeds. *Foliage:* evergreen; eaten by hoary elfin caterpillars (this butterfly now apparently extirpated from Connecticut). *Note:* Populations in some areas were depleted by historical overcollection by gardeners.

Wintergreen *(Gaultheria procumbens).* Common, low, trailing subshrub typical of well-drained acidic soils of woods and wooded roadsides. *Flowers:* white, pollinated by insects in summer. *Fruits:* red; some last through winter; provide food for ruffed grouse, turkey, and other birds and mammals. *Foliage:* evergreen; turns maroon in fall (November), then becomes green again in spring; deer browse twigs and foliage. *Note:* The fruits and foliage are pleasantly aromatic.

Black huckleberry *(Gaylussacia baccata)* (Fig. 9.70). State's most common huckleberry; shrub of hilltops or other dry woods, clearings, and boggy areas; large, low thickets form through sprouting of spreading underground rootstocks. *Flowers:* greenish, red-tinged; pollinated by insects in late spring or early summer (May to June). *Fruits:* noticeably seedy (compared to blueberries); ripen black from mid-July through August; few if any remain after September; eaten by summer resident animals such as wild turkey, ruffed grouse, thrushes, other birds,

Fig. 9.70. Huckleberry is a typical shrub of dry forests.

Fig. 9.71. Sheep laurel produces pink flowers in late spring in sand plains and wetlands. Paul J. Fusco/Connecticut DEP Wildlife Division.

small mammals, and bears. *Foliage:* deciduous; red in fall; eaten by brown elfin butterfly larvae. *Note:* With lowbush blueberry, huckleberry survives fire and rebounds quickly by sending up new sprouts.

Sheep laurel *(Kalmia angustifolia)* (Fig. 9.71). Small shrub of bogs, swamps, pond edges, rock outcrops, woods, and dry sandy soils; often forms groups of stems through sprouting of underground runners. *Flowers:* pink; pollinated by insects in late spring and early summer. *Fruits:* rounded seed capsules; mature in late summer, persist through winter; seemingly ignored by wildlife. *Foliage:* evergreen; leaves droop in frozen soil. *Note:* Sheep laurel reappears quickly and spreads after fire removes competitors.

Mountain laurel *(K. latifolia)* (Fig. 9.72). Large, common shrub; most often in acid soils of shady, wooded slopes, also cedar swamps; forms large, monotonous clones in upland areas through sprouting of underground parts; sprouts vigorously after fire. *Flowers:* usually white, sometimes pink; pollinated by bumblebees, honeybees, wasps, bee flies, and other insects, mainly in June and early July (peak usually in mid-June); individual flowers last up to 16 days, averaging a little more than a week; honey made from this plant and from rhododendrons is toxic to humans but

Fig. 9.72. Mountain laurel flowered prolifically in 2000 but only sparsely in 2001. The pollen-producing stamens are tucked into pockets in the petals.

not to bees; flowered very prolifically in 2000 and 2003, very sparsely in 2001, numerously in 2002; anthers held under tension at end of stamens in small pockets in petals, spring toward center of flower when flower is jostled, flinging sticky pollen onto body of potential insect pollinators or onto stigma; near end of flower's life, anthers may pop out of petal pockets spontaneously or when flower clusters are moved in wind, but stigma/style is too long for efficient self-pollination; in absence of pollinators, this potential self-pollination mechanism may enable at least minimal fertilization success (but some populations apparently do not self-pollinate). *Fruits:* dry capsules; contain minute, wind-dispersed seeds; some animals reportedly eat fruits and seeds, but wildlife generally ignores these low-quality foods. *Foliage:* evergreen, but old leaves turn yellow before falling after a few years; ruffed grouse sometimes consume leaves and buds; once I watched a walkingstick feed on a leaf; deer may eat foliage and twigs, but these are not preferred; fungus (*Cercospora kalmiae*) causes dark-edged dark to whitish spots, surrounded by a pale halo, to form on leaves. *Wood:* hard, strong, but brittle; moderately heavy. *Note:* Wood thrushes and hooded and worm-eating warblers nest among the foliage.

Bog laurel (*K. polifolia*). Small, pink-flowered, evergreen shrub of bogs and pond edges (OBL) in northern Connecticut.

Labrador-tea (*Ledum groenlandicum*). Shrub of boggy sites (OBL) in northern Connecticut; highly localized but sometimes numerous; not very tolerant of shade. *Flowers:* white, fragrant, bisexual; pollinated by bees and syrphid flies, mid-spring to early summer. *Fruits:* many-seeded capsules persist through winter. *Foliage:* evergreen; turns brown in fall, persists until next year; fragrant when crushed; turned-under edges, rusty-brown fuzz on underside (initially whitish). *Note:* Caterpillars of the rare Labrador-tea tentiform leaf miner (*Phyllonorycter ledella*) feed on the foliage. See bog-rosemary.

Fetter-bush or **swamp leucothoe** (*Leucothoe racemosa*). Uncommon shrub, mainly in swamps (FACW). **Maleberry** (*Lyonia ligustrina*). Shrub of swamps, pond and stream edges, and thickets (FACW). *Flowers:* white, pollinated by bees and other insects in late spring and early summer; buds for next year's flowers conspicuous in late summer. *Fruits:* dry, long-persisting capsules, release dustlike seeds to wind, not very attractive to wildlife. *Foliage:* deciduous, red in fall; sometimes eaten by deer.

Rhodora (*Rhododendron canadense*). Uncommon shrub of swamps, boggy areas, and other moist soils (FACW). *Flowers:* pink-purple, beautifully conspicuous in mid- to late spring (late April to early May) before leaves develop. *Fruits:* many-seeded, long-persisting capsules of this and other rhododendrons do not seem to be attractive to wildlife. *Foliage:* deciduous.

Great laurel (*R. maximum*) (understory in Fig. 6.6). Large shrub of a few widely scattered swampy areas, such as Atlantic white cedar swamps in Pachaug State Forest. *Flowers:* showy, white or pink; open in early summer, after bloom of exotic rhododendrons used locally for landscaping; pollinators of this and other rhododendrons include bees and other insects. *Foliage:* evergreen; new leaves grow in May, live 3 to 7 years. *Note:* See "Plants in Winter" later in this chapter for further information.

Pink azalea (pinxter-flower) (*R. periclymenoides*). Beautiful shrub of upland woods and freshwater aquatic and wetland borders. *Flowers:* usually pink, appear in spring (peak in mid- to late May) as leaves develop; bumblebees make holes at bases of flowers (Fig. 14.68) to get at nectar but in doing so do not pollinate. **Swamp azalea** (*Rhododendron viscosum*). Common shrub of freshwater wetlands and aquatic edges (FACW). *Flowers:* white, late May through July. *Foliage:* deciduous, red in fall.

Lowbush blueberries (*Vaccinium angustifolium*, *V. pallidum*). Shrubs of well-drained, sunny areas of open woods and clearings. *Flowers:* whitish to purplish; pollinated by insects in mid- to late spring, beginning in latest April and early May when leaves are not fully grown; cross-pollination often necessary for high level of fruit set, although some lowbush blueberry clones are not self-sterile. *Fruits:* ripen dark blue in mid-summer, most abundant in sunny sites; some remain into September. *Foliage:* deciduous; eaten by caterpillars of brown elfin butterfly, adults of which are conspicuous on hilltops near these plants in spring, beginning in late April and early May. *Note:* These shrubs tend to recover quickly and increase after fire. Associates are similar to those of highbush blueberry.

Highbush blueberry *(V. corymbosum)*. Common and ecologically versatile, living in swamps, along streams, and in moist or dry woods (FACW); plants can be common, but seed germination and seedling establishment appear to be rare events. *Flowers:* white (may be pink-tinged), yellowish, or greenish (Fig. 9.73); pollinated by insects mainly in mid- to late spring after leaves are well developed; nectaries are deep inside flower; bumblebees and carpenter bees commonly visit flowers and sometimes take nectar without pollinating by jabbing tongue through base of flower from outside, rather than by usual method involving hanging below flower, inserting tongue between pistil and anthers, and getting facial dusting of pollen (and effecting cross-pollination if insect has pollen on it from another flower). *Fruits:* ripen blue-black (often with whitish coating) from mid-June through September (a few remain into October); individual plants vary greatly in fruit production, which fluctuates on a two-year cycle; eaten and seeds dispersed by ruffed grouse, many summer-resident songbirds, chipmunks, white-footed mice, other small mammals, foxes, raccoons, and bears. *Foliage:* deciduous; pink/red in fall; some leaves may remain attached well into fall or early winter; eaten by brown elfin butterfly caterpillars; foliage and twigs commonly browsed by deer. *Note:* Cynipid wasps cause kidney-shaped galls (Fig. 14.114) on twigs. These galls also host small flies whose larvae feed in the wasp-induced growth. Blueberry bud mites feed in and may destroy the flower buds.

Fig. 9.74. Cranberry, a native bog plant, may have both flowers and large green fruits in July.

Large cranberry *(V. macrocarpon)* (Fig. 9.74) and **small cranberry** *(V. oxycoccos)*. Prostrate plants of acidic bogs, pond shores, and other wet areas (OBL). *Flowers:* pinkish; pollinated by mining bees, halictid bees, bumblebees, and other insects in late spring-summer; some flowers may self-pollinate; flowers and large green fruits occur together in July. *Fruits:* ripen red in late summer and early fall; may persist through winter and into early spring; occasionally eaten by various frugivorous birds and small mammals; voles and bog lemmings may open fruits and eat seeds; raised for human consumption in commercially managed "cranberry bogs." *Foliage:* evergreen; eaten in spring and early summer by caterpillars of bog copper butterfly, summer-flying adults of which never go far from cranberry plants. *Note:* Roots of cranberries have the mycorrhizal fungal association typical of members of heath family, enhancing nutrient acquisition.

Ashes, Olives, and Relatives (Oleaceae)

White ash *(Fraxinus americana)*. Moderately shade-tolerant tree; scattered individuals in deep, fertile, moist but well-drained soils; grows slowly in shade but quickly if competitors are thinned; stumps readily send up sprouts. **Black ash** *(F. nigra)*. Uncommon, small, shade-intolerant tree of boggy, swampy, or freshwater-edge areas (FACW). **Green ash** *(F. pennsylvanica)* (Fig. 9.75). Locally common, fast-growing tree of water edges, swamps, and bottomlands (FACW); wide, shallow root system tolerates extended

Fig. 9.73. Highbush blueberry ranges from swamps to dry uplands.

Ecology of Trees, Shrubs, and Vines in Connecticut | 171

Fig. 9.75. Green ash flowers appear in spring before or as the leaves emerge (pollen flowers shown).

flooding during dormant season. *Flowers:* purplish, petal-less, wind-pollinated; present mainly in May before or as leaves grow; pollen flowers and seed flowers on separate plants (except black ash, which usually has bisexual flowers or at least pollen flowers and seed flowers on same plant); some ashes produce mature fruits without fertilization or seed development. *Fruits:* winged (samaras); green ash fruits fully formed (but not necessarily mature) by July; white ash produces large crop every three to five years, green ash almost every year; clusters of samaras may persist on trees through winter; dispersed by wind or water, remain viable for years before germinating, sometimes eaten by various seed-eating birds (e.g., turkeys, purple finches) and rodents (e.g., white-footed mice), relatively unattractive to most animals. *Foliage:* deciduous; usually yellow in fall (sometimes brown, purple, or reddish), as early as late September; fallen foliage decays readily and quickly disappears among tougher leaves of oaks; ashes leaf out late, drop leaves relatively early (among the first trees to become entirely leafless in fall); leaflets of white and green ashes tend to fall independently from each leaf stalk; folklore asserts that white ash foliage is effective defense against being bitten by a rattlesnake (but I wouldn't count on it!). *Wood:* black ash wood is soft, light, rather weak but durable; with pounding, it splits into tough strips, fine for basketry and such; burls yield exquisite veneer; green ash is hard, strong, tough, heavy, elastic, splits easily, and makes good fuel; white ash wood is similar but lighter and not durable in contact with soil, but excellent for fuel, tools, sporting goods, and countless other uses. *Note:* Rabbits, beavers, and porcupines gnaw the bark. Deer and cattle relish white ash seedlings and quickly remove them from heavily grazed forests. Ashes stressed by unfavorable weather suffer increased vulnerability to pathogens, and parts or all of tree may die as a result. Since the 1980s, white ash has suffered from "ash decline" or "ash die-back," an affliction of uncertain cause that kills part or all of the tree.

Madders (Rubiaceae)

Buttonbush *(Cephalanthus occidentalis)* (Fig. 9.76). Common in shallow water of wetlands and pond edges, along margins of slow streams, and in vernal pool basins (OBL). *Flowers:* white; visited in late spring and summer by nectar- or pollen-seeking ruby-throated hummingbirds, bees, beetles, swallowtails and other butterflies, and various other insects; pollen brushes onto pollinators from tip of not-yet-mature pistils, which were anointed with pollen during flower development; later, the pistil matures and receives pollen from other flowers. *Fruits:* spherical clusters begin disintegrating in November, some persist into winter; wood ducks and various other dabbling ducks eat seeds. *Fol-*

Fig. 9.76. Shrubs of the wettest shrub wetlands: buttonbush (upper, fruits) and leatherleaf (lower, flowers).

iage: deciduous; yellowish or green in fall. *Note:* Buttonbush provides nest sites for eastern kingbirds, red-winged blackbirds, and common grackles. Plant-juice-sucking aphids sometimes are tightly packed on stalks just below the flower-heads in summer.

Partridge-berry *(Mitchella repens).* Ground-hugging, nonwoody plant of acidic woodlands. *Flowers:* white, mainly in pairs; appear in June and July. *Fruits:* ripen red in fall, some last through winter and into spring or even early summer; eaten by ruffed grouse ("partridge" in northern New England vernacular), raccoons, foxes, and small mammals. *Foliage:* evergreen.

Honeysuckles (Caprifoliaceae)

Northern bush-honeysuckle *(Diervilla lonicera).* Uncommon shrub of dry forest edges, thickets, and openings such as clearings along powerline corridors. *Flowers:* usually yellow, present in late June and July. *Fruits:* many-seeded capsules. *Foliage:* deciduous.

Japanese honeysuckle* *(Lonicera japonica)* (Fig. 9.77). Common, widespread vine; sprawls over ground and climbs into and shades out and literally strangles shrubs and trees, especially along forest edges and thickets. *Flowers:* fragrant, white-yellowish flowers; pollinated from late spring to early fall by sphinx moths and bees; fruit set requires cross-pollination between genetically different plants. *Fruits:* black in late summer and fall and into winter; eaten by various songbirds, which disperse the seeds of this and other honeysuckles. *Foliage:* deciduous, but many leaves may stay attached and green or red-green into winter; sphinx moth caterpillars eat this and other honeysuckles.

Morrow's honeysuckle* *(Lonicera morrowii)* (Fig. 9.78), and a few additional non-native honeysuckle species. Common, invasive shrubs of various disturbed areas in scattered locations. *Flowers:* white to yellowish flowers, abun-

Fig. 9.77. Japanese honeysuckle is an invasive non-native vine that can kill and stress native vegetation through shading and strangulation. On the positive side, the flowers do smell good.

Fig. 9.78. Morrow's honeysuckle is an increasingly abundant, invasive, non-native shrub.

dant by mid-May. *Fruits:* ripen red (less often yellow in some species) by early summer; catbirds, robins, and other birds frequently eat fruits and disperse seeds, resulting in quick spread of plants. *Foliage:* deciduous. *Note:* Shading by non-native honeysuckles (including also *L. maackii, L. tatarica, L. xylosteum*) can negatively impact native forest tree seedlings and herbaceous plants (see "Non-native Invasive Plants" at the end of this chapter). The problem is accentuated because the leaves of non-native honeysuckles are subject to relatively little herbivory. Connecticut's native honeysuckles are uncommon to rare and much less often encountered than the invasive species.

Common elder *(Sambucus canadensis).* Shrub of sunny or shady swamps, streamsides, wet woods, and roadsides (FACW). *Flowers* (Fig. 9.79): white, bisexual; pollinated by insects in late spring and early summer after leaves develop; produce little nectar, so pollen is the attractant; a midge *(Youngomyia umbellicola)* lays eggs in flower buds and causes gall formation. *Fruits:* ripen purple-black in late summer and early fall; eaten by thrushes and many other birds and some mammals. *Foliage:* deciduous; cecropia moth larvae eat foliage and make overwintering cocoons on twigs. *Note:* Elder borer beetle larvae bore through stems, branches, and roots, and carpenter bees and spider and potter wasps may later use these hollowed stems. Solitary wasps also hollow out stems for use as brood chambers. Blue-black and orange adult elder borer beetles feed on pollen, as do many other beetles. Tree crickets sometimes lay eggs in elder stems. Deer browse the twigs.

Red-berried elder *(S. racemosa).* Shrub of rocky woods and thickets. *Flowers:* yellow-white; pollinated by insects in spring (beginning in late April and early May) as leaves develop. *Fruits:* red (Fig. 20.59), present in early to late summer

Fig. 9.79. Common elder flowers conspicuously along roadsides in late spring.

and early fall; inedible to humans but nourish many birds. *Foliage:* deciduous.

Maple-leaved viburnum *(Viburnum acerifolium)* (Fig. 9.80). Common shrub of deciduous forest understory. *Flowers:* white, bisexual; pollinated by insects mainly in late spring. *Fruits:* purple-black; mature in late summer (beginning by late August); ignored by birds, at least initially; many fruits last through fall and winter; may serve as bird food in late winter or early spring when other foods may be scarce. *Foliage:* deciduous; pink-purple, reddish, or yellowish in fall.

Northern arrowwood *(V. dentatum).* Common shrub in moist soils of woods, clearings, swamps, and freshwater shorelines (FACW). *Flowers:* white, bisexual; pollinated by insects from mid-spring (after mid-May) to early summer. *Fruits:* ripen blue or blackish in August; many gone by end of September, but some plants retain fruits into at least late October; see nannyberry. *Foliage:* deciduous; bronze to red in fall.

Witch-hobble *(V. lantanoides).* Shrub of shady, moist forests (e.g., near ponds) and ravines, mostly in northern Connecticut. *Flowers:* white, bisexual; pollinated by insects, beginning in late April or early May; showy outer flowers sterile, less conspicuous central ones form fruits (Fig. 9.81). *Fruits:* ripen from red to dark purple or black in August and September, may persist through fall. *Foliage:* deciduous; red or orange in fall. *Note:* Deer often browse the buds and twigs.

Nannyberry *(V. lentago).* Shrub or small tree of forest edges, moist thickets, and freshwater borders. *Flowers:* white, bisexual; pollinated by insects in late spring. *Fruits:* large, sweet, blue-black when ripe in late summer (Fig. 9.82); some persist into winter; ruffed grouse, thrushes, cedar waxwings, chipmunks, white-footed mice, foxes, and sometimes other birds and omnivorous mammals eat fruits and disperse seeds of this and other viburnums. *Foliage:* deciduous; purplish-red in fall; browsed by deer.

Wild-raisin, possum haw *(V. nudum,* incl. var. *cassinoides).* Treelike shrub of swampy freshwater wetlands, aquatic edges, thickets, and mesic woods and forest margins (FACW). *Flowers:* white, bisexual; pollinated by insects from mid-spring to early summer. *Fruits:* blue/black, present from late summer into winter; see nannyberry. *Foliage:* deciduous; red in fall. *Note:* Similar **black-haw** *(V. prunifolium)* is uncommon.

Fig. 9.80. Maple-leaved viburnum is a common forest shrub with long-lasting fruits.

Fig. 9.81. The showy outer flowers of witch-hobble are sterile.

Fig. 9.82. Nannyberry fruits (mostly green here) ripen in September.

Composites (Asteraceae)

Groundsel-tree *(Baccharis halimifolia)* (Fig. 9.83). Shrub, bordering salt marshes and tidal streams, often just above marsh-elders. *Flowers:* white-yellow, appear in late summer; pollen flowers and seed flowers on different plants, the latter with numerous white pappus tufts, very conspicuous in early fall. *Fruits:* tuft-bearing seeds disperse in fall. *Foliage:* deciduous; much remains attached through fall.

Marsh-elder *(Iva frutescens)* (Fig. 3.25). Common shrub-like plant found at high-tide line of salt marshes (FACW). *Flowers:* greenish-white; pollen flowers and seed flowers in same head; present mainly in August and September. *Fruit:* small, dry, hard. *Foliage:* deciduous; some plants retain attached leaves well into fall. *Note:* See the salt marsh section of Chapter 3 for further ecological information.

Fig. 9.83. Groundsel-tree becomes conspicuous when flowering and fruiting in late summer and early fall.

Ecology of Early Spring Wildflowers

One of the joys of the end of winter and the beginning of spring is the emergence and flowering of woodland wildflowers. Established hiking trails on trap-rock ridges offer the best opportunities to see a wide variety of Connecticut's spring flora.

Production of flowers early in the season requires that the plants store resources during the previous summer(s), so nearly all of these early spring wildflowers are perennials, though some are biennials (for example, common fleabane, yellow corydalis). Plants that must grow from seed each year do not have enough time to germinate, grow, and flower before tree foliage develops in May.

As illustrated in some of the examples that follow, early flowering can be advantageous because shading makes forest-floor flowers less attractive to pollinators, and shade often reduces the plant's ability to produce mature fruits. Development of the leaf canopy can reduce the amount of solar radiation that reaches the forest floor from 50 to 80 percent in early spring to less than 5 percent by late spring. Some spring wildflowers grow in forest edges and openings where shading from woody plants is minimized. These plants in sunnier locations produce more flowers and mature fruits than do plants in less sunny sites.

Most early spring wildflowers depend on insects for pollination, but some rely on cross pollination by the wind (early meadow-rue) and others can self-pollinate (bloodroot, violets, wild ginger, hepaticas) or set seed without the fertilization of the ovules (trilliums, common fleabane). Most of the pollinators of early spring flowers are solitary bees of various types (mining bees, anthophorids, halictids, megachilids), social bees (Apidae: bumblebees, honeybees), bee flies (Bombyliidae), and beelike flies (Syrphidae). Relatively few of the early pollinators are butterflies.

Many spring wildflowers, such as hepatica, bloodroot, trillium, violet, wild ginger, Dutchman's breeches, and trailing arbutus, have ant-dispersed seeds. Ants are not attracted to the seeds per se, but they get involved because of a lipid-rich food body (elaiosome) that is attached to each seed. For example, ants attracted to bloodroot elaiosomes carry the seeds to their nest. After removing and eating the food body, the ants discard the seeds, which end up some distance from the parent plant, perhaps in a site suitable for the establishment of a new plant. There is evidence that ant-manipulated violet seeds germinate better and produce more vigorous plants. Similarly, yellowjacket wasps have been observed taking trillium seeds and removing the elaiosomes. The relationship between ants and plants appears to be mutually beneficial. Ants nourished by food bodies attached to the seeds of spring wildflowers such as bloodroot exhibit increased production of offspring.

Summer- and fall-flowering herbaceous plants, and woody plants, generally have vertebrate- or wind-dispersed seeds. These plants readily recolonize sites recovering from disturbance, whereas ant- and gravity-dispersed plants such as early spring wildflowers have limited dispersal capabilities and recolonize such sites slowly or not at all, even after several decades. Perhaps this explains in part why various early spring wildflowers are curiously absent from some seemingly suitable locations in Connecticut (where nearly all forests have been disturbed to different degrees). Active reintroduction may be necessary to restore these plants to previously disturbed forests.

In contrast to herbaceous wildflowers, many woody plants that release pollen in early spring are wind pollinated. These include pines and other conifers, cottonwoods, aspens, birches, maples (also insect pollinated), oaks, beeches, hickories, hazelnuts, hornbeams, alders, ashes, and elms. The pollen cones or flowers of most of these, often small and high in the trees, may escape notice, but their pollen can be conspicuous on cars parked outside.

Following are examples illustrating the ecology of woodland wildflowers that bloom in early spring.

Skunk cabbage *(Symplocarpus foetidus)* (Fig. 9.84) is a familiar perennial of freshwater wetlands. Its tiny flowers, clustered in a thumb-shaped spadix and mostly enclosed in a conical, tentlike spathe, appear in February or March, long before other spring flowers. The spathe has a slitlike opening on the side that admits flies and other insect pollinators.

Flies are attracted to the plant by its somewhat putrid odor. Insects also find the plant attractive due to its warmth—skunk cabbage is indeed one of the few plants known to warm itself through endogenous heat production, powered by exceptionally high levels of metabolism and reflected in an elevated rate of oxygen consumption. The heat seems to enable the plant to mature its gametes more quickly than ambient temperatures would allow, giving skunk cabbage a "jump" on competing plants. Red leaf pigments (anthocyanins), which presumably absorb more solar radiation than do green areas, perhaps also help warm the leaves on sunny days in spring when temperatures are still cold. After flowering, skunk cabbage leaves burst forth and provide most of the first new splash of brilliant green ground foliage in swampy wetlands in April.

Bloodroot *(Sanguinaria canadensis)* (Fig. 9.85) is a long-lived perennial that sometimes occurs in groups (clones) formed from vegetative reproduction. Look for it on traprock ridges and slopes, where it often persists in or colonizes disturbed areas. Its white flowers open in April and early May, and the flowers appear earlier in warmer years. Usually

Fig. 9.84. Skunk cabbage is remarkable not only for its odor but also in generating its own heat and flowering in winter.

Fig. 9.85. Bloodroot flowers make a brief appearance in spring.

there is only one flower per plant. Each flower opens for up to twelve days and closes each night.

Bloodroot flowers produce much pollen but no nectar. Within three to five days after the flower opens, the anthers may bend and touch the stigma, which becomes dusted with pollen. The few days before this self-pollination allow for cross-pollination (outcrossing), if any insects are available to do it. One study reported that the flowers last only a few days and that anther-stigma contact occurred only a few hours after the flower opens, but my observations indicate that at least some bloodroot flowers in Connecticut do not conform to this abbreviated pattern.

Potential bloodroot pollinators include honeybees*, syrphid flies, mining bees, and halictid bees. I saw bee flies visit bloodroot flowers, and they also probably pollinate. Air temperature generally must be at least 55°F for significant numbers of pollinators to be active. Syrphid flies may be active at slightly lower temperatures.

In early spring when bloodroot flowers, there is a high degree of unpredictability in the availability of pollinators because temperatures are often too cold for insect flight. Field experiments show that bloodroot can form fruits through self-pollination if insect pollinators are not present. In these experiments, some flowers were outcrossed by the experimenter (the pollen-bearing anthers were removed before they opened to prevent self-pollination). To prevent outcrossing, other flowers were enclosed in bags before the flowers opened. Both conditions resulted in successful seed set. If cold weather prevents insect activity during the short period when the bloodroot flower is open, the "back-up system" of self-pollination ensures that most flowers still will produce seed-containing fruits. Among other spring wildflowers, hepatica also has self-pollinating flowers that produce no nectar.

Bloodroot fruits open in late spring. Seed capsules open, the seeds fall out, and ants carry them away for processing (see introductory discussion).

Two closely related plants each known as **spring beauty** (*Claytonia caroliniana, C. virginica*) are early spring perennial wildflowers occurring in certain rich woods and swamps (Fig. 9.86). The pink or white flowers of spring beauty occur in groups (inflorescences) in April or early May. Each inflorescence has open flowers for an average of about three weeks. A particular flower is staminate (male) for one day, then pistillate (female) for the next one to eight days.

Spring beauty exhibits a reproductive strategy that is quite different from that of bloodroot. Unlike bloodroot, spring beauty flowers secrete nectar and require a pollen vector (only 1 of 29 bagged, unmanipulated flowers set fruit). In field experiments, artificial self-pollination usually was successful within an inflorescence but usually not

Fig. 9.86. A bee fly just departed from these spring beauty flowers in mid-April.

within individual flowers. Artificially outcrossed flowers produced fruit in 26 of 29 crosses. Successful production of fruits and seeds thus depends largely on the movement of pollen from one flower to another.

Pollination is accomplished primarily by mining bees, which during this time may collect pollen only from spring beauty, and sometimes by bee flies or rarely bumblebees. Syrphid flies also visit the flowers but have poor pollen-carrying abilities.

Spring beauty flowers open only when the temperature exceeds about 52°F, and they close when shaded in late afternoon, at night, and on dark cloudy days. Closing up when pollinators are unlikely to be active conserves pollen by protecting it from being washed away by rain. Wood anemone provides another example of a spring wildflower whose flowers close up under dark or shaded conditions.

As with bloodroot and many other spring wildflowers, the flowering time of spring beauty varies among years and is related to temperature, with earlier flowering occurring in warmer years. In contrast, leaf canopy development in the woodland habitat of this species exhibits relatively little variation. Detailed studies reveal that spring beauty flowers that appear early and late in the flowering period exhibit reduced seed production. Early flowers have poor seed set because cool temperatures limit pollinator activity. Late flowers do poorly because development of the forest canopy and consequent heavy shading reduces the wildflower's photosynthetic capability and ability to produce mature fruits. The flowering period of spring beauty seems to represent a balance between flowering late enough to ensure pollinators and early enough to produce fruit before canopy closure.

However, early and late flowering still have some potential benefit. Because of the relatively small number of open flowers during the early and late stages of the flowering period, pollinators have to travel farther to find plants with flowers. In doing so they may carry a plant's pollen (and genetic material) substantial distances. One advantage of farther dispersal of the genes is that it may reduce the probability of detrimental inbreeding.

When ripe, the fruit capsules of spring beauty rupture explosively and fling the seeds up to two feet away. The plant withers quickly and, by late spring, only the underground corms are still intact. These may be dug up and eaten by small mammals, making the successful production of seed all the more important.

Trout lily (*Erythronium americanum*) grows in moist deciduous forests and thickets, along streams, and also on dry ridge tops (Fig. 9.87). In spring, the above-ground parts of many plants consist of just a stem and one mottled leaf. Trout lily clones, produced by budding of new corms from exisiting ones, may cover large areas. Each clone tends to have a fairly distinctive mottling pattern on the leaves, so in early spring you may be able to detect which clumps represent different clonal groups.

After spring, when the tree canopy blocks most sunlight, trout lilies stop photosynthesis, leaves grow old, and the plants return to their dormant underground existence. The roots wither by late spring or early summer. New roots grow in late summer and are quickly colonized by mycorrhizal fungi, some of which may be connected to other plants, enabling transfer of nutrients.

In late April and early May, trout lilies that have two basal leaves produce a single yellow flower. Each flower lasts several days and closes at night. Nectar is produced at the base of the "petals" (actually tepals), and the anthers yield pollen.

Bees of several families and blowflies visit the flowers and collect pollen and/or nectar. Some bees are efficient pollen gatherers. A single visit by one bee may result in the removal of 25 to 90 percent of the available pollen. Clearly just a few visits may totally deplete a flower's pollen supply. Small beetles also climb on the flower parts and appear to eat pollen; perhaps they pollinate as well.

Field experiments demonstrate that fruit set and seed production are best if trout lily flowers are cross-pollinated, but self-pollinated flowers can also set fruit and produce viable seeds. Seed set is most successful if the cross

Fig. 9.87. Trout lily flowers open in April and May.

pollination occurs between plants that are relatively far from each other (for example, 130 feet apart versus 30 feet apart). This suggests that inbreeding may impose a limitation on reproduction. Because trout lily exhibits prolific vegetative reproduction, plants that are close to one another are likely to be genetically identical and thus probably not good sex partners.

Field studies also indicate that reproduction may be limited by a shortage of pollinators or at least ineffective pollination by insects. This is suggested by the fact that hand-pollinated flowers produce seed from a significantly higher percentage of their ovules than do naturally pollinated flowers. Trout lily is thus a wildflower that benefits from, but is not totally dependent on, healthy populations of bees.

In addition, it appears that it takes a trout lily plant a number of years to accumulate enough resources to attain its maximum reproductive output; percentage seed set and seed mass increase with increasing plant size.

Autumn Wildflowers

When summer ends and temperatures are steadily dropping, a large number of hardy plants are still flowering, and you may find many of these flowers in good condition well into October, even after the season's first frost. By November, the flowers of all but a few native plants are gone. Some non-natives, out of synch with the local environment, may flower into early winter. For example, I found common dandelions* (*Taraxacum officinale*) with fresh flowers in December and January. Some of the autumn-flowering species have an extended flowering season that begins in early summer whereas others do not even begin to bloom until August or September.

Like plants that flower in early spring, autumn wildflowers sometimes experience conditions in which insect pollinators are inactive due to cold temperatures. However, mild and often sunny fall weather and still-numerous insect populations make pollinator activity in autumn much more reliable than in spring. And many fall wildflowers inhabit open sunny areas that favor insect pollinator activity when the air is cool. Plants often found in cool air in shade, such as white wood aster, may be pollinated by "semi-warm-blooded" insects like bumblebees that readily fly in such habitats (see Chapter 14). Ragweeds and some other autumn-flowering plants are wind-pollinated and do not rely on insect pollinators at all.

The undisputed champion among late-flowering woody plants is **witch-hazel,** which blooms from late September or October into November (Fig. 9.88). It is all the more impressive because it is the only woody plant in which fall is the main flowering season, and even the very few woody species that flower in late summer seldom have any flowers left after September. In general, autumn flowering is not a viable option among woody plants, most of which are unable to continue fruit and seed maturation beyond the season in which pollination occurs. Witch-hazel fruits develop and mature the year after flowering. All other plants in which fruits mature in the second year, such as red oak and white pine, flower in spring and have a full growing season before winter's cold arrests fruit development.

Among native herbaceous plants, the grasses, sedges, goldenrods (*Solidago* spp.), and asters (*Aster* spp.) are emblematic of fall flora. These natives are joined by many exotic species that flourish in fields, along roadsides, and in other open disturbed areas.

Goldenrods are among the most conspicuous and well known of all autumn wildflowers, but they are sometimes misunderstood (Fig. 9.89). Contrary to popular belief,

Fig. 9.88. Witch-hazel blooms in October and November.

Fig. 9.89. Goldenrods are quintessential autumn wildflowers.

these beautiful native wildflowers do not contribute significantly to airborne pollen and thus do not normally afflict those who suffer from allergies. The heavy, sticky pollen normally requires bees, wasps, and other insects to move it from one plant to another (goldenrods generally are self-incompatible). Goldenrod fruits are small, dry, and hard, with an attached mass of fuzz that aids in wind dispersal.

Despite the tiny amount of sugar in each goldenrod floret, masses of goldenrod flowers are attractive sources of nectar and pollen for many beneficial insects in late summer and early fall. The foliage also supports a diverse insect fauna. In upstate New York, one researcher found that at least 138 species of insects fed on the foliage of one species of goldenrod (*Solidago altissima*). And there are more than 40 insect species with immature stages that specialize on goldenrods. Adult and larval goldenrod beetles (Chrysomelidae: *Trirhabda* and *Microrhopala* spp.) are often dominant herbivores on these plants. See "Galls and Gall Makers" in Chapter 14 for information on goldenrod galls.

Some goldenrods form extensive, long-lived clones that develop from new stems arising from a spreading rhizome system. Forming a large clone may be beneficial because large stands of flowering goldenrods attract more insect visitation than do isolated plants. Why? Insects likely exhibit

Autumn Wildflowers | 179

this behavior because they get more return for their effort when the flying distance between flowers is small.

With their conspicuous purple, pink, blue, and white outer rays, and yellow, red, or purple centers, **aster** flowers beautifully complement the tones of their goldenrod cohorts in the late summer and early fall flora (Fig. 9.90). Asters, represented by about 30 species in Connecticut, are perennials (rarely annual). Like goldenrods, they have some specialized associates. Flower moth caterpillars (Noctuidae: *Schinia* spp.) eat the flowers and seed capsules. I found the attractive little adults of the arcigera flower moth in mid-September. Caterpillars of the pearl crescent (spiny, black, with yellow dots and side-bands) often eat New England aster *(Aster novae-angliae)* foliage, and Harris' checkerspot larvae (spiny, orange, with black bands) depend on the foliage of flat-topped white aster *(Aster umbellatus)*. You may find feeding groups of either of these caterpillars. Adult viceroys, pearl crescents, and other butterflies visit aster flowers for nectar. Several kinds of gall midges induce galls on almost any part of the plant. Sparrows, mice, and chipmunks sometimes eat the seeds.

Plants in Winter

An early freeze in fall or a late freeze in spring can kill the leaves and flowers of even the hardiest of trees and shrubs. Yet these same plants easily survive extreme freezing temperatures in winter. For example, species such as red oak, American beech, black cherry, and sugar maple survive temperatures as low as −40 to −45°F, and basswood, quaking aspen, and paper birch survive even lower extreme temperatures.

Attainment of extreme cold tolerance requires a period of "hardening" by cold exposure in fall. Supercooling—a process in which the temperature of the plant falls below the freezing point without ice formation—is a common means of avoiding damage. For example, water in most forest trees does not freeze until temperatures drop as low as about −40°F.

Some plants of extremely cold climates (e.g., aspens, birches, willows, and alders) do not rely on supercooling but rather survive partial freezing (i.e., formation of ice in extracellular spaces). As is usual in living organisms, intracellular ice formation is lethal. Apparently, under the influence of abscissic acid, water readily moves out of the cells as ice formation proceeds extracellularly at temperatures of 28 to 23°F. The higher concentration of soluble carbohydrates within the cells inhibits intracellular freezing even as ice forms outside the cells. Ice crystals may push against the cell membranes without damaging the cells. At some point the rapid efflux of water from cells that occurs during extracellular freezing is reduced due to changes in cell wall and cell membrane properties and other mechanisms. Otherwise, cells would die from excessive dehydration.

Some herbaceous plants also undergo frost hardening when exposed to low, nonfreezing temperatures. Specific proteins accumulate and somehow confer greater frost tolerance.

Connecticut's native prickly-pear cactus *(Opuntia humifusa)* is a winter-hardy member of a family well known for its occurrence in warm climates. How does this cactus deal with freezing winter temperatures? It does this partly by avoidance; it lives only along the coastal zone where winters are relatively mild (but still subject to freezing). The cacti also cope by tolerating dehydration. As is clear from its wilted winter appearance, the cacti lose a great deal of water, even in moist soils. Apparently, water uptake effectively ceases at low root temperatures. Additionally, the cacti tolerate extracellular ice formation. This is facilitated by the accumulation of extracellular mucopolysaccharides during exposure to low temperatures. These hygroscopic molecules serve as a source of water for ice formation during extracellular freezing, protecting the cells from damage from excessive dehydration. They also facilitate ice nucleation and may inhibit ice crystal growth.

There is some evidence that at very low temperatures, intracellular fluids in plants may solidify (vitrify) without the formation of damaging ice crystals. This also would prevent lethal intracellular dessication that might occur if water kept moving out of the cell and onto growing extracellular ice.

Some plants, such as Connecticut's native rhododendron (great laurel), exhibit overt behavioral responses to freezing temperatures. Great laurel leaves droop (change leaf angle) and curl when exposed to freezing cold. (Leaves freeze when their temperature falls below 20°F.) Leaf droop, caused by a

Fig. 9.90. New England aster is one of the most stunning autumn-flowering asters.

drop in turgor pressure in the leaf stalk, protects leaf membranes from the damage that can occur when frozen or cold leaves are exposed to strong sunlight. Leaf curl evidently is related to poor mechanisms for turgor maintenance when water availability is low, not by effects of ice formation in the leaves; curling occurs at temperatures well above those required for formation of ice in the leaf tissues. Leaf curl helps protect against problems caused by high levels of light and cold temperature and may influence the rate of leaf rewarming after freezing. Thus it may also help protect against damage that can result from the frequent rapid thawing that may occur under fluctuating weather conditions. More so than biochemical adjustments, leaf movements are critical aspects of great laurel's cold tolerance; however, it turns out that this means of cold tolerance makes the plants vulnerable to water stress.

Dormant buds that will give rise to flowers and leaves the following year are conspicuous on perennial plants in winter. These buds were fully formed in mid- to late summer. Bud dormancy and cessation of stem growth are responses to shortened days and low temperatures. The end of dormancy generally is stimulated by long day length, perceived by the buds. Some plants require a period of cold in order to respond to the long-day stimulus.

Parasitic Plants

Most plants have chlorophyll and use light energy from the Sun and carbon dioxide from the air to make the organic compounds they require. But some plants (about 3,000 species worldwide) have evolved mechanisms for "stealing" organic nutrients from other plants. Others establish intimate connections with and use nutrients supplied by mycorrhizal fungi that are associated with the roots of other plants. Generally these plants do not conduct photosynthesis and, accordingly, their leaves are mere vestiges, and many of the plants are not green (about 400 species of vascular plants lack chlorophyll). Following are a few examples.

Beech-drops *(Epifagus virginiana)* (Fig. 9.91) is a small, slender, chlorophyll-lacking plant that is a benign parasite on the roots of American beech. It flowers, often in deep shade, under the beech canopy in late summer and early fall (September to October). The lower flowers stay closed and produce seeds by self-fertilization or asexually, whereas the showier upper flowers, which may be visited by bees, nevertheless are reported to be sterile (however, my observations indicate that both upper and lower flowers can produce seeds of unknown viability). In the cool, deep shade under beech trees, insect activity is minimal at best, so self-fertilization may be a good tactic. The fruits that develop split open across the top and release minute seeds, begin-

Fig. 9.91. Beech-drops is a benign parasite on the roots of American beech.

ning in October. This small plant is most conspicuous in winter against a backdrop of snow-covered ground.

Beech-drops is a member of the broomrape family (Orobanchaceae), which also includes squawroot *(Conopholus americana)*, a perennial associated with rich deciduous forests, and one-flowered cancer-root *(Orobanche uniflora)*, an annual plant of damp woods and thickets. Both are spring-flowering parasites that obtain nutrients from the roots of other plants. Squawroot often parasitizes oaks, and cancer-root attacks goldenrods and various other plants.

Dodder *(Cuscuta* spp.) (Fig. 9.92) is a parasitic annual vine with white to yellowish flowers and tiny scalelike leaves. Seeds germinate (sometimes after years of dormancy) and take root in the soil in damp areas. The seedlings grow toward light and in opposition to gravity on their own roots for up to several weeks, sometimes extending more than 12 inches. Upon contacting any of a wide variety of host plants, the dodder stem tightly wraps around it (coiling counterclockwise). At the points of contact, dodder taps the vascular system of the host through modified adventitious roots (haustoria), and eventually obtains all of its water and

Fig. 9.92. Dodder starts as a rooted plant, but later its roots wither after it attaches to and becomes a complete parasite on other plants.

nutrients from the host as its own roots wither. The host be killed or stunted to varying degrees.

Dodder flowers in summer or early fall; possible pollinators include hymenopterans, but are poorly known. Flowers yield abundant four-seeded fruit capsules. The primary mechanism of seed dispersal is undocumented. Dodder reportedly may contain some chlorophyll in the buds, fruits, and stems, but the amount of food manufactured in these tissues is inconsequential. I found thick growths of dodder with abundant fruits and flowers from August through October.

Dwarf mistletoe (*Arcuethobium pusillum*) in Connecticut is a rare, long-lived, slow-to-mature parasite on black spruce branches. The tiny plants, which contain chlorophyll, are less than 1 inch long and have scalelike leaves. Minute flowers appear in spring, with pollen flowers and seed flowers on separate plants. Insects, and perhaps the wind, carry the pollen. Cedar waxwings and thrushes may eat the fruits and presumably disperse the seeds in their droppings. Uneaten fruits open explosively to eject the seed up to several yards. The seed coat is sticky and adheres easily to branches, where germination may occur in the absence of water. Seeds may stick to other things as well, such as bird feathers, and a bird can spread the seeds to suitable germination sites when it wipes its bill on a branch after preening. Seedlings soon penetrate the host plant, tap into its vascular system, and start extracting nutrients, water, and minerals. Mistletoe can reduce host growth and cause abnormal growth patterns.

Indian pipe (*Monotropa uniflora*) (Fig. 9.93) is a very common, white, chlorophyll-lacking plant that parasitizes soil-dwelling mycorrhizal fungi associated with plant roots. A close relative, **pinesap** (*Monotropa hypopithys*), with pink multi-flowered stems, has the same parasitic habit but is much less common. Both Indian pipe and pinesap require the fungus as an intermediary in obtaining plant-produced sugars. Because these plants do not carry out photosynthesis, they can and often do grow in heavily shaded locations devoid of herbaceous plants that rely on photosynthesis. Both grow singly or in sparse to dense clusters (Indian pipe clusters have forty or more flowering stalks). I found freshly emerged Indian pipes from late June through late October. The flowers are open and fresh for about two weeks. Bumblebees visit and evidently pollinate the flowers. I found scattered clusters of pinesap stems under dense mountain laurel, oak, and hemlock trees in late September and October.

Usually Indian pipe emerges from under leaf litter in sites apart from the old, dry, blackened stalks of earlier growth. The tiny seeds reportedly are dispersed by the wind, but that does not seem to account for the dense clusters of plants in sites lacking older stalks, unless the seeds disperse in clumps. Perhaps the dense clusters of Indian pipe repre-

Fig. 9.93. Indian pipe depends on mycorrhizal fungi to supply its nutritional needs.

sent situations in which the seeds of previous plants simply dropped straight down to the ground and the old stalks decomposed prior to germination.

Wood betony *(Pedicularis canadensis)* is a spring-flowering hemiparasite that gets part of its sustenance from the roots of a great variety of other plants. It contains chloroplyll, so it can also produce its own food, but it always establishes a parasitic connection. It connects to its host as a seedling via tiny rootlets.

Understanding Orchids

Among the multitude of plant groups, orchids stand out not only as exceptionally beautiful but also ecologically interesting. Often we associate orchids with tropical environments, but many orchids live only in temperate latitudes, ranging north to Alaska, Greenland, and Scandinavia and south to Tierra del Fuego and New Zealand. Epiphytic orchids (those growing on other plants) are mainly tropical or subtropical in distribution.

The number of orchid species worldwide probably is 20,000 to 25,000 but further study may alter this estimate. Orchids are most diverse in equatorial regions, with over 8,000 species in tropical America and only about 150 in North America. Some 40 species are native to Connecticut. Orchids and composites (aster family) are the world's most species-rich plant families.

Orchids often get their nutrients through mycorrhizal fungi associated with secondary roots that arise from the orchid stem. Coral-roots *(Corallorhiza* spp.) provide an example of this relationship, which is important for good growth, though not necessarily required, especially in mature plants. Most orchids in nature cannot germinate without the fungus. Some of the orchid-associated fungi are parasitic on roots of living trees, and the fungi may serve as a vector for nutrient transport from the tree to the orchid. However, orchids are not always immediately dependent on other plants. Some orchids have tuberlike roots that store nutrients.

We may think of orchids as delicate, but actually many are tough plants that tolerate environmental stresses. Some orchids require pristine conditions whereas others, such as autumn coral-root *(Corallorhiza odontorhiza)*, lily-leaved twayblade *(Liparis liliifolia)*, and ladies'-tresses *(Spiranthes* spp.), thrive in disturbed areas and are scarce in undisturbed locations.

Orchid flowers often are intricately structured for pollination by specific insects. Pollination of orchids in Connecticut usually involves an insect, especially bumblebees and sometimes other bees. Examples of bee-pollinated orchids include lady's-slippers *(Cypripedium)*, rattlesnake-plantains *(Goodyera)*, ladies'-tresses *(Spiranthes)*, and whorled pogonias *(Isotria* spp.). Mosquitoes, syrphid flies, bee flies, tachinid flies, and butterflies may pollinate yellow-fringed orchids *(Platanthera ciliaris),* and moths pollinate white-fringed orchids *(Platanthera blephariglottis)*. Certain orchids, particularly saprophytes, self-pollinate without help. Some orchids, such as nodding ladies'-tresses *(Spiranthes cernua)*, can produce seeds without fertilization by pollen but rarely are limited to this method.

Generally, suitable pollinating insects that visit orchids end up having a mass of pollen stuck to them, and this mass detaches onto the stigma of the next visited flower. In orchids with a chamber in the flower, the pollen mass does not attach or detach from the insect until it exits the flower. Part of the stigma secretes a sticky fluid that functions in attaching pollen masses to visiting insects.

Flowers often reward pollinators with pollen and/or nectar. Orchid pollen, clumped as it is in bulky masses (pollinia), is unavailable to bees, but these insects are attracted to flowers' nectaries, which may occur on a spur on the liplike petal or elsewhere. Certain tropical orchids (none in Connecticut) have oil glands or scent glands that attract bees, and the flowers of some tropical orchids resemble female insects to such a degree that male insects attempt to copulate with the flowers!

Some orchids, including dragon's-mouth *(Arethusa)*, grass-pink *(Calopogon)* (Fig. 6.13), and rose pogonia *(Pogonia ophioglossoides)*, trick inexperienced bees by presenting yellow hairs (pseudopollen) that resemble pollen-bearing anthers but offer no reward. In grass pink, the pseudopollen-bearing lip is hinged and flips down when the bee lands on it, bumping the bee against the pollinia.

Another form of deception involves lady's-slipper orchids (Fig. 9.94). These attract bees but appear to lack any reward for them (no pollen or nectar). The bees push between the lips of the pouchlike petal but cannot easily exit this way. Instead they exit through an opening to the side and in doing so first contact the female part (stigma) and then pick up pollinia from the anther. The bees end up depositing the pollinia on the stigma of the next visited flower. Apparently the bees repeat this unrewarding activity multiple times before learning that it is wasted effort.

Individual orchid plants often do not produce flowers and fruits every year. Presumably nonflowering years in part reflect a shortage of resources available to the plant.

In contrast to the elaborate insect-dependent pollination mechanisms of orchids, seed dispersal appears to be a simple wind-driven event. The seeds are dustlike and easily become airborne.

The ecological relationships discussed in this section can serve as a model as you learn about the ecology of other Connecticut plants.

Fig. 9.94. Lady's-slippers attract bumblebee pollinators despite a lack of nectar or accessible pollen.

Conservation. Some orchids in Connecticut are relatively common, but many are naturally rare or have been depleted or extirpated by habitat destruction or by collectors and people who thoughtlessly pull them up in admiration gone wrong. The best way to appreciate orchids is observe living plants carefully on hands and knees. Record your obsevations in a notebook and with drawings or photographs, and come back year after year to see how things have changed. If the landowner is likely to be sympathetic, you might suggest cooperation with a professional conservation organization to ensure compatible management (simple protection may not be enough for long-term conservation).

Non-native Invasive Plants

One of the most important conservation issues facing Connecticut today is the invasion and proliferation of non-native plant species in natural areas (Fig. 9.95). Virtually every habitat, ranging from aquatic to wetland to upland, is being affected negatively by invasive plants. Invasive plants are detrimental in many ways. For example, they compete with and can displace native plants; they can degrade conditions for animals that depend on native plants; and they can interfere with natural processes that shape natural ecosystems.

One of the keys to the success of many invasive plants is early development of the leaves in spring. Shading by invasives can prevent flowering and normal growth of native plants, especially those adapted for blooming in the sunny

Fig. 9.95. This green ash tree along the Connecticut River is being overgrown by Asian bittersweet, which climbed an adjacent tree.

conditions that prevail before native trees and shrubs have leafed out. The invasive species also may draw pollinators and seed dispersal agents (e.g., birds) away from native plants. Some impacts are subtle. Garlic mustard *(Alliara petiolata)*, for example, may affect soils such that the competitive abilities of native plants are impaired through inhibition of mycorrhizal associations and root growth. Chapters 3 through 7 contain further discussion of invasive plants and their impacts.

For more information on this issue, visit the website of the Connecticut Invasive Plant Working Group (http://www.eeb.uconn.edu/cipwg/). For information on the biology and management of specific plants, visit The Nature Conservancy's science and stewardship website: http://nature.org/wherewework/northamerica/states/connecticut/

science. This site also describes a recent federally funded initiative aimed at removing harmful invasive species from 9,000 acres of forests of the Berkshire Taconic landscape in Connecticut, Massachusetts, and New York. The website http://www.eeb.uconn.edu/research/invasives/art_pubs.html also has information on methods for controlling invasive exotic plants.

In related activity, in early February of 1999, President Clinton signed an Invasive Species Executive Order to prevent the introduction of invasive species (including both plants and animals), provide for their control, and minimize their economic, ecological, and human health impacts. This directive outlines federal agency duties, creates a new Invasive Species Council and defines its duties, and directs creation of an Invasive Species Management Plan. For more information on this and other federal policy, see the following website: http://refuges.fws.gov.FICMNEWFiles/FICMNEWHomePage.html.

Chapter 10

Sponges through Ectoprocts

The Earth's visible biological diversity consists primarily of animals. I qualify this statement with the word "visible" because the diversity of microorganisms has not been fully explored, and they may outnumber animals in number of species. Some people are surprised that there are more species of animals than plants. In fact, worldwide there are more kinds of beetles than there are plants. The next several chapters review the impressive diversity of animals that inhabit Connecticut. Not all groups (phyla) are mentioned, but nearly all of the readily observable ones are.

Sponges

Sponges (Phylum Porifera) are simple animals that have been around for probably at least a billion years. To call sponges "animals" is a bit of an oversimplification because the evolutionary relationship between sponges and other animals is far from clear. Likewise debatable is whether a sponge is a single organism or a colony of separate but integrated individuals.

Sponges inhabit all oceans and occur in freshwater habitats of every continent except Antarctica. Worldwide, about 6,000 sponge species occur in salt water, and 150 or so inhabit fresh water (not more than about 30 species in North America). All sponges are sessile (attached) in the mature form.

In Connecticut, about 20 species of sponges occur as crusts or branched or fingerlike growths attached to rocks, shells, plants, or other objects in subtidal areas, tidepools, or the lower intertidal zone of Long Island Sound and river-mouth estuaries. In eastern Connecticut, paddlers may notice branching, fingerlike or irregular masses of the often yellowish or gold Bowerbank's halichondria *(Halichondria bowerbanki)* on the sides of tidal channels. Beach walkers may find washed-up specimens of the bright red or orange red beard sponge *(Microciona prolifera)* (Fig. 10.1). This sponge forms crusts in shallow or turbulent water but in deeper locations grows as upright branches often with squared-off tips. When dry, the bright colors fade and it turns brown.

Sponges generally do not alter the attachment substrate. However, estuarine boring sponges (*Cliona* spp.), which are not uninteresting but rather are named for their habit of making holes and tunnels in the shells of living mollusks or empty shells, may overgrow and eventually completely disintegrate the shell of their host (Fig. 10.2). The boring action is accomplished by special amoebocyte cells that dissolve the calcareous substrate along the edges of the cell, eventually removing a minute shell chip as the cell edge works its way down and under the shell chip. Many of the quahog shells I found on Connecticut beaches were riddled with holes made by *Cliona* sponges. Occasionally I found amorphous masses of *Cliona,* several inches long and with a much-pocked surface crust.

Fig. 10.1. Red beard sponge is common in Long Island Sound and often washes up on beaches.

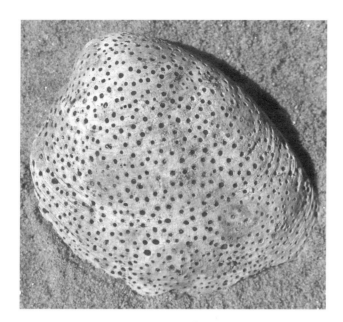

Fig. 10.2. This quahog shell is riddled with holes made by a boring sponge.

A dozen or so species of the family Spongillidae occur in Connecticut freshwater streams, lakes, and ponds, usually attached to rocks and sometimes plants, concrete dams, or other surfaces. Freshwater sponges sometimes serve as food for aquatic insects, crayfish, snails, and ducks. The four most common and widespread sponges in freshwater habitats in Connecticut are *Spongilla lacustris* (green, brown, or white; when mature, with long cylindrical branches extending from a basal crust, attached to many kinds of substrates in flowing or quiet water), *Ephydatia muelleri* (green, brown, or gray, with an irregular papillate surface), *Eunapius fragilis* (green, gray, or brown, usually forming thick half-rounded masses with large openings, living on a wide range of substrates in flowing and still water), and *Anheteromeyenia ryderi* (green, brown, or white, forming small domes in flowing water and large bumpy or branched masses in still water, often in dark sites).

All sponges have a porous structure formed by loose aggregations of cells supported by a unique "skeletal" structure consisting of needlelike calcareous or glassy spicules, and/or protein fibers. Because each sponge species may vary greatly in size and shape, sponge identification can be tricky without microscopic examination of these diagnostic spicules.

Within the structure of the sponge is an internal canal system through which water is "pumped" by the beating action of numerous minute, hairlike projections (flagella) that extend from certain sponge cells. Sponges feed on microscopic bits of organic material removed from water flowing through the canals. Digestion is intracellular; there is no "gut." The water current also brings in oxygen, removes cellular waste products, and transports sperm and larvae.

Sexual reproduction involves the release of sperm into the water. Sperm enter other sponges and fertilize eggs in situ. Embryos develop and eventually larvae are released. Sponge larvae are free-swimming at first, then settle, attach to an object, and grow into the mature form. An individual sponge may be male, female, or both.

Asexual reproduction may occur when "buds" become detached and form a new sponge. Another form of asexual reproduction, characteristic of freshwater sponges (Class Demospongiae, Order Haplosclerida) and some saltwater species, involves the production of resistant seedlike gemmules that pass the winter in a dormant state and "germinate" in spring to form an adult sponge. You may find gemmules (white specks) of haliclona (*Haliclona loosanoffi*) from late summer through winter after most of the sponge disintegrates. This tan to gold saltwater sponge, with conspicuous pores and often rounded "chimneys," attaches to hard or soft substrates, including algae and eelgrass. Gemmules also survive desiccation, such as may occur when the water level in a stream or pond drops during summer drought. In Connecticut, most sponges, including those that retain their basic mature structure year-round, are basically dormant in winter.

Sponges often harbor an assortment of small associated invertebrates that use the sponge for shelter. Sea ravens commonly deposit their eggs at the base of estuarine sponges. Some freshwater sponges are tinted green by symbiotic zoochlorellae or other single-celled algae that may provide the sponge with carbohydrates. Such sponges may contribute significantly to the productivity of ponds and small lakes.

Jellyfishes, Hydroids, and Sea Anemones

The soft-bodied aquatic animals of Phylum Cnidaria, formerly known as coelenterates, are well known for their specialized stinging structures. The miniature harpoonlike stings (nematocysts), distributed on tentacles and sometimes on other parts of the body, are used to capture the small animals upon which cnidarians feed. Relatively few species have a painful sting. The vast majority of the 9,400 species of jellyfishes, hydroids, and sea anemones are marine or estuarine. About 1,620 species occur in the United States.

Cnidarians exhibit two body types: medusa (the swimming jellyfish form) and polyp (an attached hydroid form, which is basically like an upside-down jellyfish stuck to the substrate). Depending on the group, either the medusa or polyp form is dominant. Reproduction may involve both sexual mechanisms and asexual budding.

Jellyfishes

At a safe distance or in an aquarium, jellyfishes evoke admiration for their delicate, pulsing beauty. Up close and personal along a beach in summer, more often the responses are hysterical shrieks and quick abandonment of the water.

About 200 species of jellyfishes (Class Scyphozoa) are distributed throughout the world's oceans and seas. Only a few species are common in Long Island Sound and several others are rare or occur as strays from the north or south.

Among the jellyfishes most likely to be encountered in Long Island Sound is the **moon jelly** *(Aurelia aurelia)*, a flattened, whitish-clear jelly with a fringe of short marginal tentacles, four horseshoe-shaped gonads, and four ruffled oral arms extending from the center (Fig. 10.3). Moon jellies are common in late spring and early summer and are harmless to humans.

Lion's-mane jellyfish *(Cyanea capillata)*, usually red in the center, with purple areas in large individuals, is common in spring and present through early fall (Fig. 10.4). These jellyfish have eight clusters of long stinging tentacles around the margin and four long oral arms extending from the center. The sting of this jellyfish is much feared by swimmers, but my personal experience along Connecticut beaches is puzzling. On several occasions I tried to let them sting me, in order to know the experience first-hand, but never had any success. Yet some people do get stung.

Sea nettle *(Chrysaora quinquecirrha)*, another stinging species, is whitish, up to about four inches in diameter, and has long oral arms and up to about 24 long, distinct,

Fig. 10.4. Lion's-mane jellyfish often washes up on beaches of Long Island Sound.

marginal tentacles (Fig. 10.5). It is fairly common in river-mouth estuaries in late summer and early fall. Some sea nettles host tiny hitchhiking spider crab juveniles.

Most jellyfish reproduce sexually. Fertilization of eggs by sperm occurs within the body of the jellyfish. Eventually a swimming larva is released. The larva settles to the bottom

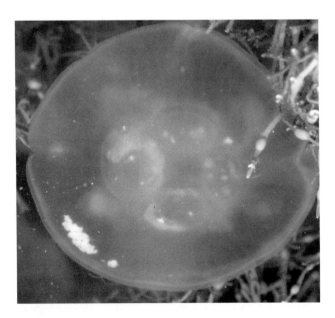

Fig. 10.3. Moon jellies, harmless to people, are numerous in Long Island Sound in late spring and early summer.

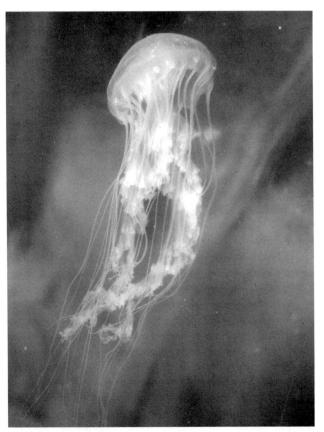

Fig. 10.5. Sea nettle is a beautiful, stinging jellyfish.

and grows into an attached polyp, which, in general, ultimately buds off a series tiny organisms that grow into medusae that are either male or female. The solitary, gelatinous medusa, shaped like a bell or umbrella and usually well supplied with stinging tentacles, is the dominant (or only) stage in the life of a jellyfish.

Jellyfishes use their tentacles to feed on larval invertebrates and other small zooplankton that swim or float in the water column. Some species rarely capture and consume animals as large as small fishes. Some fishes, unaffected by the stinging cells, may swim among the tentacles under the jellyfish.

Hydrozoans

The Class Hydrozoa includes the freshwater hydra, well known in biology classes, and the Portuguese man-of-war, a species with long stinging tentacles that make it a nemesis of swimmers (it rarely drifts into Long Island Sound). In Connecticut, there are dozens of species in addition to the few mentioned here.

Most hydrozoans are plantlike or fuzzy-looking, mostly inconspicuous animals. Various species inhabit salt water and fresh water; most species are marine/estuarine and form colonies attached to rocks, pilings, algae, or eelgrass. Under magnification, as these animals are best appreciated, each branch or stalk of the organism is tipped with a tiny anemonelike polyp, complete with tentacles. Some species are naked, whereas the stalks and polyps of others are supported by hollow, sleevelike tubes and collars secreted by the hydroid's epidermis. Members of a colony share a common digestive cavity.

Some hydrozoans are mobile even though firmly attached to a hard substrate. **Roughspined snailfur** (*Hydractinia echinata*), for example, is common as a short fuzz on shells dragged around by hermit crabs, as well as on rocks and pilings. Another snailfur hydroid, *Podocoryna carnea*, is less common.

Polyp and medusa stages are present in the life cycles of most hydrozoans, but some have no medusa stage and others omit the polyp. Reproduction may be sexual, with eggs fertilized internally or externally by sperm shed into the water by medusae, or asexual, such as in the production of medusae or the budding-off of new individuals.

The so-called **"freshwater jellyfish*"** (*Craspedacusta sowerbyi*) is an example of a hydrozoan with an easily seen, sexual medusa (Fig. 10.6). The tiny attached polyps are basically invisible in the field, but the small (less than an inch across) jellyfishlike medusa can be tremendously abundant. I saw "blooms" of many thousands of these hydrozoans in a

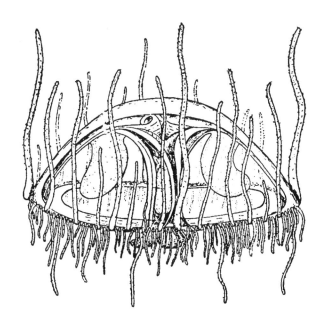

Fig. 10.6. Freshwater "jellyfish" is actually a hydrozoan that sometimes appears in enormous numbers in ponds and lakes. From Hyman (1940). The invertebrates: Protozoa and Ctenophora. Reproduced with permission of the McGraw-Hill Companies.

pond in late summer. The medusa stage consists of males and females that reproduce sexually. The polyps reproduce asexually by budding off new individuals (including medusae). Both stages feed on various small invertebrates. This species is believed to have been introduced into North America from the native range in China; it was first found in the United States in 1908.

Sea Anemones, Corals, and Relatives

Corals in Connecticut? In fact, corals do live in Long Island Sound. Northern star coral (*Astrangia poculata*, Order Scleractinia) doesn't form big reefs but rather builds little pinkish colonies only a few inches across. It inhabits clear waters on subtidal rocks, shells, and pilings, and favors cool, undiluted seawater. The polyps secrete and occupy stony, cuplike structures, with up to a few dozen per group.

The unique star coral is joined in the Sound by a dozen or so species of anemones, all of which, when fully expanded, look vaguely like aster flowers (Fig. 10.7). When disturbed, they contract into a lump with the tentacles partially or completely hidden. Anemones can exist as solitary polyps, but many form small or large colonies.

Most Connecticut anemones live at or below low tide level, so casual observers rarely see even the most abundant species. Clonal plumose anemones (*Metridium senile*), up to 4 inches tall, with countless fine frilly tentacles, are common in dark places on rocks and pilings. Lined anemones *(Edwardsiella lineata)*, topped with a few dozen whitish tentacles extending out of a brown mucus tube, live in subtidal waters, often in crevices or on the underside of rocks and sometimes in big colonies. White (ghost) anemones *(Diadumene leucolena)*, characterized by forty to sixty tentacles and pale longitudinal stripes on the trunk (up to about 1.5 inches tall), can be common on eelgrass, algae, or pilings or under rocks in shallow water (Fig. 10.7). Northern red anemones *(Urticina felina)*, a robust species up to 2 inches tall and 5 inches wide, with red-and-white ringed tentacles, occur in low abundance on solid substrates in the subtidal zone at the Sound's eastern mouth. They are more common north of Cape Cod. Several additional species burrow in muddy/sandy bottoms, with just the tentacles showing.

One species, the orangestriped green anemone* *(Diadumene lineata = Haliplanella luciae)*, thrives in the intertidal zone. This Old World native arrived in Connecticut in the late 1800s and now is common along salt marsh edges, in tidepool crevices, and on rocks, pilings, cordgrass, and algae. It is less than an inch tall, has 25 to 50 tentacles, and has many pale stripes along its blackish or dark green trunk.

The anemone with the most bizarre habitat is the so-called comb jelly anemone, often referred to as *Edwardsia leidyi*. Actually, this pink, wormlike animal is the larval state of the lined anemone. Up to an inch long, it lives within the clear gelatinous body of Leidy's comb jelly and "steals" food taken in by its host. It can cause dermatitis in humans.

Anemones can reproduce in either of two ways. Sexual reproduction results in a free-swimming planula larval stage that eventually settles and develops into the attached form. Anemones can also replicate themselves asexually by budding or fission. There is no medusa (jellyfishlike) stage.

Comb Jellies

Comb jellies (Phylum Ctenophora) resemble jellyfish, and often are mistaken for them, resulting in undue alarm among beach-goers, but comb jellies do not sting. One species has stinging structures, but it isn't harmful to humans. The gelatinous body is nearly transparent, but at close range you can see long whitish rows of comblike plates, which are brilliantly iridescent in bright sunlight.

Comb jellies swim slowly by beating microscopic hairlike cilia, but more often they are at the mercy of currents. I saw hundreds or thousands per minute flow by in some tidal channels. They feed voraciously on invertebrate larvae and other zooplankton, captured with their two sticky tentacles or by pumping water into the space inside the body. Their predators include various fishes. All comb jellies are hermaphrodites, and in all but a few fertilization occurs outside the body in sea water.

Worldwide, there are only about 100 species, with somewhat fewer than half of these occurring in U.S. waters. All are marine or estuarine. A few species occur in Long Island Sound.

Leidy's comb jelly or **sea walnut** *(Mnemiopsis leidyi)* is the most common comb jelly in the Sound (Fig. 10.8). It can be numerous in early summer and may be especially abundant in late summer and early fall. When disturbed, these 4-inch, usually clear comb jellies, recognized by their two long body lobes, emit a green glow, visible only in darkness. The feeding tentacles next to the mouth are short and not easily

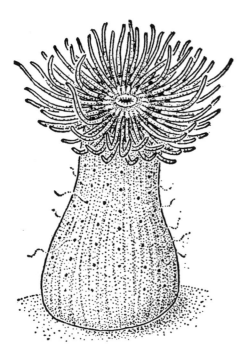

Fig. 10.7. Anemones capture prey with stinging cells on their tentacles. This is a white or ghost anemone. From Miner (1950).

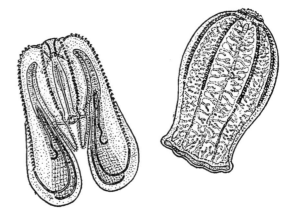

Fig. 10.8. Leidy's comb jelly (left) and Beroe's comb jelly (right) are common inhabitants of Long Island Sound. From Miner (1950).

seen. Ciliated grooves move planktonic prey toward the slit-like mouth. This species is an invasive exotic species in Europe's Black and Azov seas, where it has caused a significant decline in anchovy populations, which also consume zooplankton. The pink wormlike larva of the lined anemone commonly lives in the body of this species.

Sea gooseberry *(Pleurobrachia pileus)* is most common in winter and spring, less common in summer. It has a spherical body and two thin, branched tentacles. **Beroe's comb jelly** *(Beroe ovata)* is a common reddish species (Fig. 10.8).

Flatworms

Flatworms (Phylum Platyhelminthes) are flattened, unsegmented worms with no appendages. Most of the world's 20,000 species and the United States' 6,000 species are less than an inch long.

It is safe to say that no flatworm ranks among the state's most popular wildlife. Some, including all of the species in the classes Trematoda (flukes) and Cestoda (tapeworms), are parasitic and often have complicated life histories involving multiple hosts. For example, the trematode *(Austrobilharzia variglandis)* that causes "swimmer's itch," sometimes a nuisance in Long Island Sound, requires a mudsnail and a bird to complete its life cycle. Sometimes the bird-seeking stage bores into human skin.

Most turbellarian flatworms (Class Turbellaria), comprising a few thousand species distributed worldwide, live in salt water, but some inhabit fresh water or moist places on land. These flatworms are free living or associated with certain other animals in a nonparasitic way. Sometimes I found groups of horseshoe-crab flatworms *(Bdelloura* spp., *Syncoelidium pellucidum)*, less than an inch long, on the underside of horseshoe crabs. Turbellarians use their mouth, located on the underside of the body, to feed on various small animals or other minute organic matter.

Turbellarians reproduce sexually by cross-fertilization between different individuals. Eggs are enclosed in "cocoons," and there may or may not be a free-swimming larval stage. These flatworms also reproduce asexually by fission (splitting of one individual into two) or parthenogenesis ("virgin birth").

Ribbon Worms

The Phylum Nemertea includes some 900 species of fragile, flattened worms that live mainly on or in the bottom, or under objects, in shallow and intertidal saltwater habitats. Some live commensally with other invertebrates. They prey on annelid worms and other small living and dead invertebrates. Reproduction may be sexual, through shedding of

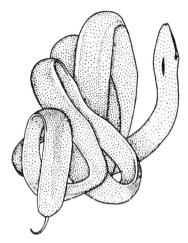

Fig. 10.9. Milky ribbon worm burrows in sandy mud or hides under intertidal rocks. From Miner (1950).

eggs and sperm by males and females or by self-fertilization, or asexual, via fragmentation of individuals into multiple pieces, followed by regeneration.

Several species occur in Long Island Sound and at least one species (the hermaphroditic *Prostoma graecense*, with two or three pairs of small distinct eyes, up to about 1.2 inches long) lives in various freshwater habitats in Connecticut, usually in quiet water among algae, plants, or detritus. It is most numerous in autumn.

Most nemerteans are less than 1.5 inches long, but one startling estuarine species, the milky ribbon worm *(Cerebratulus lacteus)*, sometimes exceeds a few feet (one meter) in length and rarely reaches 20 feet (Fig. 10.9)! This cream-colored to pink worm tends to burrow by day and may swim in the open at night.

Nematodes

Nematodes or roundworms are unsegmented, seldom-seen worms, despite the fact that they are probably the most abundant animals on earth. Most escape notice because they are tiny and secretive, but one species that parasitizes sperm whales grows up to 30 feet long. Scientists have described over 80,000 species, just a small fraction of those that exist. They live almost anywhere there is soil, mud, water, or living or dead plant or animal tissue.

Most nematodes are beneficial, with diets including microorganisms, detritus, fungi, algae, and tiny animals. But many roundworms are important parasites of animals or plants, responsible for heartworm, hookworm, lungworm, pinworm, trichinosis, elephantiasis, and other unsavory diseases. Plant parasites may cause damage to plants by sucking out their cell juices or entering the plant and feeding on living tissue, sometimes causing gall formation.

Almost all roundworms are male or female, and reproduction involves mating and internal fertilization. Some parasites have complex life cycles that involve one or two intermediate hosts in addition to the primary host. For example, the parasite may require a worm, a fish, and a mammal to complete the life cycle and attain maturity.

Horsehair Worms

Horsehair worms are long, wiry worms often found with two to several individuals tangled together in a pool of water (Fig. 10.10). Adults of both sexes are free-living in pools, mud, or moist ground, and apparently most do not feed. Horsehair worms (Phylum Nematomorpha) are represented by only about 240 species worldwide.

The life cycle is remarkably complicated and obscure. In spring, adults that overwintered in soil, under rocks, or in organic debris mate, and the females deposit enormous numbers of eggs in detritus or gravel in shallow water, in slender gelatinous strips that may become tangled in vegetation. Soon thereafter the adults die. Within a few months of egg laying, the parasitic larvae hatch. The larvae may encyst on plants or nearshore substrate and, after the water level drops, may be incidentally eaten by an insect such as a cricket, grasshopper, katydid, ground beetle, or cockroach. Or the larvae may enter and encyst in aquatic insects (e.g., crane fly larvae), snails, or earthworms; possibly grasshoppers, crickets, or beetles acquire the horsehair worms by feeding on an infested aquatic insect, snail, or earthworm. Some evidence suggests that horsehair worms must pass through an encysted stage in an invertebrate host before developing into the adult stage in an orthopteran. Horsehair worm cycts possibly survive one or two years in hosts that live that long. Whatever the situation, the larvae develop in an insect (or sometimes other invertebrates) and are nourished therein by absorbing nutrients through their body wall (they have no gut). Apparently the worms escape from the insect host only when it is in the water or wet. Presence of the horsehair worm in the host may actually cause it to seek water when the worms are ready to emerge. It may be no coincidence that I sometimes found remains of dead crickets in or next to pools that contained horsehair worms.

Ectoprocts

Ectoprocts, sometimes called bryozoans, are colonial aquatic animals that can be hard to recognize as such. They may resemble a branched tuft of seaweed (generally less than 1 foot tall), form a limy-lacy crust (Fig. 10.11) or a gelatinous, spiny, or rubbery mass (which may look like a branched sponge), or consist of short individuals distributed along an attached, rootlike network (which may resemble a hydroid). You may find them by careful search of submerged rocks, logs, shells, boat hulls, pilings, lobster traps, floats, algae, eelgrass, or mooring lines.

In most species, each individual animal (zooid) is enclosed in a case of chitin and calcium carbonate, which is secreted by the animal. The animal can withdraw into the case for protection. A ring of tentacles (lophophore) surrounds the mouth and, extended from the case, is used to procure food. Ectoprocts differ from hydroids in having ciliated tentacles and a separate digestive tract in each individual zooid.

Ectoprocts produce more of themselves in various ways. Sexual reproduction consists of fertilization of eggs by

Fig. 10.10. Horsehair worm larvae develop as parasites of insects (adults shown, Colorado).

Fig. 10.11. Lacy cases of an ectoproct colony cover this piece of algae.

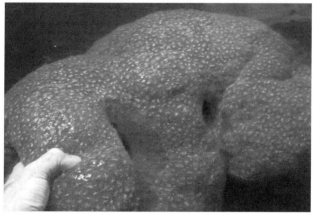

Fig. 10.12. The ectoproct pectinatella, shown among tidal marsh plants (water-celery, water-milfoil, hornwort) in late summer, often reaches football size or larger.

sperm (generally involving an interaction between different hermaphroditic individuals), followed by egg brooding and a planktonic larval stage. Ectoproct colonies often form through extensive budding of new individuals. Freshwater forms may produce new individuals by budding, fragmentation, or by the formation of resistant capsules of cells (statoblasts) that overwinter and germinate in spring, ultimately giving rise to new colonies.

This worldwide group (Phylum Ectoprocta) of 4,000 species worldwide and 900-plus species in the United States are mostly marine, but there are a fair number of freshwater species as well, including 25 in North America. Many freshwater ectoprocts favor the outflows of lakes and reservoirs.

The most noticeable ectoproct in Connecticut is a freshwater species, pectinatella *(Pectinatella magnifica)* (Fig. 10.12). Globular masses of this colonial animal attract attention because they look vaguely brainlike to some people. Pectinatella occurs throughout the state in flowing and nonflowing fresh water (including tidal areas), especially in shallow, clear, warm water and below dams. It consists of rosettes of reddish zooids embedded in firm gelatinous material (secreted by the zooids) attached to objects such as logs or the stalks of aquatic plants. Small colonies can move very slowly. Colonies increase in size throughout the warmer months and, attached to submerged logs, may grow to several feet across. I saw large colonies most often in July, August, and September. Colonies disintegrate as water temperatures drop in early fall. Pectinatella survives winter as tough flattened seedlike structures that germinate into zooids in spring. Colonies form through repeated asexual budding of new zooids. These harmless tentacled animals feed on minute organisms and organic material that they extract from the water.

An example of a saltwater species is the lacy ectoproct* *(Membranipora membrancea)*, introduced from the Pacific Ocean. It grows as a thin, lacelike crust on algae and in so doing sometimes debilitates or kills kelps. Its rectangular zooid chamber lids have a knob at each corner.

Ectoprocts can be a significant element in aquatic food webs, as consumers of phytoplankton and minute organic particles and as food for fishes and larval caddisflies. Encrusting species sometimes clog municipal water supply systems or foul boat hulls.

Chapter 11

Segmented Worms

Everyone knows what a worm is, but not all "worms" are members of this group. Earthworms and leeches that live on land or in fresh water are the most familiar examples of segmented worms (Phylum Annelida), but most of the 15,000 species are marine and estuarine polychaetes. Tapeworms, roundworms, and most other unappealing wormy species are not segmented worms.

Segmented worms occur worldwide, with 3,360 species in the United States. Many are burrowers or live in tubes that they secrete. Ecologically they are highly influential in terrestrial, wetlands, and aquatic ecosystems. The feeding and breeding habits vary greatly and are described for the different groups in the following sections.

Fig. 11.1. Clam worms are free-living polychaetes.

Polychaetes

Almost all polychaetes (Class Polychaeta) live in salt water. These worms generally have separate sexes, external fertilization, and reproduce only sexually. Some undergo a change in body form in transforming from a nonsexual form to a reproductive individual, and these may swarm at the surface of the water, shedding eggs and sperm. The water soon may contain enormous numbers of eggs, sometimes in fragile chains of polygonal jelly masses. I encountered these in summer along the coast of Rhode Island.

The following account does not do justice to this large and diverse, worldwide group of some 9,000 species. Instead it just gives a flavor of the diversity with a few examples of Long Island Sound's substantial polychaete fauna.

Mobile Polychaetes. Mobile or free-living polychaetes have a segmented, wormlike appearance. Some have strong protrusible jaws and are predators on other small invertebrates; others are scavengers or omnivores. They are important in the diets of shorebirds and benthic fishes, which may capture free-swimming worms or extract them from burrows. Clam worms (*Nereis* species), some of which reach 8 inches or more in length, are mobile polychaetes. They have a somewhat flattened body, five pairs of slender appendages on the head, a distinct fleshy flap on each side of each body segment, and protrusible jaws. Clam worms are common in mudflats, sandflats, and marshes, and also occur under rocks or shells or among algae and organisms attached to pilings or other objects. Beautifully iridescent *Nereis virens* adults spend most of their time in burrows in intertidal mudflats (Fig. 11.1). In spring, males swarm and release sperm at the water's surface on moonless nights during spring tides (females release eggs at the burrow). Clam worms take several years to become sexually mature, and they die after spawning. They prey on small animals, algae, and diatoms and are eaten by fishes and various shorebirds.

Sedentary Polychaetes. These worms vary substantially in body shape and feeding method. Some look not at all like worms but rather vaguely anemonelike. They make and live in tubes and use a tuft of tentacles and/or gills at the head end of the plain, segmented body to extract food particles suspended in the water. For example, coiled worms or hard tube worms (*Spirorbis borealis*) live in shell-like, coiled or

Fig. 11.2. Coiled worms live in limey tubes often attached to eelgrass.

curving tubes and are very common on eelgrass and algae in shallow water (Fig. 11.2). Larger carnation worms *(Hydroides dianthus)* are common in meandering limey tubes attached to shells, rocks, or sometimes sponges (Fig. 11.3). Cone worms *(Pectinaria gouldii)* live within a conical tube of sand and burrow into the substrate, ingesting sediments and digesting some of its organic content. Some sedentary polychaetes that burrow may extend their tentacles over the substrate and feed on deposited detritus. Other species live in a permanent tube, pump water through it, and extract suspended food material.

One of the most unusual polychaetes is the lobster gill worm *(Histriobdella homari)*. This colorless worm, only about 1 millimeter long, is a parasite that lives on and about the gills of lobsters.

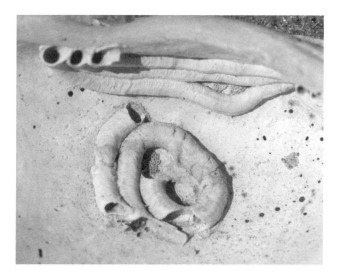

Fig. 11.3. Carnation worms commonly build their tubes on quahog shells.

Earthworms and Relatives

Worms of the Class Oligochaeta include the familiar earthworms and nightcrawlers. We may take them for granted, but as soil decomposers and aerators they are critical components of ecosystems. The earthworm is just one of over 6,000 species in this group worldwide.

Oligochaetes live not only on land in the soil but also in detritus deposits, in wetlands, and in freshwater, estuarine, and marine habitats, where they burrow in mud or crawl among debris and vegetation or under rocks. In Connecticut, relatively few species inhabit saltwater environments. Oligochaete worms can be tremendously abundant, especially in rich nonacidic soils and on the bottoms of lakes, ponds, and slow streams. Terrestrial worms are most abundant in nonacidic soils that are high in calcium. They avoid dry or freezing conditions by burrowing deeper into the soil.

Most oligochaetes feed on decaying organic matter, such as vegetable matter in the soil or water. Some aquatic species feed on living microscopic protoctistans (animal-like organisms) or flatworm larvae.

All oligochaetes are hermaphroditic. In most species, mating between two individuals results in internal fertilization and later deposition of an egg-containing cocoon, from which the young worms emerge after hatching. A few oligochaetes are characterized by self-fertilization or parthenogenesis. Asexual reproduction, usually by splitting of one worm into two or more individuals, occurs commonly in aquatic oligochaetes.

Leeches

Leeches are about as popular as mosquitoes and ticks, but their aquatic habits make them easier to avoid. Most of the world's 500 species of leeches (Class Hirudinea) occupy shallow, quiet, freshwater habitats, with smaller numbers in marine, estuarine, and tropical terrestrial environments. About 60 species inhabit fresh waters of the United States. Few species tolerate acidic environments of pH 5 or lower.

In Connecticut, leeches occur mainly in freshwater, but there are some in salt water as well (Fig. 11.4). Small leeches sometimes attached themselves to my legs when I waded in pools, but never inflicted any real harm or pain; in fact, I didn't know they were attached until I happened to see them.

Leeches do not really deserve their bad reputation. Though some leeches are blood-sucking ectoparasites on vertebrates, including humans, most live innocuous lives as scavengers or parasites or predators on small invertebrates. Leeches in Connecticut do not transmit diseases to people.

Parasitic leeches feed by attaching to a host with their mouth and slitting or otherwise penetrating the skin. They

Fig. 11.4. This leech was on a Chinese mysterysnail on a lake bottom.

then suck out body fluids, sometimes ingesting several times their own body weight. Digestion of a meal is slow, and bacteria seem to play an important role in digestion in both parasitic and predaceous species. Leeches tend to feed infrequently.

Reproduction occurs only through direct sperm transfer from one individual to another (all leeches are hermaphroditic), with internal fertilization of the eggs. Leeches deposit eggs externally, usually in a cocoon that is placed in the habitat, attached to the host, or attached to the underside of the leech. Some aquatic leeches crawl out of the water and place their cocoons in moist sites on land. Most species breed in spring and mature in one year.

Chapter 12

Mollusks

Mollusks include examples of the world's most homely animals as well as the most intricately beautiful. Some mollusks (Phylum Mollusca) are barely visible, but others are enormous; giant squids and tropical giant clams sometimes exceed 400 pounds. Their habitats are similarly varied, ranging from deserts and alpine summits to small temporary ponds to ocean abysses. Because of their shell-building requirements, mollusks tend to be most common in calcium-rich environments. The group includes at least 70,000 species worldwide and several thousand species in the United States.

Though most mollusks tend to be relatively sedentary as adults, some (for example, squids) are fast swimmers and the majority of species have free-swimming aquatic larvae. A large number of mollusks are of significant economic importance as exquisite food, feeders on garden plants, sources of jewelry (pearls), or, though their tunneling activity in wood, destroyers of ship hulls and wharves. Mollusks are important in passive recreation as well, the shells being one of the main attractions for beach walkers.

Snails and Other Gastropods

Over 50,000 species of snails and their snail-like relatives (Class Gastropoda) occur in salt water, fresh water, wetlands, and moist terrestrial habitats nearly worldwide. These diverse habitats are reflected in their breathing apparatus, which ranges from gills or gill-like structures to an air-filled lung. Most lunged snails live in fresh water or on land.

Gastropods exhibit a wide diversity of feeding habits. Depending on the species, they may filter minute organic particles from water, graze on algae, plants, and microbes, parasitize animals, scavenge dead plants and animals, or actively prey on animals (including other mollusks). Some of the predatory species use acidic secretions of the accessory boring organ and a filelike radula to make small holes in the shell of the prey (big enough to slip the mouthparts through and begin feeding). Many gastropods are important predators and prey in the estuarine ecosystem.

Gastropod reproduction generally involves cross fertilization between individuals of the opposite sex or between hermaphrodites. Eggs of most marine snails are deposited in the habitat in distinctive cases that often can be found and identified in the field.

Saltwater Gastropods

Long Island Sound supports a wide variety of gastropods, including at least a few dozen species that are relatively easy to observe, at least as beach-cast shells, plus others that are likely to be found only by scuba divers or trawlers. Examples of some of the noteworthy or easily observable species are listed here. See Chapter 3 for further information on these snails.

Filter Feeders. One of the most abundant snails along the Connecticut coast is the **common Atlantic slippersnail** *(Crepidula fornicata)*. It is often tremendously abundant, especially in stacks of its own species (Fig. 12.1) or attached to other shells, horseshoe crabs, or eelgrass. You may see live slippersnails on objects on sandy-silty bottoms if you kayak in calm, shallow waters. After storms, beaches sometimes are strewn with uncountable thousands of these mollusks, scattered singly and in stacks of several individuals. I saw sanderlings, dunlins, ruddy turnstones, and ring-billed gulls extracting the flesh from beach-cast slippersnails.

Slippersnails feed by pumping water under the shell through ciliary movements. They use their long gill filaments and a mucous sheet to filter suspended organic particles from the water.

Slippersnails have an unusual sex life and change sex from male to female as they grow. In stacks of slippersnails,

Fig. 12.1. Common Atlantic slippersnails, with large females attached to an oyster shell and smaller males atop the females. As these snails grow, they change sex from male to female.

the large bottom individual is a female that is inseminated by the male attached to her shell. Eventually the male grows larger and becomes female and mates with the male on her shell. She deposits a gelatinous egg mass in front of her foot. Larval development includes a free-swimming veliger stage.

Close relatives of the common slippersnail include **convex slippersnails** *(Crepidula convexa)* and **eastern white slippersnails** *(Crepidula plana* complex) (Figs. 12.2 and 12.9). These are most common on other shells, the latter especially inside abandoned whelk or moonsnail shells or on horseshoe crabs.

Fig. 12.2. Eastern white slippersnails (two at top), when alive, often attach inside empty whelk shells (See Fig. 12.8). A common Atlantic slippersnail is shown at the bottom.

Grazers and Scavengers. Common grazing snails along Connecticut's rocky intertidal coast include three *Littorina* periwinkles (Fig. 3.8). **Common periwinkles*** *(Littorina littorea)* are very common in summer in rocky areas and tide pools in the middle intertidal zone, and they also inhabit some bare, mud-peat exposures on the seaward side of salt marshes (Fig. 12.3). These snails tolerate regular exposure to air (and can breathe through the highly vascular mantle cavity), even on hot summer days. If necessary, they seal the shell to rock with mucus and close their operculum, greatly reducing water loss. Common periwinkles remain abundant in exposed intertidal areas through at least November, even when air temperatures fall into the 15 to 30°F range. Like wood frogs, these snails (and rough and yellow periwinkles) tolerate freezing of their extracellular body fluids. They tolerate temperatures as low as 3° to -4°F. Nevertheless, I noticed that along the Connecticut shore these snails largely vacate shallow water and spend the coldest part of winter and early spring in deeper water where they avoid freezing conditions.

Beginning in the mid-1800s and extending over a period of about a century, common periwinkles extended their range from Nova Scotia to Maryland. Possibly they arrived in Nova Scotia with Norse sailors around 1000 A.D. and did not expand there until sea conditions became more favorable. They reached Connecticut in the 1870s. Certain evidence suggests that the invading common periwinkles displaced mudsnails from most of the latter's former habitat except muddy areas.

The common periwinkle's broad diet includes algae, tiny animals, plant detritus, and carrion. Constant grazing by these abundant snails greatly reduces algal cover and alters the competitive relations among different algal species (see

Fig. 12.3. Common periwinkles and small barnacles await submergence by the next tide.

"Rocky Shores" in Chapter 3). Also, their activities reduce sediment buildup on gentle cobbly shores.

Common periwinkles may live for several years, if they evade the various crabs, fishes, predatory snails, gulls, and ducks that feed on them. They release egg packets into the water and have free-swimming planktonic larvae.

A smaller species, the **rough periwinkle** (*Littorina saxatilis*), is another common intertidal grazer that lives in two very different habitats: the high intertidal zone in rocky areas, often in crevices, and among smooth cordgrass of low salt marshes (Fig. 3.8). These snails are not only rough but tough and even more tolerant of heat, dryness, and cold than are common periwinkles*, and they can live out of water for up to a few weeks. The diet includes the black cyanobacteria that coats high intertidal rocks as well as algae and detritus on the surface of rocks, marsh plants, or peat. There is no free-swimming larval stage. Instead, the young are retained by the adult until ready to adopt the crawling lifestyle of the adult.

The third grazing periwinkle, the **yellow periwinkle** (or smooth periwinkle) (*Littorina obtusata*) is less conspicuous but nevertheless common on seaweeds in the lower intertidal zone (Figs. 3.8 and 15.5). You can find this species, as well as other snails, on lower intertidal rocks, well into fall even after temperatures drop into the 15 to 30°F range, but their position low in the intertidal zone means that their exposure to air is relatively short, and they avoid temperature extremes and desiccation under the shelter of algae. In shape and color these cryptic snails resemble the air bladders on the rockweeds on which they often occur. They have the same relationship with knotted wrack, but in Connecticut I usually found these snails on rockweed. By perching on the algae they probably avoid detection by various visual predators such as shorebirds. They also commonly graze on open rock surfaces among attached macroalgae. The eggs are laid on algae and hatch into miniature snails, with no planktonic larval stage.

In at least some parts of the range, colonization of the shoreline by predatory green crabs* has been accompanied by production of thicker, more protective shells by yellow periwinkles. The response can occur during the development of individual snails and does not require natural selection and evolutionary change.

A less evident grazer, at the southern limit of its range in Connecticut, is the **plant (tortoiseshell) limpet** (*Tectura testudinalis*). It firmly attaches its tiny, wide-bottomed, conical shell to rocks at low tide level. Limpets are known for having a relatively fixed home base, but they do move about while grazing on rock surfaces. A relative of plant limpet, the **eelgrass limpet** or **bowl limpet** (*Lottia alveus*), was formerly abundant on eelgrass from Labrador to New

Fig. 12.4. Eastern mudsnails often are abundant in shallow waters of Long Island Sound and its tidal marshes and lagoons.

York. It went extinct during a major eelgrass die-off in 1930s; apparently it could not persist in low-salinity areas where eelgrass survived.

At low tide, or while paddling in coastal shallows and channels, you may see large numbers of inch-long or smaller dark snails on mud or sand flats and channel banks in intertidal and shallow subtidal areas, often adjacent to salt marshes. These are **eastern mudsnails** (*Nassarius obsoletus*), sometimes tremendously abundant as they graze on surface-sediment algae (Fig. 12.4). They also scavenge dead organisms that they find by chemical tracking, feed on nematode worms and other invertebrates, and sometimes bore into and eat small live bivalves. During the cold winter months, they bury themselves in the mud. Mudsnails deposit groups of tiny, spiny, whitish egg-capsules on various objects.

Mudsnails are a host for the parasitic trematode worm that causes "swimmers' itch" (schistosome dermatitis). The worm requires both the snail and a bird (for example, a duck) to complete its life cycle. The eggs of the parasite are passed in the fecal droppings of a bird and the larval worm enters a snail, in which another larval form develops; this latter larva emerges from the snail and enters the skin of a bird (or a person).

Another common grazer/scavenger, a bit smaller than the mudsnail, is the **threeline mudsnail** (*Nassarius trivittatus*) (Fig. 3.9). It lives on sandy bottoms, including eelgrass meadows, in shallow subtidal waters.

The premier grazer in salt marshes in the **eastern melampus** or **salt marsh snail** (*Melampus bidentatus*) (Fig. 12.5). This small (usually less than 0.5 inches), air-breathing snail is very common in high salt marshes, especially in salt-meadow cordgrass or spike grass, also in well-drained stands

Fig. 12.5. Tiny eastern melampus snails graze on salt marsh peat at low tide.

of stunted smooth cordgrass. It grazes during low tide on algae, detritus, and carrion on the surface of salt marsh peat. In summer, when the marsh is flooded at high tide, look for these snails above the water on the stems of marsh grasses. To survive the coldest times of winter, eastern melampus snails rely on antifreeze substances that are present in their cells.

Melampus snails reproduce on a lunar rhythm. Mating and egg laying occur in pulses that begin somewhat in synchrony with spring tides. Larval hatchlings move out of the marsh and into the water with a subsequent spring tide, becoming part of the zooplankton. During a spring tide two to six weeks later, the young snails settle in the high marsh to begin a life that lasts only a few years at most.

Additional small grazers occur on eelgrass and algae in quiet, shallow waters. Northern lacunas or chink shells (*Lacuna vincta*) are less than ½ inch long and have a curved groove along one side of the shell opening. Alternate bittiums or horn snails (*Bittiolum alternatum*) have a waffle-like surface texture and grow up to ¼ inch long. Careful examination of eelgrass blades may reveal the gelatinous egg masses of these snails (doughnutlike in lacuna, coiled in bittium).

Predators. Northern moonsnails (*Euspira heros*) are the most commonly encountered (as a beach shell) of the several species of predatory moonsnails in Long Island Sound (Fig. 12.6). They inhabit sandy substrates in lower intertidal and subtidal areas. Moonsnails grow to an impressive 4 inches in diameter and are active predators on clams, mussels, and snails. They gain access to the soft tissue of their prey by patiently boring a neat round hole in the shell with their rasping mouthparts and an acid-secreting organ (Fig. 12.7). It may take two or three days to penetrate the shell. Moonsnails deposit eggs in a sand-coated collarlike structure (left on the bottom) from which tiny snails eventually emerge. On many occasions I saw herring gulls drop moonsnail shells onto intertidal rocks and extract hermit crabs from the broken shell.

Two other big predators are **channeled whelks** (*Busycotypus canaliculatus*) (Fig. 12.8) and **knobbed whelks** (*Busycon carica*) (Fig. 12.9). These massive snails occur on sand in lower intertidal and subtidal waters, where they prey on bivalves and probably scavenge on recently dead animals as well. Unlike drills and moonsnails, whelks access the soft edible tissues of their prey by using their shell and muscular foot to pry apart and chip away at the shells of bivalves. Whelks produce strings of hollow, round disks that contain the young, which emerge from the disks as tiny shelled snails.

Fig. 12.6. Northern moonsnails cruise sandy bottoms of Long Island Sound.

Fig. 12.7. A northern moonsnail bored into this common periwinkle.

Fig. 12.8. Channeled whelk shells and egg cases (with tapered edges) often wash up on Connecticut beaches.

In the middle of the predator size scale, **Atlantic oyster drills** (*Urosalpinx cinerea*) (Fig. 3.9) and **thick-lip drills** (*Eupleura caudata*) are easy to find in subtidal oyster beds and among blue mussels and barnacles on intertidal rocks, the Atlantic being more abundant in and more tolerant of waters of lower salinity. Both species bore into and kill barnacles, young oysters, mussels, and other small animals, making them a scourge to commercial shellfisheries. If you kayak rocky coastal waters, you may see breeding aggregations of these snails in spring and summer. Looking closely, you may discover the vaselike egg cases attached to rocks or other objects. The eggs hatch into tiny snails, which crawl and feed in the usual way; there is no free-swimming larval stage. Drills pass the winter in deepwater sediments.

Fig. 12.9. Knobbed whelks produce egg cases with squared-off edges. Marks inside the shell indicate where eastern white slippersnails were attached.

Similar in size to the drills, **Atlantic dogwinkles** (*Nucella lapillus*) are common in exposed rocky intertidal areas but do not inhabit areas of low salinity (Fig. 3.9). Like the drills, these predators bore holes into the shells of barnacles, blue mussels, and other bivalves. They may kill barnacles by secreting a toxic purple substance onto the prey, then inserting the mouth when the valves of the dead or drugged barnacle relax. Breeding aggregations form in late summer. Spindle-shaped egg cases adhere to rocks or other objects and, the following spring, hatch into tiny snails, which crawl and feed in the usual way. Dogwinkles appear to have responded to the invasion of their habitat by predatory green crabs* by increasing the thickness of the shell.

The smallest common predators include **lunar dovesnails** (*Astyris lunata*, to 5 mm) and other larger dovesnails (to 18 mm), which are sometimes common on eelgrass or algae in shallow quiet waters. These snails prey on attached organisms such as colonial sea squirts.

Freshwater Snails

Worldwide, freshwater habitats harbor about 5,000 snail species, and 680 species occur in the United States. The freshwater snail fauna of Connecticut includes a few dozen species.

Most of the state's freshwater snails live in lakes and ponds, and some of these quiet-water snails also occur in slow currents in streams. Impoundments on rivers generally have the same kinds of snails as do lakes of similar size and chemistry, unless subject to frequent large decreases in water level, which result in unfavorable conditions for littoral zone vegetation and associated invertebrates such as snails. Just a couple of species, the piedmont elimia and creeping ancylid (Fig. 12.10), are entirely restricted to streams. A few successfully inhabit temporary waters (see the following list for examples). These survive dry periods by estivating in mud or among vegetation.

Most freshwater snails favor substrates of plant material that has fallen into the water, detritus, or rocks, logs, and hard surfaces that are covered with algae, bacteria, diatoms, and other minute organisms on which the snails graze. Some small snails graze on live aquatic plants.

The largest numbers of species occur in productive lakes that have not been significantly altered by human activities. Acidic lakes with a pH of 5 or less generally are devoid of snails, but a diverse snail fauna can exist in acidic waters above pH 6.0. Generally snails are most abundant and diverse in waters with relatively high calcium content (calcium is an important component of the shell), though species diversity does not necessarily increase with increasingly

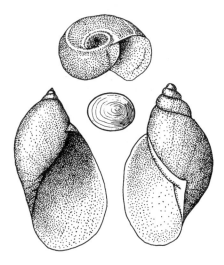

Fig. 12.10. Examples of four families of freshwater snails. Top: an orb snail, the two-ridge rams-horn *(Helisoma anceps)*. Middle: top view of a freshwater limpet, the creeping ancylid *(Ferrisia rivularis)*. Left: a pond snail, the mimic lymnaea *(Pseudosuccinea columella)*. Right: a pouch snail, the pewter physa *(Physa [Physella] heterostropha)*.

high calcium levels. Eileen Jokinen found that Connecticut lakes with exceptionally diverse snail faunas include Roseland Lake in Woodstock (17 species, pH = 6.4, Ca^{++} = 7.2 mg/l), East Twin Lake in Salisbury (15 species, pH = 7.4, Ca^{++} = 35.0 mg/l), and Lake Quonnipaug in Guilford (14 species, pH = 7.0, Ca^{++} = 9.1 mg/l). Sampled river sites had fewer (less than 10) snail species. Several species, such as the woodland pondsnail *(Stagnicola catascopium)*, marsh pondsnail *(Stanicola elodes)*, marsh rams-horn *(Planorbella trivolis)*, and turret snail *(Valvata tricarinata)*, appeared to be restricted to sites with medium-hard to hard water.

Snails are scarce or absent in areas polluted with certain heavy metals, such as copper. Copper releases from industry and from intentional applications for algae control have eliminated or reduced local snail populations in some Connecticut lakes and streams. Several species are state-listed as Special Concern. One species (boreal fossaria, *Fossaria galbana*) is believed to be extirpated. At least one exotic species (Chinese mysterysnail), is widely established as a result of accidental and intentional introductions associated with the aquarium pet trade.

There are two major groups of freshwater snails, prosobranchs and pulmonates, and they differ in several ways. **Prosobranchs** have gills, separate sexes (except turret snails, *Valvata* spp.), and a horny operculum or plate that covers the shell opening when the snail withdraws. Some have a line on the operculum that makes several spirals (multispiral operculum); others have concentric growth rings. Prosobranchs have relatively few teeth on the radula (a filelike mouthpart), and the mouth is on the end of a muscular rostrum. These gill-dependent snails seldom occur in stagnant or warm water where oxygen levels may be low. **Pulmonates,** which include physid, lymnaeid, planorbid, and ancylid snails, have a lunglike chamber and may obtain oxygen from the air, though some pulmonates have gill-like skin areas that allow oxygen uptake from the water. Pulmonates are hermaphroditic, lack an operculum, have many teeth on the radula, and have the mouth not on a snoutlike rostrum.

Most freshwater snails deposit eggs in a gelatinous mass attached to an object in the water. Mysterysnails (Viviparidae) give birth to well-developed young.

The following brief overview of Connecticut's freshwater snails includes all the families represented in the state, with specifics for a few species.

Prosobranchs (Gilled snails with an operculum). **Round-mouthed snails** or **turret snails** (Valvatidae: *Valvata* spp.). Shell less than ¼ inch wide, with low spire and round, multispiral operculum; one species has raised spiral ridges on shell; gill is external; highly localized in a few ponds and lakes.

Piedmont elimia (Virginia river snail) (Pleuroceridae: *Elimia virginica*). Survives in Connecticut locally in much-reduced range in Connecticut River north of Hartford; operculum has few spiral lines; shell about twice as long as wide.

Mysterysnails (Vivparidae). Large; operculum marked with concentric lines. Banded mysterysnail *(Viviparus georgianus)*. Shell up to about 1.4 inches tall; known from sites in western Connecticut; abundant in East Twin Lake and Lake Waramaug; has a slitlike umbilicus and reddish stripes on outside or inside of shell. Chinese mysterysnail* *(Cipangopaludina chinensis)* and Japanese mysterysnail*

Fig. 12.11. Non-native mysterysnails of Asian origin are Connecticut's largest snails.

Fig. 12.12. Pointed campeloma is common in lakes and ponds.

(C. japonica) (Fig. 12.11). Strikingly large (up to 2 inches tall); narrow slitlike umbilicus (indentation at base of shell); native to Asia; Chinese species is now widely distributed in lakes, ponds, and rivers in Connecticut; Japanese species has been found in several sites (Jay Cordiero). Pointed campeloma *(Campeloma decisum)* (Fig. 12.12). Common, mainly in large ponds and lakes and slow river areas; sandy areas on, near, or under submerged logs, or burrowed in sand or mud; shell up to 1.6 inches tall, lacks an umbilicus.

Hydrobiids (Hydrobiidae). Usually on plant material or detritus in quiet water; represented by a few small (not more than 5 mm) species with shell opening on right when coiled spire is pointed up and shell opening faces you; operculum has few to many spirals.

Pulmonates (Snails with lungs, no operculum). **Pond snails** (Lymnaeidae). Diverse habitats; spiraled, spired shell (up to an inch tall) with opening on right when you look at opening with spire pointed upward (Fig. 12.10). Pygmy fossaria *(Fossaria parva)* exploits temporary bodies of water.

Pouch snails (Physidae). Spiraled, spired shell (less 0.8 inches tall) with opening on left when you look at opening with spire pointed upward (Fig. 12.10); various habitats. Vernal physa *(Physa vernalis)* is usually found in pond or stream habitats that dry up or dry around edges in summer. Lance aplexa *(Aplexa elongata)* is another inhabitant of temporary pools.

Orb snails (Planorbidae). Wide variety of habitats, ranging from small vernal pools to larger permanent waters; flattened, coiled shell up to an inch in diameter in largest species (Fig. 12.10). Thicklip rams-horn *(Planorbula armigera)* is a common vernal pool species.

Freshwater limpets (Ancylidae). Low conical shell without spiraled whorls; largest species is up to 0.4 inches in diameter; various habitats. Creeping ancylid *(Ferrissia rivularis)* is a widely distributed stream specialist (Fig. 12.10).

Terrestrial Snails and Slugs

Connecticut's land snail and slug fauna includes several dozen species, mainly native snails but also several native slugs. Also established are a few non-native snails and a number of exotic slugs. A significant percentage of the native snail species are also native to Europe. Space limitations prevent me from presenting more than a brief discussion of the state's land snails and slugs.

Because shell production depends on calcium availability, land snails tend to be most numerous in areas with calcium-rich soils. Recent findings of declining calcium availability in New England forests, likely an outcome of acid deposition deriving from air pollution, suggest that snail populations could be negatively affected. Because land snails may be important sources of calcium needed for eggshell production by forest birds, bird populations might negatively impacted as well.

Connecticut's largest native land snail (generally up to about an inch in shell diameter) is the **whitelip** *(Neohelix albolabris)* (Fig. 12.13). It lives among logs and leaf litter in wooded areas and deposits eggs along leafy edges of logs or rocks in late spring or early summer. A few weeks later these hatch into tiny snails. Wintering sites are under logs or rocks, where the snails seal up the shell opening with a mucous membrane and wait for spring. Careful search in leaf litter and around logs may yield examples of Connecticut's several species of smaller land-dwelling snails, some of which have shells less than 0.1 inches long.

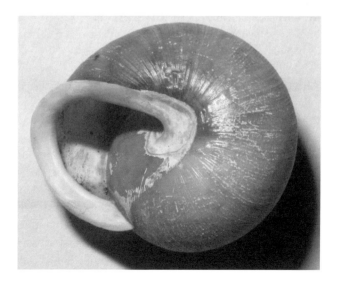

Fig. 12.13. The whitelip is a large native land snail.

Fig. 12.14. This philomycid is one of the few slug species native to Connecticut.

Slugs are represented in Connecticut by a few native species of the families Philomycidae (mantleslugs, *Philomycus, Pallifera*) (Fig. 12.14) and Limacidae, plus several exotic species of Mebeidae and Arionidae, including *Arion fasciatus* and *A. subfuscus*. Philomycids often are associated with decomposing logs on the forest floor. On my property, philomycid slugs emerged from crevices in a stone retaining wall usually on humid or wet nights but sometimes when conditions were dry. They crawled around on the smooth, bare surface of a rock ledge, evidently grazing on microscopic algae. Others crawled up the trunks of beech trees and sometimes assembled there in (mating?) groups. They usually retracted their eyestalks when first exposed to the beam of a flashlight. A few remained out until mid-morning the next day.

Clams, Mussels, and Oysters

Mollusks enclosed in two shells are known to scientists as bivalves, but to many people they are known simply as food to be relished or avoided. The Class Bivalvia or Pelycypoda includes some 10,000 species of clams, mussels, and oysters distributed in aquatic habitats worldwide.

Most bivalves are filter feeders that extract phytoplankton, zooplankton, and minute bits of organic material suspended in the water or within bottom sediments. They are thus important in clarifying the water and converting fine organic matter into larger edible morsels that fishes, birds, and other animals can eat. They feed by drawing in food-laden water through one of their two siphons. Food particles are trapped by the gill filaments and may be sorted by size as they are filtered from the water along with oxygen used for respiration. "Filtered" water is expelled (minus food particles) through the other siphon. Clams burrow and move horizontally using a muscular, protrusible "foot." Cilia on the foot gather subsurface organic matter that can then be ingested as food.

The sexes are separate in most species. Hermaphrodites occur among the shipworms, some oysters and scallops, freshwater fingernail clams, and a few freshwater mussels. Reproduction in most marine species involves external fertilization and a planktonic larval stage. In freshwater mussels, fertilization is internal, with later release of the larvae.

Freshwater Mussels

Freshwater mussels (Fig. 12.15), which are inedible, are quite different from the mussels popular as seafood. About 300 of the worldwide fauna of 1,000 species of freshwater mussels (Order Unionoida) occur in the United States and Canada. The largest number of species occurs in the Mississippi River basin. For example, the native mussel fauna of just one portion of the drainage, the Tennessee River system, includes (or, more accurately, formerly included) some 100 species.

Unfortunately, this rich fauna is not doing very well. Only about 25 percent of the mussel species in the North America are regarded as not endangered, threatened, or of significant conservation concern, and several are already extinct. Dams, channel modification, siltation, and water

Fig. 12.15. Examples of Connecticut's freshwater mussels: tidewater mucket (top left), eastern pondmussel (top right), alewife floater (center left), eastern floater (center right), eastern elliptio (bottom left), eastern lampmussel (bottom right), and dwarf wedgemussel (lower center).

pollution have destroyed or degraded much mussel habitat. Dams can alter even downstream habitat, by changing stream flow and releasing water too cold for effective growth and reproduction of mussels. Two introduced, nonindigenous bivalves, the zebra mussel and Asian clam, compete with and have been implicated in the declines of some native mussels. Overharvesting (first for the pearl button industry and later for the freshwater pearl culture) has depleted the mussel fauna in some rivers.

Some of Connecticut's 12 native species have experienced severe declines in distribution and abundance due to pollution and damming of streams. For example, the dwarf wedgemussel (*Alasmidonta heterodon*), historically known from seventy sites from New Brunswick to North Carolina, now persists in only two dozen locations, including a few sites in Connecticut (Fig. 12.15). It is included on the federal and state lists of endangered species. Most populations are declining. It lives in mud, sand, or gravel bottoms of streams with slow to moderate current, often near shore under overhanging trees. Maximum length is less than 2 inches. The yellow lampmussel (*Lampsilis cariosa*), found along the Atlantic coast from Nova Scotia to Georgia, formerly occurred in the Connecticut River, but this species is now likely extirpated from Connecticut. It has a distinctive yellow, oval shell.

Mussels generally are most abundant in water less than about 6 feet deep. They may occur abundantly in unpolluted rivers with a good current. With some exceptions, they are most common in stable substrates of gravel, sand, or sand-gravel mixtures. Freshwater mussels rarely occur in acidic or low-calcium waters, which are unfavorable for shell production and maintenance. River mussels, which are subject to a high level of physical stress, tend to have a thicker, stronger shell than do lake species. The largest local species grow to a length of about 6 inches.

In most mussel species, the sexes are (with rare exception) separate—that is, individuals are either male or female, rather than having both sexes represented in one individual as in other mollusks such as snails. However, in some populations of eastern elliptio (*Elliptio complanata*), individuals may change their sexual function from male to female as they grow (and some become hermaphrodites).

Male mussels shed their sperm into the water. In some species, eggs are fertilized in summer or early fall as females draw in sperm through the incurrent siphon. The embryos overwinter in the female, and larvae are released the following spring or summer. Others exhibit spring fertilization and release of larvae in summer of the same year.

Reproduction in freshwater mussels is closely tied to fish populations. Mussel larvae, which have hinged valves and are called glochidia (singular, glochidium), are released and become temporary and obligatory parasites on a fish (or, much less often, a salamander). Females of some mussels, including the eastern lampmussel (*Lampsilis radiata*) and yellow lampmussel, lure potential host fish with convincingly minnowlike flaps that protrude outside the shell. When a fish comes to investigate the undulating "minnow," the mussel bombards it with glochidia.

The larvae generally die within several days if they do not attach to the gills or skin of a suitable host. Some species are host-specific whereas others are able to parasitize several different species. Host fishes of the eastern floater *(Pyganodon cataracta)* (Fig. 12.15), a common, thin-shelled species found on mud or sand, sometimes gravel, of slow creeks, rivers, reservoirs, lakes, and ponds, include a wide variety of species such as threespine stickleback, white sucker, bluegill, pumpkinseed, rock bass*, common carp*, and yellow perch. The alewife floater *(Anodonta implicata)* (Fig. 12.15), locally common on sand and gravel in low-gradient streams (abundant in the Connecticut River) and stable coastal ponds, resides within the range of its primary host fish, the alewife (also uses white sucker, threespine stickleback, pumpkinseed, and white perch). Known host fishes of the eastern elliptio *(Elliptio complanata)* (Fig. 12.15) include yellow perch, banded killifish, and various centrarchids*. This common species lives in mud, sand, gravel, or cobble substrates in slow or still water of permanent streams, lakes, and ponds. Glochidia of the eastern pearlshell (*Margaritifera margaritifera*), an uncommon species of sand, gravel, or cobble substrates in creeks and small rivers of the Connecticut River system, attach to Atlantic salmon and trout. Eastern lampmussel *(Lampsilis radiata)* (Fig. 12.15), becoming uncommon on various substrates in creeks, rivers, and natural ponds, uses various centrarchids*, bluntnose minnow*, white perch, and yellow perch. Triangle floater *(Alasmidonta undulata)*, a mussel of creeks and rivers, less common in ponds and lakes, is known to parasitize blacknose and longnose dace, common shiner, fallfish, pumpkinseed, largemouth bass, slimy sculpin, and white sucker (plus additional species in other states). Brook floater *(Alasmidonta varicosa)*, a rare species generally living in coarse sand or gravel of creeks and rivers (often with aquatic vegetation), attaches to longnose and blacknose dace, golden shiner, pumpkinseed, slimy sculpin, and yellow perch. Creeper *(Strophitus undulatus)*, an uncommon, typically riverine species that often inhabits sand and fine gravel, may attract its varied hosts (dozens of fish species plus two-lined salamander larvae) by extruding gelatinous strings to which its glochidia are attached (to a fish the strands probably look like something to eat). Creepers are unusual because they can complete the life cycle without attaching to a host. However, in general, healthy and diverse fish populations are important to freshwater mussels. The possible impact of introduced fishes needs further study.

The mussel larva becomes covered by the tissues of the fish host and remains encapsulated usually for about a week to several months. Eventually, the young mussel breaks out of the cyst and settles to the bottom, where it burrows into the substrate, feeds on detritus, and develops into a mature mussel in one to several years. Generally, larval mussels do little harm to parasitized fishes.

Muskrats are the most important mammalian predator on mussels. Piles of discarded shells along shorelines indicate where muskrats have eaten. Shortnose sturgeons often swallow small mussels whole. Channel catfish* also eat freshwater mussels, so it would be prudent to learn whether this fish, which is becoming increasingly abundant in Connecticut rivers, is having an impact on native mussel populations. Mussels that escape predators and other hazards can live a long life. Individuals of the eastern pearlshell may live more than 150 years, though most other species generally live up to a decade or two.

Additional freshwater mussels in Connecticut not mentioned above include tidewater mucket (*Leptodea ochracea*) (Fig. 12.15), in various substrates in quiet tidal freshwater of the lower Connecticut River; and eastern pondmussel (*Ligumia nasuta*) (Fig. 12.15), an uncommon species that lives in various substrates slow-water areas of lowland streams and natural ponds.

Fingernail Clams

Fingernail clams (Order Veneroidea, Family Sphaeriidae) are small freshwater clams, up to 0.5 inches long but often much smaller. They live as filter feeders on or shallowly buried in bottoms of lakes, ponds, streams, and wetlands. They are so inconspicuous that many people are surprised to learn that clams inhabit a familiar body of water. About 20 species occur in Connecticut. The most common, widespread species are the pond fingernail clam (*Musculium securis*), ubiquitous peaclam (*Pisidium casertanum*), ridgebeak peaclam (*P. compressum*), rusty peaclam (*P. ferrugineum*), and grooved fingernail clam (*Sphaerium simile*).

Some species, such as the swamp fingernail clam (*Musculium partumeium*) inhabit temporary waters. These survive pond drying by sealing the shell closed while lodged in mud or debris at the bottom of the pond basin, which greatly reduces water loss until the pool refills.

In the absence of flowing water, fingernail clams lack any means of moving large distances by themselves, but they sometimes travel from pond to pond while clamped onto the toes of an amphibian or the feathers of a duck. These clams are hermaphroditic and can self-fertilize. The young develop in the adult's gills and emerge as immature but fully formed shelled clams. Thus one clam transported by a frog or duck can establish a new population in a pond.

Non-native Freshwater Clams

Asian clam (*Corbicula fluminea*), native to Asia and Africa, became established and abundant in the Connecticut River in the 1990s (Fig. 12.16). Possibly the population originated from fish bait gathered or purchased elsewhere and discarded in the river, though this is simply speculation. Regardless, they thrived in the river and are now numerous in some tributaries and have recently spread to some tributaries of the Housatonic and Thames rivers (Jay Cordeiro).

Unlike our native clams, these mollusks do not require a natural environment. In fact, these clams colonized the cooling system of the Connecticut Yankee nuclear power plant, much to the alarm of plant operators. It had been thought that the Connecticut River environment was too cold for this warm-water species. The power plant had to initiate continuous chlorination to control *Corbicula*.

Asian clams not only infest power plants but also alter ecosystem processes by filtering phytoplankton and seston from the water column and collecting food from streambeds by using cilia on the foot to gather subsurface organic matter. They may sometimes compete with native freshwater mussels for food and space, but generally they are not so detrimental as zebra mussels.

These small clams (usually less than 1 inch long in Connecticut) are prolific breeders that mature quickly and produce many offspring. Usually the sexes are separate, but

Fig. 12.16. Non-native Asian clams became abundant in the Connecticut River in the 1990s. These are from the Mattabesset River in Middletown-Cromwell.

some are hermaphrodites capable of self-fertilization. Once they become established, Asian clams are able to spread rapidly to new areas by secreting mucous strands that catch the current and move them along. In tidal rivers, they can use this method to move either upstream or downstream. At the present time, there are no predators, parasites, or diseases that have a significant limiting impact on the clam's populations. Cold water temperatures likely are the primary limiting factor.

Zebra mussel (*Dreissena polymorpha*) is the European starling of the mollusk world (Fig. 12.17). Native to Eurasia, it became established in the United States in the mid-1980s. Now it is widespread and often tremendously abundant in hard, nonacidic waters (where Ca^{2+} exceeds 12 mg/l and pH is above 7.3). The species was first found in Connecticut in 1998 in East Twin Lake in Salisbury. Surveys in 2000 and 2001 found increased numbers and size, plus a newly established population in adjacent West Twin Lake. Zebra mussels form dense aggregations that attach to nearly any solid surface, including other mollusks. In East Twin Lake I found zebra mussels attached to the outside of live Chinese mysterysnails and inside large empty shells of banded mysterysnails. Jay Cordeiro found them attached to eastern elliptios. Sometimes they attach to crayfish.

Zebra mussels do their damage by reproducing abundantly and clogging water conduits of municipal water supply systems, or by outcompeting, overwhelming, and completely extirpating other mollusks such as native freshwater mussels.

Some zebra mussel impacts are complex. In large numbers, their filtering activities can remove large amounts of material from the water and exert major changes in the ecosystem. For example, in the tidal freshwaters of the Hudson River in New York, deeper waters recently impacted by zebra mussels had reduced density of bottom-dwelling invertebrates (macrozoobenthos), perhaps resulting from a reduction in food supply caused by the feeding activity of zebra mussels. Shallow waters exhibited increased density of large bottom-dwelling invertebrates that may have been related to increased biomass of macrophytes and attached algae, which likely resulted from increased water clarity caused by zebra mussel feeding.

Common Saltwater Clams

Among the several well-known clams of Long Island Sound, the premier species is the **eastern oyster** (*Crassostrea virginica*) (Fig. 12.18). This bivalve is famous as seafood but, due to its often irregular and unclamlike shape, is often not recognized for what it is when seen in the field by "landlubbers" and the culinarily meek. Oysters grow to a length of about 7 inches.

Oysters formerly blanketed subtidal and low intertidal bottoms in virtually every river mouth and harbor around Long Island Sound. Harvest from these vast native beds (and new beds created through transplanting of seed oysters in deeper waters) boomed in Long Island Sound in the 1800s, with a peak around the end of the century. Around that time, the impacts of industrial pollutants and sewage flowing down rivers began to take their toll, wiping out some oyster beds and contaminating others. Outbreaks of typhoid fever linked to oysters, and ongoing destruction of coastal habitats, contributed to a major decline in the oyster

Fig. 12.17. Zebra mussels, here attached to a Chinese mysterysnail, recently became established in Connecticut and could have detrimental impacts on native mussels.

Fig. 12.18. Clockwise from upper left: quahog, eastern oyster, transverse ark, ribbed-mussel, blue mussel, and common jingle.

industry in Long Island Sound by the 1920s. In recent decades, with better pollution control and expansion of suitable beds through dumping of vast quantities of shell imported from the mid-Atlantic coast, oysters increased in abundance and in commercial importance in several areas in the western and central Sound, where today there are many tens of thousands of acres of productive oyster beds. However, not all is well. Most of the state's oyster beds are closed due to sewage contamination, and parasitic diseases known as MSX and dermo, which may have resulted from unusually warm water temperatures, wiped out many oysters in Long Island Sound in the late 1990s.

Other than people, the main predators are sea stars, mud crabs, and snails (oyster drills), except in some brackish water habitats that are not salty enough to support the predators. Oysters in Long Island Sound spawn in summer, from late June or early July through September.

The remainder of this section provides a brief synopsis of other common clams of Connecticut's coast and Long Island Sound. They are arranged roughly by size, beginning with the largest clams (up to 8 to 10 inches long) and ending with the small ones and a very unusual species.

Atlantic surfclam (*Spisula solidissima*) (Fig. 12.19). Most massive clam along the U.S. Atlantic coast; common on sandy bottoms offshore or in lowest intertidal zone; more common in the eastern Sound than in the west; herring gulls sometimes take clams exposed by lowest tides or washed ashore during storms, break open the shells by flying up and dropping them onto a hard surface.

Sea scallop (*Placopecten magellanicus*) (Fig. 12.20). On sandy bottoms primarily in deep water in the eastern

Fig. 12.20. In Long Island Sound, bay scallops (left) are more common that sea scallops (right).

Sound; broken shells wash up on beaches after storms; tiny inquiline snailfishes sometimes occupy the mantle cavity (see Chapter 17); shells of large adults often adorned with sponges, hydroids, bryozoans, tunicates, tube-dwelling polychaete worms, and algae; life cycle includes planktonic and crawling larval stages.

Atlantic jackknife (*Ensis directus*) (Fig. 12.21). Longest clam (10 inches); common in mud and sand bottoms in and just below intertidal zone.

Northern quahog (*Mercenaria mercenaria*). Known to seafood consumers as hard-shelled clam, cherrystone, littleneck, and round clam; recognizable by purple stain on inner surface of thick shell (Fig. 12.18); common burrower in subtidal sand or mud; beach shells often riddled with holes caused by boring sponges (Fig. 10.2), or adorned with cemented meandering tubes made by polychaete worms (Fig. 11.3).

Softshell clam (*Mya arenaria*) (Fig. 12.22). Common in sandy-muddy bottoms in intertidal and shallow subtidal zones; important commercial and recreational species, supplies market for fried clams and "steamers."

Fig. 12.19. Atlantic surfclams are most abundant in eastern Long Island Sound. These shells are from two different individuals.

Fig. 12.21. Atlantic jackknife clam is a burrower in sand or mud bottoms.

Fig. 12.22. Softshell clams supply the market for steamers and fried clams.

Ribbed-mussel *(Geukensia demissa)* (Figs. 12.18 and 12.23). Abundant, partially embedded in peat or mud at base of cordgrasses along margins of tidal creeks and coves; eaten by herring gulls, which open the hard shells by dropping them on rocks, pavement, roofs, or other hard surfaces; see Chapter 3 for further ecological information.

Blue mussel *(Mytilus edulis)* (Fig. 12.18). Locally common in dense aggregations on intertidal and subtidal rocks; clusters also on soft sand-silt bottoms in shallow subtidal lagoons; takes about a year to mature; goes through planktonic and crawling stages before settling down to sedentary lifestyle; eelgrass is important habitat of one developmental stage; attachment to rocks or shells occurs via strong byssal threads that mussel secretes (small mussels actually can move [very slowly] by attaching and pulling on these threads); threads may ensnare predatory snails; eaten by tautogs (blackfish), sea stars, lobsters, American oystercatchers, gulls, and people intent on steaming them with garlic.

Bay scallop *(Argopecten irradians)* (Fig. 12.20). Most numerous in the eastern Sound where eelgrass is abundant; bay and sea scallops react to touch, chemicals in water, and, through numerous "eyes," to passing shadows (useful in sensing potential predators); can swim quickly using jets of water ejected when shells are rapidly and repeatedly opened and closed; one-year-olds and the few that survive two years breed in late spring and early summer; larvae planktonic; juveniles attach to eelgrass or other surfaces above bottom; adults are bottom dwellers.

File yoldia *(Yoldia limatula)*. Common in bottom muds of shallow subtidal areas; deposit feeder and suspension feeder; significant food resource for flounders; smooth shell round on one end, pointed on other, less than 2.5 inches long.

False quahog *(Pitar morrhuanus)*. Up to 2 inches long; common in bottom mud of shallow waters; gray shell covering often eroded away to expose white at shell apex.

False angelwing *(Petricolaria pholadiformis)*. Uses ribbed end of shell to bore into peat or clay; fairly common as beach shell; look in beach-cast chunks of peat (Fig. 12.24).

Common jingle *(Anomia simplex)* (Fig. 12.18). Common as attractive, glossy beach-cast shell; lives attached to solid objects, including living lobsters, mainly in subtidal areas; shell that attaches has hole in it, rarely washes up on beaches.

Transverse ark *(Anadara transversa)* (Fig. 12.18). Common in sand or mud in subtidal areas.

Atlantic awningclam *(Solemya velum)*. Burrows in mudflats; up to an inch long; brown, yellow-streaked shell covering extends beyond shell as distinctive fringed edge.

Fig. 12.23. Ribbed-mussels live along the banks of tidal creeks and salt marshes.

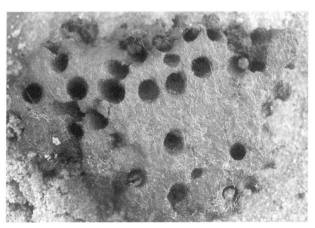

Fig. 12.24. This chunk of marsh peat, containing false angelwings and their burrows, washed up on a beach.

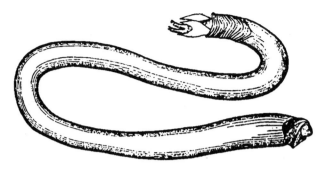

Fig. 12.25. Common shipworm. The shell, at the front end of the animal at lower right, is used to tunnel through wood. From Verril and Smith (1874).

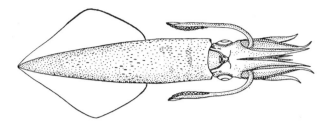

Fig. 12.26. Longfin inshore squid are seasonally common in Long Island Sound. From Miner (1950).

Dwarf surfclam *(Mulinia lateralis)*. Common in bottom muds of bays and shallow water; frequent food for various fishes and sea ducks; less than an inch long; one end of shell has a flattened area bordered by distinct ridge.

Atlantic nutclam *(Nucula proxima)*. Common deposit feeder on muddy bottoms in shallow subtidal areas; small (to 1/4 inch), shiny, triangular, greenish or grayish; has row of teeth on inside edge of shell on each side of apex.

Amethyst gemclam *(Gemma gemma)*. Common in sandy-silty bottoms of intertidal areas and subtidal shallows; small, triangular; only up to about 0.1 inches long; differs from softshell clams and quahogs by brooding young (versus having planktonic larvae); unlike nutclams, lacks rows of teeth along inside edge of shell.

The most highly modified clam is the **common (naval) shipworm** *(Teredo navalis)* (Fig. 12.25). This unusual bivalve has a long, wormlike body, and the shell is reduced to two tiny "teeth" used to bore tunnels through water-logged wood. It feeds on "sawdust" (aided by intracellular cellulose-digesting bacteria in gut) and plankton. The free-swimming larval stage attacks wood. Once a shipworm is established, it does not leave its wooden abode. Shipworms are beneficial in "recycling" wood, but they can destroy pilings and ship hulls. Shipworm abundance recently increased along the coast of northern New England, likely as a result of steadily warming sea temperatures.

Squids and Octopuses

All of the several hundred species of cephalopods (Class Cephalopoda) are marine or estuarine, and all are predators on other invertebrates and fishes, captured with the tentacled arms and torn to bits with beaklike jaws. Some immobilize their prey with salivary toxins. Most have well-developed eyes and excellent vision.

Longfin inshore squid *(Loligo pealeii)* is common throughout Long Island Sound, arriving in mid-spring and departing for deeper offshore waters in mid-fall (Fig. 12.26). Squids swim in schools and prey on various crustaceans, small fishes, and smaller squids. They live up to a year or two, spawn, and die. Spawning involves the transfer of sperm from male to female via a specialized arm as the squids embrace. "Sea mops," the masses of fingerlike jelly capsules containing the embryos, normally are attached communally to submerged objects, but sometimes I found them washed up on beaches. Juvenile abundance in the Sound peaks in late summer.

Chapter 13

Chelicerates

Phylum Chelicerata, an assemblage of some 80,000 named species of arthropods, has in common with each other six pairs of appendages (the chelicerae and pedipalps at the front of the head and four pairs of walking legs) in the adult stage. They lack mandibles and antennae. About 9,560 species occur in the United States.

Horseshoe Crab

The horseshoe crab (*Limulus polyphemus*, Class Merostomata) is a contemporary representative of an ancient lineage (Fig. 13.1). It is fairly common along the Connecticut coast. You may see it cruising over sandy-silty bottoms of shallow coastal lagoons in late spring and early summer.

This animal is not a crab or even a crustacean but rather is more closely related to scorpions, spiders, mites, and ticks, as is clear if you carefully study its anatomy. For example, crustaceans have jaws and two pairs of antennae whereas the horseshoe crab has neither.

Horseshoe crabs eat bottom-dwelling invertebrates and detritus. Look on the underside to see the spiny bases of the walking legs that form a "food mill" for grinding up hard-shelled animals.

Fig. 13.1. Horseshoe crabs spawn on the shores of Long Island Sound in late spring.

Horseshoe crabs spawn on sandy or gravelly beaches in May and June, often during a spring tide. Females plow along the beach next to the water, followed closely by one or more males, which fertilize the numerous eggs as they are laid at the water's edge. Frenzied avian activity often accompanies horseshoe crab spawning as numerous hungry birds rush in and squabble over the succulent eggs. Horseshoe crab eggs are a critical food resource for many northbound migrating shorebirds.

Hatchlings from eggs laid in the upper tidal zone emerge and disperse into shallow water two weeks later during the next spring tide. Growth is slow; up to a decade is required before an individual attains sexual maturity.

Concern over declining populations of horseshoe crabs, which are harvested for use as bait in various commercial fisheries, and the impact this may have on migratory birds, recently led the Atlantic States Marine Fisheries Commission (ASMFC) to direct states, including Connecticut, to reduce the harvest of horseshoe crabs by 25 percent. ASMFC manages the species through an Interstate Horseshoe Crab Fishery Management Plan. Harvest is illegal without a commercial license.

For effective protection of spawning horseshoe crabs and their eggs, beach cleaning (raking, grading) should be conducted before May 10 or after June 25, and then only with the required permit from the Connecticut Department of Environmental Protection.

Sea Spiders

Sea spiders (Class Pycnogonida) are slow-moving, inconspicuous, spiderlike animals that live among *Eudendrium* hydroids, ectoprocts *(Bugula turrita)*, or other attached organisms on the sea bottom or on pilings. Approximately 1,000 species are distributed worldwide in all oceans. They use their sucking or rasping mouthparts to feed on the small sedentary invertebrates among which they live. Young may

Fig. 13.2. Male sea spiders carry the egg masses. From Miner (1950).

be commensals or parasites of cnidarian medusae or echinoderms. Egg fertilization occurs sexually as a result of mating between a male and a female. Males carry and brood the eggs until they hatch into crawling larvae (Fig. 13.2). Among the several species occurring in southern New England are the lentil sea spider (*Anoplodactylus lentus*), which has brown or black eyes, and long-necked sea spider (*Callipallene brevirostris*) which has red eyes.

Spiders

Like snakes, spiders evoke aversion in many people but fascination and admiration in others. Spiders (Class Arachnida, Order Araneae) comprise a diverse, worldwide group of some 35,000 named species (plus vast numbers of undescribed species). About 3,000 species occur in the United States and Canada. Just a few of the approximately 30 families and 500 species occurring in Connecticut are discussed here.

Spiders live in nearly all kinds of terrestrial habitats, including buildings and vehicles. One summer a spider traveled around with me for several weeks within the housing of the side-view mirror of my often-used truck. Spiders also inhabit wetlands and some even dive under water. For example, **fishing spiders** (Pisauridae: *Dolomedes* spp., 6 species in Connecticut) are common along the margins of freshwater habitats. They use their legs to row or gallop across the water surface, and sometimes swim and dive. Air trapped in the hairy coat, appearing as a silvery sheen when the spider is submerged, serves as a supply of oxygen during prolonged dives. Some fishing spiders also range into woods far from water; I frequently found females, with a 3-inch legspan, perching in the open on tree trunks and other woody stems at night in late summer and fall, sometimes at near-freezing temperatures (Fig. 13.3). Fishing spiders overwinter as juveniles and mature in late spring.

All spiders are predatory, and nearly all of their victims are insects, killed or paralyzed by envenomation through two "fangs," which are used to suck up the venom-digested juices of the prey. In general, spiders prey on whatever insects happen to be available, but many reject ants, wasps, bugs, and beetles. Fishing spiders sometimes prey on small fishes and amphibian larvae. Some of the larger spiders use

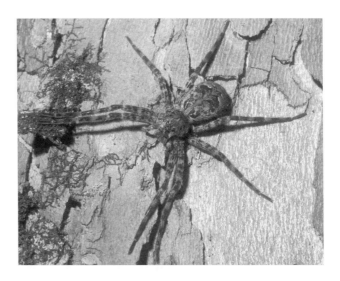

Fig. 13.3. This fishing spider, with a legspan of 3 inches, was on a stream-side tree trunk at night in late summer.

their mouthparts to mash the prey before ingesting the fluids. Ecologically, spiders function importantly in reducing insect populations. I like to have a few spiders around my windows to rid the house of any flies or mosquitoes that sneak in.

Orb-web spiders (Araneidae; 40+ species in Connecticut) are best known for their beautiful orb webs used for prey capture. The spider perches in the center of the web or in a shelter to the side. There is evidence that webs are not just passive food collectors—some insects are actually attracted to the webs, due to the webs' reflectance of light (including ultraviolet). On the other hand, a visible web might allow avoidance by some prey, particularly in daytime, so there is a tradeoff.

The **black-and-yellow garden spider** (Araneidae: *Argiope aurantia*) is the largest of the state's orb-weaving spiders (Fig. 13.4). The females are most conspicuous; at about an inch long (excluding the legs), they are three times as long as the males and spin much larger webs, usually prominently adorned with a vertical streak of dense silk near the center. I found these spiders and their showy webs in tall weeds and grass in my garden and in a red-cedar right next to my house. You might wonder why the web of this spider would have such a conspicuous white streak on it. Several explanations have been proposed, including protection from damage (conspicuousness makes webs highly visible and thus less likely to be contacted and damaged by large animals such as birds or mammals) and prey attraction (stabilimenta reflect ultraviolet light and may attract spider prey such as bees; the spider's body also reflects UV). Use of web decorations for prey attraction has its costs; sphecid wasps, the only known predators, may use the decorations to locate webs and their spider prey.

Fig. 13.4. Why do black-and-yellow garden spiders adorn their webs with a conspicuous streak of dense silk?

Not all spider webs are flat. For example, the central feature of the **filmy dome spider**'s (Linyphiidae: *Prolinyphia marginata*) web is a dome about 4 to 6 inches in diameter, constructed in bushes or under or among rocks or logs (Fig. 13.5). I became familiar with the great abundance of this spider on misty summer mornings when clinging dew made the webs conspicuous.

Spiders may respond in different ways when their webs catch few prey. Some may move and build in a new location,

Fig. 13.5. Filmy dome spiders and their distinctive webs are abundant in Connecticut forests.

whereas others may stay put and simply subsist on less, which results in weight reduction or slow growth. Web structure and location often result in the capture of prey types that the spider is best able to handle.

Prey caught in the web may be bitten immediately (or somewhat later) or sprayed with silk while the spider remains at a safe distance (stinging insects may be handled this way). Some difficult-to-handle prey may be discarded or allowed to escape.

Though spiders have silk-producing glands opening at the rear end of the body, not all spiders build webs for prey capture. Some webless species, such as **wolf spiders** (Lycosidae; about 45 species in Connecticut) are roving predators that actively hunt prey in fields, woods, marshes, and shores. You can easily detect wolf spiders at night by looking for their reflected eye-shine in the beam of a lantern. **Jumping spiders** (Salticidae; about 50 species in Connecticut), recognized by their three rows of eyes, have excellent vision and generally use catlike stalking and ambush to obtain prey. I saw jumping spiders catch skippers (butterflies) larger than themselves. One of Connecticut's oddest-looking jumping spiders, the rare *Peckhamia picata*, appears to be an ant mimic that may benefit from the ant aversion of certain birds that normally eat spiders without hesitation. **Crab spiders** (Thomasidae; about 20 species in Connecticut) obtain prey by stealth and ambush (Fig. 13.6). I frequently found crab spiders that had captured bees, wasps, flies, and other insects on goldenrods and milkweeds in late summer and fall. Crab spiders have eight eyes in two rows, and their legs are in an arrangement reminiscent of crabs, hence the common name. These spiders may gradually change color (between white and yellow) and often closely match the color of the flowers on which they perch. Because they absorb ultraviolet light, they may be inconspicuous to insects attracted to UV-reflecting flowers. Although none of these spiders build webs, many make silk-lined retreats for shelter.

Silk may be used for purposes other than web-making and housing. Some use silk simply as a support "rope." Eggs generally are enclosed in a silken sack. Hatchlings, and adult males of some species, may disperse to new areas by releasing a long strand of silk and ballooning through the air when the silk catches the wind.

Mating occurs by the male inserting one or both of his palpi (appendages behind the mouth parts) into the genital opening of the female. Prior to copulation, he transfers sperm from his genital opening to the palpus. Parthenogensis (development of eggs without fertilization) may occur in some spiders.

Spider eggs are covered with silk and may or may not be attended by the female (Fig. 13.7). You may see spider egg

Fig. 13.6. This crab spider caught a honeybee visiting milkweed flowers in August.

Fig. 13.7. I found this black widow female with her egg cocoon and a prey carcass by overturning a log.

cocoons attached to webbing under the eaves of your house or in other sheltered locations. In contrast, female wolf spiders carry their egg cocoon around with them, attached to the spinnerets at the end of the body; the hatchling spiders may climb onto the mother's body. Garden spiders *(Argiope)* construct egg cocoons in late summer. Small spiderlings overwinter in the cocoons and emerge in the spring.

Wasps probably are the primary predators of spiders, which are stung and used as food for the young. Mud dauber wasps, for example, stock their pipe-organ mud nests with spiders and lay an egg in each pipe. Upon hatching, the wasp larvae feed on the stored spiders.

Spiders tend to be short-lived. Most orb weavers live less than a year, with eggs being the only stage that survives winter. In our area, wolf spiders are among those that live the longest, but this amounts only to a few years.

Spiders of very few species ever bite humans, and even fewer species can be regarded as dangerous. Venomous **black widows** (Theridiidae: *Latrodectus* spp.) do occur throughout much of the state but are not very common. Widows build a "messy" irregular-mesh web with a funnel retreat in sheltered sites. The large, black, red-blotched females mature about three months after emerging from the egg cocoon, and some may live up to a few years. Despite the spider's common name, males usually are not eaten by the female after mating in late spring. The bite of the female is dangerous to humans but, when disturbed, these spiders generally flee rather than attack (Fig. 13.7). They are most likely to bite if disturbed while guarding their egg cocoon. The bite generally is not initially painful, but severe symptoms (cramps, swelling, anxiety) develop within an hour. With proper treatment (calcium gluconate and antivenin) symptoms disappear within a few hours.

Harvestmen

Harvestmen or "daddy-long-legs," are harmless arachnids that resemble spiders. More than 3,000 species (Class Arachnida, Order Opioliones) occur worldwide in humid temperate and tropical regions. They use their outlandishly long, slender legs, which do not regenerate if lost, to clamber over plants, ground, or rocks, and they are not averse to entering houses during their mainly nocturnal prowls. The sensitive second pair of legs functions somewhat like antennae. Individuals I observed by flashlight sometimes probed other individuals with these long legs. Some that may have been disturbed by my flashlight bobbed their body up and down. The sides of the body have scent glands that produce defensive secretions.

Harvestmen prey on or scavenge tiny insects, spiders, snails, earthworms, fruit, and fungi, They do not survive dry conditions or food and water deprivation for very long. Unlike other arachnids, males have a penis for copulation. Females use their long ovipositor to insert eggs into damp soil or humus. Eggs hatch in spring. Few adults survive the winter.

Ticks and Mites

Ticks and mites differ from spiders and scorpions in having the head, thorax, and abdomen fused into one unit (Fig. 13.8 and 13.9). This large, worldwide group (Class Arachnida, Order Acarina) includes more than 40,000 named species,

Fig. 13.8. This blood-engorged dog tick dropped off its host and will eventually lay its eggs.

most of which are mites. There are vast numbers of undescribed mite species.

Ticks (over 1,000 species worldwide) generally are larger than mites and all are parasites as both adults and young, some being important disease carriers. Females must ingest a blood meal in order to complete development of the eggs. According to the Connecticut Agricultural Experiment Station, 22 different tick species have been collected in Connecticut. Six of these are known to feed on humans and in doing so can transmit six different diseases (Lyme disease, ehrlichiosis, babesiosis, Rocky Mountain spotted fever, powassan encephalitis, and tularemia).

Black-legged tick (*Ixodes scapularis*), formerly known as **deer tick,** is one of Connecticut's most infamous animals, as a result of carrying and transmitting the spirochete bacterium (*Borrelia burgdorferi*) that causes Lyme disease. The tick is unfortunately common in grassy and wooded areas statewide.

Black-legged ticks complete their life cycle of four developmental stages (egg, larva, nymph, and adult) over a period of two years. Engorged females lay eggs on the ground in spring or early summer. Eggs hatch in summer. The minute larvae attach to a host such as a bird or small mammal, ingest a meal of blood, and drop off the host. Larval ticks rarely acquire the Lyme disease bacteria when feeding. White-footed mice are the primary reservoir of the bacteria, and this is the species from which the larvae are most likely to acquire the spirochete. Eastern chipmunks also easily transmit the spirochete to feeding ticks, whereas ground-dwelling songbirds are less effective reservoirs. Larvae that acquire bacteria are capable of transmitting the spirochete during their late nymph and adult stages.

After this single blood meal, the larvae then molt into larger nymphs, which overwinter. The next year, mainly from May to July, the nymphs seek a host and partake of a blood meal. Again, various small and large mammals and birds are likely hosts, with white-footed mice the foremost species. The nymph stage is the one most likely to bite humans. Up to about 25 to 35 percent of the nymphs can be infected with the bacteria. That fits with the percentage of times I developed a red rash after discovering an imbedded tick on my skin.

After feeding, nymphs drop off the host and molt into the adult stage, which is larger than the nymph stage but notably smaller than an adult American dog tick. That same fall, adults, of which up to 50 to 70 percent may be infected with *Borrelia,* perch on the tips of vegetation and attach to and feed on passing deer or other large mammals, but humans are not a common host for the adults. Adults mate while on the deer. Then adult females drop off, overwinter, and lay eggs the following spring. Adults may also seek medium to large mammal hosts on warm winter days or in spring. Adult males die after mating.

Black-legged ticks may crawl on the host for several hours before biting. Once the mouthparts are imbedded in the host, feeding may continue for several days. The spirochaete is transmitted when the tick is engorged and begins to return blood plasma to the host. Removal of ticks before or soon after they bite may allow one to avoid contracting Lyme disease.

First recognized in Lyme, Connecticut, in the mid-1970s, Lyme disease now has been recorded in all of the contiguous United States and on other continents. The disease, often accompanied by a red rash and later by arthritic symptoms and/or nervous system problems, generally responds well to early treatment with antibiotics. An FDA-approved human Lyme disease vaccine is available. Wild animals that carry the spirochete do not exhibit disease symptoms, but the bacterium may cause arthritis in dogs and horses.

Black-legged ticks also spread other unpleasant diseases: human babesiosis (caused by a protozoan parasite of red blood cells, *Babesia microti*) and human ehrlichiosis (caused by bacteria, *Ehrlichia* spp.), each of which produce flu-like symptoms.

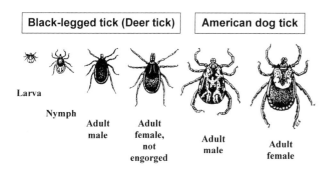

Fig. 13.9. Ticks are ectoparasites that can transmit Lyme disease and other diseases to humans. Adult black-legged ticks are only about 1 millimeter wide. Kirby C. Stafford, III, Connecticut Agricultural Experiment Station.

Fig. 13.10. Underside of a water mite.

American dog tick (*Dermacentor variabilis*) adults (large size, white marks on upper body) are common on wild and domestic mammals. These ticks can transmit the causal agent *(Rickettsia rickettsii)* of Rocky Mountain spotted fever, but this disease is quite rare in Connecticut.

Mites exhibit diverse habits. Some are parasitic on animals whereas others ingest plant juices or prey on smaller animals. Some live on animals (e.g., in human hair follicles) without producing obvious symptoms. All but a few mites are egg layers.

Water mites (Hydracnidia, Fig. 13.10) swim or crawl in streams, ponds, and lakes, especially among vegetation. They can be tremendously abundant, but you'll have to get down on your knees and look closely to see these minute animals. Most are fluid-ingesting predators or parasites active only in daytime. Mating involves transfer of a spermatophore from male to female. In most species, the usually red eggs are deposited on rocks, plants, or other objects. Larvae are parasitic on aquatic insects such as water boatmen, water scorpions, backswimmers, midges, and mosquitoes. The next stage, the nymph, generally is predatory on small aquatic organisms. However, the life histories of mites are exceptionally variable among species. Some, for instance, live their whole lives within freshwater mussels. Others occur in vernal pools and withstand the dry period buried in damp mud or debris. Water mites sometimes attach themselves to insects or birds and get flown from one place to another (Fig. 13.11).

Fig. 13.11. Hundreds of parasitic water mites are attached to the underside of this blue dasher dragonfly in late July.

Chapter 14

Insects, Centipedes, and Millipedes

If you wish to have a life filled with endless novelty and incredible (if often subtle) drama, simply watch insects and cultivate your interest in their lives. Insects also easily satisfy appetites for beauty and the bizarre.

There are far more species of insects (Phylum Arthropoda, Class Insecta) than of any other class of animals. More than 950,000 insect species have been formally described by scientists, but recent surveys of tropical insects and regional analyses of global insect diversity suggest that the total number of species may range up to 10 million. Species richness of insects far surpasses that of plants. Roughly 100,000 species have been recorded in the United States. Insects are ecologically significant if not dominant organisms in nearly every habitat, aside from the oceans.

The most speciose insect groups in the United States are the beetles (Order Coleoptera, 25,000 species), bees/wasps/ants (Order Hymenoptera, 18,100 species), flies (Order Diptera, 20,630 species), and butterflies/moths (Order Lepidoptera, 12,300 species). These groups are also the most species-rich insect groups in Connecticut, each represented by well over 1,000 species. The following pages describe the diversity and natural history of selected groups of Connecticut's insects.

Flightless Insect Relatives

The Class Entognatha includes springtails, proturans, and entotrophs, arthropod orders that formerly were regarded as insects. However, these differ from insects in the fundamental structure of the head and mandibles and so are now regarded as distinct from insects. All of these insect relatives exhibit very little external change in appearance as they grow, other than size. Except for springtails, they are unlikely to be found except by intent specialists.

Springtails (Order Collembola) are cosmopolitan and include about 6,000 species occurring in all types of habitats, including land and fresh and salty water. Some 840 species inhabit North America north of Mexico. Most are less than 0.2 inches long. Springtails are named for the taillike structure (furcula) that folds under the body of many species. Rapid extension of the furcula launches the springtail briefly into the air. Presumably, this facilitates escape from potential predators.

Most springtails eat minute or decaying organic matter; digestion may be assisted by gut-dwelling bacteria. Some are predatory. Reproduction generally involves females taking up sperm deposited on objects by the males.

Snowflea (*Hypogastrura nivicola*) is locally common in woods (Fig. 14.1). This species is most noticeable when large numbers of dark-blue individuals swarm on the snow surface on warm winter days. Sometimes they invade maple-sap-collecting buckets. Females deposit eggs on the ground in early spring.

Other common springtails are aquatic or amphibious and live on the water surface or on vegetation along the

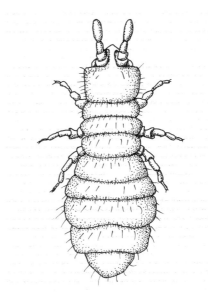

Fig. 14.1. Snowfleas sometimes swarm on snow. Redrawn from Arnett (2000).

shores of lakes, ponds, or streams, in tidepools, or among rocks, algae, or debris in salt marshes and other intertidal areas.

Mayflies

Perched beside a stream, or rising over the water, mayflies reveal the delicacy and fragility that purist anglers attempt to imitate at the end of a flyfishing line. To most people, all mayflies look rather alike, except in size, but this group (Order Ephemeroptera) is quite diverse, represented by about 2,000 species worldwide and several hundred species in North America (Fig. 14.2). More than 100 species have been recorded in Connecticut.

Mayflies live in and near freshwater lakes, ponds, streams, and wetlands, and relatively few are tolerant of pollution. They are perhaps best known for the sometimes enormous numbers of individuals that simultaneously emerge from lakes and rivers, evoking excitement among fishes and anglers alike. Aquatic immatures (naiads), generally recognizable by their three tail filaments (Fig. 14.3), often live among rocks or on vegetation, but some are mole-like burrowers in muddy bottoms or debris. Most immature mayflies are detritivores or scrape algal material from rocks; relatively few are predators.

Mayflies exhibit gradual metamorphosis to the adult form. In most species, the eggs take several months to hatch. The immature stage often lasts about three to six months. At the end of their development, the larvae molt into the subimago stage and emerge from the water (they can fly); these soon molt into the imago or true adult stage, mate, and, in most species, die within a few days (they do not feed). Females generally deposit their eggs in the water. Both the larvae and adults are important foods for fishes and other aquatic animals.

Fig. 14.3. Larval mayfly of a burrowing species, *Hexagenia bilineata*. From Needham (1920).

Dragonflies and Damselflies

If beauty and diversity in color and impressive powers of flight explain the popularity of birds and butterflies, such attractiveness may suggest why an increasing number of field naturalists are turning their attention to this group of big-eyed, four-winged insects (Figs. 14.4 and 14.5). The diversity of the Order Odonata is large but not overwhelming. Some 5,760 species of dragonflies (Suborder Anisoptera) and damselflies (Suborder Zygoptera) exist worldwide, far fewer than birds. There are about 450 species in North

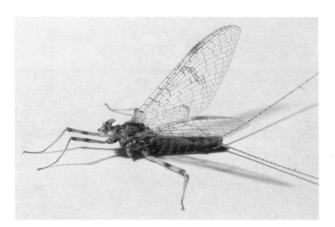

Fig. 14.2. Mayfly adults live a short life focused on mating and egg laying.

Fig. 14.4. Twelve-spotted skimmer is common in many lakes, ponds, slow streams, and freshwater wetlands.

Fig. 14.5. Ebony jewelwing (a damselfly) finishes off its prey (note insect leg protruding from mouth).

America and approximately 150 (nearly 50 damselflies and 100 dragonflies) in Connecticut.

Species differ greatly in distribution and abundance. The eastern forktail *(Ischnura verticalis)*, a damselfly that occurs in many kinds of marshy wetlands, ponds, lakes, and streams, is the state's most ubiquitous odonate. In contrast, the ringed boghaunter *(Williamsonia lintneri)* is a rare dragonfly that deposits its eggs in pools in fishless, sphagnum bogs and fens. This species is known from just a few sites in eastern Connecticut, and it is rare at each site. Most of the colonies of this species are in Rhode Island and Massachusetts.

The immature stages of most dragonflies and damselflies develop in marshes, ponds, lakes, or slow streams. Some, such as the common whitetail *(Libellula lydia)*, tolerate less than pristine conditions with organic pollution and low oxygen levels.

Other species are more sensitive and specialized. For example, rare tiger spiketails *(Cordulegaster erronea)* prefer perennially flowing cold springs, including small rivulets. Midland clubtail *(Gomphus fraternus)*, cobra clubtail *(G. vastus)*, skillet clubtail *(G. ventricosus)*, and umber shadowdragon *(Neurocordulia obsoleta)* are rarities along the sandy shores of the Connecticut River from Middletown north. Seaside dragonlet *(Erythrodiplax berenice)* nymphs develop only in salt marshes and brackish marshes, where the adults fly all summer. A rare spring flier, the beaverpond clubtail *(G. borealis)*, inhabits beaver-dammed streams.

Most species breed in permanent bodies of water. Wandering glider *(Pantala flavescens)* and spot-winged gliders *(P. hymenaea)*, both common in Connecticut, depart from this pattern. These strong-flying, pioneering insects use temporary and newly formed pools in open areas as the larval habitat. The wandering glider is an extraordinary colonist, so much so that the species occurs on all continents except Europe. Familiar bluet *(Enallagma civile)* damselflies also readily colonize temporary pools and newly created ponds. I found common green darner *(Anax junius)* larvae on the floor of a pond that dried up in summer. They survived the dry phase in moist sites under rocks or wood. I found that other dragonfly larvae do the same thing in intermittent streams.

Odonate mating is somewhat complicated but easy to observe. Male odonates are unusual in that the genital opening from which sperm is released is separate from the penis, so mating involves an extra step. The male first transfers sperm from his primary genital opening to the sperm vesicle at the base of his abdomen, then approaches a female and clasps the area behind her head with his caudal appendages. The female, hanging onto the male with her legs, curls her abdomen forward to the sperm vesicle immediately adjacent to his penis and obtains his sperm (Fig. 14.6). Females store the sperm until ready to lay their eggs. Males often guard the female with which they mated, sometimes remaining attached to her. For example, males of most spread-winged (Lestidae) and narrow-winged (Coenagrionidae) damselflies, various common skimmers (Libellulidae), and common green darners attempt to prevent other males from copulating with the female, thus maximizing the probability that the eggs will be fertilized by their own sperm.

Most dragonflies drop the eggs into the water while hovering and may or may not touch the tip of the abdomen to the water as the eggs are released. A male may clasp the female during hovering egg-laying episodes (e.g., members of the common skimmer family, Libellulidae). Damselflies and some dragonflies (e.g., common green darner) insert the eggs into algae, moss, decaying leaves, or the stems of herbaceous or woody plants, above or below water. Stream-dwelling clubtail dragonflies (Gomphidae) may place the

Fig. 14.6. A female (bottom) slaty skimmer obtains sperm from a male (top). Stephen P. Broker.

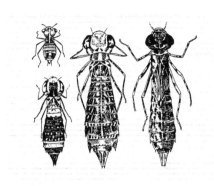

Fig. 14.7. These are nymphs (different ages) of the common green darner. From Kellogg (1905).

eggs in the bottom substrate. Various broad-winged (Calopterygidae) and narrow-winged (Coenagrionidae) damselflies may submerge completely when laying eggs in algae or plant material.

Hatching occurs usually in two to four weeks, though some species overwinter and spend several months in the egg stage. Species with overwintering eggs include spread-winged damselflies *(Lestes)* and various darners *(Aeshna)*, emeralds *(Somatochlora)*, and meadowhawks *(Sympetrum)*. These may also overwinter as larvae.

Depending on the species, the nymphs or naiads may burrow into the bottom sediments or detritus (e.g., clubtail and spiketail dragonflies, families Gomphidae and Cordulegastridae), hide under bottom debris, cling to or climb on submerged objects, or simply perch on the bottom (Figs. 14.7 and 14.8).

Dragonfly nymphs obtain oxygen by pumping water to the gills inside the end of the abdomen. Strong pumping ac-

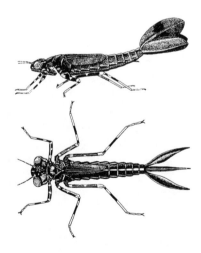

Fig. 14.8. Narrow-winged damselfly nymph *(Argia)*, showing three leaflike gills at rear end of body. From Kennedy (1915).

tion can also be used for propulsion. Damselflies also do some rectal pumping, but they have at the end of the body three platelike gills that probably are the major sites of gaseous exchange.

The length of the larval period varies among local species from a few months to multiple years. During this time the nymph goes through roughly a dozen molts and gradually grows and progresses toward the adult form. Presumably all species in Connecticut spend at least one winter in the aquatic (egg or larval) stage. Because many common species overwinter as larvae, a dip net raked over a pond bottom in the water in late fall, winter, and early spring often yields well-developed larvae.

Eventually larvae emerge from the water, the nymphal exoskeleton breaks open, and the adult emerges, expands and dries its wings, and flies off. Newly emerged adults have a soft integument, incompletely developed coloration, and fly weakly. They are full grown but not immediately sexually mature. The aerial adult stage lasts a few weeks to several months. Adults do not survive the winter in Connecticut (some migrate).

For most species, activity of flying adults begins in mid- to late spring and is maximal on sunny, warm summer days, and many fly and mate through early fall. However, some have restricted or unusual flight seasons and times. For example, rare ringed boghaunters, on the wing from late April to early June, are among the first dragonfly species to fly each year. I saw common green darners patrolling marshes by late April. Early season activity of adult emerald spreadwings *(Lestes dryas)* reflects their use of waters (temporary pools, marshes, or wooded ponds) that may dry up in summer. Eastern forktail damselflies sometimes fly as early as early April.

Some odonates fly late in the year. Abundant cherry-faced meadowhawks *(Sympetrum internum)* and yellow-legged meadowhawks *(Sympetrum vicinum)* fly all summer and well into fall, and I found active cherryfaces as late as early December. A few common green darners may fly on mild days into early December.

You may see some odonates flying even as darkness approaches. Typical dusk fliers include vesper bluet *(Enallagma vesperum)*, shadow darner *(Aeshna umbrosa)*, fawn darner *(Boyeria vinosa)*, swamp darner *(Epiaeschna heros)*, prince baskettail *(Epitheca princeps)*, and shadowdragons *(Neurocordulia* spp.*)*.

Some dragonflies engage in extensive movements or migrations. Common green darners, recognized by the green head and thorax and "bulls-eye" mark on the forehead, are large (about 3 inches long), very common, summer-flying pond dragonflies that make northward flights in spring. Migrants arrive beginning in late April, coincident with the passage of many migratory songbirds. Southward movements by the next generation occur in late summer. The southward migrations sometimes involve huge numbers of individuals that may be evident along the coast, usually peaking in September. Additional migratory species include other large darners, plus wandering glider, spot-winged glider, and saddlebags (*Tramea* spp.).

All odonates are predatory, both as immature aquatic nymphs and as flying adults. Adults use their huge eyes and excellent vision to detect prey. With quick, maneuverable flight (forward, backward, sideways, upside-down) and spiny forelegs that form a "basket," they readily nab flying insects, including such human annoyances as mosquitoes and deer flies. Such items are reduced to pulp in the dragonfly's or damselfly's strong jaws. Some species take relatively large prey of nearly their own size. Odonate nymphs often lie in ambush and lash out with their raptorial, retractable lower lip to prey on aquatic insect larvae, crustaceans, worms, and sometimes small fishes or amphibian larvae that they detect by sight or touch.

Fishes are the primary predators of the aquatic nymphs, and various birds, including flycatchers and kingbirds, and sometimes leaping fishes or frogs, catch and eat the adults.

Adult odonates fly or perch but do not walk; the legs may be used to hold captured prey. Dragonfly nymphs walk or zoom along by pumping water out of their rectum. Damselflies crawl on the bottom or on submerged objects, or they swim by somewhat fishlike, side-to-side undulations. Some odonate larvae habitually burrow into bottom sediments.

Odonate Conservation

Odonatologists David Wagner and Michael Thomas recently identified several conservation needs for the state's damselflies and dragonflies. These included: protection of cool spring runs and seepage areas that could support populations of tiger spiketails (*Cordulegaster erronea*) and gray petaltails (*Tachopteryx thoreyi*) (the latter not yet found in Connecticut); increased availability of free-flowing streams with sandy bottoms and sunny openings suitable for clubtail dragonflies; increased availability of shallow, sandy coastal plain ponds, which could be created in abandoned sand and gravel pits; protection of sandy shores and sand bars in the Connecticut River from heavy recreational use and large boat waves; and maintenance and enhancement of the availability of freshwater streams, ponds, and small lakes that are free of introduced warm-water fishes (many odonate populations do poorly when exposed to predation by fishes other than native brook trout).

Stoneflies

Stoneflies (Order Plecoptera) are characteristic inhabitants of Connecticut streams and, as nymphs, are indeed associated with stones (Fig. 14.9). These insects occur primarily in the cooler regions of the world. Described species number about 1,550 worldwide and 610 in the United States and Canada. At least 75 species have been found in Connecticut.

The immature stages, called nymphs or naiads, of nearly all species are aquatic, and the vast majority occur under or among rocks in cool, clean water (Fig. 14.10). An abundant stonefly population is an excellent indicator of an unpolluted stream.

Most stonefly nymphs eat detritus, decaying plant material, diatoms, or algae. Large nymphs of the families Perlidae and Perlodidae generally eat small aquatic invertebrates. Nymphs are most active at night.

Adults of most species can fly, and they perch on objects near water. The short-lived, weak-flying adult stage occurs in spring in most species, but several species fly in winter well before the vernal equinox and others can be found in fall well after cold weather has ended the active season for most other insects. I always look forward to the beginning of the stonefly flight season in December, when *Allocapnia recta* adults emerge from permanent and intermittent streams. Adults of this species, which may be active at freezing temperatures, can be found throughout winter and into early spring. Several other species in the genera *Allocapnia, Taeniopteryx,* and *Brachytera* first emerge as adults in February. For naturalists attuned to details, these stoneflies make the winter come alive, just when all seems dead. Just a few stonefly species fly in the warmer months after early summer.

Adult stoneflies do little or no feeding. The diet, if any, consists of algae, lichens, pollen, nectar, water, or other plant material.

Some stonefly males attract females by drumming their abdomen against the substrate. The eggs are deposited in

Fig. 14.9. This is a large adult stonefly.

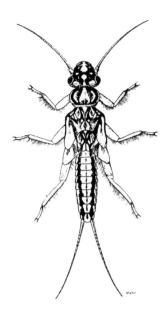

Fig. 14.10. Stonefly larvae are sensitive to water pollution. From Hitchcock (1974).

the water and in most species hatch in three to four weeks. The immature stage generally lasts one or two years. Nymphs gradually metamorphose toward the adult form. Adult and larval stoneflies are important in the diet of trout and other fishes.

Walkingsticks

Walkingsticks are harmless, twiglike, leaf-eating insects that live well-hidden on or near oaks and other deciduous woody plants. They are most active at night. Walkingsticks often perch with the front legs extended straight forward, in line with the long antennae, with the rear end of the body raised and the ends of the front legs and antennae in contact with the substrate. This tilted posture helps obscure their insect symmetry and enhances their resemblance to a twig. The group (Order Phasmatodea) includes over 2,000 species found mostly in the tropics, though 29 species occur in the United States, mainly in the southern states.

Fig. 14.11. Northern walkingsticks are well camouflaged when they climb on twiggy vegetation.

Females mate and lay eggs, or they produce eggs without mating (males are very rare or absent in many species). The eggs are simply expelled, a few each day, and drop to the ground, where they spend the winter before hatching in spring. The nymphs gradually metamorphose into the adult form. I found adult common walkingsticks (*Diapheromera femorata*) in August, September, and October (Fig. 14.11).

Grasshoppers, Crickets, Katydids, and Relatives

Just as birds dominate the sounds of spring, songs of katydids and crickets are one of the distinctive features of late summer and early autumn, even in cities where the incessant cacophony of mechanical and electronic devices competes with the insect symphony. Orthoptera offer a nightly chorus, thoroughly spectacular in its magnitude and variety.

Grasshoppers, crickets, and katydids (Order Orthoptera) are a dominant group of conspicuous and ecologically important insects throughout the world. The global number of species is in the low tens of thousands. About 1,225 species occur in the United States and Canada. Most are plant eaters, and a few of these interfere with human enterprise by eating crops or competing with livestock on rangelands. Some, such as adult tree crickets (Oecanthinae), prey on live insects. Many orthopterans supplement their diet by scavenging dead insects.

Reproductive habits vary somewhat among the major groups, but most reproduce prodigiously. Grasshoppers and some crickets lay eggs in the soil, whereas katydids (Tettigoniidae) and tree crickets insert their eggs into plant tissues. Northern mole crickets (Gryllotalpidae: *Neocurtilla hexadactyla*) lay their eggs in burrows. The female guards them and the young nymphs.

In most Connecticut species, the eggs overwinter and hatch in spring. Nymphs gradually metamorphose into the adult form as they grow, and the sexually mature adult stage is reached in mid-summer. Adults die in fall. In contrast, spring field crickets (*Gryllus veletis*) and some grasshoppers overwinter as nymphs and attain maturity in spring. Adult house crickets* (*Acheta domesticus*) often survive winter in houses. Northern mole crickets hibernate as nymphs or adults.

The Autumn Chorus

When katydids (Fig. 14.12) and crickets mature, generally in July or early August, their sound-making structures become functional and they begin making conspicuous sounds. The orthopteran chorus reaches full magnitude by mid-August and continues through September. Some species fall silent by October. Others survive light frosts and

Fig. 14.12. Male true katydids produce loud sounds in late summer.

Fig. 14.13. True katydids have a resonating membrane associated with the sound-making apparatus.

continue their sound-making activities until sometime in November, when hard frosts finally silence (and kill) the singers. Some hardy species persist into late fall. I heard ground crickets (Nemobiinae) and narrow-winged tree crickets *(Oecanthus niveus)* singing even as late as early December. Spring field crickets are an exception to the rule of late summer to early fall singing in orthopterans. The overwintered nymphs mature quickly, and adults sing from May to July.

Katydids and crickets sing only when air temperatures are at least 45 to 50°F. For example, in Middletown, George Zepko found that snowy tree crickets *(Oecanthus fultoni)* in his backyard raspberry patch stopped trilling whenever air temperature fell below 46°F. Higher temperatures result in high calling rates. In fact, snowy tree crickets are good thermometers; the number of trills in 13 seconds plus 40 equals the air temperature in degrees Fahrenheit.

In contrast to the vocal sounds that typify birds, Orthoptera produce sounds by rubbing together various body parts. Most grasshoppers stridulate by moving projections on the inner surface of the "thigh" across raised veins on the front wing. Crackler or cracker grasshoppers *(Trimerotropis verruculatus)*, associated with barren rocky outcrops in the highlands of northern Connecticut, produce their loud sounds by rapidly opening and closing the hind wings in flight. Katydids and crickets make sounds by rubbing a projection on one front wing over a filelike ridge on the underside of the other front wing. Some katydids produce internal heat that raises muscle temperature and allows high wing-muscle contraction rates. A membrane at the wing base of katydids acts as a sound resonator (Fig. 14.13). Vigorous and prolonged sound production is energetically costly, raising the metabolic rate 4 to 25 times above basal level.

Cricket sounds often consist of musical chirps or trills. Snowy tree crickets emit a steady pulsing series of short mellow trills, and nearby crickets often sing in unison. Black-horned tree crickets *(Oecanthus nigricornis)* and four-spotted tree crickets *(Oecanthus quadripunctatus)* produce a long, continuous trill. Narrow-winged tree crickets, often present in towns, produce loud high-pitched *"reeeee"* trills lasting one to five seconds, often two seconds, with 10 to 15 trills per minute. Groups sound like distant toads trilling (at the wrong season). Spring field crickets, fall field crickets *(Gryllus pennsylvanicus)* and sand field crickets *(Gryllus firmus)* chirp irregularly at a rate of more than one short trilled *"eee"* per second. Little, quick-hopping ground crickets (Nemobiinae: *Allonemobius, Eunemobius,* and *Neonemobius*), common in many habitats, emit buzzing chirps, trills, or tinkling notes. Burrowing northern mole crickets sing a steady rhythm of loud, low-pitched, froglike *"querr, querr"* sounds, one to three per second. Yellowish, ground-dwelling house crickets*, introduced from Europe and resident in or near buildings, produce one or fewer chirps per second.

In contrast to crickets, grasshoppers and katydids produce harsh mechanical sounds. The common true katydid's *(Pterophylla camellifolia)* rasping "zih-zih-zih, zih-zih-zih,

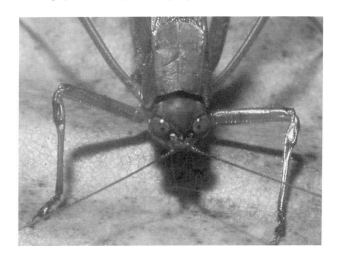

Fig. 14.14. The hearing organs of katydids consist of small innervated membranes on their front legs.

Grasshoppers, Crickets, Katydids, and Relatives | 223

zih-zih," or so-called "katy-did, katy-didn't," phrases of two to four syllables, often repeated monotonously, is a dominant sound of late summer and early fall (Fig. 14.12). The several species of bush and round-headed katydids (*Amblycorypha*, *Scudderia* spp.) generally emit various loud or soft clicks, lisps, rasps, or buzzes (Fig. 14.14). Most meadow katydids (*Conocephalus* spp. and *Orchelium* spp.) produce long series of loud buzzes, sometimes separated by ticking sounds. The wingless American shieldback (*Atlanticus americanus*), found in leaf litter in open upland woods, produces soft, high-pitched, lisping buzzes or pulses. Green sword-bearing coneheads (*Neoconocephalus ensiger*) produce a long, rapid, continuous series of lisps, easily distinguished from the continuous loud buzz of brown robust conehead (*Neoconocephalus robustus*). Late in the season, I heard the latter species buzzing weeks after sword-bearing coneheads had become silent. Coneheads are named for their pointed forehead (Fig. 14.15).

Generally each species sings from a particular type of perch, as follows: common true katydids: often high in tall trees; tree crickets: shrubs or low in trees; round-headed and bush katydids: low in trees, forest edges, shrubs, tall grass and weeds, and meadowy wetlands; meadow katydids and conehead katydids: grassy or weedy vegetation; field crickets (Fig. 14.16), ground crickets, and shieldback: ground surface. Mole crickets sing from burrows dug with their heavily clawed front limbs in marshy areas or adjacent to water. Many of these sound-makers are rather difficult to locate and observe, but a little determination and patience can yield good observations of several different species.

The chorus is most robust on warm, late summer nights. Many species, such as meadow katydids, also sing in daylight. Species that sing mainly at night often shift to more

Fig. 14.16. Female fall field crickets have a long ovipositor.

daytime singing toward the end of the season when nighttime temperatures become cool. In warmer weather, male true katydids usually begin singing at dusk and may continue through dawn. They sing in light rain but tend to be quiet when noisy downpours make singing ineffective. The rate of sound production is noticeably greater when ambient temperatures are higher. The presence of other singing males also affects singing rate; at a given temperature, a male singing solo sings faster than does a male with a singing neighbor. A male duet, if not farther apart than about 20 to 50 feet, usually alternates phrases of their songs. Late in the season, daytime singing is not unusual. In late October and even early November I heard them even after repeated light frosts down to at least 29°F. In autumn, you may find these arboreal, usually hard-to-see insects on the ground, evidently having fallen from the trees after being stunned by the first frosts.

In most species, only males produce conspicuous sounds, and the primary function of the male stridulations is to attract a sexually responsive female. Female tree crickets identify males of their own species by means of pulse rate qualities of the sounds. Female black-horned tree crickets can assess the body size of calling males by the pitch of the sound they make; they prefer the low-pitched sounds of large males (Fig. 14.17).

Males of some species produce more than one kind of sound; for example, a song that attracts females and a different sound that makes the female more receptive to his copulation attempts. When a male tree cricket detects an approaching female, he stops singing, turns, and touches her with his antennae. Then he courts her with an elaborate special song accompanied by body vibrations. Tree cricket courtship and copulatory behavior also include feeding by the female on secretions produced by a gland on the male's back.

Female bush and round-headed katydids may produce weak clicks in response to male stridulations, and the male is attracted to the calling female. Sound qualities of singing males vary with temperature, and song preferences of fe-

Fig. 14.15. The adult conehead at right has just emerged from its shed exoskeleton at left

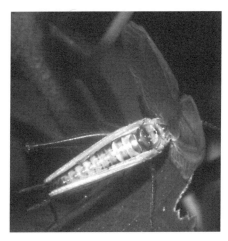

Fig. 14.17. Black-horned tree crickets contribute to the sounds of late summer and early fall. When trilling, tree crickets hold their wings vertically from the body.

males change in parallel. Certain males of some species may not sing but rather perch quietly near singing males and attempt to intercept females attracted to his neighbor's songs.

The songs of the males also may function in interactions among males as well. Males of a particular species generally are well spaced in the environment, and acoustic interactions are the primary basis for this spacing. Some species have calls that are used only or mainly in an aggressive context when another male is nearby. Male meadow katydids sometimes approach and fight with other singing males of their species; the stridulating sounds apparently trigger the approach and serve as a navigational aid. Usually the fights end with the silent withdrawal of one of the males. Through this mechanism, a dominant male may be able to maintain an exclusive area and thus increase his likelihood of attracting a mate without the interference of competing males. Some orthopteran sounds may play a role in startling enemies.

Considering the obvious importance of acoustic communication in many Orthoptera, it is to be expected that many Orthoptera have hearing organs or "eardrums." In acridid grasshoppers, eardrums usually are present on the sides of the first abdominal segment. The hearing organs (tympana and associated acoustic tracheae and spiracles) of katydids and crickets, if present, are on the tibiae of the front legs (Fig. 14.14). These organs allow the animals to recognize and locate sounds. Certain katydids can alter their best hearing frequency by opening or closing their acoustic spiracles. This allows for better detection of their songs or the sounds of natural enemies, as appropriate.

One potential problem with sound production is the possibility of attracting enemies. The high-pitched sounds of orthopterns do attract certain foliage-gleaning bats, and there is evidence that the sounds of some katydids have evolved characteristics that minimize detection. Orthopterans generally are able to detect and often move away from ultrasound, such as bat echolocation calls. Parasitoid insects such as some sarcophagid and tachinid flies also are attracted to orthopteran sounds, and associated reductions in sound production have been reported for crickets in areas frequented by these flies.

For further information on singing insects, and to hear samples of katydid and cricket sounds, visit the following website: http:buzz.ifas.ufl.edu.

The Silent Ones

Not all orthopterans are singers. Here I mention just a few of the silent ones. If you search forest floors at night with a flashlight, or look under objects on the ground in daytime, you might find a silent spotted camel cricket (Gryllacrididae: *Ceuthophilus maculatus*) (Fig. 14.18). I found adults of this wingless species most often in August and September, but sometimes they are active even into December.

Many short-horned grasshoppers (Acrididae) are also largely noiseless. Sand locusts *(Trimerotropis maritima)* are common, soundless, and beautifully cryptic on sandy seashores (Fig. 14.19). The red-legged grasshopper *(Melanoplus femurrubrum)*, an inhabitant of moist grassy areas and the

Fig. 14.18. Camel crickets live among leaf litter of the forest floor.

Grasshoppers, Crickets, Katydids, and Relatives | 225

Fig. 14.19. Sand locusts are well camouflaged on sandy beaches.

state's most common grasshopper, is another silent species. This species is an intermediate host for various parasitic worms of poultry, turkeys, and quail, and, according to one source, this grasshopper "should not be eaten unless thoroughly cooked."

Earwigs

Earwigs are easily identified by the forcepslike appendages at the end of the body (Fig. 14.20). Large earwigs can give a sharp but harmless pinch. Some species also defend themselves by releasing a repulsive secretion from glands on the top of the abdomen. This worldwide group (Order Dermaptera) includes some 2,000 species, with only about 20 species in the United States and Canada. Connecticut's earwig fauna includes a few native species in natural habitats, plus non-native European earwigs* *(Forficula auricularia)* in yards and buildings. Contrary to folklore, none is likely to crawl into a person's ear.

Earwigs generally live among vegetation and in damp or humid places on or near the ground, but once I found an errant one far from land on a lily pad in tidal freshwater. Most earwig species have wings, one pair intricately folded beneath a leathery pair, while others are wingless. Regardless of wings, local species do not fly. Activity occurs mainly at night. Most scavenge organic debris or eat tiny invertebrates or garden plants. In most species, females lay eggs in soil chambers, tend the eggs until they hatch, then guard the hatchlings for up to several weeks.

Mantids and Cockroaches

Mantids (or mantises), being large and looking distinctly extraterrestrial, always seem to command attention when encountered. So do cockroaches, but the reaction is more often revulsion rather than fascination. These insects (Order Dictyoptera) comprise some 5,500 species worldwide, with about 70 species in the United States and Canada.

Mantids live on bushes or in clumps of tall herbaceous plants, hidden by their leaflike appearance. Most of the 1,500 species of mantids occur in the tropics; only a few of the 20 species in the United States and Canada are common. Mantids feed on insects (including each other), caught with quick movements of their raptorial forelimbs (Fig. 14.21). Well fed and full grown by late summer, female mantids mate with a male (and sometimes simultaneously eat him!), then deposit their eggs in an airy styrofoamlike mass attached to a twig or stem. Nymphs emerge from the egg case in spring and gradually grow and metamorphose into the adult form. I saw adults in Connecticut from early August to early November. European (praying) mantis *(Mantis religiosus)*, a common species up to about 2.5 inches long with a distinct targetlike "bulls-eye" near its front armpit, is native to Europe and was introduced into North America in 1899. I found this species and its egg masses in meadowy habitats, including salt marshes and old fields. Chinese mantis *(Tenodera aridifolia)*, up to about 4 inches long, with sharply defined green and brown areas on each front wing, is native to China. It was introduced into the eastern United States in 1896 and is now common in many areas.

Fig. 14.20. Earwigs are harmless, ground-dwelling scavengers.

Fig. 14.21. Mantids capture prey with their raptorial front legs.

Cockroaches, including some 4,000 species, also are a mainly tropical group. Most of the 50 species in the United States and Canada occur in the southern states. Several exotic and native species inhabit Connecticut, and the exotic ones generally are pests in buildings, where they scavenge on organic debris. Most have wings but rarely if ever fly. Activity occurs at night. Females hide egg packets in the habitat or carry them externally or internally until hatching. Cockroaches may contaminate food and leave unpleasant odors and so generally are viewed as a nuisance, but they don't transmit diseases. Pennsylvania wood cockroach (*Parcoblatta pensylvanica*) is an example of a native species that occurs in natural habitats under wood, bark, or rocks in woods, and sometimes it invades buildings.

Termites

Termites are generally pale, soft-bodied insects that eat wood and other cellulose materials, but they cannot digest this food by themselves. They depend on protozoan microorganisms that live in the termite's gut.

People usually regard termites as pests, and homeowners are justly concerned about the damage they can do to wooden parts of buildings. However, in natural ecosystems, termites can be very important in the breakdown of wood and recycling of nutrients.

Termite reproduction involves swarming by winged males and females. After pairing and flying a short distance, the wings break off and a male and female construct a small mating chamber in the soil. There they raise a batch of offspring, which become male and female workers and/or large-jawed soldiers that maintain and defend the colony. Eventually a colony produces a generation of winged reproductive individuals. Adulthood is reached through gradual metamorphosis.

Fig. 14.22. Winged adult termites emerge from a rotting stump to mate in spring.

About 44 of the world's 1,900 species of termites (Order Isoptera) occur in the United States. Most of them are restricted to the southern states. Eastern subterranean termite (*Reticulitermes flavipes*) is widespread in North America. In Connecticut, I saw a mass emergence of winged termites from a rotting tree stump in late May (Fig. 14.22).

True Bugs

To many people, a bug is any kind of insect, but to a biologist a bug is but one of the many groups of insects. At least 50,000 species of bugs (Order Hemiptera) occur in terrestrial and aquatic habitats throughout the world. Some 3,850 species inhabit North America north of Mexico. About 400 North American species are aquatic or semi-aquatic in fresh water. Marine water striders are the only bugs (indeed the only insects) that habitually occur in the ocean far away from land. Hundreds of species occur in Connecticut. Bugs generally differ from Homoptera (aphids and relatives; see following section) in holding their wings, which are thickened at the base, horizontally over the body instead of angled in a tentlike orientation.

Most bugs use their piercing and sucking mouth parts to feed on plant juices, and some are significant crop pests. Certain bugs suck blood or body fluids; a few of these transmit diseases in humans. Most aquatic bugs prey on other small aquatic animals, though a few are scavengers and some in the water boatmen family feed primarily on detritus.

Reproduction involves copulation followed by egg laying in open or hidden sites. Nymphs gradually metamorphose into the adult form. Most overwinter as adults, and some of these, such as stink bugs, leaf-footed bugs, and water striders, are among the conspicuous adult insects active in early spring. Adults lay their eggs in spring, with usually one generation per year, though many water striders may produce two generations per year, and stinkbugs also may have multiple generations in summer.

When diving, aquatic bugs bring air down with them. They may carry air in the space between the body and the folded wing or as a bubble held by hairs on the underside of the body.

Terrestrial bugs generally have a strong odor that emanates from a secretion produced in glands in the abdomen or thorax. Presumably this inhibits the attacks of predators.

Ambush bugs (Phymatidae) are goldenrod specialists (Fig. 14.23). If you carefully search goldenrod flowers you may find small numbers of ambush bugs (*Phymata americana*) among the many other insects that visit these flowers. I found these insects on goldenrod from late July through October.

Fig. 14.23. Ambush bugs prey on insects on goldenrod flowers.

Ambush bugs prey on insects that visit their flowery home. They jab their piercing and sucking mouth parts into the prey, inject digestive enzymes, and suck out the liquefied results. It is not unusual for ambush bugs to capture and kill insects such as bees, wasps, moths, larger than themselves. For example, in late September I found one holding a live honeybee by the bee's head. Occasionally I saw ambush bugs with owlet moths whose body was several times larger than the bug.

Studies in New York indicate that many goldenrod-visiting insects avoid inflorescences inhabited by ambush bugs. Some insects, such as paper wasps and honeybees, spend less time on ambush bug–occupied inflorescences. When they encounter a bug they fly away.

The ambush bug's life cycle includes eggs that overwinter in leaf litter on the ground, green nymphs that hatch in spring and hide among the foliage, and yellowish adults that lurk among goldenrod flowers from mid-summer to early fall. Mating occurs among the flowers. While the male is attending a female, he feeds on insects that the larger female has caught. When not with a female, he sometimes catches his own food, which tends to be smaller insects than those caught by females.

Large milkweed bugs (Lygaeidae: *Oncopeltus fasciatus*) feed on milkweed seeds. In doing so, they ingest—but are

Fig. 14.25. This water strider was on the surface of a stream pool at night.

not harmed by—cardiac glycosides contained in the plant juices. The bugs incorporate these chemicals into their own bodies, which thus become unpalatable or toxic to potential predators. As in monarch butterflies and milkweed beetles, vivid orange and black coloration warns of the bugs' toxicity. Milkweed bugs pass the winter in the adult stage, then mate and lay eggs. Hatchlings pass through five nymphal stages before becoming adults several weeks later.

Water striders (Gerridae) are very common bugs that "skate" upon the surface of ponds, lakes, and rivers, buoyed by the surface tension of the water (Fig. 14.25). They feed on small live and dead animals. I found them active day and night. Adults, which may be winged or wingless, hibernate in winter but may emerge during warm spells; many are active by early spring. *Gerris (Aquarius) remigis* is very common. Adults copulate often and for hours at a time over the two- to three-month mating season.

Fig. 14.24. Milkweed bug adults and nymphs assemble on a milkweed pod in late September.

Fig. 14.26. This backswimmer was swimming in a stream pool at night.

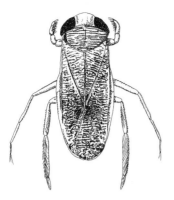

Fig. 14.27. Water boatmen are among the typical insects of freshwater wetlands.

Fig. 14.29. Giant water bugs are formidable predators of aquatic animals.

Backswimmers (Notonectidae) are common rowboat-like aquatic insects that habitually swim upside down in ponds and stream pools (Fig. 14.26). Related to this is their whitish *dorsal* color, in contrast to the paler ventral color of most two-toned animals. They fly well, presumably rightside up. Stranded on land, they hop and walk rightside up, which makes sense given the ventral location of their legs. I found them easier to observe by flashlight at night than during daytime, though daytime dip-net samples in submerged vegetation nearly always yield some. Different species exhibit differences in temperature tolerance and egg laying sites and may avoid competition by using different habitats.

Backswimmers feed on small water animals. Backswimmers can inflict a painful stab, but I carefully handled many without ever experiencing it. Adults are active through mild fall weather, and I saw them swimming under ice in winter and early spring.

Water boatmen (Corixidae) are common swimming insects (less than ½ inch long) that feed on plants or small animals (Fig. 14.27). Most occur in fresh water, where they can be tremendously abundant. I found that at least one species can be plentiful in mosquito ditches in salt marshes. Underwater, these insects carry a supply of air under the wings. *Hesperocorixa interrupta* is a common species.

Waterscorpions (Nepidae) are fairly common, slender to wide-bodied ambush predators that use their raptorial front legs to prey on small animals in marshes, ponds, and other calm fresh waters (Fig. 14.28). Submerged waterscorpions access air by penetrating the water surface with a long two-parted tube at the rear of the body. These insects may appear to be uncommon, but I often caught several of them each time I ran a dip net through submerged marshy vegetation in summer.

Giant water bugs (Belostomatidae) are powerful, strong-swimming aquatic predators that feed on other aquatic animals (Fig. 14.29). Through careless handling I learned that they can give a very painful stab, evidently releasing a quantity of the toxic saliva that quickly kills their prey. But the "sting" is not dangerous to humans, and the pain lasts for only a few minutes. These bugs spend most of their time hidden on pond bottoms or clinging to submerged plants near the water surface. Occasionally, while searching ponds at night with a flashlight, I saw swimming males carrying big loads of eggs attached to their backs. Backswimmers, water boatmen, and giant water bugs all can fly and sometimes are attracted to lights at night, though I've never found any near the lights at my house.

Cicadas, Aphids, and Relatives

The worldwide Order Homoptera includes some 32,000 species of tiny to large, land-dwelling, plant-juice sucking insects. A significant number of the 6,400 species in the United States and Canada are regarded as pests. Many are tremendously abundant. Life cycles can be quite complex and may involve parthenogenesis (see following examples).

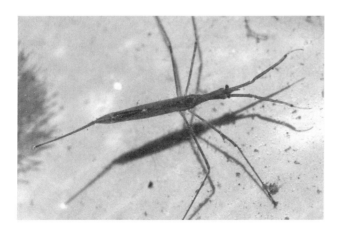

Fig. 14.28. Waterscorpions are common predators in freshwater marshes. These are insects, not true scorpions.

Fig. 14.30. Cicada adults produce loud sounds during warm weather. Paul J. Fusco/Connecticut DEP Wildlife Division.

Nymphs gradually metamorphose into the adult form, and there may be a distinct pupal stage. Following are some examples of this group.

Cicadas (Cicadidae) are well known summer soundmakers and have interesting life histories (Figs. 14.30 and 14.31). Their life cycle includes long-lived burrowing nymphs and short-lived noisy adults.

Periodical cicada (*Magicicada* species complex) is famous for its astoundingly long life cycle. In a particular location, it emerges above ground only once every seventeen years. Locally, the last emergence was in 1996, so the next one will be in 2013. Major broods of this cicada emerge in different parts of geographic range in different years. Twelve of the seventeen potential year classes of this species currently exist; the others are extinct and some perhaps never existed. Past deforestation eliminated many Connecticut colonies of seventeen-year cicadas. The remaining colonies are in the central part of the state.

Females lay their eggs in slits that they make in twigs high in the trees. After hatching, the young cicadas drop to the ground and dig down through the soil to reach tree roots. There they attach and suck out their diet of plant juices.

After seventeen years of subterranean life, they finally emerge above ground, climb up onto a tree trunk or similar object, shed the nymphal skin, and fly off to reproduce. Emergence occurs toward the end of May. The buzzing cicadas are most conspicuous in early June. Only male cicadas have sound-producing organs, located where the abdomen meets the thorax. Muscles vibrate the stiff, sound-making membranes (tymbals), and the sound resonates in large air sacs in the base of the abdomen and adjacent thorax. The hearing organs lie just behind and below the tymbals. Large numbers of cicadas emerging synchronously evidently satiates their predators (e.g., birds, shrews) and ensures that many will survive and reproduce (most or all off-year emergers likely would be eaten). By early July, most have completed their short adult life and are dead.

Fig. 14.31. Dog-day cicadas emerge from the soil every year, leaving nymphal exoskeletons attached to trees and other objects.

Periodical cicada species in the southern United States have 13-year life cycles. The long life cycles of periodical cicadas (both based on prime numbers) apparently evolved during the cold periods of the last Ice Age, when many summers may have been too cool for cicadas to emerge and mate, but the precise evolutionary mechanisms are far from clear.

Dog-day cicada (*Tibicen cancularis*), sometimes incorrectly called a locust, produces its familiar loud buzzing in summer and into early fall (early or mid-July to early October). On warm summer evenings (early August, 80°F), I heard cicadas buzzing as late as 8:20 P.M. In September and October, Eric Mosher determined that calls usually last 4 to 80 seconds, with most less than 20 seconds (average 15 seconds). Calling in fall peaked in mid-afternoon and did not occur before 9:30 A.M. or after 5:36 P.M. Cicadas buzzed under sunny conditions when air temperature was above 62°F, most frequently at temperatures above 70°F. Calling ceased after October 1, even though warm days occurred after that date. The immature stage of this cicada lasts at least a few years, but broods are staggered over time so adults appear every year.

Aphids (Aphididae), represented by numerous tiny species, are among the most abundant of insects, usually seen in dense groups feeding on new plant growth (Fig. 14.32). Feeding aphids void excess food as a sticky sweet "honeydew" and are tended by ants, which feed on the secretion and may actually defend the aphids, "herd" them, transport them to suitable host plants, or even raise them as a sort of insect "cattle." Aphid life cycles often are complex and may include parthenogenetic (asexual) and sexual reproduction, wingless and winged generations, and two different plant hosts. Aphids winter in the egg stage. Certain aphids cause plants to form galls (see "Galls and Gall Makers," pp. 267–71).

Fig. 14.32. This group of aphids fed on a milkweed in late summer.

Woolly alder aphids *(Prociphilus tessallatus)* live and reproduce on maples for a few months in spring and early summer, then produce a winged generation that flies to alder, where they feed and reproduce through the end of summer (Fig. 14.33). I found conspicuous, dense aggregations of these aphids, and careful examination revealed that they were closely tended by large ants. By early fall, another winged generation is produced, and generally these fly back to maple, where they mate. The female lays an egg on the bark, where it stays until hatching in spring. Asexual reproduction by females occurs on both maples and alders.

Beech blight aphids *(Fagiphagus* or *Grylloprociphilus imbricator)* are small plant-juice-sucking insects that are adorned with a fuzzy covering of waxy white filaments. They have a life cycle similar to that of the woolly alder aphid, except that the aphids alternate between the branches and leaves of beech trees and the roots of other plants, usually conifers (probably eastern hemlock in Connecticut). These aphids are most evident on beeches in late summer and early fall. Below dense aphid aggregations you may find large, black spongy masses of sooty mold fungus on the branches and leaves of beech trees. In central Connecticut, many beech trees were heavily laden with these dark fungal masses in the fall of 1992. Evidently the fungus is nourished by the honeydew excreted in droplets by the aphids. Sometimes the branches are simply covered by the honeydew and appear to be wet.

Adelgids (Adelgidae) are represented in Connecticut most notably by the exotic **hemlock woolly adelgid** *(Adelges tsugae)*, a small insect that is native to eastern Asia (Fig. 14.34). It arrived in western North America no later than the 1920s and in eastern North America in the early 1950s. In 1985, it appeared for the first time in Connecticut, apparently blown in from Long Island with the strong winds of Hurricane Gloria in September. Wind, birds, deer, and other mammals may have played a role in the subsequent spread of the adelgid within Connecticut. Today the adelgid is a major pest of eastern hemlock trees and has killed large numbers of hemlocks in southern and central Connecticut. Hemlocks in northern Connecticut and/or in cooler, moister sites have not suffered as much. This pest now ranges from Massachusetts to North Carolina.

Adelgids attack hemlocks by inserting their piercing and sucking mouth parts into young branches, sucking the sap, and injecting toxic saliva. This desiccates the needles and kills them, and the needles fall from the tree. The buds on the end of infested branches cease growth, and the branch

Fig. 14.33. Ants tend a small group of woolly alder aphids in late summer.

Fig. 14.34. Non-native, invasive hemlock woolly adelgids have decimated eastern hemlocks in many areas.

may die in one or two years. A large infestation of feeding adelgids can kill a tree within a year, though usually it takes several years. I noticed that in some areas the adelgids appear to come and go, and the hemlocks live on for years, though with noticeably sparse foliage.

The life cycle is as follows: Adult females that overwintered lay eggs in woolly sacs on young twigs beginning in mid-February. "Crawler" nymphs hatch from these eggs in April and May. The nymphs feed on the trees and develop into the adult form by mid-June. A winged form of adult dies off without reproducing (its required host is not available in the United States). Wingless adults lay egg masses that hatch around early July. The nymphs that hatch from these eggs settle on new growth. Soon they become dormant until October, when they start feeding again. They feed and and develop through winter and mature by spring, when they lay eggs and start the cycle again.

Elimination of adelgids from isolated, ornamental hemlocks can be accomplished with various sprayable, injectable, or implantable pesticides, but such methods are not possible, practical, or appropriate for forest stands. An imported lady beetle *(Pseudoscymnus tsugae)* that preys on the woolly adelgids shows promise as a biological control agent, but so far it looks like the beetle is not going to solve the adelgid problem. Severe winter cold helps reduce adelgid populations and slow their spread, so the recent trend toward relatively mild winters is actually facilitating the adelgid onslaught. However, cold winters are not a reliable solution to the adelgid problem because these prolific insects can quickly rebound from high winter mortality. As the adelgids weaken trees, reduce new hemlock growth, and kill off their host trees, they create poor conditions for subsequent generations of adelgids, and the adelgid population may plummet on previously infested (and seriously weakened) hemlocks.

Other adelgids feed on other conifers, and some produce the swollen galls commonly seen at the tips of spruce branches (see "Galls and Gall Makers").

Dobsonflies, Antlions, Lacewings, and Relatives

The rather diverse assemblage of about 4,700 species that make up the Order Neuroptera are distributed worldwide, from subpolar to equatorial latitudes, with 350 species in the United States and Canada. Most species are terrestrial (only about 300 aquatic species).

The lifestyle of these insects varies among the major groups. The larval stage tends to be predatory on invertebrates. Examples include terrestrial, aphid-eating lacewings; pit-dwelling antlions; and large stream-bottom-dwelling hellgrammites (larval stage of the dobsonflies; Corydali-

Fig. 14.35. This fishfly larva was under a rock in a dried-up creekbed in late August.

dae). Adults may be predatory (lacewings, snakeflies), feed on nectar or pollen (antlions), or may not feed at all (dobsonflies). Spongillaflies live near lakes, and the larvae feed on freshwater sponges. Egg, larva, pupa, and adult stages are represented in the neuropteran life cycle.

Eastern dobsonfly (Corydalidae: *Corydalus cornutus*) and **fishfly** (Corydalidae: *Chauliodes* and various other genera) larvae are impressively large and locally fairly common among rocks in flowing rivers or large creeks (Fig. 14.35). The short-lived adult stage sometimes can be found on plants (Fig. 14.36). Adult fishflies flew outside my lighted windows from mid-May to mid-August; in a nearby intermittently flowing stream, I found large, vigorous larvae under rocks when the stream was dry. The dobsonfly life cycle lasts two to five years, dominated by the large, strong-jawed larval stage (hellgrammite), a bottom-dwelling predator on small insects. Flying adults lay eggs near water. While kayaking in a strong-flowing stream in early summer, I found white patches of dobsonfly eggs on steep rocks in the shade above water, a typical location. Newly hatched larvae drop directly into the water. After completing their development, full-grown larvae crawl out of the water in spring and pupate in the ground under rocks or wood.

Antlions (Myrmeleontidae) are better known as larvae (Fig. 14.37) than as adults, which resemble weak-flying damselflies with clubbed antennae. The larvae hide under objects on the ground or at the bottom of a self-made cone-shaped sand pit that traps passing insects. The pits

Fig. 14.36. Fishfly adults may appear at lighted windows in spring and summer.

Fig. 14.37. Antlion larvae are well-equipped predators.

Fig. 14.38. Common shore tiger beetle is common on the beaches of Long Island Sound.

usually are in fine sand under a protective overhang that keeps them dry in the rain. In recent years, hundreds of antlion larvae took up residence along the foundation of my house (protected under wide, overhanging eaves), and before long ants completely disappeared from my kitchen. (In appreciation, I now pull out any plants that threaten the antlions' dusty habitat.) I found groups of several dozen pits per 10 square feet.

When an ant or other small insect enters the pit, the antlion larva starts flicking sand out from the bottom, causing a miniature landslide as the loose sand on the slopes of the pit slides downward, which often prevents the ant from climbing out. Rapid sand flicking by the antlion continues the landslide and eventually brings the ant to the bottom of the pit, where it is grabbed by the long jaws of the antlion.

Antlions spend up to two winters as fossorial larvae before pupating for a month and finally emerging as winged adults. After mating, adult females deposit eggs in fine sandy ground. The adults feed on flowers or not at all and live about one month. Adults appeared at my lighted windows on warm nights in late July. After hibernation, larvae renew their pits with the return of warm dry weather by about mid-April. The common antlion species in Connecticut is *Myrmeleon immaculatus*.

Beetles

Among all animals, no group is represented by more variety than the beetles (Order Coleoptera). This diverse group of flyers, swimmers, runners, burrowers, borers, and diggers globally includes some 290,000 described species, surpassing the number of vascular plant species worldwide. The United States and Canada host 25,030 species. Beetles occur throughout the world in essentially all terrestrial and freshwater habitats. Some of the more notable beetles in Connecticut are mentioned below. There are far too many to mention more than a small fraction of the many families found in the state.

Beetle diets range from living and dead plant material, including pollen, nectar, leaves, stems, flowers, fruits, and seeds, to living and dead invertebrates, dung, and vertebrate carcasses. A few species are parasitic on other animals. Some beetles damage forest trees or growing or stored crops.

Beetle life history includes egg, larva, pupa, and adult stages. Pheromones (air-borne sex attractants), or quiet, high-pitched sounds, play a role in the mating sequence in some species.

Tiger beetles (Cicindelidae, or Carabidae: Cicindelinae) are attractive, big-eyed, big-jawed insects (Figs. 14.38 and

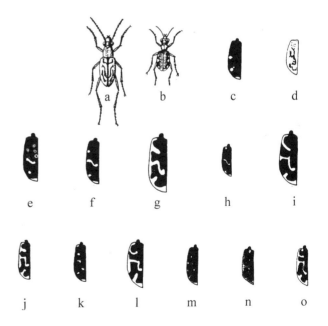

Fig. 14.39. Wing-cover patterns of tiger beetles: a = northeastern beach, b = smooth, c = six-spotted, d = dune ghost, e = cowpath, f = clay bank, g = big sand, h = red-bellied, i = oblique-lined, j = common shore, k = twelve-spotted, l = beach-dune, m = punctured, n = salt marsh, o = puritan. After Kellogg (1905).

14.39). Most occur in open sandy habitats, but some Connecticut species favor barren hard-packed soils (punctured tiger beetle, *Cincindela punctulata*), granitic or basalt outcrops (red-bellied tiger beetle, *C. rufiventris*), muddy coastal shores (salt marsh tiger beetle, *C. marginata*), or agricultural land (cowpath tiger beetle, *C. purpurea*). The six-spotted tiger beetle *(C. sexguttata),* green with light dots, active in spring, is solitary in forests. Nearly 190 of the world's 2,300 tiger beetle species occur in North America.

Fifteen species historically inhabited Connecticut, but many, including the northeastern beach *(Cicindela dorsalis),* big sand *(C. formosa),* dune ghost *(C. lepida),* puritan *(C. puritana),* cowpath *(C. purpurea),* and oblique-lined tiger beetles *(C. tranquebarica),* are rare, highly localized, or extirpated. The puritan tiger beetle is restricted to the Chesapeake Bay region and a few sites along the banks of the Connecticut River in Connecticut and Massachusetts (and formerly New Hampshire), where adults are active between mid-June and mid-August. Its habitat has been degraded or destroyed by dams, reservoirs, and shoreline alterations. A stable population of several hundred adults exists in the Portland-Cromwell area in Connecticut. The northeastern beach tiger beetle, which apparently requires clean, untrammeled sandy beaches (a rarity today), apparently is extirpated. The cowpath tiger beetle, also apparently gone, likely declined with post-agricultural reforestation. Because of the precarious conservation status of most of the state's tiger beetles, you must have a permit from the Connecticut Department of Environmental Protection in order to collect any species of tiger beetle. Federal law prohibits collection of the puritan and northeastern beach tiger beetles.

Both the adults and fossorial larvae are predatory on other insects. Larval burrows range in depth from just a few inches (common shore tiger beetle, *C. repanda*) to 1 to 2 feet (punctured tiger beetle). Larvae in ambush stay well anchored in their burrows, with the top of the head plugging the opening, and partially emerge to nab prey. Adults are fast runners and fliers and are active on warm, sunny days.

Females distribute their eggs singly in the ground. In one life-cycle scenario, adults emerge in late summer or fall, overwinter, then mate, lay eggs, and die; larvae may live a year or more before pupating. Usually this life cycle takes two years.

Another life-cycle pattern involves mating and egg laying in mid-summer. Larvae overwinter once or twice, depending on the species, and pupate in spring, with emergence of the adults in early or mid-summer. By fall, all adults are dead.

Ground beetles (Carabidae), represented in Connecticut by approximately 360 species, live mainly on the ground, but some do occur up in the canopy of forest trees. In one

Fig. 14.40. This ground beetle emitted a series of soft, high-pitched buzzing or hissing sounds as I handled and photographed it.

study by Krinsky and Godwin, canopy samples from Connecticut taken over five years yielded 693 carabids of 27 species. All but a few species were present in low numbers.

Most ground beetles are black and somewhat flattened (Fig. 14.40). Larvae and adults of most species prey on other small invertebrates. Lifespan is up to a few years. Bombardier beetles (*Brachinus* spp.) occur near streams or ponds or under objects in the woods. They are notable for their ability to spray a defensive gas from their anal glands at attacking predators. Larvae parasitize pupae of aquatic beetles. For unknown reasons, many of the state's 11 species have become very scarce in Connecticut.

Whirligig beetles (Gyrinidae) swim in groups on the surface of lakes, ponds, and streams (Fig. 14.41). They may dive underwater if threatened. These odorous beetles are predatory, both as adults and as larvae. The eyes of adults are divided, with one part functioning in aerial vision and the other part allowing underwater viewing. When fully grown, the larvae leave the water and pupate in shelters they make near water. Adults are active over a long season; I saw them from early April to late November. Adults seen in early spring survived winter. The larger whirligigs (9 to 15 mm) are members of the genus *Dineutus;* the smaller ones (4 to 7 mm) are *Gyrinus*.

Fig. 14.41. Whirligig beetles often form large aggregations on the surface of lakes, ponds, and streams.

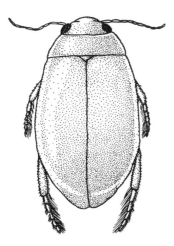

Fig. 14.42. Adults of some predaceous diving beetles are well over an inch long.

Predaceous diving beetles (Dytiscidae) have a smooth oval body, strong swimming legs (stroked simultaneously), and relatively long antennae (Fig. 14.42). They are very common in freshwater habitats, including temporary pools that the flying adults readily colonize. Adults secrete fluids that may give them some protection against predatory fishes.

Adult diving beetles are active predators or scavengers. They may dive with a supply of air under their wing covers. Females deposit eggs on or in water plants or among roots or decaying plants in moist soil. Larvae (Fig. 14.43), which take in air through openings at the rear of the body, are predatory on small aquatic animals, which are drained of body fluids through the beetles' long, grooved jaws.

Diving beetle larvae pupate in moist sites near the water's edge. Adult diving beetles may survive the winter at the bottom of a body of water.

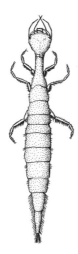

Fig. 14.43. The large size and sharp jaws of predaceous diving beetle larvae allow them to capture prey as large as tadpoles.

Water scavenger beetles (Hydrophilidae) are, as streamlined adults, mainly scavengers in quiet areas of ponds and lakes, though some may prey on invertebrates. Adults resemble diving beetles but swim with alternating strokes of their hind legs. When diving, they carry a sheet of air on the underside of the body. The larvae, which unlike predaceous diving beetle larvae, have a single claw at the tip of each limb, prey on small aquatic animals.

Carrion beetles (Silphidae), generally black with red or orange markings are attracted to carcasses, which provide food for the adults and larvae of most species. They bury small carcasses (mice, small birds) by excavating soil from under the carcass; burial may exclude competing fly maggots. I sometimes found these beetles under dead mice and shrews on trails. American burying beetle *(Nicrophorus americanus)*, a large species 1.2 to 1.4 inches long, formerly was common but now is quite rare or extirpated. It has not been found recently in Connecticut, but it does occur on Block Island, the only remaining natural occurrence east of the Mississippi River. The historical range included 35 states in the eastern and central United States and 3 Canadian provinces. By 1989, populations could be found only in Nebraska, Oklahoma, Arkansas, and Rhode Island. More recently this beetle was rediscovered in South Dakota and Kansas. The Block Island population has been used as a source for captive breeding and local introductions in New England.

Rove beetles (Staphylinidae) are quick-moving beetles with short wing covers that leave most of the abdomen exposed (Fig. 14.44). Rove beetles that I found at small mammal carcasses curled the abdomen upward as they ran away from my probing fingers. Habitats are extremely varied; many occur on carrion, other decaying organic matter, or fungi. The larger, more observable rove beetles have strong jaws and prey on maggots and other insects, mites, and small worms, as do the larvae.

Scarab beetles (Scarabaeidae), recognizable by antennae that may form a club or expand at the end into separate

Fig. 14.44. I found this rove beetle next to a shrew carcass.

Fig. 14.45. This "June bug" is actually a scarab beetle that appears outside lighted windows by early May.

Fig. 14.47. This eyed click beetle was on a pile of rotting logs (the larval habitat).

flattened plates, comprise a large number of species with diverse feeding habits. Eastern tumblebug *(Canthon pilularius)* adults, active in late summer, make, roll, and bury balls of dung that serve as the egg-laying site and source of food for the larva. May beetles or June bugs *(Phyllophaga* spp.), common at lighted windows in late spring, eat roots during a larval period that may last up to a few years (Fig. 14.45). Adults feed on tree leaves. Japanese beetle *(Popillia japonica)* is a major pest that was introduced accidentally in the United States in the early 1900s. Larvae eat roots, adults eat foliage or fruits of many plants.

Water penny beetles (Psephenidae: *Psephenus* spp.) occur in and adjacent to streams. The short-lived adult stage hides under logs or rocks near water; females enter water to lay their eggs. The flattened, rounded larvae, for which the group is named, cling to rocks in swiftly flowing water and graze on periphyton (Fig. 14.46).

Click beetles (Elateridae) are able to abruptly snap their body parts and produce a clicking sound. This action can launch them into the air, which is useful in evading or loosening the hold of predators or righting themselves should they be overturned. Eyed click beetle *(Alaus oculatus)* is a gigantic (1.6 inches) beetle with two large black "eye" spots on the thorax of adults, which reside under loose bark (Fig. 14.47). Larvae, firm-bodied "wireworms," live in soil and rotting wood and eat roots and small soil animals.

Firefly beetles, better known as fireflies or lightning bugs, are not flies or bugs but rather beetles of the family Lampyridae, which includes about 125 species in the United States and Canada. Some species produce a cold, greenish-yellow light that makes them instantly recognizable at night, but they are otherwise rather nondescript and often not recognized when seen in daylight. The light is produced in one or more of the abdominal segments (Fig. 14.48). Males have more light-producing segments than do females, which in many species are wingless and resemble larvae. Each species emits its light in a unique pattern of single or double flashes. The glittering display of a group of fireflies on a dark, early summer night is truly a beautiful sight.

The function of the light is to attract the opposite sex, though females of *Photuris pennsylvanicus* may flash their light in a mimicking, deceptive way that lures males of

Fig. 14.46. Water penny larvae live in streams. From Kellogg (1905).

Fig. 14.48. Light-producing segments are visible at the end of the abdomen of this adult firefly beetle.

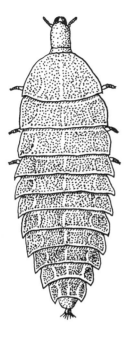

Fig. 14.49. Luminescent firefly beetle larvae ("glowworms") are ground-dwelling predators. They can retract the head into the thorax and may appear to be headless.

Photinus scintillans close enough to be captured and eaten. Apart from these sly predators, adults do not feed at all. In central Connecticut, I saw the flashing lights of these beetles from May to September, with the largest numbers in June and early July. I found day-active adults as early as April.

The larvae of firefly beetles, known as glowworms, also produce light (Fig. 14.49). They hatch in spring and spend one or two years using their long jaws to feed on snails, slugs, and other soft-bodied invertebrates. Winter is spent inactive in the soil. Light production by the larvae serves no known function other than probably as a warning to potential predators, which may thus more quickly learn that these animals are distasteful (they contain defensive steroids similar to those found in toads and certain plants). *Photuris* females cannot synthesize these defensive chemicals, but they obtain them by eating *Photinus* males. I readily found glowworms by searching the ground at night without a flashlight in August, September, and October.

Lady beetles (Coccinellidae), or "ladybugs," are well known as shiny, rounded, often red or orange, black-spotted insects (Fig. 14.50). In the animal world such bold (aposematic) coloration is associated with defensive chemicals and facilitates avoidance learning by predators, and at least some coccinellids are indeed distasteful or toxic. Adults of most species and larvae prey on small invertebrates and insect eggs, including those of many pest species such as aphids. In early November, I found several hundred lady beetles of several species flying around and walking on a high rock outcrop in deciduous forest. On the same day, hundreds flew to and perched on the south side of my house. Many seek hibernation sites inside buildings. A recent field survey of lady beetles in the northeastern United States came up with 13 native species and four introduced ones. The seven-spotted beetle* (*Coccinella septempunctata*) was introduced in the United States for pest control. It has become established in Connecticut and may have caused the disappearance of the once common native nine-spotted lady beetle (*C. novemnotata*).

Long-horned beetles (Cerambycidae) are represented in Connecticut by several dozen species, some impressively large as adults and larvae. They have long antennae that arise from next to the partially surrounding eyes. Adults feed on plant matter, if anything. The larvae bore through and may damage woody plants. Examples include the following:

Giant root borer (*Prionis* spp.) larvae initially feed on root bark then bore through and feed on the roots of woody plants for three to five years. Both the adults and fully grown larvae (the latter can be found in or under rotting logs or stumps) are astonishing in their size (2 to 3 inches long). I

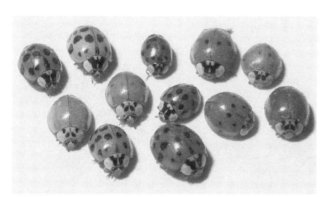

Fig. 14.50. This is a small sample of the lady beetles that appeared on the side of my house in early November.

Fig. 14.51. Adult giant root borers are among the most massive of Connecticut's insects.

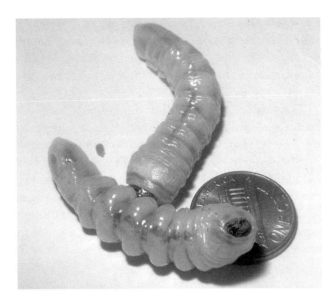

Fig. 14.52. Round-headed borer larvae live in rotting wood.

Fig. 14.54. Water-lily leaf beetles are common on lily pads in summer.

found adults, sometimes in buzzing flight at night outside lighted windows, from early July to early August (Fig. 14.51). Some adults that I handled made a high squeaky sound by rubbing a hind leg against the edge of a wing cover.

Round-headed borer (*Saperda* spp., and others) larvae (Fig. 14.52) tunnel through the wood of various living, dying, or dead trees and shrubs for one to three years, pupate, then emerge as adults in late spring or early summer.

Red milkweed beetle (*Tetraopes tetraophthalmus*) is one of an array of insects specialized for life on milkweeds (Fig. 14.53). Females lay eggs on milkweed stems near the ground surface. Larvae bore into the stems and overwinter in the roots. I found adults from late June to mid-August. When disturbed, they readily let go of the plant and drop to the ground.

Asian long-horned beetle* (*Anoplophora glabripennis*) has been found in New York and may soon appear in Connecticut. This large (1 to 1.5 inches), black, white-spotted beetle with long black and white antennae is an undesirable that may kill various deciduous trees. Signs of this beetle include oval or round holes up to one-half inch in diameter in tree trunks or branches and associated piles of coarse "sawdust." Observations of this beetle should be reported to the Connecticut Agricultural Experiment Station (203-974-8474).

Leaf beetles (Chrysomelidae) feed on plants as larvae and as adults. Many (e.g., asparagus beetle, potato beetles, cucumber beetles, etc.) are well known as pests of crops. Water-lily leaf beetles (*Donacia* spp.) live on water-lilies, pickerelweed, and similar plants (Fig. 14.54). The little summer-flying adults, easily found by observant canoeists and kayakers, eat foliage and pollen, whereas larvae feed on submerged plant parts. Most species have a two-year life cycle.

Fig. 14.53. Red and black milkweed beetles specialize on milkweed. Paul J. Fusco/ Connecticut DEP Wildlife Division.

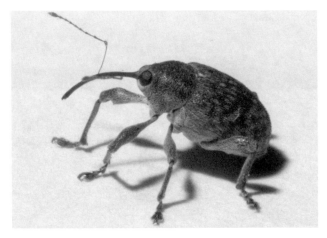

Fig. 14.55. Some weevils have a spectacular snout used to bore into nuts.

Fig. 14.56. Bark beetles make intricate patterns under tree bark. The central tunnel in each array is the brood gallery made by the adult beetle. The multiple branches off the brood galleries were made by feeding larvae. E. Kaston, Connecticut Agricultural Experiment Station.

Snout beetles and **weevils** (Curculionidae), generally have a long snout bearing "elbowed" antennae (Fig. 14.55). The group includes a multitude of species with diverse habits, but adults and larvae generally feed on plant material of one sort or another.

Northern pine weevil *(Pissodes approximatus)* females cut holes and insert eggs in the bark of leader stems of usually sun-exposed pines. The larvae feed on and often kill the leader, causing a fork trunk or crooked growth (a side branch later takes over the leader role).

Acorn weevil *(Curculio* spp.) females bore a hole into an acorn and deposit an egg inside. The larvae hatch, feed on the acorn meat, and when full-grown emerge from the acorns through a tiny hole in late summer and early fall (collect acorns and keep them on your desk to see this). They pupate on the ground. Adults emerge and are active in summer Other *Curculio* species attack hickory nuts, hazelnuts, and chestnuts. For further information, see "Acorn Ecology" in Chapter 7.

Bark beetles (Curculionidae: Scolytinae) are small stout beetles. Some bore under tree bark and feed on the inner bark, sometimes engraving the wood with attractive patterns (Fig. 14.56). Others tunnel into the heartwood of trees, inoculate the tunnels with fungal spores that they carry, and feed on the resulting fungal growth. Females lay eggs in galleries under the bark. The larvae feed and pupate inside the tree.

Wasps, Bees, and Ants

Order Hymenoptera. an enormous, worldwide group of insects, includes about 103,000 species, with some 18,100 in the United States and Canada. The life cycle includes distinct egg, larval, pupal, and adult stages. The following includes only a small selection of this diverse group.

Ichneumons (Ichneumonidae)

Ichneumonids are a very large group of wasps represented by several thousand species in the United States and Canada and a few hundred species in Connecticut. The long, animated antennae often have white or yellow markings, and in many the abdomen is long and slender, and the egg-laying structure (ovipositor) sometimes is very long. The larvae of all are parasitoids and serve as important biological controls on populations of other insects.

Giant ichneumons *(Megarhyssa)* are the most impressive members of the family (Fig. 14.57). These wasps have long slender legs, long antennae, and an exceptionally long ovipositor, which, fortunately, is not used aggressively as a stinger! In Connecticut, I saw adults flying in late August and September.

These wasps have sensitive powers of detection. Adult males explore the surfaces of logs and can determine the site from which a female will emerge several hours before she appears. Males aggregate at such sites, tapping their antennae against the log surface. Evidently chewing sounds or vibrations attract the males. Males mate with the newly emerged females. Interestingly, only these emerging females are attractive to males. Perhaps already emerged females usually have mated and somehow the males can detect this.

Females also have excellent sensory systems and can detect organisms hidden in rotting logs or trunks of diseased trees. This is critical because she must lay her eggs on or very near wood-boring insect larvae (or pupae). By late summer

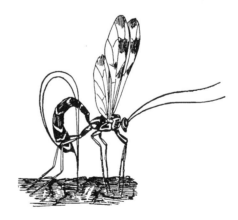

Fig. 14.57. This female giant ichneumon is laying an egg on a horntail wasp larva hidden within a log. From Kellogg (1905), after Comstock.

or early fall, mating has occurred and the female searches logs for potential host larvae. She lands on a log (usually a maple, elm, or oak) and examines its surface with her antennae. If she detects a larva, she inserts her long ovipositor vertically downward into the wood, with the basal part of the ovipositor arched high above her head. The distal end of the ovipositor remains straight and does not curve as it penetrates 3 or more inches into the wood. She deposits an egg on the larva, usually a horntail wasp (Siricidae: *Tremex*), which may be a little more than an inch long. After laying a single egg, she flies off in search of another host. Field observations that involved breaking into host logs have confirmed that the females are quite accurate in placing the egg on the horntail larva. After the ichneumon larvae hatches, it penetrates the host larva and slowly feeds on it from within. Eventually the ichneumon larva reaches full size and pupates, and later the adult wasp emerges from the log. Adults do not eat.

Gall Wasps (Cynipidae)

Gall wasps (Fig. 14.58) are tiny, hump-backed, "potbellied" wasps that lay their eggs in plant tissue, inducing the formation of a distinctive gall, within which the larva feeds. Some lay eggs in already-formed galls. Adults feed on nectar, water, or nothing at all. For further information, see "Galls and Gall Makers," near the end of this chapter.

Pelecinids

American pelecinid (Pelecinidae: *Pelecinus polyturator*) females are sometimes locally common in deciduous forests, but males are very rare (Fig. 14.59). Adults feed on nectar. Females thrust their abdomen into the soil or decaying wood and deposit eggs on hidden June beetle larvae, one egg on each larva. The pelecinid larva burrows into and eventually kills the host, scavenges on the carcass, then pupates. I saw active adults in August and September.

Fig. 14.58. This adult cynipid wasp was inside an oak gall in early fall.

Fig. 14.59. American pelecinid wasps fly in late summer.

Ants (Formicidae)

Ants are ubiquitous social hymenopterans with some 8,800 species worldwide and 700 species in the United States and Canada. Connecticut has a fairly diverse ant fauna. For example, ant surveys in small areas on the Thimble Islands and adjacent coastal mainland yielded 35 species, with as many as 21 species on Horse Island and 31 species on the mainland. Areas with the greatest microhabitat diversity had the largest number of ant species.

Ants may range widely when foraging but are based in nests located underground, under rocks or logs, or in wood. All ants live in societies that generally include wingless sterile female workers, which normally forage, build the nest, and feed the colony's young, plus periodic swarms of temporarily winged fertile males and females. The males die after mating, but the females (queens) attempt to establish new colonies. Ants generally are predators, scavengers, seed gatherers, and/or eat honeydew voided by aphids.

The small size of most ants, and a widespread perception of them as pests, belies their tremendous ecological importance. Some are involved in mutualistic (mutually beneficial) interactions with plants. For example, *Aphaenogaster* ants disperse the seeds of bloodroot, wild ginger, Dutchman's breeches, and a large number of other woodland wildflowers, allowing seeds to colonize new areas, reach suitable sites for germination and seedling establishment, or avoid competition with the parent plants. This comes about as a result of the ants' attraction to nutritious food bodies (elaiosomes) that are attached to the seeds. The ants carry the seeds to their nests, where the food body is removed; the seed is discarded or stored, sometimes a good distance from the source plant.

Ants also may effectively protect plants from leaf-eating animals and thereby foster their own food sources. For example, certain *Formica* ants tend and protect plant-juice-sucking treehoppers that provide the ants with food in the

Fig. 14.60. Black carpenter ants are among the largest North American ants.

Fig. 14.61. Mounds of the ant *Formica exsectoides* sometimes are very large. From Viereck et al. (1916).

form of "honeydew" that the treehoppers excrete. The treehoppers (Membracidae) live on goldenrods that are subject to defoliation by leaf-eating chrysomelid beetles. The ants attack beetle larvae and deter them from feeding, which ensures vigorous plants that provide a good food source for the treehoppers. The ant-protected goldenrods grow taller, have more foliage, and produce numerous seeds, whereas some beetle-eaten goldenrods lacking ants are so debilitated that they fail to flower or produce seeds. For another example of an ant-plant relationship, see also the black cherry account in Chapter 9.

Black carpenter ants (*Camponotus pensylvanicus*) excavate extensive nesting galleries in dead wood, including trees and buildings (Fig. 14.60). They bite off bits of wood but do not eat it; their diet consists of insects and sweets. Carpenter ants can bite but do not sting. Colonies may include several hundred individuals. Winged males and females emerge and mate in early summer. Each mated female ("queen") starts a new colony. Carpenter ants are big; I found queens that were nearly 0.75 inches long. Pileated woodpeckers often dig into rotting tree trunks to feed on these ants (Fig. 20.33).

"Slave-making" ants (*Polyergus lucidus*) are bright red ants equipped with slender, pointed jaws. Workers of this species do not engage in typical ant-worker acitivites but rather raid the nests of other ants (*Formica* spp.), kill resisting workers of the other species, and carry home their live pupae. Upon hatching, the kidnapped ants become ordinary workers in their new home and raise the young of the *Polyergus* queen. The queen starts a colony by invading a *Formica* nest and killing the resident queen. Usually the *Formica* workers adopt her as their new queen and raise her *Polyergus* young. *Polyergus* nests are under rocks or in similar sites. In other slave-makers (e.g., *Protomognathus americanus*), the queen enters a host colony (e.g., *Leptothorax* spp.) containing larvae and pupae and forces the adults to abscond. Shortly thereafter the newly emerged adult workers of the host accept the slave-maker queen and tend her young, which later engage in slave raids on other nests.

Allegheny mound ant (*Formica exsectoides*) is a common ant that constructs a mounded nest. Mounds are usually up to 1 to 2 feet in diameter and several inches high, but see figure 14.61.

Potter Wasps, Yellowjackets, Hornets, and Paper Wasps (Vespidae)

Wasps and hornets are well-known stingers that live solitarily or in large colonies in paper nests that they build. What is the difference between a wasp and a hornet? Nothing, really. "Wasp" is a general term that refers to hornets, yellowjackets, and a large number of similar insects, most of which are harmless or even beneficial to humans. Thus yellowjackets and hornets are kinds of wasps.

Fig. 14.62. This paper wasp *(Polistes)* nest was inside an empty tubular bird feeder.

Paper wasps *(Polistes)* are more slender than the yellowjackets and hornets and in flight show long dangling legs. They often place their open-faced, single-comb nests under the eaves of a building (Fig. 14.62). I found them also in bird nest boxes, empty tubular bird feeders, on window frames, and in other similar sites. Use of nest boxes by paper wasps, including the European paper wasp *(Polistes dominulus)*, a non-native species that has been present in the Northeast since 1980, may deter nesting by native eastern bluebirds. Paper wasps are important ecologically as effective predators on caterpillars.

Bald-faced hornet *(Dolichovespula maculata)* is a black wasp with a yellowish-white face, shoulder stripes, and spots on the rear end of the body. More well known than the hornets themselves are their large globular paper nests that become conspicuous in fall after trees or bushes drop their leaves (Fig. 14.63).

Fig. 14.63. This bald-faced hornet nest, opened on one side, shows several tiers of brood-rearing cells and the multi-layered outside covering. A nest is used for just one season.

Life of the Yellowjacket

Yellowjacket is the name used for several species of black and yellow cross-banded wasps (*Vespula* spp.) that usually nest underground, in logs, in rotten stumps, or in buildings (Fig. 14.64). In summer, yellowjackets are infamous as unwelcome guests at cookouts and as visitors to open garbage cans.

Winter. As any picnicker knows, yellowjackets can be distractingly abundant in summer and early fall (some nests are active into November), but where are the hungry hordes in winter? Nearly all of them are dead, victims of starvation and cold in late fall. The only ones to survive the winter are certain large females—the queens. Queens mate in late summer or early fall, then crawl into a crevice or underground cavity to pass the winter. With her antennae tucked behind her head, wings folded beneath her body, and mouthparts clamped onto some object, each queen waits out the winter, protected from freezing by antifreeze compounds in her body fluids.

Spring Nest Building. With the warming days of spring, yellowjacket queens emerge from their winter retreats and fly to early blooming flowers to sip nectar. Once the weather is consistently warm, each queen chooses an underground (usually) cavity and begins to build her paper nest. Nest sites are sometimes in the walls or under the eaves of buildings.

Finding a source of soft or decayed wood, or even a piece of old cardboard, she tears off a chunk and chews it up, mixing it with saliva to form a sort of paper pulp. Then she flies to her chosen nest chamber and sticks a bit of the paper to a root or stone in the ceiling. Repeating the process hundreds of times, the queen eventually fashions a hanging nest enclosed (except for a small entrance) by two or three globular paper envelopes. Inside is a horizontal layer of six-sided tubes or cells resembling a honeycomb in shape.

Fig. 14.64. The wings of this male yellowjacket did not develop to full size.

Eggs and Larvae. In each cell, the queen lays one egg, which eventually hatches into a small grub or larva. Upon hatching of the larvae, the queen becomes an aggressive predator, for she must feed her brood. She pounces on nearby insects, bites them to death (yellowjackets do not sting their victims), and chews them up into a paste. She feeds this insect hamburger to the larvae. When full grown, the larvae enter a brief resting (pupal) phase. Eventually an adult yellowjacket emerges from each cell.

Sex Determination. All of the yellowjackets that emerge from the new nest are females, noticeably smaller than their queen mother. In these wasps, sex is determined by whether or not the egg has been fertilized by male sperm. Fertilized eggs become females whereas unfertilized eggs develop into males. The queen obtains sperm by mating with a male during the autumn. The sperm is stored in her body until needed the following spring.

Nest Expansion. Soon after emergence, the queen's first brood starts remodeling the nest, adding more cells to each comb, adding additional combs beneath the original one, and enlarging the paper envelope. The queen lays more fertilized eggs in the new cells and in old ones that have been reconditioned. When the second batch of eggs hatch, the females of the first brood assume the task of catching insects to feed to their developing younger sisters. At the same time, they continue to enlarge the nest. The queen stays in the nest and restricts her activity to egg laying.

When the queen's second brood emerges, they join their older sisters in further expanding the nest and feeding their larval siblings. If necessary, they make more room in the underground nest chamber by removing soil one bit at a time, first moistening it with drops of water if it is too hard and dry. Because of their seemingly never-ending jobs, these female offspring of the queen are referred to as workers. The queen and her daughter workers raise several successive broods of more workers throughout the warm season, and the colony and nest continually grow. This is why yellowjackets are most abundant and bothersome in late summer, when a single nest may be home for several hundred workers.

Queen Dominance. Although all of the workers in a yellowjacket colony are females like their queen mother, they usually do not lay eggs of their own. Why not? One explanation is that the yellowjacket queens emit a special chemical (pheromone) that somehow prevents the workers from producing eggs. Queens may also attack and maul workers that attempt to lay eggs. If a queen is killed or removed from the nest some of the workers start laying eggs several days later. Because they are virgins, the workers lay unfertilized eggs that develop into males. Only the fertilized queen can produce more workers.

Food and Foraging. The workers are quite opportunistic in gathering food to feed to the protein-hungry larvae. Besides live flies, spiders, and various insect larvae, workers take advantage of other convenient sources of protein, including bees at hives, hunks of meat carved from discarded chicken bones, neglected hamburgers, and dead animals. They also detach the lipid-rich appendage from trillium seeds and may feed it to the larvae.

The workers themselves do not eat meat. Instead they sip nectar, fruit juice, secretions from various plant-juice-sucking insects, and other sweet liquids. This explains their habit of crawling down into soft-drink cans. Workers also drink a sweet fluid secreted from the mouth of the larvae. This larval liquid may be important on days when bad weather prevents the workers from flying out in search of food. Because she does not leave the nest once the workers have emerged, the queen depends on this liquid and also on drops of sugary liquid that the workers regurgitate for her and each other.

It is amazing how quickly a swarm of yellowjackets can appear when a good source of food (such as a picnic lunch) becomes available. When a worker returns to the nest from such a bonanza, its excited behavior and/or odor stimulates other workers to go in search of food, but the wasp does not communicate exactly where the food is or lead others to it. Yellowjackets use both vision and smell in finding food. Their vision is not especially good—when hunting insects they often mistakenly pounce on spots on leaves or on other objects that vaguely resemble a small insect.

New Queens and Males. In late summer or early fall, yellowjacket workers start building larger cells in the nest. The queen lays fertilized eggs in some of the large cells. Larvae in these cells are fed extra meals and grow larger than workers, and these large females develop into queens.

At the same time, males are produced from unfertilized eggs laid in other large cells. The male eggs are laid by the queen and, in some kinds of yellowjackets, by some of the workers. As the large cells become more numerous, production of new workers ceases.

Colony Demise. The appearance of males and new queens marks the beginning of the end for the yellowjacket colony. Newly emerged queens and males loiter in the nest for several days, feeding on sweet droplets regurgitated for them by the larvae and workers. Then they fly away and mate. Soon after mating, the young queens find a secluded spot and begin hibernation, completing the cycle begun by their mother one year earlier. Males remain active, sipping nectar from goldenrod and other fall flowers as long as the supply lasts. They do not have the instinct to hibernate, so eventually they starve to death or freeze. But their only essential function in yellowjacket society—fertilizing the queens—has been fulfilled.

The workers gradually die off while raising the males and new queens. When there are not enough workers left to feed all the hungry larvae, the efficient yellowjacket "factory" quickly deteriorates. The workers may pull larvae and pupae from their cells and feed them to other larvae or simply throw them out. Soon the workers abandon the nest and, along with their queen mother, die. In the cold-winter climate of the northeastern United States, only the young, newly fertilized queens survive to begin the cycle anew.

Comparison with Other Social Wasps. The life cycle and behavior of the bald-faced hornet and the paper wasps are much the same as those described for yellowjackets. One difference is that the nests of paper wasps may be started by more than one foundress queen. Eventually one of the queens becomes dominant, whereupon the other queens begin behaving as workers, then depart (apparently of their own volition) after the emergence of the first brood of workers. In addition to their egg-laying duties, paper wasp queens occasionally gather pulp, build cells, and feed larvae even after the workers have emerged.

Nest Usurpers. A few species of yellowjackets are unusual in that they do not bother to make their own nests. Instead the queen invades the nest of a different species of yellowjacket, kills or drives off the resident queen, and lets the deposed queen's workers raise her offspring, which may be all queens-to-be and males. Some kinds of nest parasites are capable of building their own nest and only occasionally take over the nests of other species. A few parasites know no other way of life; they have not the foggiest idea of how to build a nest.

Wasp Stings and Venom. Admittedly, it is difficult to appreciate the fascinating intricacies of wasp life without knowing something about the sting and how to avoid it. Located at the rear end of the body, the stinger is a modified part of the wasp's egg-laying mechanism, so only females (workers and queens) can sting. The nonstinging males, common on flowers in late summer and fall, can be recognized by the relatively long, curly-ended antennae and light-colored face. Though a male wasp poses absolutely no danger, catching one by hand can be a thrilling experience if you have any doubt about the wasp's true gender.

Wasp venom, contained in a pouch attached to the stinger, is a mixture of enzymes and amino acid derivatives. Some people are very sensitive to wasp or bee venom and, if stung, may die from anaphylactic shock unless antivenin is quickly administered (venom chemicals start a chain reaction that causes certain cells in the blood and connective tissue to release substances that constrict the breathing passages, resulting in convulsions and suffocation).

How does one avoid being stung? *Do not disturb wasp nests.* Wasps are most protective of their nests and immediately defend it against any disturbance. Using chemical and/or acoustic signals, yellowjacket and hornet workers signal an alarm to their sisters inside the nest at the slightest disturbance. On several occasions I was stung after running over a yellowjacket ground nest with a lawnmower or as a result of breaking open a hidden nest while looking under logs for salamanders. If you accidentally disturb a nest and are attacked, immediately run away from the site and you will suffer fewer stings, since the wasps do not pursue very far.

Nests can be knocked down relatively safely at night, but early in the season the wasps may quickly build a new nest in the same location. Though wasps are defensive about their nests, they can be approached and watched quite closely as long as they are not physically disturbed and the entrance is not obstructed.

Do not swat at wasps. Yellowjackets, hornets, and paper wasps sting only as a defensive measure and swatting at one may be interpreted as an attack. Wasps flying around the picnic table or even landing on people are interested only in getting food. Assuming you did not disturb a nest, a wasp that simply lands on you will not sting unless you press it against your skin or somehow trap it against your skin under your clothing. At a picnic, it may be helpful to set a plate of food scraps at some distance from where you are eating and let the wasps help themselves. This may reduce the number of wasps flying around you and your plate of food.

A wasp in the car can be extremely hazardous, mainly because of the distraction or needless panic it may cause in the driver. Numerous accidents have been caused by distracted drivers swatting at wasps. The best thing to do is pull over and park, then gently assist the wasp out the window or door. Letting the insect crawl onto the end of some object, such as a piece of paper or umbrella that can be removed from the car works well. Trying to kill the wasp will only increase the likelihood of a sting.

A wasp almost always employs its stinger in response to some provocative action by the stingee. The danger posed by wasps is more imagined than real. While there are circumstances that make it prudent to avoid wasps or destroy their nests, the best wasp protection is knowledge of their behavior.

Spider Wasps (Pompilidae)

Spider wasps are fidgety, black or blue, solitary wasps that can sting. In most species, females dig burrows in the ground and stock them with paralyzed spiders to which an egg is attached. The wasp larva feeds on the spider. Adults feed on nectar and may be active through early fall.

Digger Wasps and Mud Daubers (Sphecidae)

The species of digger and mud dauber wasps in North America are solitary and can sting. Adult females hunt insects and spiders and paralyze or kill them by stinging. These they provision for their young, which may be reared in underground burrows, in mud cells made by the wasp, or in other sites (Fig. 14.65). Some species bring food several times to their growing young, others provide only one meal large enough for all the larva's needs. Adults feed on nectar, prey body fluids, or aphid honeydew.

By far the largest species in Connecticut is the **cicada killer** *(Sphecius speciosus)*, females of which are nearly 2 inches long. This is a ground-nesting cicada hunter. More than one female may dig a common nest tunnel in a sunny location, but each female provisions her own nest chambers at the end of the tunnel. Females sting (and thus paralyze) cicadas and drag them back to the nest. Sometimes the wasp drags the cicada up a tree to a perch from which it can launch itself for a flight back to the nest with its heavy cargo.

Inside the nest, the wasp deposits an egg on the cicada, with one egg and usually one or two cicadas per nest chamber. After the wasp larvae hatch, they feed on the still-living cicada. Larvae overwinter before pupating. Adults emerge and mate the next summer. On several occasions I found active burrows among sparse grass of city lawns in September. Females disturbed at their burrows may produce an alarm buzz but tend not to attack (the sting is weak).

A smaller sphecid that nests in "colonies" in sandy ground is the black and yellow **common sand wasp** *(Bembix americana)*. This wasp stocks its nests with flies. Prepupal larvae overwinter in the nests, and adults emerge in summer. I found a colony in a bare area on a broad, sandy levee along the Connecticut River.

Mining Bees (Andrenidae) and Halictid Bees (Halictidae)

Mining bees are often black or dark brown, fuzzy, and have darkened wings. The more noticeable halictid bee species are bright metallic green. Both nest in tunnels in the ground, sometimes in large groups. Adults drink nectar and pollinate wildflowers. They provision their nests with nectar and pollen.

Carpenter Bees (Anthophoridae)

Carpenter bees (e.g., *Xylocopa virginica*) resemble bumblebees but have a largely naked upper abdomen (which may be fuzzy at the end) (Fig. 14.66). Carpenter bees emerge from hibernation in early spring and are among the pollinators of early spring wildflowers. Males, recognized by their yellowish-white face patch, conspicuously patrol nest sites and chase each other beginning in late April. Males and females soon mate and are quite active around nesting sites into June

Fig. 14.65. A carpenter bee tunnel contained these wasp-stunned katydids, plus four wasp larvae (at top). The adult wasp plugged the tunnel with the plant material at right center.

Fig. 14.66. Carpenter bees nest in tunnels excavated in wood. The opened tunnel contains two large carpenter bee larvae. The adult bee constructed the partitions.

or early July. Females nest in long tunnels that they excavate in solid, dead wood, including buildings. Sometimes they lengthen old tunnels. They provision a mixture of nectar and pollen in a linear series of brood cells each containing an egg; the developing larvae feed on this food. Females reportedly die soon after nest completion, but I found a lively adult female in a tunnel that I opened in late July (the tunnel contained two large larvae). Might females guard their nests? The bees are inconspicuous until the next generation of adults emerges, beginning by mid-August and extending into September. Some newly emerged adult bees may disperse to new locations, whereas others collect and store pollen in their natal tunnels. I often found them on goldenrod flowers in late summer. They hibernate when the weather turns cold in fall.

Some larvae are killed by larvae of the tiger bee fly (*Anthrax tigrinus*). I found adults of this large bee fly near carpenter bee tunnels in mid-August.

Sphecid wasps sometimes nest in carpenter bee tunnels. A tunnel that I opened in late July contained 14 small, intact katydids, many katydid body fragments, and 4 wasp larvae. The intact katydids were stunned but alive (evidently stung by the wasp), and the adult wasp had plugged the tunnel opening with a mass of dry plant material. The wasp larvae clearly had been feeding on the stockpiled katydids.

Bumblebees (Apidae)

Bumblebees (*Bombus* spp.) are fuzzy, robust, black and yellow bees (Fig. 14.67). They are a cold-tolerant group of insects. Some species range to the high arctic. Because of their internal heat production, good insulation, and large body mass favorable for heat retention, they can fly at low air temperatures that incapacitate many insects. In Connecticut, you can see newly emerged queens foraging on and pollinating early spring flowers on cool days in April.

Spring Emergence and Nest Building. Fertilized queens emerge from hibernation in spring. In central Connecticut, over a period of nine years, I saw the first active queens April 9 to 19.

A newly emerged queen begins a new nest in a ready-made hollow such as an abandoned mouse nest underground. The queen arranges a nest of grass or similar plant material, builds an egg cell of wax and pollen, primes it with nectar-moistened pollen, lays some eggs, then closes the cell. The wax comes from glands between her abdominal segments. She also makes a wax honeypot and fills it with nectar, which is used for food during periods of inclement weather.

Hatched larvae feed on the stored nectar-pollen, supplemented by meals of nectar and pollen supplied by the queen (provided through openings in the larval cells). The larvae make a cocoon and pupate after about 10 days. The queen broods the cocoons and feeds herself from the honeypot.

Foraging Behavior. The bumblebee workers (females) emerge in a week or two and begin work after a few days. I saw recently emerged workers in late May.

Workers forage and provide food (a nectar-pollen mixture) to the larvae of the next brood. After the first group of workers begins to forage, the queen generally stays in the nest. Each nest yields one or more broods of workers.

Workers begin their foraging career by visiting various species and concentrating on the most rewarding ones. They tend to switch among different kinds of flowers more so than do honeybees, but generally they focus on just one or two species on a single foraging trip. Bumblebees have no known means of communicating to each other information about forage quality.

The tongue of a bumblebee is not long enough to reach nectar at the base of long-tubed flowers or at the tip of a nectar spur. They may overcome this by puncturing the base of the flower or spur (Fig. 14.68). I saw them do this to Dutchman's breeches flowers in late April and pink azalea flowers in May. Some may forage at holes made by other species. This kind of nectar "robbing" generally does not result in pollination.

Despite frequent nectar thievery, bumblebees do serve as important pollinators of various native wildflowers, and they also efficiently pollinate alfalfa, fruit trees, cotton, and many other crops. Queens often visit the pollen flowers of red maples and willows in mid- to late April.

Fig. 14.67. This pollen-coated worker bumblebee is sipping goldenrod nectar.

Fig. 14.68. Look carefully and you can see where a queen bumblebee "stole" nectar from this pink azalea by puncturing the base of the flower.

Reproduction. New queens and males (the latter from unfertilized eggs) appear in early summer in some species, in late summer or fall in most species. Earlier in the season, the queen suppresses egg production by her female workers through pheromones (chemical emissions). Removal of the founder queen results in egg laying by one or more of her orphaned workers, no longer suppressed by the pheromones or other effects of the queen. When there are many workers per larva, some large females, the new queens, are produced.

Winter. Only new young queens, which mate in autumn, survive through winter (a few workers are active through October). They hibernate underground or in other protected sites, usually far from their natal nest in most species. In fall, the new queens subsist on the nectar they have in their crop and on their fat reserves. They use glycogen reserves for their minimal energy needs in winter.

Honeybees (Apidae)

Honeybees* *(Apis mellifera)* were introduced in North America from the Old World (Fig. 13.6). They nest in tree hollows or in hives and store large volumes of food (honey) in beeswax combs in the nest. As in bumblebees, the workers are sterile females that have been raised on a diet dominated by a honey-pollen mixture. Males are short-lived and are involved only in mating with new queens.

New honeybee colonies are started by old queens that depart with a swarm of workers. Newly produced queens, which have been fed a continuous diet of "royal jelly," a white pasty secretion of the workers, emerge from their cells, fly out to mate, then return to the nest to rebuild the colony. Queens live usually two to three years.

Colonies may include tens of thousands of workers. Workers communicate the direction and distance of good supplies of pollen/nectar by means of a "dance" they perform inside the hive. The kind of flower involved is made known by the odor of the bee or the honey it carries.

In the mid-1990s, parasitic mite syndrome, caused by tracheal mites, resulted in a die-off of many honeybee colonies in the northeastern United States.

Caddisflies

Adult caddisflies resemble moths but lack coiled mouthparts, and the wings usually are covered by fine "hairs" rather than scales (Fig. 14.69). Caddisflies (Order Trichoptera) occur in and near freshwater habitats throughout most of the world. Most species are associated with cool, flowing waters, and relatively few species inhabit temporary waters. About 1,400 of the 10,770 species occur in the United States and Canada.

Adults are active mostly at night (they were regular visitors to my lighted kitchen window), though I sometimes found many newly emerged adults active in daytime along streams. Their short lives focus on reproduction. They eat little or nothing. Females lay eggs in or near the water or in sites that later hold water.

Caddisflies are best known for their often conspicuous larval cases (Fig. 14.70). Larvae of many species construct and reside in silk-bound cases of plant material or sand grains, with each genus making a distinctive type of case. Larvae anchor themselves inside the case by hooklike appendages at the end of the body. Some carry their house around with them, giving the startling impression of animated plant material. To find these, just look into calm, clear water at night with a flashlight and watch for movement as the larvae, with their head and forelegs protruding, slowly drag their cases over the bottom or climb on plants. Other caddisfly larvae live in a fixed abode that in some instances may consist only of a silk tube. The silk is secreted through an opening near the mouth. The larvae of some

Fig. 14.69. Look for caddisfly adults near streams, ponds, and freshwater wetlands.

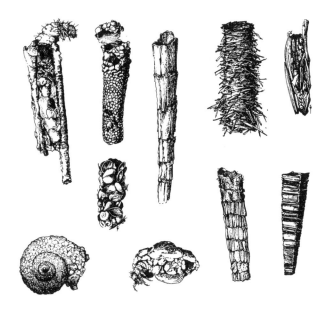

Fig. 14.70. Many caddisfly larvae build and reside in portable cases, usually of plant material or sand grains. Here are examples of several types of cases. From Betten (1934).

caddisflies range freely without a case. The aquatic larvae of most species have gill tufts on the abdomen.

Cases likely serve as camouflage and physical protection against predators and as a desiccation barrier when habitats dry up. They also increase respiratory efficiency because body undulations of the larvae easily create a ventilation current of water through the case. Additionally, cases may provide streamlining, ballast, buoyancy, and water resistance.

Larval diet varies among species and often changes as the larvae grow. Some larvae filter bits of organic material from the water, sometimes employing a silk net adjacent to the larva's fixed shelter, whereas others feed on diatoms, algae, or detritus, or prey on other small invertebrates.

Caddisflies pupate in the water, but pupae may crawl out of the water before the adult emerges. Most species have a one-year life cycle, but some require two years. Adults may emerge from the water simultaneously in large numbers. At such times they may become important food items for various fishes, land animals, and birds. The larvae also can be important in fish diets.

Butterflies and Moths

Butterflies are notable in that they are admired even by persons who find most insects suitable subjects for a rolled-up newspaper or a quick trip down the household plumbing system. Beauty, delicacy, and inoffensive behavior do have their advantages.

Butterflies and moths (Order Lepidoptera) occur worldwide wherever there is at least some vegetation, and several species range into the high arctic. Most of the world's 17,500 species of butterflies occur in the tropics, but there is a good diversity in North America (around 620 species) and in southern New England (around 125 species). Moths are represented by about 100,000 species worldwide. David Wagner has compiled a list of more than 2,300 moth species that have been found in Connecticut. Approximately 12,260 species of lepidopterans occur in the United States. The owlet moths, represented by more than 3,000 species in the United States, comprise the largest family.

Butterflies and moths are distinguished from other insects by having wings that are covered or partially covered with tiny, often colorful, easily-rubbed-off scales, and mouthparts that form a long, coiled tube. The larvae, also known as caterpillars, resemble certain larval hymenopterans, but the lepidopterans have pairs of leglike appendages (prolegs) on abdominal segments three through six and on the last segment, whereas hymenopterans have prolegs on at least six abdominal segments.

The lepidopteran life cycle includes four major stages: egg (often laid on the larval host plant, detected by the female through sensors on her prolegs or antennae), larvae (caterpillar), pupa (in moths, surrounded a silk case, or cocoon, made by the caterpillar), and adult. The duration of each stage varies among different species, but in Connecticut none of the four stages lasts for more than eight or nine months.

Adult butterflies (Figs. 14.71 to 14.75) are active in daytime, whereas most moths fly at night. Being day-active and colorful, it may be no surprise that many butterflies employ visual phenomena in their social and reproductive behavior. In addition, males of some species have special sex-attractant scent organs on their wings, legs, or body. Ready-to-mate female moths of many species attract males with

Fig. 14.71. Eastern tiger swallowtails are one of Connecticut's most common large butterflies.

Fig. 14.72. Brown elfins are common in spring on hilltops near blueberry shrubs.

Fig. 14.73. Mourning cloaks overwinter as adults and are among the earliest spring butterflies.

Fig. 14.74. American lady butterflies migrate into Connecticut each year.

Fig. 14.75. Great spangled fritillaries are common in meadows. Larvae feed on violets.

chemical secretions (pheromones) that the males easily detect with their sensitive antennae—an effective method for the darkness of night.

With a few exceptions, the larvae feed on the foliage and other parts of flowering plants. Adult diet consists of nectar, sap, water, fluids from carrion or dung, or not much at all.

Butterflies visit and may pollinate flowers of a great many colors and shapes, though many butterfly flowers are shaped so as to provide a convenient flattened landing platform. Fragrant, white, tubular flowers that stay open at night tend to be pollinated by moths. Most flowers pollinated by butterflies or moths are on herbaceous plants; few are trees or shrubs.

Butterfly Status, Ecology, and Life History

Each of Connecticut's butterfly species has a unique set of distributional, ecological, and life history attributes, yet general patterns are evident (Table 14.1). Certain butterflies fly only during restricted periods (i.e., early, mid-season, or late), whereas other species have a long flight season. Species with short seasons generally have only one generation per year and a short adult life span (most butterfly adults live no more than one or two weeks, with a good number of exceptions). Those with longer seasons may have multiple generations (the first generation of adults produces offspring that fly during the same calendar year), or the adults may simply be relatively long-lived.

A good way to see many of the earliest flying spring butterflies (Table 14.1) is to walk through oak woodlands to a ridge top on a warm sunny day in late April. Early summer is the best time to see the highest diversity of butterflies (Fig. 14.76). In summer and early fall, a visit to a lush, flowery meadow often yields many butterflies. Too often, though, meadows are mowed, leaving a depauperate field of cropped grass where a rich wildlife habitat could have been.

Knowing the food plant will help you determine where to look for specific butterflies and caterpillars (note that butterflies often sip nectar from plants that are not food for the larvae). The larvae of some butterfly groups are dependent on a restricted group of plants (for example, skippers and grasses/sedges, fritillaries and violets) whereas other butterfly groups and individual species have a wide array of larval food plants (Table 14.1).

If you compare the flight seasons with the stage in which each species spends the winter you can detect some general trends. Butterflies that fly early in the year tend to be those

Butterflies and Moths | 249

Table 14.1 Butterfly Status, Ecology, and Life History

Species	Status	J	F	M	A	M	J	J	A	S	O	N	D	Larval food plant	Wintering stage
Pipevine swallowtail (*Battus philenor*)	U/L					X	X	X	X	X				Aristolochia	Pupa
Black swallowtail (*Papilio polyxenes*)	C				X	X	X	X	X	X				Carrot Family	Pupa
Eastern tiger swallowtail (*Papilio glaucus*)	C				X	X	X	X	X	X				Black cherry, tulip-tree, aspen, birch	Pupa
Spicebush swallowtail (*Papilio troilus*)	C				X	X	X	X	X	X				Sassafras, spicebush	Pupa
West Virginia white (*Pieris virginiensis*)	U/L				X	X								Toothworts	Pupa
Cabbage white (*Pieris rapae*)	C			X	X	X	X	X	X	X	X	X		Mustard family	Pupa
Falcate orangetip (*Anthocharis midea*)	U/L				X	X								Rock-cresses	Pupa
Clouded sulphur (*Colias philodice*)	C				X	X	X	X	X	X	X			White clover, other clovers	Larva, pupa
Orange sulphur (*Colias eurytheme*)	C				X	X	X	X	X	X	X	X		Pea family	Larva, pupa
Cloudless sulphur (*Phoebis sennae*)	V									X	X			Sennas	—
Little yellow (*Eurema lisa*)	V								X	X	X			Sennas	—
Harvester (*Feniseca tarquinius*)	U/L					X	X	X	X	X				Woolly aphids on alder, beech	Pupa
American copper (*Lycaena phlaeas*)	C				X	X	X	X	X	X	X			Docks, sheep sorrel	Larva
Bronze copper (*Lycaena hyllus*)	U/L						X	X	X					Water dock, curly dock	Egg
Bog copper (*Lycaena epixanthe*)	U/L						X	X						Cranberries	Egg
Coral hairstreak (*Satyrium titus*)	C						X	X						Wild cherry, wild plum	Egg
Acadian hairstreak (*Satyrium acadica*)	U/L						X	X						Willows	Egg
Edwards' hairstreak (*Satyrium edwardsii*)	U/L						X	X						Small oaks	Egg
Banded hairstreak (*Satyrium calanus*)	C						X	X						Oaks, hickories	Egg
Hickory hairstreak (*Satyrium caryaevorum*)	U/L						X	X						Hickories	Egg
Striped hairstreak (*Satyrium liparops*)	U/L						X	X						Cherry, blueberry, etc.	Egg
Oak hairstreak (*Satyrium favonius*)	U/L						X	X						Oaks	Egg
Brown elfin (*Callophrys augustinus*)	U/L				X	X								Blueberries, other heaths	Pupa
Frosted elfin (*Callophrys irus*)	U/L				X	X								Wild indigo	Pupa
Henry's elfin (*Callophrys henrici*)	U/L				X	X								Hollies, buckthorn, blueberries, others	Pupa
Eastern pine elfin (*Callophrys niphon*)	U/L				X	X	X							Pines	Pupa
Juniper/olive hairstreak (*Callophrys gryneus*)	U/L				X	X			X					Red-cedar	Pupa
Hessel's hairstreak (*Callophrys hesseli*)	U/L				X	X								Atlantic white cedar	Pupa
White M hairstreak (*Parrhasius m-album*)	U/L				X	X			X	X				Oaks	Pupa
Gray hairstreak (*Strymon melinus*)	U/L				X	X	X	X	X	X	X			Various	Pupa
Eastern tailed-blue (*Everes comyntas*)	C				X	X	X	X	X	X	X			Pea family	Larva
Spring azure complex (*Celastrina ladon*) and relatives	C				X	X	X	X	X	X				Various, often dogwood, viburnum	Pupa
Northern metalmark (*Calephelis borealis*)	U/L						X	X						Round-leaved ragwort	Larva
American snout (*Libytheana carinenta*)	V						X	X	X	X				Hackberry	—
Variegated fritillary (*Euptoieta claudia*)	V						X	X	X	X	X	X		Violets	—

Table 14.1 (continued)

Species	Status	\multicolumn{12}{c}{Main flight season in Connecticut}	Larval food plant	Wintering stage
		J	F	M	A	M	J	J	A	S	O	N	D		
Great spangled fritillary (*Speyeria cybele*)	C						X	X	X	X				Violets	Larva
Aphrodite fritillary (*Speyeria aphrodite*)	U/L						X	X	X	X				Violets	Larva
Atlantis fritillary (*Speyeria atlantis*)	U/L						X	X	X					Violets	Larva
Silver-bordered fritillary (*Boloria selene*)	U/L					X	X	X	X	X				Violets	Larva
Meadow fritillary (*Boloria bellona*)	U/L				X	X	X	X	X	X				Violets	Larva
Harris' checkerspot (*Chlosyne harrisii*)	U/L						X	X	X					Flat-topped white aster	Larva
Pearl crescent (*Phyciodes tharos*)	C					X	X	X	X	X	X			Asters	Larva
Baltimore checkerspot (*Euphydryas phaeton*)	U/L						X	X						Turtlehead, English plantain	Larva
Question mark (*Polygonia interrogationis*)	U/L					X	X	X	X	X	X			Nettles, elms, hackberry	Adult*
Eastern comma (*Polygonia comma*)	U/L			X	X	X	X	X	X	X	X			Elms, nettles	Adult
Compton tortoiseshell (*Nymphalis vaualbum*)	U/L			X	X						X			Birches, willows	Adult
Mourning cloak (*Nymphalis antiopa*)	C			X	X	X	X	X	X	X	X			Willows, aspen, cottonwoods, etc.	Adult
Milbert's tortoiseshell (*Nymphalis milberti*)	V			X	X	X	X	X	X	X	X			Nettles	Adult
American lady (*Vanessa virginiensis*)	C				X	X	X	X	X	X	X			Sunflower family	Adult*, pupa
Painted lady (*Vanessa cardui*)	U/L				X	X	X	X	X	X	X			Thistles and others	Adult*
Red admiral (*Vanessa atalanta*)	C				X	X	X	X	X	X	X			Nettles	Adult*
Common buckeye (*Junonia coenia*)	U/L					X	X	X	X	X	X			Gerardias, toadflax, plantain	Adult*
Red-spotted admiral (*Limenitis arthemis*)	C					X	X	X	X	X				Cherry, birch, poplar and others	Larva
Viceroy (*Limenitis archippus*)	C					X	X	X	X	X	X	X		Willows	Larva
Hackberry emperor (*Asterocampa celtis*)	U/L						X	X	X	X				Hackberry	Larva
Tawny emperor (*Asterocampa clyton*)	U/L							X	X	X				Hackberry	Larva
Northern pearly-eye (*Enodia anthedon*)	U/L						X	X	X					Grasses	Larva
Eyed brown (*Satyrodes eurydice*)	U/L						X	X	X					Sedges	Larva, Pupa?
Appalachian brown (*Satyrodes appalachia*)	U/L						X	X	X					Sedges	Larva, pupa?
Little wood-satyr (*Megisto cymela*)	C					X	X	X						Grasses	Larva
Common ringlet (*Coenonympha tullia*)	C					X	X			X				Grasses	Larva
Common wood-nymph (*Cercyonis pegala*)	C							X	X	X				Grasses	Larva
Monarch (*Danaus plexippus*)	C					X	X	X	X	X	X			Milkweeds	Adult*
Silver-spotted skipper (*Epargyreus clarus*)	C					X	X	X	X	X				Black locust, other pea family	Pupa
Long-tailed skipper (*Urbanus proteus*)	V									X				Legumes	—
Hoary edge (*Achalarus lyciades*)	U/L						X	X						Tick-trefoils, bush-clover, false indigo	Larva
Southern cloudywing (*Thorybes bathyllus*)	U/L						X	X						Legumes (pea family)	Larva
Northern cloudywing (*Thorybes pylades*)	U/L				X	X	X							Legumes (pea family)	Larva
Dreamy duskywing (*Erynnis icelus*)	U/L					X	X							Usually willows, poplars; birches	Larva
Sleepy duskywing (*Erynnis brizo*)	U/L				X	X								Scrub oak	Larva

Table 14.1 (continued)

Species	Status	Main flight season in Connecticut												Larval food plant	Wintering stage
		J	F	M	A	M	J	J	A	S	O	N	D		
Juvenal's duskywing (*Erynnis juvenalis*)	C				X	X	X							Oaks	Larva
Horace's duskywing (*Erynnis horatius*)	U/L				X	X	X	X	X					Oaks	Larva
Columbine duskywing (*Erynnis lucilius*)	U/L					X	X							Columbine	Larva
Wild indigo duskywing (*Erynnis baptisiae*)	C					X	X	X	X	X				Crown vetch, wild indigo	Larva
Persius duskywing (*Erynnis persius*)	U/L					X	X							Lupines, wild indigo	Larva
Common checkered-skipper (*Pyrgus communis*)	V							X	X	X	X			Mallow family	Larva
Common sootywing (*Pholisora catullus*)	C					X	X	X	X	X				Lamb's-quarters, amaranths, ragweed	Larva
Arctic skipper (*Carterocephalus palaemon*)	V													Grasses	—
Swarthy skipper (*Nastra lherminier*)	U/L						X	X	X					Little bluestem	Larva
Least skipper (*Ancyloxypha numitor*)	C						X	X	X	X				Grasses	Larva
European skipper (*Thymelicus lineola*)	C						X	X						Timothy	Egg
Fiery skipper (*Hylephila phyleus*)	V									X				Bermuda grass	—
Leonard's skipper (*Hesperia leonardus*)	U/L								X	X				Grasses	Larva
Cobweb skipper (*Hesperia metea*)	U/L					X	X							Bluestem grasses	Larva
Indian skipper (*Hesperia sassacus*)	U/L					X	X							Grasses	Larva
Peck's skipper (*Polites peckius*)	C					X	X	X	X	X				Grasses	Larva, pupa
Tawny-edged skipper (*Polites themistocles*)	C					X	X	X	X	X				Grasses	Pupa
Crossline skipper (*Polites origenes*)	U/L						X	X						Grasses	Larva
Long dash (*Polites mystic*)	C						X	X						Grasses	Larva
Northern broken-dash (*Wallengrenia egeremet*)	U/L						X	X	X					Panic grasses	Larva?
Little glassywing (*Pompeius verna*)	C						X	X	X					Purple top, probably other grasses	Larva?
Delaware skipper (*Anatrytone logan*)	C							X	X					Grasses	Larva?
Mulberry wing (*Poanes massasoit*)	U/L							X	X					Sedges	Larva?
Hobomok skipper (*Poanes hobomok*)	C					X	X	X						Grasses	Egg, larva, pupa
Zabulon skipper (*Poanes zabulon*)	C					X	X			X				Grasses	Larva?
Broad-winged skipper (*Poanes viator*)	U/L							X	X	X				Phragmites, wild rice, sedges	Larva?
Dion skipper (*Euphyes dion*)	U/L						X	X	X					Sedges	Larva?
Black dash (*Euphyes conspicua*)	U/L							X	X					Tussock sedge	Larva?
Dun skipper (*Euphyes vestris*)	C						X	X	X					Sedges	Larva
Dusted skipper (*Atrytonopsis hianna*)	U/L					X	X							Bluestem grass	Larva
Pepper and salt skipper (*Amblyscirtes hegon*)	U/L					X	X							Grasses	Larva
Ocola skipper (*Panoquina ocola*)	V										X	X		Unknown locally (likely grasses)	—

*Most or all adults winter south of Connecticut; some adults (not monarch) may survive winter locally. Adults are migratory.

C = Common, found in a wide variety of habitats; U/L = Uncommon and/or geographically restricted or habitat-specific; V = Vagrant, not known to breed in Connecticut, migrates to this area, can be common some years. List does not include certain rarities or several historically recorded species not recently seen in Connecticut. See text for further explanation.

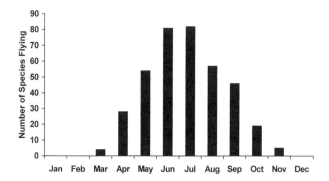

Fig. 14.76. Annual pattern of adult butterfly activity in Connecticut.

that overwinter locally as adults or pupae. Early fliers do most or all of their development during the preceding year. Species that do not emerge as adults until later in the summer generally overwinter as eggs or minimally developed larvae. For these, the spring and early summer are devoted to feeding by the caterpillars, which may not have much to eat until foliage fully develops in late spring.

The flight seasons indicated in Table 14.1 are generalized and not absolute. Some of the butterflies listed for a particular month fly only during the early or latter periods of that month, and some fly in the early or latter parts of months outside the indicated primary flight season. Some do not fly in the indicated months every year—the flight period varies from year to year depending on weather conditions. Also, the flight season varies among different regions and different elevations in Connecticut, with warmer regions exhibiting earlier flights than cooler areas.

Connecticut Butterfly Atlas Project

A multi-year project to document the distribution and status of Connecticut butterfly species was initiated in 1995 and was nearing completion in early 2003. For more information, check the Connecticut Butterfly Atlas Project website (see bibliography).

Life of the Monarch

Distribution and Uniqueness. Perhaps the most widely recognized of Connecticut's butterflies, the monarch breeds throughout most of the United States and southern Canada (Fig. 14.77). It is a year-round resident in California and most of tropical America, though California also hosts migrants. During the 1800s, it colonized and became resident on various Pacific and Atlantic islands.

Fig. 14.77. Monarchs make heavy use of seaside goldenrods in late summer and early fall.

The unique aspects of the monarch's life history are its long-distance migrations and the immense colonies it forms in winter. In areas where winters are cold, most insects spend the winter in the egg stage or as a mummylike pupa. In certain other species, the adults overwinter in an inactive state in a place protected from the most severe weather or at least in a site where they are not readily available to predators. The group of butterflies to which the monarch belongs (family Danaidae) evolved in tropical climates, and no stage of the life cycle (egg, larva, pupa, adult) can tolerate prolonged freezing. The presence of monarchs in the northern United States and Canada depends on highly evolved migratory behavior. Of the danaids, the monarch is the only species that is able to exploit high latitude environments.

In a certain sense, the behavior of the monarch is similar to that of many birds. They migrate south for the winter and return north the following spring. However, the migrations are not exactly like those of birds, for an individual monarch from the northern states does not make the entire round-trip journey. The following account should clarify this potentially confusing statement.

Late Summer and Fall Migration. As day length shortens and temperatures decline in late summer, the population of 100 to 500 million monarchs in eastern North America ceases reproductive activity and begins to fly southward. By mid-September, or as early as late August, and through mid-October, large numbers can be seen in flight. They stop each day before sundown and rest until the next morning, often forming aggregations in roosting areas. Some can be seen flying in Connecticut through early November. Migrating monarchs are easy to observe because they fly during daylight hours and often stay within 15 to 20 feet of the ground, though while hawk-watching through binoculars I often saw monarchs higher, as much as a few hundred feet above ground. Sometimes I saw them flying westward over Long Island Sound, hundreds of yards from shore.

By November, many reach the Gulf Coast, where some of them spend the winter. Others continue west and south to wintering sites in Mexico. This long trip takes some monarchs up to 2,500 miles from the field in which they hatched. Though they begin the journey with enough stored energy (fat) reserves to fly for about 100 hours without feeding, they pause frequently to sip nectar from goldenrod, thistle, and other flowers, ensuring that they arrive in wintering sites with fully stocked fat reserves.

Winter. On the wintering grounds, migrant monarchs create one of the most spectacular displays in the animal world. In some areas, they gather into small to immense flocks, covering trees with a shimmering blanket of orange. Some of the best known of these aggregations occur in California (more than a hundred aggregation areas) and have been famous tourist attractions for many years. The largest assemblages in California include about 170,000 monarchs. The largest, most impressive winter colonies, including tens of millions of monarchs blanketing trees in an area of only a few acres, only recently (mid-1970s) became known to scientists. They are in cool coniferous forests in some thirty scattered sites on a dozen volcanic mountain massifs west of Mexico City in south-central Mexico. Conservationists have been working to stop the ongoing loss and degradation of the forests that support these spectacular and unique aggregations.

The monarchs stay in the wintering areas for several months, living off their fat reserves. Cool weather restricts their activity on most days (monarchs cannot fly if their thoracic temperature is less than about 55°F), but they may warm up enough on sunny days to fly out to drink water. Wintering monarchs are preyed upon by grosbeaks, orioles, mice, and sometimes other animals.

Northward Movement. The northward exodus from the wintering sites begins in February. By March, monarchs from the Mexican colonies have mated and have begun laying eggs in the southeastern United States. Not until May do large numbers arrive in Connecticut and other northern states and Canada, though some may arrive in the latter half of April. In some years I did not see the first monarchs until early July. These monarchs, however, are not the ones that overwintered in Mexico. Monarchs that migrate into Connecticut represent successive broods of migrants that are the offspring or grand-offspring (or great-grand-offspring) of the initial overwintering migrants that laid their eggs in areas to the south. Very few (if any) monarchs make a complete round trip between the northernmost breeding areas and the wintering sites. (In contrast, migratory birds breeding in northern latitudes usually return to their previous breeding areas.).

By mid-summer, all of the tattered, travel-worn migrant monarchs from the Mexican wintering sites have died. When the next migration begins, it is their third- or fourth-generation descendants that make the journey south. In some years, severe winter weather in Mexico kills large numbers of monarchs, and the number of north-bound migrants is greatly reduced. The small number of monarchs in Connecticut in 1992 resulted from a massive die-off the previous winter.

Navigation. How do millions of monarchs, coming from many different areas, manage to travel thousands of miles over totally unfamiliar terrain to a specific traditional wintering site? Simple learning, or "following the leader," cannot be involved because none of the monarchs in the southern wintering sites has ever been there before. Migration for each monarch is a once-in-a-lifetime event. Perhaps instinctive responses to such factors as day length, the daily path of the Sun through the sky, the Earth's magnetic field, and/or certain landscape features are involved. Recent research indicates that monarchs can use a magnetic compass for navigation. Overall, we can say only that monarch migration is a complex, inherited behavior.

Courtship and Mating. Breeding activity among monarchs begins as males watch for and chase passing females. Actually, the easily stimulated males will pursue just about any monarch-size object, including blowing leaves, birds, and butterflies of other species—their vision is not particularly sharp.

Upon locating and reaching a female monarch, the male may flutter near her and extrude a feathery "hair pencil" from the tip of his abdomen, filling the air with a flowery-sweet aroma. The pungent fragrance may induce the female to land, where she is joined by the male (an unreceptive female eludes her pursuer with a rapid zig-zag flight). Her initial anti-social response overcome by the sex "perfume," the female perches close-winged while the male struts nearby, slowly beating his wings. The fragrance derives from a tiny black glandular pouch (androconial organ) on each of his hind wings; the hair pencil is anointed by being dipped into the glandular wing pouch. Males obtain the chemical needed to produce the fragrance from plants. The pouches continue to release the sex fragrance when the male withdraws his hair pencils, which he must do before mating. However, these pheromones and related behaviors are modestly developed in the monarch, and forced copulation without preliminary courtship is common.

Monarchs may remain locked together in copulation for over an hour as the male deposits a packet of sperm in the female's abdomen. Finally they fly their separate ways. There is evidence that males effectively transfer not only sperm but also lipid nutrients to the females. This may enhance female survival and reproductive success, especially in wintering aggregations where females may participate in multiple matings.

Eggs, Larvae, and Pupae. Each female lays a few hundred eggs almost exclusively on the young, tender leaves of milkweed. She deposits a single pin-head-sized egg on the underside of a leaf, then flies to another plant before laying the next egg.

After several days, a tiny black-headed caterpillar chews its way out of the egg and begins eating the underside of the leaf. Though drab at first, the caterpillar soon takes on a striking appearance, with a pair of black threadlike structures extending from each end of a yellow, black, and white banded body (Fig. 14.78). If disturbed, the caterpillar flails the harmless filaments.

Gorging itself on whole milkweed leaves, the gaudy caterpillar grows quickly. To accommodate this growth, the caterpillar replaces its tightening skin four times. After its ten- to twenty-day feeding binge, the caterpillar abandons the milkweed and crawls to some nearby object, such as a tree or bush. On the underside of a branch, it attaches a small bundle of silk, grasps it with its rearmost "legs," and hangs head downward. Soon the caterpillar's skin splits open and slides off, revealing a pale green, gold-studded capsule lightly sculptured with certain butterfly features

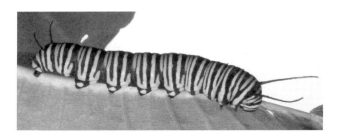

Fig. 14.78. Monarch caterpillars subsist on milkweed foliage.

Fig. 14.79. This monarch chrysalis was on a red maple next to a small patch of milkweed.

(Fig. 14.79). In this pendant chrysalis the monarch completes its metamorphosis.

Emergence. The covering of the chrysalis gradually becomes transparent, and the butterfly within can be seen clearly after several days. Then, in the morning or early afternoon, the chrysalis cracks open and a monarch butterfly crawls out. Hanging downward, it pumps fluid into its crumpled wings. In 10 to 20 minutes the wings flatten out to full size. After the wings dry and harden for several hours, the monarch flies off in search of its first sip of nectar.

Caterpillar Diet. The sweet nectar diet of the monarch is typical of butterflies, but the monarch is the only butterfly in which the caterpillar feeds on milkweeds. Milkweeds are the only plants on which monarchs successfully develop.

Few other animals, herbivorous mammals included, can eat these plants. The milky sap of many species of milkweed contains distasteful and toxic chemicals (cardiac glycosides or cardenolides) that in large amounts cause vomiting and abnormal heart pumping in birds and mammals. The monarch is unaffected by these chemicals and thus has the advantage of eating a food that is avoided by other plant eaters.

The milkweed toxins not only help protect the monarch caterpillar's food supply, but, stored in the monarch's body, also provide a chemical anti-predator defense mechanism for the caterpillar and later the butterfly. Laboratory experiments have shown that eating a single toxin-laden adult monarch is enough to make a blue jay refuse to attack another monarch. The jays evidently learn to associate the bold color of the monarch with its distastefulness; the birds not only refuse to attack monarchs but also monarch look-alikes such as the viceroy (at least some of which are also distasteful).

However, not all milkweeds contain cardiac glycosides and, as determined first-hand by one very dedicated biologist, monarchs raised on nontoxic plants can be eaten without ill effect. Birds seem to be able to detect this variation in toxicity, for field observations have shown that various birds do attack and eat monarchs, sometimes discarding ones that presumably are distasteful. Along the Gulf Coast of the United States, the milkweeds eaten by monarch caterpillars in early spring contain high levels of toxins that become sequestered in the monarchs' bodies. In contrast, monarchs that develop in the north tend to have low cardenolide concentrations, but they may be somewhat protected if predators learn to avoid eating monarchs after their encounters with the toxic individuals that arrived earlier.

Monarch body odors may play a role in warning potential predators of the monarch's toxins. Bitter and toxic pyrrolizidine alkaloids, obtained from plants by adults, also may serve as predator protection.

Memorable Moths

Most moths are small and drab, so they are not especially distinctive to most persons. Their cryptic coloration challenges the skills of the many visually oriented diurnal predators that would eat them if only they could find them. The typical muted colors of nocturnal moths also reflect the lack of importance of color in their behavioral interactions, contrasting sharply with diurnal butterflies. In mate attraction, most moths depend on odors rather than visual displays. Some moths do have colors and patterns that are bright and bold in a display case, but sometimes these are actually cryptic in the natural habitat. A minority of species (e.g., tiger moths) resemble butterflies in being brightly colored and active in daytime.

Some moths have bright flash colors or eyespots, usually hidden, that may confuse or startle a would-be predator. Bright flash colors, such as those conspicuously exposed by underwing moths in rapid flight (as when disturbed from their daytime resting spot), draw the attention of predators. When the moth suddenly lands, the colors are quickly hidden, and to the predator, focused on the moving patch of bright color, the moth simply vanishes. Sudden exposure of big eye spots may momentarily cause a predator to pause and think that it has encountered something quite a bit bigger than it actually is. That delay may be all it takes for the moth to escape.

Here I list just a very few of Connecticut's larger, more colorful species, plus a selected few of the many species that have remarkable caterpillars or attractive adults or that are of economic importance—the species most likely to attract the attention of someone who is not a moth specialist. The sequence of families is random.

For photographs and information on the known distribution of moths in Connecticut, visit the following U.S. Geological Survey website: www.npsc.nbs.gov/resource/dist/lepid/moths/ct/toc.htm.

Slug Caterpillar Moths (Limacodidae)

The unusual naked or hairy caterpillars of these moths glide sluglike on their bellies. The thoracic legs are very short and the abdominal prolegs are replaced by suckers. The skiff moth (*Prolimacodes badia*) caterpillar is naked, but larvae of many species, such as the green, red-marked crowned slug (*Isa textula*), have numerous lobes or fringes adorned with fine stinging hairs (Fig. 14.80). Limacodid caterpillars feed on foliage of a wide variety of woody plants. They winter in a loose cocoon. The small, handsomely patterned adults fly in late spring and summer. I found spiny oak slug moths (*Euclea delphinii*), attractively marked with green patches on the brown forewings, in early July.

Fig. 14.80. These unusual caterpillars are the larvae of the skiff moth (upper) and crowned slug (lower).

Fig. 14.81. Tulip-tree beauty is a large inchworm moth, common near sassafras and tulip-trees.

Inchworm or Geometer Moths (Geometridae)

The geometers include many moths that perch with wings outspread. The distinctive caterpillars are variously known an inchworms, measuringworms, spanworms, or loopers. Many are convincing twig mimics. One of the largest species, with a 2-inch wingspan, is the tulip-tree beauty *(Epimecis hortaria)* (Figure 14.81). Its caterpillars feed on the foliage of tulip-trees, sassafras, and other deciduous trees. I found adults from mid-April through August.

In autumn, look for Bruce's spanworm *(Operophtera bruceata)*, fluttering slowly among the barren tree trunks and branches of Connecticut's forests long after the leaves have fallen. Unnoticed by most persons, these drab brown, slim-bodied moths are sometimes common in the autumn woods. Bruce's spanworm is unusual in that it commonly flies in daylight, as well as at night. Most remarkable is that this and some other inchworm moths can fly at temperatures at and even slightly below freezing, conditions that incapacitate nearly all other insects (see "Insects in Winter" at end of chapter) (Fig. 14.82). In the case of Bruce's spanworm moths, all of the flying moths are males; the females are wingless and simply perch on the ground or plants. The males find the females by using their sensitive antennae to follow the scent signals released into the air by the adult females. Neither the male or female adults do any feeding. They depend on the resources that they stored in their bodies when they were foliage-feeding caterpillars. These moths provide interesting autumn nature watching, but usually I have to point out to my companions that something impressive is happening in their presence.

Tent Caterpillar and Lappet Moths (Lasiocampidae)

The most observable representative of the lasiocampid family is the eastern tent caterpillar *(Malacosoma americana)* (Fig. 14.83). The caterpillars, with a light dorsal stripe, make a conspicuous communal silk tent, occupied by inch-long larvae as early as April, and forage out from the tent to defoliate various trees. I found that small black cherry trees are a favorite food; several hundred large caterpillars may completely defoliate the trees by the first of May. Pupation generally occurs away from the larval host tree. The brownish adults, with two parallel streaks on each forewing, fly in late spring and early summer. Females deposit dark brown egg masses that overwinter as frothy collars on twigs.

Fig. 14.82. This inchworm moth flew to my lighted kitchen window at midnight at an air temperature of 31.5°F in early December.

Fig. 14.83. Large communal groups of eastern tent caterpillars often defoliate cherry trees in spring.

Giant Silkworm and Royal Moths (Saturniidae)

The huge, colorful members of the silkworm moth family are the most memorable of all moths. Unfortunately most are now simply a memory. Populations throughout the northeastern United States have declined greatly over the past few decades, evidently as a result of attacks on the caterpillars by the parasitoid tachinid fly *Compsilura concinnata*. The fly was introduced into the United States on several occasions from 1906 to 1986 by the U.S. Department of Agriculture, in attempts to control gypsy moths and other pests. Two native species (regal moth and imperial moth) apparently have been extirpated from Connecticut. Silkworm caterpillars, often adorned with prominent knobs, horns, or spines, can be as eye-catching as the adults. They pupate in large cocoons of silk. This section mentions just a few of the surviving species.

Most of the big, showy silkworm moths fly in late spring or early summer (often beginning in mid- to late May). The caterpillars tend to eat a wide range of deciduous trees and shrubs. They overwinter as pupae in silk cocoons. The cocoon may be hidden in leaf litter on the ground, as in luna moths *(Actias luna)* (Fig. 14.84), or attached to a twig and readily visible in winter when trees are leafless, as characteristic of cecropia silkmoths *(Hyalophora cecropia)* (Fig. 14.85). Luna moths fly at night, as do cecropia moths, but I saw several of the latter species flying in daylight as well.

Not all members of the family fit this pattern. For example, the eastern buckmoth *(Hemileuca maia* complex) (Fig. 14.86), a rare, fast-flying species with a highly localized known distribution in central Connecticut, flies only in daytime and specializes on oaks (often scrub oak) as the larval food source, at least initially. After feeding in spring and early summer, the larvae, adorned with stinging spines, pupate in leaf litter or soft soil. Adults fly in late summer and early fall. Larvae emerge in spring from collarlike egg masses on twigs.

Fig. 14.85. Cecropia moths mating in May. The male, attracted to the female's odor, flew to her in daytime.

Sphinx or Hawk Moths (Sphingidae)

A few dozen species of sphinx moths inhabit Connecticut. These robust, fast-flying moths generally have narrow wings and a tapered abdomen (Fig. 14.87). Depending on the species, they fly at night, at twilight, or in daytime. Larva are essentially naked and nearly all have a dorsal horn on the "rump." They defend themselves with vigorously lateral thrashing and sometimes by regurgitating fluid.

Larvae of many sphinx moths are restricted to a fairly narrow range of food plants. For example, the five-spotted hawk moth *(Manduca quinquemaculata)* and the similar Carolina sphinx *(Manduca sexta),* more familiar as the tomato hornworm and tobacco hornworm, respectively, both consume tomato, potato, and tobacco foliage. Gardeners sometimes find the pupae, which winter in the soil (Fig.

Fig. 14.84. This luna moth appeared at my lighted kitchen window in mid-May.

Fig. 14.86. Eastern buckmoth is a rare species most likely to be seen in Connecticut near scrub oaks.

Fig. 14.87. This azalea sphinx moth (*Darapsa pholus*) came to my lighted kitchen window in August.

Fig. 14.89. The false eye on the rump of Abbott's sphinx moth caterpillars may startle predators or deflect predator attacks away from the caterpillar's head.

14.88). The tomato hornworm has eight L-shaped lines on each side and a black-edged green horn. The tobacco hornworm has seven diagonal lines and a red-tipped horn. Another food specialist is Abbot's sphinx (*Sphecodina abbottii*), whose caterpillars, adorned with a startlingly eyelike bump on the last body segment (Fig. 14.89), feed on foliage of grapes and other members of the grape family, such as Virginia creeper.

Tussock Moths (Lymantriidae)

The most notorious of all tussock moths is the **gypsy moth*** (*Lymantria dispar*). This moth was introduced in Massachusetts in 1869 for attempted silk production, escaped, and is now a major, periodically abundant pest throughout the northeastern United States. Adult females are whitish with dark wavy lines and so heavy bodied that they cannot fly. After mating, females deposit eggs in fuzzy brown masses that overwinter. The caterpillars hatch in spring, feed voraciously (and audibly), and are full grown by late June (Fig. 14.90). Major outbreaks that defoliated large tracts of forest occurred in 1953 and most recently in 1981. Larvae pupate in early summer, and adults emerge in July and August. I netted (with difficulty) the mostly brown, fast-flying males in late July.

Gypsy moth caterpillars feed on many kinds of trees and shrubs but favor oak foliage (especially white oak and chestnut oak), and gypsy moth populations do not fare as well on the leaves of maples. A preference for oak may seem surprising since tannins and other phenols in oak leaves reduce caterpillar size and egg production by the adult moths, which may reduce caterpillar populations. However, these phenols somehow disable a wilt (nucleopolyhedrosus) virus that readily kills the caterpillars, so oak-leaf-eating caterpillars are not so vulnerable to the virus, which may explain the caterpillar's predilection for oak leaves. The virus causes major mortality only when there are high densities of caterpillars.

Recently, in 1989 and 1990, a fungus (*Entomophaga maimaiga*) native to Japan caused die-offs of gypsy moth caterpillars in Connecticut and several other states. Abundant rain in May seems to favor the fungus. The fungus was introduced in the United States as a possible caterpillar control agent in the early 1900s and in the mid-1980s, but at those times it did not cause any die-offs. The strain that caused the 1989 die-off likely was recently introduced. The fungus has been a significant mortality factor ever since. The spores can survive several years in forest leaf litter. On tree trunks on my property in June 2001, I was delighted to find hundreds of large dead caterpillars that clearly succumbed to this fungus

Fig. 14.88. I found this tobacco hornworm pupa in my garden soil.

Fig. 14.90. During outbreaks of gypsy moth caterpillars, forest trees may be completely defoliated.

or some other pathogenic microbe. Unfortunately they consumed much oak foliage before they died. Fungus-killed caterpillars hang vertically with the head down. Viral mortality often leaves afflicted caterpillars hanging in an inverted-V shape.

When gypsy moth populations are low, birds (e.g., cuckoos) and small mammals (white-footed mouse, shrews) may help keep them in check by eating the caterpillars and pupae, respectively. Such predators are ineffective at preventing outbreaks, but large populations of white-footed mice, resulting from abundant acorn production the previous year, may have a significant impact on gypsy moth populations. Because acorn production is severely reduced during the one or two years following a major defoliation event, gypsy moth caterpillars can, in turn, reduce mouse populations. Among the other predators of gypsy moths are dermestid beetles, which eat the eggs, and *Formica* ants, which attack larvae. Caterpillars may avoid bird predators by hiding in bark crevices or leaf litter in daytime and ascending into the canopy to feed at night, when mice and shrews are active on the ground.

Human attempts to control gypsy moth caterpillars lately have involved applications of the microbial pesticide *Bacillus thuringiensis* (Bt). However, this agent can have detrimental impact on nontarget Lepidoptera and so must be used with discretion. Some research indicates that insecticide-reduced lepidopteran caterpillar populations do not appear to reduce the reproductive success of warblers that eat caterpillars, but this certainly depends on there being an adequate supply of alternative prey.

Another biological control agent used against gypsy moths may actually be a biological disaster. A parasitic fly introduced to control gypsy moths appears to be involved in the decline of silkworm moths throughout the northeastern United States (discussed on pages 258 and 265).

Since female gypsy moths do not fly, how does natural dispersal occur? Large caterpillars are limited to crawling and generally do not move very far. However, newly hatched larvae are light and fuzzy and capable of being blown aloft. They also spin silk strands that may catch the wind. Airborne larvae occasionally may travel many miles before alighting and feeding.

Tiger Moths (Arctiidae)

The most prominent example of the Arctiidae family is the Isabella tiger moth *(Pyrrharctica isabella)* (Fig. 14.91). Few people are familiar with the rather nondescript adult stage, but the larvae are famous as the so-called "woollybear," a fairly large, mainly black, bristly caterpillar with a

Fig. 14.91. Upper: Woollybear, larva of the Isabella tiger moth. Lower: Banded woollybear moth (Isabella tiger moth) is the adult of the woollybear.

reddish band around the middle. The red band becomes wider as the caterpillars molt and grow. The caterpillars are conspicuous in early fall as they crawl along the ground, presumably in search of a suitable place to spend the winter. I found active woollybears during mild weather throughout autumn and even in mid-winter, and some wintered under stacked firewood. In summer, the caterpillars eat the foliage of many kinds of plants. The ones that survive the winter pupate in spring. I found newly emerged adults around the vernal equinox. This moth is also known as the banded woollybear moth or blackened bear.

Some arctiid moths produce ultrasonic clicks that warn bats of the moths' unpalatability.

Owlet or Noctuid Moths (Noctuidae)

Owlets comprise the most species-rich family of Lepidoptera. Their abundance, diversity, and typical larval diet of plant foliage make them one of the most ecologically and economically influential insect groups. At rest, most of these usually drab, gray-brown moths hold the wings tent-like over the body (Fig. 14.92).

Some adult owlets that overwinter do so in leaf litter, crevices, or similar protected sites. Owlets generally fly in early spring when breeding occurs, but you may find some fluttering outside lighted windows even in mid-winter (see "Insects in Winter" at end of chapter).

The caterpillars of owlet moths, as well as those of many other kinds of butterflies and moths, are common hosts for

Fig. 14.92. This owlet (Morrison's sallow) appeared at my lighted window in mid-winter at an air temperature of 41°F.

Fig. 14.94. The larvae of this underwing moth (known as "The Bride," *Catocala neogama*) feed on the foliage of hickories, walnut, or butternut.

parasitoid flies (Tachinidae) and wasps whose larvae feed on the caterpillars (Fig. 14.93).

Underwing moths (*Catocala* spp.) are an interesting group of large attractive moths. Adults are well camouflaged when perched on tree bark, but most have vividly patterned hindwings that they expose when disturbed. Individuals that I inadvertently flushed in deciduous forest flew rapidly, with bright colors flashing, then suddenly perched on a tree trunk and "disappeared," concealing the flash colors. Many species showed up at my lighted kitchen window on warm summer nights, particularly from mid-July to mid-August but also until mid-September (Fig. 14.94). Depending on the species, larvae feed on the foliage of hickories, oaks, poplars, willows, blueberries, and other trees and shrubs.

The so-called large (or greater) yellow underwing* (*Noctua pronuba*) is similar to the true underwing moths. This Old World species has orange or orange-yellow hind wings, each with a single black band. It was first detected in North America in Nova Scotia in the 1970s and is now rapidly spreading across the continent. I first saw it in central Connecticut in the mid-1990s, and it is now common in at least that part of the state. The flight season begins in June and extends through summer. The caterpillars are "cutworms" that feed on a wide range of weedy and cultivated herbaceous plants.

Borer moths (*Papaipema* spp.) are often colorful moths whose highly specialized larvae bore into the stems, roots, and rhizomes of ferns, pitcher plants, and various other herbaceous plants, with different species generally using different host plants. I found large larvae in ostrich ferns in mid-July (Fig. 14.95). Adults fly in late summer and early fall.

Conservation

Like butterflies, some moths, particularly the giant silkworm and royal moths, evoke admiration for their color and size. But many moths have been targeted for eradication because of the damage the caterpillars inflict on garden plants, field crops, orchards, forest and yard trees, and stored clothing. Even bug-huggers like me look askance at caterpillars in the vegetable garden. However, it is important

Fig. 14.93. The holes in this moth caterpillar (probably American dagger moth) indicate where parasitic wasps (Chalcididae?) emerged. The wasp larvae fed and developed inside the caterpillar, eventually killing it.

Fig. 14.95. This borer moth *(Papaipema)* larva was extracted from an ostrich fern.

to realize that caterpillars of native moth species play important roles as herbivores and prey in most terrestrial and wetland ecosystems. They warrant as much positive consideration as do birds and bees. Caterpillar control measures should be carefully targeted such that innocuous and beneficial insects, including ones that may feed on the pest, are not obliterated.

For further information on conservation of butterflies, moths, and other invertebrates, visit the website of the Xerces Society, an international organization dedicated to protecting biological diversity through invertebrate conservation: www.xerces.org.

Common Scorpionflies

This small insect group (Order Mecoptera: Family Panorpidae) is notable for the long snout and the male's scorpion-like genitalia (Figure 14.96). Adults, which do not bite or sting, eat insect carrion, rotting fruit, and nectar. I found adults in late June. Caterpillarlike larvae live in the soil; they scavenge and prey on insects.

True Flies

To the chagrin of many people, flies are diverse in species and often abundant. About 20,630 of the world's 100,000 species of flies (Order Diptera) occur in the United States. Well over a thousand species inhabit Connecticut. Adults of most groups generally are recognizable by their single pair of membranous veined wings, though some are fairly convincing mimics of bees or wasps.

This major insect group is one of the most significant in terms of ecological and economic importance. Many are major predators or parasites of other insects, numerous species play a role in the breakdown of dead plant or animal matter, some damage crops, and others spread diseases such as malaria and typhoid to humans.

Dipterans occur in almost any terrestrial habitat, lakes, ponds, streams, ephemeral pools, and freshwater and saltwater wetlands. The life cycle includes eggs, larvae, pupae, and adults.

I found it impossible to do justice to Connecticut's highly diverse fly fauna in the space available in this book. Only a relatively few of the many fly families in Connecticut are characterized below.

Crane fly (Tipulidae) larvae generally live in mud or plant debris in ponds, stream backwaters, or wetlands and feed on detritus or rotting vegetation. Several dozen species inhabit Connecticut. The harmless adults (Fig. 14.97), sometimes mistaken for giant mosquitoes, live in damp areas with much vegetation, eat little or nothing, and commonly fly into houses through open doors or windows. Some are impressively large: one adult that I found in mid-August had, in its normal perched posture, a front-to-rear legspan of 4 inches.

Phantom midges (Chaoboridae). The mostly transparent aquatic larvae (Fig. 14.98) are more distinctive than the adults. The larvae can be abundant, major predators on small insects and crustaceans in vernal pools, ponds, lakes, and stream pools.

Mosquitoes (Culicidae) are all too familiar to everyone. Adult females have long, piercing mouthparts that can inflict an itchy welt on human skin. Nearly 50 mosquito species occur in Connecticut, but "only" a dozen of these are regarded as pests to humans or livestock.

Adult males, with conspicuously fuzzy antennae, are harmless and feed on nectar and sap, if anything. Females of most species also may sip nectar or sap but must suck up a blood meal from a mammal, bird, reptile, or amphibian in order to produce eggs (though *Culex pipiens molestus* can produce its first batch of eggs without a blood meal). Mosquitoes find their hosts through odors (carbon dioxide, octenol, lactic acid) or warm, moist convection currents emanating from the breath or body surface, or through visual detection of host movement.

Fig. 14.96. Common scorpionflies fly in early summer.

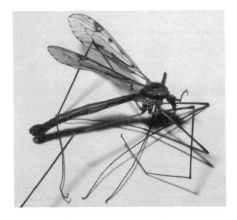

Fig. 14.97. Adult crane flies such as this often enter houses.

Fig. 14.98. Phantom midge larvae are aquatic predators. From Johannsen (1934).

Eggs may be laid in water or on land in areas likely to be flooded (Fig. 14.99). Larvae are aquatic and occupy virtually any pooled water, including that in tree holes, the vaselike leaves of pitcher plants *(Sarracenia purpurea)*, or human artifacts. They eat detritus and microscopic algae and aquatic organisms and get oxygen from the air at the water surface or by tapping plant stems.

Adults emerge after a short pupal stage. Most of them are eaten by birds, dragonflies, spiders, and other predators, or they succumb to drought, heavy rains, or strong winds. Survivors may live a few months or as long as six to eight months if they hibernate. Some mosquitoes winter only as cold-hardy eggs containing live embryos. Adults generally fly from early spring to mid-fall and are most numerous in the warmer months.

Recently, two mosquito-related public health issues have developed: eastern equine encephalitus (EEE) and West Nile virus (also referred to as West Nile fever). EEE is a rare but serious viral infection that may spread to humans only if a person has been bitten by a mosquito that has bitten an infected bird. Symptoms include fever, headache, and/or brain swelling. Normally the virus is restricted to birds and to mosquitoes that do not bite people. Most sites where EEE in mosquitoes has been found are near freshwater wetlands and wetland-forest margins.

West Nile fever is a viral infection that can cause inflammation of the brain. Like EEE, it is spread to humans by the bite of a mosquito (usually the common house mosquito, *Culex pipiens*) that has bitten an infected bird. Birds that acquire the virus have it in their blood for a week or so, then either die from the infection or survive by developing antibodies that kill the virus. In humans, feverlike symptoms generally appear 5 to 15 days following the bite of an infected mosquito (very few mosquitoes are infected).

If you are bitten by a mosquito, there is only a very remote chance of contracting either infection. As of early 2001, there were no documented human cases of EEE in Connecticut. An outbreak of West Nile virus occurred in New York in 1999, and since then human cases of the infection have been recorded in Connecticut. Human infections are most likely to occur from mosquito bites in August and early September. In Connecticut, the state has implemented coast-wide mosquito testing and ground spraying of a pyrethroid-based insecticide (Resmethrin) to reduce the probability of transmission of mosquito-borne viruses. As alternatives to chemical toxins, insect growth regulators (e.g., methoprene, which prevent larvae from molting and developing into adults) and bacterial agents specific to mosquito larvae have been used to control mosquito populations. An ecologically benign method of mosquito management that has been implemented in salt marshes in recent years is known as open marsh water management, which creates ponds that support small fishes that feed extensively on mosquito larvae. These ponds also provide improved habitat for waterbirds.

Current emphasis is on public education concerning personal protection and eliminating mosquito breeding sites (standing water) in urban and residential areas. Because effective mosquito control seems unlikely, given the millions of potential breeding sites in small containers that hold rain water, the future of the disease and its impact on humans probably will reflect weather patterns that either foster or discourage mosquito breeding and population recruitment. A long season of wet weather and abundant mosquitoes means that more birds will acquire the West Nile virus, so spread of the virus (through bird movements) and transmission to humans becomes more likely.

Black flies (Simuliidae) are small, stout, hump-backed flies. The adult females of many species bite humans and sometimes are an annoyance during springtime and early summer walks in the woods. The onslaught usually intensifies in the latter third of April in central Connecticut. Blackflies are most effective at biting a nonmoving target, as I learned whenever I stopped to take a photograph or write some notes. Male and female adults feed opportunistically on homopteran honeydew and flower nectar. Females lay eggs in or near water. Larvae (Fig. 14.100), less than ½ inch long, use a ring of minute hooks at the rear end of the body to cling to silk pads that they attach to rocks and other objects in flowing stream waters; they feed on nearby detritus

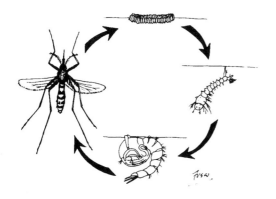

Fig. 14.99. Mosquito eggs (top), larva (right), pupa (bottom), and adult (left). Paul J. Fusco/Connecticut DEP Wildlife Division.

True Flies | 263

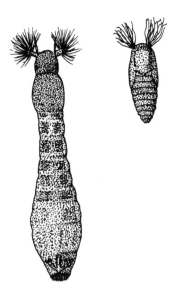

Fig. 14.100. Black fly larva (left) and pupa (right, removed from its pocketlike cocoon), dorsal view.

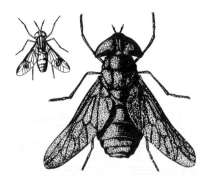

Fig. 14.102. Deer fly (upper left) and black horse fly (right) are aggressive biters. After Arkansas Agricultural Experiment Station.

or on microscopic life obtained by filtering the water with their head filaments.

Midges (Chironomidae) are small, mosquito-like but harmless, short-lived dipterans that often swarm near water (Fig. 14.101). Larvae generally are bottom-dwelling scavengers or mini-predators in ponds, streams, or wetlands. Red "bloodworms" (*Glyptotendripes paripes*) are common in polluted water. Larval midges often are very abundant and are important food items for fishes and other aquatic predators. Similarly, adults can be abundant near certain lakes and streams throughout much of the year, and these often incur predatory attacks by dragonflies.

Horse fly and **deer fly** (Tabanidae) females are aggressive, big-eyed biters of deer, livestock, humans, other mammals, and birds. Males, whose eyes are especially large and meet dorsally, feed on nectar and pollen. Larvae are bottom-dwelling predators in wetlands and shallow water; they hatch from eggs attached to vegetation near or above water. Deer flies are significant transmitters of disease in some parts of the world. Deer fly (*Chrysops*) adults (Fig. 14.102) can be locally numerous and persistently annoying, circling around your head as you walk in wooded areas near wetlands in late spring and early summer. Generally they have brown patches in the wings. Horseflies (*Tabanus* and other genera) can be larger (Fig. 14.102). Occasionally, massive, inch-long reddish-brown horseflies circled and landed on my bare back as I worked outdoors. Greenhead fly (*Tabanus nigrovittatus*), half an inch long with green eyes, is common along the coast, where the larvae live in marshes and adults pester summer sunbathers. Perhaps surprisingly, several horse flies are listed by the state as Endangered, Threatened, or Special Concern. Presumably they declined with decreases in populations of large livestock as Connecticut's economy shifted from agriculture to industry.

Robber flies (Asilidae) include some of most attractive flies, especially those that resemble bumblebees or wasps and don't look like flies (Fig. 14.103). Robber flies frequent sunny, dry, open areas. Adults have excellent vision and are common pierce-and-suck feeders that aggressively chase

Fig. 14.101. Adult midges resemble mosquitoes. Males generally have fuzzy antennae. From Kellogg (1905).

Fig. 14.103. This robber fly, photographed in Colorado, resembles bumblebee-like robberflies that occur in Connecticut.

Fig. 14.104. I found this tiger bee fly near the tunnel of a carpenter bee, the larvae of which serve as food for the bee fly larvae.

and attack other insects and small spiders. The immature stages generally are predators on insect larvae in soil or rotting wood. It takes more than one year to complete the life cycle.

The two largest Connecticut species are each about 1.4 inches long. A slender species *(Proctacanthus philadelphicus)* makes a buzzing sound in flight, favors sandy pastures, and does not hesitate to attack able-bodied insects such as yellowjackets and hornets as well as bees, flies, beetles, bugs, butterflies, and small grasshoppers. Adults fly mainly from early July to early September. The other large species *(Laphria grossa)* is heavy bodied and resembles a bumblebee or carpenter bee and attacks these large bees, Japanese beetles, and other slow insects. The best time to search for this uncommon species is in July and August. I regret that I never saw either species.

Bee fly (Bombyliidae) adults, superb hoverers, consume nectar and pollen. Bee flies with long, needlelike but harmless mouthparts, and other larger, furry species that resemble bumblebees, are conspicuous among the pollinating insects active in early spring. Larvae generally consume the eggs and young of various ground insects, including those in bee nests. Larvae of the tiger bee fly *(Anthrax tigrinus)* feed on larval carpenter bees. I found newly emerged adults near carpenter bee tunnels in mid-August (Fig. 14.104). The

Fig. 14.105. Many syrphid flies resemble bees or wasps but are harmless.

Fig. 14.106. The tachinid fly *Compsilura concinnata* inserts its larvae into caterpillars. The introduced fly appears to be responsible for a large decline in silkworm moth populations. U.S. Forest Service, Northeastern Forest Experimental Station.

black-mottled wings of the adults of this large fly span more than 1 inch.

Flower flies or **hover flies** (Syrphidae), as adults, often resemble wasps or bees, but they have a thick waist (rather than slender as in wasps) and do not bite or sting (Fig. 14.105). They may hover close to humans, sometimes evoking unwarranted concern, between their visits to flowers for nectar and pollen. Adults are common on goldenrods in late September and early October. Larval habits vary quite a bit among different species, ranging in diet from aphids to rotting plants.

Parasitic flies (Tachinidae) are significant controls on certain pests of trees and crops (Fig. 14.106). The larvae generally are internal parasitoids of insects, especially caterpillars, such as the gypsy moth. The adult flies lay eggs or larvae on or in the host, or the host inadvertently ingests an egg that was laid on a plant. The larvae slowly consume the host, focusing first on nonvital tissues, thus keeping it alive until they are fully developed. Adults are robust, strong fliers; many have conspicuous bristles on the end of the short abdomen. The tachinid fly *Compsilura concinnata*, introduced in attempts to control gypsy moths, is believed to be responsible for a drastic decline in populations of several species of beautiful silkworm moths. In spring and summer, this fly hunts for caterpillars and injects its larvae into them. It survives winter as an immature inside a live butterfly or moth pupa or caterpillar. The fully grown larvae emerge and pupate outside the host (e.g., in soil or tree bark crevices).

Fleas

Fleas (Order Siphonaptera) are wingless, compressed, blood-sucking parasites of mammals and birds. They occur worldwide (2,300 species) and are represented in the United States and Canada by about 325 species. Eggs yield larvae, which pupate prior to emergence of the adults. Only the adults feed on blood; the secretive larvae feed on organic debris generally in the nest or den of the host.

Generally each flea species parasitizes only a specific group of hosts (e.g., cats, rodents, etc.). Females require a blood meal in order to mature their eggs. Fleas can be significant transmitters of disease (e.g., typhus, plague) and parasites (e.g., tapeworms). The cat flea (Pulicidae: *Ctenocephalides felis*) and dog flea (*C. canis*) are common on pets. Various additional species occur on other carnivores, rodents, shrews, and other mammals and birds. Adults of some species can jump exceptionally well.

Insects in Winter

Mourning cloak butterflies flying in the woods during winter thaws vividly illustrate the cold tolerance of many insects and other arthropods. Tucked away in a tree hole or crevice, these butterflies endure and easily survive the hardest freezes, and they sometimes emerge and fly around during mild respites from winter's cold.

Supercooling and Freeze Tolerance. Invertebrates generally withstand freezing temperatures much better than do vertebrates, yet most insects are killed if ice forms in their body tissues. As a general rule, formation of an ice crystal within a cell of any organism, plant or animal, kills that cell. Nonetheless, freeze-intolerant insects can survive in cold climates. They do this by supercooling—lowering the body temperature, without ice formation, to levels below that at which freezing normally occurs.

Supercooling is a phenomenon that is widespread among all life stages of insects. Eggs often are exposed to temperature extremes and, not unexpectedly, generally have a lower supercooling point (SCP) than do other stages of the life cycle. The SCP is the temperature at which spontaneous freezing occurs during gradual cooling. For example, overwintering eggs of the eastern tent caterpillar (*Malacosoma americanum*) supercool to –31°F. Eggs of other insects may reach – 40°F or lower before freezing.

The immature stages of insects also have a low SCP. For example, the SCP is –79°F in larvae of the willow cone gall midge (*Rhabdophaga*) in northern Canada. Insect pupae supercool to about – 4 to –22°F.

The SCP of adult freeze-intolerant insects and spiders varies widely from 25 to –33°F. Experiments have demonstrated that insects readily survive supercooling for brief periods, but not much is known about the effects of long-term exposure to temperatures close to the SCP.

Freeze-intolerant insects are protected from freezing at very low temperatures by having in their body fluids various cryoprotectants ("antifreeze"), of which glycerol is the most common. Other cryoprotectants include other polyols, sugars, and proteins, the last of which occur in species such as lady beetles, milkweed bugs, and spiders. In most species, cryoprotectants are produced only in response to low temperatures, so freeze protection disappears in summer.

Generally, freeze-intolerant insects that overwinter as adults stop feeding prior to the onset of freezing temperatures. This may facilitate survival because food residues and mineral dust particles that may be present in the food apparently promote the formation of ice at relatively high subfreezing temperatures. These insects also tend to select dry wintering sites that are less likely to expose them to ice crystals that could trigger ice formation in the body. Additionally, they undergo controlled dehydration and incorporation of water into other molecules, which reduce the amount of freezable body water.

Some insects tolerate the formation of ice within the body. The SCP of many or most of these freeze-tolerant species is relatively high (mostly 23° to 14°F). Woollybear caterpillars survive 23° to 14°F and readily survive 23°F for up to at least a week, but they often select wintering sites that rarely fall below their SCP.

Surprisingly, most freeze-tolerant insects have substances (termed nucleating substances, probably proteins or peptides) in the body fluids that actually promote the formation of ice. The nucleating agents assure that ice formation occurs at a relatively high temperature and that freezing is extracellular (outside the cells). Presumably the lipid portion of the cell membrane prevents penetration of ice into the interior of the cell. Extracellular freezing evidently increases the concentration of extracellular solutes, which results in water outflux from the cells. Consequently, increased solute concentration inside the cells results in greater resistance to intracellular freezing (just as salty water freezes at a lower temperature than does fresh water). At least some freeze-tolerant insects, such as woolly bears, have cryoprotectants (e.g., glycerol) that prevent intracellular freezing.

Various freeze-tolerant insects have a very low SCP of –40° to –58°F, due to high cryoprotectant concentrations. These species lack nucleating substances and generally never experience ice formation.

Activity in the Cold. Some insects ward off the cold and warm up enough to fly by generating their own body heat through contractions of the wing muscles, evident as low amplitude wing vibrations. Once airborne, these self-heating insects can maintain body temperatures well above

100°F when flying at air temperatures of 37 to 41°F. Some insects, such as bumblebees and owlet moths, have an insulating cover of "fuzz" or dense, hair-like scales that may help them retain heat.

Bald-faced hornets are notable for their activity even when temperatures are quite low. Using a tiny thermometer, biologist Bernd Heinrich found that the hornets stinging him had body temperatures of up to 108°F though the air temperature was near freezing. Many of the flying insects that hornets prey upon are numbed by cold, and on cold mornings they cannot fly well enough to elude the warm-bodied hornets.

Internal heat production may be an essential mechanism allowing bumblebees to inhabit alpine and high arctic areas and to fly under conditions that incapacitate most insects (I saw bumblebees flying in the shade at an air temperature of 41.5°F, when no other insects were evident). Warming through internal processes not only allows the bees to remain active when ambient temperatures are low, but also enables the queen to accelerate the development of her brood by increasing its temperature (she presses her warm abdomen against the brood cluster, warming it to around 86°F, even at ambient temperatures as low as 37°F).

Like bumblebees and hornets, "furry," heavy-bodied owlet moths (Fig. 14.92) may use muscular contractions (shivering) to warm up before they fly. Certain species regularly fly in mid-winter if temperatures rise above freezing. Some owlets are genuinely adapted for cool conditions. At "room temperature" they are unable to fly for long before succumbing to heat prostration. As you might predict for an animal that expends a lot of energy to warm itself, these moths feed as adults, taking advantage of sap flowing from tree wounds and flower nectar.

Not all moths that fly at cold temperatures raise their body temperatures through muscular contractions before taking flight. Some inchworm moths (Fig. 14.82), for example, have enzymes, muscles, and a nervous system that function well at temperatures near freezing. The moths can fly (though not fast!) despite being no warmer than the cold air around them.

Some other groups of insects are predictably active in cold weather. Examples include adults of several small stonefly species, which are remarkable in normally emerging from their larval stream habitat to mate in winter, and certain springtails or "snowfleas." These insects can be found in winter flying or crawling about when temperatures are quite low (but usually above freezing). Winter crane flies are another group of insects that commonly fly when temperatures are cold. Like cold-weather inchworm moths, none of these warm up much before activity; they can move well even at body temperatures near freezing. Occasionally I saw spiders crawling in the open, sometimes even crossing snow, on mild winter days.

Migration and Immigration. A few butterflies that sometimes or always fail to survive Connecticut's cold winters as egg, larvae, pupa, or adult are nevertheless present in the state every summer. How so? Adults from areas to the south fly northward in spring and colonize Connecticut each year. They may even reproduce, but their progeny perish in winter's cold, though some may survive mild winters.

Examples of these yearly colonists include the red admiral and common buckeye. Question mark and American lady butterflies migrate southward in late summer and fall, and some of these same individuals migrate north in spring to recolonize New England. And of course monarch butterflies migrate far south each fall, and the offspring and later generations recolonize Connecticut the next year. Several other butterflies irregularly or rarely immigrate from the south into Connecticut in summer. Among other insects with similar behavior, a good example is the common green darner, a dragonfly that regularly migrates southward for winter and returns in spring.

Galls and Gall Makers

What Is a Gall?

In late autumn, winter, and early spring, when most plants are barren of leaves, you may notice various smooth, spiny, or scaly swellings or projections on twigs or still-attached leaves. Later in spring and in early summer, conspicuous lumps and odd protuberances appear on newly grown leaves. Most of these growths are insect-induced galls, though sometimes mites, other invertebrates, or microorganisms cause them. This chapter focuses on insects, but non-insect galls are discussed here for the sake of continuity.

Galls are abnormal plant growths in the sense that they do not occur in the absence of the outside stimulus. Basically a gall is a source of food and shelter that the animal or microorganism induced the plant to produce. The microorganism or larval or immature stage of the animal develops within and feeds upon the gall. The inner layers of a gall that are eaten by animal gall-makers generally are composed of abnormal (for example, multinucleate) cells. The cells in the outer part of a gall are less altered and often form a hard protective cover.

Galls provide no benefit to the plant. A heavy gall infestation can be detrimental by reducing the photosynthetic area or by consuming resources that could have been used for reproduction or normal growth.

Galls range from rather nondescript swellings to elaborate flowerlike structures. The galls induced by the different species of gall-inducing organisms are unique and generally more distinctive than are the organisms themselves. Each gall-inducing species is consistent in the type of plant that it uses and in the particular part of the plant on which the gall forms.

The specific stimulus that causes formation of the gall evidently consists of chemical substances secreted by the larval insect or microorganism. This idea is supported by a few experiments that successfully produced characteristic galls by injecting plants with a chemical extract of the gall insect larvae. Gall growth generally depends on presence of a living gallmaker. It is believed that in some cases irritation caused by the feeding of the animal also may play a role in gall formation. Most galls form in spring when plant tissues are tender and actively growing.

The principal gall inducers are plant mites, aphids (so-called plant lice) and their aphidlike relatives, gall midges, and gall wasps. Certain kinds of thrips and weevils also cause galls to form. Fungi and other microorganisms also may induce growths that can be called galls. Plants commonly afflicted by galls include members of the rose family, aster/sunflower family, and oak/chestnut family.

Plant Mites

Plant mites (Order Acarina) are microscopic or barely visible animals having only four legs. They occur on a wide variety of plants. About 165 species of gall-making mites occur in North America.

The feeding of certain of these mites causes the formation of "hairy" patches on leaves or globular pocket- or pouch-shaped galls, such as those on the lower surface of silver maple leaves, caused by *Vasates quadripedes*. This latter species feeds and reproduces in the gall and spends the winter in bark crevices of the host. The long, slender galls and bumps on the upper surface of cherry leaves, appearing in mid-May, are caused by mites (Fig. 14.107). Eriophyid mites (*Cecidophytes betulae*) cause the formation of dense masses of bud tissue on birch twigs. Tiny mites also live in galls on the leaves of chokeberry.

Homopterans

Aphids, adelgids, psyllids, and phylloxera (Homoptera: Aphididae, Adelgidae, Psyllidae, and Phylloxeridae) are tiny soft-bodied insects with sucking mouthparts. In North America, a few dozen species produce galls. Generally these galls are globular and have a small opening to the outside. The more easily observed of the galls caused by these animals include spruce galls (swellings at the end of branches); many of the galls on hickory leaves; the spiny bud gall and conical leaf galls of witch-hazel; globular swellings on the leaf stalks of aspens, cottonwoods, and poplars; and nipplelike galls on hackberry leaves. As the following examples illustrate, many of these insects have complicated life histories.

Eastern spruce gall adelgid* (*Adelges abietis*) attacks planted Norway spruces (and sometimes black, white, and red spruces) and always remains on spruce. This insect was introduced into the United States sometime before 1900.

Fig. 14.107. Mites caused these galls on cherry leaves.

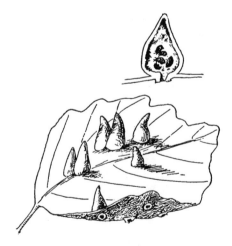

Fig. 14.108. These galls were caused by the witch-hazel leaf gall aphid. Upper figure shows aphids within an opened gall. Drawing by T. Pergande, from USDA.

The female nymphs overwinter on the trees, often at the base of the buds. After maturing in spring, the aphids lay eggs that hatch at about the time the new needles are emerging. The feeding activity of the newly hatched adelgids among the new needles causes a conelike gall to form around them. The gall protects the insects from parasites, predators, and severe weather. The nymphs live inside the gall until it turns brown and opens in July or August. They then leave the gall, molt and develop wings, and eventually lay eggs that yield the overwintering nymphs.

Witch-hazel leaf gall aphid (*Hormaphis hamamelidis*) causes conical galls on the leaves of witch-hazel (Fig. 14.108), and it also uses birch, the latter being colonized by later generations of the population that caused the witch-hazel galls. The galls on witch-hazel leaves form as the nymph begins to feed in May soon after the leaves have grown. Soon the aphid is contained inside the gall, and there it produces offspring. These mature and emerge from the opening on the underside of the leaf. Some fly to birches in late spring or early summer, but do not cause gall formation; others (appearing scalelike) remain on witch-hazel and can be found on leaf undersides in late summer. Multiple generations of aphids develop in summer, and in late summer or fall, a winged generation flies from birches back to witch-hazel. There these lay eggs that overwinter.

Another homopteran, the **witch-hazel bud gall aphid** (*Hamamelistes spinosus*), also alternates between witch-hazel and birches. It winters as an egg on witch-hazel twigs, or as a dormant female on birch. Eggs hatch in spring, the nymphs begin to feed on flower buds, and, as a result, a spiny gall forms and encloses the aphid. The aphid nymph matures and produces offspring in the gall. These mature, exit the gall via an opening in its base, and fly to birch. Dormant females that overwintered on birch move to new leaves in spring and give birth to young, which may become very numerous on the undersides of the leaves. Feeding by the aphids causes corrugations on the leaf. By the end of June, some aphids mature and develop wings, fly to witch-hazel, and lay eggs.

Poplar petioleball aphids (*Pemphigus* spp.) cause globular swellings on the leaf stalks of cottonwoods (Fig. 14.109). Eggs are laid under bark in fall, overwinter, and hatch in spring. Nymphs feed on developing leaf stalks and soon become enclosed in expanding plant tissue. The enclosed nymph gives birth parthenogenetically to many young inside the hollow ball. I found large balls crowded with young and their mother in October. The young develop wings and escape the ball through a slit that develops in summer. The young establish themselves on the roots of plants in the mustard or aster family and produce new generations, one of which flies back to cottonwoods, mates, lays eggs, and dies.

Hackberry nipple gall psyllid (*Pachypsylla celtidismamma*) makes nipplelike galls on the underside of hackberry leaves. Adults overwinter, then mate and lay eggs in spring as leaves unfold. Nymphs feed on leaves and become enclosed in a globular gall through spring and summer. Adults emerge in late summer or early fall. I found wide-bodied nymphs with small wing buds still in the galls in mid-October. Within thirty minutes of being exposed, they molted into a large-winged form. Psyllid nymphs may be parasitized by chalcid wasps.

Gall Midges

Gall midges or gall gnats (Diptera: Cecidomyiidae) are small (rarely more than 3 mm), delicate flies with long, threadlike antennae; they somewhat resemble mosquitoes. The many species in this group have varied life histories; the larvae of many of them (more than 5,000 species) develop in galls. The gall-inducing species (nearly 700 in North America) usually occur on willows, other deciduous trees, and on herbaceous plants.

Some of the more common galls caused by gall midges include those caused by the **willow cone gall midge** (*Rhabdophaga strobiloides*) on the tips of willow branches (Fig. 14.110). In this species, adult midges lay eggs in terminal buds of willows in spring; the larvae and budlike galls grow through summer. The larvae overwinter in the gall and pupate in early spring. Soon thereafter the adults emerge. The larvae of other midge species may develop in old willow cone galls, feeding on decaying gall tissue.

Willows are also often affected by the **willow beaked-gall midge** (*Mayetiola rigidae*), which causes a bud to turn into a beaked swelling on a twig. Adults emerge in spring and lay eggs on buds. Newly hatched larvae burrow into the bud. By

Fig. 14.109. This cut-open gall on the stalk of a cottonwood leaf was home to many aphids. Similar galls may form on aspens.

Fig. 14.110. A willow cone gall midge caused this budlike gall at the tip of a willow branch.

early summer the larva and gall have grown large. The larva overwinters in the gall, pupates there in late winter, and the adult emerges in spring.

Other galls caused by gall midges include dense globular masses of small leaves ("cabbage galls") at the top of goldenrod stems (Fig. 14.111), galls in catkin scales of hazelnuts, spiny rounded galls at twig tips and podlike galls on leaves of spiraea (meadow-sweet), globular galls at the ends of gray dogwood and black cherry twigs, and small lumps on maple leaves.

Other kinds of flies also may cause galls. **Goldenrod gall fly** (Diptera: Tephritidae: *Eurosta solidaginis*) produces bulbous galls in goldenrod stems (Fig. 14.111). The fly larvae feed in the goldenrod stem, reach full size in September, and spend the winter in the gall. In May, adults emerge from the galls using a tunnel cut almost to the surface by the larvae in late summer and early fall. Adults mate, and females deposit eggs in the leaves surrounding the apical bud of goldenrod ramets. After five to eight days, the larvae hatches and then chews its way down into the stem, where the gall begins to become evident a few weeks after egg laying. The gall reaches maximum size about three weeks later. Downy woodpeckers and black-capped chickadees commonly peck into goldenrod galls and feed on their inhabitants in winter when other food is scarce. Gray squirrels sometimes open galls and eat the insects. The gall fly larvae also are attacked by beetle (*Mordellistena*) larvae and by the larvae of parasitoid chalcid wasps (*Eurytoma* spp.) that deposit eggs in gall fly eggs or larvae in goldenrod buds or in fully formed galls.

Some gall midges are reproductively unusual in that larvae may give rise to more larvae without an intervening adult stage. The daughter larvae grow inside the mother larva, consume her from within, then emerge and become free living.

Gall Wasps

Together with the gall midges, gall wasps (Hymenoptera: Cynipidae) are the most common gall makers in North America (several hundred species, with over 10,000 species worldwide). Adults are small (mostly 6 to 8 mm) and do not sting (Fig. 14.58). Life cycles of gall wasps often alternate between sexual reproduction and production of viable eggs by virgin females. Females insert each egg inside the tissue of a plant. A whitish, legless larva emerges from the egg and soon becomes surrounded by gall tissue. The larva feeds and pupates within the gall, and eventually the adult wasp chews its way out or uses an exit route chewed by the larva before it pupated.

Fig. 14.111. These galls on goldenrod were caused by the goldenrod gall moth (left), goldenrod gall fly (center), and a gall midge (right).

Fig. 14.112. I cut open this oak apple gall in September and found a single cynipid wasp larva (fat, curved) and five unknown larvae (three visible at left center).

Fig. 14.113. These oak twig galls were caused by cynipid wasps.

The plants utilized by gall wasps primarily are oaks (Figs. 14.112 and 14.113), but the spiny rose gall wasp causes the growth of spherical, long-spined galls on rose leaf primordia, in which the larvae overwinter, and other species attack blueberry (Fig. 14.114). Most of the galls are formed in spring when newly emerged adults lay eggs in woody tissue or on leaves. Fresh, pea-sized, growing oak apple galls can be seen on new oak leaves in late April, and these are evident as brown, papery globes (about an inch in diameter) that in autumn may be still attached to dry leaves or fall onto the ground. I found green, red-dotted oak apple galls, recently detached from red oak trees, in late summer and early fall. By late September, these usually contained single fat, C-shaped, segmented-looking larvae about 4 millimeters long. Some contained a single large larva plus multiple smaller and thinner larvae in the central chamber, likely representing the wasp and a secondary gall user or perhaps a wasp parasite.

Probably the best time to see an adult cynipid wasp is in fall. Carefully open the small, woody, spherical galls that occur in clusters on small twigs of white oak or chestnut oak. In fall, adult wasps resting inside seedlike pupal cases

Fig. 14.114. This gall on a blueberry twig contained larval cynipid wasps that emerged through the holes.

in the galls become quite lively when removed. These galls have a small opening to the outside near their attachment to the twig, providing an exit for the adult wasp. Apparently, the adults overwinter in the galls and emerge and lay their eggs in fresh growth in spring, but I did not confirm this suspicion.

Some gall wasps do not produce their own galls. Instead, the larvae develop in galls caused by other organisms. Swellings on shoots and leaf petioles of willows may be induced by hymenopteran sawflies of the genus *Euura*.

Gall Moths

Tapered enlargements on goldenrod stems are caused by goldenrod gall moths (Gelechiidae: *Gnorimoschema* spp) (Fig. 14.111). Females lay their eggs on old goldenrods (mainly *Solidago altissima* and *S. gigantea*) in late summer or early fall. Larvae hatch the following spring and burrow and feed within new shoots, causing the gall to form. Upon reaching full size, larvae chew a tunnel to the surface of the gall at its upper end, cover the opening with silk, then pupate. Adult moths emerge from the escape tunnel in late summer.

Not all gall moths are successful in this method of reproduction. Ichneumon wasps sometimes lay an egg next to the moth pupa, and the wasp larva feeds on and eventually kills the pupa.

Studying Galls

The best way to observe gall makers is to collect galls and store them in a finely screened container until the adults emerge. However, because of the potential presence of insects that exploit the galls formed by other species, including chalcid wasps and other species that attack the gall-inducing insect, one cannot always assume that the insect that emerges from the gall is the one that caused its formation.

Centipedes and Millipedes

Centipedes and millipedes (Class Myriapoda) sometimes are recognized as separate classes of mandibulate arthropods. **Centipedes** (Order Chilopoda) include about 3,000 species occurring worldwide in temperate and tropical regions (Fig. 14.115). One may find them in soil or humus or under leaf litter, rocks, or logs on the ground. Presumably they roam in the open only at night if at all. Centipedes avoid light and if exposed run quickly to cover. In trying to photograph centipedes, I found that these energetic animals can run continuously for several minutes without tiring.

Fig. 14.115. Centipedes are predators that avoid light.

Fig. 14.116. Connecticut's largest millipede may reach a length of 4 inches.

Most centipedes in Connecticut are only an inch or two long. Each body segment has a single pair of legs and, in some species, defensive repugnatorial glands. A pair of fanglike appendages, each fang having an associated venom gland, is positioned just behind the head. These venom claws are used in preying on the small invertebrates that share their soil habitat. All of the centipedes in Connecticut are harmless to humans.

Centipede reproduction generally involves courtship interactions, deposition of sperm on a silk web by the male, picking up of the sperm packet by the female, egg deposition in soil or decayed wood, and, in some species, brooding of the eggs by the female. After hatching, it may take up to several years for the centipede to reach maturity.

Millipedes (Order Diplopoda) are a worldwide group comprising several thousand species of secretive, slow-moving, land-dwelling animals (Fig. 14.116). They generally avoid light (photoreceptors are present in the integument of species lacking eyes) and spend most of their time in soil or leaf mold, under tree bark, or under objects on the ground, but sometimes I found them crawling in the open in broad daylight. The exoskeleton lacks a waxy coating or other effective waterproofing, so millipedes are vulnerable to dehydration and quickly die in dry environments.

The diagnostic feature of millipedes is their two pairs of legs on each of the numerous body segments. Some millipedes have a pair of repugnatorial glands on each or most of the body segments. The glands secrete a substance that may be toxic or repellent to potential predators. Millipedes may writhe, curl, and squirm in response to probing or handling. Most millipedes feed primarily on decomposing vegetation.

Millipede reproduction involves transfer of sperm from male to female, fertilization at the time of egg laying in soil or humus, and, in many millipedes, brooding of the eggs by the female. Development is characterized by hatching generally several weeks after egg laying, growth that includes the addition of body segments with each molt, and a potential lifespan of several years. The large millipede *Narceus americanus* (Order Spirobolida) is fairly common in wooded habitats in Connecticut. I found vast numbers of small millipedes (up to ½ inch long) under objects on the ground in every season.

Chapter 15

Crustaceans

Familiar examples of crustaceans include crabs, lobsters, and barnacles. A far greater number are relatively obscure animals, including fairy shrimps, clam shrimps, cladocerans, ostracods, copepods, isopods, amphipods, and many others. The familiar species can be so big and feisty that you wouldn't want to hold one, whereas others are so small and innocuous that you likely wouldn't know it if you *were* holding one. This group of arthropods (Phylum Crustacea) includes some 45,000 species worldwide and about 10,000 species in the United States. Most species inhabit salt water, but crustaceans also occur in fresh water and in moist places on land.

Fairy Shrimps and Clam Shrimps

Fairy shrimps (Order Anostraca) are one of four major groups of branchiopod crustaceans (Fig. 15.1). These shrimps are delicate, defenseless crustaceans that live in vernal pools and other fishless freshwater habitats. They swim upside down, propelled by beating movements of the legs. Their food consists of microscopic organic particles removed from the water and accumulated in a ventral groove on the body by the leg movements.

The sexes are separate, but parthenogenesis, in which females produce viable offspring without mating with males, is common in the group. To mate, the male (in dorsal position) clasps the female with his second pair of antennae.

Fig. 15.1. Fairy shrimp are one of the species unique to vernal pools.

The eggs are carried by the female, then are dropped or sink to the bottom when the female dies. Of the two kinds of eggs, thin-shelled ones hatch almost immediately. Thick-shelled resting eggs resist heat, cold, and desiccation and allow the shrimp population to get through dry summers and icy winters; the resting eggs hatch in late winter or early spring. A common species is *Eubranchipus vernalis,* which can be found, sometimes very abundantly, in vernal pools from late February through April.

Clam shrimps (Orders Spinicaudata and Laevicaudata) somewhat resemble small clams but actively swim using their numerous appendages. The animals are enclosed in a pair of shell-like valves that may or may not have concentric growth lines. Like fairy shrimp, these crustaceans thrive in temporary or semi-permanent ponds. One rare coastal species, *Limnadia agassizii,* develops very quickly in rain pools that form in summer, with adults (carapace has growth lines) appearing less than 10 days after the drought-resistant eggs hatch. Another more common clam shrimp, *Lynceus brachyurus,* whose carapace lacks growth lines, also develops quickly. In a semi-permanent pond, I found this species (about 0.1 inches long) to be very abundant in mid-May and much less numerous but still present in mid-July of the same year.

Cladocerans

Cladocerans are small, barely visible but often abundant branchiopod crustaceans (Fig. 15.2). Recent classifications divide the order Cladocera into multiple orders, but here I group them all together. Approximately 150 freshwater species inhabit the United States.

These so-called "water fleas" are a plentiful component of freshwater zooplankton of ponds, lakes, and rivers, especially following the pulse of spring reproduction. Their abundance contributes to their importance as a food source for many fishes and larval pond-dwelling salamanders.

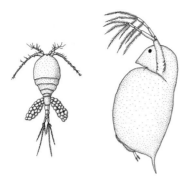

Fig. 15.2. Freshwater zooplankton: cyclopoid copepod (left, carrying two sacs full of fertilized eggs) and cladoceran (right). These animals are about 1 to 2 millimeters long.

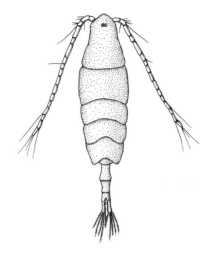

Fig. 15.3. This copepod *(Acartia)* is common in Long Island Sound.

Cladocerans move by stroking their large antennae. They often migrate upward toward the water surface as darkness falls and return to greater depths around dawn. These movements may reduce their exposure to visually oriented predators. They obtain food usually by filtering microorganisms and other organic particles from the water via movements of the legs, which lie between the two sides of the shell-like carapace that encloses most of the body.

Reproduction is by parthenogensis for most of the year—females produce more females without the involvement of males. Periodically females produce eggs that yield females and males, enabling sexual reproduction. Short daylengths stimulate females to produce resting eggs that survive the winter.

Copepods

Copepods are a diverse group of crustaceans, totalling some 8,000 to 10,000 species worldwide. If you pull a fine-mesh plankton net through a large body of salt water or fresh water, you'll be able to appreciate that copepods are often abundant members of the zooplankton community (Figs. 15.2 and 15.3). They are scarce or absent in small headwater streams. Some crawl on the substrate and others, the ones likely to end up in your net, are good swimmers.

Some copepods, such as the harpacticoids, are bottom deposit feeders. Others graze on phytoplankton and sometimes are predatory on other small animals. Some are highly specialized parasites attached to the skin or gills of fishes.

Copepods often are important elements in aquatic food webs, serving as common fare in the diet of small fishes and larval amphibians. They also serve as intermediate hosts of parasites (e.g., tapeworms, flukes, nematodes) that attack various fishes, amphibians, and birds.

Reproduction generally involves mating of a male and female, females carrying egg sacs until the larvae hatch, and several immature stages prior to sexual maturity. Several generations may be produced each year. Both activity and reproduction are reduced during the colder months. Females may produce thick-walled, dessication-resistant eggs in response to the drying of vernal pools. Such conditions, and possibly anaerobic conditions as well, also may cause adults to form protective, resistant cysts.

Similar to cladocerans, some copepods exhibit cyclic vertical migrations, moving toward the water surface with darkness and returning to the bottom when daylight returns. In addition to movements powered by their own efforts, some copepods also sometimes move about while attached to flying aquatic insects or waterfowl, allowing them to colonize new habitats.

Ostracods

Ostracods, also known as "seed shrimps," occur in saltwater and freshwater habitats throughout the world. At least a few hundred species inhabit the United States. Many species have huge, multicontinent distributions; others have never been found outside their original discovery site.

These tiny, specklike crustaceans superficially resemble small clams, enclosed as they are in two "shells" (Fig. 15.4). Crawling or swimming are accomplished by stroking of the antennae (lacking in clams) and other appendages between the shells.

Ostracods generally live on or about objects or the bottom in still or flowing water. In such situations they can be tremendously abundant. Some vernal pools that I sampled were teeming with ostracods. Some ostracods are specialized

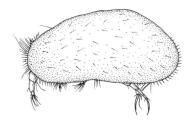

Fig. 15.4. Ostracods are crustaceans that resemble tiny clams.

for life on the gills of crayfishes. Food consists of minute organisms (bacteria, fungi, etc.), detritus, or sometimes dead or living animals.

Depending on the species, reproduction may involve parthenogenesis or mating between males and females. In permanent waters, ostracods may complete two or three generations each year whereas in vernal pools a single generation per year is the norm.

Barnacles

Judging from appearance, it's hard to believe that barnacles are related to crabs and other crustaceans (Fig. 15.5). Firmly glued to rock, wood, living or dead shellfishes, or other solid objects, and enclosed in a pyramidal limey case, they look more like mollusks. But they are genuine crustaceans that happen to be highly modified for a sedentary existence. All of the world's 900 species live in salt water. Away from their typical habitat, I saw groups of barnacles (species unknown), actively beating their feeding appendages, attached to logs floating in the quiet water of a brackish marsh.

Fig. 15.5. Barnacles, here closed at low tide, feed on organic matter suspended in the waters of Long Island Sound. The snail is a yellow periwinkle.

Barnacles use their feathery "legs" to strain small bits of organic material out of the water. Breeding in local species, a late summer or autumn event, occurs through cross-fertilization via extension of a sperm-containing appendage between neighboring hermaphrodites. Diatom-eating larvae are released into the water in winter. By early spring, the larvae attach to rocks, secrete a little "house" of calcified plates, and begin their struggle for survival. Atlantic dogwinkles are the primary predator. Barnacles that escape dogwinkles or dislodging by storms may live up to three to five years.

How do barnacles cope with cold temperatures when exposed at low tide in winter? They simply tolerate partial freezing of their body water. They survive unless the temperature drops below about −2°F.

Northern rock (acorn) barnacle *(Semibalanus balanoides)* is abundant on intertidal and upper subtidal rocks, and it commonly attaches to boat hulls and other submerged artifacts. Little gray barnacle *(Chthamalus fragilis)* occurs on higher rocks, outcompeted (at high population densities) at lower levels by *Semibalanus*. Several additional species of barnacles occur in Long Island Sound.

Isopods

Isopods are small (usually less than an inch long), often grayish crustaceans represented by 4,000-plus species worldwide. Most are marine, a good number are completely aquatic in fresh water, and others are completely terrestrial and in fact comprise the only large group of land-dwelling crustaceans.

The vast majority of isopods are scavengers or nibble on algae, fungi, moss, bark, decaying organic matter, and the like; these isopods may contain symbiotic bacteria that digest cellulose in plant matter. Some isopods are predatory and others are parasites (see end of section).

Reproduction may involve copulation of a male and a female, or copulation of hermaphrodites. Eggs are retained and internally brooded until they hatch, then the young depart and fend for themselves.

Terrestrial isopods, known colloquially as pillbugs, rolypolys, woodlice, and potato bugs, are represented in Connecticut by both native and introduced species (Fig. 15.6). For example, the woodlouse *Oniscus asellus* (up to 0.6 inches, shiny, brownish, light and dark spots along side of body, antennae end with three short segments), introduced from Europe, is common and lives side by side with the native wood louse *Porcellio scaber* (up to 0.6 inches, dull gray with numerous tubercles, antennae end with two short segments) in natural areas and around buildings. It is easy to take these common animals for granted until you reflect on how few crustaceans live completely on land.

Fig. 15.6. These two isopod species are land-dwelling scavengers.

Terrestrial isopods are secretive, generally active at night, and avoid warm, dry conditions. They lack the waxy cuticle that helps insects reduce water loss, and die quickly if left exposed to dry air. During daylight, they hide under logs, rocks, leaf litter, or other objects on the ground. Sometimes in winter I found hundreds congregated under logs that were frozen to the ground. When disturbed, some species defend themselves by rolling up into a round ball with only their armorlike dorsal side exposed. Isopods have poor vision but active, sensitive antennae. Odor plays a significant role in their social behavior, and defensive secretions help protect them from attacks by small predators.

Freshwater isopods (*Caecidotea* spp.) live among the decaying vegetable matter in aquatic habitats. They can be tremendously abundant among the leaf litter at the bottom of some vernal pools, whereas in other pools they are curiously scarce or absent. They stay hidden during daylight hours, but at night a flashlight reveals them walking about on submerged vegetation. Many aquatic isopods crawl and swim, and some can burrow into the bottom. Each habitat generally supports only one of the few species in Connecticut.

Estuarine isopods of several species (e.g., *Philoscia vittata*, 8 mm, two wide brownish bands along back) are common under salt marsh debris; on sandflats and mudflats; on eelgrass, algae, and sponges; under and on rocks; on pilings; and on carrion (Fig. 15.7). Some saltwater isopods are parasitic: aegid and cymothoid isopods are ectoparasites on various estuarine fishes, whereas bopyrid isopods ectoparasitize shrimp and crabs.

Amphipods

Amphipods, sometimes called scuds or sideswimmers, comprise nearly 5,000 species worldwide. They are common on the bottom of all kinds of freshwater and saltwater aquatic habitats, especially among living or decaying vegetation. Most are gray or brown, flattened from side to side, less than an inch long, and somewhat shrimplike. They move by hopping, crawling, swimming (on the side), or burrowing, but usually stay hidden during daylight.

Life styles of estuarine amphipods are diverse and include free-living scavengers, jellyfish parasites (*Hyperia galba*), and others that live in close association with sponges, peanut worms, polychaetes, or sea squirts. Four-eyed amphipods (Ampeliscidae) are common filter feeders on muddy or sandy bottoms of Long Island Sound.

Reproduction involves transfer of sperm from male to female. In some common species, the male carries the female on his back for days before transfering sperm to her. Females carry the fertilized eggs in a brood chamber, which shelters the young for up to a week after hatching.

Amphipods are a food source for many kinds of larger animals. They are an intermediate host for tapeworm and roundworm parasites of various fishes, amphibians, and birds.

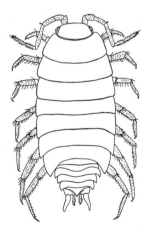

Fig. 15.7. The estuarine isopod *Philoscia vittata* is abundant under debris in high salt marshes. From Kunkel (1918).

Fig. 15.8. Beach hoppers emerge at night on sandy beaches.

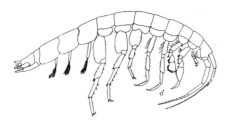

Fig. 15.9. The eyeless, unpigmented, freshwater amphipod *Stygobromus tenuis* is restricted to underground waters and spring-fed streams. From Kunkel (1918).

Beach fleas *(Orchestia)* and larger **beach hoppers** *(Talorchestia)* are common amphipods on beaches of Long Island Sound (Fig. 15.8). These terrestrial amphipods are most active at night. During daylight you can find them burrowed in damp sand or hiding under algae (lift large masses of beach-cast algae to expose huge numbers of hopping beach fleas). They are important foods for many shorebirds.

Marsh amphipods *(Orchestia grillus)* are abundant in detritus of high salt marshes. *Hyallela azteca* and *Gammarus* species are common scavengers in freshwater or brackish water habitats. Among the fewer than 10 freshwater species in Connecticut, *Crangonyx pseudogracilis, Gammarus fasciatus,* and *Hyalella azteca* are most common and widespread.

Some of the specialized, uncommon amphipods in Connecticut include: Mystic Valley amphipod (*Crangonyx aberrans*), a species of freshwater swamps and marshes; Piedmont groundwater amphipod (*Stygobromus tenuis tenuis*) (Fig. 15.9), a rarity restricted to a few subterranean waters and spring-fed streams; and coastal pond amphipod (*Synurella chamberlaini*), which inhabits ponds, swamps, and their feeder streams in its small range from eastern Connecticut to southeastern Massachusetts.

Crabs and Relatives

Crabs, lobsters, and shrimps (Order Decapoda) are the largest and most familiar crustaceans inhabiting the subtidal and intertidal areas of Long Island Sound. They are important food items for many fishes. Several species are valuable commercial and recreational resources.

Most decapods prey on small living animals and scavenge recently dead ones. Crabs and lobsters detect prey by sight or through sensors on the antennae. Powerful enlarged claws allow predation on clams, snails and other shelled prey—by chipping away a part of the shell or crushing it outright. Some decapods eat plants and detritus. Peak activity for most decapods occurs at night (fiddler crabs are an exception).

Crabs and lobsters produce enormous numbers of eggs (carried by the females) that hatch into minute larvae that drift in the water with other zooplankton before adopting the usual benthic (bottom-dwelling) life style. In many crabs, and in the lobster, females must be recently molted (softshell) in order for mating to occur. Generally, crabs live only two to five years at most.

Saltwater Decapods

American lobster. Long Island Sound's most famous decapod is the American lobster *(Homarus americanus)*. This esteemed crustacean lives on various types of bottoms, almost exclusively in subtidal areas. I rarely found small lobsters stranded in tide pools at low tide. They are most abundant in the western half of the Sound. Lobsters occur wherever there are rocks or other objects that can be used as cover, or where they can burrow into firm sediments (as in the western part of the Sound). They tend to stay in their shelters in daytime and are most active at night. Because lobsters are nearly always submerged, their shells often harbor encrusting bryozoans, barnacles, jingles, tunicates, or other estuarine organisms that attach to just about any available bare surface.

Lobsters may not look well suited to travel, but they can make long-distance movements. In the eastern Sound, some move large distances between offshore wintering areas and coastal summer habitats, while lobsters farther west in the Sound are more sedentary.

Lobsters eat various small marine animals that are caught alive, and they scavenge carrion from the sea floor. Scavenging is what gets lobsters into trouble in baited lobster traps. Lobsters use their impressive front claws for feeding and defense. The left claw is a powerful crusher whereas the right one is good for tearing or cutting.

It takes about five years for lobsters in the Sound to reach legal harvest size, at which time they have reproduced at least once. After mating in late spring or summer, females carry sperm for a year or more before using it, and they carry their eggs for almost a year before the hatchlings emerge.

In the late 1990s, after years of exceptionally high abundance, the lobster population in Long Island Sound underwent a large decline, especially in the western portion where fall landings declined over 90 percent. As of 2003, the lobster population and fishery had not recovered. One abnormality associated with the decline was premature molting by egg-bearing females. Some have speculated that this may have resulted from the unintended effects of a larvicide (methoprene, or "altocid") applied in coastal areas in attempts to control mosquito populations. The decline also was associated with high abundance of a protozoan parasite (*Paramoeba* sp.), but the parasite increase could have been a result of poor lobster health that was caused by something else, such as the unusually warm temperatures of 1999. High levels of hydrogen sulfide and ammonia in the Sound's bottom waters also may have played a role in the decline. As part of the continuing effort to further understand the conditions associated with the lobster die-off in 1999, the DEP, in cooperation with the New York Department of Environmental Conservation, National Marine Fisheries Service, the University of Connecticut, and Connecticut and New York Sea Grant, developed a comprehensive research and monitoring program. Management of this species is also conducted through the American Lobster Fishery Management Plan of the Atlantic States Marine Fisheries Commission.

Big Crabs. The larger crabs in Long Island Sound range from gangly, slow-moving, cryptic species to sleek, fast swimmers. Foremost among the first category are the **portly spider crab** (*Libinia emarginata*) and **longnose spider crab** (*L. dubia*) (Fig. 15.10). These lethargic scavengers and predators range over various substrates but are most common on soft muddy bottoms. In fall, they aggregate on the floor of the Sound in big piles of molting individuals. They spend the winter buried in bottom sediments. Egg-bearing females can be observed in late spring and summer. Young spider crabs have some interesting behaviors. Larvae may attach themselves to sea nettles, and tiny juveniles live on and in the jellyfish for awhile before dropping to the bottom. After adopting their bottom-dwelling life style, small spider crabs actively adorn their upper shell with algae and sessile animals such as sponges, hydroids, ectoprocts, and tunicates, which undoubtedly helps camouflage them from predators.

Blue crabs (*Callinectes sapidus*) are the most impressive of the fast swimmers (Fig. 15.11). Their last pair of legs is flat like paddles, allowing the crabs to swim with excellent speed and maneuverability. These legs are also superb for digging in the sand, so blue crabs can quickly bury themselves (rear end first). These streamlined crabs, famous on seafood menus in the Cheasapeake Bay region, are widespread in the Sound but most abundant in the eastern end. Blue crabs winter in bottom sediments of deeper water and move into shallow waters in the warmer months, where recreational crabbers catch many in the lower parts of tidal rivers and in tidal channels of salt marshes. Mated females move out to deeper water, overwinter, then in spring extrude and carry their developing eggs. Larvae emerge from the eggs and live as zooplankton, and eventually the young make their way into estuaries. Early juveniles feed primarily on zooplankton, then switch to an animal diet supported by cordgrass or eelgrass productivity. Hence, marshes and eelgrass meadows are good developmental habitats for the young. Blue crabs live up to three years.

Another fast crab with flat swimming and digging legs is the **lady crab** (*Ovalipes ocellatus*) (Fig. 15.10), most common on sandy bottoms. Its life history is similar to that of the blue crab. Encounters with this highly mobile crab generally consist of a pinched toe as one wades out from the beach for a summer swim. The crabs aren't aggressive, but they defend themselves if stepped on.

More powerful are **Atlantic rock crabs** (*Cancer irroratus*) (Fig. 15.10). These are heavy-bodied and lack the rear-leg

Fig. 15.10. Crab carapaces: clockwise from top left: lady crab, spider crab, green crab, and rock crab.

Fig. 15.11. Blue crabs move into marshes and tidal streams in summer.

swimming paddles of the blue and lady crabs, and so move more slowly, but they have more massive claws that can crush thicker-shelled prey. Rock crabs live mainly in deep water in summer and occur in rocky shoreline areas primarily in the cooler months, although I found some in tidal creeks in salt marshes in summer. They breed offshore in late summer and fall. Young rock crabs are important in the diets of fishes of the cod family, dogfishes, tautogs, and many other fishes. The similar **Jonah crab** (*C. borealis*) tends to be in deeper water.

Small Crabs. One of the most common of the smaller crabs is a non-native species, the **green crab*** (*Carcinus maenas*) (Fig. 15.10). It arrived with ships from Europe over two hundred years ago and has successfully colonized the east coast of North America. In the warmer months, this hardy generalist occurs in a wide variety of shallow-water habitats. Sometimes juveniles can be found in the lower intertidal zone in winter; adults winter in deeper waters. Breeding occurs in summer. Life span is three to five years. Herring gulls catch and eat many green crabs, which nevertheless remain abundant.

Green crabs can be a nuisance by preying on young oysters, scallops, and softshell clams, and they also have a strong ecological impact on rocky shore communities. In hopes of reducing the threat to commerical and recreational shellfisheries, a bounty actually has been paid on green crabs on Martha's Vineyard in Massachusetts.

If you search muddy or shelly bottoms with good populations of oysters and clams, you might find any of the several species of **mud crabs** (Xanthidae) that live along the Connecticut coast (Fig. 15.12). They also live in rocky intertidal habitats and among algae and other organisms attached to rocks and pilings. Mud crabs eat bivalves, barnacles, polychaete worms, algae, and detritus. The shells of small mollusks are no match for the crabs' strong claws. Mud crabs are solitary and socialize only for mating purposes.

Three species of **fiddler crabs** are common along the Connecticut coast. Atlantic marsh fiddlers (*Uca pugnax*) can be tremendously abundant on mud banks of tidal creeks bordering salt marshes (Fig. 15.13). Atlantic sand fiddlers (*U. pugilator*) prefer sandy, sheltered, coastal shores and glasswort beds. They burrow at or above the high tide level in the zone between open water and marsh vegetation. I found red-jointed fiddlers (*U. minax*) to be locally common at the base of reeds and cattails along water edges of low-salinity coastal marshes.

All fiddlers live in colonies and dig burrows for shelter. They are very social and use sounds and visual displays of the one enlarged claw (males only) to communicate with one another. Males sometimes use their enlarged claw to "fence" with each other. (Captured by hand, fiddler crabs may pinch with the large claw, which may detach as the animal escapes. A few molts later, the crab will have regrown a new fully formed claw.)

Most activity occurs when tide waters drop below the openings of the burrows, especially on bright sunny days. I saw marsh fiddlers perched just outside their submerged burrows during high tide.

Adults feed on decaying organic detritus. Females feed with both claws and can profitably feed near the safety of their burrows. In contrast, males can feed with only one claw (the small one) and so they feed only half as fast as the females do. As a result, males tend to forage in sites with high food levels, which may take them far from their burrows. This is why males are much easier to catch.

Atlantic sand fiddlers forage in groups. In response to a predatory threat, they may bunch up and move away as a

Fig. 15.12. Mud crabs are small, strong-clawed predators in the shallows of Long Island Sound.

Fig. 15.13. Fiddler crabs are abundant along tidal creeks.

densely packed unit. This likely makes it difficult for the attacker to single out and capture any particular crab among the swarm.

Fiddlers breed in mid- to late summer. Females carry the eggs for about two weeks, then release the hatching larvae into the water during high tide. After about a month, the planktonic larvae complete development and settle along shorelines. The best way to see fiddler crabs is to drift quietly in a small boat in tidal creeks at low tide.

The preceding crabs are relatively easy to find, but not so tiny **squatter pea crabs** *(Pinnotheres maculatus)* and **oyster pea crabs** *(P. ostreum)* (Fig. 15.14). Females of these species live commensally or parasitically in the shells of various living bivalves, especially mussels and oysters; sometimes they associate with other invertebrates. Inside the shell, the crab "steals" algae captured by the bivalve's feeding apparatus. These crabs leave the bivalve shell in order to breed. Occasionally I found one in the shell of a steamed mussel I was about to eat.

Among the true crab species of Long Island Sound, one of the most notable is a newcomer, the **Asian shore crab*** *(Hemigrapsus sanguineus)* (Fig. 15.15). Native to eastern Asia, it appeared on the U.S. Atlantic coast in the late 1980s in New Jersey, evidently arriving in ballast water discharged from a ship. The crab has since spread to Maine and North Carolina. I found these crabs to be common to extremely abundant along rocky intertidal shores all along the Connecticut coast. It is impressive and frightening how quickly an exotic species can go from totally absent to abundant. The crabs prey indiscriminately on juvenile bivalves (clams and relatives), periwinkles, small crabs, barnacles, polychaete worms, fish larvae, algae, eelgrass and surely are hav-

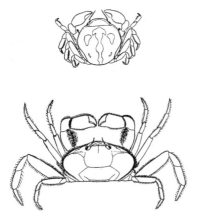

Fig. 15.14. Pea crabs, *Pinnotheres* (top) and *Pinnixa* (bottom), live in close association with oysters, mussels, and burrowing polychaete worms. From Verrill and Smith (1874).

Fig. 15.15. Asian shore crabs became tremendously abundant along the Connecticut shore in the 1990s.

ing a strong impact on rocky intertidal ecosystems. For example, as an apparent result of predation on juveniles, green crab* populations have declined in areas where shore crabs are now abundant, and probably other species have been impacted as well. Asian shore crabs are greenish to purplish, small (up to 1.5 inches across the carapace), and have three spines on each side of the roughly rectangular carapace.

If you watch the bottom in shallow water, you may see a snail shell moving a bit too fast for a gastropod. Likely the shell is occupied by a **hermit crab.** Hermits are not true crabs, but rather are examples of unusual decapods that evolved from a different ancestor. Immense numbers of longwrist hermit crabs summer in shallow water on sandy or rocky bottoms of Long Island Sound. They spend the winter burrowed shallowly in the bottom in water several yards deep and move back into shallower water in spring. Other hermits tend to stay in deeper water. Scavenged organic material is the primary food for all.

Hermit crabs are famous for their use of mollusk shells as portable protective homes. Evolution of this behavior also resulted in loss of the crab's own hard abdominal shell. Now these crustaceans are totally dependent on mollusk shells for protection. Hermit crabs change shells as they grow larger. Each species has its favorite shells. Longwrist hermit *(Pagurus longicarpus)* (Fig. 15.16), most common in calm, shallow water of high salinity, uses shells of eastern mudsnails, threeline mudsnails, oyster drills, periwinkles, and others. Acadian hermit *(P. acadianus)* and flatclaw hermit *(P. pollicaris)* generally live in deeper waters and use the big shells of moonsnails and whelks. Banded hermit *(P. annulipes)* prefers shells of greedy dovesnails.

Shells occupied by hermit crabs often have colonies of attached hydroids *(Hydractinia* spp.), anemones, ectoprocts, polychaete worms, snails, or barnacles. Some of these relationships may be mutualistic; the crab benefits through camouflage and reduced predation while the attached organisms

Fig. 15.16. This hermit crab and its eggs were removed from an eastern mudsnail shell.

Fig. 15.18. Grass shrimp are abundant in salt marshes.

benefit by being continually moved to new feeding areas. However, for each participant there are costs to the relationship as well, such as increased shell weight for the crab and possible damage to the attached organisms when the crab moves the shell.

Whelk shells inhabited by hermit crabs sometimes also contain zebra flatworms *(Stylochus zebra),* which feed on hermit crab embryos and slippersnails *(Crepidula plana)* that attach inside the opening of the whelk shell, or polychaete scale worms (obligate symbiont). Some shells occupied by hermit crabs host a small bright-red polychaete worm *(Polydora commensalis)* that bores into the shell.

Hermit crab reproduction occurs in late spring and summer. The planktonic larval stage lasts a few weeks and the crabs are full grown in about a year.

Another small "crab" that is not a true crab is the **Atlantic sand crab** or **mole crab** *(Emerita talpoida)* (Fig. 15.17).

These compact, somewhat egg-shaped crustaceans live on wave-washed sandy shores in summer and move to deeper water for winter. They are efficient burrowers, moving quickly backwards into the sand if exposed. They feed in the surf zone, straining minute algae from the water with their bushy antennae as waves recede. Reproduction, by individuals that survived their first winter, occurs during the warmer months; thereafter the adults soon die. The larval stage is planktonic. Because of their specialized feeding behavior, sand crabs are more common on open ocean beaches with big surf than they are along the relatively quiet shores of Long Island Sound.

Shrimps. It usually doesn't take much effort to see that **grass shrimp** or shore shrimp *(Palaemonetes* spp.) can be extremely abundant (Fig. 15.18). Dip-net samples in beds of eelgrass or algae, in mosquito ditches and tidal creeks, or in waters of salt marshes often contain hundreds. Grass shrimp eat algae, detritus, and sometimes various small animals.

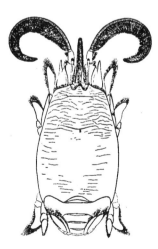

Fig. 15.17. Atlantic sand (mole) crabs are most common on wave-swept shores. From Verrill and Smith (1874).

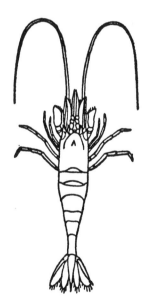

Fig. 15.19. Sand shrimp abound on sandy bottoms of Long Island Sound. From Miner (1950).

Fig. 15.20. Crayfish are secretive inhabitants of lakes, ponds, streams, and freshwater wetlands.

They mature quickly and rarely live more than a year. A distinctive feature is the long, saw-toothed rostrum.

Another abundant shrimp is the **sand shrimp** or seven-spine bay shrimp *(Crangon septemspinosa)* (Fig. 15.19). Sweep a net over a sandy bottom or in an eelgrass flat and you're likely to catch many, especially in late summer and fall. Sand shrimp occur offshore and depend on inshore waters for spawning and as a nursery area for the young. They spawn mainly from spring to early fall. The diet consists of tiny invertebrates (e.g., opossum shrimp, amphipods, polychaetes), larval fishes, diatoms, and detritus. I found many of these shrimps in the stomachs of flounders that I caught, and they are an important food for many other sport fishes.

Freshwater Decapods

Crayfish are freshwater decapod crustaceans (Family Cambaridae) variously known as crawfish or crawdads (Fig. 15.20). They inhabit permanent streams, lakes, and ponds. The approximately 400 crayfish species in North America represent nearly 70 percent of the world's total.

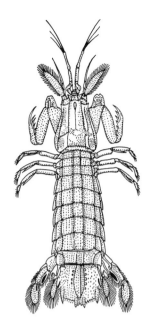

Fig. 15.21. Mantis shrimp burrow in tidal mud flats. From Miner (1950).

Alarmingly, more than 30 percent of North American species warrant conservation concern, reflecting human abuses of rivers and lakes.

Connecticut's crayfish fauna includes a mixture of several introduced and native species. Some species undoubtedly were introduced through use as fishing bait. As omnivores, scavengers, and sometimes predators, introduced crayfish can cause large alterations in aquatic communities and, through aggressive interactions and hybridization with native species, often displace or extirpate native crayfish.

Stomatopoda

Mantis shrimp *(Squilla empusa)* is a burrowing denizen of intertidal mud flats, especially in western Long Island Sound (Fig. 15.21). This summer-breeding crustacean is shrimplike but not a true decapod crustacean. It uses its raptorial limbs to seize small animals that venture close to its burrow opening. Handle mantis shrimp with care to avoid being stabbed as they flip the sharp-spined "tail."

Chapter 16

Echinoderms and Tunicates

This chapter covers two groups that share certain highly distinctive aspects of their embryonic development. In these features they parallel the vertebrates but differ from all of the other invertebrates discussed in this book.

Echinoderms

Phylum Echinodermata, a strictly marine and estuarine group, includes 7,000 species found throughout the world's oceans. Nearly all are bottom or rock dwellers. In contrast to most mobile animals, the vast majority have no obvious head or even head end and no real brain.

A diagnostic feature of echinoderms is their "water vascular system," which controls the actions of the arms of sea stars and the numerous suckerlike tube feet that are used for locomotion and/or obtaining food. Another unique feature on the surface of some sea stars and sea urchins are numerous small mobile pincers that inhibit algae and various larval animals from attaching and establishing residency.

Sea Stars

Sea stars (Class Stelleroidea), widely known as "starfish," are the most recognizable group of echinoderms. The common sea star *(Asterias forbesi)* (Fig. 16.1) is the most famliar species in Long Island Sound. It is locally fairly common to abundant in rocky intertidal and subtidal habitats, shellfish beds, and on some sand bottoms. Like other sea stars, these stars move only very slowly but nevertheless tend to migrate to deeper waters for the colder months. They prey on clams, mussels, barnacles, and many other invertebrates. As opportunistic predators, sea stars sometimes take advantage of the concentrated food resources in shellfish beds, to the chagrin of commercial interests. Through persistent pulling with their tube feet and release of digestive juices, common sea stars pry apart the shells of live bivalves to get at the flesh

Fig. 16.1. Common and northern sea stars (underside shown) are ecologically influential predators in intertidal and subtidal waters of Long Island Sound. From Coe (1912).

inside. They can locate prey through chemical means and may dig out clams buried in bottom sediments. Predation by common sea stars often exerts a major impact on the composition and functioning of rocky intertidal communities (see Chapter 3). As in human society, being "in charge" does not necessarily require an impressive brain.

Sea stars reproduce sexually by releasing eggs and sperm into the water in summer. The larvae drift in the plankton for a few weeks before changing into a crawling star.

Spider crabs are among the few sea star enemies, though their effect is relatively benign. The crabs sometimes use their claws to cut into the sea star's arm. This induces the star to cast off the limb, which is confiscated as a meal by the crab. Gulls sometimes make an awkward meal of a sea star. Human beach-goers too often collect or displace sea stars to unsuitable locations, with fatal results (to the sea star).

Less obvious than common sea stars are the brittle stars. The little brittle star *(Axiognathus squamatus)* (Fig. 16.2) is the most common and widely distributed brittle star along the Connecticut coast. It is indeed little (arms less than an inch long) and occurs on eelgrass, in old mollusk shells, or among rocks. Brittle stars eat detritus, worms, small crustaceans, and the like. Several additional sea star species inhabit Long Island Sound, but these are much less likely to be observed.

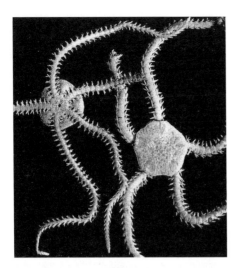

Fig. 16.2. Little brittle stars live on eelgrass or among rocks or shells. From Coe (1912).

Sea Urchins and Sand Dollars

The globular or disclike echinoderms of the Class Echinoidea lack arms. They have an attractive, limey internal skeleton that often becomes a take-home curio for beach-strollers. The body of the living animal is adorned with movable spines that, with the tube feet, may be used in locomotion (and also serve as protection).

Sea urchins have a mouth on the underside, and they graze on algae, small sessile invertebrates, or carrion. Sand dollars, rare in Long Island Sound, have a covering of short spines, burrow in bottom sands, and use cilia to collect their diet of fine organic matter.

Sea urchins and sand dollars are prey for various fishes such as oyster toadfishes, searobins, northern puffers, cod, flounders, and haddock. In most species, reproduction involves external fertilization after shedding of eggs and sperm into the water by females and males, followed by a planktonic larval stage.

Purple sea urchin (*Arbacia punctulata*) and green sea urchin (*Strongylocentrotus droebachiensis*) occur on rocky or shelly bottoms, mainly in subtidal areas. The purple is variably common and widely distributed in the Sound whereas the green is most numerous and reaches larger sizes in the eastern end of the Sound and on the coast north of Cape Cod. Both are sensitive to pollution and seldom numerous except in the cleanest water.

Sea Cucumbers

Sea cucumbers (Class Holothuroidea), bottom denizens of Long Island Sound, have a wormlike or baglike body with

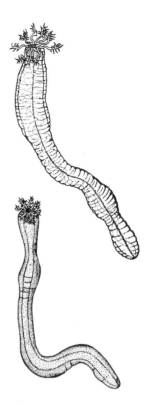

Fig. 16.3. Sea cucumbers, including the white synapta (top) and pink synapta (bottom), are sluggish benthic detritus eaters. From Miner (1950).

a circle of tentacles around the mouth end of the animal (Fig. 16.3). Tube feet may form a thick covering or occur in distinct rows, or they may be totally absent. These sluggish echinoderms feed on bottom detritus. Sea cucumbers have separate sexes or are hermaphroditic, with fertilization occurring externally in the water.

White synapta (*Leptosynapta tenuis*, often 4 to 6 inches long) and pink synapta (*Epitomapta roseola*, usually less than 4 inches long), with a fragile body, branched tentacles at the mouth end, and no tube feet, are common burrowers in intertidal and subtidal sand or mud. Their burrows are marked by small mounds of fine sand with a small hole in the center. Hairy sea cucumber (*Sclerodactyla briareus*), a species covered with numerous tube feet, may be common in muddy bottoms and eelgrass beds in shallow water.

Tunicates

Tunicates or sea squirts (Phylum Urochordata) are pouch-shaped saltwater animals. As adults, they live singly or in colonies, usually attached to a firm surface, including living shellfishes. The brief larval stage is tadpole-shaped and planktonic. Most tunicates are filter feeders. Worldwide, there are some 1,400 species, of which a few hundred occur

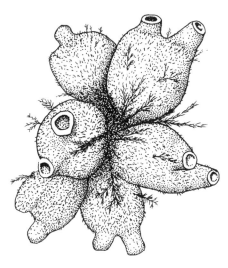

Fig. 16.4. Sea grapes are Long Island Sound's most common tunicate. These smooth, globular animals are often covered with algae and debris. Modified from G. H. Childs, in Miner (1950).

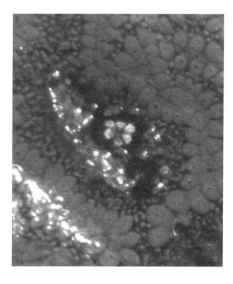

Fig. 16.5. Curving rows of the invasive Pacific colonial tunicate surround a small patch of the native golden star tunicate (flowerlike structure at center).

in the United States. The most conspicuous tunicates in Long Island Sound are the following:

Sea grape *(Mogula manhattensis)* (Fig. 16.4), an often colonial species, is locally common on any solid substrate in the lower intertidal and shallow subtidal zones, including some brackish water situations.

Golden star tunicate *(Botryllus schlosseri)* is common in summer as a soft mat or lump up to a few inches in diameter. It attaches to various solid objects, including algae and eelgrass, in subtidal areas. Its zooids are arranged in starburstlike or flowerlike clusters (Fig. 16.5).

Pacific colonial tunicate* *(Botrylloides diegensis)*, a Japanese species released in the late 1970s in Massachusetts via California is now common in southern New England. It forms a thin, dense layer of tiny, usually orange (or whitish to red) zooids (Fig. 16.5), often in curving rows, on algae, eelgrass, pilings, rocks, and other solid objects in shallow subtidal areas and the lowest part of the intertidal zone.

Pacific rough sea squirt* *(Styela clava)*, native to the southwestern Pacific Ocean, first appeared in southern New England in the mid-1970s. It is now common throughout the region, attached to low intertidal and subtidal rocks, pilings, lobster traps, and other objects. This long, baglike, yellowish to brown tunicate has a conspicuous stalk, two siphons, and a rough, lumpy surface; it grows to a length of 6 inches.

Chapter 17

Fishes

Fishes comprise several distinct groups of vertebrate animals. Those inhabiting Connecticut include lampreys (Class Cephalaspidomorphi); sharks, skates, and rays (Class Elasmobranchii); and ray-finned fishes (Class Actinopterygii). Fishes may be smooth-skinned or scaly, and they range in appearance from ordinary to quite bizarre. Worldwide, fishes range in adult length from about 0.4 inches (a goby) to 50 feet (whale shark). Note: I use "fish" for one fish or as an adjective, and "fishes" as the plural of fish.

The known worldwide fish fauna includes approximately 25,000 species. Habitats range from desert hot springs, super-salty lakes, and frigid polar waters to river torrents, lightless caves, and the dark floor of deep oceans. Worldwide, about 60 percent of species are marine, 40 percent live in fresh water, and 1 percent migrate between salt water and fresh water. About 78 percent of the marine species occur in the relatively shallow waters along the edges of land masses. Some 2,500 species inhabit Canada and the United States, including their contiguous shore waters. About 800 species occur in freshwater habitats of Canada and the United States.

The following annotated list includes a brief synopsis of the lives of fishes most likely to be encountered in freshwater and saltwater habitats in Connecticut. Not included are certain species, such as released aquarium and bait fishes, that have been collected in the state but which are currently not known to be represented by established, reproducing populations. Also not included are a few apparently native species that have not been seen in Connecticut in recent years, some non-native freshwater fishes represented by not more than a few localized populations, and various game fishes that are stocked for "put and take" fishing but which are not represented by self-sustaining populations (see Whitworth 1996). I've also excluded many species of saltwater fishes that rarely enter Long Island Sound or that are present in very small numbers. Rare strays generally are warm-water species that range northward in late summer. Due to frequent unauthorized introductions by anglers and fish hobbyists, range expansions and additions to the state's fish fauna can be expected. Virtually all of the species mentioned as occurring in Long Island Sound ("the Sound") also sometimes occur in the lower portions of coastal rivers.

Lampreys (Class Cephalaspidomorphi)

Lampreys (Order Petromyzontiformes, Family Petromyzontidae)

Lampreys include 41 species of boneless, anadromous and freshwater fishes lacking jaws, paired appendages, and true bone. Nearly half of the species are parasitic, and all die soon after spawning.

American brook lamprey (*Lampetra appendix*). Known from a few streams in north-central Connecticut; adults not parasitic, do not feed or migrate; larvae filter-feed on stream bottoms for several years; spawns in spring in gravel nests on stream bottom.

Sea lamprey *(Petromyzon marinus)* (Fig. 17.1). Parasitic adults inhabit salt water for two to three years; attach to other fishes with mouth, rasp through skin with horny toothlike structures, suck out body fluids; feeding often fatal to host; migrates up streams when ready to spawn, beginning in early spring; deposits eggs in gravelly streambed nests in spring or early summer (I found adult lampreys swimming in Connecticut River tributaries from May to July); filter-feeding larvae live in stream bottoms for several years before metamorphosing and moving out to estuarine or marine waters to continue growth. *Note:* Over the last several years, about 20,000 to 100,000 sea lampreys annually migrated up the Connecticut River as far as the Holyoke Dam in Massachusetts, and at least several hundred more entered rivers (e.g., the Salmon River) south of the dam. Historically, sea lampreys in the Connecticut River were a significant human food resource. The larvae are an important food resource for other fishes.

Fig. 17.1. The mouth of the sea lamprey is specialized for attaching to and rasping through the body wall of other fishes.

Cartilaginous Fishes (Class Elasmobranchii)

Houndsharks (Order Carchariniformes, Family Triakidae)

Smooth dogfish (*Mustelus canis*) (Fig. 17.2). Common throughout the Sound at various depths from late spring through fall; abundance declines in colder months; actively roams in groups over mud or sand bottoms; crabs dominate diet, also eats lobsters, other crustaceans, mollusks, and small fishes; crushes food with numerous small teeth; young born from mid-spring to early summer; young-of-the-year become common in trawl samples by July.

Requiem Sharks (Order Carchariniformes, Family Carcharhinidae)

Sandbar shark (*Carcharhinus plumbeus*). Juveniles, and females about to give birth, relatively common in summer in shallow waters of the Sound near shore; other adults mostly farther offshore; both adults and juveniles migrate to ocean waters off southeastern United States for the winter; primary foods are bottom-dwelling fishes and marine mollusks and crustaceans; poses no threat to humans; young born in late spring. *Note:* Excessive human exploitation, exacerbated by the shark's low reproductive rate (they take 12 to 15 years to mature, and individual females produce small numbers of young at intervals of two years or more) has resulted in a large population decline along the U.S. Atlantic coast.

Dogfish Sharks (Order Squaliformes, Family Squalidae)

Spiny dogfish (*Squalus acanthias*). Highly migratory, one of the world's most widely distributed fishes in temperate and boreal waters; may be mainly a spring and fall transient in Connecticut; most winter in deeper waters offshore; most common in trawl samples in May and June in deeper waters of eastern part of the Sound, with lower numbers in summer, followed by maximal abundance in fall, when widespread throughout the Sound; feeds voraciously on various fishes and invertebrates; slow-growing, long-lived, schooling; live-bearing, with births over extended season, mostly offshore. *Note:* Concern about overharvest of this species recently has resulted in periodic closures of the commercial fishery along the entire U.S. Atlantic coast. A venomous spine at the front of each dorsal fin makes them hazardous to handle.

Skates (Order Rajiformes, Family Rajidae)

Bottom-dwelling invertebrates (crabs, lobsters, shrimps, mollusks, worms), squids, and fishes are the primary foods. Eggs, which may be deposited almost any time of year, are enclosed in a horny case with four slender appendages. These fishes often are harvested for use as lobster bait.

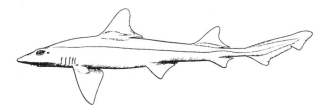

Fig. 17.2. Smooth dogfishes are common bottom feeders in Long Island Sound. E. N. Fisher, from Bigelow and Schroeder (1953).

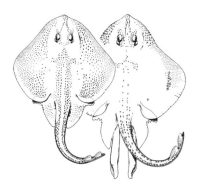

Fig. 17.3. Little skates are most numerous in the Sound during the cooler months. E. N. Fisher, from Bigelow and Schroeder (1953).

Little skate (*Leucoraja erinacea*) (Fig. 17.3). Common in the Sound in spring and fall at various depths on sand, mud-sand, or gravel bottoms; trawl samples indicate low abundance in summer; diet includes many *Leptocheirus* amphipods.

Winter skate (*L. ocellata*). Sand or gravel shoals in Long Island Sound, mainly from fall to spring.

Ray-finned Fishes (Class Actinopterygii)

Sturgeons (Order Acipenseriformes, Family Acipenseridae)

Sturgeons are long-lived, slow-maturing, and include the world's largest fishes found in freshwater.

Shortnose sturgeon (*Acipenser brevirostrum*) (Fig. 17.4). Special fish of the Connecticut River; spawns in Massachusetts portion of the river over rocky bottoms in late April or early May at water temperatures of 50 to 59°F; females first spawn when 8 to 14 years old (23 inches fork length), average several years between spawnings (can live three or four decades or more); usually stays above the saltwater intrusion zone (Essex to south of Haddam) in summer; uses a wide variety of habitats, often in river bends with sand or cobble substrate in summer in Massachusetts; in winter, tends to be sedentary in small groups on bottom in low-velocity water; feeds on bivalve mollusks, including freshwater mussels, plus amphipods, polychaetes, and insects. *Note:* The adult sturgeon population between the river mouth and Turners Falls, Massachusetts, at river mile 124 is about 1,450 individuals. The population below Holyoke Dam is sustained by downstream dispersal of individuals from the reproducing population above the dam. Populations throughout the range were decimated by the effects of historical overharvest, dams, and pollution, but now the species is relatively common in several Atlantic coast drainages.

Atlantic sturgeon (*A. oxyrinchus*). Gigantic bottom feeder commonly more than 6 feet long; does not now spawn in Connecticut, but juveniles and subadults from other populations occur in deeper waters of Long Island Sound and the lower Connecticut, Thames, and Housatonic rivers; juveniles that were tagged in the Connecticut River or Long Island Sound have been recovered in various coastal locations from New Hampshire to Maryland; in Hudson River (New York), spawns in fresh water from May to early July, spawners at least nine or ten years old; feeds on invertebrates and fishes on bottom. *Note:* Most populations throughout the range are mere vestiges of their former selves, due to the effects of overfishing, blockage of spawning habitat and alteration of stream flow by dams, and water pollution. In view of the severely depleted stocks, the Atlantic States Marine Fisheries Commission closed all U.S. sturgeon fisheries coastwide in 1996.

Freshwater Eels (Order Anguilliformes, Family Anguillidae)

American eel (*Anguilla rostrata*). Extraordinarily unusual life history; spawns in winter or spring in Sargasso Sea south of Bermuda; adults die after spawning; young drift in ocean for a year or two, eventually migrate into the Sound and river mouths; some move far up streams (including full length of Connecticut River), squirming over or around small dams and waterfalls; common in streams, spends several years feeding and growing before moving back to sea to spawn; most active at night; feeds opportunistically on insects, crustaceans, fishes, and various other small animals, including fish eggs and terrestrial insects that fall into water. *Note:* For reasons unknown, populations appear to have declined along the Atlantic coast in recent years.

Anchovies (Order Clupeiformes, Family Engraulidae)

Bay anchovy (*Anchoa mitchilli*). Pelagic schools cruise sandy or muddy shores of the Sound and river mouths in summer; probably moves to deeper waters for winter; important forage fish for many larger fishes; eats mainly zooplankton and shrimps; spawns (mainly one-year-olds) abundantly in shallow waters of western and central Sound in late spring and summer; larvae numerically dominant in ichthyoplankton of Long Island Sound.

Herrings (Order Clupeiformes, Family Clupeidae)

Most herrings swim in schools near the surface and feed on plankton. They tend to move with zooplankton and swim closer to the surface at night than during daytime. Many species support valuable commercial fisheries, and they have long been used for food, bait, and fertilizer.

Fig. 17.4. The shortnose sturgeon, an endangered species, resides in the lower Connecticut River. From Collette and Klein-MacPhee (2002), with permission.

Blueback herring *(Alosa aestivalis)*. Lives in the sea for a few years, then migrates to fresh water in spring to spawn, most often in swift, deep stretches of streams; common in Long Island Sound, often along shore but also in deeper water in summer; spawns in distinct runs over long season from May into early September, at warmer temperatures (and thus later) than alewives; life cycle similar to that of alewife and American shad; eats mainly zooplankton, plus some shrimps and larval fishes. *Note:* The Connecticut River has the only major spawning population in the state, and it has drastically declined. In 1985, 630,000 individuals migrated as far northward as the Holyoke Dam in Massachusetts. By 1996, the number dropped to about 55,000 individuals, with at least 1,300 entering the Farmington River in Connecticut. By the early 2000s, the count was down to 2,000 to 11,000 at Holyoke. Additionally, production of juveniles has dropped by 50 percent over the past decade. The declines are thought to result from predation by the increasingly abundant striped bass population. To help rebuild dwindling stocks, the DEP recently enacted an emergency prohibition on the taking of all river herring (anadromous blueback herring and alewife) from all Connecticut waters. This may help stocks to rebuild in years when striped bass populations are low. Landlocked alewife can be taken from a dozen specified lakes.

Hickory shad *(A. mediocris)*. Formerly uncommon, has increased since the mid-1990s, with decline of bluefish; appears to move between deeper saltwater habitats in summer (and probably winter) and shallower waters in spring and fall; seasonally enters Connecticut's major rivers, where it is caught by recreational anglers; feeds mainly on small fishes, augmented by squids, crustaceans, and fish eggs; spawns in freshwater in spring, with limited spawning in Connecticut; in general, life history is poorly known.

Alewife *(A. pseudoharengus)*. Generally a schooling fish; spends considerable time at sea or near river mouths; common in Long Island Sound; beginning in early spring, alewives at least a few years old migrate to tidal lagoons or up streams to ponds or other sluggish to flowing water to spawn in natal area; often associates with similar-looking blueback herrings, but spawns at lower temperatures and thus earlier; after breeding, many adults die on spawning grounds, but a substantial proportion survive to spawn in subsequent years; nutrients derived from adult carcasses in spawning areas may stimulate increased production and result in enhanced populations of insects and benthic invertebrates that are food for young fishes; young migrate out to sea by early fall, leaving the Connecticut River about a month before blueback herrings depart; permanent freshwater populations occur in some areas; migratory adults do not feed in fresh water until spawning is completed; diet includes mainly zooplankton, plus shrimps, amphipods, small fishes, fish eggs, and water-borne terrestrial insects; selects zooplankton individually or nonselectively by filter-feeding with gill rakers; selective feeding by non-native landlocked form can affect composition and characteristics of lake zooplankton, especially by reducing abundance of large forms. *Note:* Populations have been reduced by degradation of spawning habitat, blockage of spawning areas by dams, and possibly overfishing. Efforts are being made to restore alewife access to historical spawning areas. Currently, the annual alewife run up Connecticut rivers includes several tens of thousands of individuals.

American shad *(A. sapidissima)* (Fig. 17.5). Spends most of life in large schools at sea, ranging along much of U.S. Atlantic coast; best known for migration up rivers; the premier fish of Connecticut River, also occurs in Housatonic and Thames rivers; begins migration up river in late March or early April, spawns in its natal river in late April, May or early June; by end of June, most adults have departed the river, moved through Long Island Sound, and returned to the ocean; survivors may return to spawn in other years; spawns in slow water, often in sandy or gravelly shallows below obstructions; eggs drift awhile before hatching; young feed and grow throughout summer and by mid-September attain a length of 3 to 5 inches, then begin migration out to sea; young common in Long Island Sound during much of the year; immatures range widely in western North Atlantic, where they spend three to five years before returning as adults to spawn; during ocean residency, feeds opportunistically on zooplankton, shrimps, other crustaceans, occasional other invertebrates, fish eggs, and occasionally small fishes. *Note:* In recent years, a few hundred thousand individuals have migrated up the Connecticut River toward the main spawning areas in Massachusetts (375,000 in 2002). Historically, populations were decimated by pollution and dams that blocked access to spawning areas. Improved water quality and construction of fishways at dams have allowed populations to recover. Fish passages to be constructed at the Taftville and Tunnel dams will provide shad with access to additional spawning habitat in the Thames River basin. Many sport anglers go after shad with artificial

Fig. 17.5. The American shad is the Connecticut River's premier migratory fish.

lures, primarily below Enfield Dam. Commercial gill netters in the lower Connecticut River supply local markets with a seasonal delicacy.

Atlantic menhaden (*Brevoortia tyrannus*). Abundant in the Sound and lower tidal streams; high abundance near New Haven Harbor; populations increase in abundance from spring through fall; reaches the Sound through migrations from deep-water wintering areas mainly south of Cape Hatteras; swims open-mouthed in large pelagic schools, uses comblike gill rakers to filter feed on phytoplankton (unusual among adult fishes in general) and zooplankton; large schools can greatly reduce zooplankton populations; important food source for predatory sport fishes; spawns buoyant eggs throughout the Sound from late spring to early fall; spawners are at least a few years old; coastal waters and wetlands are important nursery habitats (I saw vast, dense schools of juveniles feeding at the edge of brackish marshes and in tidal creeks in salt marshes in September and October) (Fig. 3.31). *Note:* Menhaden are important to humans for bait, fish meal, oil, and fertilizer. In fact, at an annual average of 1.86 billion pounds, U.S. commercial landings of this species on the United States Atlantic and Gulf coasts in the late 1990s far exceeded that of any other Atlantic species. Commercial harvest of this fish on the east coast is managed by the Atlantic States Marine Fisheries Commission through the Atlantic Menhaden Fishery Management Plan.

Atlantic herring (*Clupea harengus*). Migratory, open-water, schooling fish; adults scarce in the Sound in summer and early fall, young juveniles abundant in summer; travels in large schools (mainly immatures); preys selectively or by filter feeding primarily on zooplankton, plus shrimps and amphipods; predatory fishes eat herrings as available, and squids consume the young; clumps of eggs, deposited in late summer and fall in southern New England, stick to objects on rocky, gravelly, shelly, or algae-covered bottoms; attains maturity at three to five years. *Note:* Populations along the northwestern western Atlantic coast are economically valuable and have been heavily exploited (e.g., as "Maine sardines," kipper snacks, fillets, etc.) and depleted in recent years.

Gizzard shad (*Dorosoma cepedianum*). Schooling, open-water species; began appearing in the Connecticut River in the 1980s, now well established, with increasing numbers of adults migrating through the fish elevator at Holyoke Dam in Massachusetts (about 1,000 in 1996 and 1998, increasing to 35,000 in 1999, down to about 3,000 in 2002); adults also have been found in Housatonic and Thames rivers; favors quiet water; filter feeds on phytoplankton and zooplankton, also extracts food items from bottom; spawns probably over sandy or rocky bottoms in shallow, quiet fresh water in late spring or early summer.

Minnows and Carps (Order Cypriniformes, Family Cyprinidae)

Most Connecticut cyprinids spawn in late spring or early summer and eat various invertebrates (especially insects and small crustaceans) and some algae (exceptions are noted below). They are important sources of food for various mammals, birds, reptiles, fishes, and invertebrates.

Common carp* (*Cyprinus carpio*). Native to Eurasia; common in larger streams, river backwaters and marshes, some lakes and reservoirs, and even brackish river segments; very tolerant of adverse conditions, such as high temperatures and low oxygen levels; adults root about on bottom when feeding, stir up sediments, uproot vegetation, and increase water turbidity; may degrade habitat for native fishes and waterfowl; spawns in vegetated shallows.

Cutlips minnow (*Exoglossum maxillingua*). Clear rocky streams from Housatonic River west; apparently introduced in Farmington River; spawns on pebble mound-nests (made by male) in flowing water; when fish density is high, sometimes uses peculiar lower jaw (three-lobed, with narrow, bony, central part) to gouge out eyes from other fishes, sometimes eats extracted eye.

Common shiner (*Luxilus cornutus*). Common in small to medium-sized streams; prefers cool, clear, weedless water and gravelly or rocky bottom; spawns over nests of other minnow species or over gravel or sand prepared by male common shiner.

Golden shiner (*Notemigonus crysoleucas*) (Fig. 17.6). Common in slow streams, lakes, and ponds, especially in loose schools among vegetation in quiet, clear water; tolerant of warm, murky water with low oxygen content; eats mostly plankton, especially mid-water zooplankton, plus typical minnow foods; spawns over vegetation.

Bridle shiner (*Notropis bifrenatus*). Slow streams, ponds, and lakes, usually among submerged vegetation; eggs sink and stick to vegetation. *Note:* Distribution and abundance may have declined over the past couple of decades.

Fig. 17.6. Golden shiners are common in quiet freshwater habitats.

Spottail shiner (*N. hudsonius*). Often abundant in quiet shallows in rivers and some impounded rivers; spawns over sandy bottoms.

Mimic shiner* (*N. volucellus*). Established in the Connecticut River system; deposits eggs over aquatic vegetation.

Bluntnose minnow* (*Pimephales notatus*). Established in Housatonic River drainage, has been found in Byram River drainage; streams and ponds, especially among vegetation; tolerates eutrophic conditions; spawns in cavity cleared by male under sheltering object; males guard nests.

Fathead minnow* (*P. promelas*). Introduced and recently established in many streams; tolerates various conditions, including ponds and turbid water; territorial males may clear and guard nesting sites on bottom; females deposit eggs on underside of sheltering objects.

Blacknose dace (*Rhinichthys atratulus*) (Fig. 17.7). Among commonest fishes in flowing water and pools of cool, clear, gravelly or rocky headwaters, creeks, and small rivers of high to moderate gradient; may shelter on bottom under or beside stones, or under banks in deepest water in winter; spawns over gravel in shallow riffles, sometimes over fallfish nests.

Longnose dace (*R. cataractae*). Common in medium-sized, gravel- or boulder-bottomed clear streams (I caught several by dislodging rocks in flowing water just upstream from my dip net); spawns probably in gravelly riffles.

Creek chub (*Semotilus atromaculatus*). Native to streams in western Connecticut; some populations from Connecticut River eastward apparently introduced; spawns in smooth water near gravelly riffle, where males make pit in bottom, cover eggs with stones, and may stand guard.

Fallfish (*Semotilus corporalis*). Common in clear, medium-sized streams, also in lakes; males build large pebble nests on bottom in stream pools and lake shoal areas.

Tench* (*Tinca tinca*). Recently established in Bantam Lake.

Suckers (Order Cypriniformes, Family Catostomidae)

Longnose sucker (*Catostomus catostomus*). Uncommon in clear stream waters of the upper Housatonic River system; spawns over gravel in spring.

Fig. 17.7. Blacknose dace are stream-dwelling fishes.

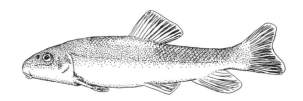

Fig. 17.8. White suckers migrate seasonally between lakes and spawning streams.

White sucker (*C. commersoni*) (Fig. 17.8). Common in streams; migrates up tributary streams to spawn in riffles (or sometimes in shoal areas of lakes) in spring (peak as early as late April in some areas; May or June in the Farmington and Salmon rivers, respectively); inhabits larger streams and lakes at other times; mature individuals in Connecticut commonly reach total lengths of 15 to 20 inches and live up to 20 to 25 years; flexible diet includes both zooplankton and large bottom-dwelling invertebrates.

Creek chubsucker (*Erimyzon oblongus*). Clear rivers and lakes; migrates upstream in spring to spawn; eats invertebrates and plant material.

North American Freshwater Catfishes (Order Siluriformes, Family Ictaluridae)

Catfishes spawn in late spring or early summer and lay masses of eggs in sheltered nests on the bottom. One or both parents guard the schools of young. Catfishes feed mostly at night and eat bottom invertebrates, fish eggs, and small fishes. Some species have toxic skin associated with the stout dorsal and pectoral spines; a prick may be painful but is not dangerous. Most species can live 10 years or more.

White catfish* (*Ameiurus catus*). Introduced and established in rivers (including brackish water), lakes, and ponds.

Yellow bullhead* (*A. natalis*). Recently introduced and established in ponds and streams.

Fig. 17.9. The brown bullhead is Connecticut's only native catfish.

Brown bullhead (*A. nebulosus*) (Fig. 17.9). Native to rivers, lakes, and ponds; I often encountered dense schools of juveniles in July.

Channel catfish* (*Ictalurus punctatus*). Recently introduced and now abundant in the Connecticut River; has been found in other rivers, probably will continue to colonize suitable habitat; recognized by scattered dark spots and forked tail fin.

Pikes (Order Esociformes, Family Esocidae)

Pikes, fast-growing, elongate, toothy fishes, are voracious predators on invertebrates, fishes, and amphibians. Spawning females scatter their eggs in shallow water in early spring. Spawned eggs sink and stick to vegetation or the bottom.

Redfin pickerel (*Esox americanus*). Common, mainly in quiet water among vegetation in small lowland streams, also lakes and ponds; spawns in thick vegetation in flooded areas or water-edge shallows.

Northern pike* (*E. lucius*). Not native to Connecticut, but present since at least the 1850s; pike introduced into Connecticut River in Massachusetts probably moved south into Connecticut; now resides in state's major rivers, several lakes, and reservoirs; spends winter in deep pools of rivers, moves into marshes and shrub swamps to spawn. *Note:* Marshes at Haddam Meadows State Park, Bantam Lake, and Mansfield Hollow Reservoir are managed as spawning areas. At Haddam Meadows, pike are trapped along the Connecticut River and placed in the marsh to spawn, then are removed using gill nets. In early summer, marsh waters are drawn down and the fingerling pike are trapped and stocked in other sites. What is the impact of managing these wetlands to propagate an exotic species?

Chain pickerel (*E. niger*) (Fig. 17.10). Vegetated areas of rivers, lakes, reservoirs, and ponds; tolerates warm water, high acidity, and brackish water.

Fig. 17.10. Chain pickerels occur widely in vegetated fresh waters.

Mudminnows (Order Esociformes, Family Umbridae)

Central mudminnow* (*Umbra limi*). Introduced and established in the Connecticut River drainage; spawns in spring in submerged areas of floodplains and at mouths of small tributary streams; lives on or near bottom, favors slow water with dense vegetation and soft bottoms; tolerates warm water and low oxygen; eats benthic aquatic insects and crustaceans (sometimes small fishes).

Smelts (Order Osmeriformes, Family Osmeridae)

Rainbow smelt (*Osmerus mordax*). Schools mainly in coastal shoal waters and river mouths; trawl samples in Long Island Sound contained this fish most often near river mouths in April; anadromous populations migrate from the Sound to spawning streams in late winter or early spring; spawns in fresh or brackish water in April and May in Connecticut River, Salmon River, and Eight Mile River; eggs sink and stick to objects on bottom; introduced populations exist in some lakes, spawn over bottoms of tributary streams; diet includes small crustaceans, polychaetes, and other invertebrates and, to a lesser extent, small fishes; excellent food fish.

Salmonids (Order Salmoniformes, Family Salmonidae)

Most salmonids are important in commercial and recreational fisheries. Hundreds of thousands of catchable-size trout (mainly brown but also rainbow and brook) are stocked in approximately 300 streams and 80 lakes and ponds in Connecticut each year, mainly from March through May. In addition, hundreds of thousands of kokanee fry are stocked each year in a few lakes. And many Atlantic salmon fry and smolts are released as part of a large multi-state restoration effort. The traditional opening day of the trout season is the third Saturday in April beginning at 6 A.M.

Rainbow trout* (*Oncorhynchus mykiss*). Numerous introductions in Connecticut, with minimally reproducing populations established in a few relatively deep lakes with suitable tributary spawning streams; spawns in early spring; zooplankton, midge larvae, other invertebrates, and small fishes are primary foods.

Sockeye salmon* ("kokanee") (*O. nerka*). Cool-water fish, stocked each year in East Twin Lake, Lake Wononscopomuc, and West Hill Pond; in East Twin Lake, kokanee spawned on gravel bottoms in lake shallows and in a tributary stream mouth, mid-October through November; fine gill rakers used for filter feeding on zooplankton; also may eat midge larvae and pupae and other insects. *Note:* Illegal introduction of plankton-eating alewives into East Twin and Wononscopomuc lakes in the early 1990s quickly led to the disappearance of kokanee from those waters. The alewives outcompeted the kokanee for food.

Round whitefish* *(Prosopium cylindraceum).* Introduced into East Twin Lake, may be self-perpetuating there; spawns in lakes over rocks or gravel in fall; bottom feeder, eats invertebrates, fishes, and fish eggs.

Atlantic salmon *(Salmo salar).* Adults leave the ocean and migrate up natal rivers in early spring, some not until early fall; spawns in late October or November, followed by hatching in early spring; young spend one or two years feeding on insects in streams before migrating to the ocean; spends usually two winters in the ocean eating fishes and crustaceans before returning to its natal stream to spawn; salmon from New England migrate to ocean areas off the coasts of Canada and Greenland, where they feed on fishes and crustaceans; migrating adults do not feed when in streams, and, unlike Pacific salmon, they may return to spawn in subsequent years. *Note:* Atlantic salmon historically migrated up Connecticut rivers to spawning streams, but the species was extirpated through the effects of dams and pollution. Over the past few decades, a restoration effort has been underway in the Connecticut River system (and to a limited extent in the Thames River drainage), but as yet the number of salmon returning to the Connecticut River and its tributaries has been small. The total number of salmon reaching fishway counting stations was 239 in 1996, 153 in 1999, and only about 40 to 50 in 2001 and 2002. Most of these salmon are removed for use in an effort to develop a population that will exhibit migratory fidelity to this drainage system. The Kensington Hatchery produces 800,000 to 900,000 Atlantic salmon fry each year. The fry are stocked into streams in the Farmington River and Salmon River watersheds. Broodstock salmon that are no longer needed for egg production are released into the Naugatuck River and Shetucket River for recreational fishing. In the headwaters of the Farmington River system, some stocked fry actually mature as "parr" and sometimes successfully spawn with female brown trout, producing hybrid offspring.

Brown trout* *(S. trutta).* Introduced and established resident in streams, with a few anadromous populations also established in coastal streams; prefers deeper, slower, more fertile segments of rivers but cannot tolerate temperatures above about 81°F; spawns over gravel in fall or early winter; female buries eggs; eats insects, crustaceans, fishes, and amphibians.

Brook trout *(Salvelinus fontinalis)* (Fig. 17.11). Common in cool to cold, small to medium-sized streams; readily re-

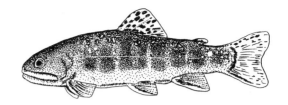

Fig. 17.11. Brook trout are native to cool streams.

colonizes and occupies intermittently flowing stream segments that dry up in summer, surviving in pools; spawns in fall; females deposit eggs in pits dug in gravel; prey includes aquatic and terrestrial insects and sometimes fishes and other small vertebrates.

Lizardfishes (Order Aulopiformes, Family Synodontidae)

Inshore lizardfish *(Synodus foetens).* Sometimes present in coastal waters, generally in summer or early fall; uses sharp teeth as ambush predator on small fishes.

Cusk-Eels (Order Ophidiiformes, Family Ophidiidae)

Striped cusk-eel *(Ophidion marginatum).* Burrowing, nocturnal, warm-water fish; uncommon in the Sound during the warmer months; adult females produce buoyant egg masses over extended spawning season; diet includes fishes and crustaceans.

Silver Hakes (Order Gadiformes, Family Merlucciidae)

Silver hake (whiting) *(Merluccius bilinearis)* (Fig. 17.12). Fast swimmer, sometimes in loose schools; year-round resident, primarily in western and central Sound, often over mud bottoms in deep water; abundance in trawl samples peaks in late spring and early summer, minimal in late summer; feeds mainly at night on invertebrates (mainly shrimps and other crustaceans), fishes, and squid; spawns offshore

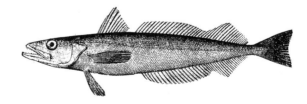

Fig. 17.12. Silver hake occurs year round in Long Island Sound. H. L. Todd, from Collette and Klein-MacPhee (2002), with permission.

in the ocean (e.g., south of Long Island and Block Island) in late spring and summer.

Cods (Order Gadiformes, Family Gadidae)

Fourbeard rockling (*Enchelyopus cimbrius*). Bottom fish, rests in burrows in daytime; usually on mud bottoms in shallow or deep water in western and central Sound; peak abundance in trawl samples in spring and summer; spawns mainly from February through May; pelagic eggs and small larvae; eats mainly crustaceans.

Burbot[*?] (*Lota lota*). Known from the Connecticut and Housatonic river drainages; status poorly known; inhabits lakes or rivers with good cover; spawns in winter, even under ice; feeds on various invertebrates and fishes.

Atlantic tomcod (*Microgadus tomcod*). Hardy fish, close to shore in the Sound; rare in trawl samples, with most captures in April and May; spawns over bottom in brackish water in late fall and winter; eggs sink and stick to objects on or near bottom; short-lived; uses sensitive chin barbel and pelvic fins to find bottom-dwelling shrimps, amphipods, mollusks, polychaetes, and fish fry. *Note:* Tomcod in polluted estuaries may contain high levels of PCBs, pesticides, and toxic heavy metals.

Pollock (*Pollachius virens*). Rare in Long Island Sound; larvae have been netted in the Sound in early spring, juveniles in summer.

Red hake (squirrel hake, ling) (*Urophycis chuss*). Adults occur seasonally over soft bottoms of the Sound, particularly in deeper water; juveniles up to about two years old may be resident; avoids warm water, so uncommon in summer and early fall; avoids shallows in winter; spawns offshore out in the ocean, apparently from late spring to fall; juveniles drift at surface, then move into mantle cavities of sea scallops for a few months, later become free living; primary foods include various crustaceans, polychaetes, and small fishes, taken mainly at night.

Spotted hake (*U. regia*). Seasonal visitor in the Sound; tends to avoid shallow warm-water areas; most common in trawl samples in deeper waters in late summer and early fall; often associated with bottom cover; crustaceans, fishes, and squids dominate diet; spawns offshore from late summer to early spring.

Toadfishes (Order Batrachoidiformes, Family Batrachoididae)

Oyster toadfish (*Opsanus tau*) (Fig. 17.13). Bottom dweller among eelgrass, rock or shell reefs, or other cover in

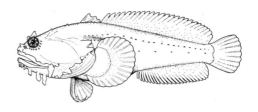

Fig. 17.13. Oyster toadfishes feed by ambushing crustaceans and fishes. Louella E. Cable, from Bigelow and Schroeder (1953).

the Sound; may move to deeper water for winter, but rare there in warmer months; in late spring or early summer, male guards sheets of eggs that female lays under objects such as shells or human artifacts; short-range ambush predator on various invertebrates (especially small crabs and shrimps) and small fishes. *Note:* Toadfishes ambush prey via a short, sudden lunge and quick gape and suck of the powerful jaws (handle carefully, they can inflict a painful bite). They produce loud sounds; both sexes grunt, and males whistle, especially during the mating season.

Goosefishes (Order Lophiiformes, Family Lophiidae)

Goosefish (*Lophius americanus*). Rare in Long Island Sound, represented primarily by juveniles on mud and mud-sand bottoms of central basin in spring, with fewer present in summer and fall; spawns outside Long Island Sound in late spring; females produce long (many feet) floating mat of purplish-brown eggs; uses huge mouth to engulf fishes and various invertebrates; fleshy "lure" on first dorsal spine, extended above mouth, entices fishes to their demise; adults can ingest huge meals that may include one or more waterbirds.

Mullets (Order Muguliformes, Family Mugilidae)

Striped mullet (*Mugil cephalus*). Uncommon in Long Island Sound in summer and fall; spawns offshore in large groups in winter, from Cape Hatteras southward; mullets have long gut, reflecting the largely herbivorous (algae, detritus) diet.

Fig. 17.14. Atlantic silverside. H. L. Todd, from Collette and Klein-MacPhee (2002), with permission.

White mullet *(M. curema)*. Uncommon in Long Island Sound in summer and fall; spawns offshore south of Cape Hatteras in spring and summer.

Silversides (Order Antheriniformes, Family Atherinopsidae)

Inland silverside *(Menidia beryllina)*. Most abundant in brackish water, also occurs in freshwater portions of some coastal streams; spawns among vegetation or debris in shallow waters of upper estuaries in spring and summer; eats primarily crustaceans and fish eggs and larvae; serves as food for striped bass, bluefish, and probably herons and various other water birds.

Atlantic silverside *(M. menidia)* (Fig. 17.14). Very common schooling fish year round in coastal shoreline waters; usually the most abundant fish in my seine samples from coastal beach shallows; some winter in deeper water; various small invertebrates (mainly crustaceans, including zooplankton) dominate the diet; spawns in grassy shallows during daytime high tides in late spring and early summer; eggs adhere to cordgrass stems or algae; temperature of developing embryo affects sex of fish, with warmer temperatures producing relatively more males; short-lived, most survive a year or less; important food for many larger sport fishes and seabirds (terns). *Note:* Compared to other abundant shoreline fishes such as mummichogs and killifishes, silversides are relatively sensitive and often quickly perish in a seine or bait bucket.

Needlefishes (Order Beloniformes, Family Belonidae)

Atlantic needlefish *(Strongylura marina)*. Inshore near surface in Long Island Sound, mainly in summer and fall; marsh edges and vegetated shallows serve as nurseries for young; juveniles prey on crustaceans; adults eat small fishes; spawns probably in late spring or early summer; eggs sink and attach to vegetation.

Killifishes (Order Cyprinodontiformes, Family Fundulidae)

Killifishes are relatively small fishes that are extremely tolerant of high temperatures, high salinity, and low oxygen. They easily survive extremes that kill other fishes. Some can survive out of water for at least 90 minutes.

Banded killifish *(Fundulus diaphanus)* (Fig. 17.15). Widespread in shallows of ponds, lakes, and streams, including

Fig. 17.15. Banded killifishes can be plentiful in tidal freshwater streams.

some brackish river mouths; I caught many by seining over muddy freshwater tidal mud flats in Connecticut River tributaries; distribution in state has expanded through stocking as forage fish; spawns in late spring and summer; eggs become entangled in vegetation; diet includes small invertebrates.

Mummichog *(F. heteroclitus)* (Fig. 17.16). Abundant in schools in shallow coastal shoreline waters, salt marshes, low-salinity stream mouths, and fresh streams draining into salt marshes; in winter, may burrow into bottom mud of marsh pools or move to deeper water near shore; eats mainly small invertebrates and fish eggs, often ingests detritus; females deposit sticky eggs in high marsh near high-tide level, among algae, shells, leaves, or sand, with reproductive peaks at highest tide levels in late spring and summer; eggs hatch two weeks later during next spring tide; important as food for herons, egrets, terns, other seabirds, crabs, and various larger fishes; life span normally not more than a few years; commonly used as bait.

Striped killifish *(F. majalis)*. Abundant in schools in shallow coastal shoreline waters and salt marsh creeks; habitat may shift to somewhat deeper waters in winter, rarely occupies fresh waters of lower parts of rivers during warmer months; primary foods include small crustaceans, snails, clams, and polychaetes (on a rising tide in late summer, I saw many plunging their long snouts into and apparently ingesting prey buried in soft sediments of tidal shallows); spawns in late spring and summer, perhaps with peaks

Fig. 17.16. Mummichogs are abundant in salt marshes and associated tidal creeks.

during spring tides; females bury eggs in bottom substrate near low tide level or in intertidal pools; common food item for many coastal birds and larger fishes.

Rainwater killifish (*Luciana parva*). Shallow coastal shoreline waters (sometimes lower parts of streams), usually among vegetation; travels in schools; spawns in late spring or summer; eats small invertebrates.

Pupfishes (Order Cyprinodontiformes, Family Cyprinodontidae)

Like killifishes, cyprinodontids are extremely tolerant of high temperatures, high salinity, and low oxygen.

Sheepshead minnow (*Cyprinodon variegatus*) (Fig. 17.17). Common in shallow shoreline waters of Long Island Sound and adjoining marshes; may enter fresh water during warmer months; often in large schools over mud bottoms, impressively tolerant of warm, stagnant water; may burrow into muddy bottoms in winter; in spring and summer, females deposit eggs in shallow-water nest depressions guarded by adult male; omnivorous, eats small animals and algae; serves as food for many larger predatory fishes, wading birds, and plunge-diving least terns, thus making marsh nutrients available to the more "glamorous" parts of the ecosystem.

Sticklebacks (Order Gasterosteiformes, Family Gasterosteidae)

Sticklebacks are well known for interesting courtship behavior and elaborate plant-material nests made by territorial males. Females deposit eggs in the nest, and males guard the eggs and (in most species) the young. Small invertebrates, larval fishes, and fish eggs dominate the diet. Lifespan is short, only one to three years at most.

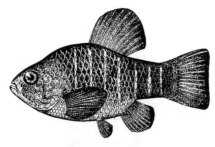

Fig. 17.17. Sheepshead minnows exploit the shallowest marsh waters of Long Island Sound. A. H. Baldwin, from D. S. Jordan and B. W. Evermann (1896–1900). The fishes of Middle and North America: A descriptive catalogue of the species of fish-like vertebrates found in the waters of North America, north of the Isthmus of Panama. *United States National Museum Bulletin 47*.

Fig. 17.18. Fourspine sticklebacks live in salt marshes and associated creeks and coves. A. H. Baldwin, from D. S. Jordan and B. W. Evermann (1896–1900). The fishes of North and Middle America. *United States National Museum Bulletin 47*.

Fourspine stickleback (*Apeltes quadracus*) (Fig. 17.18). Common among clumps of algae in shallow coastal shoreline waters and edges of adjoining marshes and lower parts of coastal streams; spawns in fresh and brackish water in late spring and early summer.

Threespine stickleback (*Gasterosteus aculeatus*). Shallow coastal waters, tidal marsh pools, and lower parts of coastal streams; usually close to vegetation or floating masses of algae or eelgrass, also in open ocean waters; enters vegetated stream mouths in spring to spawn.

Blackspotted stickleback (*G. wheatlandi*). Rare in estuaries (bays, salt marshes); small range from Newfoundland to New York; spawns in late spring or early summer.

Ninespine stickleback (*Pungitius pungitius*). Generally in dense vegetation in shallow coastal estuarine waters, tidal marsh pools, and adjacent freshwater habitats, including some freshwater lakes and ponds; perhaps catadromous—in spring, may move from fresh water to estuarine waters to spawn in quiet vegetated areas.

Pipefishes and Seahorses (Order Gasterosteiformes, Family Syngnathidae)

These are cryptic, bottom-dwelling ambush predators that use their small toothless mouth to suck in zooplankton, benthic invertebrates, and fish eggs and fry. Males have a brood pouch into which the female deposits her eggs. The male carries the eggs and tiny hatchlings for up to several weeks.

Lined seahorse (*Hippocampus erectus*). Uncommon in summer among algae or eelgrass in nearshore waters of Long Island Sound.

Fig. 17.19. Northern pipefishes generally stay hidden among eelgrass or algae in Long Island Sound. From Bigelow and Schroeder (1953), after Bigelow and Welsh (1925).

Northern pipefish *(Syngnathus fuscus)* (Fig. 17.19). Somewhat resembles a straightened-out seahorse; I netted pipefish among eelgrass and algae in shallow shoreline waters of Long Island Sound in summer and early fall; winters in eelgrass shallows and in deeper waters; spawns in spring and summer.

Searobins (Order Scorpaeniformes, Family Triglidae)

Northern searobin *(Prionotus carolinus)*. Common bottom fish, on sand in shallow and deep water of Long Island Sound, sometimes swims higher in water column; can bury itself in sand, with just eyes projecting; present mainly in warmer months of late spring to fall; moves to deeper water offshore and to south for winter; feeds on various bottom invertebrates (e.g., crabs, polychaetes) and small fishes; "walks" and probes bottom for food using sensitive, pectoral fin rays that are free from huge winglike fins; spawns throughout Long Island Sound in late spring and summer (June to August); eggs float; often hooked by beach anglers casting bait for bluefish or striped bass.

Striped searobin *(P. evolans)*. Common on sand, mud, or rock bottoms of Long Island Sound, with peak abundance in trawl samples in western and central Long Island Sound from late spring through fall; young juveniles appear in trawl samples in August; habits similar to those of northern searobin, though prey tend to be larger and include more fishes.

Sculpins (Order Scorpaeniformes, Family Cottidae)

Sculpins have few or no scales but tend to be prickly, and they generally feed on available bottom invertebrates and sometimes small fishes and carrion.

Slimy sculpin *(Cottus cognatus)*. Cool, clear, gravelly streams, mainly in northern part of state; in spring, females deposit eggs on undersides of rocks or logs, where each nest is guarded by a territorial male; male may tend clutches of multiple females.

Fig. 17.20. Grubbies employ their large mouth in capturing invertebrates in the shallows of Long Island Sound.

Grubby *(Myoxocephalus aeneus)* (Fig. 17.20). Common on bottom among algae in shallow coastal shoreline waters; scarce in deep waters, at least from spring to fall; in fall at night I found many among or near clumps of algae in large sand-bottomed tidepools; spawns in winter; eggs sink and stick to objects.

Longhorn sculpin *(M. octodecemspinosus)*. Bottom fish at various depths in Long Island Sound; like shorthorn sculpin, avoids warmest shallow waters in summer, though less of a cold-water fish than shorthorn; adults may move up brackish coastal rivers from fall to spring; spawns in estuaries in late fall and winter; eggs sink and stick to various objects.

Shorthorn sculpin *(M. scorpius)*. On bottom in Long Island Sound; avoids warm water, so absent from shallow water in summer; winters away from shallows, at least in coldest locations; spawns on the bottom in late fall and winter; males guard the eggs until they hatch.

Sea Ravens (Order Scorpaeniformes, Family Hemitripteridae)

Sea raven *(Hemitripterus americanus)*. Rocky or otherwise firm bottoms of Long Island Sound; avoids warm water, moves into shallow waters only during colder months, with peak numbers from late fall to early spring; spawns in fall, often deposits eggs among sponges; eats available bottom invertebrates and fishes.

Lumpfishes (Order Scorpaeniformes, Family Cyclopteridae)

Lumpfish *(Cyclopterus lumpus)*. Adults live over hard bottoms of Long Island Sound, such as among algae-covered rocks; sometimes they and juveniles inhabit masses of floating seaweed; attaches to rocks and objects such as lobster traps by means of ventral sucker; diet includes various invertebrates, including crustaceans, comb jellies, and jellyfish, and small fishes; spawns in inshore waters in late spring and early summer; males establish nests, guard and aerate eggs, which clump together and stick to bottom; predators include harbor seals.

Temperate Basses (Order Perciformes, Family Moronidae)

White perch *(Morone americana)*. Schools in coastal shoreline waters, lagoons, and brackish river mouths, and in freshwater streams, ponds, and lakes, often over bottoms of sand, silt, mud, or clay; some populations migrate seasonally between saltwater and freshwater spawning areas; distribution in state has been expanded through intentional introductions; feeds opportunistically on various invertebrates (often midges in lakes) and, as they grow larger, small fishes; takes a few years to mature; spawns between mid-spring and early summer (May to June in the Connnecticut River, about a month later in tributaries and coves) in fresh or brackish water, including streams, lake inlets, and along lake shores; spawning in Connecticut River occurs mainly between Cromwell and Enfield Dam; in winter, most are inactive in deepest parts of bays or streams.

Striped bass (*M. saxatilis*) (Fig. 17.21). Common in Long Island Sound, often close to land off rocky shores, in spring and fall; scarce in summer; some enter lower parts of major coastal rivers and many go far up Connecticut River; most winter along coast south of New England, but some winter in Long Island Sound or in major rivers; migrants that enter the Sound generally are at least a few years old; impressive predator, diet includes various invertebrates (especially crustaceans) and fishes (for example, silversides, menhaden, and grass shrimp); in recent decades, Long Island Sound anglers have caught individuals weighing up to 75 or 76 pounds. *Note:* Populations along the Atlantic coast have rebounded from low levels in the 1970s and 1980s, when populations may have suffered from overfishing, pollution, and habitat degradation. In 2002, around 1,100 stripers made it as far as the Holyoke Dam fishlift in Massachusetts. Striped bass in Long Island Sound spawn in late spring in brackish or fresh water of the Hudson River (New York), the breeding area closest to Connecticut, and southward to Chesapeake Bay and elsewhere. Most spawning females are at least five years old. Management of this species is conducted through the Interstate Fishery Management Plan for Atlantic Striped Bass, administered by the Atlantic States Marine Fisheries Commission.

Sea Basses and Groupers (Order Perciformes, Family Serranidae)

Black sea bass (*Centropristis striata*). Adults occur in Long Island Sound over rocky bottoms and around pilings and shipwrecks, mainly in warmer months; juvenile habitat includes eelgrass beds, shelly bottoms, wharves, and margins of shoreline jetties; winters in deeper waters offshore and to south; eats various benthic invertebrates, especially crustaceans and mollusks, and small fishes; spawns in late spring and summer; young-of-the-year have been caught in trawl nets in fall; most individuals are protogynous hermaphrodites—they change sexual function as they age and grow, beginning as female and later changing to male; by an age of two to three years, most are mature males. *Note:* Along the Atlantic coast, this fish is jointly managed by the Atlantic States Marine Fisheries Commission and the Mid-Atlantic Fishery Management Council, which set commercial quotas to ensure that populations are not overharvested.

Sunfishes (Order Perciformes, Family Centrarchidae)

Sunfishes often spawn in easily seen colonies. Adult males generally clear fine sediments from saucerlike nesting areas on the bottom in shallow water, where females deposit eggs between mid-spring and mid-summer. Males guard the eggs and young. Invertebrates are important in the diet, and some ingest plant material, whereas large basses often focus their attacks on crayfishes, fishes, and amphibians.

Rock bass* (*Ambloplites rupestris*). Introduced and established in ponds, lakes, and especially rivers with rocky bottom, aquatic vegetation, and good flow.

Banded sunfish (*Enneacanthus obesus*). Coastal plain species, primarily in sluggish, well-vegetated streams, backwaters, and lowland lakes in central and eastern Connecticut; rare here, ecologically poorly known; builds nests in vegetation rather than in sandy or gravelly bottoms.

Fig. 17.21. Striped bass are top predators in Long Island Sound and major rivers.

Fig. 17.22. Redbreast sunfish favor quiet water of lakes and slow streams.

Fig. 17.23. Pumpkinseeds (juvenile shown) are native, widespread, and common in lakes, ponds, and slow streams statewide.

Redbreast sunfish (*Lepomis auritus*) (Fig. 17.22). Lakes and slow currents in streams; spawns in vegetated shallows.

Green sunfish* (*L. cyanellus*). Introduced and established in ponds and slow streams with sheltering aquatic vegetation; more tolerant of turbid water than other sunfishes.

Pumpkinseed (*L. gibbosus*) (Fig. 17.23). Common in ponds, lakes, and slow areas of streams, most often among aquatic vegetation; may nest in loose colonies.

Bluegill* (*L. macrochirus*). Introduced and established in ponds, lakes, and slow currents of streams; commonly nests in dense colonies (Fig. 17.24).

Smallmouth bass* (*Micropterus dolomieu*). Introduced and established in lakes with rocky shoals and cool streams with deep pools, good cover, and rocky substrate; I caught some in water-celery beds in silty-sandy tidal river shallows.

Largemouth bass* (*M. salmoides*). Introduced and established in ponds, lakes, and slow areas of streams, especially in warm, vegetated waters; nests often in deeper water (but generally less than 5 feet) than that selected by breeding sunfishes (*Lepomis* spp.); nests well spaced (usually at least 30 feet apart); "sit-and-wait" foraging often occurs in shallows near vegetation, but in reservoirs lacking vegetated littoral zone may switch to active foraging on crayfish and open-water fishes.

White crappie* (*Pomoxis annularis*). Introduced and apparently established in Connecticut River drainage; lakes, ponds, and slow streams, often in turbid water.

Black crappie* (*P. nigromaculatus*). Introduced and established in ponds, lakes, and slow streams; prefers clear water and aquatic vegetation.

Perches and Darters (Order Perciformes, Family Percidae)

Swamp darter (*Etheostoma fusiforme*). Vegetated slow streams, lakes, and ponds east of mainstream Quinebaug River in eastern Connecticut (a few occurrences just west of Quinebaug may represent introductions); prefers ponds with vegetation and detrital sediments; small stream populations probably depend on pond populations for recruitment; tolerates low oxygen levels and high acidity; eggs laid among submerged plants in spring; eats mainly invertebrates including small zooplankton; important food items in Connecticut include midges and Cladocera; few live longer than a year. *Note:* This darter may have entered New England via the eastern tip of Long Island very late in postglacial time; the distribution in Connecticut may be a result of dispersal through former (postglacial) and current stream systems.

Tessellated darter (*E. olmstedi*) (Fig. 17.25). Common in sandy shallows of streams, also occurs in a few lakes and ponds; in late spring, females lay eggs on underside or sides of submerged objects; male guards eggs, sometimes of multiple females; eats invertebrates such as small crustaceans and insects (e.g., fly larvae, caddisfly larvae).

Yellow perch (*Perca flavescens*) (Fig. 17.26). Common in shallows and deep water of ponds, lakes, reservoirs, and slow areas of large streams; schooling, mostly day-active species; spawns in shallows in April and May in ponds, lakes, and streams and sometimes as early as March in coves along Connecticut River; eggs enclosed in large, conspicuous, accordianlike masses of jelly that expand and may drift

Fig. 17.24. Bluegill clear fine sediments from nest sites, which are often in dense colonies in shallow water.

Fig. 17.25. Tessellated darters inhabit stream bottoms.

Ray-finned Fishes | 299

Fig. 17.26. Yellow perch is a native inhabitant of quiet fresh water. Allan Smits.

and become tangled in shoreline vegetation (Fig. 17.27); diet includes zooplankton, other invertebrates, and small fishes; may feed in winter, so often caught by ice anglers.

Walleye* (*Stizostedion vitreum*). Introduced and established in Connecticut and Housatonic rivers and a few lakes; stays near bottom in daytime, moves into shallows to feed at night; big-eyed sight feeder, may feed in daytime in turbid water and at twilight in clear lakes; small fishes, aquatic insects, and crayfish are important adult foods; spawns in moving water of rocky-gravelly bars and shoals in early spring. *Note:* The Department of Environmental Protection recently has stocked fingerling walleyes in a number of lakes, including Gardner Lake, Rogers Lake (discontinued), and Squantz Pond. Lake Saltonstall and Saugatuck Reservoir recently have been stocked by regional water companies.

Bluefishes (Order Perciformes, Family Pomatomidae)

Bluefish (*Pomatomus saltatrix*) (Fig. 17.28). Warm-water fish, migrates into Long Island Sound in spring (not in numbers until June), stays until fall; juveniles show up in July, may enter lower parts of coastal rivers in warmer months; usually occur in schools; most likely to come close to shore and into river mouths during twilight hours; winters mainly in ocean off southeastern United States; young eat zooplankton, shrimp, and small fishes; adults eat fishes, squids, and crabs; overall, fishes, including larval forms, are the primary prey; predatory attacks often result in schooling prey leaping out of water in "boiling" masses easily seen from shore; not much known about spawning, probably occurs offshore in spring and summer; eggs drift in ocean currents. *Note:* Bluefish abundance decreased significantly during the 1990s.

Jacks (Order Perciformes, Family Carangidae)

Jacks are warm-water fishes that eat small fishes and various invertebrates. Adults of all of the following are generally scarce in Long Island Sound, but juveniles can be fairly common in late summer and early fall: yellow jack (*Caranx bartholomaei*), crevalle jack (*C. hippos*), mackerel scad (*Decapterus macarellus*), bigeye scad (*Selar crumenophthalmus*), Atlantic moonfish (*Selene setapinnis*), lookdown (*S. vomer*), banded rudderfish (*Seriola zonata*), and rough scad (*Trachurus lathami*).

Porgies (Order Perciformes, Family Sparidae)

Scup (porgy) (*Stenotomus chrysops*). Abundant in Long Island Sound; adults migrate, usually in schools, into shallow waters of Long Island Sound for warmer months of spring to fall; juveniles sometimes overwinter in Long Island Sound, but adults of this cold-sensitive species tend to move to deep water offshore; spawns buoyant eggs in late spring or early summer (peak in June) along southern New En-

Fig. 17.27. Yellow perch egg masses appear in vegetated shallows in late winter and early spring.

Fig. 17.28. Bluefish are well-equipped as migratory predators in Long Island Sound.

Summer flounder (fluke) *(Paralichthys dentatus)*. Present in Long Island Sound on sandy or muddy bottoms, mainly from June through October; tends to avoid shallow water during warmest months; moves offshore for winter; diet includes fishes and various invertebrates, especially crustaceans (e.g., sand shrimp); big-mouthed visual predator, often ambushes prey while partially buried in sandy bottom, also actively pursues prey throughout water column; spawns in ocean, beginning in late summer and continuing into winter in deeper waters; eggs float in current; juveniles move into estuaries from fall through spring. *Note:* Concern about overharvest of this commerically and recreationally important fish prompted recent fishing restrictions by NOAA and the Atlantic States Marine Fisheries Commission. For example, in 2001 the summer flounder commercial quota for Connecticut was harvested by mid-August, and further commerical landings of this species in Connecticut were prohibited for the remainder of the calendar year.

Fourspot flounder *(P. oblongus)*. Common on mud to sand bottoms in Long Island Sound, mainly in western and central parts; abundance in trawl samples peaks in June and is low from fall to early spring; spawns apparently offshore from late spring to early fall; eats mainly squids, crustaceans, and small fishes.

Righteye Flounders (Order Pleuronectiformes, Family Pleuronectidae)

Yellowtail flounder *(Limanda ferruginea)*. Sand-mud bottoms of Long Island Sound, but this generally offshore fish avoids shallow, warm water in summer; small-mouthed, eats crustaceans, mollusks, worms, and occasionally small fishes; spawns in late winter or early spring. *Note:* Populations in the region have declined over the past few decades.

Winter flounder *(Pleuronectes americanus)* (Fig. 17.31). Year-round day-active resident of Long Island Sound, most often on soft, muddy or sandy bottoms; adults are present in largest numbers during cooler months from

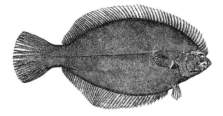

Fig. 17.31. Winter flounder is the most common bottom fish in Long Island Sound. H. L. Todd, from Collette and Klein-MacPhee (2002), with permission.

Fig. 17.32. This juvenile hogchoker was in the Connecticut River in Haddam in late July.

November to May; nestles into bottom sediments, may change color to match background; opportunistic, small-mouthed, sight feeder, eats various invertebrates, such as amphipods, polychaetes, smaller mollusks, or whatever is readily available, and sometimes small fishes; spawns in Long Island Sound in shallow water over sand or algae in winter and early spring; eggs sink to bottom; larvae are common prey for jellyfishlike hydromedusa stage of clapper hydroid *(Sarsia tubulosa)*. *Note:* This economically valuable fish has declined in abundance.

American Soles (Order Pleuronectiformes, Family Achiridae)

Hogchoker *(Trinectes maculatus)* (Fig. 17.32). Resident of Long Island Sound on bottom in nearshore waters in warmer months, in deeper waters in winter; also in brackish river mouths; preys mainly on crustaceans, worms, and clam siphons; spawns in brackish water, with peaks in late spring and early summer; in summer, larvae move upstream to low-salinity nursery areas near saltwater-freshwater contact zone; I found juveniles in tidal freshwater sections of the Connecticut River in Haddam; eyes on right side.

Leatherjackets (Order Tetraodontiformes, Superfamily Balistoidea)

These fishes have long faces and strong teeth that allow them to nip or crush hard-shelled, spiny, or clawed invertebrates such as sea urchins and crabs without exposing their eyes to damage from the prey. All of the following are uncommon to rare warm-season visitors to Long Island Sound: orange filefish *(Aluterus schoepfi)*, gray triggerfish *(Balistes capriscus)*, and planehead filefish *(Monacanthus hispidus)*.

Puffers (Order Tetraodontiformes, Family Tetraodontidae)

Northern puffer (*Sphaeroides maculatus*) (Fig. 17.33). Shallow waters of Long Island Sound, primarily in late summer and early fall; probably winters in deeper ocean waters; eats various small invertebrates (mostly crustaceans), taken mostly from bottom; spawns in late spring and summer, attaches eggs to bottom; renowned for ability to inflate with water or air. *Note:* The internal organs of puffers, and sometimes the flesh, may contain a highly potent neurotoxin.

Fig. 17.33. Northern puffers move into the Sound in late summer and early fall. H. S. Haines, from D. S. Jordan and B. W. Evermann (1896–1900). The fishes of North and Middle America. *United States National Museum Bulletin* 47.

Habitat

In Connecticut, fish diversity is greatest in estuarine habitats of Long Island Sound, even when rarities are excluded (Fig. 17.34). The state's rivers and creeks harbor more fish species than do lakes and ponds. The native freshwater fish fauna includes just a few dozen species. All of these inhabit streams; only half of the strictly freshwater species are also characteristic of ponds or lakes.

The state's native freshwater fish fauna is matched by a nearly equivalent number non-native freshwater species that are established in at least one site. Lakes have a relatively high proportion of exotic species. More than any other group of animals, our freshwater fish fauna has been highly altered by humans.

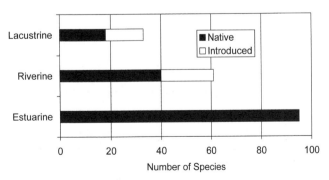

Fig. 17.34. Habitat associations of Connecticut fishes.

Ten native fish species migrate between salt water and required freshwater habitats. Dozens of primarily estuarine species that do not require fresh water nevertheless sometimes enter the lower portions of streams near Long Island Sound; these excursions generally are not mentioned in the preceding species accounts. Many additional marine species avoid fresh water altogether. In contrast to the freshwater fauna, the fish fauna of Long Island Sound consists virtually exclusively of native species, though the abundances of some of them have been greatly altered by human activities.

Food and Feeding

Most fishes feed opportunistically on available invertebrates and/or small fishes, with the frequency of fish in the diet increasing as individuals grow larger. But even small fishes may eat many tiny fish fry. Crustaceans often figure prominently in the diets of both estuarine and freshwater species. Examples of exceptions to these generalizations include sea lamprey (a fish parasite); herrings, shads, and bay anchovy (plankton eaters); and carp and killifishes (whose diet includes significant amounts of plant material and detritus).

Connecticut fishes also vary in foraging sites and methods. Sturgeons, skates, flounders, suckers, and catfishes tend to be bottom feeders. These often have sensitive barbels that project from the chin or snout (e.g., catfishes, sturgeons); suckers have a downward-facing, sucking mouth. Herrings, menhaden, and anchovies are open-water filter feeders or plankton selectors. Tautogs use their strong jaws to remove shellfish attached to rocks or pilings. Minnows, trout, and sunfishes are clear-water sight feeders. Killifishes move in with the tide and forage in very shallow waters of tidal marshes. Sharks, mackerels, bluefish, and jacks make quick-approach attacks that "panic" and disaggregate open-water schooling fishes, allowing the predators to pick off individuals. Any fish may deviate from its usual feeding mode when abundant prey (e.g., newly emerged mayflies) are available in atypical feeding locations. Turbid water may cause trout to abandon their usual habit of picking off drifting or low-flying insects and shift to benthic feeding.

Reproduction

Connecticut fishes exhibit a wide array of reproductive strategies. Some fishes are *nonguarders*—species that do not protect their eggs or young after spawning. Included are pelagic spawners, benthic spawners, and brood hiders. Pelagic spawners include schooling species that release large numbers of buoyant eggs in open waters, often near the surface (and near the bottom in shallow water). Shad, herrings, and anchovies are pelagic spawners. Benthic spawners, including stream fishes such as suckers, minnows, sturgeons, and smelts, as well as carp and pikes, may spawn in groups on the bottom or on aquatic plants. Eggs stick to the substrate or vegetation or fall into crevices among stones. Brood hiders, such as salmon, trouts, many minnows, and killifishes, may build nests and/or bury their eggs in the bottom substrate or under other cover, then abandon them.

Guarders go further than brood hiders and protect their eggs and embryos until they hatch. These fishes generally are territorial and have complex courtship behaviors. Many build simple to elaborate nests. Males are the guarders; they repel predators and, by fin and body movements, keep the eggs well-oxygenated and free of debris. Examples of guarders include sunfishes, largemouth and smallmouth basses, sticklebacks, sculpins, oyster toadfish, certain minnows, catfish, and darters.

Bearers carry their embryos either inside the reproductive tract in internal bearers (e.g., sharks) or in a special pouch in external bearers (pipefishes and seahorses). In internal bearers, egg fertilization often occurs within the female's body, and males of these fishes have the anal fin or pelvic fins modified into a copulatory organ. Requiem sharks have highly developed viviparous reproduction, with a placentalike connection between the mother and developing offspring. These fishes produce relatively small numbers of young.

Most freshwater fishes spawn in late spring or early summer. Some stream fishes, such as brook trout and Atlantic salmon, spawn in fall, as do certain saltwater fishes such as sea raven and ocean pout.

Numerous saltwater fishes of Long Island Sound spawn in late spring or early summer, but many others spawn in the colder winter months. Cold-season spawners include American eel, American smelt, rainbow smelt, Atlantic tomcod, pollock, spotted hake, grubby, longhorn sculpin, shorthorn sculpin, spot, striped mullet, Atlantic sand lance, Spanish mackerel, summer flounder, winter flounder, and yellowtail flounder. Some of these spawn offshore in areas south of Connecticut.

Reproduction sometimes is closely tied to migrations between different habitats. *Anadromous* species are those that use marine or estuarine waters for feeding and growth and spawn in freshwater. Anadromous fishes in Connecticut include sea lamprey, Atlantic sturgeon (which currently does not breed in freshwater in Connecticut), shortnose sturgeon, American shad, blueback herring, alewife, hickory shad, rainbow smelt (not obligately anadromous), Atlantic salmon, striped bass (visits but evidently does not now breed in freshwater in Connecticut), and white perch (not obligately anadromous). The American eel is a *catadromous* species, feeding and growing in fresh water and spawning in the ocean. Ninespine sticklebacks, another apparently catadromous species, also may move from freshwater to salt water to spawn.

Fishes in Winter

Overall, fishes experience more benign conditions in winter than do land-dwelling animals. Fishes can always find water that is above freezing, whereas land animals may have to cope with extremely low air temperatures that can be lethal.

Most fishes do not produce their own body heat and in fact comfortably exist at low body temperatures essentially equal to the surrounding water temperature. Their cold-adapted physiology allows them to avoid the incapacitating effects that cold water quickly exerts on humans. Of course, some fish species are more cold tolerant than others, but Connecticut's native fishes acclimate seasonally to temperature variations, and none of the winter residents are harmed by low temperatures. However, water temperature, particularly in summer, is not inconsequential. Small temperature changes of a few degrees, such as may result from water releases from an upstream impoundment, removal of streamside vegetation, or changes in climate or oceanographic conditions, can reduce or increase fish growth rates, delay, halt, or stimulate reproduction, or result in large changes in estuarine fish faunas as fish distributions shift with changing conditions.

Most fishes respond to cold temperatures in winter by reducing activity, and some species move from their shallow summer habitats to deeper water or farther south along the Atlantic coast where conditions are more stable or warmer. Some warm-water fishes do remain active and feed opportunistically in winter. Ice anglers take advantage of this and commonly hook yellow perch, bluegill, and pumpkinseed through the ice of frozen lakes.

Conservation

Several species of saltwater fishes (and shellfish) are managed by the Atlantic States Marine Fisheries Commission, which was formed by fifteen Atlantic coast states in 1942 to assist in managing and conserving their shared coastal fishery resources. Each state is represented by three commissioners, including the director for the state's marine fisheries management agency, a state legislator, and an individual representing fishery interests, appointed by the state governor.

The Commission's Interstate Fisheries Management Program (ISFMP) promotes the cooperative management of marine, estuarine, and anadromous fisheries through the development of interstate fishery management plans. In December 1993, the actions of the ISFMP were expanded further with the passage of the Atlantic Coastal Fisheries Cooperative Management Act, which provides a mechanism to ensure state compliance with mandated conservation measures in the Commission-approved fishery management plans. All member states that have a declared interest in a fishery must comply with certain conservation provisions of the plan, or the Secretary of Commerce may impose a moratorium in that state's waters for the harvest of the species in question.

The Commission's Sport Fish Restoration Program is aimed at improving fishery conservation and wise utilization of critical sport fisheries resources by promoting interstate and state/federal cooperation. These activities are coordinated through the Commission's Sport Fish Restoration Committee.

The Atlantic salmon has been a focal point for fish conservation activities in Connecticut. These efforts, though so far relatively nonproductive in reestablishing salmon, have greatly benefited other migratory fishes, particularly American shad and other river herrings, which use the fishways built to allow salmon to pass around dams that block access to spawning areas in the Connecticut River drainage. Recently constructed fishways in other drainages now allow shad, alewife, and blueback herring to bypass dams and access upstream spawning habitat. Restoration of populations of anadromous fishes has become an important conservation goal in Connecticut. Further projects aiding migratory fish passage are now in progress.

Other conservation efforts aimed at fishes largely focus on research on the distribution, abundance, and productivity of saltwater fishes; collection of harvest data; research on shortnose sturgeon; improvement of water quality; and establishment of regulations (size and creel limits, and seasonal restrictions) to prevent overexploitation of estuarine and freshwater fishes, as exemplified by the recent statewide ban on harvest of river herring. State-administered freshwater fisheries projects often are designed to meet the demand for fishing opportunities and deal primarily with non-native sport fishes.

Efforts to reduce water pollution have improved conditions for fishes in many streams and lakes. Despite pollution abatement, fish kills still occur, but the causes often are obscure. You can help conserve fish populations or contribute to increasing our understanding of the problem of fish kills by reporting die-offs as soon as possible so that causes can be remedied or at least investigated. During normal business hours (8:00 A.M.–4:00 P.M.), you may report a fish kill by calling any of the following DEP offices: Hartford: 860-424-3474; Marlborough: 860-295-9523; Harwinton: 860-485-0226; Old Lyme: 860-434-6043; and Litchfield: 860-567-8998. At other times you may call DEP Emergency Dispatch at 860-424-3333. Be prepared to provide your name and phone number and to describe the location of the kill, the number of fish observed, the type of fish observed (if possible), and anything else that seems out of the ordinary.

Though our fish fauna is highly altered by introductions of exotic species, there is evidence that fish faunas in North America are undersaturated in species and sometimes can accommodate additions without much harm to native species. However, many ecological disasters have occurred as a result of introductions of exotic fishes, particularly in rivers and lakes of the western United States and in the Great Lakes. Every effort should be made to exclude fishes from any body of water in which they are not naturally present, unless careful studies indicate that an introduction will not be ecologically harmful. In the 1980s, the state began permitting the introduction of sterile grass carp* (*Ctenopharyngodon idella*) for aquatic vegetation control in private ponds. These non-reproductive fishes should be useful for pond management without running amuck in our aquatic ecosystems.

Further information on fish conservation is included in the conservation sections of chapters 4 and 5.

Chapter 18

Amphibians

Amphibians (Class Amphibia) occur throughout the tropical and temperate regions of the world, primarily in freshwater aquatic and wetland habitats and in upland environments. Worldwide there are about 5,500 species, with the greatest diversity in the tropics. About 275 species occur in the United States, and 12 species of salamanders and 10 species of frogs and toads inhabit Connecticut. Amphibians have moist, glandular, nonscaly skin. Females lay their jelly-enclosed eggs in water or in moist sites on land.

Salamanders

Salamanders (Order Caudata) occur primarily in the Northern Hemisphere, with one family extending southward into South America. Of the world's 500 species, about 180 inhabit North America north of Mexico. Terrestrial salamanders use their sticky tongue to procure food. Aquatic salamanders (including larvae) use a "gape and suck" feeding mechanism.

Mudpuppies and Waterdogs (Family Proteidae)

Mudpuppy *(Necturus maculosus)* (Fig. 18.1). Totally aquatic, bottom-dwelling, generally stays under cover; permanent external gills; known to occur locally only in the Connecticut River, formerly south to Middletown, with recent records south only to Hartford; feeds opportunistically on small aquatic animals; courts and mates in fall; female lays eggs in May or June on underside of rocks, logs, or other cover, attends them until hatching in mid- to late summer. *Note:* Mudpuppies are occasionally hooked by anglers and reported in newspapers as a bizarre lifeform. This species may be native to Connecticut but perhaps was introduced through release of laboratory animals.

Mole Salamanders (Family Ambystomatidae)

Adult mole salamanders usually are terrestrial, secretive, and nocturnal; they eat small terrestrial invertebrates All

Fig. 18.1. Little is known about the mudpuppy population that inhabits the Connecticut River north of Hartford. Paul J. Fusco/Connecticut DEP Wildlife Division.

Fig. 18.2. Spotted salamanders require vernal pools and adjacent forests.

Connecticut species breed in ponded water; all but the marbled salamander breed in March and April. Larvae feed on zooplankton and other small aquatic invertebrates. These salamanders have noxious, toxic, and sticky skin secretions that chemically repel predators or mechanically interfere with their attacks.

Spotted salamander *(Ambystoma maculatum)* (Fig. 18.2). Locally common; except during breeding season, metamorphosed individuals are generally underground or under wood or rocks in wooded areas up to several hundred yards from breeding sites; females attach globular, clear or milky egg masses to submerged twigs, leaves, or wood in shallow water. *Note:* Sometimes I found eggs in shallow water of permanent ponds and lakes or in small puddles or bogs, but successful larval development occurs only in vernal pools containing water from at least March through most of summer and in other aquatic habitats lacking predatory fishes. In each pool, the egg-laying period often extends over a period of two weeks or more. Metamorphosis occurs as early as July or as late as October; larvae rarely overwinter.

Blue-spotted salamander *(A. laterale)* (Fig. 18.3) and **Jefferson salamander** *(A. jeffersonianum)* adults (dark with blue spotting) inhabit forested areas. They comprise a complex of populations each of which generally includes "pure" forms of one or the other species and diploid, triploid, or tetraploid individuals with mixed genetic material resulting from hybridization between the two species. Females take up sperm deposited by males, but the male's genetic material may or may not be incorporated into resulting offspring. Females attach eggs singly or in small clusters to submerged objects. Metamorphosis occurs from July through September. Metamorphosed individuals of both species spend most of their time secluded in wooded habitats surrounding and up to several hundred yards from breeding sites. Jefferson salamanders and hybridized populations dominated by their genome are locally common

Fig. 18.4. Marbled salamanders deposit eggs in dry vernal pool basins in late summer.

west of the Connecticut River; they generally breed in upland vernal pools. Blue-spotted salamander and hybridized populations dominated by the blue-spotted genome are represented by substantial populations in and near lowland swampy areas.

Marbled salamander *(A. opacum)* (Fig. 18.4). Locally common in wooded habitats in vicinity of vernal pools and swamps. *Note:* Marbled salamanders are unusual in courting, mating, and laying their eggs in late summer in dry vernal pool basins (rather than in water in spring as in most other salamanders). In September and October, I found many females attending clusters of nonadherent eggs under rocks, wood, or leaves in basins and depressions that filled with water in October and November. The female stays with the eggs until the site is flooded, then she heads into her summer and winter habitat in the woods. Males always live on land. The larvae hatch, feed, and grow in fall, rest on bottom in winter, and resume feeding in late winter or early spring, when they feast on newly hatched spotted salamander larvae. Not infrequently, breeding ponds dry up by mid-summer, soon after newly metamorphosed young depart. Juveniles may disperse to a half mile or more from their natal pool.

Lungless Salamanders (Family Plethodontidae)

All Connecticut members of this family feed opportunistically on small invertebrates.

Northern dusky salamander *(Desmognathus fuscus)*. Locally common along small streams; I frequently found many in a barely flowing rivulet below a spring, and other large populations inhabited permanent creeks with strong flow; females attach clusters of eggs by a single stalk to underside of rock or log very close to flowing water, mostly in July; larvae metamorphose from about May through July of the next year.

Fig. 18.3. Blue-spotted salamanders frequent swampy areas.

Fig. 18.5. Two-lined salamanders live in and along small streams.

Fig. 18.7. Four-toed salamanders are often associated with mossy swamps. These salamanders have an easily detachable tail that helps them survive encounters with predators.

Two-lined salamander *(Eurycea bislineata)* (Fig. 18.5). Common in and along small streams; occasionally I found larvae in ponded water along streams; in spring, females attach groups of eggs singly on underside of submerged rocks in flowing water; can be inconspicuously very abundant in small streams lacking brook trout; in North Branford I found hundreds of larvae in a stream segment only 10 feet long; larvae metamorphose after probably two years.

Spring salamander *(Gyrinophilus porphyriticus)* (Fig. 18.6). Uncommon in and along cold streams and seeps in northwestern and northeastern Connecticut; in spring, females attach groups of eggs singly on underside of submerged rock in flowing water; larval stage lasts a few years, so larvae of various sizes are present in water throughout year; eats invertebrates and smaller salamanders such as two-lined and dusky salamanders.

Four-toed salamander *(Hemidactylium scutatum)* (Fig. 18.7). Locally common (but secretive and rarely seen); generally associated with swampy areas with abundant cover of sphagnum moss, also inhabits forests near swamps; moves to breeding sites in late winter and early spring; in spring, females deposit clusters of adherent eggs in pockets in moss or other cover just above water; sometimes nests communally; white belly with black dots. *Note:* This salamander readily loses its tail when attacked. The wiggling detached tail may distract the predator, allowing the salamander to escape. It takes several months for the tail to regrow, so the process likely curtails an individual's body growth and reproduction. I had good luck finding adults and juveniles under wood near breeding pools in late summer and early fall, when adults are actively courting and mating and juveniles that metamorphosed earlier in summer are still active.

Redback salamander *(Plethodon cinereus)* (Fig. 7.14). Abundant in wooded habitats; completely terrestrial, spends much time under logs, rocks, or leaves, or in small spaces underground; in late spring, females lay eggs in small clusters with a common stalk, attached to underside of rock (Fig. 18.8) or in or under rotting wood on land; individual females produce eggs usually every other year; does not require standing or flowing water to complete life cycle; salamander

Fig. 18.6. Spring salamanders require cold streams and seeps. This and other salamander larvae that live in well oxygenated streams have relatively short gills.

Fig. 18.8. Redback salamanders attach their eggs under rocks or logs.

Fig. 18.9. Slimy salamanders in Connecticut occur only in the western margin of the state.

Fig. 18.10. Juvenile eastern newts ("red efts") live on land for at least a few years before returning to water.

that emerges from egg capsule is essentially a miniature version of adult (but not sexually mature). *Note:* This species sometimes attains densities of several hundred individuals per acre, far outnumbering any other terrestrial vertebrate. The red-striped coloration typical of most individuals may benefit them by inhibiting attacks of ground-feeding birds that happen to expose a salamander while foraging among leaves. The bird's reluctance to attack is due to the redback's color resemblance to highly toxic "red eft" (eastern newt), which birds quickly learn to avoid.

Northern slimy salamander *(Plethodon glutinosus)* (Fig. 18.9). Steeply hilly, rocky, mature forest in localized areas along western edge of state; more common and widespread in areas west of Connecticut; basic habits and life history are similar to those of redback salamanders.

Newts (Family Salamandridae)

Eastern newt *(Notophthalmus viridescens).* Common in still or slow-flowing water and in surrounding uplands; eats small invertebrates; relatively complex life cycle; courtship occurs in spring or fall; in spring, females attach eggs singly to submerged vegetation in ponded water; larval stage generally is brief and metamorphosis to terrestrial stage occurs in late summer or early fall; terrestrial "red eft" stage (actually more orange than red) lasts a few to several years; efts eventually transform into adults and renew aquatic life style; depending on local conditions, adults may remain permanently aquatic (e.g., where land conditions are excessively dry) or make seasonal migrations between upland and aquatic habitats; I found adults active around openings in ice-covered ponds in winter. *Note:* Red efts often walk around conspicuously in the woods in broad daylight, especially in warm, damp weather in summer, protected from many bird and mammal predators by their highly toxic skin (Fig. 18.10). Their vivid orange coloration makes it easier for predators to remember previous unpleasant meals and learn not to attack. These salamanders can use Earth's magnetic field and polarized light to orient themselves in the landscape.

Frogs and Toads

Frogs and toads (Order Anura) are distributed nearly worldwide in the tropical and temperate zones, with the highest diversity in the tropics. Only about 100 of the world's 4,840 species occur in North America north of Mexico.

Invertebrates are the primary food for adults of all species. Terrestrial stages use their sticky tongue to capture prey. Aquatic frogs may simply engulf some food items. Larvae (tadpoles) have beaklike jaws and rows of tiny "teeth." They feed in the water on algae, diatoms, detritus, plant tissue, carrion, and/or other minute organic material suspended in the water or attached to submerged objects. The larvae of some species, including wood frogs and bullfrogs, may also eat small live animals or amphibian eggs or hatchlings.

Spadefoot Toads (Family Pelobatidae)

Eastern spadefoot *(Scaphiopus holbrooki)* (Fig. 18.11). Rare; breeds during brief episodes in spring or early summer after heavy rains form temporary pools; breeding call consists of explosive nasal grunts; clasped by male, female attaches eggs in irregular groups or short strands to submerged vegetation; spends most of life burrowed in soil; local populations do not breed in years when rains are insufficient to form breeding pools; burrows into soil rear-end first, using a hard, sharp-edged "spade" on each hind foot. *Note:* This toad was formerly locally common in sandy

Fig. 18.11. Eastern spadefoot is the only Connecticut frog or toad with a vertically elliptical pupil.

areas, but its habitat has been destroyed by commercial and residential development in some areas, and now it is rare and occurs in Connecticut in only a few remnant habitats in the eastern and central parts of the state.

True Toads (Family Bufonidae)

True toads live mainly on land and are efficient burrowers, employing digging tubercles on their hind feet. Both Connecticut species are most common in areas with soft soils and tend to breed in semi-permanent or temporary waters. They deposit long strings of eggs that generally become tangled in vegetation in shallow water. At breeding sites, females are secretive, but males are very easy to observe (at night). The "warts" and lumps in the skin of a toad are aggregations of poison glands that help protect them from certain predators.

American toad *(Bufo americanus)* (Fig. 18.12). Common on land in habitats ranging from floodplains to hilly forests; occasionally inhabits edges of salt marshes where there is freshwater influx; brief breeding season begins

Fig. 18.12. Trills of American toads signal the return of warm weather in April.

Fig. 18.13. The calls of gray treefrogs in trees are often mistaken for those of red-bellied woodpeckers.

when warm, wet weather ensues; breeding call is a long ringing trill; I found breeding assemblages most often from mid-April to early May and newly metamorphosed young as early as late June.

Fowler's toad *(B. fowleri)*. Locally common in sandy areas, including coastal dunes and the vicinity of rain pools that may form in depressions near dunes and salt marshes; land dweller except during brief breeding season, which occurs in spring, later than that of American toad; breeding call is a short nasal "waaaah"; differs from American toad by having usually three or more (versus one or two) warts in each dark blotch on the back.

Treefrogs (Family Hylidae)

Gray treefrog *(Hyla versicolor)* (Fig. 18.13). Common in wooded areas adjacent to shrub-bordered ponds and marshes favored for breeding; breeding call is a loud, short, low trill; in mid-spring, females attach loose clumps of eggs to vegetation in permanent or semi-permanent waters; adults move to uplands after breeding. *Note:* Gray treefrogs are excellent climbers and often use tree cavities as shelter. Adults are beautifully cryptic on lichen-covered tree bark. They often vocalize from trees during warm, humid weather. Novice naturalists sometimes mistake the calls as those of red-bellied woodpeckers.

Spring peeper *(Pseudacris crucifer)* (Fig. 18.14). Adults and juveniles common on the ground in forested areas during moist periods in summer and fall (as late as early December); breeding call consists of loud peeps; males direct trilled peeps to intruding males; may call from woods in late summer and fall; in early spring, females attach single or small clusters of eggs to submerged vegetation in shallow, permanent or semi-permanent pooled water, especially in

Frogs and Toads | 311

Fig. 18.14. Spring peepers breed in spring in marshy shallows and forage on forest floors in summer and early fall.

Fig. 18.16. Pickerel frogs, often mistaken for leopard frogs, are common in ponds, wetlands, and nearby uplands.

open sunny locations. *Note:* With a flashlight and patient searching, you may be able to find the tiny, loud-peeping males among plants at the water surface.

True Frogs (Family Ranidae)

Bullfrog *(Rana catesbeiana).* Common in and at edge of deep, still waters, usually in marshy sites; breeds late, mainly in June or July, though I heard loud bellowing by males as early as late April and early May, likely indicating the beginning of territorial interactions; sheets of eggs initially float at surface of permanent water; larvae metamorphose mainly in June, July, or August; large individuals attack and ingest just about any animal, from insect to mammal, that will fit into their wide mouth; grows larger than the green frog and does not have ridge extending posteriorly along each side of back. *Note:* Because the larvae require at least a full year to metamorphose, bullfrog (and green frog) populations depend on permanent or semi-permanent water.

Green frog *(R. clamitans)* (Fig. 18.15). Common, usually seen or heard within 100 yards of a lake, pond, swamp, marsh, or stream, including stream-pools in coastal forests immediately adjacent to salt marshes; I found juveniles in just about every sort of freshwater pool or stream; breeds in late spring and early summer; breeding call sounds like a loose banjo string being plucked; sheets of eggs initially float at surface of shallow, still, generally permanent water, then sink; larvae overwinter once before metamorphosing; wintering sites usually are in flowing water of streams or seeps up to several hundred meters from breeding pond. *Note:* This frog is exceptionally unwary, often allows close approach, and makes half-hearted escape attempts.

Pickerel frog *(R. palustris)* (Fig. 18.16). Common near water; often ranges into nearby damp terrestrial habitat; breeding call is a low snore; in spring, females attach globular egg masses to submerged stems or twigs, generally in permanent or semi-permanent waters; I found many newly metamorphosed pickerel frogs in early summer. *Note:* Pickerel frogs are often mistaken for northern leopard frogs, which are far less common and more localized in occurrence.

Northern leopard frog *(R. pipiens)* (Fig. 18.17). Locally fairly common where meadows are adjacent to ponds and lakes and along floodplains of streams, primarily in Connecticut, Farmington, and Housatonic river drainages; often breeds in depressions inundated by flood waters along streams; breeding call consists of a low snore often followed by a series of grunts; females attach globular egg masses to submerged stems or twigs in shallow water in early spring; larval period lasts only a few months; I found metamorphosing leopard frogs in July in central Connecticut. *Note:* This frog is very wary and often escapes with quick and/or erratic leaps.

Fig. 18.15. Green frogs are ubiquitous, unwary inhabitants of swamps, marshy ponds, and most other freshwater habitats.

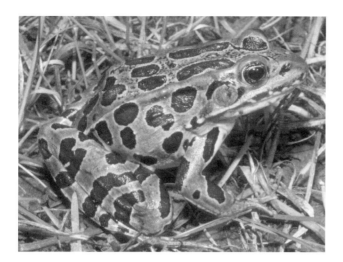

Fig. 18.17. Northern leopard frogs live along river floodplains and in marshy pond or meadow habitats.

Wood frog *(R. sylvatica)* (Fig. 18.18). Common, especially near vernal pools; land-dwelling except during brief breeding season, common on forest floors in summer and early fall; breeding call consists of ducklike quacks; breeds in heavily shaded or exposed vernal pools in late winter or early spring; eggs may hatch within two weeks; tadpoles metamorphose by early summer. *Note:* In each vernal pool, up to hundreds of females each lay a single globular egg mass on vegetation in shallow water within a small area, sometimes over a period of just a few days (Fig. 6.17). However, cold spells often interrupt breeding such that egg-laying episodes are up to a couple weeks apart. Over a ten-year period in one pool, I found 100 to 200 tightly grouped egg masses in the same place every year, then they inexplicably became very scarce. Sometimes I found these frogs moving about during mild rainy thaws in mid-winter.

Habitats

Connecticut salamanders can be grouped ecologically into three categories: pond salamanders, stream salamanders, and completely terrestrial salamanders. Pond salamanders (four-toed salamander, eastern newt, members of the mole salamander family) are associated primarily with wooded inland wetlands, including swamps and vernal pools. All of these pond species also depend on adjacent wooded uplands. These habitats additionally support a relatively high diversity of frogs and toads (Fig. 18.19). Various kinds of frogs and toads frequent temporary bodies of water (such as rain pools and floodplain pools) in addition to woodland vernal pools. Only eastern newts and bullfrog larvae are regular inhabitants of the open waters of lakes and deep ponds. Marshy habitats, including the shallow edges of lakes and ponds, harbor relatively few kinds of salamanders but a much more diverse frog and toad fauna.

Streams have fewer species, but four salamanders (mudpuppy; northern dusky, two-lined, and spring salamanders) are essentially restricted to streams and stream edges. Most frogs and toads tend to be relatively scarce in streams because of the relatively cool water and low productivity of tadpole food resources. But juvenile green frogs are common in small streams.

Salty habitats are devoid of salamanders, but a few frogs and toads sometimes occur at the edges of salt or brackish marshes where there is a significant influx of fresh water.

Fig. 18.18. Wood frogs breed in vernal pools and feed and hibernate on adjacent forest floors.

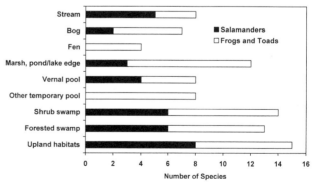

Fig. 18.19. Habitats of Connecticut amphibians. Uplands, swamps, and marshy areas harbor the largest numbers of amphibian species in Connecticut. Categorizing species by their major habitat occurrences is somewhat subjective; the values indicated here are not intended to be absolutely precise.

Only two species of Connecticut amphibians (redback salamander, slimy salamander) are strictly terrestrial. On land, these salamanders, and the land-dwelling stages of other species, use rocks, logs, leaf litter, other surface objects, and burrows for protection. Toads dig their own burrows in soft soils.

Reproduction

Reproduction by Connecticut salamanders involves stereotyped courtship behavior (nudging, rubbing, and following) and deposition of spermatophores (small mounds of jelly with sperm at the top) by males. Females pick up the sperm into their cloaca and later deposit the internally fertilized eggs in streams (a few species), vernal pools or ponds (most species), or on land (redback and slimy salamanders). Courtship and mating may occur in spring and/or fall, depending on the species.

Male frogs and toads emit a unique advertisement call during the breeding season. The call's function is to attract females and/or repel other males. Some calling may occur outside the breeding season.

All of Connecticut's frog and toad species lay their eggs in water. The male sheds sperm over the eggs as the female deposits them. Breeding sites include water with little or no current, usually ponds, swamps, lake edges, or marshes. Most species breed in sites with ephemeral or semi-permanent water. This reflects the high vulnerability of the larvae to fish predation, though bullfrog larvae common coexist with fishes.

Female frogs, toads, eastern newts, and most members of the mole salamander family abandon their eggs right after they lay them, while mudpuppies, marbled salamanders, and all Connecticut members of the lungless salamander family stay with their eggs until hatching.

The amphibian egg-laying season in Connecticut begins in March and April (Table 18.1). The amphibians that breed the earliest use vernal pools. Most species have bred by the end of May, when you can find unhatched eggs of the vast majority of the state's amphibians. Some species, such as northern dusky salamanders, green frogs, and bullfrogs, deposit eggs in early summer. The latest breeder is the marbled salamander, a vernal pool user. It is the only species whose eggs you can find after September.

Most Connecticut amphibian species have relatively fast development—eggs laid in late winter or spring yield metamorphosed young in late spring or summer of the same year. However, several salamander species and two frogs have a prolonged larval period that extends through winter and into the next year (Table 18.1). Mudpuppies, two-lined salamanders, and spring salamanders take more than a year to complete larval development. The slow development of these salamanders is related in part to their stream habitat, which tends to be much cooler than that of pond-breeding

Table 18.1. Reproduction Calendar for Amphibians in Connecticut.

	Jan	Feb	Mar	Apr	May	Jun	Jul	Aug	Sep	Oct	Nov	Dec
Mudpuppy	L	L	L	L	EL	EL	EL	EL	EL	L	L	L
Spotted salamander			E	E	EL	L	L	L	L	L		
Jefferson salamander			E	E	EL	L	L	L	L			
Blue-spotted salamander			E	E	EL	L	L	L	L			
Marbled salamander	L	L	L	L	L	L	L		E	EL	EL	L
Northern dusky salamander	L	L	L	L	L	EL	EL	EL	EL	L	L	L
Two-lined salamander	L	L	L	EL	EL	EL	EL	EL	L	L	L	L
Spring salamander	L	L	L	EL	EL	EL	EL	EL	EL	L	L	L
Four-toed salamander				E	EL	EL	L					
Redback salamander						E	E	E				
Slimy salamander						E	E	E				
Eastern newt			E	EL	EL	EL	L	L	L	L		
Eastern spadefoot				EL	EL	EL	EL	EL				
American toad			EL	EL	EL	EL	L	L				
Fowler's toad					EL	EL	EL	L				
Gray treefrog					EL	EL	L	L				
Spring peeper			EL	EL	EL	L	L	L				
Bullfrog	L	L	L	L	EL	EL	EL	L	L	L	L	L
Green frog	L	L	L	L	EL	EL	EL	EL	L	L	L	L
Pickerel frog				EL	EL	L	L	L	L			
Northern leopard frog				EL	EL	L	L	L				
Wood frog			EL	EL	L	L	L					

Letters indicate when eggs or embryos (E) and larvae (L) can be found in the field. Some atypical occurrences are not indicated.

amphibians. After metamorphosis, it generally takes another one to three years for an individual amphibian to attain sexual maturity.

Amphibians in Winter

Connecticut's amphibians vary in the life history stages that experience winter and in their characteristic wintering sites:

- **Aquatic larvae present in winter:** spotted salamander (rarely), marbled salamander, two-lined salamander, spring salamander, dusky salamander, mudpuppy, bullfrog, green frog.
- **Adults overwinter in aquatic/semi-aquatic sites:** dusky salamander, two-lined salamander, spring salamander, mudpuppy, eastern newt (also on land), bullfrog, green frog (often in flowing streams), pickerel frog, northern leopard frog.
- **Adults overwinter on land (underground):** Jefferson salamander, blue-spotted salamander, spotted salamander, marbled salamander, four-toed salamander, redback salamander, northern slimy salamander, eastern newt, American toad, Fowler's toad, gray treefrog, spring peeper, wood frog, eastern spadefoot.

Most amphibians are dormant in winter, though salamander larvae perhaps do some feeding during warmer periods of winter. Eastern newts often remain active in ponds in winter, and they even can feed and digest food (albeit slowly) at winter pond temperatures.

Many salamanders and wood frogs are active in March when freezing temperatures are not uncommon at night. These amphibians generally breed in aquatic habitats and can avoid freezing by submerging in the water, as do amphibians that overwinter under water. But some of these amphibians also can survive freezing weather even when they are in relatively exposed sites on land, such as during migrations to breeding sites.

Frozen Frogs. Gray treefrogs, spring peepers, and wood frogs (all widespread in Connecticut and eastern North America) overwinter on land under leaf litter, logs, and burrows. In some of these sites, they are subject to freezing temperatures throughout winter and in early spring when they begin activity. All three species can survive freezing (i.e., formation of ice within the body) of up to about 50 to 66 percent of their total body water content. Apparently the blood and extracellular fluids freeze completely whereas intracellular water remains unfrozen; hence cells are not damaged. Freezing of their body fluids occurs at about 30 to 28°F. Ice formation occurs rapidly, and the temperature of ice formation is highest when external ice is present. Wood frogs survive ice formation most readily when cooling proceeds slowly. Ice-nucleating bacteria may be responsible for initiating spontaneous extracellular freezing in these freeze-tolerant frogs.

How long can the frogs withstand being frozen? Gray treefrogs, spring peepers, and wood frogs tolerate freezing at 25 to 16°F for 5 to 7 days. Spring peepers exhibit about 50 percent survival after a week at 29.3°F, which freezes 45 percent of their body water. A gray treefrog survived experimental freezing at 27.5°F for 14 days. The frogs also can survive repeated freeze-thaw cycles.

After the frogs experience freezing temperatures, their tissues and urine contain elevated levels of cryoprotectants (antifreeze compounds, glycerol, and/or glucose), which are released from the liver. Heart beat and breathing eventually stop in partially frozen frogs, but heart rate accelerates temporarily as a result of the release of heat when body fluids crystallize. This slows the rate of cooling (rapid cooling is detrimental to survival) and helps circulate cryoprotectants.

In the same experiments discussed above, mink frogs (*Rana septentrionalis,* a species that occurs in northern New England) and northern leopard frogs, both species that overwinter in aquatic sites, froze at lower temperatures (27°F) than did the frogs that overwinter on land, but the aquatic frogs died if body freezing occurred. Freeze tolerance is not necessary for highly aquatic frogs such as these, which hibernate in water bodies that do not freeze solid.

Wood frog egg masses often are at the surface of the water, or protrude slightly above it, so the embryos may be exposed to below-freezing temperatures. The embryos supercool to 25°F without freezing and, like the adults, can survive freezing of their body fluids.

Conservation

The mysterious decline in amphibian populations that has gotten so much press has not occurred in Connecticut. That is, though some local populations have declined, the reason is no mystery. Habitat destruction, degradation, and fragmentation are the major factors. Where suitable, unfragmented habitat remains, so do the amphibians.

Forest fragmentation does not equally affect all species. For example, research in southern Connecticut by James Gibbs indicates that sedentary redback salamanders and habitat-flexible spring peepers often maintain populations in fragmented habitats, whereas vagile species that are mostly confined to forest, such as wood frogs and spotted salamanders, tend to be eliminated when forest cover in the region falls below 30 percent. Eastern newts, which are also wide-ranging forest amphibians, are particularly sensitive to fragmentation and disappear when forest cover falls below

50 percent. It seems that forest species that tend to disperse long distances from their breeding sites are most sensitive to fragmentation. Apparently good dispersal capability works well in continuous forest by facilitating colonization of vacant or newly available habitats, but in fragmented landscapes good dispersers are likely to end up in unsuitable habitat and may fail to survive or return to breeding pools.

Vernal pool amphibians in particular have experienced significant habitat loss and degradation in Connecticut (see the Conservation section of Chapter 6). This is part of a larger problem facing amphibians and other wildlife that favor small or shallow wetlands. Because they may look (but most definitely are not) inconsequential, these wetlands frequently have been drained and filled in conjunction with land development. Some have been made larger and deeper through dredging, impoundment, inadequate culverts under roads, or other hydrological alterations. Increasing the size and permanancy of a vernal pool may lead to the introduction of predatory fishes, which are incompatible with vernal pool amphibians (and fairy shrimp).

Thousands of amphibians are killed each year as they attempt to cross roads that divide the wetland and upland portions of their home ranges. Roads thus negatively impact salamander abundance in roadside habitats. Roads also appear to act as partial barriers to movement for certain salamanders species (e.g., eastern newts) that may be reluctant to traverse barren, open habitats. Wetlands commissions should enlist the aid of the public in identifying areas where large numbers of amphibians cross roads in their towns, and these roads should be evaluated for possible modification (e.g., installation of suitable underpass tunnels) to allow small animals to traverse the roadway corridor unharmed. New roads should be located far way from any wetlands.

Road runoff has detrimental effects on amphibians and other wetland and aquatic wildlife. For example, de-icing salts used on roads can contaminate nearby vernal pools and streams and cause mortality of amphibian embryos.

Flood control measures, especially the many dams that now impound stormwater and snowmelt runoff, have reduced the availability of breeding sites and detrimentally impacted some populations of northern leopard frogs in Connecticut. High flows and flooding are critical in scouring out the floodplain depressions that leopard frogs use for breeding. Not only is such habitat renewal now much reduced, but existing floodplain lowlands that formerly were seasonally flooded by high flows and provided breeding sites for leopard frogs now are flooded only under exceptional circumstances. Floodplain populations of American toads may be similarly affected. In the upper portions of watersheds, some leopard frog habitats have been permanently inundated by reservoirs.

Among amphibians, land development in Connecticut has been most costly to eastern spadefoot toads. This toad has disappeared from about half of its historical locations in Connecticut as a result of urbanization in New Haven County, where all former populations have been extirpated. Probably it was widespread in the sandy lowlands of central Connecticut in Hartford County prior to extensive and intensive urbanization, but now it is essentially extirpated there as well.

Selective timber harvest on lands that otherwise retain most of their natural character presently is not a severe problem for Connecticut amphibians on a state-wide scale, but intensive timber harvesting practices that greatly reduce canopy closure or understory vegetation, compact forest litter, or reduce the amount of decomposing coarse woody debris on the ground can result in local declines in salamander abundance. Negative impacts of intensive timber harvesting on salamander populations may extend at least a few dozen yards into uncut forest, suggesting that salamanders in logged areas do not always simply move into adjacent uncut forest. However, the fate of salamanders in clear-cuts is largely unknown and in need of further study.

Existing northern slimy salamander populations, which are highly localized and dependent on mature forest, warrant special efforts toward maintaining and enhancing mature forest conditions. Additionally, watersheds containing spring salamanders should be protected against high impact timber harvest or other land uses that degrade streams by increasing water temperatures, reducing oxygen levels, or introducing sediments or contaminants.

Chapter 19

Reptiles

In contrast to amphibians, reptiles (Class Reptilia) generally have dry, scaly skin, rather than moist, glandular skin. And the embryos are always contained in a membrane-bounded, fluid-filled chamber, as is typical of birds and mammals, and sometimes are encased in a shell, instead of being enveloped only in desiccation-prone jelly. Scaly skin and shelled, fluid-filled eggs make reptiles more tolerant of dry conditions than are amphibians.

Reptiles differ from mammals and birds in deriving almost all of their body heat from their surroundings rather than generating it through internal metabolic processes. They regulate their body temperature within tolerable limits through behavior, by making hourly, daily, and seasonal movements among microhabitats that differ in temperature and exposure to sunlight. In spring and fall, they often select warm sites, whereas during hot summer weather they stay in the shade or are active at twilight or night. In winter, when opportunities for attaining warm temperatures are minimal or absent, they abandon activity and hibernate (see "Reptiles in Winter" at the end of this chapter).

Reptiles occur throughout the world, including the oceans, except for polar regions and adjacent cold, high latitudes. Worldwide there are about 8,150 species. The largest number of species occurs in tropical regions. The resident reptilian fauna of Connecticut consists of 8 turtles, 1 lizard, and 14 snakes, for a total of 23 species. In addition, three species of sea turtles (green turtle, loggerhead, and Kemp's ridley) are regular, seasonal visitors in Long Island Sound, and one additional species (leatherback) may occasionally enter the eastern end of the Sound.

In all Connecticut reptiles, reproduction involves copulation, internal fertilization of the eggs, and deposition of eggs on land or live birth. Unless otherwise noted, the following reptiles are egg layers.

Turtles

The 300 species of turtles (Order Testudines) occur throughout most of the world excluding Antarctica and the high latitudes of the Northern Hemisphere. Habitats range from desert to oceans. All of the 57 North American turtle species lay shelled eggs on land, most often deposited in holes dug with the hind feet in soft soil. In the majority of species that have been studied, the sex of hatchlings is determined by the temperature an egg experiences during its development; generally, relatively warm temperatures result in females and cooler temperatures yield males. Turtles generally take several years to attain sexual maturity, and most are capable of living at least a few decades.

Snapping Turtles (Family Chelydridae)

Snapping turtle *(Chelydra serpentina)* (Fig. 19.1). Common in permanent, usually soft-bottomed river coves, lakes, ponds, and wetlands; also in some estuarine marshes; active day or night; flexible diet includes aquatic plants and small aquatic animals; very defensive (and exhibit why they are so named) when captured or closely approached on land; timid in water, flees from humans by swimming away or

Fig. 19.1. Snapping turtles are highly aquatic, but females roam on land during the late spring and early summer nesting season.

digging down into bottom mud. *Note:* In late spring or early summer, adult females often attract attention when they appear (and are frequently killed) on roads far from water as they move between aquatic habitat and terrestrial nesting areas. The abundance of this turtle in favorable habitat is indicated by the following statement from a state report describing former (now discontinued) predator control activities in Connecticut: ". . . at the Shade Swamp Sanctuary in Farmington . . . more than three thousand pounds of snapping turtles were caught during a period of two months in the summer of 1928."

Marine Turtles (Family Cheloniidae)

With rare exceptions, marine turtles come ashore only to nest on sandy beaches. The basic pattern is for an adult female to lay several large batches of eggs at intervals of one to three years, depending on the species.

Marine turtles in Long Island Sound generally appear in June or July and remain into October, when they migrate southward to overwintering areas. Some cold-stunned individuals are found in late fall and winter. No marine turtles nest in Connecticut.

All species in this family, plus the only other marine turtle, the leatherback, have suffered population declines due to human activities (degradation and disturbance of nesting beaches, harvest of adults and eggs, mortality in shrimping and fishing nets).

Green turtle (*Chelonia mydas*). Migratory; small numbers (juveniles only) in Long Island Sound in warm months; closest regular nesting area is in Florida; herbivorous, diet in New York includes algae and eelgrass.

Loggerhead (*Caretta caretta*). Migratory; most common sea turtle in Long Island Sound, but numbers relatively low; mostly juveniles; sometimes nests as far north as mid-Atlantic states, but most nesting is from Georgia southward; diet in New York dominated by benthic crustaceans.

Kemp's ridley (*Lepidochelys kempii*). Low numbers of juveniles seasonally inhabit Long Island Sound; eats benthic crustaceans such as rock crabs and spider crabs; globally rare. *Note:* Although nearly all nesting occurs on a single beach in the western Gulf of Mexico, Long Island Sound appears to function as a significant developmental habitat for juveniles of this migratory species.

Leatherback (Family Dermochelyidae)

Leatherback (*Dermochelys coriacea*). Oceanic, highly migratory; capable of diving to depths of thousands of feet; rare in Long Island Sound, known primarily from carcasses that occasionally wash up on beaches (a dead leatherback that I found in Rhode Island evidently had been hit by a boat propeller); fairly common in the Atlantic Ocean off New England and southeastern Canada; specialized diet, dominated by jellyfish. *Note:* This is one of few reptiles that is at least partially endothermic or self-warming. The large body mass and fat deposits in the skin allow retention of heat generated by metabolism or muscle contractions while swimming. Body temperatures may be several degrees warmer than surrounding water temperature, allowing use of high-latitude waters. An increasing number of carcasses have been found with the digestive tract jammed with clear plastic bags and similar material, which evidently the turtles mistake for jellyfish.

Musk and Mud Turtles (Family Kinosternidae)

Stinkpot (*Sternotherus odoratus*) (Fig. 19.2). Locally common in lowlands on bottoms of slow streams and reservoirs, particularly those with soft bottom and abundant submerged vegetation; highly aquatic, secretive; most active at night, but occasionally I found stinkpots on land in daytime, apparently heading back to water after laying eggs, and others I found while kayaking were moving in the open in sunny, shallow water; eats mainly aquatic invertebrates and carrion, sometimes considerable plant material. *Note:* When captured, stinkpots may scratch with the claws, bite, and/or release a foul-smelling secretion from two glands present on each side in the skin between the upper and lower shells, but sometimes they simply withdraw into the shell. Species is known also as the common musk turtle, but I prefer the more picturesque name.

Fig. 19.2. Stinkpots can be common in lakes and ponds with thick aquatic vegetation.

Pond and Marsh Turtles (Family Emydidae)

Painted turtle *(Chrysemys picta)* (Fig. 19.3). Abundant in many kinds of freshwater aquatic and wetland habitats, especially where there is little or no current and bottom is muddy; low numbers in channels through brackish marshes; active in daytime; diet includes assortment of plant and animal material. *Note:* Turtles seen basking in numbers on logs in ponds are always this species. Painted turtles can be extremely abundant in favorable habitats. For example, between mid-March and early May, in 151-acre Silver Lake in Berlin-Meriden, Vicki Cummings captured 226 individuals and marked them with conspicuous numbers on the shell. Later that same May she did a one-day lake-wide count of basking turtles and found 246 individuals, of which only three were marked. Even if some marked individuals were not seen clearly and recorded as unmarked, this very low "recapture" rate suggests that the lake was inhabited by at least several thousand individuals.

Spotted turtle *(Clemmys guttata)* (Fig. 19.4). Locally fairly common in ponds, marshes, swamps, and vernal pools, primarily in small, shallow waters with sedgy or shrubby hummocks (and similar areas at lake edges), including the tops of some trap-rock ridges; I found them also in pools along streams in coastal forests immediately adjacent to salt marshes; basks on logs, sedge tussocks, and shores; active in daytime; diet is dominated by invertebrates captured usually in water; females deposit eggs in sunny sites, in sedge tussocks, mossy hummocks, or near vegetation up to several hundred feet from water or wetland. *Note:* Spotted turtles are easiest to find in spring and early summer, from just after hibernation in March until temperatures become consistently hot and water levels drop in July. They hibernate in spaces beneath mossy hummocks, muskrat burrows, and similar sites in shrubby or forested swamps. In late March, many move to upland vernal pools

Fig. 19.4. Spotted turtles inhabit vernal pools and pond and lake shallows.

for up to four or five months, then leave the pools and spend several weeks in secluded terrestrial sites up to several hundred yards from permanent wetlands. Then they walk back to swamps to hibernate. Individual turtles may travel overland among multiple wetlands within a single year.

Bog turtle *(Glyptemys muhlenbergii)* (Fig. 6.14). Uncommon and local in calcareous fens and wet meadows in western Connecticut; individuals sometimes cross uplands to reach nearby wetlands; active in daylight, most observable in late spring and early summer; nest sites include tops of sedge tussocks; diet includes various terrestrial and wetland invertebrates. *Note:* This is a highly endangered species in Connecticut (see conservation section at end of chapter).

Wood turtle *(G. insculpta)* (Fig. 19.5). Uncommon or locally fairly common along rivers and smaller perennial streams and associated wooded or open riparian corridors; sometimes in and along tidal freshwater streams; scarce and highly localized in coastal and extreme eastern Connecticut; hibernates in streams, most likely to be seen in water or on streambanks in spring and late summer; forages on land on

Fig. 19.3. Painted turtles are Connecticut's most abundant turtle.

Fig. 19.5. Wood turtles live in and along streams.

Fig. 19.6. Diamondback terrapins are locally common along tidal creeks in coastal salt marshes in western and central Connecticut.

a wide variety of plant and animal material. *Note:* See conservation section at end of chapter.

Diamondback terrapin (*Malaclemys terrapin*) (Fig. 19.6). Locally common in tidal marshes, especially in tidal creeks that wind through coastal marshes in western and central Connecticut; once I counted 71 terrapins along a short stretch of tidal stream; eats various crustaceans, mollusks, worms, and dead fishes; generally ignores abundant mudsnails, evidently because this mollusk's shell is too hard to crush; females deposit eggs in nests dug in sandy upper beaches or dunes, often at high tide. *Note:* Raccoons and other mammalian egg predators destroy many nests. A study in Madison found 33 nests in July, and raccoons destroyed all of the 316 eggs in those nests.

Eastern box turtle (*Terrapene carolina*) (Fig. 19.7). Generally uncommon; usually near forest-field edges, often close to water, mostly at elevations below 500 feet; throughout most of state except Northwest Highlands and most of upland eastern Connecticut; active in daytime; forages on land for insects and other invertebrates, small fruits, fungi, and carrion. *Note:* Box turtle longevity is comparable to that of people, if the turtle can manage to avoid encounters with the latter. For about the first 15 or 20 years of the turtle's life, you can estimate the age by counting the number of layers present on the shell scutes. This also works for most other turtles with layered scutes.

Lizards

Lizards (Order Squamata, Suborder Lacertilia) are a nearly worldwide group of about 4,675 species, with 115 species in North America north of Mexico.

Skinks (Family Scincidae)

Five-lined skink (*Eumeces fasciatus*) (Fig. 19.8). Known from several scattered, rocky, semi-wooded or shrubby habitats west of Connecticut River; spends much time under or among rocks or logs; females nest in these microhabitats and stay with eggs until after they hatch; diet includes mainly various small invertebrates.

Fig. 19.8. Five-lined skinks, the state's only lizard, have a localized distribution in western Connecticut. Paul J. Fusco/ Connecticut DEP Wildlife Division.

Fig. 19.7. Eastern box turtles inhabit deciduous forests and adjacent fields.

Snakes

Nearly everyone has feelings about snakes, whether it be love, aversion, fear, or simple curiosity. These legless, sinuous reptiles (Order Squamata, Suborder Serpentes) occur worldwide in the tropical and temperate zones, with a few species ranging north of the Arctic Circle in Eurasia. Most of the nearly 3,000 species are tropical or subtropical; only one-half of snake families occur in the temperate zone. North America north of Mexico harbors about 140 species.

As indicated in the following paragraphs, snakes vary widely in their physical characteristics and behavior. Unifying all snakes is their lack of movable eyelids. The eyes are protected by a fixed transparent scale that gives snakes their unblinking gaze.

Another constant and characteristic snake feature is their pair of cloacal sacs. When physically disturbed or attacked, snakes usually void the contents of these sacs, which contain odorous fluids that may repulse certain predators.

Typical Snakes (Family Colubridae)

Some of Connecticut's colubrid snakes are egg layers; others give birth to young as noted below. When disturbed, many of these harmless snakes vibrate the tail and sometimes are mistaken for rattlesnakes.

Small Snakes that Eat Mainly Invertebrates. These snakes tend to stay hidden. When handled, they seldom if ever bite.

Worm snake *(Carphophis amoenus)*. Secretive burrower, uncommon in soft, well-drained soils.

Ringneck snake *(Diadophis punctatus)* (Fig. 19.9). Locally common in many habitats; often under debris around

Fig. 19.9. Ringneck snakes are small, secretive, and locally common.

Fig. 19.10. Smooth green snakes live in lush meadowy habitats.

abandoned cabins during mild, moist weather in spring and summer; occasionally I found individuals out in the open in daytime; diet commonly includes small salamanders as well as invertebrates.

Smooth green snake *(Liochlorophis vernalis)* (Fig. 19.10). Uncommon and localized in open areas with lush herbaceous vegetation.

Brown snake *(Storeria dekayi)*. Locally common in various habitats; I found it most often in areas highly altered by humans, including residential areas in towns; live-bearing.

Redbelly snake *(S. occipitomaculata)*. Locally common under wood and debris in moist, higher-elevation areas of northern Connecticut; live-bearing.

Larger Snakes with Generalized Feeding Habits. These snakes eat vertebrates and invertebrates. They are often seen out in the open and readily bite when handled.

Racer *(Coluber constrictor)* (Fig. 19.11). Common in fields, shrubby places, wetland borders, and wooded areas with sunny openings; often hibernates in small groups near sunny rock ledges; despite scientific name, does not constrict prey. *Note:* If approached, racers may move quickly to cover, climb into vegetation, remain immobile, vibrate the tail, assume a striking posture, or continue slow locomotion. In hand, they

Fig. 19.11. Racers are large, warmth-loving, fast-moving snakes that are especially common on and near trap-rock ridges and other rocky hills.

Fig. 19.12. Northern water snakes are harmless denizens of wetlands and their shores.

put up a vigorous defense by twisting, thrashing, and biting. In spring and fall, many that I captured by hand had a warm head and neck and a cool body, a result of differential exposure to the Sun.

Northern water snake *(Nerodia sipedon)* (Fig. 19.12). Common along streams, in marshy ponds, and in shrub swamps that contain abundant fishes or amphibians, its main prey; often basks at edge of water, quickly flees into it when disturbed; females, which grow much larger than males, give birth in late summer; hibernates in uplands adjacent to summer wetland habitat. *Note:* This harmless snakes is often mistaken for a "water moccasin."

Eastern ribbon snake *(Thamnophis sauritus)* (Fig. 19.13). Locally common in semi-open shrubby or sedgy margins of ponds, streams, and wetlands; in spring and fall, I found some on adjacent upland slopes and ridges where they winter; a population I studied ate mainly frogs and salamanders, including toxic eastern newts; live-bearing.

Fig. 19.14. Eastern hognose snakes are dietary specialists that feed mainly on toads.

Common garter snake *(T. sirtalis)*. Common in both uplands and wetlands; sometimes hibernates underwater; five individuals found underwater by John King in a partially ice-covered vernal pool in early March may have hibernated there; live-bearing.

Toad Specialist. This snake primarily eats toads, but its diet also includes frogs and other animals to a much lesser degree.

Eastern hognose snake *(Heterodon platirhinos)* (Fig. 19.14). Locally fairly common in areas with sandy soils and sunny openings; inactive periods spent underground in sandy areas or on nearby rocky slopes; may use spadelike snout to burrow or dig out toads buried in soil. *Note:* Enlarged teeth at the rear of the upper jaw may assist in swallowing toads with inflated lungs or may introduce saliva toxins that may narcotize the prey.

Constrictors. Before ingesting their prey, these muscular snakes first kill it by tightly wrapping body coils around it.

Fig. 19.13. Eastern ribbon snakes inhabit shrubby margins of freshwater wetlands.

Fig. 19.15. Eastern rat snakes are adept tree climbers. In rural areas they commonly enter old houses and help rid them of rodents.

Fig. 19.16. Milk snakes, like eastern rat snakes, kill large prey by constriction.

Fig. 19.17. Copperheads are venomous but reclusive.

Eastern rat snake *(Elaphe obsoleta)* (Fig. 19.15). Uncommon, mainly in wooded, often rocky habitats, primarily in southern and central Connecticut; excellent climber; I found them in trees, in cellars and attics of abandoned or occupied buildings and old houses, and on ground on forest paths or rural roads; eats birds and mammals; juveniles have blotched dorsal pattern.

Milk snake *(Lampropeltis triangulum)* (Fig. 19.16). Fairly common in diverse upland habitats, including woods, fields, talus slopes, and old buildings; I found milk snakes fully exposed on ground in daytime on many occasions; eats mice and various other small vertebrates, sometimes insects.

Vipers (Family Viperidae)

These dangerously venomous snakes have venom glands connected to hollow, needlelike fangs used for killing prey and defense. New World representatives (the pit vipers) have an infrared-sensing organ, used in locating prey and assessing attackers, in a pit on each side of the face. Rodents are the primary prey for the two Connecticut species. Both the copperhead and timber rattlesnake give birth to their young.

Copperhead *(Agkistrodon contortrix)* (Fig. 19.17). Locally fairly common, but secretive, in various wooded, shrubby, old field, and rocky habitats in southern and central Connecticut; locally common on and near talus slopes of traprock ridges, which serve as hibernacula. *Note:* Like timber rattlesnakes, copperheads are passive and reclusive. They never aggressively attack a person, but will strike defensively if molested.

Timber rattlesnake *(Crotalus horridus)* (Figs. 19.18 and 19.19). Uncommon to locally fairly common in a few scattered locations, mostly in the Glastonbury-Portland-East Hampton area of central Connecticut and hills of the northwestern and west-central parts of the state; active April to October in thick forest, shrub swamps, rocky areas, and sunny clearings; hibernates communally underground, often on steep, southerly facing rocky slopes; eats mainly white-footed mice and chipmunks; mates in summer, mainly in July, August, and early September; most give birth in September under rocks, in hollow logs, or on leaf litter. Today each den hosts several dozen rattlesnakes; formerly some dens sheltered hundreds of snakes. *Note:* In summer, individuals that my research team radio-tracked often ranged 1 to 2 miles from their winter den and returned to hibernate in the same den every winter. Rattlesnakes sometimes ambush prey by coiling next to a log, with the chin resting on the log (Fig. 19.20). They wait patiently and strike passing rodents

Fig. 19.18. Timber rattlesnakes are unaggressive and unlikely to strike at a person unless molested. This is the black morph.

Fig. 19.19. Timber rattlesnakes have a sensitive infrared radiation receptor in a pit on each side of the face.

that use the log as a runway. They can detect prey by vibrations, infrared radiation, odor, and vision. After a strike, the rattlesnake scent-tracks the prey to its deathplace, usually just a few feet away. The rattle consists of a series of loosely interlocking segments that produce sound when vibrated. One new segment is added to the rattle each time the snake sheds (typically twice each year). Rattle segments frequently break off, so aging a snake by counting the number of segments and dividing by two potentially works only if the rattle is complete, but rattlesnakes sometimes shed only once per year, so that too can confound rattle-based age estimates. Timber rattlesnakes are extremely passive and rarely rattle or strike unless harassed or startled. They rely on effective cryptic coloration that blends with leaf litter and most often remain coiled quietly if a person passes nearby. Unprovoked snakes do not attack humans, though a molested rattlesnake may put up a vigorous defense. A snake that is stepped on may bite defensively, so watch where you step when in rattlesnake habitat. See the conservation section at the end of this chapter for further information.

Fig. 19.20. Timber rattlesnakes often ambush rodents that use logs as paths. This is the yellow morph.

Habitats

Connecticut's reptiles inhabit a diverse array of habitats (Fig. 19.21). A large proportion of the turtle species are habitat specialists. For example, diamondback terrapins inhabit only coastal marshes, bog turtles occur only in calcareous fens, wood turtles are stream specialists that also roam adjacent woods and meadows, and eastern box turtles occur primarily in upland woods and fields. A few species of sea turtles occur in low numbers in Long Island Sound. Four of the eight nonmarine turtle species occur in freshwater ponds and marshes.

Of the 14 snake species and 1 lizard species, 13 occur primarily in upland habitats, especially those that are semi-wooded. Only two species (northern water snake, eastern ribbon snake) are strongly associated with water in ponds, streams, marshes, or swamps. Most live on the ground, but three species (eastern rat snake, racer, eastern ribbon snake) commonly climb into trees or shrubs.

Reproduction and Activity Calendar

March

- Brown snakes, common garter snakes, and aquatic turtles begin activity; other species may be active on warm days.
- Spotted turtles, wood turtles, and painted turtles court and mate.

April

- Snapping turtle mating begins (through fall).
- Painted turtles begin frequent basking in groups.
- Spotted turtles, wood turtles, painted turtles, racers, and common garter snakes mate.
- Most species are now active regularly.
- Painted turtle hatchlings move from nests to water.

May

- Spotted turtles, wood turtles, painted turtles, racers, and milk snakes court and mate.
- Painted turtle hatchlings move from nests to water.
- Bog turtle activity peaks.
- Ringneck snakes, eastern hognose snakes, smooth green snakes, and worm snakes begin their primary activity season (sometimes in April)
- Snapping turtles, painted turtles, spotted turtles, wood turtles, bog turtles, and stinkpots may deposit eggs at end of month.

June

- Eastern box turtle nesting peaks.
- Snapping turtles, painted turtles, spotted turtles, wood turtles, bog turtles, stinkpots, and diamondback terrapins deposit eggs.
- Bog turtle activity peaks.
- Spotted turtle main activity period ends.
- Five-lined skink egg laying may begin.
- Most snake species begin egg laying (through July).
- Timber rattlesnakes shed their skin (first of usually two sheddings each year).

July

- Spotted turtles become increasingly difficult to find.
- Snapping turtle, painted turtle, spotted turtle, wood turtle, bog turtle, stinkpot, diamondback terrapin, and snake egg laying ends.
- Timber rattlesnake mating begins (continues through mid-September).

August

- Common garter snake, ribbon snake, northern water snake, brown snake, redbelly snake, copperhead, and earliest timber rattlesnake births occur.
- Turtle egg hatching begins (late August).
- Wood turtles may court and mate.
- Five-lined skink hatchlings emerge (through September).
- Eastern hognose snake hatchlings first appear (and into September).

September

- Turtle egg hatching occurs.
- Snapping turtle hatchlings commonly observed crossing roads.
- Common garter snake, ribbon snake, northern water snake, brown snake, redbelly snake, copperhead, and most timber rattlesnake births occur.
- Racer hatchlings are common.
- Eastern rat snake hatchlings first appear.
- Common garter snakes court and mate.
- Activity becomes less frequent for many species.

October

- Snake road kills are common in early October, but activity of most reptiles ends by mid- to late October.
- Common garter snakes may court and mate.
- Spotted turtle hatchlings may leave nests.

November

- Occasional activity by a few species (e.g., painted turtle, common garter snake, racer, timber rattlesnake) when sunny, mild weather prevails.

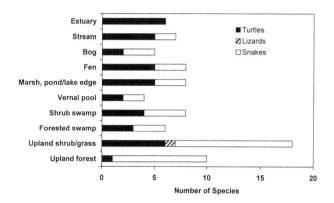

Fig. 19.21. Habitats of Connecticut reptiles. Reptiles use many types of wetlands and aquatic habitats, but open upland habitats support the largest number of species and provide nesting sites for several aquatic turtles. Categorizing species by their major habitat occurrences is somewhat subjective; the values indicated here are not intended to be absolutely precise.

Reptiles in Winter

Most reptiles, as well as amphibians, insects, and other invertebrates, have no effective internal mechanism for producing significant amounts of body heat. Near-freezing weather generally results in low body temperature, which greatly reduces the metabolic rate and capacity for activity, and feeding and reproduction become difficult or impossible. Most high latitude and high elevation species spend cold periods (up to several months) inactive in a secluded site where they are minimally vulnerable to predation. Some species select sites that remain above freezing, whereas others use sites that may be subject to freezing temperatures.

Hibernation Sites. Many species overwinter underground where ambient temperatures remain at least a few degrees above freezing. However, hibernating reptiles avoid conditions that are truly warm, since warm temperatures result in increased metabolism and could lead to exhaustion of energy reserves before conditions are suitable for feeding again in spring.

Most turtles and some snakes (for example, sometimes the common garter snake) spend the winter underwater. Garter snakes that hibernate in air are more vulnerable to mortality from dehydration. Turtles winter in freeze-protected sites such as chambers in mossy hummocks or rock caverns or on pond bottoms. Other reptiles such as box turtles take refuge near the ground surface where they may be exposed periodically to temperatures below 32°F.

- **Reptiles that winter underwater:** snapping turtle, painted turtle, spotted turtle, wood turtle, bog turtle, diamondback terrapin, common musk turtle, common garter snake.
- **Reptiles that winter on land (underground):** eastern box turtle, five-lined skink, worm snake, racer, ringneck snakes, eastern rat snake, eastern hognose snake, milk snake, northern water snake, smooth green snake, brown snake, redbelly snake, ribbon snake, common garter snake, copperhead, timber rattlesnake.
- **Reptiles whose hatchlings overwinter on land in the nest:** snapping turtle (but usually not), painted turtle, spotted turtle (sometimes), wood turtle (sometimes), eastern box turtle (sometimes).

No species has viable unhatched eggs that overwinter. In no species do the females normally overwinter in advanced stages of pregnancy.

Supercooling. Exposure to freezing temperatures is not necessarily harmful to reptiles. Almost all reptiles can supercool without freezing (that is, without ice formation) to 30 or 28°F, and common garter snakes can supercool to 22°F. Some amount of supercooling should not be too surprising because water also supercools by about the same amount (or more) before turning to ice. (The so-called freezing point of water—32°F—is actually its melting point, the highest temperature at which ice can exist in a solution.) Cryoprotectants that may enhance supercooling in reptiles include glucose, glycerol, lactate, and free amino acids. Because supercooling often does not extend to temperatures much below 32°F and may be of short duration (e.g., only an hour or two), the effectiveness of supercooling as a strategy for freeze avoidance is limited.

Freeze Tolerance. Actual freezing (i.e., internal ice-formation) is lethal to most reptiles, but recent studies indicate that at least some species can survive partial freezing for up to a few hours (common garter snake) or a few days (box turtle) or at least 11 days (painted turtle hatchlings). Freezing of extracellular fluids such as the blood is tolerated, but intracellular freezing is lethal. Special ice-nucleating proteins, made in the liver, facilitate extracellular freezing in such a way that the ice crystals remain small and do not damage delicate tissues. In box turtles, the heart stops beating only when body ice content is quite high. Freeze tolerance in box turtles and painted turtle hatchlings is clearly related to their use of shallow terrestrial hibernation sites.

Metabolic Adjustments. Winter-dormant reptiles at low temperatures exhibit very low metabolic rates, only a small fraction of the rate at the same temperature in animals tested in summer. This phenomenon has been demonstrated in snapping turtles, garter snakes, and water snakes. Reduced metabolic rate has obvious energy-saving advantages where winters are long and food is not available. Some reptiles increase metabolic rate after exposure to low temperatures. This response may occur in environments that rarely become very cold and in which food is available throughout the year, so it is not to be expected in Connecticut.

Role of Stored Fat. In Connecticut and elsewhere in the northern United States, the common garter snake spends much of the cold season (about five months) dormant at body temperatures probably not above 41 to 50°F. Prior to winter, these snakes develop large fat bodies in the abdominal cavity, but fat-body lipids provide little or no energy during dormancy. Instead, glycogen and proteins from muscle and liver tissues provide nearly all the energy used during the winter.

The importance of the fat bodies is that they provide the energy needed during the initial spring activity period when males search for and court females. In females, much of the energy stored in the fat bodies goes into the production of offspring. Some reptiles that are intermittently active during winter or that are dormant at relatively high temperatures do extract energy from the fat bodies during winter.

The Problem of Anoxia. Aquatic turtles in some habitats have been thought to spend most or all of the winter buried in soft mud at the bottom of a body of water where the minimum temperature may be 39 to 41°F. Turtles in such sites would easily avoid freezing.

A turtle submerged in mud has a thermally favorable overwintering site, but the location imposes a different problem—anoxia. Turtles that remain buried in the mud may have no access to oxygen, which is consumed by decomposition of organic material in the stagnant water. Access to air may be prevented by ice covering the water. Turtle metabolism under such conditions is anaerobic. The turtles' energy source under these conditions is glycogen stored prior to hibernation.

Anaerobic metabolism results in the production of lactate, which cannot be eliminated without the presence of oxygen. As lactate accumulates, the blood becomes more acidic. Respiratory acidosis also contributes to this problem. Death due to acidosis normally occurs at about 1 pH unit below normal.

Northern populations of painted turtles avoid harmful levels of acidosis by minimizing the production of lactate (through minimal activity and low metabolic rate) and by increasing the concentrations of calcium and magnesium in the blood (calcium and presumably magnesium chemically offset the lactate-induced acidosis). Experiments done with painted turtles that were held in cold (37°F) anoxic water (no access to air) demonstrate that they can survive anoxia and avoid lethal acidosis for periods of up to five months or more. When given access to air, the turtles hyperventilate, rapidly restore blood pH, and more slowly lower blood lactate and correct ionic balance.

Because turtles can extract oxygen from the water through their vascularized buccal cavity or cloacal bursae, acidification of the blood during hibernation is less of a problem in oxygenated water than it is in anoxic conditions. Why then would a turtle select an anoxic site for overwintering? Possibly this is related to avoidance of predation during periods when cold-induced "grogginess" makes it difficult for them to defend themselves or escape an attack. A turtle buried in mud may be overlooked by a hungry otter.

Actually, it appears that turtles often do not hibernate buried in mud. Instead, they frequently hibernate in sheltered sites on the bottom above the mud where the water contains substantial oxygen that can be obtained through extrapulmonary mechanisms, and some spend the winter in partially flooded chambers that may allow access to air. They do not always experience pronounced changes in blood acidity or lactate concentration and may not need to employ special physiological adjustments, particularly in mild winters when a short duration of ice cover facilitates oxygenation of the water and access to air. Wintering sites are more often in shallow areas rather than in the deepest area. This may allow a quicker response to spring thaw, since ice melts away from the edges first.

Conservation

Habitat loss, Degradation, and Fragmentation

Virtually all upland reptiles have been detrimentally affected by habitat alterations. Among the most vulnerable species in Connecticut are bog turtles, wood turtles, and timber rattlesnakes.

Turtles. Bog turtle populations have been greatly reduced by destruction and fragmentation of habitat, habitat alterations caused by flooding and dewatering, and successional habitat change from open wetland to closed-canopy red maple swamp. Some of these habitat changes are attributable to human activities whereas others are "natural." However, even so-called natural changes that negatively affect bog turtles are in part human-caused. For example, by preventing flooding by beavers, which kills red maples and eventually restores open wetland conditions, people interfere with the natural processes that replace habitats temporarily lost to vegetation succession. And even if flooding occurs, human land uses may be such that there is no suitable habitat nearby to allow the local turtle population to survive flooding of their current habitat. Bog turtle populations do best in a network of interconnected wetlands that allow natural vegetation dynamics to proceed such that as some patches become unsuitable others become newly available. Light grazing by carefully managed cattle provides one method by which open wetlands can be maintained in areas where bog turtle habitat is shrinking as a result of encroachment of trees (the cattle eat colonizing tree seedlings).

Wood turtle populations also have suffered from habitat alterations and fragmentation. Local populations are highly vulnerable to persistent, excessive mortality on roads that parallel and cross rivers. Keeping development and traffic away from riparian corridors is an important aspect of wood turtle conservation. In fact, most turtle populations in Connecticut suffer from road mortality. What makes this particularly damaging is that the individuals killed often are gravid females moving to or from nesting sites.

Land disturbance by people has facilitated the establishment of exotic plants that have degraded habitats required by bog and wood turtles. Exotic plants degrade habitats by shading, changing soil properties (temperature, moisture, friability), acting as a physical barrier to movement, and likely in other ways as well.

Timber Rattlesnake. Recent housing developments, and developments now planned or under construction, threaten the remaining timber rattlesnake populations in central Connecticut. Residential housing developments in Glastonbury have usurped the snakes' foraging habitat, and they interfere with snake migrations between winter dens and summer foraging areas. Developments not only remove habitat and disrupt migrations but also expose the snakes to danger or death when they end up in the yards of the new houses or as the snakes cross newly constructed roads or existing roads with increased traffic. Yet planning and zoning commissions continue to approve destructive developments in rattlesnake habitat. Endangered species laws need to be strengthened to better protect habitat of disappearing wildlife.

Rattlesnakes make long migrations and require large areas of habitat. Through radio-tagging, my research team determined that individual snakes in central Connecticut move up to 2.2 miles (in any direction) from their winter den (they return to the same den each fall), so the minimum habitat size required for a population using just one den is more than 4 miles across. Most populations are represented by more than one den, so the area used is even larger. Protection of the remaining undeveloped tracts of forest in central Connecticut is a critical conservation need not only for rattlesnakes but also for the multitudes of other native plants and animals that inhabit rattlesnake habitat.

Rattlesnakes tolerate timber cutting in small patches as long as most of the landscape remains mature forest. In fact, gravid female rattlesnakes sometimes are attracted to logged openings for basking, especially for a few years when shrubby regrowth is mixed with open areas and piles of sun-exposed logging debris. However, to avoid causing road mortality of snakes as they move among forest and open habitats, timber cuts should never be located near public roads or on the opposite side of a road from a rattlesnake den. Despite use of logged clearings by gravid females, effective rattlesnake conservation requires protection of large areas of mature, uncut forest, which is essential as the primary foraging habitat for juveniles, males, and nongravid females. Fallen trees and logs, which serve as cover and foraging sites, are an important component of rattlesnake habitat that should be maintained throughout forested and logged areas. Logging should be done from November to March, when the snakes are hibernating.

Collecting, Vandalism, and Incidental Mortality

Turtles. Bog turtle and wood turtle populations have been hard hit by now-illegal collection for the pet trade. Bog turtles are particularly attractive to pet enthusiasts because they are small, which simplifies housing. Because of their rarity, the turtles command high prices, and this has put pressure of the remaining populations. Vigilant protection of these populations from collection is an absolute necessity.

Researchers in New Haven County found that two formerly stable wood turtle populations declined drastically after a protected drinking water supply area was opened to recreational use. Presumably most of the turtles that disappeared were taken by people.

Mortality incidental to seemingly innocuous human activities is a problem for some turtles. Road kills were mentioned previously in conjunction with habitat alterations and fragmentation. Diamondback terrapins are vulnerable to drowning in submerged crab and lobster traps. In intensively trapped areas, local terrapin populations rapidly decline to rarity and can be extirpated in just three or four years. Sea turtles in Long Island Sound incur mortality through collisions with boats and incidental capture or entanglement in fishing gear.

The demographic characteristics of turtles (slow to mature, low fecundity, naturally low survival of eggs and young due to predation) make populations highly vulnerable to excessive mortality of adults. In fact, persistence of turtle populations depends in large part on high levels of adult survival. Basically, any increase in the removal of adults from a population is likely to cause a population decline. Thus effective turtle conservation generally entails not only the protection and maintenance of habitat but also protection from excessive collecting and road mortality of adults. Efforts to increase reproductive success (e.g., through captive breeding or artificial incubation) usually are less effective. Giving the long-lived adults enough opportunities (years) to breed often is sufficient to ensure that each adult will produce at least one offspring that will reach reproductive age.

Timber Rattlesnake. Populations of this endangered snake are extremely vulnerable to collection and killing because they aggregate at communal hibernation sites. Through repeated visits as the snakes emerge from their hibernacula in spring or return in fall, intent rattlesnake hunters have decimated or extirpated many local populations. In fact, recent surveys by Hank Gruner indicate that 85 percent of the state's rattlesnake populations have been lost.

Populations of timber rattlesnakes, like those of turtles, cannot withstand any significant increase in the death rate of adults. Rattlesnakes take several years to mature, females generally reproduce not more than once every three years and produce small litters of young, and the young experience high mortality. As a result, only a few new individuals are added to the breeding population each year. So rattlesnake populations plummet when substantial numbers of adults are killed. In contrast to deer and certain other game species, rattlesnakes and turtles do not exhibit a significant compensatory increase in reproductive rate when adult populations are reduced.

Chapter 20

Birds

More than any other vertebrates, birds (Class Aves) have a truly worldwide distribution and are familiar to people everywhere. There is almost no habitable place on Earth where one would not be able to see at least a few of the world's approximately 10,000 species. The Neotropical region, extending from southern Mexico into South America, has more bird species than anywhere else; the avifauna of Colombia alone includes more than 1,500 bird species, over twice as many as in Canada and the United States combined.

Connecticut, despite its small size, hosts a substantial bird fauna. Counting all bird species known to have occurred in Connecticut without human assistance within historical time, including nonbreeders such as migrants, winter visitors, and vagrants, as well as breeding species and recently extinct and extirpated species, the state total comes to around 410 species, of which about 280 species regularly occur. The groups represented by the largest number of species are the warblers (Parulidae); swans, ducks, and geese (Anatidae); sandpipers (Scolopacidae); and the sparrows, towhees, longspurs, and relatives (Emberizidae).

The Atlas of Breeding Birds of Connecticut (Bevier 1994) listed 173 species as confirmed breeders in Connecticut with an additional 12 species probably also breeding during the atlas survey period (1982 to 1986). About 30 or so of the confirmed breeders are rare, very local, or irregular nesters. Additional species nested in Connecticut historically but have not done so in recent years. Seven introduced species are included among the confirmed regular breeders. The latest assessment puts the total for regular breeders at approximately 150 species.

Intensive field surveys of the "Summer Bird Count" (actually conducted primarily in June before the summer solstice and thus in late spring) from 1992 to 2001 detected up to 201 species in Connecticut. This number includes primarily locally nesting species but also quite a few species that do not breed in the state. The survey encompasses over 1,642 square miles (31 percent of the state).

For each species recorded by the breeding bird atlas, breeding status (confirmed, probable, or possible breeder, or no evidence of nesting), was recorded for each of 596 atlas survey areas (blocks) distributed statewide. Each block was one-sixth of a standard United States Geological Survey topographic map, encompassing approximately 10 square miles. Thirty-five blocks had 100 or more breeding species; most of these were in the northwestern Connecticut, conforming with the general pattern of greater species density in blocks in that part of the state.

Species recorded as breeding in nearly all of the blocks (at least 585) included the song sparrow, American robin, gray catbird, black-capped chickadee, American crow, mourning dove, northern cardinal, blue jay, northern flicker, common grackle, house finch*, European starling*, common yellowthroat, house sparrow*, and chipping sparrow. The rarest breeders, those confirmed in ten or fewer blocks, included the cattle egret, long-eared owl, evening grosbeak, king rail, black skimmer, little blue heron, upland sandpiper, roseate tern, olive-sided flycatcher, glossy ibis, vesper sparrow, grasshopper sparrow, yellow-crowned night-heron, American oystercatcher, willet, and common moorhen.

The following annotated species list includes most of the bird species that have been adequately documented as occurring in Connecticut. However, I did not include all species that occur as nonbreeding rarities or unpredictably, and bird occurrences that are very rare in a particular place, habitat, or time of year generally are not mentioned. For historical interest, I did list several extinct species. *Connecticut Birds* (Zeranski and Baptist 1990), *The Atlas of Breeding Birds of Connecticut* (Bevier 1994), and recent issues of *The Connecticut Warbler* (journal of the Connecticut Ornithological Association) are excellent sources of information for these less common birds and occurrences, as well as for further information on common species. You can call the state's Rare Bird Alert at 203-254-3665 to learn about recent occurrences of rarities.

The sequence of families and species content of each family, as well as the scientific and English names, follow the American Ornithologists' Union (1998), as updated by supplements in 2000, 2002, and 2003.

Spring, summer, fall, and winter as used here refer to the seasons as defined by the solar equinoxes and solstices. Thus early September is late summer (not fall), and early June is late spring (not summer).

Loons, Grebes, Storm-petrels, Gannets, and Cormorants

Loons (Order Gaviiformes, Family Gaviidae)

Loons in North America breed primarily along the shores of lakes and ponds in Canada, Alaska, and other northern states. They spend winter primarily in coastal areas. Loons feed mainly on small fishes, crustaceans, and other small animals obtained by diving underwater from the surface, using their feet for propulsion.

Red-throated loon *(Gavia stellata)*. Fairly common, mainly in coastal waters and mouths of major rivers, during spring and fall migration; uncommon in winter, when most numerous in eastern Long Island Sound; sometimes 100 to 200 individuals gather at single locations in November; many depart Connecticut by early winter (species then most abundant along North Carolina coast).

Common loon *(G. immer)* (Fig. 20.1). Historically a rare nester along lake shores in Connecticut; no recent nesting (does nest in Massachusetts); fairly common spring and fall migrant in coastal waters, large lakes, and reservoirs; winter sightings most likely in eastern Long Island Sound; subadults and nonbreeding adults sometimes present in summer. *Note:* This loon no longer nests in many areas of its former range in the northeastern United States, due to loss of food resources from lake acidification, reduced reproductive success resulting from mercury poisoning (coal-fired power plants are a source of atmospheric mercury pollution), and excessive disturbance by humans and lakeshore development. Successful nesting is also inhibited by raccoon predation (whose populations are large due to human-augmented food resources) and fluctuating water levels in water supply reservoirs.

Grebes (Order Podicipediformes, Family Podicipedidae)

Grebes generally breed in lakes and ponds and winter in inland and coastal waters. These aquatic birds feed mainly on small fishes obtained by diving underwater from the surface.

Pied-billed grebe *(Podilymbus podiceps)* (Fig. 20.2). Rare breeder in freshwater marshes and marshy ponds and lakes in western Connecticut; nests generally float in shallow water; widespread but usually uncommon spring and fall migrant and winter visitor in lakes, ponds, marshes, and coastal waters.

Horned grebe *(Podiceps auritus)*. Variably numerous spring and fall migrant, less common in winter; usually in coastal waters of Long Island Sound, sometimes on large inland lakes and rivers.

Red-necked grebe *(P. grisegena)*. Usually uncommon during spring and fall migration, rare in winter; mainly coastal waters of Long Island Sound, sometimes on inland lakes, ponds, and rivers.

Fig. 20.1. This immature common loon was resting on a coastal beach in mid-July.

Fig. 20.2. A hatchling pied-billed grebe is riding on the back of a parent, while unhatched eggs remain in the nest.

Storm-Petrels (Order Procellariiformes, Family Hydrobatidae)

This group includes 21 species that together range nearly worldwide throughout the oceans. Storm-petrels feed on small marine animals captured at the surface of the water. They nest in colonies in borrows or crevices.

Wilson's storm-petrel *(Oceanites oceanicus).* Occurs rarely yet somewhat regularly in summer in Long Island Sound, particularly the eastern end. This far-ranging bird nests on islands off southern South America, on the Antarctic Peninsula, and on islands around Antarctica.

Gannets and Boobies (Order Pelecaniformes, Family Sulidae)

Northern gannet *(Sula bassanus).* Rare spring and fall migrant; seldom occurs in mid-winter, though fairly regular in December; mostly in eastern Long Island Sound; autumn groups sometimes include a few dozen individuals; feeds primarily on fishes obtained by spectacular dives into water from high in the air; nests in colonies on sea cliffs or islands in eastern Canada.

Cormorants (Order Pelecaniformes, Family Phalacrocoracidae)

Cormorants generally nest in colonies on coastal islands or in trees standing in inland waters. They feed mainly on fishes obtained by diving underwater from the surface.

Double-crested cormorant *(Phalacrocorax auritus)* (Fig. 20.3). Rare migrant a century ago, now common from spring to fall; nests in trees or on open ground on some coastal islands, sporadically on inland lakes and river islands; much less common (though increasing) into early winter; mainly along coast, major rivers, and some inland lakes; large groups roost communally at night in favored large riparian trees. *Note:* The nesting population has increased in recent decades and included around 1,000 pairs in the late 1990s. There is concern that expanding nesting colonies of cormorants may displace other colonial nesting birds whose populations are less robust. Fishing interests sometimes complain that these birds eat too many fishes, and control measures have been undertaken in New York and southern New England. In contrast to the multi-decade increasing trend, Summer Bird Counts in the most recent years suggest an ongoing decline in the number of cormorants present in Connecticut in June.

Fig. 20.3. Connecticut's population of double-crested cormorants increased over the past several decades. Paul J. Fusco/Connecticut DEP Wildlife Division.

Great cormorant *(P. carbo).* Fairly common in coastal waters from fall through spring, with smaller numbers on the Connecticut River and other large rivers and lakes; single pairs have nested on coastal Massachusetts islands; primary breeding areas are to the north in Canada.

Wading Birds and Vultures

Herons, Bitterns, and Allies (Order Ciconiiformes, Family Ardeidae)

These birds often forage by wading in shallow water and catching small animals with a quick stabbing motion of the head. Most species nest in colonies in trees or shrubs, though bitterns and green herons are generally solitary.

American bittern *(Botaurus lentiginosus).* Uncommon or rare, but widespread, throughout year; rare nester in emergent vegetation in scattered freshwater marshes; mostly in large marshes in winter and summer; migrants may use small marshes. *Note:* Nesting numbers probably declined as common reed invaded cattail marshes.

Least bittern *(Ixobrychus exilis).* Rare in fresh and brackish marshes, spring to fall; nests in large marshes of cattails or other emergent plants, usually in marshes with a roughly equal mixture of vegetation and open water.

Great blue heron *(Ardea herodias)* (Fig. 20.4). Familiar bird of marshes, ponds, lakes, and streams, spring through fall; localized nester, but increasing in recent decades, primarily away from coast; nests in large trees in or near water, including trees killed by beaver-caused flooding; uncommon (but increasing) in winter.

Fig. 20.4. Great blue herons are adept predators of fishes and frogs. Paul J. Fusco/Connecticut DEP Wildlife Division.

Fig. 20.5. Snowy egrets nest on a few islands in Long Island Sound.

Great egret *(A. alba)*. Fairly common in coastal marshes and waters, spring through fall (rarely into early winter), less common inland; approximately 100 pairs nest in woody vegetation on small islands along the coast (e.g., Great Captain Island off Greenwich, Charles Island near Milford Harbor, and Duck Island at Westbrook); increasing numbers recorded on recent Summer Bird Counts; sometimes one hundred or more aggregate in coastal salt marshes on late summer evenings; dozens may join night roosts with other water birds in riparian trees. *Note:* This and other egrets historically were decimated by plume hunters but recovered with protection in the 1900s.

Snowy egret *(Egretta thula)* (Fig. 20.5). Fairly common in coastal marshes and waters, spring through fall; dozens aggregate in good feeding areas in coastal tide pools and salt marshes in summer; less common inland; many may roost at night with great egrets and cormorants in certain favored riparian trees; nests in woody vegetation on small coastal islands (e.g., Great Captain Island off Greenwich, Charles Island near Milford Harbor, and Duck Island at Westbrook). *Note:* The population nesting on islands in Long Island Sound (including New York) declined somewhat from 1,650 pairs in 1977 to 1,390 in 1998.

Little blue heron *(E. caerulea)*. Uncommon in coastal marshes and waters, spring through early fall; seen most often seen in mid- to late summer and early fall when immatures invade state from nesting areas to south; has nested in woody vegetation on coastal islands (e.g., Charles Island, Great Captain Island). *Note:* The nesting range expanded northward along the U.S. Atlantic coast in the 1900s.

Tricolored heron *(E. tricolor)*. Rare in coastal marshes and waters, spring and summer; occasionally has nested in woody vegetation on coastal islands at Norwalk. *Note:* This heron increased in abundance in southern New England and New York in the mid-1900s.

Cattle egret *(Bulbulcus ibis)*. Rare in coastal marshes and waters, spring and summer; has nested periodically in woody vegetation on coastal islands at Norwalk. *Note:* This Old World egret showed up in South America in the 1880s and spread to Florida by the early 1940s. The range expanded northward along the U.S. Atlantic coast in the mid-1900s.

Green heron *(Butorides virescens)*. Fairly common, usually at low densities, in marshes and swamps, spring through fall; nests generally in shrubs near (often over) water. *Note:* Loss and degradation of marshes probably caused a decline in the 1900s.

Black-crowned night-heron *(Nycticorax nycticorax)*. Fairly common along coast and Connecticut River, spring through fall, rare in winter; disperses inland along Connecticut and Housatonic rivers in late summer; nests on coastal islands (e.g., Great Captain Island off Greenwich, Charles Island near Milford Harbor, and Duck Island at Westbrook), sometimes in other locations near water. *Note:* Around 500 pairs nest in the state. Abundance and nesting distribution declined in the northeastern United States during the 1900s, probably due to human disturbance and pesticides. The population nesting on islands in Long Island Sound (including New York) decreased from 2,400 pairs in 1977 to 1,390 in 1998.

Yellow-crowned night-heron *(N. violaceus)*. Rare in coastal marshes from spring to early fall (mainly western Connecticut); rare nester on islands and a few other coastal sites.

Fig. 20.6. These juvenile glossy ibises are products of one of Connecticut's few nesting sites. Paul J. Fusco/Connecticut DEP Wildlife Division.

Fig. 20.7. Turkey vultures have become increasingly common in recent years. Paul J. Fusco/Connecticut DEP Wildlife Division.

Ibises and Spoonbills (Order Ciconiiformes, Family Threskiornithidae)

Glossy ibis *(Plegadis falcinellus)* (Fig. 20.6) A few dozen pairs nest in shrubs on coastal islands near Mystic, Westbrook, Milford, and Norwalk; gregarious; small flocks fairly easy to find in coastal marshes in spring and summer; feeds on small aquatic animals in shallow water. *Note:* Ibis abundance increased locally in the mid-1900s.

Vultures and Condors (Order Ciconiiformes, Family Cathartidae)

Genetic evidence indicates that vultures, long grouped with hawks, actually are more closely related to storks. Vultures tend to be gregarious around large carcasses and when roosting.

Black vulture *(Coragyps atratus)*. Rare but increasingly numerous; mainly inland along Housatonic and Naugatuck river valleys and adjacent ridges. *Note:* The first documented nesting in Connecticut occurred in 2002.

Turkey vulture *(Cathartes aura)* (Fig. 20.7). Low numbers nest along cliffs and ledges; common migrant, primarily inland, in late winter and early spring and late summer to fall; numbers increasing in winter, when roosting aggregations sometimes include more than a hundred individuals; well-developed olfactory system, able to locate carrion through odor. *Note:* Turkey vultures reportedly were rare in Connecticut from the mid-1800s through the early 1900s, following an earlier period of greater abundance. Then the nesting and wintering ranges expanded northward, and these birds have become increasingly common in Connecticut since the 1920s, possibly due to warming climate and/or increased availability of food (road kills).

Swans, Geese, and Ducks

Waterfowl (Order Anseriformes, Family Anatidae) include some of the heaviest birds that fly. Some (particularly swans and geese) exhibit long-lasting, sometimes life-long pair bonds. Ducks tend to be relatively short-lived; about 33 to 50 percent of adults die each year. The reproductive biology of waterfowl is notable in that males are among the few birds that have a phallus.

Swans and geese in Connecticut primarily are grazers on wetland and aquatic plants and algae, though Canada geese often graze on land. Most ducks feed in the water on algae, seeds, tubers, foliage, insects, crustaceans, mollusks, worms, or amphibian larvae. Diving ducks tend to eat more invertebrates than do dabbling ducks, but nesting and juvenile dabblers eat many insects, and canvasbacks eat submerged algae and plants. Mergansers use their serrated bill to nab fishes.

Snow goose *(Chen caerulescens)*. Variably numerous as spring and especially fall migrant (some flocks include a few hundred birds); most seen in Connecticut are migrating between arctic breeding areas and wintering sites along U.S. mid-Atlantic coast; some winter along coast.

Canada goose *(Branta canadensis)*. Nonmigratory population present all year, nests on freshwater islets and marshy shores; paired adults generally stay together for life; families stay together for nearly a year before parents and young separate; different families sometimes form cooperative or "gang" broods. *Note:* The nesting population in Connecticut, not present before 1950, descended from captive stocks that through successive generations in confinement lost their migratory tradition. This introduced population has increased to many thousands and now occurs throughout the state, except when snow and ice force them to the coast. They are especially common on golf courses and other large lawns with adjacent water. Among the state's waterfowl species, only the mallard nests in Connecticut in larger numbers. Numbers are reduced somewhat by predators, including coyotes, red foxes, raccoons, snapping turtles, and dogs. Prior to the 1950s, only migratory populations occurred in Connecticut, and wintering birds were rare. Migratory Canada geese still pass through the state in spring and fall. These migrants nest on the Ungava Peninsula in northern Quebec (a seriously declining population) and in Canadian Atlantic Maritime Provinces (a stable population) and winter in Atlantic coastal states south of New England. Recently it was learned that an expanding goose population nesting in southwestern Greenland visits Connecticut mostly in late October and November; most of these geese continue on to wintering areas on Long Island and in southeastern Pennsylvania. Goose hunting seasons in Connecticut are timed to protect these migratory populations and focus the harvest on the large nonmigratory population.

Brant *(B. bernicla)* (Fig. 20.8). Spring and fall migrant along coast, occasionally on inland water; flocks of several hundred in October and November; sometimes common in winter (flocks of hundreds or thousands), but most are southward along mid-Atlantic states; makes heavy use of sea lettuce along Connecticut shores; nests on arctic tundra.

Mute swan* *(Cygnus olor)* (Fig. 20.9). Introduced into

Fig. 20.9. Mute swans are non-native birds that may have detrimental impacts on native waterfowl.

North America from Europe; established with large releases of birds in 1910 through 1912 in New York; expanded to Connecticut by 1930s; local breeding began in 1940s or 1950s; now very common and present year-round in coastal waters, also inland on rivers and lakes when not frozen; nests on freshwater and brackish shores mainly in southern half of state; aggressive, able to deter predation by skunks, raccoons, and opossums. *Note:* There is concern that the state's growing population of mute swans may detrimentally affect native waterfowl through aggressive interactions and effects on food resources (swans eat algae and aquatic plants). Efforts have been made to curtail the mute swan population (statewide total minimally 1,500 to 2,000) by inhibiting reproduction through "egg shaking." Shaking the eggs and returning them to the nest makes the eggs inviable and does not stimulate the female to lay a new clutch.

Tundra swan *(C. columbianus).* Arctic breeder, migrates through Connecticut in small numbers in spring and fall; mainly quiet coastal waters; most of east coast population winters farther south.

Ducks that Nest in Connecticut. All but the mergansers are dabblers, that is, they feed at the surface or by submerging the head while floating.

Wood duck *(Aix sponsa)* (Fig. 20.10). Fairly common in inland marshes and swamps, spring to fall; nests in natural tree cavities or nest boxes in or near water (once I found a female with downy young walking in the forest almost a mile from the nearest pond); single ponds or marshes sometimes host 100 to 500 or more individuals during September to October migrations; rare in winter. *Note:* After the mallard, this is Connecticut's most common nesting duck, with several thousand nests each year. Resurgence of the formerly depleted populations of this duck probably is due to

Fig. 20.8. Brant migrate through Connecticut, and some winter in Long Island Sound.

Fig. 20.10. Wood ducks are native waterfowl that nest in tree cavities and nest boxes. Paul J. Fusco/Connecticut DEP Wildlife Division.

increased habitat produced by the increasing beaver population, forest maturation and increased availability of large natural tree cavities, erection of nesting boxes in many marshes and swamps, and improved hunting regulations. Now that wood duck populations have rebounded, a problem has arisen. Where multiple nest boxes have been erected out in the open over water, the easily discovered nests often are subject to egg "dumping" by females nesting nearby, resulting in excessively large clutches and nest failure. One such nest I examined contained four dead downy young and 16 unhatched eggs (typical clutch size is 6 to 10 eggs).

American black duck (*Anas rubripes*). Common year-round in coastal and inland waters and wetlands; nests on ground in vegetation near water. *Note:* The black duck is far less numerous now than before the 1960s when a major decline began, due probably to loss of suitable habitat and negative effects (including hybridization) of the increasing mallard population. The winter population has fluctuated over the past few decades; populations in the early 1990s were near the low point of the recent fluctuation cycle.

Mallard (*A. platyrhynchos*). Common all year in coastal and especially inland waters and wetlands; Connecticut's most abundant nesting duck; nests on ground in vegetation near water; winter ice on inland waters may push mallards into coastal waters, where they compete with or displace black ducks, but competition probably is reduced somewhat by dietary differences (mallard prefers seeds, black focuses on submerged plants). *Note:* The Connecticut population includes both "wild" migratory birds and a population of relatively tame birds descended from introduced stock. Populations have increased in recent decades, due to introductions and favorable habitat changes (e.g., increase in suburban ponds).

Blue-winged teal (*A. discors*). Coastal and inland waters and wetlands; has nested sporadically in marshes; nests on ground in vegetation near fresh or brackish water; more common as early migrant in late summer to early fall than in spring; rare or absent in winter.

Gadwall (*A. strepera*). Present all year, usually uncommon; small but increasing numbers nest along coast, on ground in vegetation near fresh or brackish water.

Green-winged teal (*A. crecca*). Inland waters and coastal marshes; rare nester (on ground in vegetation near water); uncommon to common spring and fall migrant; coastal sites harbor gatherings of up to a few hundred during spring migration; uncommon in winter.

Hooded merganser (*Lophodytes cucullatus*) (Fig. 20.11). Diver in coastal and inland waters (mainly the former in winter); nests in tree cavities and nest boxes in swampy forests; look for females attended by young in June; usually uncommon from fall to spring, flocks of up to at least several dozen occur in late fall. *Note:* Populations in New England and New York increased over the past few decades, probably due to forest maturation and increased availability of suitable nest cavity trees, plus augmented wooded pond habitat created by increasing beaver populations. Several hundred pairs reportedly nest in Connecticut.

Common merganser (*Mergus merganser*). Diver, mainly in large rivers and lakes; localized nester, usually in tree cavities near water; fairly common fall through spring. *Note:* Sometimes assemblages of more than a thousand occur on certain lakes in early spring (e.g., Lake Waramaug in New Preston) or early winter (Lake Zoar in Southbury). This duck has been increasing southward as a nester in recent decades.

Ducks that Do Not Nest in Connecticut. These birds occur primarily during spring and fall migration or in winter.

Fig. 20.11. Connecticut's nesting population of hooded mergansers has increased in recent years. Paul J. Fusco/Connecticut DEP Wildlife Division.

The first two species are dabblers whereas the others are divers, that is, they forage while completely submerged.

Northern pintail *(Anas acuta)*. Mainly a migrant in coastal and inland waters; a few winter along coast; possibly a rare, sporadic breeder.

American wigeon *(A. americana)*. Fairly common spring and fall migrant, coastal and inland waters; can be locally common along coast in winter.

Canvasback *(Aythya valisineria)*. Fairly common to locally numerous during migration, late winter to early spring and mid- to late fall; generally uncommon in winter; mainly coastal (eats much sea lettuce), much less common inland. *Note:* Canvasback numbers in Connecticut are much lower now than they were before the 1990s.

Ring-necked duck *(A. collaris)*. Fairly common migrant, late winter to early spring and fall; rivers, lakes, and ponds; sometimes in groups of up to 200 to 300 or more; rare or uncommon in winter.

Greater scaup *(A. marila)*. Locally common (but declining), fall through early spring; mainly coastal waters. *Note:* The wintering population along the Connecticut coast declined from about 40,000 in the 1950s to just a few thousand in the 1990s and early 2000s, though flocks of a few thousand sometimes can still be found along the coast in late winter, especially in the area from West Haven to Fairfield. The decline may be due to poor nutrition, the effects of contaminants, and reduced habitat quality. High numbers during the 2000/2001 Christmas Bird Count hopefully signal an eventual reversal of recent declines.

Lesser scaup *(A. affinis)*. Uncommon in late winter to early spring and fall migration, rare in winter; primarily coastal waters, sometimes lakes and ponds.

Labrador duck *(Camptorhynchus labradorius)*. Extinct since 1870s or soon thereafter; occurred primarily along coast of southeastern Canada and northeastern United States, including Long Island Sound; may have succumbed to overharvest by early seafaring people.

Surf scoter *(Melanitta perspicillata)*. Variably numerous, fall through spring; coastal waters, rare inland; groups of several hundred sometimes seen in fall.

Black scoter *(M. nigra)*. Uncommon to rare, fall through spring; coastal waters, rare on inland lakes and reservoirs (though groups of a few dozen may occur during migration).

White-winged scoter *(M. fusca)*. Locally common, fall through spring; coastal waters, rare inland. *Note:* This and the other scoters seem to be less abundant in recent decades than in previous years.

Long-tailed duck (oldsquaw) *(Clangula hyemalis)*. Variable abundance, fall through spring; mainly coastal waters, rare inland; groups of up to several hundred sometimes

Fig. 20.12. Migrating and wintering common goldeneyes use large bodies of water.

occur at coastal sites around the vernal equinox; far more common in oceanic waters (e.g., 50,000 or more winter around Nantucket, Massachusetts).

Common goldeneye *(Bucephala clangula)* (Fig. 20.12). Locally common, fall through spring; coastal waters and rivers, sometimes inland on large lakes; coastal coves may harbor hundreds in late winter. *Note:* I usually see a few when I kayak on the Connecticut River in Haddam in winter.

Bufflehead *(B. albeola)*. Fairly common, fall through spring; coastal waters, rivers, and lakes.

Red-breasted merganser *(Mergus serrator)*. Common, fall through spring; primarily coastal waters, can be seen in winter almost anywhere along the shore of Long Island Sound; rare inland on large bodies of water. *Note:* Nesting at Milford Point was suspected but not confirmed in 2000.

Ruddy duck *(Oxyura jamaicensis)*. Generally uncommon migrant, late winter to early spring and fall, and usually rare or uncommon in winter; coastal and inland waters. *Note:* Despite the general scarcity of this duck, single lakes sometimes harbor groups of 100 to 250 migrants in October and November, and nearly 900 were counted on Candlewood Lake on New Year's Day in 2000.

Eagles, Hawks, and Falcons

Osprey, Eagles, and Hawks (Order Falconiformes, Family Accipitridae)

Hawks and eagles capture prey with their long, sharp talons. Unless otherwise specified, the diet includes mainly various small vertebrates.

Osprey *(Pandion haliaetus)* (Fig. 20.13). Locally common, spring through fall; fairly common nester, mainly on platforms erected along central and eastern coast; migrations peak in early to mid-spring and late summer to early

Fig. 20.13. These osprey fledglings symbolize the dramatic comeback of this bird over the past few decades. Paul J. Fusco/Connecticut DEP Wildlife Division.

fall; many adults migrate in late summer and fall prior to subsequent peak in juvenile migration; feeds almost exclusively on medium-sized fishes such as flounders, caught by aerial dive and feet-first plunge into water. *Note:* The osprey has rebounded, with assistance from humans, from its former rarity that was caused by reproductive failure (due to pesticide contamination), nest disturbance, and shooting. The number of active nests in Connecticut increased from just 9 in 1974 to 162 (producing 315 fledglings) in 1999. The nesting range has expanded westward along the coast and inland up the Connecticut and Quinnipiac rivers. At Great Island at the mouth of the Connecticut River, low productivity in the mid-1990s may have been due to nest predation by raccoons that climbed nesting platform poles before barriers were installed on the poles. Today it is amazing to consider that many ospreys nested successfully on the ground in the days when raccoons were scarce.

Bald eagle *(Haliaeetus leucocephalus).* Nests locally in tops of large trees along rivers and lakes; incubation begins as early as March; young have fledged in early to mid-July; approximately 50 to 130 (usually around 60 or 70) present in winter in recent years, mainly along major rivers; feeds mainly on unhealthy or dead fishes and waterfowl. *Note:* As a result of pesticide-caused reproductive failure, shooting, and disturbance by humans in the early and middle 1900s, bald eagle populations throughout much of North America plummeted, and nesting of bald eagles in the state dwindled to zero during the 1930s to 1950s. After the early 1970s, with reduced pesticide use, better protection, and reintroductions of young eagles, North American populations strongly rebounded. Bald eagles began attempting to nest again in Connecticut in 1990 and were successful in 1992 (one pair at Barkhamsted Reservoir). Since then, the number has risen to several nesting pairs, and the nesting range has expanded to additional sites across the state. Six pairs fledged ten chicks in 2003. Eagles hatched in Connecticut have been observed in subsequent years in Massachusetts, New York, and Connecticut. A female that hatched in Connecticut in 1994 nested successfully along the Hudson River in New York in 2000, 2001, and 2002.

A special eagle observation area has been established along the Housatonic River. To make a reservation to visit the Shepaug Bald Eagle Viewing Area (9:00 A.M. to 1:00 P.M., Wednesdays, Saturdays, and Sundays, December to mid-March), call 1-800-368-8954 (Tuesday through Saturday, 10:00 A.M. to 4:00 P.M.), or check out www.nu.com/environmental/steward/shepaug.asp. You also might try searching from any of the boat launches along the wooded portions of the Connecticut River.

Northern harrier *(Circus cyaneus).* Long known as marsh hawk; nests (on ground) at Great Meadows in Stratford, perhaps also in a few other scattered marshes; uncommon to common during late summer to fall and spring migrations; a few winter; mainly in marshes and fields; small mammals such as voles dominate diet, occasionally preys on ducks, songbird nestlings, or eggs. *Note:* Nesting was more common in the 1800s. Habitat loss, pesticide poisoning, increased populations of predatory mammals (resulting from human-augmented food sources), and disturbance by humans during the nesting season likely were factors causing its decline in the 1900s.

Sharp-shinned hawk *(Accipiter striatus)* (Fig. 20.14). Rare nester in well-hidden sites near top of conifer trees (often white pine) in conifer-dominated stands; most common during spring and especially fall migrations; sometimes lingers into winter (I regularly see them attempting to capture birds near my feeders); diet dominated by small songbirds, plus occasional mice *(Peromyscus, Microtus)* and shrews *(Blarina).* *Note:* On a good day in September at

Fig. 20.14. This immature sharp-shinned hawk is eating a songbird it caught at a bird feeder on a snowy morning.

Lighthouse Point Park in New Haven, one can see hundreds of sharp-shins flying westward along the coast. In late summer and fall, thousands funnel through this area from nesting areas extending far to the north and east. This species was a common nester in Connecticut until at least the early 1900s, but it declined to apparent extirpation as a breeder by the late 1950s. More recently, breeding observations have increased, and now small numbers nest again in the hilly parts of northern Connecticut.

Cooper's hawk *(A. cooperii)*. Nests in small numbers in large trees in northern and central uplands, primarily in west; uncommon migrant in late summer to fall and spring; rare in winter; eats mainly birds; like all *Accipiter* hawks, sometimes takes nestling songbirds.

Northern goshawk *(A. gentilis)*. Nests in small numbers in large trees in northern and western uplands, primarily in live trees near water; uncommon migrant in late summer to fall and late winter to spring; rare in winter; main foods are birds and mammals the size of grouse and squirrels. *Note:* The nesting population appears to be at least stable and likely gradually increasing.

Red-shouldered hawk *(Buteo lineatus)*. Uncommon spring and fall migrant and nester in large deciduous trees in undisturbed forested wetlands and other forests near water or wetlands; usually rare in winter.

Broad-winged hawk *(B. platypterus)*. Fairly common nester in large deciduous trees near water in forested areas; large numbers pass through in September and October and April to mid-May, when migrating to and from wintering areas in South America; absent in winter; feeds opportunistically on various small vertebrates, sometimes including bird nestlings and eggs. *Note:* On occasions at Quaker Ridge in Greenwich, over 30,000 migrants have been seen on a single day, though a few thousand is more typical for a good day at a single count area.

Red-tailed hawk *(B. jamaicensis)* (Fig. 20.15). Present year-round; widespread nester in large trees near open areas; fairly common winter resident and late winter to spring and late summer to fall migrant. *Note:* This is the most commonly seen soaring hawk and is conspicuous in winter as it perches in leafless trees along forest edges and highways.

Falcons and Caracaras (Order Falconiformes, Family Falconidae)

Falcons prey on small animals, often captured at the end of a high-speed aerial dive. Birds often figure prominently in the diet, though kestrels eat mainly small mammals and insects.

Fig. 20.15. Red-tailed hawks often perch in trees along highways in winter. Paul J. Fusco/Connecticut DEP Wildlife Division.

American kestrel *(Falco sparverius)*. Rare nester in tree cavities or nest boxes in open areas with scattered trees and along forest edges; most common during spring and fall migrations; uncommon at best in winter. *Note:* Loss of open habitat and nesting cavities has resulted in a significant decline in the abundance of kestrels in the northeastern United States over the past couple of decades.

Merlin *(F. columbarius)*. Uncommon to rare spring and fall migrant (seen mainly in fall); rare in winter; nests mainly in Canada and Alaska; does not breed in Connecticut.

Peregrine falcon *(Falco peregrinus)* (Fig. 20.16). Rare nester, spring and late summer to fall migrant, and winter visitor; mainly along coast (preys on shorebirds). *Note:* This falcon historically nested in Connecticut (until the 1940s) but was extirpated through uncontrolled shooting and egg collecting. After World War II, widespread use of the pesticide DDT resulted in disastrous eggshell thinning and reproductive failure. The nesting population dropped from at least 350 pairs east of the Rocky Mountains prior to the 1940 to zero by the mid-1960s. The banning of DDT use in the United states and release of hundreds of young falcons in the northeastern United States since the mid-1970s, together with a virtual elimination of egg collecting and shooting, allowed recovery. The population in northern New York and northern New England was up to 42 pairs in natural nest sites by 1996. Today a few pairs nest each year in Connecticut (four successful nests in 2003), generally on ledges of tall buildings (e.g., the Travelers Tower in Hartford) or under bridges (e.g., P. T. Barnum Bridge in Bridgeport, where the

Fig. 20.16. Connecticut has a small nesting population of peregrine falcons. Paul J. Fusco/Connecticut DEP Wildlife Division.

Fig. 20.17. Ruffled grouse populations may have declined in recent years. Paul J. Fusco/Connecticut DEP Wildlife Division.

Wildlife Division and Department of Transportation have installed nest structures). These city-associated birds have helped rebuild populations in natural habitats. For example, a chick hatched on the Travelers Tower in 1997 showed up as an adult in Maine in 2000 and nested successfully on a cliff ledge in New Hampshire in 2001. Nesting on structures is typical in most of the eastern United States, except in northern New York and northern New England, where cliff nesting is the rule. Prior to the DDT era, almost all nests in the eastern United States were on cliffs, including several traprock ridges in Connecticut. With recovery of peregrine populations, we're beginning to see these falcons visiting some of their historical nesting ledges.

Pheasants, Grouse, Turkeys, and Quail

Pheasants, Grouse, Turkeys, and Old World Quails (Order Galliformes, Family Phasianidae)

The Phasianids are primarily ground-dwelling birds that eat buds, fruits, nuts, seeds, leaves, flowers, and insects.

Ring-necked pheasant* *(Phasianus colchicus).* Native to Asia; uncommon resident in fields and other grassy/brushy areas; some feral birds nest, primarily in the southwestern quarter of the state; nests in vegetation on ground. *Note:* Most of the pheasants that one sees are pen-raised birds that have been released for hunting. In recent years the state has released more than 20,000 birds each fall. Few of these birds survive very long, even if they avoid being shot. They are ill-equipped for survival, and a state such as Connecticut, where much of the land is unsuitable forest and where many areas of suitable farmland habitat recently have been lost to residential development, does not provide especially favorable conditions for these grassland-adapted birds. They do, however, sometimes take advantage of food spilled from bird feeders.

Ruffed grouse *(Bonasa umbellus)* (Fig. 20.17). Forest and forest-edge dweller; resident year-round; nests on ground among forest vegetation; beginning in April, listen for unbirdlike accelerating thumping sounds made by males in efforts to attract females. *Note:* Ruffed grouse were scarce in the 1800s when the state was largely deforested, but they subsequently rebounded with reforestation. However, populations seem to have declined in recent years, perhaps due to renewed loss of forest habitat and aging of remaining forests, which provide fewer food resources than do multi-age forest mosaics in which this bird thrives.

Greater prairie-chicken or "heath hen" *(Tympanuchus cupido).* Extirpated from the northeastern United States; formerly occurred in open scrub oak and pine barrens from Massachusetts to Virginia, including Long Island (New York) until 1840s and Martha's Vineyard (Massachusetts) until 1932; probably occurred in southern Connecticut prior to the mid-1800s. *Note:* Habitat change, likely including exclusion of fire by humans, and excessive shooting probably caused the demise of the heath hen.

Wild turkey *(Meleagris gallopavo)* (Fig. 20.18). Common year-round resident in deciduous forest and adjacent old fields, meadows, and agricultural lands, mostly away from coast and central lowlands; winter roosts often in large white pines near water (provide good shelter, openly spaced limbs for easy access and group roosting, and convenient source of drinking water); nests on ground among dense forest understory vegetation, with incubation beginning in May; vast majority of females nest each year, but most are unsuccessful; many females succumb to mammal predation during nesting season. *Note:* This relatively huge bird was widespread and abundant in the state during initial

Fig. 20.18. Wild turkeys, formerly extirpated from Connecticut, now are common statewide. Paul J. Fusco/Connecticut DEP Wildlife Division.

colonial times but was totally extirpated by the early 1800s as a result of excessive hunting and elimination of its forest habitat. Beginning in 1975 and continuing through the late 1970s and 1980s, after much of the state was reforested following the decline of farming, reintroductions of wild-caught birds from New York and later translocations within Connecticut, followed by natural population growth in the now favorable habitat, resulted in the reestablishment of the turkey as a common Connecticut resident. Turkeys (more than thirty thousand birds in 2000) have now repopulated virtually all suitable habitat within the state, including all of the state's 169 towns. Large flocks, rarely including 200 to 300 birds, are now a common sight along country roads. The statewide population is large enough to have allowed transplants from Connecticut to other states attempting to restore this species. The burgeoning population also supports spring and fall hunting seasons, though the birds' wariness limits the kill. In 2001 and 2002, hunters annually harvested approximately 2,000 wild turkeys in Connecticut in spring and around 250 to 360 birds in fall.

New World Quails (Order Galliformes, Family Odontophoridae)

Northern bobwhite *(Colinus virginianus)*. Uncommon, year-round, ground-nesting resident; most numerous east of Connecticut River; released pen-raised birds appear almost anywhere; major foods include buds, fruits, nuts, seeds, leaves, flowers, and insects. *Note:* Probably the scarcity of wild bobwhite in the state is due mainly to the scarcity of suitable field-thicket habitat, severe winter conditions for which this species is not well adapted, and the deleterious effects of cross-breeding with less-hardy birds that were imported from southern states and released in Connecticut.

Rails, Gallinules, and Coots

Most rails (Order Gruiformes, Family Rallidae) are mud treaders and waders that eat mainly invertebrates, seeds, and other plant material picked from wet ground or shallow water. Gallinules and coots also swim and dive for food.

Black rail *(Laterallus jamaicensis)*. Rare marsh dweller, late spring through early fall; possibly a very rare nester but apparently does not nest in Connecticut in most years.

Clapper rail *(Rallus longirostris)* (Fig. 20.19). Fairly common nester in coastal saltwater and brackish-water marshes; most migrate out of Connecticut for winter.

King rail *(R. elegans)*. Close relative of clapper rail (the two species sometimes hybridize); rare nester in freshwater or brackish marshes; generally absent in winter.

Virginia rail *(R. limicola)*. Uncommon during most of year; nests on ground in many freshwater marshes; may occur in coastal marshes during migration; favorable sites harbor a dozen or more during peak migration periods; rare in winter. *Note:* I saw an adult with two downy young in a freshwater marsh at the end of July.

Sora *(Porzana carolina)*. Uncommon to rare nester; uncommon spring and fall migrant; mainly in inland marshes. *Note:* In late May, while slogging through the marshy margin of a beaver pond, I inadvertently flushed a sora off a nest that contained ten eggs. In late summer and early fall, soras can be plentiful in freshwater tidal marshes of the lower Connecticut River valley, feeding on wild rice.

Common moorhen *(Gallinula chloropus)*. Also known as common gallinule; uncommon at best from spring

Fig. 20.19. Clapper rails nest in Connecticut's salt marshes. Paul J. Fusco/Connecticut DEP Wildlife Division.

through fall; nests sparsely in scattered freshwater marshes and marshy ponds; generally absent in winter;

American coot *(Fulica americana).* Inland marshes, ponds, lakes, and coastal waters; generally uncommon during spring and fall migration (especially the latter), rare at other seasons; some lakes and large ponds host two hundred or more during November migrations; sometimes locally common in mild winters, especially on Bantam Lake; no recent records of nesting in Connecticut.

Shorebirds

Plovers and Lapwings (Order Charadriiformes, Family Charadriidae)

Plovers tend to pick food items from the ground surface rather than probe into it.

Piping plover *(Charadrius melodus)* (Fig. 20.20). Mainly coastal; uncommon from early spring through early fall; nests on sandy, open or partially vegetated flats above adjacent beach and tidal flat feeding areas; absent in winter. *Note:* In 1996 to 2002, the population along the Connecticut coast was relatively stable at 21 to 32 pairs nesting at 9 to 10 sites, producing about 20 to 60 fledglings annually. This species receives careful protection from human disturbance by a network of volunteers, state wildlife biologists, and private conservation organizations such as The Nature Conservancy (55 High Street, Middletown CT 06457, 860-344-0716), but nests sometimes are destroyed or abandoned as a result of predation or flooding by high tides.

Killdeer *(Charadrius vociferus).* Present year-round, least common in winter; nests on open ground, often in open gravelly or shelly areas.

Black-bellied plover *(Pluvialis squatarola).* Most common from mid-summer through fall, less numerous in spring (but aggregations of several hundred may occur at favorable sites in late May); uncommon in winter; beaches, tidal flats, sandbars, and short-grass fields along coast; does not nest in Connecticut; eats polychaete worms, plus small clams and other small prey.

American golden-plover *(P. dominica).* Uncommon migrant in late summer to early fall, rare in spring; coastal beaches, tidal flats, sandbars, and short-grass fields; may accompany black-bellied plovers.

Semipalmated plover *(Charadrius semipalmatus).* Fairly common to locally abundant migrant on coastal beaches, tidal flats, and sandbars in late spring and mid-summer to early fall.

Oystercatchers (Family Haematopodidae)

American oystercatcher *(Haematopus palliatus)* (Fig. 20.21). Coastal beaches and tidal flats; present nearly year-round; nests in low numbers on open upper beaches at several coastal sites; least numerous in winter; uses long, straight, compressed bill to feed on mollusks and other marine invertebrates. *Note:* Populations have been increasing in recent years, and the loud whistles of this attractive shorebird are now a regular feature of Connecticut's coastal shores.

Sandpipers, Phalaropes, and Allies (Order Charadriiformes, Family Scolopacidae)

Most sandpipers gregarious wading birds that pick and probe in shallow water and mud of shores, wetlands, or shal-

Fig. 20.20. Successful nesting by Connecticut's piping plovers depends on careful protection from human beach-goers and predators. Paul J. Fusco/Connecticut DEP Wildlife Division.

Fig. 20.21. American oystercatchers expanded their nesting range into Connecticut in recent years. Paul J. Fusco/Connecticut DEP Wildlife Division.

low waters to obtain mollusks, crustaceans, worms, insects, and other invertebrates. The majority make long-distance migrations.

Sandpipers that Nest in Connecticut. All of these birds are scarce or absent in winter.

Willet (*Catoptrophorus semipalmatus*). Present spring through fall; local nester on ground in vegetation in or near salt marshes; forages in marshes and on beaches and tidal flats. *Note:* The loud calls and flashy wings of this bird make it conspicuous in summer nesting areas.

Spotted sandpiper (*Actitus macularia*): Nests widely but sparsely, on dry ground, usually near freshwater lakes, ponds, and streams, sometimes near tidal creeks; occasionally nests in areas with regular human activity; fairly common migrant in spring and late summer to early fall along nearly any tidal or nontidal shore. *Note:* Unlike most sandpipers, males incubate eggs and tend young.

Upland sandpiper (*Bartramia longicauda*). Favors large fields, such as those associated with Bradley International Airport; rare nester among vegetation on ground in north-central Connecticut; rare spring and late summer migrant.

Wilson's snipe (*Gallinago delicata*). Uncommon in marshes and wet grassy areas along coast and inland, spring and late summer to fall; rare in winter; rarely may nest (in vegetation on ground) in some inland marshes, though recent records are lacking.

American woodcock (*Scolopax minor*) (Fig. 20.22). Uncommon inland ground-nester in moist thickets and woods; uncommon spring and fall migrant in coastal and inland wetlands; generally rare in winter, though "spring" arrivals may begin in February. *Note:* Habitat loss through residential and commercial development, and natural reforestation of shrubby old fields, have caused a steady decline in the woodcock population in Connecticut and elsewhere in the northeastern United States.

Sandpipers that Do Not Nest in Connecticut. These species occur primarily during migration. Many of them migrate between arctic breeding areas and South American wintering grounds; most occur mainly along the coast in marshes and on beaches and tidal flats

Greater yellowlegs (*Tringa melanoleuca*). Common in spring and from mid-summer to early fall; sometimes assembling in groups of a hundred or more during peak migration periods; generally rare at other seasons; inland and coastal shores and shallows; often emits loud calls in flight.

Lesser yellowlegs (*T. flavipes*). Uncommon from mid-summer to early fall, less common in spring; coastal and inland shores and shallows.

Solitary sandpiper (*T. solitaria*). Uncommon in spring and late summer along inland streams and ponds.

Eskimo curlew (*Numenius borealis*). Historically common, now probably extinct as a result of market hunting; formerly a late summer to early fall migrant along Connecticut coast; last appeared in Connecticut in late 1800s; not reliably documented anywhere since the 1980s.

Whimbrel (*N. phaeopus*). Rare along coastal shores and wetlands in spring and uncommon in mid- to late summer.

Ruddy turnstone (*Arenaria interpres*) (Fig. 20.23). Common along coastal shores, tidal flats, and sandbars in spring and from mid-summer to early fall; uncommon in other seasons.

Red knot (*Calidris canutus*). Generally uncommon along coastal shores in spring and from mid-summer to early fall; very rare in other seasons.

Sanderling (*C. alba*). Fairly common along coastal beaches in spring and mid-summer through fall; small

Fig. 20.22. Woodcocks are unusual "shorebirds" that depend on old fields and woods. Paul J. Fusco/Connecticut DEP Wildlife Division.

Fig. 20.23. Ruddy turnstones migrate and winter along the Connecticut coastline. Paul J. Fusco/Connecticut DEP Wildlife Division.

flocks here and there in winter; sometimes present in summer when most are on arctic breeding grounds.

Semipalmated sandpiper (*C. pusilla*). Common in late spring and from mid-summer to early fall, mainly along coastal shores and wetlands; a few at inland bodies of water; Connecticut's most abundant migrant shorebird; aggregations may include several thousand individuals; nonbreeders often linger in summer.

Western sandpiper (*C. mauri*). Generally uncommon along coastal beaches and tidal flats from mid-summer to early fall; rare in spring.

Least sandpiper (*C. minutilla*). Fairly common in late spring and from mid-summer to early fall; rocky and muddy coastal intertidal areas, tidal mudflats along rivers, and other wet, open areas inland; few present during late spring to early summer arctic nesting season.

White-rumped sandpiper (*C. fuscicollis*). Rare or uncommon in late summer and early fall and in late spring, mainly along coastal shores and tidal flats, sometimes on inland shores and mudflats.

Pectoral sandpiper (*C. melanotos*). Uncommon in late summer and early fall and in spring, in coastal and inland marshes and wet fields.

Purple sandpiper (*C. maritima*) (Fig. 3.10). Generally uncommon along rocky coastal areas and jetties, mid-fall through mid-spring; can be locally numerous.

Dunlin (*C. alpina*). Often the most common sandpiper on coastal tidal flats, muddy shores, and sandbars in fall and spring; sometimes common in winter as well.

Short-billed dowitcher (*Limnodromus griseus*). Common in shallow water of muddy intertidal areas from mid- to late summer (flocks of up to several hundred assemble at favorable coastal sites); less common in spring; sometimes present in early summer when others are on arctic breeding grounds.

Gulls, Terns, and Skimmers

Gulls and terns (Order Charadriidae, Family Laridae) nest in colonies, usually on the ground. Gulls feed opportunistically on whatever small animals or human refuse is available in or near water and sometimes on land far from water. Terns feed primarily on small fishes, which they catch by plunging headfirst into the water after an air dive of several yards.

Gulls that Breed in Connecticut. These opportunistic gulls have benefited from human wastefulness.

Herring gull (*Larus argentatus*). Common all year, mainly along coast, locally inland near large bodies of water and especially at garbage dumps; nests on ground in scat-

Fig. 20.24. Great black-backed gull populations increased in Connecticut during the 1900s. Paul J. Fusco/Connecticut DEP Wildlife Division.

tered colonies on coastal islands and locally on large flat rooftops. *Note:* This gull, present only from fall through spring a century ago, first nested in Connecticut in the 1940s; now about 1,200 pairs nest. The nesting population has declined in recent decades. Versatile herring gulls employ many methods of obtaining food, ranging from simple scavenging to skillful predation and food processing. Countless times I watched adult and immature herring gulls break open shellfish (moonsnails, ribbed-mussels) by flying up and dropping them onto rocks, roads, parking lots, or roofs. They often search for crabs while floating in shallow coastal waters. They catch the crabs by plunging and nearly submerging head-first, sometimes using a little leap off the water to gain extra momentum for their brief dive.

Great black-backed gull (*L. marinus*) (Fig. 20.24). Fairly common all year along the coast and Connecticut River, less common inland at other large bodies of water; nests on ground along the coast, usually on islands. *Note:* In the early 1900s, this species was present in Connecticut only as a nonbreeder from fall through spring. Populations have been increasing in recent decades, with the nesting population numbering more than five hundred pairs.

Gulls that Do Not Breed in Connecticut. These species may be present during the breeding season.

Laughing gull (*L. atricilla*). Common along coast in spring and late summer to fall; nonbreeders present when others are nesting elsewhere; late summer assemblages include up to several hundred individuals; rare in winter.

Bonaparte's gull (*L. philadelphia*). Fairly common along coast, fall through mid-spring; few range inland along Connecticut River; rare but perhaps increasing in summer on Long Island Sound.

Ring-billed gull (*L. delawarensis*). Common all year along coast and to a lesser extent inland; most common inland gull. *Note:* The flocks of gulls that often occur in large ballfields generally are this species, as are the gulls that frequent parking lots of shopping centers and fast-food restaurants. This species, formerly a rare migrant, has increased in abundance in the region. For example, it was first recorded as nesting in New York in 1936 and now that state hosts more than a hundred thousand nesting pairs. Likely some of these visit Connecticut.

Roseate tern *(Sterna dougallii).* Uncommon to rare along coast, mid-spring through summer; localized colonial nester; nests in colonies on islands along central coast; most nest on Falkner Island (also known as Faulkner's Island); nests on ground under low cover; on Long Island (New York), nests very infrequently in salt marshes, but these colonies never persist. *Note:* Research by the Falkner Island Tern project, headed by Jeff Spendelow, found that the nesting colony of Falkner Island steadily declined from 150 pairs in 1997 to 70 pairs in 2002. The terns produced only 13 fledglings in 2002, which was due mostly to heavy predation by black-crowned night-herons. Some adult terns travel more than 12 miles from their colony to foraging areas. Roseate terns hatched on Falkner Island have been captured in winter south of the equator in coastal Brazil. A tern banded as a chick in Massachusetts in 1971 was captured in Brazil in 1997, setting the record for the oldest known roseate tern.

Common tern *(S. hirundo)* (Fig. 20.25). Indeed common along coast from mid-spring through early fall; thousands gather in late summer assemblages in coastal mainland sites; several thousand pairs nest in colonies at some twenty coastal sites (roughly 3,000 to 4,000 pairs on Falkner Island in recent years); nests usually on ground in open areas on islands, but historically nested in some Connecticut salt marshes, including those at the mouth of the Hammonasset River. *Note:* Common terns hatched on Falkner Island have been captured in winter south of the equator in coastal Brazil.

Forster's tern *(S. forsteri).* Uncommon near coastal marshes, late summer and fall; rare in spring; groups of two to three dozen appear along coast in September and October.

Fig. 20.25. These two hungry common tern chicks confronted their parent on Falkner Island.

Least tern *(S. antillarum)* (Fig. 3.4). Uncommon to locally common along coast (sometimes at near-coastal lakes), mid-spring through summer; nests on ground, usually in open areas of sandy upper mainland beaches. *Note:* A century ago, least terns were very rare in Connecticut, following a period when they were abundant as migrants. The state population increased dramatically from the 1960s to the 1980s, due in large part to influxes of birds that hatched in colonies in New York. In recent years, the nesting population in Connecticut, distributed in scattered small colonies, has steadily declined, down from over 1,000 nesting pairs in 1987 and 1988 to just 175 to 240 pairs in the early 2000s. The largest colony in Connecticut is at Sandy Point in West Haven. The number of pairs nesting regionally in Connecticut, Massachusetts, New York, and Rhode Island has been stable at about 7,000 in recent years. Connecticut nesting success lately has been low, with only a few dozen chicks fledged each year since the mid-1990s. Poor reproductive success has been due to predation on eggs and chicks (by black-crowned night-herons, herring gulls, American crows, small mammals, cats, and others), flooding of nests by high tides and heavy rains, and excessive disturbance by humans.

Black skimmer *(Rynchops niger).* Present from mid-spring to early fall; rare nester on ground in open areas of upper coastal mainland beaches or islands; roosts on sandy shores and bars; feeds on fishes obtained by flying with lower mandible immersed. *Note:* In the late 1990s, the first successful nestings were documented at Morse Point and Sandy Point in West Haven.

Doves, Parakeets, and Cuckoos

Pigeons and Doves (Order Columbiformes, Family Columbidae)

Columbids eat seeds, fruits, and some insects, picked up from the ground or plucked from vegetation. Both parents feed young nestlings a milky secretion produced in the adult's crop.

Rock pigeon* *(Columba livia).* Formerly known as rock dove; introduced into North America from Eurasia; common nester, particularly in vicinity of towns, bridges, and dairy farms, sometimes on cliffs; nests usually on ledges under cover in human-made structures.

Mourning dove *(Zenaida macroura)* (Fig. 20.26). Nests commonly in trees or large shrubs in various wooded habitats, including residential areas; many pass through during early spring and early fall migrations (flocks may include up to a few hundred individuals); feeds almost exclusively on seeds on ground in wooded areas and fields. *Note:* Winter populations have been increasing, probably because milder

Fig. 20.26. Mourning doves are seed eaters that sometimes occur in large flocks. Paul J. Fusco/Connecticut DEP Wildlife Division.

Fig. 20.27. Non-native monk parakeets lay their eggs in large communal nests. Paul J. Fusco/Connecticut DEP Wildlife Division.

winters and augmented food resources provided by backyard bird feeders reduce the need to migrate.

Passenger pigeon *(Ectopistes migratorius)*. Historically nesting species present in Connecticut primarily from spring to fall; extinct. *Note:* This species was formerly one of the most abundant birds on Earth, with a population of perhaps two billion, but it became rare in Connecticut by the 1870s and 1880s. The last large flock in Connecticut, including perhaps 100,000 birds, was seen (and heavily hunted) in September 1876 at the mouth of the Thames River. The species disappeared from the wild around the turn of the century, and was extinct by 1914. It declined and disappeared as settlers cleared the land of mature deciduous forest (source of the pigeons' food). Excessive shooting (for food and sport) and human disturbance of nesting colonies probably were additional important factors in the decline. The passenger pigeon, a colonial nester, may have required large flock sizes to stimulate nesting, and as the birds became scarce they may have stopped breeding. Possibly this species went extinct because of its strict social requirements for nesting, whereas other less social forest birds were able to persist in remnant forests and bounce back when forests regrew.

Parrots and Parakeets (Order Psittaciformes, Family Psittacidae)

Monk parakeet* *(Myiopsitta monachus)* (Fig. 20.27). Fruit- and seed-eating native of South America; local nester in towns along coast; multiple pairs build single large stick nest, each pair has own nest compartment; may depend in part on bird feeders to survive winter; recent statewide count yielded at least several hundred individuals.

Carolina parakeet *(Conuropsis carolinensis)*. Extinct species of southeastern United States (not confirmed alive since 1918); occasionally may have wandered north to Connecticut, but no definite evidence of this.

Cuckoos, Roadrunners, and Anis (Order Cuculiformes, Family Cuculidae)

Both Connecticut cuckoo species are attracted to areas with caterpillar infestations. Local cuckoo populations often increase dramatically when gypsy moth larvae or tent caterpillars are numerous, then plummet to rarity when the outbreak is over. Breeding Bird Survey data indicate that the abundance of both cuckoos has decreased dramatically in Connecticut over the past few decades. When food is superabundant, Connecticut cuckoos sometimes lay additional eggs in the nests of other cuckoos (either species) or other birds. Both Connecticut cuckoos winter in South America.

Black-billed cuckoo *(Coccyzus erythropthalmus)*. Uncommon nester in shrubs or small trees in brushy forest edges, mid-spring through late summer or early fall.

Yellow-billed cuckoo *(C. americanus)*. Uncommon nester in shrubs or trees in brushy forest edges or overgrown fields or orchards, from mid-spring through late summer or early fall.

Owls, Nighthawks, and Nightjars

Barn Owls (Order Strigiformes, Family Tytonidae)

Barn owl *(Tyto alba)*. Rare, declining; mainly in vicinity of nesting areas and winter roost sites in coastal region and inland lowlands; nests mainly in old buildings (e.g., barns, dilapidated water towers, attics) with easy access from outside; tolerant of human activity around nesting sites but requires nearby unwooded habitats with sufficient food resources; based on several hundred prey remains, I found that small mammals, mainly meadow voles, plus occasional star-nosed moles and short-tailed shrews, are primary foods in semi-rural areas of central Connecticut, whereas Norway rats dominate diet in urban areas; acute hearing allows prey localization by sound alone. *Note:* Many of the grasslands that formerly served as foraging habitat have been lost to reforestation and residential development. In the 1990s, to offset the loss of old farm buildings historically used for nesting, the DEP Wildlife Division began installing nest boxes for barn owls (for information, contact the Nonharvested Wildlife Program, Franklin, CT).

Typical Owls (Order Strigiformes, Family Strigidae)

Typical owls are active primarily or, in some cases, exclusively at night. Connecticut species often depend primarily on small mammals for food but may also take birds and other small animals as available.

Eastern screech-owl *(Megascops asio).* Uncommon but widespread year-round resident in wooded areas, including suburban areas with mature trees; nests in tree cavities and nest boxes.

Great horned owl *(Bubo virginianus)* (Fig. 20.28). Uncommon but widespread nesting resident in wooded areas, including small groves in landscapes fragmented by residential and commercial development; nests generally consist of old stick platforms made by other birds or squirrels (red-tailed hawk, American crow, gray squirrel, etc.) in large trees, often living oaks, red maples, and conifers (hemlock, white pine, Norway spruce); highly opportunistic predator; samples of prey remains from coastal Connecticut indicate diet dominated by Norway rats and wide assortment of birds, including waterbirds and songbirds. *Note:* These owls are long-lived. A dead great horned owl recently found in western Connecticut had been banded as a nestling in the same region nearly twenty years earlier.

Snowy owl *(Bubo scandiacus).* Rare in winter; mainly in wide open spaces of the coast that resemble owl's usual tundra habitat; some years, such as winter of 2001/2002, bring major influxes of owls and yield many scattered sightings.

Barred owl *(Strix varia).* Uncommon but widespread nesting resident in forested wetlands and other wooded areas; nests in hollows of large standing trees, stick platforms made by other birds, or in suitably large nest boxes.

Long-eared owl *(Asio otus).* Rare migrant and winter visitor, mid-fall to early spring, mainly along coast; rare nester at best (no definite records in recent years); nests usually in old stick platforms made in trees by crows or raptors, usually along streams; winter roosts often in conifers.

Short-eared owl *(A. flammeus).* Rare in large marshes and fields, fall to early spring; often flies in daylight and preys on voles; not known to have nested in Connecticut in more than a hundred years.

Northern saw-whet owl *(Aegolius acadicus).* Rare from mid-fall to mid-spring; irruptive migrant (e.g., large numbers in 1995 and 1999); migrants generally arrive beginning in mid- to late October; rare nester in tree cavities or nest boxes in forests, with most known or suspected nesting occurrences in northwestern part of state; winter roosts generally in conifers.

Nighthawks and Nightjars (Order Caprimulgiformes, Family Caprimulgidae)

Nighthawks nest on bare or leafy ground and feed on flying insects captured mainly at night or in the morning or evening. They have a small bill but a huge mouth.

Fig. 20.28. Defensive behavior of a juvenile great horned owl (late April) makes it appear larger than it actually is. Stephen P. Broker.

Fig. 20.29. Whip-poor-wills are beautifully cryptic on leaf litter. Paul J. Fusco/Connecticut DEP Wildlife Division.

Whip-poor-will *(Caprimulgus vociferus)* (Fig. 20.29). Uncommon nester in openly wooded regions; present from mid-spring through early fall; usually heard (migrants most often in May) and not seen. *Note:* Populations of this ground nester probably have been reduced through disturbance and predation by the relatively high numbers of cats, dogs, and carnivorous mammals that roam our woodlands. Populations of the native carnivores are in many areas artificially high due to abundant food resources associated with humans.

Common nighthawk *(Chordeiles minor)*. Rare nester, primarily on flat gravel roofs of buildings (undoubtedly subject to far less disturbance and predation than are natural bare ground sites); fairly common long-distance migrant (winters in South America), late summer and late spring; usually seen in high flight; notable for loud calls and/or wing sounds; sometimes hundreds move through an area in a single day.

Swifts, Hummingbirds, and Kingfishers

Swifts (Order Apodiformes, Family Apodidae)

Chimney swift *(Chaetura pelagica)* (Fig. 20.30). Flies almost all day, feeding on flying insects; uses saliva to glue nests in chimneys and similar sites; formerly nested in natural tree cavities but apparently never does so now; most arrive in mid-spring, depart by late summer to early fall; many migrants evident around autumnal equinox; migrating flocks of several hundred may roost in large chimneys at night in late summer. *Note:* The air seems charged with excitement when the first twittering migrants arrive from their South American wintering grounds in late April or early May.

Hummingbirds (Order Trochiliformes, Family Trochilidae)

Ruby-throated hummingbird *(Archilochus colubris)* (Fig. 20.31). Present mid-spring through late summer; nests

Fig. 20.30. The chimney swift, seen here in silhouette, nests in Connecticut and winters in South America.

Fig. 20.31. Ruby-throated hummingbirds are long-distance migrants that nest in Connecticut. Paul J. Fusco/Connecticut DEP Wildlife Division.

sparsely on tree branches in openly wooded areas (most commonly in Northwest Highlands); surprisingly scarce in many suburban and rural residential gardens that seemingly (but evidently not actually) provide sufficient food and nesting sites; much more common in Connecticut in 1800s; feeds on nectar and small insects associated with flowers, plus sap from wells drilled in trees by sapsuckers. *Note:* The flowers favored by hummingbirds tend to be long, red, and have sucrose-rich nectar. In contrast, flowers pollinated by insects or birds other than hummingbirds tend to have mixtures of sucrose, glucose, and fructose, or at least they are low in sucrose. However, hummingbirds sometimes visit exotic flowers that are low in sucrose but have high rates of nectar flow. Hummingbirds digest and assimilate sucrose, which is a more complex sugar than its constituent components, glucose and fructose, more efficiently than do other birds. The ruby-throat is reported to be the only pollinator for cardinal flower *(Lobelia cardinalis)*, but I saw swallowtails as well as hummingbirds visiting those flowers.

Kingfishers (Order Coraciiformes, Family Alcedinidae)

Belted kingfisher *(Ceryle alcyon)*. Remains around open water all year, though habitat becomes restricted by ice cover in winter; nests in low numbers in steep, bare earthen banks suitable for digging nesting burrows (I observed one attempted nesting in a low bank just above the high-tide line of Long Island Sound); feeds mainly on small fishes obtained by diving headfirst into water after hovering or from a perch.

Woodpeckers

Most woodpeckers (Order Piciformes, Family Picidae) feed on insects obtained by pecking the bark and wood of living and dead trees, and they also commonly eat small fruits. Their exceptionally long tongue aids in extracting insects from holes and crevices. Woodpeckers nest in tree cavities and generally excavate a new cavity for nesting each year; many other animals use abandoned cavities.

Red-headed woodpecker *(Melanerpes erythrocephalus).* Rare throughout the year; sporadic and very rare nester in deciduous woods; excavates nest cavities usually in dead wood of trees or in snags. *Note:* This woodpecker has declined in the region and was much more common in the 1800s.

Red-bellied woodpecker *(M. carolinus).* Year-round resident; fairly common nester in deciduous woods; regular visitor to bird feeders; excavates nesting cavities in dead limbs of trees; European starlings commonly usurp nest cavities. *Note:* This bird was very rare in the early 1900s. It increased in abundance and distribution in Connecticut beginning in the 1950s and 1960s.

Yellow-bellied sapsucker *(Sphyrapicus varius).* Increasing as nester in deciduous and mixed forest throughout Northwest Highlands and expanding elsewhere; often excavates nesting cavities in fungus-weakened aspen trees; uncommon spring and late summer to fall migrant in various wooded habitats; occurs sparsely in winter, mainly along coast; drills sap wells in trees, consumes the sap (as do hummingbirds, squirrels, and various insects).

Fig. 20.33. A pileated woodpecker tore open the side of this tree while feeding on carpenter ants.

Downy woodpecker *(Picoides pubescens).* Common nester and year-round resident in various wooded habitats; excavates nesting cavities in snags or dead wood of trees; able to nest in relatively small-diameter trees, so more common than other woodpeckers in young woods.

Hairy woodpecker *(P. villosus).* Year-round resident and nester; usually less abundant than downy woodpecker, generally prefers stands of larger trees; excavates nesting cavities in trunks or large limbs of living or dead trees.

Northern flicker *(Colaptes auratus)* (Fig. 20.32). Common nester; year-round resident, but most of population appears to move south of Connecticut for winter; nests in various habitats with mixture of trees and open areas; excavates nesting cavities in dead wood of trees, sometimes in live trees or other sites; sometimes reuses cavities; unlike most woodpeckers, obtains much of its food (particularly ants) from the ground. *Note:* Breeding abundance has declined in recent decades, probably due to competition with European starlings for nest cavities.

Pileated woodpecker *(Dryocopus pileatus).* State's largest woodpecker; uncommon, permanent resident in mature forest; nests in greatest numbers in Northwest Highlands; excavates nesting cavities usually in dead wood of large snags; versatile feeding behavior: tears open sides of snags to get at carpenter ants (Fig. 20.33), pecks into downed logs for beetle grubs, and plucks fruits (e.g., wild grapes).

Fig. 20.32. Northern flickers mate for life, and they sometimes consume thousands of ants in a single meal.

Flycatchers

The primary feeding technique of flycatchers (Order Passeriformes, Family Tyrannidae) involves flying out from a perch to catch a flying insect and then quickly landing on the same or another perch. Most of Connecticut's flycatchers are long-distance migrants that winter primarily in Central and/or South America.

Olive-sided flycatcher *(Contopus borealis)*. Nests rarely in early summer in far northern part of western Connecticut (no documented nestings in recent years); nests usually on branches of conifers (sometimes deciduous trees) in semi-wooded areas with snags; usually detected as uncommon migrant in May to June and mid- to late summer; in late summer, I saw single migrant individuals foraging for long periods of time from the dead top of a large hemlock tree. *Note:* Curiously, this species has declined in southern New England and New York despite an ample supply of apparently suitable breeding habitat (e.g., areas with standing dead trees in beaver-flooded areas).

Eastern wood-pewee *(C. virens)*. Fairly common nester on tree branches in wooded areas; present mainly May to October.

Yellow-bellied flycatcher *(Empidonax flaviventris)*. Uncommon migrant in late spring and late summer in coniferous and mixed forests; nests farther north.

Acadian flycatcher *(E. virescens)*. Localized nester, often on tree branches along streams bordered by overhanging conifers and deciduous trees; present mainly May to September. *Note:* This flycatcher historically was a localized nester, then the northern end of its nesting range shrank southward in the 1950s, followed by recolonization of northeastern U.S. nesting areas in the 1960s and subsequently. Reasons for the fluctuation are unknown. This bird often uses hemlock stands and is vulnerable to impacts caused by hemlock woolly adelgid*.

Alder flycatcher *(E. alnorum)*. Sparse nester in shrubs of wetland thickets, most abundantly in Northwest Highlands; present mainly May to September.

Willow flycatcher *(E. traillii)*. Variably numerous inland and coastal nester in wetland thickets, often in large shrubby willows; present mainly May to September.

Least flycatcher *(E. minimus)*. Fairly common nester in trees in wooded areas, mainly away from coastal region and central lowlands; favors early and mid-successional forests, declines in abundance as forests mature; present mainly May to September.

Eastern phoebe *(Sayornis phoebe)* (Fig. 20.34). Common nester, usually on or under structures built by humans, exceptionally tolerant of human proximity (one active nest was on the ferry that historically ran between Middletown and Portland); natural sites include niches under rock ledges and among uplifted roots of fallen trees; pair often produces two broods in one season; among flycatchers, earliest to arrive, due to close proximity of wintering areas in

Fig. 20.34. Eastern phoebes spend more time in Connecticut than any other flycatcher. Paul J. Fusco/Connecticut DEP Wildlife Division.

southeastern United States; migrants commonly arrive before vernal equinox; few linger into winter and may switch from eating insects to fruits.

Great crested flycatcher *(Myiarchus crinitus)*. Fairly common nester in natural or woodpecker-dug tree cavities (and suitable nest boxes) in mature deciduous or mixed forests; present mainly May to September.

Eastern kingbird *(Tyrannus tyrannus)*. Common nester in trees or shrubs, primarily along edges of open areas such as water, marsh, or field; present mainly May to September.

Songbirds

Shrikes (Order Passeriformes, Family Laniidae)

Shrikes are predators that feed on large insects and small vertebrates and are known for impaling and storing prey on thorns and thornlike objects.

Northern shrike *(Lanius excubitor)*. Rare, irregular in late fall and winter; more common farther north; most often seen in open areas with patchy thickets, but in most years not seen in Connecticut at all; winter of 1995/1996 produced large numbers in the region.

Loggerhead shrike *(L. ludovicianus)*. Rare migrant, late summer to mid-fall, mainly in mostly open areas near coast; very rare at other times; last nesting record was over a hundred years ago. *Note:* Prior to European settlement, this shrike likely was rare or absent from New England. It expanded into the region with deforestation and agriculture, and it became fairly common and is known to have nested widely in New York. Now it nests there in very small numbers if at all. Its decline in the northeastern United States likely is related to reforestation and reduction in agricultural landscapes.

Vireos (Order Passeriformes, Family Vireonidae)

Vireos in Connecticut are migratory and feed mainly on insects and small fruits obtained from trees or shrubs. Their easily recognized cup-shaped nests are suspended between a fork near the end of a branch.

White-eyed vireo *(Vireo griseus)*. Localized nester in shrubs in thickets; present mainly May to October.

Blue-headed vireo *(V. solitarius)*. Localized nester in coniferous and mixed forests, primarily in northern half of state; nests in small branch forks of large shrubs or low in trees; present mainly late April to October.

Yellow-throated vireo *(V. flavifrons)*. Fairly common nester in stands of mature deciduous trees, especially near water; nests in fork near end of tree branch; present mainly May to September.

Warbling vireo *(V. gilvus)*. Fairly common nester in groves of deciduous trees, often in riparian and floodplain situations; nests in fork near end of tree branch; present mainly May to September.

Red-eyed vireo *(V. olivaceus)* (Fig. 20.35). Common nester in deciduous and mixed forests; nests in fork in shrub or near end of low tree branch; present mainly May to October.

Crows and Jays (Order Passeriformes, Family Corvidae)

Corvids are opportunistic feeders that eat a wide assortment of plant and animal matter, with scavenging common among crows and ravens.

Blue jay *(Cyanocitta cristata)* (Fig. 20.36). Common nesting resident in wooded areas; nests usually in trees; in some years, substantial migration occurs, late summer to mid-fall and early to mid-spring. *Note:* Migrations are most evident when acorn crops are low (e.g., 1987, 1996). In years with abundant acorns, most blue jays do not migrate, and they spend a great deal of time storing vast numbers of acorns in the ground. They make use of these through the fall and winter. Unretrieved acorns may sprout and contribute to oak regeneration. Blue jays, like other members of the corvid family, exhibit good learning abilities, though I have evidence that not all jays are equal in this regard. In a color-banded group of jays that I studied in Middletown, one individual was adept at repeatedly removing peanuts from a funnel trap without being captured, whereas all of the many other jays that entered the trap could not find their way out.

American crow *(Corvus brachyrhynchos)*. Common resident; nests in top of trees in wooded areas; limited migration, late summer to mid-fall and late winter to mid-May, but not well documented; social, often feeds and perches in groups; large roosting aggregations of several hundred to several thousand crows form in some areas in winter (several tens of thousands roost next to Interstate Highway 84 near the Hartford/West Hartford town line). *Note:* This bird is a game species with a specified (late October to early March) hunting season, with no limit. This isn't because they're sought for food but rather they're a limited holdover from the agricultural era, when crows were persecuted year-round because of their damage to crops.

Fig. 20.35. An adult red-eyed vireo feeds a brown-headed cowbird nestling (vireo nestling at right). Paul J. Fusco/Connecticut DEP Wildlife Division.

Fig. 20.36. Blue jays are spectacularly beautiful and ecologically important. Paul J. Fusco/Connecticut DEP Wildlife Division.

Fish crow *(C. ossifragus)*. Increased in the region over the past century, now a fairly common resident along coast and major rivers (and at garbage dumps in winter); nests locally in conifer groves or sometimes deciduous trees.

Common raven *(C. corax)*. Recently colonized (mainly northwestern) Connecticut as nesting resident; expansion has continued; nests on rock ledges or occasionally in trees. *Note:* Lately I've been excited to hear the croaking voices of ravens in central and southern Connecticut, where they now nest in low numbers.

Larks (Order Passeriformes, Family Alaudidae)

Horned lark *(Eremophila alpestris)* (Fig. 20.37). Rare breeder in extensive areas of short grass and/or barren ground near coast and around airports and agricultural fields; nests are usually on the shaded north side of a grass clump, bush, or rock; common migrant and winter visitor, most often near coast; walks (rather than hops) on ground, feeds on seeds and insects. *Note:* This bird expanded its range in eastern North America with nineteenth-century land clearing and agricultural expansion, but reforestation and urbanization have resulted in a subsequent decline.

Swallows (Order Passeriformes, Family Hirundinidae)

Swallows spend most of their daylight hours flying in open areas, catching insects on the wing. All of the species listed below nest in Connecticut, and most wintering occurs in Central America and/or South America.

Purple martin *(Progne subis)*. Uncommon, localized breeder, mainly along coast; nests in boxes (martin houses) adjacent to large open areas, usually including large body of water; present mainly late April to Septem-

Fig. 20.37. Horned larks depend on large expanses of sparse vegetation. Paul J. Fusco/Connecticut DEP Wildlife Division.

Fig. 20.38. Tree swallows descend on a coastal bayberry thicket in September.

ber. *Note:* Martins are unusual among Connecticut birds in being totally dependent on nest boxes for reproduction. Apparently they ignore tree cavities and cliff crevices that are sometimes used in the western United States and formerly were used in the east. The abundance of this bird in the state has fluctuated with changes in the amount of open habitat, availability of suitable nest boxes, populations of nest competitors such as house sparrows and European starlings, and possibly other factors. In summer, you can readily see nesting purple martins at Hammonasset Beach State Park.

Tree swallow *(Tachycineta bicolor)* (Fig. 20.38). Common nester in nest boxes and woodpecker-excavated holes in trees killed by flooding caused by beaver dams; present mainly from late March or April to October, rarely into early winter; flocks of many thousands swarm over and roost in coastal common reed marshes beginning in August; dense flocks descend on coastal bayberry shrubs to consume fruits in late summer. *Note:* This swallow has benefited from nest boxes put up for bluebirds and from increased beaver populations. In late summer, I frequently observed large flocks bathing by briefly but repeatedly splashing down into the water as they flew low over a pond. Tree swallows often engage in extra-pair copulations; genetic research indicates that 50 to 80 percent of nests have eggs fathered by more than one male.

Northern rough-winged swallow *(Stelgidopteryx serripennis)*. Nests at low densities in holes in vertical banks or, more commonly, in holes or small pipes in artificial, vertical banklike structures, normally near water; fairly common migrant, spring and mid- to late summer.

Bank swallow *(Riparia riparia)*. Localized colonial nester, usually away from coast; nests in steep earthen banks bordering water or some distance away (Fig. 20.39); present mainly April to September.

Fig. 20.39. Bank swallows nest in burrows in steep banks, as do belted kingfishers.

Cliff swallow *(Hirundo pyrrhonota)*. Very localized, colonial, inland nester, primarily on bridges and dams along Housatonic River in western Connecticut; nests in gourdlike structures made of mud, generally attached beneath overhangs on concrete bridges and dams; present mainly May to September.

Barn swallow *(H. rustica)*. Common nester in partially open wooden buildings, or under bridges, in open areas; present mainly April to October.

Titmice and Chickadees (Order Passeriformes, Family Paridae)

Tits are gregarious, mostly nonmigratory, cavity-nesting birds that feed on insects and seeds obtained in trees and shrubs.

Black-capped chickadee *(Poecile atricapillus)*. Common, generally sedentary resident in wooded areas and near bird feeders; weak excavator, often digs own nesting cavity in dead wood of trees, sometimes uses existing cavities and nest boxes. *Note:* In some years, such as those with low seed crops in areas north of Connecticut, southward migrations of mostly immature birds bring numbers of chickadees from the north into southern New England.

Tufted titmouse *(Baeolophus bicolor)* (Fig. 20.40). Common resident in wooded areas and near bird feeders; nests in natural and woodpecker-dug tree cavities. *Note:* This species was very rare in Connecticut in the early 1900s. It colonized and expanded across Connecticut in the 1940s, 1950s, and 1960s. Today it is hard to imagine that it ever was rare. The increase in range and abundance presumably reflect increased food resources (bird seed at feeders) and perhaps milder winters.

Fig. 20.40. Not so many years ago, tufted titmice, now common at feeders, were a rarity in Connecticut. Paul J. Fusco/Connecticut DEP Wildlife Division.

Nuthatches (Order Passeriformes, Family Sittidae)

Nuthatches feed on insects, seeds, and small fruits obtained while climbing up or down over the trunks and larger branches of trees. The exceptionally long claw on the hind toe aids their head-first, downward movements on tree trunks.

Red-breasted nuthatch *(Sitta canadensis)*. Local nester primarily in Northwest Highlands; strongly associated with conifer trees, the seeds of which are an important winter food resource; often excavates own nesting cavity in trees, sometimes uses nest boxes, anointing opening with conifer pitch. *Note:* In some years major southward migratory movements in late summer and fall bring relatively large numbers into the state for the winter; these migrations appear to coincide with conifer cone-crop failures in the north.

White-breasted nuthatch *(S. carolinensis)* (Fig. 20.41). Common nesting resident in areas with mature deciduous trees; nests in natural or woodpecker-excavated cavities in trees.

Creepers (Order Passeriformes, Family Certhiidae)

Brown creeper *(Certhia americana)* (Fig. 20.42). Uncommon, most frequent in Northwest Highlands; present both as

Fig. 20.41. White-breasted nuthatches are year-round residents that nest in tree cavities. Paul J. Fusco/Connecticut DEP Wildlife Division.

Fig. 20.43. This marsh wren nest is in a brackish marsh.

nester and in winter, some migration occurs in spring and fall; nests usually behind loose bark of large living or dead trees in mature forests or swamps; ascends tree trunks, feeds on insects obtained by probing into bark crevices.

Wrens (Order Passeriformes, Family Troglodytidae)

Carolina wren (*Thryothorus ludovicianus*). Fairly common nesting resident in thickets, mainly in southern half of state; adaptable nester, uses many different kinds of sites, both natural and "artificial," including cavities in trees and various secluded niches (a pair nested successfully among plants in a hanging pot outside my kitchen window); spring and fall migrations may occur. *Note:* In the early 1900s this wren was initially rare, then increased greatly. Subsequently large population declines were reported and attributed to severe winters. Today, severe winters probably are less likely to result in a big decline because of the abundant food resources available at thousands of bird feeders.

House wren (*Troglodytes aedon*). Nests commonly in wooded areas, often in tree cavities, also in nest boxes; present mainly late April to October. *Note:* Band recoveries suggest that many house wrens raised in Connecticut winter in Florida, as do many people.

Winter wren (*T. troglodytes*). Uncommon, local breeder, primarily in Northwest Highlands; most often encountered in mesic, shady habitats with abundant woody cover near ground; nests in various hollows, generally not far above ground; present mainly March to December, usually uncommon in winter.

Sedge wren (*Cistothorus platensis*). Formerly a rare nester in grassy wetlands with scattered clumps of tall dry grass and shrubs; no documented nestings in recent decades, but may rarely nest; very rare migrant, May to June and September to October. *Note:* This wren has declined in the Northeast, apparently due to loss and degradation of suitable habitat from mowing, grazing, and development.

Marsh wren (*C. palustris*) (Fig. 20.43). Locally common nester in herbaceous emergent wetlands (tall vegetation of estuarine and freshwater marshes, especially cattails), primarily along coast and Connecticut River and tributaries; present mainly late April to November, rare into early winter.

Fig. 20.42. As they forage, brown creepers move upward on tree trunks and large branches.

Kinglets (Order Passeriformes, Family Regulidae)

Golden-crowned kinglet (*Regulus satrapa*). Rare nester in mature forests dominated by conifers, including planted stands; nests near ends of branches of conifer trees; fairly common migrant in wooded areas, spring and early to mid-fall; usually uncommon in winter.

Ruby-crowned kinglet (*R. calendula*). Fairly common migrant in wooded areas, mainly April to May and September to November; some migrants linger until early winter; no evidence of nesting in Connecticut; both kinglets feed mainly on insects obtained in trees and shrubs.

Gnatcatchers and Old World Warblers (Order Passeriformes, Family Sylviidae)

Blue-gray gnatcatcher (*Polioptila caerulea*). Uncommon nester in areas with deciduous trees, often near water; nests on branches of trees or shrubs; present mainly late April to September; forages on insects in wooded and shrubby areas. *Note:* Gnatcatchers increased in abundance in Connecticut during the mid-1900s.

Thrushes (Order Passeriformes, Family Turdidae)

Most thrushes are migratory. Invertebrates and small fruits dominate the diet, and most species frequently feed on the ground.

Eastern bluebird (*Sialia sialis*) (Fig. 20.44). Locally common nester in tree cavities and nest boxes around edges of open areas; locally common migrant, March to April and September to October; increasingly numerous in winter. *Note:* Provision of nearly 40,000 nest boxes has allowed bluebirds to maintain their numbers or even increase despite competition with introduced European starlings for natural nest sites and a reduction in the extent of suitable habitat (due to reforestation and residential/commercial expansion). Nest boxes installed for bluebirds may be used by tree swallows (more often than by bluebirds), house wrens, house sparrows*, and sometimes chickadees and other species (e.g., white-footed mouse). For information on bluebirds and nest boxes, contact the DEP's Wildlife Diversity Program, or the North American Bluebird Society (see bibliography).

Veery (*Catharus fuscescens*). Common nester in vegetation near or on ground in mature and early successional forests with thick undergrowth; most common near swamps in large tracts of forest; present mainly May to October.

Swainson's thrush (*C. ustulatus*). Uncommon migrant in wooded areas, May to June and September to October; not yet known to nest in Connecticut but may do so in coniferous or mixed forests in Northwest Highlands; nests in small or shrubby trees, often conifers.

Hermit thrush (*C. guttatus*). Uncommon or localized nester in ground vegetation in wooded areas, primarily in large tracts of forest in northern half of state; present mainly April to November, rare in winter.

Wood thrush (*Hylocichla mustelina*). Nests commonly in shrubs or small trees in deciduous or mixed forests with thick undergrowth; favors mid-successional forests, declines in abundance as forests mature; present mainly May to October. *Note:* Near my home over three consecutive years I heard a peculiar song type that I never heard elsewhere; probably the same male nested in the same territory each year.

American robin (*Turdus migratorius*). Common nester in trees and shrubs in various (especially semi-wooded) habitats; present mainly March to December, usually uncommon or rare in winter (large numbers present in recent mild winters); migrants conspicuous in spring and fall.

Catbirds, Mockingbirds, and Thrashers (Order Passeriformes, Family Mimidae)

The mimids have an omnivorous diet of insects, small fruits, and seeds.

Gray catbird (*Dumetella carolinensis*). Common nester in shrubs or small trees in wooded thickets; present mainly April to December, rare in winter.

Northern mockingbird (*Mimus polyglottos*). Common nester in shrubs and small trees in partly open areas with patches of woody vegetation, almost always in areas altered by humans; present all year, fairly common in winter (especially along coast and in large river valleys). *Note:* Though this bird is common now, it was a rarity several decades ago. Per-

Fig. 20.44. Eastern bluebirds, displaced from tree-cavity sites by non-native starlings, now depend on nest boxes. Paul J. Fusco/Connecticut DEP Wildlife Division.

haps its increase has been due at least in part to the introduction and increase of multiflora rose, which provides excellent nesting cover and fruits favored as a winter food resource.

Brown thrasher (*Toxostoma rufum*). Nests uncommonly in shrubs or on ground under shrubby vegetation in woody thickets; present mainly April to October, generally rare in winter. *Note:* This species formerly was a common summer resident. Populations apparently declined in recent decades, possibly as a result of a reduction in the amount of brushy habitat. However, thrashers are now scarce even where suitable habitat remains, and birds with similar ecologies (mockingbirds, catbirds) are still common. These other birds may compete with thrashers, but whether they have displaced them is an open question.

Starlings (Order Passeriformes, Family Sturnidae)

European starling* (*Sturnus vulgaris*). State's most numerous bird, but not native to North America; introduced in New York in 1890s, subsequently spread throughout most of North America; very common all year; nests in tree holes and cavities in human-built structures; diverse diet dominated by insects, seeds, and fruits. *Note:* Starlings form large, noisy roosting aggregations in fall and winter, releasing showers of fecal material. Other unfavorable habits include the consumption of grain at dairy feedlots and the usurpation of nest sites from native cavity-nesting birds such as eastern bluebirds. Christmas Bird Counts and Breeding Bird Surveys indicate that starling abundance in Connecticut steadily declined over the past two decades, perhaps as a result of a reduction in farming and other land use changes.

Pipits and Wagtails (Order Passeriformes, Family Motacillidae)

This family includes 65 species occurring nearly worldwide. Most species are migratory, mainly insectivorus, and inhabit open, mainly unwooded areas.

American pipit (*Anthus rubescens*). Rare to uncommon migrant, spring and fall; rare in mid-winter; open expanses of short grass or low sparse vegetation or relatively barren fields or shores; does not nest in Connecticut; walks on ground when foraging for invertebrates and seeds.

Waxwings (Order Passeriformes, Family Bombycillidae)

Cedar waxwing (*Bombycilla cedrorum*) (Fig. 20.45). Fairly common year-round resident in wide array of habi-

Fig. 20.45. Cedar waxwings make heavy use of small fruits. Larry Master.

tats, generally near fruiting woody plants; nests in trees primarily in young forest, most often in mid-summer, well after other songbirds (except American goldfinch); some apparent migration in spring and fall, but movements mainly erratic; diet dominated by small fruits and (seasonally) insects caught in flight.

Wood Warblers (Order Passeriformes, Family Parulidae)

Nearly all of these warblers are colorful insect-eating birds that migrate to the Neotropics for winter.

Warblers that Nest in Connecticut. The state hosts a rich assemblage of breeding warblers.

Blue-winged warbler (*Vermivora pinus*). Common nester among low plants at ground level in thickets and shrubby forest edges; present mainly late April to September. *Note:* This warbler expanded its range across Connecticut during the early to mid-1900s, encountering and hybridizing with the golden-winged warbler in the process. Over the past few decades, the Connecticut population has declined (likely due to reduction in habitat), raising concern for the species, since a significant percentage of the total population of this species nests in Connecticut.

Golden-winged warbler (*V. chrysoptera*). Increasingly rare ground- or near-ground nester in shrubby, early successional habitats and forest edges, mainly in Northwest Highlands; present mainly May to September. *Note:* The abundance and distribution of this species have declined greatly in southern New England in recent decades, at least in part as a result of the reduction in grassy/shrubby habitat as abandoned farms became forested or were converted to residential use. Hybrids between blue-winged and golden-winged warblers (so-called Brewster's warbler) also nest in Connecticut.

Nashville warbler *(V. ruficapilla)*. Rare nester on ground in shrubby young woods, primarily in Northwest Highlands; present mainly May to October.

Northern parula *(Parula americana)*. Formerly nested at least locally in trees in swampy woods, but nesting not confirmed in Connecticut in recent decades; present mainly May to October. *Note:* The decline in breeding paralleled the extirpation of the beard lichen *(Usnea)*, which grew on trees and was used for nest construction. The reason why the lichen disappeared is not known, but declining air quality was probably involved.

Yellow warbler *(Dendroica petechia)* (Fig. 20.46). Nests commonly in shrubs or low in trees in wetland thickets and open uplands with tallish, thick herbaceous vegetation; present mainly late April to September. *Note:* This is one of the state's most easily observed warblers.

Chestnut-sided warbler *(D. pensylvanica)*. Locally common nester (most numerous in Northwest Highlands) in shrubs or low in small trees or vines in relatively dry brushy woods and forest edges; favors early and mid-successional forests, declines in abundance as forests mature; present mainly May to September.

Magnolia warbler *(D. magnolia)*. Locally fairly common nester in shrubby or small-tree conifers in open, relatively young, coniferous or mixed forest, primarily in Northwest Highlands; present mainly May to October.

Black-throated blue warbler *(D. caerulescens)*. Locally fairly common nester in shrubs and small conifers in

Fig. 20.46. Yellow warblers often nest in wetland thickets. Paul J. Fusco/Connecticut DEP Wildlife Division.

Fig. 20.47. Yellow-rumped warblers are one of the few birds that can profitably feed on bayberry fruits. Paul J. Fusco/Connecticut DEP Wildlife Division.

shrubby woods in northern Connecticut, primarily in Northwest Highlands; present mainly May to October.

Yellow-rumped warbler *(D. coronata)* (Fig. 20.47). Rare, local nester in stands of coniferous trees in northern Connecticut, most abundant in Northwest Highlands; common migrant in various wooded and brushy habitats, April to May and September to October; huge numbers migrate through coastal sites in October. *Note:* This species is unusual among warblers in that it inhabits Connecticut (in low numbers) in winter, mainly along the coast and riparian lowlands. These birds can winter in the north because they readily shift from an insect diet to one dominated by bayberry fruits when insects become scare in the colder months. Yellow-rumped warblers are among the few birds that have digestive system traits allowing efficient assimilation of the saturated fatty acids occurring in waxy bayberry fruits.

Black-throated green warbler *(D. virens)*. Locally common nester at various heights in conifer trees and sometimes other plants in uplands, primarily away from coast in forests of hemlock or hemlock-hardwoods; present mainly late April to October. *Note:* The ongoing decimation of hemlocks in Connecticut by the woolly adelgid, an insect pest native to Asia, poses a substantial threat to this warbler. If the adelgid continues to spread, the black-throated green warbler could be eliminated from much of its breeding range in the state.

Blackburnian warbler *(D. fusca)*. Locally common nester in tall conifers, primarily in Northwest Highlands, to lesser extent in northeast; present mainly May to October. *Note:* Like the black-throated green warbler, this species is potentially threatened by habitat loss caused by the introduced hemlock woolly adelgid.

Yellow-throated warbler *(D. dominica)*. Very rare migrant in wooded areas in April to May, rare and irregular in fall; has nested in one site (Kent) in multiple years; singing males seen elsewhere in summer with no nesting evident; nests on tree branches.

Pine warbler *(D. pinus)* Nests locally in stands of white pine or pitch pine; present mainly April to November, rarely lingers until early winter.

Prairie warbler *(D. discolor).* Local nester, primarily in shrubby plants such as those in deforested areas under major power lines; present mainly late April to September. *Note:* Populations have declined in recent decades with reductions in old-field habitat.

Cerulean warbler *(D. cerulea).* Rare and local nester, primarily in tall trees of mature, deciduous, riparian forest, such as along Housatonic and lower Connecticut rivers; rare migrant in May and June, very rare in late summer. *Note:* Available population data indicate that overall this species has declined over the past few decades, though this may not be the case in Connecticut. The small winter range in the Andes foothills in South America has been extensively logged and cultivated.

Black-and-white warbler *(Mniotilta varia).* Common nester on ground, often at base of tree, shrub, log, or rock, in deciduous and mixed forests; present mainly late April to October.

American redstart *(Setophaga ruticilla).* Fairly common nester in shrubs, small trees, or among vines in deciduous and mixed forests and shrubby areas; favors early and mid-successional forests, declines in abundance as forests mature; fairly common migrant, mid- to late spring and mid-summer to early fall.

Worm-eating warbler *(Helmitheros vermivorus).* Uncommon to fairly common nester on ground among leaves, often at base of a shrub or small tree, in deciduous and mixed forests, mainly in southern half of state, usually on steep slopes in relatively large tracts of forest; present mainly May to October.

Ovenbird *(Seiurus aurocapilla).* Common ground-nester in deciduous and mixed forests; present mainly May to October (rarely remaining into early winter); forages as it walks on ground.

Northern waterthrush *(S. noveboracensis).* Nests uncommonly in stream-bank crannies and among fallen-tree roots in shrubby swamps, usually with hemlock trees, with largest numbers in Northwest Highlands; present mainly late April to September.

Louisiana waterthrush *(S. motacilla).* Locally fairly common nester, primarily in stream-bank crannies or pockets among roots of fallen trees along flowing woodland streams in noncoastal uplands; present mainly late April to August.

Kentucky warbler *(Oporornis formosus).* Very localized nester in moist, shrubby deciduous forest, on or just above ground, in or under overhanging plant material; present mainly May to September. *Note:* This bird seems to do best in older forests than are far from agricultural areas and the impacts of cowbirds.

Common yellowthroat *(Geothlypis trichas).* Common nester in wetland and upland thickets; nests just above ground or water in thick grassy-weedy or shrubby vegetation; present mainly late April to October, with rare occurrences into late fall or early winter.

Hooded warbler *(Wilsonia citrina).* Uncommon nester in shrubs, small trees, or vines in moist deciduous forests with thick undergrowth, primarily in southern half of state; present mainly May to September.

Canada warbler *(W. canadensis).* Uncommon nester in thicker undergrowth of moist deciduous forest, primarily near creeks or wetlands in noncoastal uplands; nests concealed, often by overhanging plants, in various sites on or near ground; present mainly May to September.

Yellow-breasted chat *(Icteria virens).* Very rare (or former), localized nester (primarily in southern Connecticut) in deciduous shrubs and vines in large, low thickets or shrubby fields; present mainly May to October, with a few birds remaining through winter. *Note:* This warbler, formerly a common nester, has declined in the state over the past several decades, coincident with a decrease in the amount of suitable habitat, which was more abundant earlier in the 1900s when many abandoned farm fields passed through a shrubby stage prior to reforestation. Genetic data suggest that this bird may be more closely related to certain members of the blackbird family than to typical warblers.

Warblers that Occur in Connecticut During Migration But Do Not Nest in the State. These birds pass through the state on their way to and from more northern nesting areas.

Tennessee warbler *(Vermivora peregrina).* Uncommon in wooded and brushy areas, May and June and late summer to early fall.

Cape May warbler *(Dendroica tigrina).* Uncommon at best, May and late summer to early fall; prefers spruce trees, but migrants use other trees as well.

Palm warbler *(D. palmarum).* Uncommon in April and May and September to November, rarely remaining until early winter; ground-dweller, favors forest edges, open shrubby areas, and marshy thickets.

Bay-breasted warbler *(D. castanea).* Uncommon in wooded areas, mid-spring and late summer-early fall.

Blackpoll warbler *(D. striata).* Fairly common in wooded areas in May and late summer to early fall.

Wilson's warbler *(Wilsonia pusilla).* Uncommon at best in thickets, May to June and late summer to early fall.

Tanagers (Order Passeriformes, Family Thraupidae)

Scarlet tanager (*Piranga olivacea*) (Fig. 20.48). Common nester in deciduous and coniferous trees in large deciduous and mixed forests; present mainly May to October; arboreal feeder on fruits, flowers, insects, and/or nectar; winters in northwestern South America.

Emberizids (Order Passeriformes, Family Emberizidae)

Emberizid diets generally are dominated by seeds and insects obtained from the ground or low plants.

Eastern towhee (*Pipilo erythrophthalmus*). Common nester on ground in shrubby wooded areas; present mainly April to December, rare to uncommon (but increasingly numerous) in winter, mainly along coast and near bird feeders.

American tree sparrow (*Spizella arborea*) (Fig. 20.49). Fairly common in shrubby open areas in migration and winter, mainly October to April; does not nest in Connecticut.

Chipping sparrow (*S. passerina*). Common nester, often in coniferous trees bordering short grass or bare ground; present mainly April to November, very rare in winter, in partly open areas.

Field sparrow (*S. pusilla*). Locally fairly common nester on ground or in low shrubs in open brushy areas (including powerline corridors); most numerous March to November, uncommon in winter. *Note:* Populations have declined in recent decades with reductions in old field habitat.

Vesper sparrow (*Pooecetes gramineus*). Uncommon migrant in April and May and September to December, very rare in winter, mostly in open areas; rarely has nested in re-

Fig. 20.48. Scarlet tanagers nest in Connecticut forests and winter in South America. Paul J. Fusco/Connecticut DEP Wildlife Division.

Fig. 20.49. American tree sparrows migrate southward to Connecticut for winter. Paul J. Fusco/Connecticut DEP Wildlife Division.

cent decades, usually on ground in dry fields. *Note:* Populations in the Northeast have been declining, due probably to forest regrowth and changes in hay-cutting schedules.

Savannah sparrow (*Passerculus sandwichensis*). Nests uncommonly among vegetation on ground in large grassy fields (sometimes smaller ones), if not disrupted by early summer mowing; fairly common migrant in spring and late summer to late fall, uncommon in winter, in grassy/shrubby/open habitats; favorable sites may host more than a hundred individuals during October migrations.

Grasshopper sparrow (*Ammodramus savannarum*). Highly localized nester among vegetation on ground in extensive grassland with patches of open ground, in north-central (Connecticut River valley) and northeastern Connecticut; present mainly late April–October. *Note:* In 2000, the state's largest known colony, represented by 35 singing males, was discovered in East Hartford, unfortunately on a site slated for construction of a University of Connecticut football stadium and other development. Early summer mowing reduces nesting success, as does predation by the now abnormally high populations (due to human-augmented food resources) of predators such as raccoons and striped skunks.

Henslow's sparrow (*A. henslowii*). Very rare migrant in grassy/weedy areas in May and late September to early October. *Note:* This sparrow formerly nested in weedy fields in Connecticut but no longer does so, in conjunction with a widespread population decline in the Northeast. The decline was associated with a reduction in recently post-agricultural lands as forests regrew and suburban development increased.

Nelson's sharp-tailed sparrow (*A. nelsoni*). Likely an uncommon migrant in coastal marshes. A recent extensive breeding-season survey did not detect this species in any of forty-four Connecticut marshes.

Fig. 20.50. This saltmarsh sharp-tailed sparrow nest is in high salt marsh.

Saltmarsh sharp-tailed sparrow *(A. caudacutus)* (Fig. 20.50). Locally common ground nester in large expanses of high salt marsh, though marsh destruction has extirpated some populations; present mainly late April through November, very few remaining into winter. *Note:* Nests commonly are destroyed by flooding during high tides. Studies in Rhode Island show that individuals often return to breeding sites they used the previous year, and even juveniles exhibit relatively strong fidelity to their natal sites.

Seaside sparrow *(A. maritimus)*. Locally fairly common near-ground nester in tall grasses of large coastal salt marshes; present mainly April to November. *Note:* In 2002, researchers began an intensive study of seaside and sharp-tailed sparrows in Connecticut (http://dep.state.ct.us/burnatr/wildlife/special/smspar.htm).

Fox sparrow *(Passerella iliaca)*. Uncommon migrant and winter visitor, mainly October to April, in brushy areas.

Song sparrow *(Melospiza melodia)*. Common ground and shrub nester in open shrubby areas; present all year, with significant migration in early spring and mid-fall.

Swamp sparrow *(M. georgiana)*. Locally common nester in emergent marshy vegetation or in shrubs in or adjacent to water in shrubby-marshy wetlands; present all year, with significant spring and fall migration; uncommon in winter, mainly along coast.

White-throated sparrow *(Zonotrichia albicollis)*. Nests on ground, hidden in vegetation, in swamp-edge or boggy habitats, mainly in Northwest Highlands and in scattered locations elsewhere; present all year, with significant spring and fall migration (usually in small flocks) in shrubby areas.

White-crowned sparrow *(Z. leucophrys)*. Uncommon migrant in shrubby areas in May and fall; rare in winter.

Dark-eyed junco *(Junco hyemalis)*. Uncommon nester among ground vegetation, leaves, or other low cover in coniferous forest, primarily in northwestern and northeastern parts of state; present all year, with significant migration in spring and fall; common in winter, in forest edge and other habitats.

Lapland longspur *(Calcarius lapponicus)*. Present in small numbers, mid-fall to early spring, usually on upper beaches, open dunes, or in fields, often with horned larks and snow buntings; nests in arctic tundra.

Snow bunting *(Plectrophenax nivalis)*. Uncommon or locally common, mid-fall through mid-winter, usually in flocks of up to a few hundred on short-grass fields, open weedy areas, open dunes, or upper beaches; nests far to north in arctic regions of Alaska, Canada, and Greenland.

Cardinals and Allies (Order Passeriformes, Family Cardinalidae)

Cardinals, grosbeaks, and their relatives are strong-billed birds that eat seeds, small fruits, and insects.

Northern cardinal *(Cardinalis cardinalis)*. Common nesting resident in shrubs, thick vines, or small trees in and near woody thickets; pairs may stay together year-round, can produce multiple broods in single nesting season; very sedentary, no evidence of migration. *Note:* Formerly very rare, this species increased in abundance over the past several decades, probably due to the increased food resources now available at bird feeders, aided historically by an increase in shrubby nesting cover in the post-agricultural era.

Rose-breasted grosbeak *(Pheuticus ludovicianus)*. Fairly common nester in shrubs or small trees in deciduous and mixed woods; favors mid-successional forests; present mainly May to October. *Note:* This species appears to have declined in abundance in recent decades as many Connecticut forests have matured.

Blue grosbeak *(Passerina caerulea)*. Rare migrant, April to May and September to October, in field-edge thickets. *Note:* The state's first known nesting occurred in 1996 along the impounded Farmington River in Windsor. This bird nests in shrubby vegetation.

Indigo bunting *(P. cyanea)*. Uncommon nester in shrubs, small trees, or vine thickets in shrubby areas bordering deciduous woods; present mainly May to October.

Dickcissel *(Spiza americana)*. Rare fall migrant and winter visitor (mainly at bird feeders or in grassy or weedy fields). *Note:* Dickcissels were locally common nesters in Connecticut before 1840. Land-use changes may have led to their disappearance as a nesting species.

Blackbirds (Order Passeriformes, Family Icteridae)

Blackbirds and their kin eat seeds, small fruits, and insects.

Bobolink *(Dolichonyx oryzivorus)* (Fig. 20.51). Uncommon nester on ground in moist meadows and hayfields; present mainly May to October. *Note:* Nesting success is reduced when fields are mowed in June before the young have fledged. Residential development has destroyed much habitat.

Red-winged blackbird *(Agelaius phoeniceus)*. Common nester on ground or low in tall herbaceous plants or shrubby growth in marshy wetlands and grassy-weedy fields; present all year, with conspicuous migrating flocks, late winter through early spring and late summer through mid-fall; generally uncommon in winter.

Eastern meadowlark *(Stunella magna)*. Now an uncommon to rare nester on ground in large grassy fields; present mainly March to November, generally scarce in winter. *Note:* Meadowlarks are becoming increasingly scarce as a nester. The decline probably is related to loss of pastureland and a shift to cutting of hayfields during the nesting season.

Rusty blackbird *(Euphagus carolinus)*. Generally uncommon migrant, March to April and October to December, but groups of 100 to 300 or more appear in fall at favorable shrubby or wooded swamps (e.g., White Memorial Foundation); rare in winter; nests in areas north of Connecticut. *Note:* This blackbird has undergone a steep (about 90 percent), range-wide population decline in abundance since the mid-1800s. The decline has continued in recent decades and may be due to human-caused changes in their habitat.

Boat-tailed grackle *(Quiscalus major)*. More typical of habitats farther south; rare and erratic visitor, but recently regular, in coastal marshes. *Note:* Since 1995, this species has nested at Great Meadows Marsh in Stratford. Nests generally are in trees, shrubs, or tall robust herbage over or adjacent to water.

Common grackle *(Q. quiscula)*. Common nester, often in small groups, in conifers, shrubs, residential shade trees, or various other sites, often near water or wetlands; present all year, with large migrating flocks in fields and wooded areas, February to May and September to November; locally common in winter, mostly along coast. *Note:* The air is always charged with excitement whenever a big noisy flock of these birds arrives, or when hundreds stir up forest leaf litter looking for food.

Brown-headed cowbird *(Molothrus ater)* (Fig. 20.52). Common breeder in forest edge and partly wooded areas; present all year, but generally uncommon in winter, except around dairy farms and feed lots.

Note: Cowbirds always deposit their eggs in nests of other birds (the list includes over 200 species), and they never build a nest or incubate. A female lays 30 to 40 eggs per season, usually one egg per nest. Under ideal captive conditions, a female can lay 77 eggs per season (one laid an egg a day for 67 consecutive days). The female commonly removes an egg from the nest shortly before or after she lays her own egg. Sometimes the egg that is removed is eaten. Cowbird eggs often hatch before or at the same time as host eggs, despite often being much larger. This seems to be due mainly to relatively rapid development of the cowbird

Fig. 20.51. Bobolinks need unmowed grassy fields for nesting. Paul J. Fusco/Connecticut DEP Wildlife Division.

Fig. 20.52. Brown-headed cowbirds (male shown) lay their eggs in the nests of other songbirds. Paul J. Fusco/Connecticut DEP Wildlife Division.

embryo plus disruption of incubation of the host eggs caused by the presence of the cowbird egg(s).

Brood parasitism by cowbirds often reduces the host bird's reproductive success. Having a big, hungry cowbird nestling in the nest means that a significant amount of the food delivered to the nest by the host birds goes to the cowbird rather than to the host's young. Recent studies indicate that cowbird young sometimes even eject host nestlings from the nest, and an adult female cowbird was videotaped as she killed and removed six nestling blue-winged warblers from their nest in New London County.

Various warblers, vireos, indigo buntings, and some others readily accept cowbird eggs and incubate them as their own. Common hosts in Connecticut include red-eyed vireo (Fig. 20.35), common yellowthroat, and ovenbird. Red-winged blackbirds reduce cowbird parasitism by being aggressive toward cowbirds but, if parasitized, can raise their own young along with the cowbirds because blackbird young are larger than the cowbirds and compete well for food.

Yellow warblers also can raise their young with cowbird young because the warblers have a short incubation period and rapid nestling development, so the cowbird nestling is less able to outcompete the host's young. The warblers may desert a parasitized nest or bury the eggs under nest material and start a new clutch. They sometimes tolerate removal of two or more of their own eggs without deserting the nest.

American robins and gray catbirds combat cowbird brood parasitism by removing cowbird eggs from the nest. Catbirds learn to recognize their own egg by seeing it in the nest, then they innately reject other eggs.

Formerly a rarity, the brown-headed cowbird has undergone a population explosion in the eastern United States over the past several decades. Extensive forest fragmentation, in conjunction with augmented food supplies at dairy farms, allowed greater numbers of cowbirds to penetrate farther into forested areas. Now they may be negatively affecting forest interior birds that formerly did not experience cowbird brood parasitism.

Orchard oriole *(Icterus spurius)*. Localized nester in groves of mature deciduous trees, often near water, mainly in southern half of state and in major river valleys; present mainly May to July. *Note:* The southward migration is very early; most have disappeared from nesting areas by the end of July. Local nesting may be sporadic; near my home along the Connecticut River, this oriole nested once but not in several other years, despite stable habitat conditions.

Baltimore oriole *(I. galbula)* (Fig. 20.53). Nests commonly in mature trees of deciduous groves; present mainly May to October, very rare in winter (as is its western relative, Bullock's oriole). *Note:* Near my home I heard a unique song variation over a period of three years, suggesting that the

Fig. 20.53. Baltimore orioles nest in groves of mature deciduous trees. Larry Master.

same male nested in the same territory over those three years. I never heard this particular vocalization among many other singing orioles.

Cardueline Finches and Allies (Order Passeriformes, Family Fringillidae)

Finch diets often are dominated by seeds and also may include small fruits and insects. Most of the species that occur in Connecticut move into the area for winter, in largest numbers in years when seed crops are low in areas to the north (Canada).

Pine grosbeak *(Pinicola enucleator)*. Irregular visitor, mainly October to April; apparently completely absent in many years; generally associated with coniferous trees.

Purple finch *(Carpodacus purpureus)*. Uncommon nester in mature coniferous trees, primarily in Northwest Highlands; present all year, variably numerous in winter. *Note:* In the fall of 2001, I saw a purple finch that was apparently afflicted with conjunctivitis (see house finch).

House finch* *(C. mexicanus)*. Common breeder and year-round resident; tends to nest on vine-covered walls and in conifers near buildings; generally occurs in human-modified habitats. *Note:* This finch of the western United States was introduced in New York in the 1940s. The population expanded greatly (reaching Connecticut in the early 1950s) and now covers most of the eastern United States. Since the early 1990s, populations have been infected with infectious conjunctivitis that causes swollen and runny or crusted tissue around the eyes. This contagious disease, caused by a common respiratory pathogen of domestic poultry, peaked in the northeastern United States in 1995 and since has fluctuated in recurring cycles, with peaks in late fall and winter when birds are crowded at bird feeders. It is estimated that tens of millions of house finches in eastern North America have died from the disease, and populations have markedly declined in the northeast.

Red crossbill *(Loxia curvirostra)*. Irregular visitor, mainly November to April; absent in some years and sometimes fairly common, in areas with mature coniferous trees; rare nester, when and where conifer seeds abound. *Note:* This species may actually consist of multiple species, each adapted to feeding on the cones of different kinds of conifers. Crossbills use their peculiar crossed mandibles to pry apart conifer cone scales so that the seed can be extracted with the tongue.

White-winged crossbill *(L. leucoptera)*. Irregular visitor, mainly November to April; absent in some years and sometimes fairly common; rare nester almost any time of year in habitats with mature coniferous trees (especially spruce, fir, or larch), which provide nest sites and food (seeds).

Common redpoll *(Carduelis flammea)*. Irregular visitor, mainly November to April; absent or rare in most years, common in other years; frequents weedy/brushy areas, places where birch, alder, or willow seeds are available, and yards with bird seed scattered on ground; nests far to the north in arctic and subarctic latitudes.

Pine siskin *(C. pinus)*. Nests sparsely in conifers, mostly in Northwest Highlands, sometimes in human-modified habitats; variably and unpredictably numerous in various situations in winter and during migration, late winter through early spring and mid- and late fall.

American goldfinch *(C. tristis)* (Fig. 20.54). Common nester in trees or shrubs in semi-open brushy areas in summer when thistles and other weedy plants provide abundant nesting material and seeds that adults feed to nestlings; nests contain eggs mainly in July and August; present all year, variably numerous in winter.

Evening grosbeak *(Coccothraustes vespertinus)*. Very rare nester in trees in northeastern hills; recently rare in wooded and semi-wooded areas in winter (sometimes mostly absent); if present, mainly September to May. *Note:* This species in the past has been an erratically numerous cold-season visitor. Christmas Bird Counts in the 1970s and 1980s yielded up to 4,545 evening grosbeaks in Connecticut, with an average of about 1,800 birds. The 1990s and early 2000s saw dramatic declines—over the past decade, most years have produced only 0 to 3 birds.

Old World Sparrows (Order Passeriformes, Family Passeridae)

House sparrow *(Passer domesticus)*. Locally common breeding resident near farms and in cities; nests in various cavities, crevices, and vine-covered walls, particularly those associated with buildings; sometimes usurps nest boxes from eastern bluebirds; eats seeds, insects, and small fruits. *Note:* This bird was first introduced in the United States in Brooklyn, New York, in 1851 to 1853. Following a population explosion, it declined substantially in the latter half of the 1900s, probably due to loss of nesting site nooks in modern architecture, less food (fewer horses depositing seed-laden dung), and competition from European starlings* and house finches*.

Seasonal Pattern of Bird Diversity

The number of bird species present in Connecticut varies seasonally (Fig. 20.55). April to May and September to October are the times of maximum diversity, owing to the simultaneous occurrence of breeding birds, migrants passing through, and some remaining or early-arriving winter visitors. Some of these species are present only in low numbers and/or in localized areas, so any particular area will have fewer species than indicated in figure 20.55. Christmas Bird

Fig. 20.54. American goldfinches subsist on seeds. Paul J. Fusco/Connecticut DEP Wildlife Division.

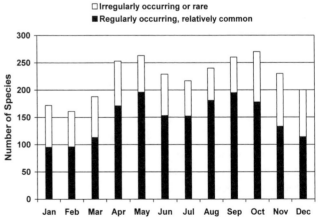

Fig. 20.55. Monthly number of bird species occurring in Connecticut.

Counts conducted from mid-December to early January generally yield about 155 to 170 species, whereas June counts in the same areas find about 190 species in an average year.

Habitats

Upland (terrestrial) habitats are used by 85 percent of Connecticut's breeding bird species (Fig. 20.56). Almost half of the species use thickly wooded habitats and almost half use shrubby, semi-open, or forest edge habitats. About 30 percent of the species use fields and other areas with sparse or only grassy or weedy vegetation. This high level of use of terrestrial habitats might be expected, since the vast majority of the state consists of woods, fields, and other upland habitats.

Despite the relatively small extent of wetlands in the state, these habitats are relatively heavily used by breeding birds—over half of the species use freshwater marshes and swamps and almost one-fourth use intertidal estuarine habitats. Habitats of lesser importance from the perspective of species usage include rivers, lakes, and the open waters of Long Island Sound. However, species using these habitats tend to be fully dependent on them.

A relatively small percentage of species, including swallows, chimney swifts, and common nighthawks, can be classified as making specialized use of the aerial habitat. These birds obtain all or most of their food from the air. To these could be added several species of flycatchers and a few other birds that also feed on flying insects, but these I omitted from the aerial category in figure 20.56 because they are more strictly associated with vegetated habitats rather than being aerial specialists per se.

The number of species cannot be used as the sole indicator of a habitat's importance. Most habitats support a large number of species that occur in no other kind of habitat, and thus each habitat, even those with relatively few species, contributes uniquely to the state's avifauna. For example, virtually no grassland or marsh birds occur in forests, and vice versa. Generally habitats with similar vegetation structure (i.e., size and spacing of plants) support similar bird species, though of course some birds select deciduous trees over conifers or vice versa.

Food and Feeding

Invertebrates, particularly insects, dominate the summer diet of the vast majority of Connecticut's breeding bird species (Fig. 20.57; the flying insect category includes only species that specialize on this food source). Hence factors that affect insect abundance (e.g., weather patterns, pesticide applications) are likely to affect birds as well. Most invertebrate-eating birds feed on the ground or in woody vegetation.

Among plant material, seeds, fruits, and nuts are the items eaten most often (see also "Acorn Ecology" in chapter 7). Relatively few species eat fishes, other vertebrates, or foliage.

Of course, many species fall into two or more diet categories. Others are restricted, and some food categories are critical even if supporting few species. For instance, although fishes are not consumed by many breeding birds, they are the only diet for ospreys.

Bird diets vary with the birds' seasonal status. The Connecticut diet of the vast majority of birds that migrate south to warmer climates consists of invertebrates such as various land insects, aquatic and intertidal invertebrates, or wetland plant material, though many migratory breeding species also feed heavily on fleshy fruits in late summer and fall (see following section). Since invertebrates and wetland plant material are scarce or unavailable in winter, it makes sense for the birds to migrate to areas where these foods are still

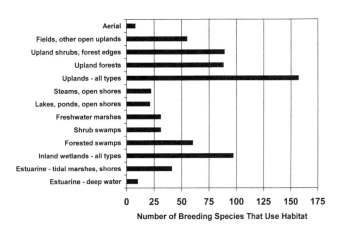

Fig. 20.56. Habitats of Connecticut breeding birds.

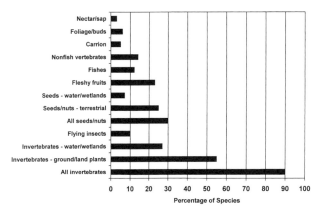

Fig. 20.57. Summer diets of Connecticut breeding birds.

accessible. Some year-round residents also eat these foods but are flexible enough to be able to switch to other sustenance in winter. Species occurring only in winter, and year-round residents in winter, eat estuarine or riverine fishes, small mammals or birds, carrion, seeds, or small fruits—foods that are readily available even during freezing weather. Even on snow-covered ground, wintering sparrows and other small seed-eaters often can find seeds of birches and other plants that release their seeds gradually throughout the winter.

Diet is only one aspect of a bird's feeding ecology. Species that eat similar foods may obtain them in different locations, at different times, or by using different methods, thus reducing their ecological similarity and the potential for competition. For example, among the many species of small birds that feed on insects associated with tree branches and foliage, red-eyed vireos often glean insects found by slow, careful visual searching whereas American redstarts move more quickly, flash their colorful wing and tail patches, and pursue insects flushed into movement. Flycatchers perch quietly and watch for airborne insects, then quickly fly out to nab them.

Fruits and Fruit-Eating Birds

Connecticut has a rich assortment of wild fruit-producing plants and a diverse array of birds and other animals that depend on these natural food sources. In the sense used here, fruits are structures consisting of fleshy, nutritious, edible pulp surrounding or attached to one or more seeds or in which seeds are embedded. The seed contains the plant embryo and the resources it needs to become a seedling.

Fruits vary in how they form and what they include. They may be *simple* (deriving from a single ovary, as in viburnums and spicebush; Fig. 20.58), *aggregate* (formed from a group of ovaries in a single flower, as in raspberries), *multiple* (composed of ovaries of a group of flowers borne on the same flowering stalk, as in mulberries, figs, and pineapples), or *accessory* (including one or more ovaries plus additional tissues such as the sepals and/or receptacle, as in strawberries, apples, and pears.

In Connecticut, dozens of species of plants with fleshy fruits depend on birds to disperse their seeds. Many of these plants are in the families Rosaceae (e.g., rose, bramble, chokeberry, cherry, hawthorn, shadbush, strawberry), Caprifoliaceae (honeysuckle, viburnum, elderberry; Fig. 20.59), Ericaceae (huckleberry, blueberry; Fig. 20.60), Vitaceae (grape, Virginia creeper; Fig. 20.61), Berberidaceae (barberry family; Fig. 20.62), Anacardiaceae (sumac, poison ivy), Cornaceae (dogwoods), and Aquifoliaceae (hollies). Most of the plants that depend on frugivorous birds for seed dispersal are shrubs and vines; relatively few are trees or herbs. The majority of wild fruits are small, usually 0.2 to 0.4 inches (5 to 10 mm) in diameter.

Fig. 20.59. Elder (elderberry) fruits are relatively low in nutrient content.

Fig. 20.58. Spicebush fruits, here beginning to ripen in late August, are relatively high in fat and protein content. Seed dispersers include migratory birds.

Fruit Function. Why would a plant invest significant resources in producing a colorful, fleshy fruit? Fleshy pulp is most important as a dietary enticement to potential seed dispersers. Seeds contained within fruits eaten by a bird, mammal, or box turtle may be moved by the animals to a location suitable for successful germination and growth. In contrast, seeds that drop to the ground beneath the parent plant may encounter poor conditions for survival due to nutrient competition, shading, or other factors. Hence it is advantageous for a plant to produce fruits that are attractive to frugivores (fruit-eaters) that move the seeds to another location. Fruit eaters generally digest the pulp but void the seed(s) intact, though some mice may destroy the seeds in the fruits they eat. Deer may eat certain fruits, but in doing so they grind up and destroy the seeds rather than disperse them.

Fig. 20.60. Blueberry fruits are available in early to mid-summer, so the seeds are dispersed mainly by summer-resident birds and mammals.

Fig. 20.62. Many birds eat Japanese barberry fruits and disperse the seeds of this non-native invasive plant. *Paul J. Fusco/Connecticut DEP Wildlife Division.*

Another function of the fleshy part of the fruit is to protect the enclosed seed(s). The flesh of a fruit, through simple mechanical interference, may inhibit the attacks of certain seed predators. Some fruits may contain chemical substances that deter seed predators or inhibit consumption by less effective dispersal agents. Green, unripe fruits often contain distasteful compounds that probably are effective in discouraging frugivores from removing the fruits before the seeds have ripened. The saponins and tannins in unripe American holly fruits may function in this way.

Fruit Color. A fruit's color also has important functions. Upon ripening, bird-dispersed fruits generally change color from green to reddish, blackish, purple, dark blue, or whitish. In some species, the fruits go through a series of color changes. These color changes may signal impending ripening to potential seed dispersers and make the fruits conspicuous to visually searching fruit eaters such as birds.

Green fruits often contain distasteful or digestively upsetting chemicals, which inhibit fruit eaters from eating them. Thus, from the plant's perspective, color change helps ensure that only mature seeds that are capable of germinating will be taken. If the fruits did not signal ripeness through color change, birds likely would have to test for ripeness, and much of the plant's investment of resources into its seeds and fruits could be lost as birds plucked and discarded unripe fruits.

Foliar Flagging. Many plants undergo a change to brightly colored foliage at about the time the fruits ripen. Many of these have nutritious fruits that do not last very long. The conspicuous foliage probably facilitates discovery of the fruit by birds and helps ensure that the fruits are consumed and dispersed before they rot or fall to the ground and become less available to good dispersers (though fallen and low-growing fruits may be found, eaten, and dispersed by various mammals such as foxes and raccoons).

Seasonality. Most bird-dispersed plants display mature fruits in late summer and fall, when bird populations temporarily are high and substantial migration is occurring. This is the major season of fruit-eating for most frugivorous birds, many of which eat insects in spring and summer.

Fewer species produce mature fruits before or after the peak bird migration season. Some plants that produce fruit relatively early (mid-summer) include blueberries, huckleberries, strawberries, mulberries, and blackberries. The fruits of these mostly low-growing plants have small seeds and tend to grow within reach of various mammals such as foxes, raccoons, and bears, as well as birds. Examples of early fruits that are large and large seeded include cherries and plums. Early fruiting plants tend to have sweet fruit (high in water and carbohydrates) and rely on nonmigratory birds and mammals for seed dispersal. Although raccoons disperse

Fig. 20.61. This raccoon scat is filled with grape skins and seeds.

Food and Feeding | 365

many seeds, rodents often eat and destroy seeds deposited in fecal latrines by raccoons (Fig. 20.61). Examples of birds that eat fruit during the breeding season include waxwings, thrushes, catbirds, mockingbirds, and towhees.

Many fruit eaters remain in Connecticut during winter and thus there is considerable potential for dispersal of seeds in winter for plants that retain fruits during the cold season. Examples of plants on which fleshy fruits may still be available in winter include common juniper, red-cedar, greenbriars, hackberry, common barberry*, Japanese barberry*, American mountain-ash, multiflora rose*, black chokeberry, hawthorn, crabapples, poison ivy, sumacs, American holly, winterberry, inkberry, American bittersweet, common buckthorn, autumn-olive*, bearberry, wintergreen, cranberry, partridge-berry, snowberry, and maple-leaved viburnum. These join nonfleshy nut or seed sources such as balsam fir, red pine, pitch pine, eastern hemlock, bayberry, American hazelnut, beaked hazelnut, birches, American beech, oaks, tulip-tree, witch-hazel, box elder, buttonbush, and ashes. These fruits tend to be eaten by fall migrant and winter resident birds and mammals. Some may not be eaten until spring. Fruits that remain attached through winter and are eaten in spring tend to have relatively low nutritional quality and contain chemicals that defend against fungal rot.

Fruit Quality. The fruits of different plants may vary greatly in their nutritional content. Fruits of gray dogwood, flowering dogwood, spicebush, sassafras, black gum, arrow-wood, Virginia creeper, and poison ivy are relatively high in lipid (fat) content. These fruits generally ripen in late summer and early fall and are especially attractive to migratory birds such as thrushes and catbirds (but not to American robins, which absorb digested lipids relatively poorly). Fruits with high protein levels include Solomon's-seal, American bittersweet, and spicebush. Black cherry, pin cherry, shadbush, raspberries, highbush blueberry, Solomon's-seal, pokeweed, spicebush, grapes, and multiflora rose* are high in carbohydrate content.

Some fruits, including common elder, red baneberry, maple-leaved viburnum, choke cherry, greenbriar, roses, poison ivy, winterberry, sumacs, bayberry, and junipers are relatively low in nutrients (on a per fruit basis). Fruits that are low in nutritional content tend to stay on the plants for a long time, and some of them are not eaten until periods of food scarcity in winter or sometimes not until the following spring when north-bound migrants pass through and feed.

Despite this variation in the nutritional content of different fruits, available evidence indicates that birds may or may not forage selectively on the more rewarding fruits (e.g., those with proportionally more pulp and/or relatively high fat content). Cedar waxwings feed almost exclusively on sugar-rich fruits, whereas thrushes often eat fruits with high fat content. Birds sometimes seem to forage selectively on the fruits of certain individual plants, but the factors that account for this selectivity are poorly known. Fruit selection by birds undoubtedly is influenced by the various toxins present in fruit pulp or skins. In general, birds do not specialize on certain species of fruiting plants—each bird species eats and disperses fruits of several species of plants, and birds may need to consume a mixture of sugary fruits and insects (or protein-rich plant material) to attain nutritional balance. Some birds, such as American robins, preferentially select fruits that exist in high abundance, which are more accessible and require lower search costs. Robins also prefer larger fruits that yield more total resources per unit of foraging effort. Birds such as cedar waxwings, which inefficiently digest sucrose, learn to avoid sucrose-rich fruits and prefer those containing glucose or fructose, perhaps because eating sucrose leaves them feeling hungry.

Fruit-Eating Birds. In Connecticut, birds that eat and disperse fruits include bobwhite, ruffed grouse, woodpeckers, flycatchers, American crow, thrushes (including American robin), northern mockingbird, gray catbird, brown thrasher, black-capped chickadee, European starling*, cedar waxwings, finches, sparrows, vireos, warblers, Baltimore oriole, scarlet tanager, and others. Some of these fruit eaters (for example, bobwhite, rose-breasted grosbeak, and some finches) may be seed predators—that is, they digest and destroy at least some seeds. Some fruit eaters, such as scarlet tanager, sometimes rose-breasted grosbeak, and Baltimore oriole, are poor fruit dispersers because they may eat the fruit and/or fruit juice but leave the seeds behind (Fig. 20.63).

Fig. 20.63. Sometimes fruit eaters do not disperse seeds. Here a white-throated sparrow removed the fleshy part of two flowering dogwood fruits, leaving the seeds behind (upper left).

Frugivory and Seed Handling. Some fruit-eating birds are able to ingest large numbers of fruits at one time. Waxwings, for example, have an expandable esophagus that allows them to ingest relatively large meals. American robins may swallow ten blueberries whole, one right after another.

Seeds contained within the fruits ingested by frugivorous birds may be defecated or regurgitated. Cedar waxwings usually defecate all seeds. Red-eyed vireos defecate only the smallest seeds, such as those of common elder, and regurgitate some small seeds as well as all larger ones. Thrushes regurgitate most seeds. American robins defecate seeds as large as black cherry and sometimes regurgitate seeds smaller than dogwood. Overall, regurgitated seeds generally spend less time in a bird than do defecated seeds. For example, hermit thrushes generally regurgitate seeds about 10 to 11 minutes after eating the fruit whereas seed defecation usually takes about 30 minutes. For some fruits, regurgitation and defecation times do not differ. In most frugivorous birds, food remains are defecated approximately an hour or less after the material was eaten.

Effect on Germination. Contrary to what one might expect, fruit pulp generally does not serve as fertilizer for germinating seeds, and attached pulp actually can inhibit germination. Removal of fruit pulp from seeds, such as occurs in the gut of frugivorous birds, may enhance germination of seeds of some plants, especially those with lipid-rich fruits such as spicebush and some viburnums. In contrast, germination rate of blueberry seeds is reduced by about 20 percent after passing through the digestive systems of birds and mammals. This may be a small price to pay for getting the seeds dispersed.

Competition for Fruits. Limited information suggests that there may be competition among fruit-eating birds for fruits when fruits are scarce. For example, mockingbirds may defend fruiting plants from other frugivorous birds (including their own and other species) in winter. It is possible that birds and mammals sometimes compete for fruits, but this has not been investigated much.

Mutual use and possible competition for fruits may occur between birds and various invertebrates, but little is known. One interaction involves birds, a small fly, and hollies. American holly fruits infested with the larvae of the holly berry midge (Diptera: Cecidomyiidae) remain green whereas uninfested fruits turn red as is normal. During winter, birds eat the green infested fruits to a lesser degree than they eat the normal red fruits. By interfering with the development of the red color that indicates ripeness, the midge larvae reduce the probability of predation on the fruit and thus increase their probability of survival until spring when the adult midge emerges from the fruit. Many other similarly interesting but as yet little known relationships undoubtedly exist among fruits, vertebrate frugivores, and invertebrates.

Unanswered Questions. As the above example suggests, the relations between fruits and fruit consumers may be complex and often may be difficult to interpret because it is likely that many seed and fruit traits are affected by factors other than dispersal agents. And we should not forget that perplexing fruiting phenomena may result from the impacts of humans. For example, what fruit characteristics evolved in response to the formerly enormously abundant passenger pigeon? How are the numerous exotic fruiting shrubs and vines now established in the state affecting seed dispersal and populations of native species? And finally, why is it that fruits in the supermarket can look so perfect but taste so bad or rot before they ripen?

Reproduction

Seasons

Singing generally heralds the onset of the breeding season, but males of some species sing well before actual nest building and egg laying occur. Singing functions in mate attraction and establishment of breeding territories. Common birds that begin singing in March include mourning dove (which may coo as early as late January or early February), eastern phoebe, tufted titmouse, black-capped chickadee, Carolina wren, northern mockingbird, American robin, northern cardinal, chipping sparrow, song sparrow, red-winged blackbird, and house finch*. Various owls vocalize regularly in February and March, and woodpeckers may begin drumming in March.

In Connecticut, most birds begin nesting in May. However, great horned owls, among the earliest of nesters, often begin nesting in February, and several additional species may lay eggs in March, including black vulture, bald eagle, red-tailed hawk (Fig. 20.64), red-shouldered hawk, northern goshawk, wood duck, hooded merganser, mourning dove, barn owl, eastern screech-owl, barred owl, northern saw-whet owl, long-eared owl, woodcock, horned lark, common raven, eastern bluebird, house finch*, and pine siskin. These may be preceded by crossbills, which sometimes nest as early as January. Nesting by rock pigeons* extends year-round, and house sparrow* nests may have eggs from February through September.

The overall pattern is that the early breeders tend to be carnivorous and omnivorous species and/or permanent residents of Connecticut. Many of the permanent residents

Fig. 20.64. Upper: Juvenile red-tailed hawks in a nest in a black birch. Stephen P. Broker. Lower: American goldfinches build compact, durable nests. This one, photographed in early spring (in a silky dogwood), was used the previous summer.

have a flexible diet that includes a large component of seeds or fruits in the colder months. Thus early nesters do not have to wait until insects are readily available in May. Additionally, they tend not to depend on deciduous foliage to conceal their nests. An early breeding flycatcher, the eastern phoebe, differs from other small insect eaters in wintering in areas not far from Connecticut (southeastern United States and Mexico) and in not being dependent on foliage for nest cover (it nests mainly in buildings and other structures or in nooks of sheltering rock outcrops).

Late-nesting birds commonly or sometimes do not lay eggs until June. Many late nesters, such as the common nighthawk and willow flycatcher, are long-distance migrants that may not arrive in nesting areas until well into May. Others, such as clapper rails, marsh wrens, and seaside sparrows, arrive earlier but may wait until adequate nesting cover has developed. American goldfinches, which feed seeds to their nestlings, often do not nest until mid-July; late nesting ensures a good food supply (and a source of thistle fluff for their nests) (Fig. 20.64).

Nest Sites

Connecticut's birds nest in a wide variety of microhabitats (Fig. 20.65). Not unexpectedly, a large number of species nest in trees. Most of these attach their nests to branches or among foliage. Examples of these include most hawks, great horned owl, hummingbirds, vireos, tanagers, gnatcatchers, jays, some warblers, orioles, grosbeaks, and many others. The tree nesters category includes nine species, such as the pine warbler, that nearly always nest in conifer trees.

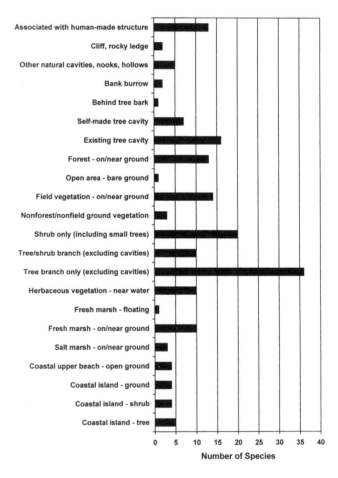

Fig. 20.65. Nest sites of Connecticut breeding birds. I assigned each species to only one primary nest-site category, so the species totals for some categories are somewhat conservative. Included are virtually all species that have nested in Connecticut since 1990.

About two dozen bird species nest in tree cavities. Red-headed woodpecker, red-bellied woodpecker, yellow-bellied sapsucker, downy woodpecker, hairy woodpecker, northern flicker, and pileated woodpecker usually excavate their own nesting cavities. Among songbirds, black-capped chickadees and red-breasted nuthatches also are cavity excavators. Other species, including wood ducks, hooded mergansers, common mergansers, American kestrels, eastern screech-owls, northern saw-whet owls, great-crested flycatchers, purple martins (formerly), tree swallows, tufted titmice, white-breasted nuthatches, brown creepers, house wrens, winter wrens, Carolina wrens, and eastern bluebirds, use naturally formed cavities or those excavated by woodpeckers. Cavities excavated by the pileated woodpecker may be important sites for nesting wood ducks and other wildlife requiring large cavities. The European starling is an introduced species that often usurps nesting cavities from native birds. Forestry practices that remove snags and diseased or injured trees that often serve as cavity trees are likely to be detrimental to populations of these birds (see "Forestry Practices and Missing Resources" in chapter 7 for further discussion of this topic).

Not all forest-dwelling birds nest in trees or shrubs. More than a dozen species (e.g., wild turkey, ruffed grouse, woodcock, hermit thrush, chipping sparrow, eastern towhee, black-and-white warbler, and ovenbird) nest on the ground. And, across all habitats, there are almost as many ground- or near-ground-nesting birds (more than 60 species) in Connecticut as there are tree nesters. This suggests the importance of maintaining natural ground cover in both forested and nonforested habitats and preventing the establishment of non-native, ground-dwelling predatory mammals (as well as unnaturally high populations of native species).

Migration

For many birds that nest in Connecticut, cold temperatures and the short period of daylight in winter result in low food availability, impose severe physical stresses, and make it difficult to maintain a favorable energy balance. Some birds, such as chickadees, titmice, nuthatches, and crows, deal effectively with these challenges (see following section, "Birds in Winter") and stay in Connecticut year-round, but most, including warblers, vireos, swifts, and many others, avoid winter's problems and migrate to more favorable locations.

A few dozen bird species (e.g., various warblers and sandpipers) that you may see in Connecticut do not nest or winter in the state, but occur only as migrants between nesting areas to the north and wintering sites southward. About 50 species of northern breeders show up in Connecticut only as nonbreeders in the colder months (e.g., various waterfowl and sandpipers, Bonapart's gull, fox sparrow, tree sparrow, etc.).

Migration Examples

Following are some examples of the migrations of birds that occur in Connecticut:
- Broad-winged hawks, chimney swifts, and common nighthawks fly overland to South America and back.
- Loons, grebes, diving ducks, and geese may fly from inland areas in Canada and Alaska to the North American coast then proceed coastally southward.
- Many shorebirds fly between the arctic and South America, following a coastal route, inland route, or transoceanic route.
- Many passerines migrate between North America and South America by flying over the Gulf of Mexico (radar studies and moon watching indicate that up to 50,000 birds per mile of front per day enter southern Louisiana from the Gulf of Mexico in late April and early May).
- Possibly millions of individual passerines fly from southeastern Canada and New England to the West Indies and South America by way of a southeastward then southwestward flight over the Atlantic Ocean, taking advantage of favorable winds along their journey. These migrations may involve a nonstop flight of 80 to 90 hours. Connecticut birds that winter in South America include, in addition to those mentioned previously, cuckoos, olive-sided flycatcher, eastern wood-pewee, eastern kingbird, purple martin, bank swallow, barn swallow, veery, Canada warbler, Connecticut warbler, Baltimore oriole, and many others. Swainson's thrush, rose-breasted grosbeak, orchard oriole, and others winter primarily in Central or South America.
- "Leapfrog" migration occurs in some species, such as the peregrine falcon and fox sparrow, with the most northerly breeding populations migrating farther south than do more southerly breeding populations.

Spring Migration

In the northern part of North America, spring brings a softening of physical conditions and favorable feeding and nesting opportunities, and birds respond accordingly. Spring migration is unified among nearly all birds in that movements are basically northward, but different groups of birds differ in their peak migration periods.

Spring migration of waterfowl in Connecticut occurs mainly in March and April for most species. Large numbers of raptors migrate in April, especially the latter half of the month when winds are out of the southwest, but some species that do not winter very far to the south, such as the red-tailed hawk, red-shouldered hawk, and turkey vulture (not actually a raptor), may migrate much earlier (as early as mid-winter) and other species, such as the broad-winged hawk, commonly migrate through Connecticut well into May. The late migration of the broad-winged hawk is related to its long return from wintering areas in South America. Spring migration of raptors is not so concentrated or evident as in autumn, so less is known about it.

Shorebird migration is concentrated mainly in April and May. Many of these birds breed in the high arctic and winter in South America. The migrations of many shorebird species (for example, ruddy turnstones, semipalmated sandpipers, dunlins, and several others) are timed to coincide with the late-May spawning of horseshoe crabs, the eggs of which are a critical food resource used for replenishing energy stores depleted by the long flight from South America and for ensuring that the birds are in good condition for their continued migration to breeding areas in the high arctic. Good places to see migrating shorebirds include Milford Point, Sandy Point in West Haven, Hammonasset Beach State Park in Madison, Griswold Point-Great Island in Old Lyme, and Barn Island Wildlife Management Area in Stonington.

Small migratory landbirds pass through Connecticut in largest numbers in May, though the songbird migration is well underway in Connecticut by the last week of April. Migration coincides with the period when tree and shrub foliage emerges and grows to full size. These migrants include many long-distance travelers that winter primarily in South America (for example, cuckoos, common nighthawk, chimney swift, eastern kingbird) and insect eaters that forage primarily among foliage (for example, vireos, warblers, scarlet tanager, orioles). Among the warblers and other species, the species that winter the farthest south tend to be the ones that arrive latest in migration. The early-returning species, however, do not necessarily begin to nest any earlier than do late returners. Most of the seed eaters such as sparrows, finches, and blackbirds have protracted migration periods that begin in late winter or early spring; their early migration is possible because they are not so dependent on insects.

Autumn Migration

Autumn migration, of course, reverses the spring pattern and brings a vast southward movement of bird populations. And as in spring, the migration schedules vary among different kinds of birds.

Water Birds. Southward migration of waterfowl extends from late summer through late fall or early winter. The earliest migrants in August and September include blue-winged teal, green-winged teal, northern shoveler (rare), mallard, gadwall, and American wigeon. Most other species migrate through Connecticut beginning in October or early November, though several species, including canvasback, common goldeneye, common merganser, and red-breasted merganser, do not arrive in large numbers until December. Substantial southward migration of shorebirds gets underway in mid-July, rapidly increases through late July and early August, and continues through October.

Raptors. Late summer to early fall is a good time to observe raptor migration in Connecticut. Especially notable are the large numbers of migrating broad-winged hawks and sharp-shinned hawks. For example, on 14 September 1986, 30,535 broad-winged hawks were counted from Quaker Ridge in Greenwich; on 15 September 1995, 31,988 broad-wings were counted from the same location. However, a few thousand on a good day in southern New England is more typical.

Broad-winged hawk migration in Connecticut peaks in mid-September, but during the latter half of September and early October, you can see large numbers of other raptors (mostly sharp-shinned hawk, osprey, American kestrel, Cooper's hawk, northern harrier, and merlin). Migrating red-tailed hawks increase in abundance through October, peaking with south-moving cold fronts in late October and early November.

Among the birds counted from all of the hawk watch locations in Connecticut, the broad-winged hawk is always the most numerous (sometimes over 100,000), followed by the sharp-shinned hawk (often around 15,000 to 20,000), American kestrel (several thousand), osprey (a few thousand), northern harrier (1,000 or somewhat more or fewer), red-tailed hawk (around 1,000 or somewhat fewer), Cooper's hawk, turkey vulture, and merlin (hundreds of each), and several other species all represented by fewer than 100 counted. Of course, many raptors migrate through the state uncounted, so these figures indicate only relative and minimum numbers.

Lighthouse Point in New Haven is one of the best places in New England to observe migrating raptors. Its coastal location concentrates raptors that are averse to crossing large expanses of open water, and it is far enough west to be along the flight path of migrants from most of New England and southeastern Canada. In some years, 25,000 to 30,000 raptors have been counted during September and October. Most of these are species that tend to migrate along the coast, including northern harriers (especially adult females and immatures), sharp-shinned hawks (especially immatures), Cooper's hawks, ospreys, American kestrels, merlins, and peregrine falcons.

The numbers of raptors observed can vary greatly, from only a few per hour to a few hundred per hour; numbers may vary by two orders of magnitude from one day to the next. These variations are, in large part, the result of weather conditions, especially the daily pattern of wind direction and speed. For example, a prolonged period of strong winds out of the northwest in mid-September may result in a buildup of southward-bound broad-winged hawks along the Connecticut coast, yielding high counts at Lighthouse Point. Under other wind conditions, most broad-wings migrate farther inland, and sharp-shinned hawks and American kestrels are the dominant species at the Point. During such times, migrating broad-winged hawks (as well as other raptors) can be observed from many peaks and ridges in southern New England and southeastern New York. Other typical inland migrants include the red-tailed hawk and turkey vulture.

Large hawks tend to fly mainly when rising masses of warm air (thermals) are available, such as in mid-morning and afternoon when the Sun's angle is likely to heat certain cliffs, hillsides, or other topographic features. By "riding" these air currents (and other kinds of updrafts, such as those occurring where winds are deflected upward by a ridge or hill), hawks can soar more and flap less and thus reduce energy expenditure.

In Connecticut, big September migrations occur after skies clear following the passage of a cold front, especially following a period of cloudy or rainy weather and when winds blow from the north, northwest, or west, which facilitate the development of thermals and deflective updrafts. Smaller hawks, such as the sharp-shinned, do not depend as much on thermals and often can be observed in flight relatively early in the morning and at other times when conditions are unfavorable for the occurrence of updrafts.

Most raptor species that migrate through Connecticut spend the winter in the area extending from the central and southern latitudes of the eastern United States south to the West Indies and Caribbean region and through Mexico to Central America. Cooper's hawks, goshawks, and red-tailed hawks mostly go no farther south than southern United States and Mexico. Broad-winged hawks winter principally in South America (some in southern Florida).

Over the past couple decades, counts of migrant hawks from hawk lookouts in northeastern North America indicate increasing numbers of bald eagles, peregrine falcons, merlins, ospreys, and Cooper's hawks. Counts of northern harriers suggest an increasing trend whereas those of broad-winged hawks suggest a decline. No consistent trends are evident for other hawks.

Certain owls also migrate. The three most common migrant owls are the northern saw-whet owl, long-eared owl, and barn owl. All of these owls essentially complete their fall migrations in our area before the end of November.

Songbirds. Some songbirds begin moving southward by mid-summer, but the greatest numbers migrate in late summer and early fall. Departure of warblers from nesting areas begins as early as late July and increases through August. Very large numbers of migrants pass through coastal sites such as Bluff Point in Groton in September and into mid-October; under optimal conditions, one might tally 3,000 to 4,000 warblers of 20 or more species in one day.

Periodic Migrations of Boreal Seed-Eating Birds

When there are ample seed crops of high latitude trees, most individuals of certain boreal seed-eating birds remain in the north in winter and are scarce in Connecticut. When seed crops are poor, the birds migrate southward. In these winters, seed eaters such as red-breasted nuthatch, evening grosbeak, purple finch, pine grosbeak, common redpoll, pine siskin, red crossbill, and white-winged crossbill may be synchronously common in Connecticut and areas farther south. These birds (though not all species) were numerous during the winter of 1997/1998, extremely rare or absent in 1998/1999, and plentiful again in 1999/2000. Recent analyses indicate that food is not the only factor that plays an important role in causing and synchronizing the movements of these birds, but the additional factor or factors remain unknown.

Birds in Winter

Dealing with Cold Temperatures

The primary winter adaptations in birds are as follows:
- Increased plumage insulation (thicker coat of feathers), especially in larger birds.
- Heat production through low-amplitude shivering, powered by energy stored as body fat. This increases the time that normal body temperature can be maintained at low ambient temperature. Even small birds such as chickadees have greater heat production capability and higher metabolic rates in winter relative to summer.
- Selection of roosting sites that minimize heat loss. For example, many species roost under plant cover, grouse

may roost in snow cavities or under cover of evergreens, and chickadees, titmice, nuthatches, and woodpeckers roost in tree cavities.
- Communal roosting. Several individuals of certain species such as bluebirds and creepers may roost in a single tree cavity and benefit from the increased warmth.
- Reduced body temperature at night (a drop of up to 18 to 22°F in chickadees), which reduces energy requirements at a season when food may be scarce.
- Reduced foot temperature. In cold weather, birds reduce blood flow to the feet and allow them to cool. This reduces heat loss to cold air and when walking on snow (Fig. 20.66). Even small arboreal songbirds that do not walk on snow nevertheless have cold feet in winter. I verified this by immediately recovering birds killed after colliding with windows; the birds' feathered upper legs were warm but their feet were very cold.
- Reduced extraneous activity, which also reduces energy expenditure and thus decreases the amount of food needed.
- A high percentage of the activity period spent foraging.
- Food storage. For example, chickadees store excess food under bark and in similar nooks and retrieve it up to several weeks later.

Frequent feeding is an important activity for small birds, which have enough energy reserves to get them through only about two to three days. Some birds, of course, avoid the cold and scarcity of food by migrating to warmer climates with greater food availability.

Winter Birds at the Feeder

For several winters I maintained a feeding station at my home in Higganum along the Connecticut River in central Connecticut. The feeding station included three hanging tubular feeders, two filled with black oil sunflower seeds and the other containing thistle seeds, and a wire feeder that I periodically stocked with blocks of suet. In addition, I scattered sunflower seeds and mixed bird seed on the ground every few days. Here is what I saw:

Species Observed Virtually Every Day. Mourning dove, downy woodpecker, blue jay, black-capped chickadee, tufted titmouse, white-breasted nuthatch, northern cardinal, house finch*, dark-eyed junco, American tree sparrow, white-throated sparrow, song sparrow, and American goldfinch.

Present Every Winter, Sometimes Daily. Red-bellied woodpecker, eastern towhee, Carolina wren, and field sparrow.

Present Irregularly. Sharp-shinned hawk, ring-necked pheasant*, hairy woodpecker, European starling*, purple finch, common redpoll, pine siskin, red-breasted nuthatch, fox sparrow, red-winged blackbird, and common grackle.

These lists include only the species that actually fed on the seed provided or on the birds attracted to the seeds. They are representative of what most bird feeders attract in winter, though the frequency of occurrence of the various species varies among different feeding stations, depending on location and surrounding habitat. Birds not listed here but commonly occurring at feeders in winter include the house sparrow*, rock pigeon*, and American crow. In addition, evening grosbeaks can be expected at feeders in some years, though I never detected this species in my yard. Wild turkeys also are becoming more frequent at bird feeders.

To participate in a project that gathers data on birds at feeders, visit the Cornell Laboratory of Ornithology's Project Feeder Watch website: http://birds.cornell.edu/pfw.

Conservation

Forest Fragmentation, Predation, and Cowbird Parasitism

A major problem facing certain forest birds in Connecticut and elsewhere is the ongoing fragmentation of forests

Fig. 20.66. A wild turkey passed this way (ruler is 6 inches long).

into smaller and smaller pieces and the general increase in the amount of forest edge and the simultaneous shrinkage of the forest interior (i.e., forest habitat that is far from an edge) (Fig. 20.67). Sometimes you will hear that creating forest edge is good because it increases biodiversity. However, the vast majority of forest-edge birds and other wildlife are resilient, common species that have plenty of habitat and are not experiencing population declines as a result of human activity. Some are even increasing in abundance and range. In contrast, several bird species characteristic of forest interiors and large tracts of forest with minimal edge appear to be experiencing significant population declines. The extent of the problem is not clear, due to somewhat conflicting results from different analyses.

Examples of birds that require large continuous tracts of forest habitat for nesting include red-shouldered hawk, barred owl, whip-poor-will, hairy woodpecker, pileated woodpecker, Acadian flycatcher, red-eyed vireo, wood thrush, ovenbird, worm-eating warbler, Louisiana waterthrush, hooded warbler, scarlet tanager, and others. These large-forest birds do not necessarily avoid the edges of occupied forest tracts. Many are long-distance migrants that nest in North America and winter in the neotropics.

The birds that appear to be declining generally do not breed in small tracts of forest, or at least they are relatively scarce in such habitats. Since small tracts of forest are becoming more common and large tracts more rare, a logical conclusion is that the populations of birds that favor forest interiors must be declining. Data from the North American Breeding Bird Survey and other research also suggest that significant declines are occurring in certain of these species.

Why does fragmentation result in population declines or local extirpations of forest birds (and other forest species)? Increased predation is one possibility. Egg and nestling destruction by edge-prowling mammals, birds, and snakes may reduce breeding success and detrimentally impact populations of birds in fragmented forests that have a relatively large amount of edge. Examples of forest-edge predators that may destroy eggs or young include raccoons, fishers, gray squirrels, eastern chipmunks, flying squirrels, white-footed mice, opossums, Baltimore orioles, gray catbirds, blue jays, rat snakes, racers, milk snakes, and other animals. Increased mammalian predator pressure along edges seems to be most pronounced in agricultural landscapes. However, some predators may be less numerous near edges than in forest interiors, and research results do not consistently indicate that predation is greater along forest edges, so there is debate about the significance of this factor.

Close proximity of forest edge and open areas also may expose forest-interior birds to the detrimental effects of brood parasitism by the brown-headed cowbird, a for-

Fig. 20.67. Small forest fragments, such as those surrounded by residential, agricultural, or commercial development, do not support the full diversity of forest wildlife.

merly scarce but now abundant species that thrives in open country and forest-edge habitats. Cowbirds lay eggs in the nests of other bird species, which may raise the young cowbirds and experience reduced production of their own young (Fig. 20.35). Recent research indicates that while cowbird parasitism may detrimentally affect populations of certain bird species that nest in shrubby habitats in various regions of North America, populations of forest interior birds in the eastern United States, even if sometimes brood-parasitized by cowbirds, are not necessarily reduced significantly by this factor.

Stable populations in a local area do not necessarily indicate that all is well. Such populations may be reproductively unsuccessful and simply absorb dispersing individuals produced in extensive forests elsewhere. However, research on worm-eating warblers in Connecticut indicates that in largely forested landscapes reproductive success is not reduced in small forest patches (averaging 96 acres), though patches smaller than about 50 acres tend to be devoid of nesting worm-eating warblers.

Overall, available data indicate that predation and cowbird parasitism in small forest fragments can, but do not always, detrimentally impact songbird abundance and reproduction. It is important to recognize that some forest fragments do support productive bird populations and may be valuable in areas where conservation of large tracts of forest is not possible. Furthermore, even nonproductive bird populations play ecologically important roles as predators, prey, and dispersal agents.

Non-native plants might possibly play a small role in increasing predation along forest edges. A large number of exotic shrubs are well established in Connecticut in forests and along their edges. These provide suitable branch structure for supporting bird nests and they do indeed attract nesting birds. Shrubby exotic honeysuckles leaf out early, and this early cover may be attractive to birds exploring potential nest sites at the beginning of the nesting season. Yet recent research indicates that birds nesting in the exotic shrubs such as shrubby honeysuckles and buckthorns may experience greater nest predation than do the same species nesting in native plants. This may be because nests in the exotic shrubs tend to be lower and thus more accessible. Further study is needed to clarify whether this impact of exotic plants might contribute significantly to bird population declines.

Another explanation of bird population declines in fragmented forests is that forest birds may prefer habitats that can support many individuals of their species such that mating opportunities are maximized (many songbirds are not strictly monogamous with respect to copulation). Small patches of forest can support relatively few birds and thus may limit mating opportunities; such areas may not be very attractive to the birds even if habitat conditions are favorable. At high population levels, competition may force birds into small forest patches. When populations are low, whatever the cause, the marginal conditions afforded by small tracts of forest may be seldom used.

Small populations in small habitat patches are more prone to extirpation due to the effects of random or cyclic environmental fluctuations or random changes in reproduction and survival than are larger populations in larger habitat patches. Clearly, a small population reduced from 100 to 5 generally is in much greater danger of disappearing than is a larger one reduced from 1,000 to 50. This phenomenon applies mostly to sedentary species with limited ability to colonize isolated patches of habitat and so is not a significant factor in forest bird declines.

Some biologists have suggested that the disappearance and declines of forest interior birds in smaller forest tracts is not a result of predation, cowbird impact, or other breeding ground factors but rather is a symptom of a decline caused by deforestation of winter habitat in the tropics. Perhaps populations of birds reduced by wintering-ground impacts are simply using the best remaining breeding habitat. Large tracts of forest probably are most likely to provide favorable nesting and feeding habitat; small tracts inevitably have less habitat variety and resources.

In reality, declines of forest birds likely are due to multiple factors, which may vary with the species, geographic region, and habitat. Simple explanations seldom apply in the natural world.

To enhance forest habitats, conservationists suggest the following guidelines:

- Avoid fragmenting large, contiguous tracts of forest and attempt to maintain or create forested tracts that are at least thousands of acres in size.
- Plan for future forests by scheduling timber harvests such that contiguous forest cover is maximized throughout a cutting cycle.
- Place buildings, roads, and other disturbances around the margins of a forest rather than in the interior.
- Manage forests by maximizing the amount of interior and minimizing the amount of edge—avoid restricting forest to narrow swaths. Edges do create local diversity, but in general edges are not in short supply whereas interior forest habitat is a declining habitat that should be enhanced through long-term regional planning.
- Attempt to maintain or restore forested connections between isolated or partially isolated forests. These may be important in facilitating wildlife movements among (and maintaining wildlife populations in) the connected forest tracts.
- Manage forests for maximum structural diversity, including large trees, small trees, shrubs, and herbaceous vegetation. This maximizes the microhabitats available to wildlife. Reduction of deer populations may be necessary in some areas.
- Maintain continuous availability of patches of old growth and decadent trees throughout the forest. These create small openings (when old trees fall) and provide cavities and coarse woody debris favored by various wildlife.
- Maintain continuous availability of large, nut-producing trees of various species. These produce important wildlife food resources.
- Maintain and manage for high tree diversity rather than monocultures than offer fewer resources.
- Remove invasive exotic plants that commonly thrive on disturbance and invade deforested areas or fragmented forests. These may outcompete native plants and offer fewer usable resources for native wildlife.

While habitat fragmentation clearly affects the species composition and abundance of forest bird populations, it is important to recognize that bird populations in both fragmented and unfragmented forests also change as habitat conditions change through natural ecological processes that affect vegetation structure. Declines and disappearances of bird species in forest fragments are not necessarily due to predators, cowbird parasitism, or outside influences.

Grassland and Open-country Birds

The populations of many bird species in Connecticut have fluctuated greatly over the long term as a result of vegetation changes associated with clearing of the land for agricultural expansion during the colonial period, followed by reforestation when farms were abandoned as the human economy became increasingly industrial. Populations of forest birds were much reduced during the agricultural peak, and grassland and grass-and-shrub-associated birds expanded and became more abundant. With reforestation, forest birds made a big comeback whereas the open-country birds declined. Grassland and thicket species continue to decline as many remaining fields become forested, are converted to human residential use, or are inappropriately managed. At least three species of grassland birds have been extirpated as breeders in Connecticut.

Today, specific conservation actions are aimed at maintaining populations of open-country birds, such as the bobolink, eastern meadowlark, upland sandpiper, grasshopper sparrow, and barn owl, as interesting and unique members of the biota. For example, destruction of the state's largest population of nesting grasshopper sparrows on the site of the new University of Connecticut football stadium and other development is being mitigated by establishment of a 200-acre grassland habitat on state property in Somers.

One might argue that under "natural" conditions, with relatively little human alteration of the landscape, these grassland birds would be scarce at best in this naturally forested state and that their presence is largely an artifact of human activity. But humans have been a part of the landscape throughout Connecticut's post-glacial history, and there is good evidence that humans have maintained large openings in the forest (through burning) for millennia. Also, beaver activity, which leads to the development of open meadows after ponds are abandoned (beavers abandon ponds when they deplete the local food supply), has also made available grassy-sedgy openings for thousands of years. With respect to biodiversity concerns, grasslands are thus a legitimate component of the Connecticut landscape.

Open-country birds are most attracted to and do best in large patches of habitat. Assurance of long-term viability of the state's grassland bird populations probably will require continuous availability of at least a few suitably managed grassland patches each encompassing a few hundred acres or more. A basic management protocol for providing favorable habitat conditions for these birds involves periodic mowing or burning to prevent the invasion of woody plants and conducting such activities outside the bird nesting and brood-rearing season. Unfortunately, many fields around the state are mowed during the nesting season, destroying nests and eliminating the necessary protective vegetation. Mowing also degrades conditions for meadow voles, which are a major food resource for grassland-foraging barn owls, red-tailed hawks, and other predators. Mowing should be avoided from April through July.

Ironically, the best populations of grassland birds in the Northeast tend to be in areas with relatively low, sparse cover dominated by non-native cool-season grasses, whereas native warm-season grasses may form tall, dense stands that provide poor conditions for grassland birds of concern. Hence, anthropogenic habitats with non-native plants may be essential for effective conservation of grassland birds in the region.

Coastal and Riverine Birds

Heavy summer use of coastal beaches by humans poses problems for birds that attempt to nest there. Most of these birds, such as the piping plover, least tern, black skimmer, and American oystercatcher, nest on upper beaches where frequent foot traffic may prevent the birds from nesting, even if it does not actually crush eggs.

People and their dogs often displace adult birds and expose eggs or young to extreme temperatures and predatory gulls. Eggs and chicks on beaches are vulnerable not only to gulls but also to burgeoning populations of free-ranging predatory mammals (raccoon, opossum, skunk, red fox, feral cat), all of whose numbers are much larger than they used to be because of enhanced food resources made available by humans. Recreational impacts are not limited to the nesting season. For example, migrating shorebirds attempting to rest or feed on coastal beaches also face constant disturbance from people and roaming dogs.

Several coastal sites are actively managed for birds. Protection of the coastal nesting areas of piping plovers and least terns in Connecticut is an ongoing activity of the Department of Environmental Protection, The Nature Conservancy, and many volunteers. Temporary fencing and on-site education are the primary conservation tools (Fig. 20.68). At Milford Point, state and federal agencies, together with private conservation organizations, recently restored nesting habitat for the plover and tern. These protection efforts have had unexpected side benefits—in 1998, the state's first-ever successful nesting by black skimmers occurred in an area that had been fenced to provide a disturbance-free area for terns and plovers. On Falkner Island, old tires and boxes carefully placed on the ground provide secluded nesting sites for roseate terns (Fig. 20.69).

Fig. 20.68. Exclosures protect piping plover nests from predators and increase nesting success.

Fig. 20.70. Sandpipers feast on horseshoe crab eggs in preparation for their spring migration to arctic nesting grounds.

Another concern about coastal birds is the impact of declining populations of horseshoe crabs. Many kinds of shorebirds depend on the eggs of the horseshoe crabs to fuel the birds' long migration to their arctic nesting grounds (Fig. 20.70). Excessive harvest of horseshoe crabs along the Atlantic coast has reduced their abundance, cutting into the food supply for the migrating shorebirds. Recent harvest restrictions hopefully will allow the horseshoe crab populations to rebound.

Coastal islands in Long Island Sound are very important for bird conservation. These islands are relatively isolated from heavy impacts by mainland predatory mammals and provide critical nesting refuges for colonies of wading birds, such as great egrets, snowy egrets, black-crowned night-herons, and little blue herons, and common and roseate terns. In fact, coastal islands are the only places in the state that support significant nesting populations of these birds. Unfortunately, thoughtless disturbance by people has reduced nesting success on some of these islands, despite attempts to protect the bird colonies through educational signs, fencing, and seasonal closures. Enhanced public education and better on-site protection are needed. Predation by raccoons has also caused water birds to abandon certain coastal islands for nesting. Raccoon populations often are higher than normal because of easy availability of human-augmented food resources.

Chemical Contaminants. Birds that feed on organisms that live in coastal waters, rivers, wetlands, and other habitats contaminated with pesticides and industrial chemicals have suffered reduced reproductive success and population declines. Banning of the use of the pesticide DDT in the United States in 1972 was an important step in the recovery of populations of the peregrine falcon, bald eagle, osprey, and probably various fish-eating sea birds and wading birds. DDT's metabolite, DDE, accumulated in the birds' tissues as they ate contaminated prey. Contamination ultimately was derived from applications of DDT in agricultural situations, sometimes far from some populations of affected birds. Runoff carried the toxin into streams and eventually to the sea. DDE interfered with calcium metabolism and caused eggshell thinning and reproductive failure, exacerbating problems the birds faced from human disturbance, shooting, and egg collecting.

Today, eggshells are thicker and populations have at least partially recovered, but problems related to this long-persisting pesticide linger in some areas of the United States. Migrants from North America continue to ingest DDT in areas in Latin America where this pesticide is still or was recently in use.

Along the Hudson River in New York, just west of Connecticut, tree swallows feed on insects that emerge from the river and accumulate high concentrations of polychlorinated biphenyls (PCBs) that have entered the river for over

Fig. 20.69. Bird management on Falkner Island includes installation of old tires that provide cover attractive to nesting roseate terns (two eggs are visible at center).

three decades from two capacitor manufacturing plants. The swallows exhibit reduced reproductive success. Likely similar impacts have occurred along rivers in and downstream of industrial areas of Connecticut.

Effects of Food Augmentation and Climate Change

Some native birds have benefited from changing conditions in Connecticut. Certain birds that thrive in forest-edge habitats, and/or that take advantage of the many tons of high-quality food now available at bird feeders throughout the state, have undergone population increases or range expansions. Milder winters in recent years have facilitated these increases and led to a general northward expansion of the range of certain resident songbirds in eastern North America. Examples of birds that have become more common in Connecticut at least in part as a result of these factors include red-bellied woodpecker, American crow, Carolina wren, tufted titmouse, northern cardinal, and eastern towhee.

Additionally, opportunistic birds such as crows and most gulls have become more numerous as a result of augmented food resources available at garbage dumps. Increased gull populations can cause problems for terns and other coastal water birds by displacing them from choice nesting areas (above flood level) and by preying on their eggs and young.

In the 1990s, a previously undocumented disease (evidently bacterial) affecting the eyes of birds, especially house finches*, appeared in Connecticut. Apparently birds at feeders contract the disease through crowding or contact with feeder parts that have been used by infected birds. Periodic thorough cleaning of bird feeders may help reduce disease transmission and is especially appropriate if you notice birds with closed, swollen, or crusty eyelids.

Cats

Domestic and feral cats have been blamed for killing large numbers of birds, but the magnitude and significance of the problem sometimes have been exaggerated as a result of inappropriate extrapolation from atypical killing sprees of particular cats or other nonrepresentative data. Many cats rarely if ever catch and kill birds. Observant cat owners know that rodents and shrews are the usual victims; these are killed or, if alive, are regarded as unworthy of a trip to a wildlife rehabilitator and thus do not show up in statistics.

Birds killed by cats most often are plentiful, yard-dwelling species rather than birds undergoing population declines. And cat predation does not necessarily cause a significant decline in the overall population of affected bird species. The basic principles of population biology indicate that even if a factor causes significant mortality in a prey species it does not necessarily follow that the long-term population of the prey will be significantly changed—often one mortality factor may simply substitute for another.

We should also bear in mind that losses attributable to cats are miniscule compared to, say, the permanent wildlife population losses that result from habitat destruction and degradation by humans, or the daily slaughter of wildlife by humans in automobiles. And it has been documented that a single tall communications tower can kill more than 20,000 migratory birds in a single night. Also, I suspect that bird deaths resulting from collisions with windows near bird feeders may outnumber cat kills, and birds stunned by such collisions certainly are easier for cats to capture. Pointing a finger at cats and cat owners leads to a general misunderstanding of bird conservation and may detract from important conservation needs such as habitat protection and ecologically sound land management.

That said, do restrict your pet's outdoor time if it is a habitual killer or harasser of wildlife, and help prevent cats and other exotic species from occupying natural habitats. This applies to dogs, too. Sharp, hooked claws make cats more effective at catching small animals than are dogs, but dogs habitually chase and harass larger wildlife. I keep my dog indoors when I see wild turkeys or ruffed grouse outside, but the dog's frequent outdoor activity surely restricts the freedom of native wildlife.

It may be that predation by the state's burgeoning population of coyotes has helped reduce any "cat problem" that may have existed.

Exotic Birds

Several non-native bird species are established within the state. The two species that have the greatest negative impact are the European starling and mute swan. The starling surely is now the state's most abundant bird species. It is an aggressive cavity nester that usurps the tree-cavity nest sites of native birds. Eastern bluebirds have lost most of their natural nesting sites due to the presence of starlings. For further information, see the species accounts earlier in this chapter.

The abundance of the introduced mute swan increased greatly to about 2,000 birds by the mid-1990s. Swans consume large quantities of wetland and aquatic plants, and today's large and growing swan population threatens to reduce the food supplies of native waterfowl. Also, the aggressive mute swan may detrimentally affect native waterfowl by displacing them from otherwise suitable habitat. A population management technique applied to mute swans involves shaking the eggs and replacing them in the nest. This kills the embryos but keep the birds sitting on the now nonviable eggs. Eventually they abandon the nest, but often not before it is too late to lay a new clutch and successfully raise a new brood.

The Canada goose population in the state has exploded, a result of the introduction and establishment of non-native, nonmigratory populations coupled with the excellent habitat provided by extensive lawns with human-made ponds. Grazing and fecal deposition by large numbers of non-native Canada geese can be a nuisance on golf courses and parks and may cause economic damage in certain agricultural situations. These geese also can have significant impacts on native wetland vegetation (see the bird section of Chapter 4).

Important Bird Areas Initiative

One project underway in the state, the Important Bird Areas initiative, seeks to identify key bird habitats so that these areas receive adequate consideration by open space planners. Leading this program are the National Audubon Society and the Connecticut Ornithological Association; additional cooperators include state and federal wildlife agencies and various private conservation organizations.

Chapter 21

Mammals

Mammals are the only animals that have hair, and they are the only animals in which the young are suckled on milk produced by mammary glands. The words "mammals" and "animals" sometimes are used as if they meant the same thing, but mammals (Class Mammalia) are just one of the many groups of animals, which include birds, reptiles, amphibians, fishes, insects, crustaceans, mollusks, and many others.

Mammals comprise some 4,800 species and occur throughout the world, on land, in rivers and lakes, and throughout the oceans. The greatest diversity of species occurs in the tropics (1,100 species in the Neotropical region). In North America north of Mexico, the largest numbers of mammal species occur in the vicinity of the mountain ranges of the western United States. North America north of Mexico hosts about 430 native species, plus nearly 30 established exotics. The wild mammal fauna of Connecticut includes about 60 native species (exact number depends on how many rarely occurring marine mammals are counted), plus a few introduced ones. Most species are secretive and/or active primarily at night, so only a small proportion of the state's mammals are ever seen alive by most persons.

Surprisingly, species richness of mammals (excluding marine species) in Connecticut is comparable to that of Florida, despite the fact that Connecticut is much smaller, recently glaciated, and farther north where overall animal biodiversity is much lower. Connecticut makes up for these biodiversity limiters by being within the range of a large number of mammals associated with the Appalachian Mountains and boreal North America.

The annotated species list in this chapter focuses on the mammals that have or had established populations in Connecticut. Excluded are various transients and rare visitors, such as belugas, West Indian manatees, and other marine mammals that have occasionally appeared in Long Island Sound.

Opossums

Opposums (Order Didelphimorphia, Family Didelphidae) were combined with kangaroos and other pouched mammals in the Order Marsupialia until recently, but that diverse group is now appropriately regarded as comprising seven distinct orders.

Opossum *(Didelphis virginiana)* (Fig. 21.1). Common in various habitats, including wooded areas, farmland, the drier portions of wetlands, and rural residential areas; colonized Connecticut from south in early 1900s; general warming trend in North America, plus abundant human-associated food resources (crops, garbage), likely facilitated northward range expansion; active mainly at night; (once, while studying mice, I inadvertently captured two juvenile opossums in small live traps and kept them on my porch for a few days. They slept all day and through early evening, then erupted into vigorous activity around 11:00 P.M., raising enough of a ruckus to awaken me); inactive during coldest winter weather, though hunger may stimulate activity in daytime; spends much time searching for food; eats almost anything edible, including carrion; spends inactive periods in underground burrows (often those abandoned by woodchucks), tree hollows, or similar secluded sites; females with young generally use burrow dens; natural predators include various mammalian carnivores and large raptors. *Note:* Opossum reproduction differs greatly from that of other local mammals. The placental connection between mother and fetus is not well developed, and the gestation period is very short. Births occur only 12 to 13 days after copulation, in late winter or spring. Newborn young are smaller than honeybees yet crawl to the mother's pouch, where they suckle for more than two months (Fig. 21.2). The long nursing period compensates for the short gestation. The young finally emerge and cling to the mother's fur. Some females produce two litters in a single season.

Fig. 21.1. Young opossums travel with their mother as they cling to her hair.

Opossums are short-lived and generally live less than two years. Survival is particularly low during the coldest winters. Thousands die in Connecticut each year as a result of collisions with automobiles. Opossums are basically unaffected by venom of copperheads, rattlesnakes, and other pit vipers, although these snakes rarely if ever attack opossums in Connecticut, or vice versa. But perhaps venom resistance is significant in the opossum's Neotropical habitat where pit vipers get very big and may be more likely to attack an opossum. Death feigning ("playing possum") is highly developed in these mammals. In response to a predator attack, the opossum falls over and lies immobile on its side. Because heart rate and brain activity remain normal, death feigning appears to be a voluntary behavior. It probably saves opossums from being killed by half-hungry predators whose killing instincts are most stimulated by struggling or fleeing prey.

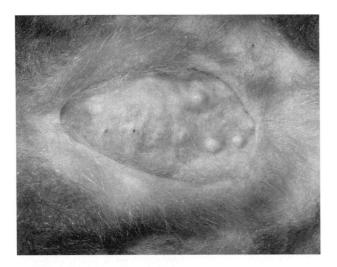

Fig. 21.2. Opossum young develop in a pouch on the mother's belly.

Shrews and Moles

Shrews (Order Lipotyphla, Family Soricidae)

Shrews are small mammals that eat mainly invertebrates and sometimes plant material. Activity occurs day and night throughout the year, though nocturnal activity predominates and substantial periods of inactivity may occur in winter. Shrews eat frequently, but their rate and amount of consumption generally are exaggerated in popular literature. They have small eyes and poor vision, and a long, pointed snout. At least some species are able to orient using echolocation (a form of ultrasonic sonar that is most highly developed in bats) to gather information about their surroundings.

Short-tailed shrew (*Blarina brevicauda*) (Fig. 21.3). Very common in any habitat with leafy or grassy ground cover (mainly wooded areas); burrows through leaf litter and soft soils, also uses mole tunnels; most numerous when invertebrate food resources are abundant, such as after massive emergences of periodical cicadas; commonly eaten by owls. *Note:* While walking quietly along forest paths, I often detected hidden shrews by the rustling they made when moving under leaf litter. Sometimes they emitted soft, brief squeaks as they moved. These shrews can incapacitate prey with their toxic saliva (from submaxillary gland ducts that open at the base of the lower incisors) and store it for later use when prey is scarce. The bite may cause considerable pain and swelling in humans. Short-tailed shrews are often captured but usually not eaten by cats, perhaps due in part to the shrew's odoriferous secretions, which emanate from a gland on each flank and one on the underside. The odor may function in the shrew's sexual behavior. Over the years my cats brought many of these shrews into the house. Even in darkness, I identified this species by the pungent odor emanating from the shrew as the cat hopped onto the bed with a muffled "meow."

Smoky shrew (*Sorex fumeus*) and the similar **masked shrew** (*S. cinereus*). Common in various habitats, especially moist and shady sites; in most areas, smoky shrew seems more common than masked shrew; smoky shrews are a bit larger, can take larger prey, and may use wider range of foraging microhabitats. *Note:* I seldom saw live *Sorex* shrews. My encounters were mainly with dead specimens found along trails or brought home by my cat.

Water shrew (*S. palustris*). Uncommon along edges of upland streams and sometimes lakes, ponds, marshes, or bogs; hair-fringed hind feet allow it to skitter over water surface; water-resistant fur allows underwater activity; feeds underwater and in water-edge situations.

Fig. 21.3. Short-tailed shrews are common under forest leaf litter and among thick, low vegetation.

Fig. 21.4. Eastern moles have powerful equipment for tunneling through soils of forests and fields.

Least shrew *(Cryptotis parva)*. Rare in Connecticut, known from a few grassy areas along coast; small, less than 3.5 inches (90 mm) long, brownish, with short tail (up to 20 mm long, versus up to 30 mm in grayish short-tailed shrews); all other Connecticut shrews have much longer tail (at least half total length of shrew).

Moles (Order Lipotyphla, Family Talpidae)

Moles are larger than shrews and, as burrowing mammals, have massive front feet and powerful forelimbs useful for tunneling. External ears are absent. Like shrews, they have poor vision, eat mostly invertebrates, and may be active day or night throughout the year. Moles are solitary and territorial.

Eastern mole *(Scalopus aquaticus)* (Fig. 21.4). Common, mainly in well-drained open areas such as fields and lawns but also in wooded areas; relatively rare at higher elevations in northern Connecticut; fossorial, seldom ventures above ground; makes surface tunnels with upraised ridges especially after rains have softened soil, used for foraging for earthworms, insects, and some vegetable matter; in winter, spends more time in deeper permanent tunnels; individuals range over a few acres or less; females give birth in underground nests in early spring; other small mammals, such as voles and shrews, sometimes search for food or travel in mole tunnels. *Note:* These moles are "obsessive" about repairing holes in their tunnels. One autumn when I lived in Old Lyme, I was curious about how often a mole used a particular tunnel, so I made a small hole in the roof of a mole's tunnel once or twice each day for several consecutive weeks and found that the mole plugged up the hole each day or night, usually within 12 hours. Sometimes people ask me what they can do about the moles in their lawns. I usually suggest watering during summer dry spells, to make it easier for the moles to tunnel and feed. There's nothing more satisfying to see in the suburbs than a lawn supporting native wildlife.

Star-nosed mole *(Condylura cristata)* (Fig. 21.5). Common in wet soils near water, sometimes in moist upland areas; good swimmer, often forages underwater; not as strong a digger as eastern mole; nests underground or in stump, above water level. *Note:* At night, star-nosed moles evidently are active on the ground surface much more so than is the eastern mole. Among 583 food items that I found in a sample of barn owl pellets in central Connecticut, 24 were star-nosed moles. Barn owls normally feed on small mammals that are active above ground. No eastern mole remains were present, though they likely were common in the owls' foraging areas. The remarkable, fleshy, starlike structure of twenty-two rays on the end of the mole's snout, looking like an invention of Dr. Seuss, may be involved in

Fig. 21.5. Star-nosed moles employ their unusual snout in detecting prey.

food-finding through detection of electrical fields generated by moving prey.

Hairy-tailed mole (*Parascalops breweri*). Fairly common in well-drained soils in northwestern Connecticut; life resembles that of eastern mole.

Bats

Bats (Order Chiroptera, Family Vespertilionidae) are the only vertebrates besides birds that fly (some fishes, amphibians, reptiles, and mammals are adept at gliding). With nearly 1,000 species worldwide, bats are one of the major groups of mammals, comprising one-fifth of all species.

Evening bats use sonar (echoes of their own rapid series of ultrasonic vocalizations) to detect and hone in on their flying insect prey. The bats can identify the location, movement, size, shape, and texture of their prey, and the detailed structure of their surroundings, solely through echoes. The pulse duration shortens and pulse rate increases as the bat closes in on its prey. Using ultrasonic bat detectors, researchers have determined that different kinds of bats generally produce recognizably different echolocation sounds.

Mating occurs mainly in late summer and early fall. Females store the sperm internally over winter. Fertilization and development of the young occur in spring, and most births are in June or July. Among Connecticut species, the little brown bat and northern bat give birth to single young, the eastern pipistrelle, silver-haired bat, big brown bat, and hoary bat commonly produce twins, and the red bat usually gives birth to three to four young.

Bats sometimes carry the rabies virus and should not be handled with bare hands (despite Fig. 21.6).

Bats that Occupy Buildings. Generally these are in old buildings with unscreened openings to the outside.

Little brown bat (*Myotis lucifugus*). Common, widespread; breeding colonies often in warm attics, barns, and similar sites.

Fig. 21.6. Bat wings consist of thin skin extending between the limbs, fingers, and body. This is a male northern bat in breeding condition (note penis).

382 | 21. Mammals

Big brown bat (*Eptesicus fuscus*). Breeds and winters in Connecticut; frequently hibernates in buildings and sometimes uses caves; summer roosts often are in attics, behind shutters, or in similar crevices; any bat flying outside in Connecticut in winter probably is this one.

Bats that Roost in Trees in Summer. Among local bats, only the first three species migrate south for the winter

Silver-haired bat (*Lasionycteris noctivagans*) (Fig. 21.7). Uncommon; most likely to be encountered near water; typical roost sites include tree foliage, tree hollows, and crevices behind loose bark, though migrating individuals use various available nooks.

Red bat (*Lasiurus borealis*). Mostly at lower elevations; normally roosts singly among foliage of deciduous trees in semi-open situations. *Note:* Once, in mid-October, I found a red bat sleeping in the top of a tree; it spent just one day there and probably was migrating through the area.

Hoary bat (*L. cinereus*). Roosts alone or as mother-pup family groups in trees in semiforested habitats.

Northern bat (*Myotis septentrionalis*). In summer, generally roosts in deciduous trees, in cavities or behind loose bark, sometimes in groups of up to a few dozen adults.

Bats that Hibernate in Caves and Tunnels. Of these bats, the little brown bat is by far the most common.

Eastern small-footed bat (*Myotis leibii*). Likely always scarce; believed to be extirpated in Connecticut.

Little brown bat (see above). In a water conduit tunnel that I periodically check, a few hundred little browns are

Fig. 21.7. Silver-haired bats migrate and spend the winter south of Connecticut. Paul J. Fusco/Connecticut DEP Wildlife Division.

present each winter, hanging from the ceiling, mostly in tight-knit clusters of up to several dozen individuals.

Northern bat (see above) (Fig. 21.6). When hibernating, generally hangs alone, not in contact with other individuals, often in nooks and crevices.

Indiana bat (*M. sodalis*). Only one individual has been found in the state over the past several decades.

Eastern pipistrelle (*Pipistrellus subflavus*). Roosts in summer in rock outcroppings, deciduous trees, or old buildings; in hibernation, it roosts singly or in small loose groups.

Great Apes

The family Hominidae of the Order Primates includes orangutans, gorillas, chimpanzees, and humans. Genetically, humans and chimpanzees are nearly identical; this emphasizes both our common evolutionary heritage and how small genetic differences can result in enormous phenotypic differences, such as those related to mental ability.

Humans (*Homo sapiens*). Colonized what is now Connecticut about 10,000 years ago, after Pleistocene ice sheets melted and vegetation and game animals reappeared; for many centuries, populations probably remained at not more than a few thousand; large increase occurred after settlement by Europeans began in 1600s; total population about 3.3 million, and very abundant throughout year, though significant seasonal migration occurs between Connecticut and Florida; population in Connecticut not locally sustainable; vast majority of food, clothing, building materials, and energy resources needed to support population are imported from elsewhere; bulk of population involved in nonessential (but surprisingly well paid) activities.

Rabbits and Hares

Rabbits and hares (Order Lagomorpha, Family Leporidae) are active throughout the year, mainly at dusk and early in the morning. Most species occur along the open edges of shrubby or wooded areas. They take cover in dense vegetation, and cottontails sometimes use burrows made by woodchucks. Leafy forbs dominate the diet in summer, whereas woody material (twigs, bark) becomes important in late fall, winter, and early spring.

Rabbits defecate two kinds of fecal pellets. Pellets containing partially digested material from the caecum (a blind pouch off the intestine) are reingested and further nutrients are extracted.

Individual females produce one or more litters of young in spring and summer. Offspring in a single litter may be fathered by more than one male. Nests generally are on the surface of the ground, often under cover in a small depression. Cottontail babies are naked and helpless at birth and require a few weeks of suckling before they are ready to leave the nest. Newborn hares are fully furred, precocial, and generally leave the birth site within two to three days. Rabbits and hares are short-lived; only a small percentage of the young survive their first winter, and most that do survive do not make it through their second winter.

Eastern cottontail* (*Sylvilagus floridanus*) (Fig. 21.8). Introduced into Connecticut from south; now the state's most common rabbit; shrubby to open habitats, often occurs in disturbed areas. *Note:* A good place to watch rabbits is Hammonasset Beach State Park, where you may see as many as 30 to 40 individuals along the entrance road in the early evening.

New England cottontail (*S. transitionalis*). Native cottontail, now relatively uncommon and replaced in most areas by introduced eastern cottontail (which it closely resembles but has a larger number of chromosomes); most recent records from west-central and southeastern Connecticut; usually in shrubby wetlands and forests with dense understory. *Note:* The decline of this species may be due to loss of adequately large tracts of suitable cover and associated increases in predation, combined with competition with the non-native eastern cottontail. To help in an ongoing study of cottontail distribution and abundance in Connecticut, contact or deliver specimens to the Franklin Wildlife Management Area (860-642-7239) or Sessions Woods WMA (860-675-8130).

Snowshoe hare (*Lepus canadensis*). Locally common in northern part of state, primarily in areas with dense thickets; fur changes from brown to white in fall, and back to brown in the spring; young weaned and on their own within about one month; populations in Connecticut do not exhibit

Fig. 21.8. Eastern cottontail, an introduced species, is now more common than the native New England cottontail. Paul J. Fusco/Connecticut DEP Wildlife Division.

dramatic cycles in abundance that are characteristic of populations farther north in boreal Canada.

European hare* (*L. europaeus*). Old World species, introduced in Connecticut, but now rare or extirpated; requires wide open spaces.

Rodents

Squirrels (Order Rodentia, Family Sciuridae)

Most members of the squirrel family are active in daytime, but flying squirrels are nocturnal. Connecticut species, aside from the woodchuck, have a varied diet of nuts, seeds, fungi, insects, and occasional other animal matter such as bird eggs and nestlings.

Woodchuck (*Marmota monax*) (Fig. 21.9). Also known as groundhog; common in fields, pastures, and other areas with thick herbaceous vegetation; also uses well-drained wooded areas as hibernation and birthing sites (once I saw one enter a burrow at the edge of a tidal river); most active in daytime, sometimes feeds at night; mainly a terrestrial foliage and fruit eater, but sometimes climbs into trees and shrubs to feed; relishes many cultivated plants; basically solitary and territorial; excellent burrower; individuals may occupy several burrow systems thoughout home range; periodically refurbishes underground nest with dry leaves carried to burrow in mouth; abandoned burrows used as shelter by many other mammals; soil disturbances from burrowing, combined with herbivory, lead to increased local diversity of flora and fauna; hibernates in colder months (some may stay active into December), living off stored body fat (see "Mammals in Winter"); resumes activity in late winter, when males begin seeking sexually receptive females; young are born in April or May. *Note:* A good fence is the best way to discourage woodchucks from eating garden plants, though they may climb over fences that do not have a sharply angled overhang at the top. Once I found a woodchuck inside my garden fence devouring my plants. Acting on instinct, I managed to corner it, pin it under my boot, then capture it by hand by grabbing the nape skin. I managed to get the struggling animal into a barrel. Woodchucks are strong and have sharp teeth, and they can (but rarely) carry rabies—I wouldn't recommend this capture method!

Gray squirrel (*Sciurus carolinensis*). By far the state's most frequently observed mammal; abundant and active in daytime throughout year (unfortunately, the number seen dead on roads often seems to outnumber the live ones); favors forested habitats with mature nut-bearing trees such as oaks, hickories, and beeches (or well-stocked bird feeders); cavities in large trees are important den sites, but in absence of suitable cavities squirrels construct spherical shelters of sticks and leaves high in trees; nests may be occupied by one or more individuals; communal sleeping is most common among females; births occur in early spring and/or late summer; in fall, stores large numbers of nuts, one at a time, in small holes in ground, later uses memory and smell to retrieve stored food, which is important in winter and early spring. *Note:* Nuts stored but uneaten by squirrels may germinate far from the source tree, aiding tree dispersal and facilitating reforestation. I often found hickory saplings far from any mature hickory trees. Through seed predation and seed dispersal, these squirrels have major impacts on forest ecology (see "Acorn Ecology" in Chapter 7). Young squirrels may establish themselves in or near the natal area if food is plentiful, but when food is scarce, aggression from resident adults may cause them to disperse to areas with few squirrels. Adults also move in response to food shortages. Long-distance movements of large numbers of squirrels sometimes occur in late summer or early fall. These movements include swims across major rivers and lakes. Such movements seem to have been more common in the past. Gray squirrels sometimes take up residence in the attic, eaves, or within wall spaces of occupied buildings. Because of the damage they can do (e.g., to electrical wiring), it is best to keep squirrels out. This may require careful sealing with strong wire mesh any holes under eaves or attic ventilation openings, making sure that no squirrels remain inside (an adult that exits may leave young inside). Keeping trees and vines away from the building also helps. A squirrel in the chimney can be dealt with by inserting a thick rope or long length of coarse fabric that the animal can use to climb out. Various commercially available squirrel guards are available for bird feeders, but I find it easier simply to put food on the ground so that the squirrels leave the bird feeders alone. That way I get to enjoy both birds and squirrels, though at considerable expense in sunflower seeds!

Red squirrel (*Tamiasciurus hudsonicus*). Usually associated with and fairly common (usually fewer than one per

Fig. 21.9. This woodchuck was active in December, completing its preparations for winter.

acre) in areas with stands of mature conifers such as white pine or eastern hemlock; conifer seeds are a primary food, obtained by biting off cone scales; also lives among deciduous trees and eats their nuts and seeds; nests in tree cavity or in underground chamber, or makes spherical nest in tree crotch; often breeds twice annually, generally in February to March and June to July; active in daytime year-round, except during stormy weather. *Note:* Red squirrels may store and aggressively defend large numbers of conifer cones and other foods for use during colder months, or they may establish many small caches and not be so protective. Survival and reproductive success tend to decline when conifer seeds are scarce. In late winter or early spring, they sometimes start sap flowing from sugar maples by biting into the bark; later they return to lick the sugary residue.

Southern flying squirrel *(Glaucomys volans).* Locally common where productive nut trees and nesting cavities are available, often along streams or near wetlands. **Northern flying squirrel** *(G. sabrinus)* (Fig. 21.10). Known from just a few areas in northern part of state; generally associated with old-growth forests, warrants conservation attention throughout range in eastern United States. Both species active at night; in addition to nuts (dietary staple), commonly eat fruits, insects, and fungi (including truffles) (see note); nest primarily in trees in abandoned woodpecker holes and natural cavities, but other cavities and even old buildings may be used (I found flying squirrels nesting in cavities in a swamp-edge yellow birch, in a dead red maple on a rocky hillside far from water, and in bird nest boxes along a small forested stream on my property). *Note:* Flying squirrels can evict small birds from desirable cavities, but larger cavity-nesting birds may oust the squirrels. In winter, several individuals may share a nest; this cuts heat loss and reduces the amount of food that is needed. These squirrels are generally quiet, but on several occasions I evoked vocalizations from southern flying squirrels at night by imitating calls of great horned owls. Flying squirrels do not actually fly but rather use their extensive skin membranes extending between the limbs to glide from one tree to a lower point on another tree. Glides usually are less than 80 feet and rarely exceed 150 feet in horizontal distance. Truffles, an important food, are the fruiting bodies of underground ectomycorrhizal fungi, which help trees absorb water and nutrients from forest soils. They are a seasonally important food resource. Squirrels locate the truffles by smell and by searching near decaying logs that often are associated with truffles. Squirrels and other truffle-eating mammals aid the truffles by dispersing the spores in their fecal droppings. See the gray squirrel account for information on how to deal with squirrels that invade human residences.

Eastern chipmunk *(Tamias striatus).* Common in wooded habitats; active on ground surface but commonly

Fig. 21.10. Flying squirrels can be fairly common but are seldom seen due to their nocturnal habits. Larry Master.

forages in trees and shrubs as well; solitary and territorial (except mothers and young, or during mating season); active much of year but stays in underground burrows most of winter; in autumn, stores food in burrow for use in winter (see "Mammals in Winter"); transports food items in large internal cheek pouches; most common where oaks and acorns are numerous ; prefers low-tannin white oak acorns over high-tannin red oak acorns (see "Diet" at end of chapter); food supply appears to affect abundance (populations were very high in 1999 after large acorn crops of 1997 and 1998); breeds in spring and summer, when males move to female territories to mate; juveniles emerge above ground in late spring; yearling females that did not breed earlier may produce a litter in summer; vocalizations include chips (when terrestrial predators are present), chucks (aerial predators), or trills (given close to the burrow at the end of a predator chase, probably serve mainly to warn kin). *Note:* I noticed that when under attack by a domestic cat, chipmunks may go into a hump-backed, low-hopping form of locomotion that looks semi-paralytic. This is not necessarily the result of injury because later they may dash off at full speed. I saw gray squirrels employ this same behavior under similar circumstances.

Beavers (Order Rodentia, Family Castoridae)

Beaver *(Castor canadensis)* (Fig. 21.11). Largest North American rodent; common in small to large low-gradient

Fig. 21.11. Beavers, formerly extirpated from Connecticut, were reintroduced and now have recolonized all of the state's suitable habitat.

Fig. 21.12. Beavers may build a lodge surrounded by water or excavate a burrow in the bank of a river or lake. A drop in water level in this large pond likely caused the beavers to build a new lodge in deeper water next to the old lodge.

streams (including tidal sections of lower Connecticut River), lakes, and other permanent waters deep enough to avoid freezing to bottom in winter; most common where favorite food plants (for example, aspen, birch, willow, cottonwood, and soft aquatic plants) abound; also cuts, eats, and builds with dozens of other kinds of woody plants; tail-slapping warns family members of danger and may frighten undesirable intruders.

Note: Beavers often create or augment their own habitat by building dams across small streams, forming "beaver ponds," but they also commonly inhabit rivers too big and strong to be dammed. Dams on some streams last only a short time before being washed out by high flows. Beavers also build lodges that protect them from temperature extremes and predators. Lodges are hollowed out within large piles of cut tree branches and mud, and they may be out in the beaver pond or along the shoreline. Sometimes beavers simply burrow into a river bank without adding much wood.

In some areas, winter survival depends on food stored in underwater caches near the lodge. Food caches are piles of branches that generally have preferred food items at the bottom where they are readily available when the pond freezes. One function of a beaver dam is to deepen the water enough to allow development of these underwater food storage areas.

Beavers live in family groups that occupy one or more lodges (Fig. 21.12). Families may maintain multiple dams. They mark their territory with mounds of mud anointed with secretions from their castor and anal glands. On average, young stay with their parents until two to three years old, then disperse to new sites, often going downstream after spring thaw.

Beavers were common in Connecticut before Europeans arrived, but excessive trapping for their luxuriant fur led to their extirpation from the state by around 1842. Reintroductions in the early 1900s, followed by subsequent natural population increase and range expansion, resulted into today's burgeoning population. The beaver population in Connecticut has increased to the point where an annual trapping harvest of approximately 500 to 1,000 (record high of 1,224 in 2001/2002 trapping season) is sustained without depleting the population. The total population was about 5,000 to 8,000 beavers statewide in 2000.

Today's robust beaver populations are having a profound impact on bottomland ecosystems and hydrology. Although beavers create habitats that benefit numerous other wildlife (see "Beaver-influenced Wetlands" in Chapter 4), they have become unpopular with some people because of their often indiscriminant tree cutting and the flooding they cause, which may damage wells, septic systems, or houses placed too close to streams. Flooding is sometimes of concern to conservationists because it may detrimentally affect certain rare wetland communities or plants that today exist in just a few locations. Locally, beaver damage to trees can be discouraged by surrounding trees with heavy wire mesh. Properly designed wire exclosures can be effective at keeping beavers from plugging culverts. Flooding may be prevented by installing in a beaver dam one or more PVC drain pipes that keep the water level from rising above the level of the pipe (periodic maintenance is needed). Anyone wishing to alter wetlands or water flow in this manner must first obtain a permit from their town wetlands commission.

Mice, Rats, Voles, Lemmings, Hamsters, and Gerbils (Order Rodentia, Family Muridae)

White-footed mouse *(Peromyscus leucopus)* (Fig. 21.13). Very common in wooded areas and especially forest edges, particularly habitats with abundant high-quality food resources (acorns, beech nuts, and other large seeds); terrestrial and arboreal; on ground, tends to use logs and branches as runways, probably to avoid producing predator-attracting noise that results from scampering over leaf litter;

Fig. 21.13. White-footed mice are excellent climbers and often nest in tree cavities, abandoned bird nests, and nest boxes created for birds. Larry Master.

nests are in tree cavities, underground burrows, abandoned bird nests (roofed over by the mice), or bird nesting boxes; active at night throughout year but regularly torpid in colder months (see "Mammals in Winter"). *Note:* This woodland mouse, familiar to many rural Connecticut residents as the mouse that invades houses when the weather turns cold in fall, sometimes ranges into unwooded areas in summer and early fall when food and cover are available. It can cross open areas between woodlots up to at least 1 1/4 miles apart. Researchers in Massachusetts documented amazingly long-distance mouse movements of 4.3 miles and 9.2 miles through an extensive forest, likely triggered by high mouse population density and low food (acorn) availability. For several years, I monitored the use of three nest boxes that I attached to trees specifically for white-footed mice. Communal occupancy of nest cavities was common in the colder months. The nests inside the boxes had a thick roof in the colder months but no roof at all in summer. In fall, I found up to fifteen mice in a single box; most of these were two generations of young plus multiple adults. The young stayed in the nest box day and night until they were a little more than two weeks old.

Deer mouse *(P. maniculatus).* Restricted to northern part of state; habits like those of white-footed mouse, but more likely to be active in early morning hours and has greater tendency to use arboreal nest sites.

Allegheny woodrat *(Neotoma magister).* Formerly occurred in one site in western Connecticut, now extirpated. *Note:* This native rat has disappeared from many areas in the northeastern United States. In some areas, it apparently succumbed to effects of roundworm (nematode, *Baylisascaris procyonis*) parasitism transmitted by raccoon population, which has been abnormally high in recent decades due to superabundance of food (e.g., human garbage and garden crops). Woodrats may ingest roundworm eggs when foraging on undigested seeds present in raccoon feces, in which the eggs remain infective for years. The roundworms may cause severe or fatal neurologic disease in woodrats but do not harm raccoons.

Red-backed vole *(Clethrionomys gapperi).* Locally common, primarily in forested areas (especially coniferous); I live-trapped many in forested areas with abundant ground cover such as logs and rocks (including old stone walls); nests in holes in stumps or under logs; diet includes leaves and stems of small plants, small fruits, seeds, fungi, and sometimes insects and other small invertebrates; active day or night throughout year; long breeding season; individual females may produce multiple litters in single year. *Note:* This vole is sensitive to large-scale timber cutting. It disappears from large, clear-cut areas but regularly occupies smaller patch cuts.

Meadow vole *(Microtus pennsylvanicus)* (Fig. 21.14). Often abundant in meadows, pastures, marshes, and other areas of thick, unmowed grasses or sedges; digs underground burrows and, in thick vegetation, often makes extensive networks of cleared runways on surface of ground; in wetlands, I saw voles dive underwater, probably to enter a submerged burrow opening; nests underground, under objects, or in thick grass on ground surface; active day or night all year; active under snow in winter (while cross-country skiing, I frequently found the tunnels they make to the snow surface); eats mostly herbaceous plants, seedlings of trees and shrubs, and sometimes bark, insects, and seeds; important as food for various owls, hawks, carnivorous mammals, and snakes; prolific; individual females mature sexually within several weeks, may produce up to several litters of young during short lifetimes of usually much less than a year; local populations tend to fluctuate in size, from a few to hundreds of voles per hectare (2.5 acres), with peaks every three or four years. *Note:* Feeding by these voles can

Fig. 21.14. Meadow voles can be abundant in fields with thick grass.

delay reforestation and probably has significant effects on the overall vegetation dynamics of grassy fields.

Woodland vole *(M. pinetorum)*. Locally common, mainly in partly wooded upland habitats; spends much of time in burrow system dug in well-drained soils, active day or night all year; diet includes all kinds of plant material, fungi, and sometimes small invertebrates; despite highly fossorial habits, occasionally taken by predatory birds, mammals, and snakes; low reproductive rate compared to meadow vole; lives in extended family groups that may include mature nonbreeders; females generally produce only a single small litter during their short life. *Note:* On many occasions I watched one or two of these voles emerge from their burrows to gather seeds in daytime from among the low ground cover beneath my bird feeder. Unfortunately, my cat soon presented me with several woodland voles from this population before he was "grounded."

Muskrat *(Ondatra zibethicus)*. Common in lakes, ponds, slow-moving streams, canals, swamps, and marshes; nests in burrow in stream bank or along shore of a lake or pond, or constructs house of cattails or other herbaceous wetland plants in shallow marshy areas; active year-round, day or night; prolific breeder with short lifespan; populations tend to vary with food availability; eats wetland and aquatic plants and, in some areas, freshwater mussels (Fig. 21.15) and other aquatic animals.

Southern bog lemming *(Synaptomys cooperi)*. Highly localized in systems of underground burrows and surface runways among succulent herbaceous plants, generally in bog-edge habitats of sphagnum moss but also sometimes in shady upland habitats with thick humus soils; active day or night throughout year; feeds primarily on leaves, stems, and seeds of grasses and sedges.

House mouse* *(Mus musculus)*. Common in cities and farms; introduced from Europe; highly associated with humans and croplands; feeds unfastidiously on just about any kind of food; soils buildings with urine and feces; gnaws cable and wire insulation, which can lead to fires. *Note:* Even rodent aficionados like myself find nothing enchanting about these mice.

Norway rat* *(Rattus norvegicus)*. Like house mouse, arrived as stowaway on ships coming from Europe; locally common where there is easily obtained food in cities, farms, waterfront areas, garbage dumps, and streamside habitats; nests in underground burrows, buildings, or other structures; good swimmer, readily enters water; active year-round, mainly at night but also in daylight; diet includes anything that could be regarded as at least marginally digestible. *Note:* I saw rats casually foraging among intertidal rocks along the coast and around garbage dumpsters in towns, and I found a few rats dead on roads in rural residential areas with old scattered farm buildings. I remember one enormous road-kill that I initially thought was an opossum. Steve Broker and I found abundant skeletal remains of these rodents in barn owl pellets below owl roosts in a semi-open building next to the New Haven landfill.

Jumping Mice (Order Rodentia, Family Dipodidae, Subfamily Zapodinae)

These mice are indeed good jumpers and, initially, are easily mistaken for a leaping frog should you inadvertently flush one from its resting spot. Both Connecticut species hibernate in underground nests from early fall through early or mid-spring. Activity reportedly occurs mainly at night, but I encountered them several times in daylight. Compared to most mice, jumping mice are relatively long lived and reproductively are not very prolific. Nests may be underground or, in summer, in an decaying stump or clump of grass. They eat seeds, small fruits, low foliage, fungi, insects, and sometimes small invertebrates.

Meadow jumping mouse *(Zapus hudsonicus)* (Fig. 21.16). Locally fairly common, most often in damp areas with thick herbaceous vegetation, including meadows, shrubby areas, old fields, and forest edges, often near water or wet-

Fig. 21.15. After eating freshwater mussels, muskrats leave piles of shells on shore.

Fig. 21.16. Meadow jumping mice are hibernators that inhabit meadows or lush streamside growth. Larry Master.

lands; sometimes establishes well-used runways through vegetation.

Woodland jumping mouse *(Napaeozapus insignis)*. Locally fairly common in moist forested or shrubby situations, generally along streams; if startled, this superb jumper may flee with leaps of 6 to 8 feet. *Note:* In Middletown, I found woodland and meadow jumping mice in the same area along the edge of a moist riparian woodland.

New World Porcupines (Order Rodentia, Family Erethizontidae)

North American porcupine *(Erethizon dorsatum)* (Fig. 21.17). Generally uncommon in forested areas in northern Connecticut, most often in mixed forests including eastern hemlock; uses secluded den sites such as crevices or holes among rocks or under ledges, tree hollows, underground burrows, or old buildings; usually dens singly but occasionally two may share den; sometimes simply sleeps in trees and does not use den, even in winter; active year-round, in daylight or (more often) at night, on ground and in trees; diet includes various kinds of plant material, ranging from inner bark, buds, and fruit of woody plants (for example, acorns and beechnuts) to leaves of soft herbaceous species; breeds in late summer and fall, gives birth to single pup in spring after long gestation period; may live a dozen years. *Note:* The porcupine's extensive digestive system is packed with symbiotic microorganisms that convert some of the tree roughage into usable nutrients. Porcupines facilitate digestion by selectively foraging on the most rewarding plant parts. In northwestern Connecticut, I found two porcupines side by side digging in soil at the base of a hemlock tree, likely feeding on fungi. In winter, small conifer branches dropped to the ground by arboreal-feeding porcupines may provide a bonus food supply for ground-dwelling herbivores. Porcupines are attracted to salt and rarely miss a chance to gnaw on objects that have contacted salted food or human sweat. They commonly chew on undersides of vehicles. The bold black and white color bands on the spiny, barbule-tipped quills may warn or remind potential predators of the unpleasant consequences of attacking a porcupine. This visual signal may be reinforced by a unique odor released by disturbed porcupines. Though porcupines look ungainly, they have excellent climbing abilities, aided by strong claws, large foot pads, muscular legs, and a spiny tail that can be used as a brace.

Whales and Porpoises

Whales (Order Cetacea, Family Delphinidae)

Towns on Long Island Sound and the lower Connecticut River historically boasted a productive whaling industry that peaked in the early to mid-1800s. However, they were supported by whale populations in oceans as far away as the South Atlantic or Pacific Ocean, not by the scant populations in Long Island Sound.

Long-finned pilot whale *(Globicephala melas)*. Sometimes enters Long Island Sound, but more common near Cape Cod, where well-publicized mass strandings regularly occur; carcasses rarely wash up on Connecticut beaches.

Porpoises (Order Cetacea, Family Phocoenidae)

Harbor porpoise *(Phocoena phocoena)*. Occasionally along coast of Connecticut, but currently much too rare to regard as ecologically significant part of Long Island Sound ecosystem.

Carnivores

Dogs, Wolves, Coyotes, and Foxes (Order Carnivora, Family Canidae)

Coyote *(Canis latrans)* (Fig. 21.18). First appeared in Connecticut in 1950s, following general range expansion eastward into southeastern Canada and northeastern United States; now common (for a large carnivore), with total population of probably a few thousand; howls and yip-howls of coyote groups now audible throughout

Fig. 21.17. North American porcupines move slowly on the ground but are accomplished climbers.

much of state, including both remote forests and semi-developed areas in towns; research in Vermont and Connecticut indicates an average adult home range size of less than 10 square miles; state Wildlife Division biologists have determined that Connecticut coyotes mate in February and produce litters averaging about seven pups in April; nearly all adult females breed every year; both parents raise pups and tend them until late fall or winter; highly opportunistic feeder, eats almost any available animals or plant material, often including deer, rabbits, voles, mice, woodchucks, squirrels, turkeys, grasshoppers, and fruits; there is at least one documented instance of a Connecticut coyote depredating mute swan* eggs; in winter, coyotes need the equivalent of about three rabbits every two days; prefers larger animals that offer the most food for the least effort; most active at twilight or at night but I occasionally saw them abroad in daylight as well. *Note:* Coyotes in the northeastern United States are larger than coyotes farther west. It is possible that this is a result of interbreeding with wolves as coyotes expanded eastward, but we need more information on coyote genetics and on the nutritional dependence of their growth characteristics before other explanations can be ruled out. For example, larger size in the east could be adaptive for taking larger prey (such as white-tailed deer) or could be associated with better nutrition afforded by more diverse food resources. Coyotes readily prey on domestic cats and small dogs left unattended and so have become a nuisance to pet owners. Attacks on humans are extremely rare, but in 1998 a three-year-old child was attacked on Cape Cod. The state Wildife Division suggests that homeowners take the following steps to discourage coyote (and fox) activity near their homes: eliminate food attractants, such as outdoor pet food, table scraps on compost piles, or dropped fruit beneath fruit trees; attempt to harass or frighten away coyotes and foxes with loud noises; spray strong, unnatural odors (e.g., deodorant soap) to discourage bold coyotes and foxes from using the sprayed areas; trim low branches of conifer trees and remove low brush cover in or near yards that may serve as cover; and install yard lights or lights with motion detectors. Pets can be protected by limiting their time outdoors, especially at night, by supervising pets while outdoors; and by excluding coyotes and foxes from pet areas with fencing or a kennel. Both coyotes and foxes can carry rabies, but this viral disease, which affects the central nervous system, is relatively rare in these mammals in Connecticut. The Wildlife Division recommends that anyone who observes a fox or coyote that is exhibiting obvious signs of rabies (uncoordinated movements, aggressiveness, seizures, or extreme lethargy) should contact local police or animal control authorities

Fig. 21.18. Coyotes colonized Connecticut about fifty years ago and now occur statewide. Paul Rego/Connecticut DEP Wildlife Division.

or, if they cannot be reached or are unable to respond, contact the DEP at 860-424-3333.

Gray wolf *(Canis lupus)*. No longer occurs in Connecticut; deliberately exterminated by early settlers and negatively affected by decimation of its food supply (mainly deer) in 1800s; now basically replaced in ecosystem by coyote. *Note:* According to some taxonomists, the wolf that historically inhabited Connecticut and the Northeast likely was the so-called eastern Canadian wolf *(Canis lycaon)* rather than the gray wolf, which may be restricted to Eurasia and western North America. *Canis lycaon* and the red wolf *(C. rufus)* have been proposed as forming a single species *(C. lycaon)*, but another hypothesis is that *lycaon* is a subspecies of the gray wolf that hybridized with the red wolf.

Red fox *(Vulpes vulpes)*. Native to New England, but probably interbred with red foxes introduced from Europe in past years; common in habitat mosaics that include fields and forest edges; tends to be absent from areas regularly used by coyotes, but may persist in areas between coyote territories; versatile diet includes meadow voles, rabbits, and other small animals and various fruits; hunts mainly at twilight and at night, but daytime activity not uncommon; dens in woods or fields, frequently in abandoned woodchuck burrows or in holes among rocks. *Note:* I used to see foxes regularly at night in a large weedy/grassy field associated with a dairy farm, but now that area is populated by houses and mowed lawns and the foxes have to go elsewhere to find food in what remains of the shrinking habitat of grassland and old field wildlife. I found red fox pups outside their den at the base of a trap-rock ridge on April 8. Another group of pups was active outside their den on a rocky wooded hillside on May 12. Red foxes often occur in rural residential areas and in farmland, but they pose virtually no danger to humans and are not a primary carrier of rabies in this region. They may kill small livestock or free-ranging cats, but this can be prevented by proper pet supervision, fencing, and by otherwise preventing access to pet food, fallen fruit, or compost piles that might attract foxes and lead to predation on pets or livestock.

Gray fox *(Urocyon cinereoargenteus)*. Fairly common, but likely less abundant than red fox; does not stray far from thick cover, so seen far less often than red fox; dens in holes among rocks or in tree hollows; eats small animals and fruits; active day or night. *Note:* This fox seems to have become more common in the northeastern states in recent decades, possibly because of a warming trend that may favor this primarily southern species.

Bears (Order Carnivora, Family Ursidae)

Black bear *(Ursus americanus)*. Rare in most of Connecticut, but becoming fairly common in Litchfield and Hartford counties; primary habitat needs include a suitable winter den site, such as a chamber among rocks, under tree roots, or under a fallen tree, and ample food resources, particularly hard mast (for example, acorns, beech nuts, and formerly chestnuts) and fruits; these foods provide calories needed for fattening prior to winter dormancy, when feeding ceases and bears spend two to three months "sleeping" (see "Mammals in Winter"). *Note:* Bear sightings have increased in recent years as bear populations have expanded gradually southward from core habitat in the state's northwestern hills and as human residential areas keep expanding farther into forest. The total Connecticut population in 2002 was probably more than one hundred and growing. State Wildlife Division biologists have captured males weighing up to 450 pounds. A study in western Massachusetts indicated that the survival rate of the young is low, especially among males; only 2 of 21 male cubs survived to adulthood. The state Wildlife Division suggests that homeowners prevent bears from becoming a problem by following these guidelines: after winter, remove bird feeders or hang them out of a bear's reach (at least 8 feet above the ground and 6 feet from tree trunks or vertical poles); store garbage in secure containers or indoors, and periodically clean garbage cans to reduce residual odor; avoid continuously leaving pet food outdoors; clean your barbecue grill of grease and store it inside a garage or shed; do not put leftover meat scraps or sweet foods in compost piles; and use electric fencing to protect beehives, livestock, and berry bushes. Anyone who observes a black bear in Connecticut should notify the DEP's Sessions Woods office (860-675-8130, M–F, 8:30 A.M.–4:30 P.M.) or the DEP 24-hour dispatch line at 860-424-3333.

Raccoons and Relatives (Order Carnivora, Family Procyonidae)

Raccoon *(Procyon lotor)*. Common near lakes, ponds, marshes, and streams; most active at night, but also forages under dark cloudy skies during day; tends to stay in dens during periods of deep snow or very cold weather; readily climbs trees; feeds on whatever plant or animal food is readily available, including aquatic and wetland vertebrates and invertebrates (for example, frogs, crayfish, turtle eggs), many kinds of small fruits (Fig. 20.61), and high-quality garbage we provide in ample supply (as I wrote this paragraph, at 10:42 P.M. on 3 March, air temperature 23°F, a very fat raccoon foraged on sunflower seeds under my bird feeder); sleeps and bears young in underground burrow, large tree cavity, and similar site; prefers tree hollows for maternal dens; births occur in April or May. *Note:* A rabies epidemic swept through the raccoon population in Connecticut in the early 1990s and killed perhaps up to three-quarters of the population, yet these adaptable mammals remain common in some areas and the state population is still probably a few hundred thousand. Raccoon rabies remains present at a low level in Connecticut, with a couple hundred confirmed cases per year, involving skunks and cats as well as raccoons. Avoid any mammals that behave abnormally, such as those that are overly aggressive or unwary, disoriented, shaking or convulsive, or paralytic, and do not handle mammals dead from unknown causes. Report these cases to the police or animal control authorities.

Weasels, Otters, and Skunks (Order Carnivora, Families Mustelidae and Mephitidae)

The mustelids and skunks are carnivores that feed mostly on small vertebrates and large invertebrates. All species are active year-round, mainly at night and twilight but also sometimes in daylight. All have anal scent glands that produce powerful odors, which are put to effective defensive use by skunks.

River otter *(Lontra canadensis)* (Fig. 21.19). Formerly scarce, now fairly common in many lakes and large ponds; takes shelter in shoreline burrows; feeds on fishes, also frogs, crayfish, and other aquatic animals. *Note:* One morning in late May in Salisbury, a group of otters circled my canoe and looked at me with seeming curiosity. Along the banks of the pond I found areas of flattened herbaceous vegetation where the otters evidently had been resting and lolling, and there were scattered piles of scats consisting largely of fish scales and bones. On another morning in late June, I saw two juveniles running through a forest a quarter mile from a

Fig. 21.19. River otters depend on fishes and other aquatic animals in lakes, ponds, and rivers. Stephen P. Broker.

nearby reservoir. Studies of harvested and road-killed otters in Connecticut indicate that about 75 percent of females more than one year old breed; a smaller percentage of yearlings breed.

Mink *(Mustela vison)* (Fig. 21.20). Fairly common in streams, ponds, lakes, and marshes; in contrast to fish-eating river otters, minks forage along shorelines and eat a wide assortment of terrestrial and aquatic animals; sleeps and rears young in den in bank, under log or tree roots, or similar site. *Note:* Occasionally, while kayaking, I had fleeting glimpses of minks along river and lake shores. In winter, as I sat quietly at the edge of a river, I watched a mink investigate spaces under logs and tree roots, dive into the water, and emerge onto the open ice that fringed the river. Once I observed a juvenile cavorting in a marsh as I stood only 15 feet away. Sometimes I detected mink presence near wetlands by their tracks in the snow. A population of exceptionally large minks (sea mink) may have occurred along the coast of Connecticut in the 1800s, but these large minks are now extinct.

Long-tailed weasel *(M. frenata)* and **ermine** (short-tailed weasel) *(M. erminea)*. Fairly common ground-dwelling predators in woods and thickets and along old stone walls; apparently most common along water courses, based on tracks I saw in snow; den among rocks, under tree roots, in burrows or brushpiles, or in similar secluded sites. *Note:* These two long, slim animals minimize competition with each other by taking prey of different sizes and in different locations. Thus larger long-tailed weasels may prey on larger prey such as rabbits whereas smaller ermines, especially females, can pursue voles and other small mammals along their narrow runways and tunnel systems. Ermines change color from brown in summer to white in winter; seasonal color change might appear to be related to camouflage in an environment that changes from leafy to snowy, but instead (or in addition) may reflect the better insulative qualities of hollow white fur compared to pigmented brown fur.

American marten *(Martes americana)*. Status uncertain; one recent road-kill in New Hartford was the first hard evidence of this species' occurrence in Connecticut; studies in Maine indicate that marten populations do best in areas with large blocks of forest with minimal fragmentation; average male home range (1,112 acres) included more than 600 acres of unfragmented forest.

Fisher *(M. pennanti)* (Fig. 21.21). Denizen of large tracts of thick forest; depends on large tree cavities for maternal den sites; probably most often sleeps in burrows in winter and up in trees in summer; winter diet in Connecticut dominated by gray squirrels; also eats various smaller mammals; male fishers, which are much larger than females, commonly take cottontails, raccoons, and sometimes porcupines. *Note:* Fishers were extirpated from southern New England as a result of historical land clearing for agricultural purposes. After being absent from Connecticut for more than a century, fishers became reestablished as a result of natural range expansion allowed by reforestation and inten-

Fig. 21.20. Mink are fairly common but infrequently seen along Connecticut waterways. Larry Master.

Fig. 21.21. Fishers are repopulating the state's forests following reforestation, reintroductions, and natural range expansion. Larry Master.

tional reintroductions of fishers obtained from New Hampshire and Vermont in 1989 through 1991. Fishers are now well established in northern Connecticut (particularly the northwest) and have expanded into large forests in south-central Connecticut. These generally solitary, wide-ranging mammals have home ranges that may encompass thousands of acres. On average, populations normally consist of not more than one fisher per several hundred acres.

Striped skunk *(Mephitis mephitis)*. Common in diverse habitat, often attracted to improperly discarded garbage in towns; one of the mammals most often seen dead on roads in Connecticut (with opossums, raccoons, and gray squirrels); dens in underground burrows (for example, abandoned woodchuck holes), chambers under rocks or stumps, or beneath buildings; active mainly at night throughout year, but stays in dens during the coldest periods or deepest snows of winter; feeds mostly on small vertebrates and large invertebrates obtained on ground surface or dug out of ground; also eats available fruits and garbage. *Note:* Of course, skunks are notorious for their chemical defense system. Unassisted by the wind, the scent can be sprayed a distance of about 9 or 10 feet. The skunk's bold color pattern undoubtedly is intimately related to its chemical defenses. Potential predators and harassers are likely to remember and avoid such a conspicuous animal after an odoriferous encounter. Skunks are relatively unwary and fairly easy to approach, suggesting that they do not feel especially threatened by much larger mammals. However, small skunks may be easy prey for great horned owls, which do not seem to be repelled by the strong odor. To avoid problems with skunks around your house, seal off access around the base of buildings and porches (be sure that no skunks, including young, are within!) and keep edible trash out of reach. Skunks may carry rabies, so avoid any that behave abnormally (that is, that display lack of coordination or that are not oblivious or shy around humans).

Cats (Order Carnivora, Family Felidae)

Bobcat *(Felis rufus)* (Fig. 21.22). Sparsely distributed in vicinity of thickets and patchy woods in less-developed regions of state, especially Northwest Highlands; commonly preys on cottontails, snowshoe hare, deer, gray squirrels, voles, shrews, mice, and birds; dens in tree hollows, spaces among rocks, or in thickets; most births in late spring or early summer. *Note:* Bobcats tend to be scarce in areas where coyotes are numerous (due to direct predation and/or competition for food), so perhaps the large increase in the coyote population in Connecticut over the past several decades caused a decline in the bobcat population. How-

Fig. 21.22. Bobcats are uncommon residents of Connecticut's less-developed landscapes. Paul J. Fusco/Connecticut DEP Wildlife Division.

ever, the bobcat population trend in Connecticut is unknown, and the two species can coexist to some degree. In a freak incident, a rabid bobcat attacked a man in Plainville, Connecticut, in 2003.

Lynx *(Lynx canadensis)*. Formerly may have ranged occasionally into Connecticut, but historically this essentially northern mammal apparently was not a permanent resident of the state.

Eastern cougar or **mountain lion** *(Felis [Puma] concolor)*. Formerly perhaps an uncommon resident of hilly sections of northern Connecticut; no firm evidence that a wild population currently exists in Connecticut; some reported sightings of cougars in the region probably were based on escaped or released individuals or more likely were simply the product of inaccurate observation and an active imagination. However, it is possible that someday wild cougars might expand their breeding range into New England, where deer (their primary prey) are abundant.

Seals (Order Carnivora, Family Phocidae)

Harbor seal *(Phoca vitulina)*. The only marine mammal that is a regular inhabitant of Connecticut; coastal waters, primarily in eastern part of state but also westward (not uncommon in the vicinity of Hammonasset Beach State Park); at least several hundred seals inhabit waters at eastern end of Long Island Sound in colder months of late fall through mid-spring; rests on isolated rocks and ledges; feeds primarily on fishes. *Note:* Historically harbor seals may have been permanent breeding residents of southern New England, but sealers likely put an end to that. They migrate northward to coastal Maine for the summer, paralleling the behavior of a fair number of humans. There is evidence that seals are arriving in the Sound earlier and departing later. Perhaps they eventually will reestablish themselves as year-round breeding residents in Long Island Sound, but the high level of human activity in coastal areas in summer may not be compatible with the maternal needs of these disturbance-sensitive mammals.

Gray seal *(Halichoerus grypus)* occasionally appears in Long Island Sound, but primary habitat is in more northern locales.

Hoofed Mammals

Deer (Order Artiodactyla, Family Cervidae)

White-tailed deer *(Odocoileus virginianus)* (Fig. 21.23). Common in Connecticut's ideal mixture of young and mature forests, fields, wetlands, and edible cultivated plants; breeding peaks in November; spotted fawns become evident in June, with well-fed females annually producing twins and sometimes triplets; annual home range size for individuals in southern Connecticut is about 100 to 200 acres, and some may range over areas of up to 500 acres.

Fig. 21.23. A white-tailed deer, with snow on its head and shoulder, browses on eastern hemlock in mid-winter.

Note: The statewide population is substantial and growing, due to an increasing area of rural residential lands suitable for deer but not amenable to hunting, forestry practices that create a mosaic of young and mature growth, milder winters, and reduced or eradicated populations of large predators. As of 2000, the total state population was minimally about 76,350, up from around 20,000 in the 1970s and fewer than 20 in the whole state at the close of the 1800s, when habitat loss and excessive hunting virtually eliminated deer from Connecticut. Winter density ranges up to about 40 per square mile in southwestern Connecticut, with a statewide mean of 21 per square mile. The annual hunter harvest in recent years was between 10,000 and 14,000. The reported road kill in 2000 was just over 3,000 (probably 5,000 to 8,000 here hit by motorists).

Male deer grow antlers, and older, well-fed individuals grow the largest antlers. Antler growth begins in spring and ends in late summer or early fall, prior to the mid-fall breeding season. Antlers function in male-to-male combat and affect access to females for breeding. Generally antlers drop off around the first of the year as testosterone levels plummet.

In late summer or early fall, deer rub their fully grown antlers and glandular forehead on small trees. The function of this behavior is partly to remove the dried "velvet" and more importantly to produce visual and chemical signals that may function in establishing dominance relationships among males. These rubbed trees may also attract and affect the reproductive condition of females. Scraping of the ground by the hooves, and urination in the scraped area, usually done by older males in autumn after leaves fall, may function in a similar way.

Deer eat a wide variety of plant material, ranging from soft wetland plants to the tips and sprouts of woody plants to acorns. Researchers monitoring bird nests with video cameras recently were shocked to learn that deer sometimes eat nestlings of ground-nesting birds!

Through their selective feeding on seedlings, young trees, tree sprouts, and other low vegetation (Fig. 21.24), deer can have a strong impact on vegetation development and the species composition of forests, sometimes leading to degraded conditions and reduced biodiversity. In some places, intense browsing and grazing by the large deer population is having a significant impact on forest undergrowth, inhibiting tree and shrub regeneration (Fig. 21.25) and eliminating or reducing populations of woodland wildflowers. Deer-altered understory vegetation may result in changes in the abundance and composition of bird communities. For example, ovenbird populations may be reduced or eliminated where heavy deer browsing supresses the density and diversity of woody understory vegetation. Some browse-resistant non-native plants, such as Japanese barberry*,

Fig. 21.24. Deer browsing on red-cedars in winter often leave the plants devoid of their lower branches and foliage.

Asian bittersweet*, and winged spindle-tree* may increase as more palatable native competitors decline. Deer also damage orchard trees by browsing and rubbing. Feeding by deer on landscape plantings, especially in winter and early spring, has resulted in human-deer conflicts in some residential areas, especially where hunter access is restricted.

One goal of deer management is to maintain populations below levels at which they cause unacceptable alterations in the ecosystem or damage to cultivated plants or property. Hunting is the usual method of control. Hunting as a means of population control can be especially effective when localized populations are targeted. However, under most circumstances, a primary effect of such deer harvests is to increase the health and survival of the surviving individuals (due to more resources available per deer with a smaller population) and stimulate increased reproduction (more young per female), ensuring that the population will increase and the perceived deer problem will arise again. If hunting is the primary management tool, then usually it must be continued at least periodically in perpetuity because total elimination of deer is seldom feasible or desirable. One reason that hunting has not stopped continued deer population growth in Connecticut is that much deer habitat is highly intermingled with low-density residential areas, making many areas unsuitable for hunting despite the presence of many deer.

Experimentation with immunocontraceptives (chemicals that temporarily block egg fertilization) has been conducted at Mumford Cove in Groton, with the hope of developing an effective alternative population control method in situations where hunting is not possible or desirable. Unfortunately, such procedures are costly, labor-intensive, and, as yet, not very effective.

Damage to evergreen landscaping around houses can be minimized by shielding deer-favored shrubbery with burlap in winter, if it's worth the time and effort to you. A better alternative is to replace deer-favored plants such as yews, junipers, arbor vitae, hosta, lilies, sunflowers, tulips, and such with andromeda and other plants less attractive to deer (contact a good nursery or the Connecticut Agricultural Experiment Station in New Haven). You can obtain copies of a booklet *Managing Urban Deer in Connecticut* from the Connecticut Department of Environmental Protection.

Safety considerations dictate that hikers be aware of deer hunting seasons. Hunting with firearms occurs throughout November and December, the exact dates depending on whether the land is private or public. On state lands, the season begins in mid-November and extends into the first week of December. Bowhunting begins as early as mid-September in certain areas. Check with the DEP or consult the DEP's current Hunting and Trapping Field Guide for specific regulations.

Moose *(Alces alces).* Has wandered into Connecticut from Massachusetts with increasing frequency over the past decade; now becoming a rare resident breeding species along parts of Massachusetts-Connecticut border; first documented reproduction (a female with two calves) in Connecticut occurred in 2000. *Note:* In addition to habitat limitations, increase of the moose population in Connecticut might be limited by transmission of brainworm (a nematode) from deer to moose. Deer host the worm without getting sick, but for moose brainworm infection is fatal. Deer pass brainworm larvae in their feces, then land snails feed on deer feces and ingest the larvae, which develop further in the snail. Next the snail may be accidently ingested by deer feeding on plants, and the brainworm completes its life cycle by maturing and laying eggs which hatch within the deer. A moose striding across a highway in Connecticut

Fig. 21.25. Deer browsing on stump sprouts and seedlings can inhibit reforestation. Here dozens of sprouts have been nipped off.

is always a source of considerable excitement, but such events pose severe danger to the moose and to motorists. Moose sightings should be reported to the DEP (Franklin Wildlife Management Area, 860-642-7239; 24-hour DEP phone line is 860-424-3333).

Habitats

In Connecticut, wooded and partly wooded terrestrial habitats and wooded inland wetlands, which provide a wide range of microhabitats and food resources, harbor the largest number of mammal species (Fig. 21.26). Some habitats, such as streams and lakes, tend to have relatively few mammals, yet these habitats contribute importantly to mammalian biodiversity because the species that use them (e.g., beaver, river otter) tend to be highly specialized for their restricted environment and unable to live anywhere else.

Many mammals use burrows for shelter. Woodchucks are the primary diggers of large burrows in Connecticut, and other mammals often use (and often modify) abandoned woodchuck burrows. Small mammals generally dig their own burrows. Caves and large tunnels are used mostly by bats.

Cavities in standing dead and living trees provide important den and nest sites for many mammals, such as opossums, northern bats, silver-haired bats, gray squirrels, red squirrels, southern and northern flying squirrels, deer mice, white-footed mice, porcupines, gray foxes, raccoons, and fishers. Bears sometimes use tree hollows, if they can find one sufficiently large. Several additional mammals use hollows in fallen logs. Some of the mammals that use tree cavities or log hollows also nest in underground burrows. See "Forestry Practices and Missing Resources" in Chapter 7 for a discussion of habitat management for cavity-using animals.

Food and Feeding

Connecticut's mammals exhibit a wide range of food habits (Fig. 21.27). Land-dwelling invertebrates, fleshy fruits, seeds, and nuts are important foods for the largest percentage of species. A perhaps surprisingly high percentage of mammals eat significant amounts of carrion and fungi. These items generally are highly nutritious and readily digestible. A smaller but significant percentage of Connecticut's mammal fauna subsists on plant foliage, a diet that poses some physiological challenges for mammals and other herbivores, as discussed in the following section.

Problems for Plant Eaters

This section describes some of the challenges of an herbivorous diet. The next section reviews the ways in which mammals successfully deal with the problems of a foliage-based diet. Where appropriate, I expand the discussion to include herbivorous insects.

In spring and early summer, growing plants contain large amounts of potassium but are low in sodium. Mammals have a greater need for sodium than for potassium. In fact, a high intake of potassium can disrupt a mammal's sodium balance because in ridding the body of potassium the kidneys cannot help but excrete into the urine some of the limited sodium as well.

Plants often contain compounds that reduce food digestibility or are toxic. Cellulose, a glucose polymer, makes up 80 to 90 percent of the dry weight of most plant parts. The problem is that cellulose cannot be directly digested by vertebrates, and only a few insects lacking symbiotic gastrointestinal microorganisms secrete cellulase and can digest cellulose.

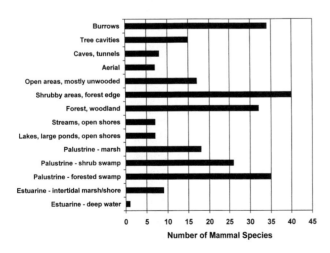

Fig. 21.26. Habitats of Connecticut mammals. Among marine mammals, only the harbor seal was included in this summary.

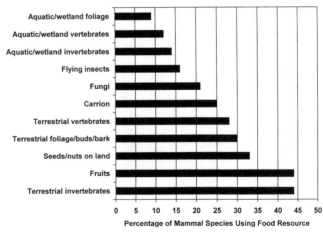

Fig. 21.27. Diets of Connecticut's native mammals (includes four species of marine mammals).

However, as we shall see, vertebrates and insects do have ways of processing and obtaining nutrients from cellulose.

Another plant compound, lignin, is resistant to normal enzymatic and acid hydrolysis and so is basically indigestible. It interferes with digestion by binding to both carbohydrates and digestive enzymes. Lignin is a high molecular weight, aromatic, nonsaccharide phenolic polymer that increases the rigidity of the cell wall, which may be up to 15 to 20 percent lignin.

Tannins are another phenolic polymer. Tannins interfere with digestion by binding with proteins and deactivating digestive enzymes. They may even be toxic at high levels. Tannins also effectively inhibit digestion by fungi and bacteria. Phenolic chemicals in aromatic spicebush and sassafras similarly interefere with digestion.

Salicin is an anti-herbivore chemical present in large amounts in aspen bark and shoots. Beavers nevertheless are able to eat large amounts of aspen without any problem.

Some plants contain inorganic crystals that discourage herbivory or inhibit digestion. Silica is an indigestible and abrasive crystal present in grasses and horsetails. It limits the digestibility of these plants. Skunk cabbage, arrow-arum, and related plants contain caustic calcium oxalate crystals that cause a burning sensation in the mouth if eaten. These plants generally show little damage from herbivores.

Alkaloids and certain other nitrogen-containing compounds have many toxic effects and are most common in herbaceous plants. Nicotine, morphine, and caffeine are example of alkaloids.

Terpenoids are another class of defensive compounds in plants. Examples of terpenoids include pyrethrin (present in chrysanthemums), peel oils of citrus, and aromatic oils of sagebrush and evergreens. Many common species contain terpenoids. Terpenoids are insecticidal and antimicrobial, and they account for the resistance of cedar wood to insect attack. Ecdysones are terpenoids that mimic insect hormones and interfere with the normal development of plant-eating insects by preventing molting or preventing the gonads from maturing.

Some plants contain cyanogenic glycosides that release hydrogen cyanide, a potentially lethal poison, when the plant tissue is damaged.

Mammal Reponses

Despite these defenses, some mammals and other herbivores frequently consume plant foliage, and microbes also attack leaves and other plant tissues. My students and I determined that many common trees in central Connecticut incur significant herbivore damage on 15 to 55 percent of their leaves, and microbial damage typically affects 25 to 50 percent of the leaves.

Mammals and other herbivores cope with plant defenses in various ways. In most mammals that feed extensively on fibrous plant matter, the cheek teeth are crowned with folded ridges of enamel that extend down the length of the tooth and stay sharp even as the tooth gradually wears down. The cheek teeth of meadow voles, which eat an abrasive diet of grasses, keep growing from below the gums as the top of the tooth wears away, so their teeth never wear out (though voles generally live less than one year).

In some mammals, such as deer, the teeth do eventually degenerate. After several years, the sharp ridges of deer teeth eventually wear down, and the deer finds it harder and harder to adequately chew its fibrous food. Poor nutrition or starvation may result.

The digestive tracts of meat eaters tend to be simple and short, in accordance with their easily digested diet, but the guts of strict herbivores are extensive, sometimes huge, and always include specialized pouches or outpocketings, such as the rumen of deer and the cecum of porcupines, that serve as fermentation chambers housing large populations of microbial symbionts such as protozoans and bacteria that have the ability to break down cellulose and make it available to the host animal. For example, hoofed herbivores digest 50 to 70 percent of the alfalfa fiber they eat, compared to only 10 percent in omnivorous humans. In addition to these extracellular symbionts, some insects have intracellular bacterial symbionts in special cells called mycetocytes in gut outpocketings or in special organs called mycetotomes. These symbiotic organisms in herbivore guts not only break down foods but also may synthesize vitamins and degrade plant toxins. However, even with these symbionts, herbivores still have to process large volumes of food daily to meet nutritional needs, reflecting the relatively low concentration of available nutrients in their plant diet. This can profoundly affect their ability to conduct other activities.

Despite their elaborate digestive systems, large plant-eaters such as deer and porcupines do poorly on the low-quality diet available in winter, and their body mass and condition decline throughout the colder months. To survive the winter, large herbivores must accumulate within their bodies sufficient nutrients in autumn.

Some herbivores compensate for inefficient digestion of plant material by ingesting their own fecal pellets and recycling the material through their system. This phenomenon, known as coprophagy, is typical of rabbits, which reingest soft pellets that are still high in protein but do not eat fibrous pellets with less protein.

Mammals may overcome the nutritional shortcomings and chemical defenses of plants by selective feeding. Some mammals, such as meadow voles, if given a choice, select plants that are relatively low in potassium. Other mammals seek out sodium sources. Deer may be attracted to wetland plants for their relatively high sodium content, and they commonly visit natural mineral licks and roadside salt deposits that result from winter salting of highways. Porcupines seek out sources of sodium, including the salty undersides of vehicles and tool handles, cabin furniture, and other objects that have come in contact with human sweat. Porcupines also go out of their way to visit trees with low levels of organic acids. These acids may interfere with their ability to excrete potassium.

Mammals tend not to eat large quantities of plants containing phenolic chemicals, but they don't avoid them altogether. Mixed with other foods, deer may eat these plants with relatively little digestive upset.

Many herbivores have in the digestive system mixed-function oxidases, membrane-bound enzymes that detoxify secondary plant compounds. Thus the chemical defenses of plants are met with biochemical responses in the plant eaters. But detoxification processes can be energetically expensive, so it is no surprise that herbivores have evolved mechanisms, including sensory systems for detecting toxins and digestion inhibitors and associated selective feeding, for minimizing the need for them.

Probably most plant-eating mammals are deterred from eating high-tannin plants by the astringent taste. In summer, porcupines preferentially forage on species that are low in tannin. Plant eaters such as deer, moose, bears, and beavers produce tannin-binding salivary proteins that minimize tannin absorption and toxicity.

In addition to the chemicals in plants, one might suppose that the diets of mammals might be restricted also by mechanical plant defenses, such as thorns and spines. To a small degree they are, but most mammals, including big-mouthed species such as deer, eat thorny plants with seeming impunity. Accordingly, only a minority of plant species have thorns or spines.

Some nonmammalian herbivores employ external mechanisms to deal with hard-to-digest plants. For example, tropical leaf-cutter ants and certain termites maintain external fungal gardens in their nests and rely on the fungi to break down the plant material and convert it into a form they can use. Bark beetles may do the same in wood.

Reproduction

The mating systems of mammals range from monogamous (rare) to promiscuous. The vast majority of species have a promiscuous or polygynous (one male mates with multiple females) mating system.

Most Connecticut mammals begin producing offspring in early spring. The mammal that gives birth earliest in the calendar year is the black bear (January). Generally not much birthing happens in February, but by the vernal equinox in late March, opossums, masked shrews, cottontails, gray squirrels, white-footed and deer mice, meadow and woodland voles, red foxes, and fishers may have produced young. By the end of April, newborns of most species have appeared.

The mammals that wait the longest to give birth are the hibernators and migrators, including several species of bats and meadow and woodland jumping mice. The late bat schedule, with most litters born in June or July, may ensure a good supply of food (flying insects) for lactating mothers and newly volant young. The year's earliest jumping mouse births, or the peak in births, generally is not until June. Jumping mice do not mate until after emerging from hibernation in late April or early May, so early births, such as is typical of white-footed mice, are not possible.

White-tailed deer also give birth late (peak in early June) relative to most mammals. The late timing of deer births probably is related to climate patterns and food availability. Cold, wet weather likely discourages birthing in early spring, since deer are too big to use well-protected dens or nests. Also, food resources to support milk production and fawn diets are better after mid-spring when a flush of succulent new growth appears.

Some of the rabbits and rodents that reproduce early produce additional litters through summer and sometimes into early fall. In contrast, individual female carnivores, bats, deer, and some rodents (beaver, porcupine, woodland vole, jumping mice) give birth to just one litter each year. Litter size tends to be largest in mammals that have a short life span (for example, small rodents) and smallest in long-lived mammals such as bats.

Bats in Connecticut and other north temperate regions exhibit an unusual reproductive pattern that involves delayed fertilization. Bat copulation usually occurs shortly before hibernation. Males inseminate females with stored sperm that they produced earlier (their testes are inactive when mating occurs). Females store the sperm in the uterus, and it stays there all winter. Shortly after emergence in spring, the females ovulate and egg fertilization finally occurs. Among the advantages of delayed fertilization is that the development of the young begins immediately in spring; the females do not have to wait for the males, which are in relatively poor shape after hibernation, to attain reproductive condition. So births can occur earlier (but late enough to ensure mild weather and plenty of food), and the young have more time to develop and accrue the fat deposits they need to get them through winter.

Another ecologically significant reproductive pattern—delayed implantation—is characteristic of members of the weasel family (Mustelidae) and some other mammals. Males produce sperm in summer, and mating and fertilization occur in late summer or early fall. A fertilized egg develops into a ball of cells, then stops developing and begins a period of suspended animation while it floats freely in the female's uterus. After several months, it implants in the wall of the uterus and resumes development. This pattern allows births early enough such that the young have sufficient time to develop before environmental conditions deteriorate at the end of the year.

In some rodents, implantation is delayed if the female is nursing young from a previous litter. The delay is beneficial because the mother avoids being saddled with two energy-demanding situations at the same time.

Mammals in Winter

Mammals respond to the stresses of winter's cold in various ways. Some Connecticut mammals, such as the silver-haired bat, red bat, and hoary bat, vacate the state for the winter and migrate to warmer areas in the southeastern United States. Conversely, the harbor seal is the only mammal that occurs here primarily in the colder months and rarely in summer. It migrates northward in spring and spends the warmer months along the coast of Maine. For this relatively cold-tolerant species, a winter in Connecticut may be akin to winter in Florida for humans. All other species cope with Connecticut's cold winters either by remaining active on a daily basis, remaining active except during periods of particularly cold or snowy weather, or hibernating in a protected nest.

Winter-Active Mammals

I enjoy going out after winter storms to look for signs of mammal activity (Figs. 21.28 to 21.36). This is most productive 24 hours or more after the snow has stopped falling. Careful search in appropriate habitats may yield signs of winter-active shrews, moles, cottontails, snowshoe hares, squirrels, beavers, white-footed and deer mice, voles, muskrats, southern bog lemmings, coyotes, foxes, fishers, weasels, mink, river otters, bobcats, and deer. Opossums, raccoons, skunks, and some of the other carnivores commonly remain inactive in their dens during snow storms or extreme cold, but they are active periodically in winter and do not hibernate. Chipmunks stay in their burrows most of the winter and go torpid for periods lasting up to several days, but I periodically saw them active above ground throughout the winter months.

Small mammals that are active in winter face a greater challenge in maintaining a warm body temperature than do larger mammals. Due to their relatively large surface area in proportion to body mass, small mammals do not retain heat very well. This problem is exacerbated by the fact that the fur of a small mammal cannot be as thick as that of a larger species. However, small mammals have several ways of combating winter cold.

Nest Insulation. Small mammals make and rest in heating-retaining nests made of shredded plant material. White-footed mice often construct their nests above ground, such as in tree cavities, whereas shrew nests are below ground. Nests tend to be larger and better insulated with thicker walls during cold seasons. This can be observed easily in free-ranging white-footed mice using artificial nest boxes. Nest insulation provides considerable energy savings, estimated at about 27 percent in the white-footed mouse, compared to energy needs of a mouse deprived of a nest. The value of reduced energy expenditure is that it increases the amount of time that a mouse can withstand cold temperatures in the absence of food.

Communal Nesting. White-footed mice often reduce heat loss by nesting communally. In winter, most mice rest in daytime in communal nests containing two to six individuals; communal nesting reduces energy expenditure by about 30 percent in both torpid (see below) and nontorpid mice. Voles, muskrats, and flying squirrels also often nest communally during the colder months. All evidence indicates that the short-tailed shrew is solitary in winter and does not snuggle with other shrews.

Food Storage. Short-tailed shrews commonly cache food (throughout the year); this may be critical in times of food shortage in winter. The saliva of the shrew apparently acts

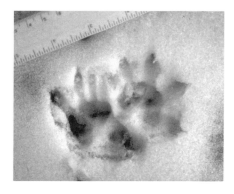

Fig. 21.28. Opossums are active in winter but generally stay in their dens during severe weather.

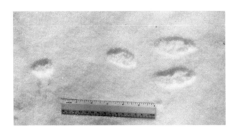

Fig. 21.29. Cottontail tracks: hind feet side by side, front feet offset, moving left to right.

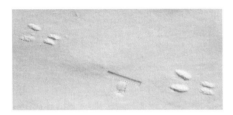

Fig. 21.30. Gray squirrel tracks: larger hind feet and smaller front feet side by side, moving fast, right to left.

Fig. 21.31. Woodchucks leave tracks in the snow when they emerge from hibernation in late winter.

Fig. 21.32. White-footed mice tracks often show an impression of the tail, hopping right to left.

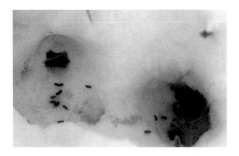

Fig. 21.33. Meadow voles tunnel to the surface of the snow, sometimes depositing green fecal pellets in their runways.

Fig. 21.34. Raccoons stay in shelter during foul weather but may leave tracks in snow or mud when conditions are less severe.

Fig. 21.35. Red foxes leave unique tracks even on hard-crusted snow.

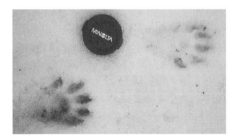

Fig. 21.36. Fisher tracks are becoming a common sight in large forests in Connecticut.

as an immobilizing agent and allows small animals to be stored alive for future use. White-footed mice also store food (for example, cherry pits and acorns), especially in fall. These energy sources can be used if food becomes scarce or foraging becomes difficult due to inclement weather. Other small mammals that store food for use in winter and early spring include chipmunks and squirrels, which may individually cache thousands of nuts each autumn (Fig. 21.37). When food is abundant, a chipmunk can store enough food in only a day or two to meet its winter needs. Chipmunks periodically awaken from winter torpor (see following) to eat food stored in or near their nest or burrow.

Subnivean Activity. Many winter-active small mammals, such as voles, avoid extremes of temperature by exploiting the subnivean (under the snow) environment. A thick layer of snow buffers temperature extremes at the ground surface and in the soil. Temperatures under the snow seldom are below 23°F, even if air temperature above the snow is less than –4°F.

Additionally, wind chill is not a factor in the subnivean environment. Movement beneath the snow is facilitated by the development of an air space at the ground surface as water vaporizes from the snow and moves upward in the snow pack.

Short-tailed shrews sometimes can avoid temperature extremes by foraging in tunnels under an insulating layer of leaves. Snow, when present, provides an additional layer of insulation.

In addition to its insulative benefits to small mammals, snow also provides a measure of protection from predators, though predatory birds and mammals with acute hearing, such as hawks, owls, and foxes, do sometimes successfully detect and capture small mammals hidden under snow.

Reduced Activity. Short-tailed shrews decrease their energy expenditure and hence food requirements by reducing their activity in winter, spending 80 to 90 percent of the day resting in the nest. Apparently the abundant, high-energy diet (mainly snails and arthropods in winter) allows the shrew to spend most of its time in a favorable microclimate. White-footed mice and other small mammals also reduce activity in winter.

Change in Body Mass. Voles and masked shrews reduce body mass in winter. The advantage of this seems to be that it reduces energy needs at a time when food is scarce (yet less mass seemingly would tend to increase the rate of heat loss due to a less favorable surface:volume ratio, everything else being equal). Short-tailed shrews, white-footed mice, and flying squirrels do not employ this method of energy conser-

Fig. 21.37. Hickory nuts supply high-quality nutrition to mammals in winter. Likely consumers of these nuts were flying squirrel (top center), white-footed mouse (top right), and eastern chipmunk (others).

vation and instead are heaviest in winter; apparently their other mechanisms of energy conservation (food hoarding, communal nesting, torpidity in mice) are sufficient.

Body Insulation. Increased pelage insulation (thicker fur) in winter has been demonstrated in short-tailed shrews, voles, and deer mice. Flying squirrels do not appear to have better insulation in the winter than they do in warmer months. Perhaps this is related to staying stream-lined for efficient gliding.

Torpor. Torpor refers to a reversible periodic lowering of body temperature. Torpor generally occurs over a period of several hours each day or at more irregular intervals.

Torpor in free-ranging white-footed mice occurs only in the colder months. On any given day, not all mice will be torpid; mice that share a nest tend to all become torpid or none of them do. Food availability apparently does not affect the incidence of torpor. Torpor occurs from early morning (prior to daylight) to midday. Minimum body temperature during torpor generally is about 64 to 72°F regardless of ambient temperature. Torpor reduces the daily energy expenditure of a mouse by an estimated 20 percent compared to nontorpid mice. Daniel Vogt determined that communal nesting, torpor, and use of a nest reduce energy expenditure 73 percent compared to nontorpid, solitary mice without a nest.

Usually chipmunks are torpid for periods of less than a week, then awaken for part of a day to eat before going torpid again. They stay torpid longer when less food is available. During winter torpor, the chipmunk's body temperature remains slightly above ambient temperatures of 37 to 61°F.

Periods of hypothermia increase in duration as ambient temperature decreases. Body temperature of active chipmunks is 97 to 104°F.

Daily torpor has not been observed in the short-tailed shrew or in voles. Body temperatures of eastern gray squirrels are only a few degrees lower in winter than in summer, with the greatest difference at night in food-deprived individuals.

Increased Heat Production. White-footed mice, voles, flying squirrels, and short-tailed shrews have a higher resting metabolic rate (increased heat-producing capacity) in winter than in summer under identical thermal-neutral conditions (there is some conflicting information for the white-footed mouse). Many small mammals, such as short-tailed shrews, masked shrews, and flying squirrels, also have an augmented capacity for heat production in winter due to nonshivering thermogenesis (NST), a mechanism that produces heat in the absence of muscular contractions. NST is mediated by the hormone norepinephrine. Brown-adipose tissue (brown fat) is an important site of NST. For various mammals, including red squirrels, brown fat appears to be present in greater amounts in winter than in warmer seasons. Brown fat can be observed readily under the skin in the shoulder-blade region in a small mammal carcass. Increased heat production is of obvious advantage in offsetting heat losses to the cold environment.

Winter-Active Large Mammals

Large mammals, too, face stresses in winter. Deer, for example, experience a shortage of high-quality food in winter and must resort to woody browse, which is low in nutritive value and digestibility. At the same time, low temperatures and snow on the ground increase their energy expenditures for keeping warm and finding food. Deer successfully cope with these conditions by increasing their food intake and putting on body fat in fall. In winter, they catabolize these fat reserves to offset the food shortage. Thus deer lose weight throughout the winter and enter early spring in relatively poor condition. But they survive winter because they are large enough to accumulate sufficient fat deposits to get them through the seasonal famine.

Beavers successfully cope with winter temperature extremes by building and residing within lodges that can be at least 36 to 54°F warmer than outside air temperature. They also exhibit reduced metabolic rate, which decreases their food requirements at a time when death of water plants and ice formation make food less available to them. Beavers further reduce winter energy expenditure by using food stored for ready use under water near their lodge. If necessary, beavers can make use of the large amount of fat they store in the body and tail prior to winter.

Many large mammals undergo an autumn molt that yields increased hair density and thus better insulation from the cold.

Hibernation and Winter Dormancy

Hibernation. Hibernation is a condition of pronounced hypothermia that is maintained for a period of days, weeks, or months during the colder months of the year. Body temperature generally remains less than about 10°F above burrow temperature. At burrow temperatures of 36°F, woodchucks maintain body temperatures of 45 to 46°F.

Hibernation differs from torpor because body temperatures fall much lower during hibernation. Also, torpor occurs daily (or at more irregular intervals) during that portion of the day when the animal is inactive and alternates with regular periods of activity. Under normal circumstances, hibernating and torpid animals retain the ability to warm themselves spontaneously using only endogenous sources of heat.

Energy Savings. Both hibernation and torpor are effective mechanisms for conserving energy (and/or water in certain species) during periods when food may not be readily available. Metabolic rate often is reduced to levels that are less than 5 percent of the nonhibernating resting rate at comparable ambient temperatures. This allows the animals to extend their energy reserves over many months without feeding.

Which Mammals Hibernate? Connecticut mammals that hibernate include the little brown bat, northern bat, eastern pipistrelle, big brown bat, woodchuck, meadow jumping mouse, and woodland jumping mouse. Hibernation ends in early spring for most species, but the woodchuck resumes activity in late winter.

Carnivores do not hibernate or exhibit pronounced torpor. During extreme winter weather they may become lethargic and sleep for extended periods. During this sleep, body temperature may fall several degrees and metabolic rate may be reduced substantially compared to active conditions, but these generally are minor changes compared to true hibernation.

Rectal temperature of the black bear during winter sleep apparently does not fall below 86°F, but heart rate may fall to as low as 8 beats per minute. Staying warm makes sense considering that the tiny young are born during the winter denning period.

Hibernation physiology. The transition into hibernation by woodchucks is characterized by a slowing of the heart rate and reduction in metabolism prior to the decrease in body temperature. Several hours are required for body temperature to stabilize at the minimum level. The cooling process may be interrupted by brief periods of warming (result of shivering) during which heart rate and oxygen consumption increase.

Heart rate during hibernation is much reduced and may be steady or erratic (the heart may not beat for over a minute). At 41°F, the heart rate of various small bats varies from about 24 to 80 beats per minute (the heart rate during activity may frequently exceed 500 beats per minutes). Woodchuck heart rate drops from about 100 to 15 beats per minute. Breathing rate also may be even or erratic, and deep hibernators may go an hour between breaths.

What happens if the hibernation den or nest begins to freeze? Hibernators normally can respond to freezing ambient temperatures by increasing metabolic rate while remaining in hibernation. This response may be stimuated by peripheral temperature sensors in the skin and/or by the preoptic hypothalamic region of the brain. If ambient temperatures drop below freezing, big brown bats arouse from hibernation and appear to seek out warmer hibernating sites. However, under natural conditions, hibernators generally select or construct dens that are not subject to freezing. Woodchuck burrows average about 45 to 46°F during the hibernation period.

Arousal. All species of hibernators that have been studied exhibit periodic, spontaneous arousal and warming. In long-term, seasonal hibernators, such as woodchucks, bouts of hibernation initially are short, then lengthen (to an average of about eight days), then shorten again as the season progresses. Woodchuck arousals average around two to three days at an average body temperature of about 97°F.

Bats have perhaps the longest uninterrupted bouts of hibernation; continuous hypothermia lasting up to nearly three months has been recorded in the little brown bat under laboratory conditions. However, hibernation bouts of one to two weeks or less are more common. Considering that arousal consumes a great deal of energy (a single arousal in a small rodent may consume as much energy as 10 days of hibernation, and a bat arousal may consume an amount of fat that would support several weeks of hibernation), one might wonder why hibernators bother to rewarm then go back into hibernation. Hypotheses proposed to explain frequent rewarming include restoration of depleted nutrients in blood, elimination of toxic substances, problems with loss of potassium from cells, and facilitation of sperm production in males. Presently there is no clear answer to the "mystery" of periodic arousal.

In rodents, vigorous muscular contraction (shivering) and brown fat deposits are the primary heat sources for warming during arousal from hibernation. Bats can warm rapidly without shivering; brown fat apparently provides the heat. Depending on the conditions and species, complete rewarming during arousal may take about one to several hours.

Stored fat. Deep hibernators do not eat during arousals; instead, they rely on deposits of stored body fat to meet their nutrient needs. Jumping mice may double their body mass during a fattening period prior to hibernation, allowing them to survive the nearly seven months they spend in hibernation. Woodchucks get very fat before winter and lose about 40 percent of their body mass during hibernation. Little brown bats start hibernation with stored fat amounting to about 30 percent of their live body mass, and they lose about 25 percent of their mass during hibernation.

Conservation

Forest Fragmentation

As with birds (see Chapter 20), forest fragmentation benefits some species and negatively affects others. Some mammals such as the fisher appear to require large tracts of forest to meet their resource needs. Mammals such as white-footed mice may thrive in forest fragments, aided by the disappearance of other species (such as food competitors and nocturnal predators) that require larger forests to meet their needs. In contrast, other small mammals such as chipmunks may experience reduced survival and have lower populations densities in forest fragments because these daytime-active mammals may face increased rates of predation due to fragmentation-caused increases in the populations of diurnal predators. While some data are consistent with this hypothesis, further study is needed to verify the effect. There is evidence that fragmentation may also negatively affect populations of gray squirrels and flying squirrels in forest fragments, though the mechanisms causing the effect need further study.

By facilitating movements and allowing population interactions, corridors of undisturbed vegetation connecting forest fragments can help reduce the impacts of forest fragmentation for some kinds of wildlife. Keeping the gaps between forests small is important. Small mammals such as chipmunks can move among tracts of forest separated by undeveloped gaps of up to a few hundred meters or so. See Chapter 20 for further discussion of the effects of forest fragmentation.

Forest Management

Forestry practices can affect mammal populations by reducing the availability of certain resources. In particular, loss of large mast-producing trees, cavity-containing trees and snags, and large logs on the forest floor reduces the food supplies and nesting and sheltering sites for a wide range of mammal species and may lead to population declines or local extirpation. To some degree, loss of cavity trees can be mitigated by putting up suitable nest boxes (if the forest remains relatively intact). See the conservation section of Chapter 7 for further discussion.

Bat Hibernacula

Hibernating bats are vulnerable to being killed or injured by vandals and, even if not directly molested, their energy reserves, needed during winter and early spring when food is scarce or unavailable, become depleted if they are aroused too often as a result of disturbance. Excessive disturbance can result in abandonment of the hibernating site.

A major project aimed at protecting hibernating bats from disturbance was initiated in 1999 at an old iron mine in Roxbury. Each winter this mine shelters at least 2,500 bats of various species. Newly installed steel gating at the Roxbury mine allows free movement of the bats while preventing unauthorized entry by humans.

Cat and Dog Impacts

Though often maligned for killing birds, domestic and feral cats are much more likely to decimate local populations of small mammals such as mice, voles, chipmunks, and shrews. You can help protect small wildlife around your yard by keeping your cat indoors if it is an habitual killer and by discouraging people who foster large populations of semi-stray cats by feeding them and letting them run wild. Likewise, free-roaming dogs often disturb, harass, injure, or kill native mammals and other wildlife and should be leashed or closely controlled when you visit parks and natural areas.

Poaching

Illegal harvest or killing of mammals or other wildlife can be highly detrimental to effective conservation. To combat this problem, the state Department of Environmental Protection has established a confidential, 24-hour toll-free telephone number (1-800-842-HELP) for reporting wildlife violations. Rewards are offered after arrests.

Additional mammal conservation topics are mentioned in the individual mammal species accounts.

Chapter 22

A Naturalist's Calendar

The following calendar includes a selection of observations I made in Connecticut over the past 20 years. One primary goal of this chapter is to give you a feel for the timing of natural events and seasonal changes in the condition and behavior of plants and animals.

Bear in mind that I recorded these observations over many years and that annual changes in weather conditions often alter the timing of events by as much as a few weeks from one year to the next. Some of that variation is evident in these notes, reflected for example in different dates for the year's first katydid sounds or birdsongs, or first flowering of a plant. These field notes also include specific examples of some of the ecological relationships discussed in this book.

January

January 1. Common evening primrose pods still contain many seeds. An American robin eats Japanese barberry* fruits. An eastern phoebe forages among emergent plants in a frozen, swampy pond.

January 2. A winter crane fly flies at an air temperature of 36°F, with no sun.

January 3. Cattail seeds are dispersing.

January 4. An American crow eats American bittersweet fruits in a vine high in a tree. A black-capped chickadee and a tufted titmouse sing repeatedly in the morning.

January 5. Seedbox fruits still contain many seeds.

January 6. A blue jay rapidly ingests forty-four black oil sunflower seeds before flying off.

January 7. A black-capped chickadee sings in the morning. A northern cardinal picks fruits from tulip-tree pods and eats the seeds.

January 8. A dead but intact woollybear is on the front steps of my house.

January 9. The ground is frozen and heavy rains cause flooding of a creek, toppling streamside mountain laurels. Near snow-covered ground, white-throated sparrows pluck winterberry fruits, eat the insides, and discard the skins.

January 10. Dark-eyed juncos repeatedly come to the bird feeder in a group of five.

January 11. Two American robins eat sumac fruits. Several of a loose flock of forty-one American robins eat winterberry fruits (ground is snow-covered).

January 12. A Carolina wren sings many times. Yellow-rumped warblers search tree bark, ignoring nearby Asian bittersweet* fruits. 10:15 P.M., 36°F: winter crane flies arrive outside my lighted kitchen window. Leaves of small bayberry plants are bright red; Japanese honeysuckle retains reddish-green leaves.

January 13. Two short-tailed shrews emerge briefly from snow tunnels and gather sunflower seeds under bird feeder. A pair of northern cardinals plucks wayfaring-tree* fruits, eats the seeds, and discards the pulp. A tufted titmouse plucks a holly fruit, holds it in its feet, and pecks at it.

January 14. The most abundant bird as I paddle along one mile of the Connecticut River in Haddam is the common merganser (110 individuals).

January 15. A male red-bellied woodpecker repeatedly takes a sunflower seed from a feeder, flies to a red maple, inserts the seed into a bark crevice, then pecks the hull open and eats the seed.

January 16. A pair of great horned owls engage in antiphonal hooting.

January 17. Small adult stoneflies are active in the Sun above a creek at 42°F, with snow and ice on the ground.

January 18. Several isopods walk across the wood floor of my living-room (they hide under the front door and enter when I open it). Snow is covered with paper birch nutlets.

January 19. An American robin forages on a frozen lawn (air temperature 13.5°F), eating a few minute, unidentifiable objects, plus a smooth-skinned, black, C-shaped, frozen, 1-inch-long caterpillar (I verified this by finding another identical one on the lawn after the robin departed).

January 20. Two crows land briefly on the Connecticut River and each takes a small food item from a floating ring-billed gull that was pecking at something. With the ground frozen and snow covered, three American robins, one northern cardinal, and three eastern bluebirds pluck and eat still-abundant winterberry fruits.

January 21. An American robin eats holly fruits.

January 22. An eastern chipmunk runs along a ledge at an air temperature of 48°F.

January 23. A northern mockingbird eats Japanese barberry* fruits. A juvenile yellow-bellied sapsucker eats suet at a feeder on a tree trunk. At 3 P.M., sunny, 36°F, with several inches of snow on the ground, a small opposum, missing the flesh at the tip of its tail (frozen off?), forages under a bird feeder, often raising its snout and sniffing the air.

January 24. Three bald eagles (immatures and adult) each ride a block of ice floating down the Connecticut River.

January 25. An adult leaf-footed bug is active in the house.

January 26. An American robin eats holly fruits. Rains and mild temperatures have softened the soil; the lawn has several new eastern mole tunnels. I fill the bird feeders this morning (they had been empty for a few days); 30 minutes go by with no birds evident, then the first birds arrive (4 titmice, 3 chickadees, 2 white-breasted nuthatches, 1 downy woodpecker), all within a few seconds of each other, so apparently most or all traveled as a mixed-species flock.

January 27. I hear the year's first cooing by mourning doves. Each of two nest boxes contains two adult white-footed mice.

January 28. A white-tailed deer browses on hemlock foliage; another eats last year's growth of shrubbery (yew*) next to the house. Sanderlings and ruddy turnstones in a mixed flock extract and eat the flesh of common Atlantic slippersnails that have washed up on the shore. One turkey vulture stands on and picks at a roadside deer carcass. Groups of turkey vultures are in flight; one group of 15, circling low over a grassy hill, and two groups of 5.

January 29. Tufted titmice sing on a sunny morning. The Connecticut River rises and spreads out into its floodplain after a thaw and rain.

January 30. I hear the year's first cooing by mourning doves.

January 31. Most winterberry fruits are now largely stripped from the plants by birds.

February

February 1. Common redpolls probe seed capsules of evening primrose. A mockingbird eats multiflora rose* fruits. An American robin eats holly fruits.

February 2. Five adult white-footed mice are in a nest box in the afternoon at an air temperature of 45°F. I hear the year's first cooing by mourning doves.

February 3. A Carolina wren sings persistently in the morning.

February 4. Over a large area in a grassy field, I find an average of 28 meadow vole snow tunnel openings per 100 square meters (1,080 ft^2).

February 5. A crow and a turkey vulture fly up from beside the road, where I find a deer carcass, completely buried in snow except for a patch of about 5 inches by 4 inches; coyote tracks are here, too.

February 6. The Connecticut River is frozen all the way across in Haddam. On a sunny morning after a major snowstorm (14 inches), I hear the year's first black-capped chickadee songs.

February 7. Three American robins are in one Japanese barberry* plant, each bird eating several fruits. Many new eastern mole tunnels are in lawn thawed by recent rain. A screech-owl trills at 6:00 P.M., with a full moon rising. I follow a new striped skunk trail in the snow over several hundred feet; the skunk walked to the entrance of but did not enter several woodchuck burrows.

February 8. I hear the year's first mourning dove cooing. With the ground frozen and snow covered, three American robins pluck and eat crab-apples*.

February 9. With light rain and record high temperature in the 50s(°F), an adult male wood frog hops on a road at 7 P.M.

February 10. Several winter crane flies fly near a light at 10:00 P.M., 35°F.

February 11. Nine bald eagles (eight immatures) are along the Connecticut River at Higganum. I hear the first thunder of the year.

February 12. Two tufted titmice on partly snow-covered ground of a rocky wooded slope repeatedly use their bills to toss sun-warmed dry oak leaves aside, apparently searching for food (invertebrates, or cached sunflower seeds?). These same birds also frequently take sunflower seeds from a feeder.

February 13. Tulip-tree fruits disperse in large numbers in stong winds, falling onto snow-covered ground.

February 14. A census of bats in a hibernaculum yields little brown bat (343 individuals), northern bat (22), eastern pipistrelle (15), and unidentified (4). An ice-covered creek flows both above and below the ice after heavy rains.

February 15. I hear the first-of-the-year tufted titmouse songs.

February 16. A woodland vole darts out of a hole in the snow and takes seeds from under a bird feeder. At 6:50 A.M., 32°F, an opossum eats seeds under the bird feeder.

February 17. I hear the year's first mourning dove cooing. Adult stoneflies emerge from along a creek.

February 18. A chipmunk is active at 36°F.

February 19. A chipmunk runs across the road. A blue jay retrieves, pecks open, and eats sunflower seeds, one by one, from bare patches at the base of mountain laurels on a steep rocky slope covered with 6 inches of new snow, while another jay perches quietly nearby for several minutes. Black-capped chickadees and tufted titmice carefully search these bare patches (near my bird feeder).

February 20. A gray squirrel with a wad of dry leaves in its mouth climbs a tree. On an exceptionally warm day, a chipmunk is active, the first this year.

February 21. I hear the first-of-the-season northern cardinal songs.

February 22. Beavers cut trees along the Connecticut River. An adult brown lacewing appears at my lighted kitchen window. A hermit thrush plucks and eats several winterberry fruits beside a small stream.

February 23. In record warmth, winter crane flies, woollybears, honeybees*, and dot-and-dash swordgrass moths are all active.

February 24. A male stonefly persistently climbs atop a female near a creek.

February 25. Tulip-tree fruits disperse in strong winds.

February 26. Winter crane flies fly and walk on the snow. Inchworm and owlet moths are outside my lighted kitchen window.

February 27. A short-eared owl flies over a marsh at the mouth of the Connecticut River at sunset.

February 28. Fresh woodchuck tracks are in the snow between two burrows.

February 29. Red maple buds are noticeably swollen.

March

March 1. At 10:42 P.M., 22.8°F, a fat raccoon is under the bird feeder.

March 2. An adult male spotted salamander walks through the woods, at an air temperature of 9°C, following heavy rains this morning.

March 3. 6:44 to 7:20 P.M.: a screech-owl trills incessantly from a single location every 15 seconds or so.

March 4. An eastern chipmunk runs across the road.

March 5. There are 109 fruit "balls" in the top 10 feet of a 65-foot-tall sycamore.

March 6. Winterberry fruits are nearly all gone; those remaining are shriveled.

March 7. A few trees are newly felled by beaver. 8:50 P.M., 41°F: A swarm of 20 inchworm moths and two owlet moths are outside my lighted kitchen window.

March 8. American goldfinches have yellow feathers coming in on the head and throat. 10:40 A.M.: cloudy, 60°F: northern cardinals sing many times. Evening temperature is above 50°F for the first time this year; I hear the season's first spring peeper choruses.

March 9. I remove wood partially frozen to the ground at the edge of the woods and find aggregations of isopods of different sizes, large and small centipedes, dozens of small millipedes, an adult rove beetle, spider eggs, and a gypsy moth* egg mass.

March 10. American robins sing many times. A large flock of red-winged blackbirds arrives. 11:00 A.M.: a hole that I made in an eastern mole tunnel yesterday afternoon is plugged when I check it this morning. Fresh digging is evident outside a woodchuck burrow.

March 11. Several American robins eat fruits of winterberry and holly.

March 12. Alder catkins open and shed pollen.

March 13. Several American robins pluck crab-apple* fruits.

March 14. Many red oak acorns are sprouting and taking root.

March 15. The first eastern phoebe of the year arrives in the yard. Tree swallows fly over coastal fields. A paper wasp is active.

March 16. About 150 to 200 wood frog egg masses are in a vernal pool, all in an area of 20 square feet.

March 17. 7:30 A.M.: an adult raccoon sleeps 50 feet up in an eastern hemlock.

March 18. American robins eat snow and Asian bittersweet* fruits. The first eastern phoebe of the year is in the yard.

March 19. Four wild turkeys fly across my bow as I paddle up the Connecticut River in Haddam.

March 20. Vernal equinox (1993); cedar waxwings eat viburnum fruits.

March 21. Small numbers of common periwinkles* are on a shallow, wide intertidal flat; most are apparently still in deeper (subtidal) water.

March 22. Hepatica flowers are almost open. Leaves of multiflora rose* and honeysuckle* are expanding.

March 23. A pair of piping plovers, recently arrived at a breeding area, feeds along the edge of sandy-cobbly tide pools. Yellow birch seeds are mostly dispersed, but some still remain attached to the conspicuous central spike of each seed cluster. Bald eagles have begun to incubate.

March 24. Compton tortoiseshells and mourning cloaks fly on a sunny day. Dozens of newly arrived robins forage on a lawn. Greenbriar still has some of last year's leaves attached. Sunset: two adult ospreys perch side by side on a nesting platform, to which they have recently added several wooden branches. An adult male red-breasted merganser surfaces with a small flatfish in its bill; two herring gulls swoop down toward the duck, which dives under water, then resurfaces a few seconds later, still holding the fish. The gulls dive at it again, causing the duck to submerge again. It soon resurfaces without the fish, and the gulls depart.

March 25. A dozen harbor seals swim off Hammonasset Beach.

March 26. A giant water bug has eggs attached to its back, swimming in a small pond at night. A vernal pool contains about 200 wood frog egg masses. Male catkins of speckled alder, hazelnut, and aspen are open and releasing pollen. Mourning cloaks, eastern commas, and spring azures fly on a sunny day. Some red maples are beginning to flower. Autumn-olive* has tiny new leaves. An American crow with a stick in its bill flies to a nest under construction.

March 27. Three-square is 1 inch tall along the shore of Connecticut River.

March 28. In a shallow pond, an adult eastern newt is perched, atypically, out of the water, perhaps in an attempt to rid itself of the leech attached to its throat. A male painted turtle courts a female; spotted turtles are mating. Cedar waxwings eat holly fruits.

March 29. A pair of eastern phoebes closely investigate a previous phoebe nest site.

March 30. I search 24 miles of back roads for amphibians on a rainy night, low 50s (°F), and find spotted salamanders (32), marbled salamander (1), red-backed salamander (1), eastern newt (1), American toads (30), bullfrogs (2), green frogs ("many"), pickerel frogs (3), and wood frogs (3). American goldfinch males are acquiring gold feathers.

March 31. Marbled salamander larvae in a vernal pool are 1.7 inches long. Marsh-marigold is flowering. Trailing arbutus has some open flowers.

April

April 1. Two great horned owl nests each contain two nestlings. False hellebore is 2 to 3 inches tall. Wood frog larvae are almost ready to hatch out of jelly covering.

April 2. The year's first red maple flowers (male) are open. Mourning dove eggs hatch. Male alder catkins are drying up and disintegrating. Black cherry leaf buds are open and leaves are beginning to grow.

April 3. A pileated woodpecker excavates wood from a cavity in a dead tree. The season's first eastern phoebe arrives in my yard, after a cold, rainy March.

April 4. Tightly twisted skunk cabbage leaves are 3 to 4 inches tall. Three common oak moths are outside lighted windows at 10 P.M. (50°F). Large numbers of American robins are migrating and feeding on lawns. Exotic plants, including Japanese barberry, autumn-olive, shrubby honeysuckles, and multiflora rose, have leaves whereas nearby native shrubs do not.

April 5. Yellow perch eggs are tangled in pond shallows. Small American beeches with bleached leaves from last year are conspicuous in the forest understory.

April 6. Skunk cabbage leaves are up to 10 inches tall; false hellebore leaves are up to 8 inches tall. American elm is done flowering, and fruits are well developed. In a colder year, American elm flowering is at its peak. A trio of American crows has been flying back and forth, sometimes carrying nest material (sticks, clumps of moss).

April 7. Many long, pleated egg masses of yellow perch are in pond shallows. Sunny, 60°F, noon: I see the first activity of carpenter bees this year; the bees investigate holes in the wood eaves of my house. A mourning dove nest in a small hemlock tree has one egg in it. Noon: Dozens of wood frogs call vigorously from several vernal pools; some other pools without calling frogs contain fresh wood frog egg masses, plus spotted salamander egg masses and many spermatophores.

April 8. Six red fox pups emerge from their den on a rocky forested slope. This is the warmest day of the year so far (low 70s °F), and I see the year's first cabbage white* in flight. Spring azures are abundant along dirt paths in deciduous forest. A tree swallow repeatedly lands on the lip of the opening of a wood duck nest box in a pond. Spicebush is beginning to flower in large numbers; not peak yet. Hazelnut male catkins are drying up. Reddish speckled dart (a moth) and an Ophionini wasp (Hymenoptera: Ichneumonidae) perch outside a lighted window at night.

April 9. Several spring azures and the year's first bumblebee fly on a sunny day. A blue jay lands in small tree, vocalizes a few times (three-part squeaks accompanied by vertical head bobbing), looks down at the ground, flies down to the ground, uses bill to toss three leaves aside, extracts an acorn from where the leaves had been, flies up and perches briefly in a tree, then flies off with the acorn. All this takes about 15 seconds as cold rain turns to sleet. A large mat of trailing arbutus has dozens of flowers, all fully open, with light snow cover. Today is the first really warm day of spring, mid-70s (°F), after only mid-40s yesterday. A lethargic carpenter bee is on my wood deck; probably it fell there from a hibernation site under the eaves of the house.

April 10. The woods resound with the thumpings of ruffed grouse. At least thirty adult eastern newts move up a narrow stream, climb a small dam, and enter a small pond. Common oak moths are outside my lighted kitchen window (41°F).

April 11. American tree sparrows are no longer at the feeder. I see 200 harbor seals in the area between Fishers Island, New York, and the Connecticut coastline.

April 12. A ribbon snake eats two spring peepers. I hear a loud evening chorus of many spring peepers, at an air temperature 35°F, but the water is warmer. Bigtooth aspen is still flowering. In 2003 (a cold March and April), acorns of red oak and scarlet oak are beginning to sprout; a patch of trailing arbutus that flowered in late March last year has not yet flowered; spicebush flower buds are still closed; hazelnut catkins are just starting to release pollen; and American elm is still flowering.

April 13. American goldfinch males are all or nearly all yellow. Today I see the year's first active bumblebees and a cabbage white* (butterfly). Black flies are becoming numerous. 3:44 P.M., 73°F: I see the year's first carpenter bee in flight. An American crow is on its newly built nest. Quaking aspen fruits are beginning to develop.

April 14. Wild lily-of-the-valley leaves are 1 to 2 inches tall. Along the Connecticut River, silver maple flowers are drying up, cottonwood catkins are beginning to extend. Four bald eagles (three adults and one immature) perch in trees along the Connecticut River at Keeney Cove.

April 15. A European starling* nest contains five eggs. An American crow flies with a stick in its bill. Timber rattlesnakes begin to emerge from dens. Common periwinkles* are not yet present on intertidal cobble-gravel flats (still in subtidal waters), but Asian shore crabs* are abundant under upper intertidal rocks; cold weather has dominated spring thus far. Alder catkins are drying up. Spicebush is beginning to flower. Skunk cabbage leaves are about 6 inches tall. Wintergreen leaves are still maroon. 9:43 P.M., 61°F: after a hot day, more than a hundred owlet moths appear outside my lighted kitchen window. Bumblebees—the season's first—visit red maple pollen flowers.

April 16. Many scarlet oak acorns are on the ground under trees (crop of 1993); black cherry, highbush blueberry, and autumn-olive* leaf buds are recently opened. With warm rains last night and this morning, American toads sing and lay eggs this afternoon in a large puddle in a dirt road on a deciduous forest hillside. Long gelatinous strands extrude from cedar-apple rust growths on red-cedar after warm rains.

April 17. The year's first bumblebees are flying. A newly built paper wasp nest is 1 inch in diameter. A dragonfly is on the wing. Sugar maples are flowering profusely, just before sassafras flower buds open.

April 18. Large gelatinous masses laid by yellow perch in lake shallows have well-developed embryos and green algae associated with the jelly.

April 19. A ribbon snake eats a spring peeper. Two large male snapping turtles wrestle and bite each other in pond shallows. In a flooded swamp, a brown creeper carries nest material to a dead tree with loose bark, and a great blue heron places a stick on a nearly completed nest in the top of a dead white pine tree. Bigtooth aspen has fuzzy seed catkins.

April 20. Black cherries have the largest leaves of any deciduous tree, about 1 inch long, but in another year the leaves are still in the buds on most trees. Eastern tent caterpillars (0.4 inches long) occupy small silk tents in pin cherry. American hornbean and eastern hop-hornbeam catkins are not yet releasing pollen.

April 21. A nest box contains two adult female white-footed mice (one apparently pregnant) plus three babies. A raccoon forages along the edge of a vernal pool on a dark cloudy morning (9:00 A.M.). Garlic mustard*, an invasive exotic plant, flowers abundantly along roadsides. Woodland sedge is abundantly flowering in oak woods.

April 22. A ruby-crowned kinglet catches flying mayflies. Migrating black-and-white warblers arrive in a deciduous forest. A group of twenty-two cormorants circle high overhead above an inland hill. Tulip-trees have small leaves. Every three or four minutes for over an hour, in a light drizzle, a male mourning dove delivers a small stick to a female in a nearly completed nest; the female stays in the nest and arranges the sticks. On the summit of a trap-rock ridge, rock-cresses are flowering and many falcate orange-tips fly and sip nectar on the rock cresses. Other flowers in abundance include woodland sedge, Dutchman's breeches, bloodroot, hepatica, and trout lily. Wild columbine flower buds are not yet open. A six-spotted tiger beetle is active on the summit of a trap-rock ridge. American toads trill nonstop all day long. It's cold, cloudy, rainy, and 43°F, but a carpenter bee is flying.

April 23. 4:49 P.M.: a mourning dove sits on the nest completed yesterday; its mate arrives at the nest, whereupon the sitting dove gets off and departs while the mate settles onto the nest. Under barn owl roosting areas in a building at the New Haven landfill Steve Broker and I find remains of pigeons* (live ones are sitting on eggs), starlings*, red-winged blackbirds, meadow voles, opossums, and Norway rats*. American beech leaves are beginning to unfurl, and many bud scales fall to the ground. A cottontail is hit by a car and killed on a road on Easter. Quaking aspens and highbush blueberry have small leaves. Multiflora rose* is well leafed out, and autumn-olive* has small leaves. Newly fallen pollen flowers litter the ground beneath red maple trees.

April 24. White-throated sparrow, field sparrow, and dark-eyed junco are no longer at the bird feeder. Silver maple leaves begin to grow. Dogbane is still releasing seeds from pods produced last year. A pipelike mud tube (one of three together) contains two 0.8-inch-long wasp pupae. A house finch* nest in a red-cedar contains at least two downy young. I recapture an adult female spotted turtle that I first captured seven years ago; it grew less than 0.1 inches during this period. Three broad-winged hawks are close together in woods at 6:00 P.M., the first arrivals of the year. A woodchuck gathers leaves and carries them in its mouth to its burrow on a rocky hillside.

April 25. The first chimney swifts of the season arrive. Skunk cabbage leaves are 1 foot tall. A beech tree still has many of last year's leaves attached. Pickerelweed leaves are above the water and unfurling. A ribbon snake eats an adult eastern newt. Lowbush blueberry, highbush blueberry, and small black birches have small leaves. A Juvenal's duskywing is flying—first of the year. A newly emerged adult mayfly perches along a creek.

April 26. Black flies are numerous in the woods. A pregnant female ribbon snake that I catch today was pregnant and 4.3 inches shorter when I caught and marked her three years ago. Noon: mixed snow and rain fall at 37°F. Male carpenter bees patrol nest sites in a wooden building and periodically chase each other.

April 27. A ribbon snake eats an adult spring peeper. Canada geese have small downy young. Attracted by a loud squealing call, I see a woodchuck chase another that is limping and has a bloody rump. Shadbush is abundantly flowering. Foliage development this week has given deciduous forests a strong green color.

April 28. A northern water snake eats an adult American toad. I bushwhack into a greenbriar-clogged swamp and find highbush blueberry already flowering. Some tulip-trees have 1-inch wide leaves. Black birch branches sag with heavy catkins. Numerous adult mayflies are active along a swampy stream. Brown elfins and spring azures, and brown elfins and Juvenal's duskywings, one of each species, engage in close aerial "combat" and brief chasing on a rocky hilltop. Flowering dogwoods on warm slopes have greenish bracts open.

April 29. A tiger swallowtail flies in a forest clearing. Chipping sparrow singing is increasing; now many are singing often. Swarms of Juvenal's duskywings sip moisture from damp soil. Most tree species in southwestern Connecticut now have small rapidly growing leaves; leaves are less developed in central and northern Connecticut. In a cool year, flowering sugar maples are now highly conspicuous, as are Norway maples that began flowering earlier. Some yellow birches have pollen catkins fully extended. A tent caterpillar nest in a black cherry with small leaves contains many half-inch-long caterpillars in it. On a balmy night, several lappet moths appear outside my lighted window. A vernal pool contains thousands of half-inch-long wood frog larvae, spotted salamander egg masses with well-developed embryos, and a pair of mallards. Many caddisflies fly low next to a small swampy stream.

April 30. American beeches shed large numbers of bud scales. A ruffed grouse nest contains at least six eggs. The year's first ovenbird is silent, perching and walking on a horizontal branch. Migrant songbirds that arrived this week in my yard included great crested flycatcher, yellow-throated vireo, warbling vireo, black-throated green warbler, black-and-white warbler, Louisiana waterthrush, hooded warbler, and worm-eating warbler. In a swamp, similar-sounding gray treefrog and red-bellied woodpecker alternate vocalizations.

May

May 1. A wood thrush, individually recognizable by its unique song, arrives on the exact same nesting territory it used last year. An adult opossum carries five large young on her back. A northern water snake eats a green frog. Lowbush blueberry flowers abundantly on a west-facing slope. Coyote walks through rocky woods at 11:45 A.M. An adult male-female pair of racers has been coiled together in the same location each afternoon for almost a week now. Recently arrived common green darners (migratory dragonflies) patrol a few feet above open marsh waters. Among the multi-species swarm of moths, midges, and mayflies outside my lighted window this balmy evening is a mating pair of arched hooktips.

May 2. The seasons' first chimney swifts arrive. As I arrive in my study area I witness someone rush forward and pummel something over and over with his heavy walking stick; I hurry ahead and discover that he has smashed the skull and killed one of the two courting racers I have been observing over the past week. Later I find another pair of racers coiled together—luckily they escape his notice. Many pin cherries are flowering.

May 3. Wood thrushes have arrived and are singing in the morning but not later in the day. I receive my first black fly bite of the season. Newly arrived migratory birds today include great crested flycatcher and chimney swifts. About 50 adult male American toads aggregate at a breeding pond, many of them frequently sing in shallow water at the edge, others swim in deeper water and sometimes grab each other. The year's first bullfrog calls begin at dusk in a large deep marsh on a warm sunny day. Many jack-in-the-pulpits are fully developed and flowering, other are less developed. Japanese barberry* is flowering. Silver maple is well leafed out and has large fruits. Three pairs of European starlings* feed vocalizing young in holes in a single silver maple tree and carry away the nestlings' fecal sacs. Red-berried elder is flowering. Tent caterpillars on black cherry are 1 to 1.5 inches long, and many saplings hosting large numbers of caterpillars are completely defoliated. Falcate orangetip, spring azure, painted lady (ragged wings), mourning cloak, eastern tiger swallowtail, and a tiny male black swallowtail fly on the summit of a trap-rock ridge. Eastern phoebe nest contains three eggs. Large numbers of mayflies emerge and fly along a semi-perennial stream.

May 4. A ribbon snake eats a spring peeper. Baltimore orioles have arrived and are singing vigorously. Black chokeberry is flowering on a rocky hilltop. Mountain laurel has new, small sticky leaves. An American toad (1.5 inches snout-to-vent length, young of 2000) forages on a rocky forest slope after three consecutive dry 90°F days. A bumblebee visits flowers of several Japanese honeysuckle* plants.

May 5. A racer eats a gray treefrog. Beech leaves are about 1.5 inches long. Oaks begin releasing massive amounts of pollen today. Beach plums profusely flower. Green ash has small leaves or none at all; flowers are present but not yet releasing pollen. Fruits of silver maple and American elm are fully formed; elm leaves are just beginning to unfurl. Canada goose incubates six eggs among leatherleaf shrubs on an island in a pond. Leatherleaf is profusely flowering, as is shadbush. Bumblebees visit leatherleaf flowers. Pickerel frogs are calling in a high-elevation pond, where yellow water-lily is beginning to flower. Witch-hobble is flowering.

May 6. A beaver felled a large cottonwood tree sometime during the last few days. Striped maple and rhodora are flowering. Trailing arbutus is nearly done flowering. Half-submerged spotted salamander egg masses in shallow water at the edge of a bog have well-developed embryos. Pickerel frog larvae are hatching in a low elevation pond.

May 7. A bumblebee in the shade shivers and pumps abdomen, then flies, at 42.8°F (6:41 A.M.). Male falcate orangetip butterflies are active this morning, with the females not flying until later in the morning when weather is warmer. Sycamores have small leaves and some fruit balls from last year still attached. American beech, red oak, and black oak have small leaves and are flowering. Many black birch catkins fall from trees—it's the end of the flowering season. Yellow birch catkins are fully extended and leaf buds begin to open at the ends of the branches. A belted kingfisher excavates a nesting burrow in vertical dirt bank.

May 8. Canada geese have downy young. American beeches shed large numbers of scales that covered the leaf buds. Spring has been warm and sunny, and forest trees are well leafed out. Piles of fallen oak flowers start to build up on the ground.

May 9. A Carolina wren carries food to young in its nest. A ribbon snake eats an adult eastern newt; two others each eat a spring peeper. Yellow birch has catkins fully extended and small leaves. A male rose-breasted grosbeak begins persistent singing in breeding territory. 11:50 P.M., 53°F: a tulip-tree beauty flutters outside my lighted window.

May 10. Seaside goldenrod leaves are several inches long, and sea-rocket is a few inches tall. An adult American woodcock with three downy young is at the edge of a deciduous forest; all sit down and "freeze" when I approach; after a minute, the adult flies weakly away and the young walk away vocalizing, with the wings (flight feathers still growing) held straight up. Huge numbers of dry cones are on the ground under large white pines (2001). Hilltop oaks and other trees have dead and dying new leaves, an apparent result of the strong drought underway. A common green darner flies vertically upward 20 feet and captures a mayfly. Some American beeches have many flowers. Pond-side highbush blueberry has many prime flowers but flowers are drying up and falling from highbush blueberries on a dry hilltop. Wild sarsaparilla begins to flower. Mountain-holly is flowering.

May 11. Piping plovers incubate eggs. Many star-flower plants have flowers. All highbush and lowbush blueberries have fully developed flowers. Wild lily-of-the-valley begins to flower. Many adult mayflies fly high above a small creek.

May 12. Baby red foxes are outside their natal den. A ribbon snake eats an adult eastern newt. Pink azalea and red baneberry are flowering. Oaks and beeches are flowering, with leaves half to nearly fully grown and still soft and delicate. Flowering dogwoods are all fully flowering. Fertile fronds of royal fern and cinnamon fern are fully grown but not mature. An adult fishfly and several adult caddisflies fly outside a lighted window at 9:30 P.M., 65°F.

May 13. A common garter snake eats an adult spotted salamander. Chokeberry is beginning to flower. Wild lily-of-the-valley is flowering abundantly. Greenbriar still has last year's fruits attached. Pink azalea and autumn-olive* are at their flowering peaks. Maple-leaved viburnum has tightly closed flower buds. Fertile fronds of cinnamon fern are full size. Noon, sunny, 65°F: an adult ringneck snake is extended and immobile on a dirt road in a shady forest. Two adult Canada geese and eight small downy young walk down a forest road to a reservoir ¼ mile away.

May 14. The wings of a newly emerged eastern tiger swallowtail go from crumpled to fully expanded in 10 to 15 minutes. Tidal flat shorebirds at midday low tide, in decreasing order of abundance, include dunlin, least sandpiper, semipalmated plover, greater yellowlegs, black-bellied plover, willet, sanderling, semipalmated sandpiper, piping plover, and short-billed dowitcher. American beech flowers fall from trees in large numbers. Nannyberry is flowering.

May 15. I count 128 eastern tent caterpillars in a tent in a small, mostly defoliated cherry tree. Bumblebees puncture the base of pink azalea flowers; small beetles and flies take nectar through holes made by the bees. Vernal pools contain many wood frog larvae. Norway maple* fruits have fully formed wings but minimally developed seeds. Black cherry and black locust* are beginning to flower. Many mature red maple fruits are on the ground, as are fewer fruits of sugar maple and silver maple. White mulberry* is flowering.

May 16. Lady's-slippers are flowering abundantly. A veery nest contains three eggs. The ground is covered with fallen oak flowers. Spherical, dark-spotted, small to fully grown oak apple galls, caused by cynipid wasps, are present on oak leaves.

May 17. I watch the moon through a spotting scope for 30 minutes tonight and see five birds fly past, going north, northeast, and east. An eastern towhee nest contains four eggs. Morrow's honeysuckle* is near its flowering peak. The drought (2001) continues; red oaks and tulip-trees are dropping new, dry, partly grown leaves; many small to medium-sized trees, including pin oaks, white oaks, hickories, and others, have many small, new, dead leaves, and on some trees all the new leaves are dead.

May 18. I watch the moon through a spotting scope for 17 minutes tonight, with no wind evident, and see 20 birds fly past, almost all going slightly north of east, though a few fly east, west, or north. Red maple fruits helicopter to the ground. Many old, weathered wild cucumber fruits, hollowed and fibrous inside, remain attached to leafless vines beside a stream. Butternut has 3-inch-long leaves and its flowers, in 2-inch catkins, are just beginning to release pollen; attached at the base of the catkins are silken cocoons completely covered with hundreds of caterpillar droppings.

May 19. European starlings* carry food to a nest. Newly emerged adult mayflies, stoneflies, and caddisflies are present along an intermittent creek. Black locust* is abundantly flowering. Black cherries in inland areas are flowering abundantly, but flower buds of coastal black cherries are still tightly closed.

May 20. Cecropia moths mate and lay eggs. White waterlily recently began flowering. Male bayberry flowers release pollen. Spore-bearing fronds of interrupted fern and cinnamon fern are fully grown. Two spotted turtles, a green frog, and a school of mummichogs share a stream pool in the edge of a coastal forest next to a salt marsh. In coastal forest, highbush blueberry, sassafras, chokeberry, and pink azalea are flowering abundantly, and glossy buckthorn* is just beginning to flower. Some black cherry leaves have many long, slender, mite-caused galls. A least tern plunges 20 feet into a salt marsh stream-pool and comes up with a sheepshead minnow. 11:10 P.M., 51°F: a male luna moth perches outside my lighted kitchen window.

May 21. Hundreds of 5-inch-long eels swim into a stream mouth from Long Island Sound. Leatherleaf is in flower. Black locust* flowers are alive with bees. 10:30 P.M., 57°F: an azalea sphinx moth is outside my lighted kitchen window, where the luna was last night.

May 22. An adult marbled salamander defecates the remains of beetles. Swamp azalea is flowering. Steady rains bring a respite from the 2001 drought.

May 23. A cat captures a live, baby cottontail at dawn. Gray treefrogs call vigorously on a balmy evening.

May 24. A sedge wren sings in tall dry grass of wet meadow. A sora flushes off its nest that contains 10 eggs. An adult starling* feeds vocalizing young in a nest box. A racer eats a hatchling painted turtle. A ribbon snake eats an adult eastern newt. A female mallard is accompanied by nine downy young.

May 25. A nest in a red-cedar contains large, spot-breasted juvenile American robins.

May 26. Piping plover adults have downy young. Beach pea is beginning to flower. More than a thousand winged termites emerge from a spruce stump. They mass on top of the stump and fly weakly and slowly up and away. The emergence lasts about half an hour and then none are to be seen at the stump.

May 27. Many gray treefrogs sing this evening. Many turtle nests next to a pond are freshly dug up, with broken shells scattered.

May 28. A sea lamprey swims up Higganum Creek. A 1.6-inch-long giant click beetle (eyed elator) walks near a stump pile. An American toad trills several times, well after the main breeding season. Black locust* flowers fall like snow from a large tree.

May 29. Blue flag and yellow iris* are in flower. Cottonwoods seeds are dispersing.

May 30. Canada goose families with downy young feed on three-square on a tidal flat along the Connecticut River. Black cherry flower petals rain to the ground. The season's first monarch arrives. Worker bumblebees have emerged.

May 31. Four river otters swim around my canoe, looking at me and snorting.

June

June 1. Pitcher-plant has red flowers.

June 2. Black flies swarm around my head in the Northwest Highlands.

June 3. Adult female ruffed grouse fans tail, hisses, and whines at me. Baltimore checkerspot caterpillars eat plantain foliage, and some are pupating.

June 4. Mountain laurel begins to flower.

June 5. I see my first clearwing sphinx (moth) of the season.

June 6. Blueberry and huckleberry have abundant green fruits.

June 7. An adult female ruffed grouse charges me, hissing, with its young nearby.

June 8. On an unusually warm night, thousands of midges swarm outside lighted windows above a creek. In mid-afternoon, a nest box contains four gray juvenile white-footed mice, huddled side by side.

June 9. A ribbon snake eats a juvenile pickerel frog. An American woodcock adult is accompanied by at least three downy young; adult flies off then quickly returns to do repeated, conspicuous double-wing flapping while walking away from young, occasionally flying a few yards to another nearby location; all the while the young sit still on the ground at my feet.

June 10. Mudsnails are extremely abundant on intertidal flats. Five adult ringneck snakes (11 to 14 inches long) are under a single piece of wood.

June 11. I smear a liquid fruit-sugar mixture on trees at the edge of the woods; at night, I find click beetles, smaller beetles, a crane fly, isopods, and a smallish moth on the concoction. A pond contains a fresh green frog egg mass and several newly metamorphosed young.

June 12. 11:15 A.M.: A large black jumping spider stalks butterflies visiting vetch flowers. 11:50 A.M.: spider captures a skipper. 12:30 P.M.: spider is again stalking butterflies. 4:11 P.M.: spider captures a European skipper. Tulip-tree is flowering; ground is strewn with the tulip-tree flower petals after a violent thunderstorm. A tulip-tree beauty (moth) perches on the side of my house near a large tulip-tree.

June 13. Mountain laurel reaches peak in flowering, with extremely prolific flowering in 2000, which is dry in early spring and wet in mid- to late spring; even some plants less than 6 inches tall have dozens of flowers. In 2001, very few mountain laurels produce flowers. Maple-leaved viburnum is flowering. Blueberries and huckleberries have green fruits, plus a few ripe highbush blueberry fruits. Female flower clusters of *Carex intumescens* are fully developed. Gray treefrogs call vigorously, after recent rains.

June 14. An eastern towhee nest contains four eggs. Fruits of black birch and sweet-fern are green but fully formed. In a shrub swamp, some fetter-bush flowers have large holes at the base of the fused petals, perhaps the result of bumblebees puncturing to obtain nectar. Pignut hickory leaflets are heavily infested with galls.

June 15. Hundreds of newly metamorphosed American toads hop along the edge of a pond. 6:20 P.M.: a large adult box turtle, at least 21 years old (based on scute layers), pauses on a paved road at the junction of woods and a field.

June 16. An adult female wood duck with at least six downy young is in woods far from water; adult "rows" across dirt road with both wings as the young run along with her. False indigo* is flowering abundantly. Shadbush fruits are ripe.

June 17. Witch-hazel fruits are about 50 percent of mature size. Numerous small wasps and bees are on abundant holly flowers. Winterberry fruits are about 1 millimeter in diameter. Blue-eyed grass is flowering. One "tall meadow-rue" plant is 7 feet tall. Large wood frog larvae surface often and quickly in a vernal pool.

June 18. A Baltimore oriole fledgling is out of the nest with its parent. For the second time this week, two shrieking blue jays roust an adult broad-winged hawk out of tall, thick shrubbery where the jays are feeding nestlings. Several times today and over the past few days a jay flies out of the nest area, lands, and emits an imitation of the broad-winged hawk call while bobbing the head up and down, but no hawk is visible to me. Under a forest log, one of two newly metamorphosed green frogs has a small leech attached to its side.

June 19. A ribbon snake eats an adult eastern newt and newly metamorphosed green frog. Partridge-berry is flowering. Common elder is beginning to flower. New white flowering stalks of Indian pipe push up through forest floor leaf litter.

June 20. White mulberries* begin to ripen. Staghorn sumac flowers begin to open; the same plants still have last year's fruit clusters attached. A tiny pool of water in a 1-inch by 2-inch by 2-inch-deep hole in an old red maple contains dozens of mosquito larvae and pupae.

June 21. Timber rattlesnakes are beginning to shed their skins. Several baby white-footed mice suckle at their mother's nipples.

June 22. A recently born white-tailed deer lies briefly in the open as my truck approaches on a dirt road.

June 23. An adult female Isabella tiger moth is active (larva is the "woollybear"). Staghorn sumac flowers attract honeybees*, bumblebees, carpenter bees, and syrphid flies. Smooth sumac flowers are not yet open. A downy woodpecker feeds loudly vocalizing young in a cavity in a red maple snag. Ringneck snakes hatched last year are only 5 inches long.

June 24. Wild blueberries and huckleberries begin to ripen. Deer flies are abundant and aggressive in the forest. Among the most observable invertebrates in a coastal lagoon behind a sand spit are comb jellies, moon jellies, grass shrimp, green crabs*, and horseshoe crabs. Fish fry are tremendously abundant among algae and adjacent to salt marshes. A single spotted fawn accompanies an adult female deer.

June 25. A freshly emerged mourning cloak butterfly takes flight. 9:25 to 10:41 A.M.: a spotted turtle lays four eggs in a hole and buries them; 7 to 14 minutes between eggs. A hognose snake eats an adult American toad. A dragonfly darts among a large concentrated swarm of midges for several minutes, picking off one every few seconds, 10 to 20 feet above ground.

June 26. 8:00 A.M., cloudy: two large juvenile river otters lope along side by side through mature forest, heading in the direction of a reservoir about ¼ mile away. Newly metamorphosed wood frogs (0.6 inches snout to vent length) hop on damp ground next to a shrub swamp. A scorpionfly is flying.

June 27. Milkweeds are covered with many red, adult milkweed beetles. Three black-legged tick nymphs are attached to me! Swarms of banded hairstreak butterflies fly rapidly around and perch on the trunks of large oak trees. Hundreds of large gypsy moth* caterpillars are dead and dying on the trunks of large oak and beech trees.

June 28. Common terns have fresh eggs, hatchlings, and large young just beginning to fly. Roseate tern nests have eggs and hatchlings. An American lady caterpillar eats dusty miller (*Artemisia*) foliage. George Zepko and I band six barn owl nestlings in each of two nests. A southern flying squirrel is eaten by an unknown predator along a forest trail, only the tail and some viscera remain.

June 29. A common garter snake eats an earthworm. Clusters of salt-marsh bulrush flowers are conspicuous along tidal channels at the high marsh-low marsh junction. Pools in high salt marsh contain 0.4-inch-long mummichogs. Common elder is the most conspicuous flowering shrub along roadsides. Several newly metamorphosed American toads are active on a forest floor. A water-lily leaf beetle flies from a lily pad.

June 30. Northern leopard frog tadpoles are about 2.5 inches long, with no externally evident front limbs yet. Two

2-inch-long cerambycid (long-horned) beetle larvae (round-headed borers, "timber worms") are in rotted wood at the base of a large dead *Viburnum* stump. Heavily shaded silky dogwood is beginning to flower, while plants in open sun are nearly done flowering. Winterberry flowers attract many insects.

July

July 1. An adult ribbon snake eats a newly metamorphosed green frog. A female wood duck accompanied by eight ducklings swims into the cover of pickerelweed. White mulberry* fruits are ripening in large numbers and are eaten by gray catbirds, American robins, cedar waxwings, and gray squirrels. A snapping turtle feeds on a mute swan* carcass in a brackish marsh.

July 2. Five ringneck snakes are together under a rock. A tiger swallowtail sips nectar from a swamp azalea flower. At sunset on a cool evening, several green frogs call continuously in a shaded pond.

July 3. A killdeer nest contains three eggs. Adult sea lampreys swim in the Salmon River. A snowy egret captures and eats a large polychaete worm in a sandy tidepool. Two downy juveniles accompany two adult piping plovers. A gray catbird and house finches* eat highbush blueberry fruits.

July 4. Natural fireworks at dusk (8:50 P.M., humid, 70°F), as dozens of firefly beetles flash while flying and perching in my yard. A vernal pool used by breeding wood frogs is dry; young wood frogs are active nearby. Pipewort is beginning to flower. Mountain laurel flowers are now nearly all gone and fruits are well developed. At dusk over the past week, a Louisiana waterthrush was been perching near the top of a tall sycamore and singing briefly but loudly. This is followed by an individually recognizable song of a wood thrush that sings until dusk.

July 5. Spider wasps carry spiders into burrows in the sand. An adult long-horned beetle (1.5 inches long) walks in deciduous forest at sunset. Newly metamorphosed spring peepers emerge from ponds, while others are still tadpoles. A female gray squirrel builds a new nest in a large sugar maple, using small leafy branches bitten off mainly within 15 feet of the nest site.

July 6. Least terns feed large young, some able to fly. Osprey nests contain large young. Purple martins feed young in nest house. Juvenile barn swallows are flying. On the mud banks of a brackish marsh, mudwort is flowering, and eastern lilaeopsis flower buds are not yet open.

July 7. A large flock of juvenile starlings* forages on a lawn. Ailanthus webworm moth* caterpillars are on ailanthus* foliage. A painted turtle lays eggs. Beaked hazelnut fruits are fully formed but not mature. Bumblebees visit American chestnut flowers. Swamp azalea and northern bush-honeysuckle are flowering. At 7:45 P.M., several common terns repeatedly dive into the water just a few feet from shore, capturing and immediately swallowing some of the thousands of juvenile fishes swarming along the shore of Long Island Sound. Least terns also plunge into the water, but most of their dives are 50 feet from shore. 10:45 A.M., sunny, 70°F: an old female box turtle (plastron length 5.5 inches), with multi-cracked shell worn smooth, crosses a forest road; she's probably much older than I am! Nine small juvenile wild turkeys, closely attended by three adult females, forage in a field along a forest edge. A coyote pup (about 35 inches in total length) is freshly dead on the road.

July 8. Water temperature in 1-inch-deep water at the edge of Salmon Cove is 100°F; no insects, amphibians, or fishes are present here. A gray treefrog jumps from my hand and lands on and clings to a horizontal string 3 to 4 feet away—impressive maneuver! Male mute swans* aggressively defend their families (females with large downy young) from intruding boaters. Swamp rose is abundantly flowering. Shallow tidal freshwaters of the Connecticut River teem with juvenile fishes. Arrowwood has small green fruits.

July 9. Grass-pink and rose pogonia orchids are flowering. My lighted office window attracts a fishfly adult. Seven Japanese beetles* feed on a single sassafras leaf. Arrowwood and maple-leaved viburnum have small green fruits. I hear the year's first dog-day cicada buzz. Butternut fruits are near full size.

July 10. An eastern kingbird sits on its pond-edge nest. Downy young killdeer are in a field with an adult. Spotted wintergreen is flowering in large numbers. Buttonbush is flowering. Groups of great spangled fritillaries visit milkweed flowers. Tiny beetles mass on meadow-sweet flowers. Wood thrushes are still vigorously singing. Deer flies are numerous and pesky.

July 11. Hazelnut fruits are well formed but not mature. Basswood is flowering. Green ash fruits are full size. Juvenile spottail shiners, about 1 inch long, are abundant in the Connecticut River. A black skimmer flys low over the water next to a salt marsh, with its lower bill slicing through the water.

July 12. Blue jays shriek at something up in the tall crabapple tree. The jay ruckus continues for over an hour, with one short break. Turns out the jays are mobbing an adult eastern rat snake draped over some branches. Air temperature is 65°F, after a relatively cool night. Evidently the snake spent the night there. It is gone at 11:00 A.M. Downy rattlesnake-plantain has well-developed flowering stalks, but flowers are not yet open. A few wild sarsaparilla plants have ripe, purple-black fruits.

July 13. A nest box contains 4 dead downy wood duck chicks and 16 unhatched (dead) eggs. White mulberry* is loaded with ripe fruit: goldfinches and rose-breasted grosbeaks nibble attached fruits; Baltimore orioles nibble attached fruits or pluck fruit, hold it against a branch with their toes, and nibble it; catbirds, juvenile and adult robins, and red-bellied woodpeckers pluck purple fruits and swallow them whole; gray squirrels nibble attached or plucked fruits, discarding most; chipmunks nibble fruits that have fallen to the ground. In a small stream pool, orange-tinted male black-nosed dace relentlessly chase and rub against females.

July 14. A marshy flood pool contains dense schools of juvenile brown bullheads. Many juvenile Fowler's toads, young of this year, forage along a lakeshore. A northern water snake eats a 6-inch redfin pickerel.

July 15. A northern water snake eats an adult green frog. Four Carolina wrens fledge from a nest in a hanging plant outside my kitchen window. A giant root borer bettle crawls across the lawn. An adult fishfly appears outside my lighted window.

July 16. Black-capped chickadees eat insects from clusters of dry, curled leaves in a small tree. The stomach of an adult male copperhead that someone killed contains an adult meadow vole. A male rose-breasted grosbeak eats part of an orange fruit of climbing nightshade*.

July 16–17. A cat goes on a rampage, captures and brings into the house six meadow voles this night.

July 17. I hear the season's first cicada buzzing. I count 71 diamondback terrapins along a short section of tidal creek. A young mink playfully and repeatedly enters and exits the water at the edge of a marsh. A newly metamorphosed wood frog (0.6 inches snout to vent length) hops across a forest path at 11:00 A.M.

July 18. Grass-pink and rose pogonia orchids are flowering. Eastern newt larvae are abundant in a shallow marsh; three larvae are 1.0 to 1.6 inches in total length. A newly metamorphosed American toadlet hops along a forest path. Pignut hickory fruits are full size. Arrow-arum fruiting stalks are pointing downward. Arrowwood fruits are small and green. Abundant Canada geese graze wild rice leaves and shoots, leaving virtually none uncut.

July 19. A ruby-throated hummingbird visits swamp azalea flowers. One square meter (10.8 ft^2) of ground has forty-one antlion burrows. Gray treefrogs are metamorphosing; some are ready to leave a pond.

July 20. Some ostrich ferns have fertile fronds, and some have *Papaipema* (borer moth) larve in the stems. Early goldenrod is flowering. A large butternut fruit is on the ground; squirrel-cut? A huge flight of midges is active along the Connecticut River. In and around a marshy pool in the meadowy floodplain of the Connecticut River, northern leopard frogs are represented by large tadpoles, metamorphosing larvae, and newly metamorphosed juveniles. The first true katydids begin singing this evening (warm summer this year).

July 21. Ailanthus webworm moth* adults are on goldenrod flowers. A recently dried vernal pool contains dead spotted salamander larvae; newly metamorphosed spring peepers are active nearby. A woodchuck climbs into a grape thicket a few feet above the ground.

July 22. I encounter four dead shrews on paths in three hours; cause of death is not evident. Piping plovers just hatched. A buzzing cicada flies vertically away and escapes from a pursuing house sparrow* over a distance of about 30 feet from the ground.

July 23. Bank swallows feed young in a colony in a partially excavated sand/gravel bank one-half mile from the Connecticut River. Small bees visit spotted wintergreen flowers.

July 24. Many gypsy moth* males fly fast and erratically along a forest edge. Hardhack is flowering. American chestnut flowers are dry and falling off the plants. Highbush and lowbush blueberry and black huckleberry have numerous ripe fruits. Goldenrod is beginning to flower abundantly. A male Fowler's toad calls several times from a rain pool beside a coastal sand dune.

July 25. American goldfinches eat timothy grass seeds. Wild rice begins to flower. I hear the year's first dog-day cicada buzz (after an exceptionally cool spring and early summer).

July 26. Young ospreys make their first flight from a nest; others flew a few days ago. Salt-meadow cordgrass and freshwater cordgrass are flowering abundantly, but smooth cordgrass is not yet flowering. Adult antlions flutter outside my lighted kitchen window at 10:00 P.M. (73°F); also an adult fishfly, several species of underwing moths, many other moth species, caddisfly adults, stoneflies, various small beetles, and other insects.

July 27. Many juvenile common terns, capable of flight, perch on a mainland beach, vocalizing loudly as adults fly in to feed them. Very large stoneflies flutter outside my lighted window near the Connecticut River.

July 28. After rain last night and this morning, vast numbers of red efts (eastern newts) walk on the forest floor near ponds this morning. An adult female dusky salamander attends a clutch of thirty-five well-developed embryos, under a rock 6 inches from a stream. Sea-rocket has both flowers and well-developed fruits.

July 29. Wild rice and many other emergent, floating, and submerged freshwater marsh plants are flowering. Fruits of maple-leaved viburnum, spicebush, and beaked hazelnut fruits are fully formed but still green. Water-celery and waterweed *(Elodea)* are flowering in a tidal freshwater marsh. A group of 120 mute swans* is in the central part of Salmon Cove. I hear the season's first oblong-winged katydids and snowy tree crickets. Conehead katydids stridulate in grassy fields.

July 30. Two downy black chicks accompany an adult Virginia rail in a freshwater marsh. American goldfinches eat chicory flowers and fruits. An adult female common merganser is accompanied by fifteen young in the Farmington River.

July 31. A copper underwing moth is active on a warm rainy night. An adult female spotted turtle, at least 18 years old, swims in a vernal pool on the top of a trap-rock ridge. 9:00 P.M., 75°F: I hear the year's first true katydid sounds.

August

August 1. Since June 26, the cat has brought to the house 18 freshly killed meadow voles, which have been abundant in a nearby grassy field. On two plants, American bladdernut pods are numerous and fully developed, but these are the only fruiting individuals evident along three miles of tidal riverbank. Ambush bug adults are active. A mink is active in tidal river shallows, sheltering in a burrow under tree roots in the riverbank.

August 2. A pregnant female ribbon snake was pregnant last year, too. An adult female pickerel frog smells like roasted nuts. Recently emerged Indian pipes include a dense cluster of 40 flowering stalks; two additional dense clusters emerge above the ground just a few inches away.

August 3. A cicada killer repeatedly enters and exits a hole in the ground. An adult redback salamander attends a cluster of 8 embryos attached to the underside of a rock. A common garter snake eats an earthworm.

August 4. Dozens of small beetles are on every sweet pepperbush flower spike. For at least a week, red oaks have been dropping partially developed acorns.

August 5. A ribbon snake eats a juvenile green frog. A ribbon snake that was pregnant on July 22 gave birth before today. A common garter snake eats an earthworm. 10:00 P.M., 74°F: a lighted window attracts a 2-inch-long fishfly adult, plus an underwing moth, at least a dozen additional species of moths, two-tailed mayflies, stoneflies, planthoppers, and a half-dozen beetle species. Several flocks of more than a hundred semipalmated plovers fly and roost along the coast at high tide. Hundreds of common terns perch and catch fish at the mouth of the Connecticut River; adults feed small fishes to their volant, vocal young perched at the shoreline.

August 6. In a drought year, very few blueberry or huckleberry fruits remain. In a wet summer, some arrowwood plants have abundant ripe fruits and others have only green fruits. Fruits of nannyberry are fully grown but green. In a small nontidal river, pondweeds, and fanwort* are flowering, bladderworts are not. A dog-day cicada emits a loud, long buzz at 8:20 P.M. (dusk) on a warm evening.

August 7. 12:44 A.M., air temperature 74°F: the night resounds with a vigorous, continuous chorus of true katydids. In a cool summer, true katydids have not yet begun to chorus. A gray squirrel bites open witch-hazel fruits and eats the seeds. On the ground below the squirrel are dozens of open seed pods, all with the seeds removed.

August 8. A large (3.5-inch) millipede crosses a dirt road on a warm humid day at 10:30 A.M. When disturbed several times from its perch in daytime in oak forest, an underwing moth flies quickly and always lands on tree trunks 6 to 12 feet above ground. Large, full-sized, and small fruits of shagbark hickory begin to appear on the forest floor.

August 9. Fruiting bodies of coral fungus appear. Rattlesnake-plantain is in full flower.

August 10. Sweet pepperbush flowers host abundant small slender beetles, bumblebees, carpenter bees, honeybees*, bald-faced hornets, and yellowjackets. In a cool summer, the year's first true katydids begin chorusing tonight.

August 11. A cluster of freshly formed galls of cynipid wasps is present on the stem of a chestnut oak sapling. Grapevine beetle adults are active. Bumblebees visit shining sumac flowers, while smooth sumac has ripe red fruits. Least, semipalmated, and spotted sandpipers, and semipalmated plovers and ruddy turnstones, forage on an algae-strewn cobble-boulder intertidal shore of Long Island Sound; the least sandpipers focus their attention on amphipods associated with the algae.

August 12. Viscid violet cort mushrooms are common in a deciduous forest. Red milkweed beetle adults are on milkweed plants. I hear the year's first true katydid sounds (cool summer this year); first heard them here on July 20 last year (hot summer).

August 13. Black cherry fruits are ripe. Beach plum fruits are beginning to ripen. Bitternut hickories in a coastal forest are heavily laden with large fruits (2001). 7:45 P.M. (dusk): tens of thousands of swallows (mainly tree swallows) swarm over marshes at the mouth of the Connecticut River and roost in common reed. Many adult blue crabs are in the lower Connecticut River.

August 14. Staghorn sumac and smooth sumac have red, mature fruits; shining sumac is just flowering. For at least 40 minutes this evening, an olive-sided flycatcher perches at the dead top of a large hemlock tree; periodically it flies up to 100 feet or more to catch a fat insect (beetle?) then returns to the treetop to eat it. Tiger bee flies are active near carpenter bee tunnels in a wooden building.

August 15. Spicebush fruits are large but still green. In tidal fresh water, water-celery and waterweed *(Elodea)* flower abundantly and many damselflies fly in tandem low over submerged plants. An adult tiger bee fly perches near a carpenter bee tunnel in a wooden building. With extended hot weather and drought, cottonwoods and aspens are dropping many leaves, and black birch leaves are turning yellow and falling; leaves of many other trees are limp or turning brown and shriveling.

August 16. Chokeberry has large green fruits. A coastal shore is alive with a spectacular concentration of birds; hundreds of noisy common terns mass near the mouth of the Connecticut River, with adults busy catching small fishes and carrying them to their volant young perched on sand bars with groups of cormorants, herring gulls, laughing gulls, black-bellied plovers, greater and lesser yellowlegs, semipalmated sandpipers, willets, semipalmated plovers, and others; in the evening before sunset, approximately 100 great egrets assemble nearby in a salt marsh.

August 17. Arrowheads, pickerelweed, and purple loosestrife* flower abundantly in a freshwater tidal marsh, and rose-mallow is flowering in brackish and freshwater tidal marshes. Animal-gnawed hickory nuts and acorns begin to appear on the ground.

August 18. Beach grass and smooth cordgrass are flowering abundantly. The water surface of tidal freshwater marsh is strewn with fallen male flowers of wild rice; red-winged blackbirds land on and investigate the not-quite-ripe rice heads; yellow flowers of water stargrass project above the water. Perched above the edge of a tidal river, a gray squirrel eats several American hornbeam nutlets, discarding the leafy scales. Carpenter bees and associated tiger bee flies are active near nesting tunnels in a building.

August 19. I flush three less-than-half-grown cottontails from a weed patch in a field. A common garter snake attacks an adult pickerel frog. A gray squirrel is in the top of a beech tree; beneath it, beechnut husks rain to the ground. At dusk, black-crowned night-herons emit "quok" calls as they fly into a salt marsh to forage, and a large flock of semipalmated plovers and fewer semipalmated sandpipers and ruddy turnstones roost in a group on a sandy coastal island at high tide.

August 20. A young white-tailed deer lacks white spots. A gray squirrel repeatedly bites off witch-hazel fruits, opens them, and eats the seeds, taking as few as five seconds to process each fruit.

August 21. Unseasonably cool weather recently; while crawling through a shrub swamp, I look up and see an adult male timber rattlesnake 5 feet above ground, coiled in partial sun on a highbush blueberry branch. Ambush bugs are paired on goldenrod flowers.

August 22. Some maple-leaved viburnum fruits are ripe. At 8:20 P.M., a common nighthawk flies and vocalizes over a brightly lit tennis stadium in New Haven, evidently feeding on light-attracted insects.

August 23. A common garter snake eats a metamorphosed two-lined salamander. Large ants tend aphids on the ends of chestnut oak branches and at a hole in a spiny gall on witch-hazel. An eastern screech-owl wails and trills after an episode of dog barking at dusk (8:00 P.M.).

August 24. Winterberry fruits are large but still green. Large ants tend clumps of woolly alder aphids on speckled alder. Fanwort* and bladderwort flowers are abundant at the surface of a pond.

August 25. A northern water snake that someone killed contains 19 fully formed babies. Hemlocks have many new cones. Glowworms are active and glowing on the forest floor.

August 26. A cat brings into the house a live, uninjured snapping turtle hatchling. A nest box contains an adult white-footed mouse and six gray-brown juveniles. An adult walkingstick perches under house eaves; it is 3 inches in body length and 5.5 inches long including the antennae.

August 27. Thousands of "freshwater jellyfish" medusae swim in a pond. 7:45 P.M., 75°F: at dusk, bats are flying before true katydids begin their nightly serenade.

August 28. A flock of about 90 tree swallows flies low over the water, and individuals repeatedly plunge briefly into it, evidently bathing; they do this when flying in one direction, then fly back from where they had come and repeat the behavior. Several newly emerged American pelecinid wasps are active in deciduous forest. 7:30 P.M., dusk, 72°F: true katydids begin chorusing. Severe winds accompanying a thunderstorm snap off the top 30 feet of a large tulip-tree.

August 29. Grape leaves have long, pointed, pink galls on them, mainly on the upper surface, and low, whitish fuzzy patches on the underside, plus a large swelling on the leaf stalk. Shining sumacs have red, ripe fruits. A crab spider captures a yellowjacket on goldenrod flowers. Ambush bugs are paired on goldenrod flowers.

August 30. Mountain laurels recently have been dropping yellow leaves. Most spicebush fruits are still green, but a few are red or orange. Black chokeberry fruits are black.

August 31. During a wet summer, young-of-the-year wood frogs are numerous on deciduous forest floor. A foggy, misty morning reveals through condensation hundreds of the domed webs of "filmy dome spiders" in the shrubby undergrowth of deciduous forest. Great spangled fritillaries sip nectar from thistle, knapweed, goldenrod, and wild carrot.

September

September 1. A timber rattlesnake gives birth to 10 young; another gives birth to 3 young. A large, laurel sphinx moth caterpillar crawls on a dirt path in deciduous forest.

September 2. Sea-lavender is still flowering. Dodder has many large fruits. Fox grapes are ripening along the edge of a pond. Adult and hatchling stinkpot turtles slowly patrol thickly vegetated lake shallows in the morning sun.

September 3. Floating-heart is still flowering. A ruby-throated hummingbird visits cardinal-flowers along a stream. Water-celery and waterweed are flowering in a freshwater tidal marsh. Eleven double-crested cormorants perch in a mostly dead cottonwood tree on the bank of the Connecticut River. Several recently dead, intact blue crabs are in shallows of the Connecticut River near the Goodspeed Opera House.

September 4. A hatchling hognose snake crawls through a forest opening. Winterberry fruits are still green. Numerous small white moths and water-lily leaf beetles fly up from lily pads as I paddle slowly through pond shallows. Frequent steady rain the past few days restored steady large volume flows to intermittent streams that had not flowed for at least two months.

September 5. Bumblebees visit abundant jewelweed flowers. New clusters of Indian pipe are still emerging from forest leaf litter. Fully developed red oak acorns fall from the trees. Predatory oyster drills are numerous on and among blue mussels, barnacles, and rockweed on an offshore coastal rock exposed at low tide. Along the shoreline, grazing common periwinkles are abundant on rocks that are devoid of mussels and oyster drills and have scant algal cover.

September 6. Two great horned owls each call two to three times per minute. A gray squirrel buries a butternut fruit. Purple- and yellow-flowering bladderworts are common in aquatic beds of large ponds. Vast dense swarms of tiny whirligig beetles occur in patches on a pond surface. I carelessly pick a giant water bug out of my dip-net and receive a painful stinglike jab; I reflexively fling the bug skyward. After two minutes, the smitten finger no longer hurts. An underwing moth flies out of the woods, over a pond, and hits the pond surface, from which it cannot take off; I investigate and find it has very worn wings. On a partially exposed sandy/muddy tidal flat being covered with water with the rising tide, the most common birds are herring gull, great black-backed gull, laughing gull, great egret, snowy egret, great blue heron, semipalmated plover, black-bellied plover, greater yellowlegs, American oystercatcher, semipalmated sandpiper, and willet.

September 7. Gray squirrels eat flowering dogwood fruits (seeds only?). Air temperature last night drops to 43°F, yet some timber rattlesnakes spend the night on the forest floor coiled on top of leaf litter. Just before sunset on a tidal flat exposed at low tide, many burrowed clams spontaneously squirt water several inches into the air. An immature herring gull floating in shallow water just off the beach in Long Island Sound looks down into the water for a few minutes, then plunges its head underwater and catches a green crab*. While floating, the gull repeatedly drops and easily recatches the struggling crab, as if trying to avoid being pinched on the face. The crab eventually escapes. In a few minutes the gull plunges again and has another crab. After several drops and recaptures, the gull swallows the crab whole. Small syrphid flies, bumblebees, paper wasps, and cabbage white* butterflies visit sea-lavender flowers in a salt marsh. Many migratory black saddlebags (dragonflies) fly over coastal salt marshes. In deciduous forest, wood thrushes (migrants?) emit "pip-pip-pip" calls at sundown.

September 8. Giant ichneumon wasps are flying. A female scarlet tanager has two hippoboscid flies on it. Bumblebees visit Indian pipe flowers. 10:00 P.M., calm, 67°F: a true katydid high in an American beech tree, one of many stridulating continuously, emits an average of 57 phrases (each two or three syllables) per minute.

September 9. A female marbled salamander attends 171 "eggs" (jelly-enclosed embryos) under wood in a dry vernal pool basin. A newborn common garter snake eats an earthworm. Leidy's comb jellies and juvenile mummichogs and silversides are trememdously abundant in tidal lagoons. Many floating eelgrass blades have patches of attached *Haliclona* sponge gemmules. Large clumps of *Halichondria* sponges festoon the submerged banks of a tidal channel. A tulip-tree beauty (moth) perches on my deck in mid-afternoon.

September 10. Sanderlings feed on amphipods and small worms along the edges of waves on sandy beaches; some individuals chase others away by running at them and vocalizing with the head lowered. Beach grass is flowering abundantly, as are all of the dominant salt marsh grasses. Calm, 71°F, 9:00 P.M.: a large underwing moth is outside a lighted window. An oak apple gall newly detached from a red oak contains one fat, kidney-shaped, 3 to 4 millimeter larva and five thinner 2 millimeter larvae in the central chamber. 11:00 P.M., 71°F: several female tree crickets and silent male bush katydids prowl screens outside lighted windows.

September 11. Some highbush blueberry plants still have a few ripe fruits. Cicada killer (wasp) adults are active at their nesting burrows. Clusters of pokeweed fruits vary from all ripe to all green.

September 12. Arrowwood fruits are abundant in coastal thickets. The ground is strewn with acorns of chestnut oak, white oak, red oak, black oak, scarlet oak, pin oak, and scrub oak, but many acorns are still attached to the plants.

September 13. A flock of several hundred tree swallows perches on a wire. Spicebush has both red and green fruits. Large numbers of migrating common green darners and black saddlebags (dragonflies) fly over coastal salt marshes.

September 14. Tonight I watch the moon through a telescope and see a heavy migration of birds across its face. Beech-drops is flowering. A yellowjacket nest in the ground was dug out and torn apart last night. A swarm of yellowjackets clings to the side of the nest that remains, but the sections of the nest that contained the larvae are gone, evidently eaten by a skunk. Nearly all winterberry fruits are still green. In a coastal forest, nearly all black gum fruits are ripe. Dense schools of menhaden feed in tidal creeks in a salt marsh. At dusk, groups of snowy egrets fly from marshes at the mouth of the Connecticut River along the coast to roosting areas somwhere to the east.

September 15. A male black-throated blue warbler, accompanied by a female, pierces a blueberry fruit, extracts much of the contents, and discards the skin. Thousands of tree swallows fly over salt marshes, perch on marsh-elders, and swarm into adjacent bayberry bushes and eat the fruits. Seaside goldenrod is flowering profusely and attracting various large and small wasps, many large carpenter bees, honeybees*, monarchs, American ladies, buckeye butterflies, orange sulphurs, and others. Eastern melampus snails climb smooth cordgrass stems as the tide is rising. Dozens of snowy egrets forage in coastal salt marshes. Some Virginia creeper leaves have turned red and so have a few of the fruits. Schools of striped killifishes and sheepshead minnows, plus a few small silversides, move out of the shallows of Long Island Sound and into adjacent marsh pools as the tide rises and floods these areas. Ospreys are conspicuously absent in a coastal ecosystem where they were common earlier in summer. Groundsel-tree is beginning to flower.

September 16. Wild rice is almost ripe. Large, colorful, "marbled spiders" (*Araneus marmoreus*) are common on their vertical orb webs between shrubs in deciduous forest. Spicebush has numerous fruits ranging from green to red. A female marbled salamander rests on top of her eggs under leaf litter next to a log in a damp vernal pool basin; a juvenile green frog is also atop some of her eggs. Ground crickets emit a continuous noonday chorus. Golden leaves of black birch are widely scattered on the ground—the beginning of seasonal leaf-fall. New growth of Indian pipe emerges from the ground.

September 17. A flying firefly beetle flashes at intervals of a few seconds; many larvae glow on the ground. Bumblebee workers visit flowers of white wood aster and silverrod. Recent rains result in the sudden appearance of many mushrooms.

September 18. Acorns of red, scarlet, black, pin, white, and chestnut oaks rain from the trees; huge crop this year (1998). Acorns are scarce this year (2000) in most areas, modest crop of red oak acorns from some trees. Wintergreen fruits are turning from white to pink.

September 19. Count of migrating hawks at Lighthouse Point: sharp-shinned hawk (1,138), American kestrel (726), osprey (443), Cooper's hawk (203), broad-winged hawk (131), northern harrier (122), unidentified (90), merlin (89), peregrine falcon (1). Good acorn crop this year—no flocks of migrating blue jays (poor acorn crop and many flocks of migrating jays last year). Bumblebees visit Indian pipe flowers. 11:00 P.M., 66°F, calm, heavy then light rain: true katydids are silent, but ground crickets trill incessantly. I watch hundreds of red-winged blackbirds and hear several soras in large stands of ripening wild rice in a freshwater tidal marsh.

September 20. A gray treefrog calls on a warm, humid afternoon. Two female marbled salamanders attend their eggs under leaf litter in a dry vernal pool. 9:30 P.M., 69°F, humid: large, gray, dorsally black-spotted and laterally striped slugs aggregate on the trunk of a beech tree; many others crawl on a dry, lichen-covered ledge; small katydids probe their ovipositors into mossy-licheny tree bark. Many chestnut oak acorns fall from trees (2000). Many adult caddisflies are active in woods near a stream, attracted to light at night.

September 21. A black-horned tree cricket lays eggs in a plant stem. 6:45 A.M., 57°F: a group of snowy tree crickets trills synchronously, 17 times in 13 seconds. Hundreds of chimney swifts migrate along the coast on a cool, windy afternoon. No flocks of migrating blue jays—fairly good acorn crop this year (1997).

September 22. Red-winged blackbirds perch on wild rice and eat the grains. 11:25 P.M., 70°F: alerted by repeated, loud, upslurred "ah-uh" calls, in a crevice in a stone wall, I discover an adult common garter snake trying to swallow a vocalizing juvenile green frog, rear-end first.

September 23. Many new white flowering stalks of Indian pipe recently pushed up through forest floor leaf litter. A gray catbird eats small summer grapes. Great spangled fritillaries visit Joe-Pye-weed flowers. A gray treefrog calls at midday. A gray squirrel bites open witch-hazel pods and eats the seeds.

September 24. An eastern phoebe nest in an abandoned cabin has been taken over and roofed (with fiberglass insulation) by a white-footed mouse. Three turkey vultures circle very low near a dead, very odorous gray squirrel that I put in the yard. Many flocks of migrating blue jays are in flight; poor acorn crop this year (1995). Winterberry fruits are red but not fully so. Dodder is still flowering and has large fruits, too.

September 25. I catch pregnant white-footed mice. A newborn ribbon snake eats a newly metamorphosed eastern newt. A northern walkingstick perches on the side of a trash can next to a house in the forest.

September 26. Ground crickets trill continuously in the rain. Tree clubmoss and common running clubmoss have full-grown "cones," but the spores are not yet being released.

September 27. Many newly fallen birch and bigtooth aspen leaves are on the ground. Much of the foliage of black huckleberry and lowbush blueberry in a hilltop oak forest has fallen. At night, many large spotted salamander larvae are in the shallows of a small, permanent, fishless pond; some have nearly completed metamorphosis, other still have large gills.

September 28. Poor acorn crop this year (1996); I see many large flocks of migrating blue jays along the coast.

Chestnut oaks and red oaks are dropping many acorns (2000); much fewer black oak and scarlet oak acorns; no white oak acorns at all. The ground is strewn with numerous red oak and white oak acorns (2002). Some chestnut oak acorns are sprouting, but most have not. Witch-hazel is beginning to flower. In an upland deciduous forest, a common garter snake (15.7 inches snout to vent length) contains a partially digested, adult female wood frog. Some vernal pools are still dry, and others have a little standing water; shrub swamps used by breeding wood frogs and marbled salamanders have extensive shallow water.

September 29. A timber rattlesnake that gave birth on September 7 eats an adult eastern chipmunk. First frost of the season last night, but some rattlesnakes still coil on the forest floor day and night.

September 30. Huge acorn crop this year (2001); no flocks of migrating blue jays. Some milkweed pods are beginning to shed seeds. Pinesap is fully grown and flowering, with red flowers and stems, under eastern hemlock, red oak, and mountain laurel. Wintergreen leaves are still green. Virginia creeper leaves are red. Winterberry and autumn-olive* fruits are fully red. Seed cones of black birch are dry and brown. Syrphid flies and bumblebees are common on goldenrod flowers. A walkingstick that I place on a mountain laurel shrub soon begins eating a leaf. Wild grapes are ripe to somewhat overripe. A large winterberry plant that fruited prolifically last year has no fruits this year.

October

October 1. A downy woodpecker pecks into a common reed stem. A fat female black-horned tree cricket perches atop the rear end of a singing male. A goldenrod crab spider captures a syrphid fly. Glasswort is bright pink-red. Thousands of Atlantic marsh fiddlers of all sizes feed on mud banks of a tidal creek in a salt marsh. Many mud crabs, ranging from tiny juveniles to adults, are in partially open, empty oyster shells in shallow water of a tidal lagoon. Swamp white oak acorns are still attached.

October 2. After summer drought (1995), shrubs have relatively few fruits. A chipmunk fills its cheek pouches under the bird feeder. Silky dogwoods still have ripe fruits attached. Buttonbush plants are loaded with fruits. Many black oak acorns are under some trees (2000). Woolly aphids are still conspicuous on alder in a swamp, but they disappeared from other alders weeks ago. Many fresh willow beaked-galls, caused by a midge, are on willows in a swamp.

Many gray squirrels and eastern chipmunks run across roads; probably gathering and storing acorns.

October 3. Chestnut oak acorns are sprouting, and weevil larvae are emerging from the acorns. An adult cynipid wasp is inside a small spherical gall on an oak twig. Each globular, red gall on the leaf stalks of eastern cottonwood contains hundreds of aphids, including winged individuals, wingless nymphs of various sizes, and one "fat" wingless individual. 5:00 P.M.: meadow katydids perch and "sing" on the top of tall goldenrods. Summer grape has many clusters of small ripe fruits. Acorns are still falling from red oak trees.

October 4. Several American chestnut burs containing fat nuts are on the ground. A nest box that I put up in a tree contains 14 white-footed mice, including three babies with closed eyes. Adult caddisflies appear at lights near an intermittent stream.

October 5. The first white-throated sparrows of the season are at my feeder. American hornbeam has a very large crop of fruits this year (2000). Small ash trees have lost most of their leaves.

October 6. Fungi fruiting bodies (mushrooms) are prolific. 6:30 P.M., calm, thick, low clouds: a school of dozens of large fishes splashes noisily at the surface as they move up the mouth of the Connecticut River.

October 7. Nighttime dip-netting yields many Atlantic silversides, grubbys, and striped killifishes, and a northern pipefish in the shallows of Long Island Sound. An adult female fork-tailed bush katydid climbs the side of a gasoline pump as I fill my tank.

October 8. Hundreds of monarchs fly along the coast. A large adult diving beetle crawls on the ground in a field next to an open shrub swamp. Many 2-millimeter-long whitish larvae are in the firm capsule in the center of an oak apple gall from this year—not the larvae of the cynipid wasp that caused the gall to form. Witch-hazel is flowering prolifically (2001). Vast numbers of chestnut oak and red oak acorns are on the ground (2001). Huge schools of juvenile menhaden feed adjacent to a brackish marsh along the lower Connecticut River.

October 9. Today produces the season's first killing frost (1988). Leaves on the ground are mainly birch, aspen, and red maple. This morning was the season's first frost (29°F, 2001). The season's first flock of dark-eyed juncos arrives.

October 10. Glasswort has full red color. Female groundsel-trees with abundant white pappus tufts are conspicuous along salt marsh edges. Large adult racers are numerous and basking near their hibernation sites in rock outcroppings. Nipple galls on a hackberry leaf each contain a single, wide-bodied psyllid nymph with small wing buds. Yellow grape foliage is conspicuous among green roadside trees and shrubs.

October 11. I watch the moon through a spotting scope for 23 minutes tonight in Middletown, with no wind evident and see 28 birds fly past, all but one heading southwest. 5:00 P.M.: A narrow-winged tree cricket sings from within a curled maple leaf 3 feet above the ground in an urban residential area. In a coastal forest, large numbers of acorns are on the ground beneath white oaks, and many nuts are still attached to pignut hickory and black oak trees. Virginia creeper and shining sumac are deep magenta and have numerous ripe fruits. Poison ivy has numerous fruits, and its leaves are orange to deep magenta. Shadbush is leafless. 11:28 P.M., 57°F: true katydids stridulate continuously; they started up around 4:45 P.M., before sunset. Several small caddisfly adults appear outside a lighted window near a flowing intermittent stream.

October 12. Last night was the seasons's first frost (1985). Many woollybears crawl across roads. Hundreds of large larval bullfrogs are in a drying puddle beside a pond; tracks of great blue heron and raccoon are beside the puddle. Tulip-tree leaves are now falling. An adult timber rattlesnake crawls right over two adult racers that are coiled side by side; neither species seems perturbed by this.

October 13. Seaside goldenrod is flowering abundantly. In another year, most seaside goldenrod plants are done flowering. Common periwinkles* are numerous on a muddy-sandy-cobbly tidal flat exposed at low tide. Woollybear caterpillars crawl around on the ground. A flock of 50 brant is in the water along a coastal shore; one individual is apart on the adjacent sand spit, busily feeding on seed heads of crabgrass*.

October 14. All but two of 17 radio-tracked timber rattlesnakes are in their winter dens. Witch-hazel is beginning to flower, with flowers on the same branches that have ripe seed pods, but most fruiting plants have no flowers at all. Sanderlings and ring-billed gulls extract the "meat" from beach-cast slippersnails.

October 15. White ash trees have dropped virtually all of their leaves; birches and sycamores are in second place in leaflessness. 2:00 P.M.: a red bat hangs in a red maple tree, closely mobbed by several tufted titmice and a white-breasted nuthatch. A skiff moth caterpillar is still active on a witch-hazel leaf. Dodder, growing on cardinal flower, has many ripe fruits and fresh flowers. At 5:00 to 6:00 P.M., several monarchs, plus an American bittern and three merlins, fly along the coast on a balmy evening. At 6:18 P.M., after sunset, a flying monarch alights in a coastal pine tree. Leaves rain down from deciduous trees.

October 16. White-throated sparrows bite the fruit off flowering dogwood fruits, leaving the seed intact on the tree. After bush-whacking through mountain laurel and walking along deer trails, I find two deer keds (blood-sucking fly larvae) on me, alerted to their presence by their bites. Yellow-rumped warblers eat poison ivy fruits. Live adult cynipid wasps are in pupal cases within clusters of thick-walled spherical galls on oak twigs.

October 17. Sprouting chestnut oak acorns that I put on the bird feeding tray several days ago have the root sprouts and apical end of the acorn bitten off (by squirrels?). Bur-cucumber has brown, fully developed fruit clusters, plus some fruits that are still small and green.

October 18. Leaves fall from trees in large numbers; ashes, hickories, and birches are already mostly leafless. A beaver felled a 1-foot-diameter cottonwood tree this past week. The season's first dark-eyed juncos of the season arrive at my feeder. No wild rice grains are left on plants.

October 19. Large hen-of-the-woods fungi are attached to the base of oak stumps. Lady beetles swarm on the side of a house. Ten orange sulphurs fly and land on flowers of yellow mustard and red clover along the coast. Hundreds of stacks of live slippersnails wash up on a beach at the high tide line.

October 20. A walkingstick walks on grass in oak woodland. Spanworm moths fly at dusk, low 30s (°F). A ribbon snake eats a spring peeper. 3:20 P.M., mid-60s: an adult ringneck snake crawls across a shaded path in a coastal forest. Ten immature herring gulls stand facing the water along a sandy beach of Long Island Sound, where waves are depositing seaweed; periodically a gull extracts a small spider crab from the seaweed and swallows it whole. Hickory nuts are falling from trees in a coastal forest, and many other are on the ground.

October 21. Yellow foliage of Asian bittersweet* is conspicuous on forest-edge trees. Large numbers of leaves fall from deciduous trees. A silky dogwood still has many fruits.

October 22. A gray catbird tugs on multiflora rose* hips. Arrow-arum fruiting pods are splitting open. Adult and juvenile Asian shore crabs* are abundant under intertidal rocks of Long Island Sound.

October 23. Red-bellied woodpeckers fly with an acorn in the bill. Thousands of common grackles fly in large flocks and move through the trees in a deciduous forest.

October 24. American robins and cedar waxwings eat flowering dogwood fruits. A herring gull catches a live 9-inch-long flounder in shallow water, flies to shore with it, and, from the underside, eats only the small area of viscera just behind the head. Other adults and immatures catch and eat live rock crabs (attacked on underside), mussels, periwinkles, and other snails; the mollusks are cracked open by being dropped on rocks from a height of 25 feet. This year (1995) has a poor crop of acorns. Large antlion larvae are still active in the bottom of their conical pits; immobile when extracted, as if playing dead, but quite animated when grasped gently with forceps, and they soon bury themselves if placed on sandy soil.

October 25. Many large aggregations of woolly beech aphids are on American beech trees. I open a bald-faced hornet nest; inside is a live but lethargic adult hornet, a dead adult, many dead larvae and pupae, a small spider, and a yellow moth. Pokeweed has abundant fruits, some drying up. Basswood trees have many fruits still attached. Pignut hickories still have some fruits attached. Witch-hazel seeds shoot out of the opened fruit pods that I brought into my office yesterday; most pods have already launched their seeds.

October 26. Cedar waxwings eat autumn-olive* fruits. An adult cynipid wasp and an opened puparium are in each of two, marble-sized, woody galls on twigs of white oak. Tulip-tree fruits drift in the wind.

October 27. A fresh growth of Indian pipes is emerging. A herring gull adult drops a mollusk onto rocks and lets a juvenile gull retrieve it and eat the contents. Beach "fleas" (amphipod crustaceans) are active under objects in high salt marsh. Some vernal pools are still dry, others are partially filled, despite the lack of rain for the past three weeks (they partially filled with heavy rains in early October).

October 28. Flowering dogwood fruits have disappeared over the past two weeks. Strong winds are detaching and scattering the leaves of oaks, beeches, and maples. Some witch-hazels are leafless, others still have many leaves attached. Maple-leaved viburnum has purple to pink leaves and abundant fruits attached. Fifty-four yellowjackets enter a nest hole in three minutes.

October 29. Clouded sulphur is flying. A survey of a bat hibernaculum yields little brown bat (368), northern bat (26), eastern pipistrelle (7), and unidentified (6). A common garter snake eats a redback salamander. We've had three mild (29 to 30°F) frosts so far this fall. Red and yellow sassafras leaves, still attached to small plants, are turning brown. At sunset, ground crickets are trilling, and an earwig prowls my kitchen sink. Leaves rain down from forest trees.

October 30. Masses of woolly alder aphids are on one branch of a small alder. A true katydid sings from a tree.

October 31. Under a log, I find freshly laid clusters of slug eggs, with two slugs nearby.

November

November 1. A woollybear caterpillar crawls across the road in the afternoon. Butter-and-eggs*, white wood aster, heart-leaved aster, goldenrod, and daisy fleabane are still flowering. Thousands of common grackles are in dense flocks moving through the trees of a moist forest adjacent to a shrub swamp. Periodically, large groups of them land on the ground and flick leaves aside with their bills—it looks like the leafy ground is boiling. When large numbers of the grackles fly up from the ground or from their perches in trees, the sound of their wings resembles the roar of a large wave crashing on an ocean shore. An adult four-toed salamander is under wood at the edge of a pond. Just before sunset on a mild evening, a true katydid calls from the canopy of a deciduous forest. Ground crickets trill continuously. Red oak acorns continue to fall from the trees.

November 2. Tulip-trees still have many attached leaves, but many leaves are falling. Indian tobacco is still flowering.

November 3. Eleven wild turkeys walk single file out of the woods and down my driveway. Cattail seeds are dispersing.

November 4. A few mosquitoes are still flying in the woods. Tulip-tree pods begin to loosen and release their winged fruits.

November 5. Late afternoon: as I scratch on the trunk of a yellow birch tree adjacent to a swamp, a flying squirrel emerges from a hole, perches motionless, and watches us for several minutes. A flock of 10 snow buntings forages on a low sand dune. A herring gull pecks at a rock crab on a cobbly beach. A sanderling spends a minute eating flesh from a rock crab carcass; pulling off strands of flesh with rapid "sewing machine" movements of the bill. Then it resumes probing in intertidal sand as it walks along the exposed shore. A lion's-mane jellyfish (16 inches in diameter) is stranded in shallow water of a flat intertidal shore. I run my fingers across the tentacles several times without being stung. Adult and juvenile Asian shore crabs* are still abundant under intertidal rocks. Common periwinkles* are now scarce in intertidal areas where they were numerous two weeks ago. At low tide, the most common birds along a cobbly-sandy shore of Long Island Sound are brant, great black-backed gull, herring gull, ring-billed gull, sanderling, dunlin, black-bellied plover, American oystercatcher, and greater yellowlegs.

November 6. Yellow foliage of aspens is still conspicuous along some roadsides. Black cherry trees have greenish, amber, and red leaves still attached.

November 7. A daisy fleabane still has several unweathered flowers. Among hundreds of seaside goldenrods that have set seed, one plant has a full set of fresh flowers. Windy and 45°F, but sunny; two orange sulphurs fly well when flushed from the ground. A black-capped chickadee eats poison ivy fruits on a windswept high beach between salt marsh and Long Island Sound. 12:50 P.M., sunny, 57°F: thirteen yellowjackets enter a nest in three minutes.

November 8. A woollybear walks across a driveway. Narrow-winged tree crickets, ground crickets, and meadow katydids are still singing. Birches are loaded with fruiting catkins. Six black-capped chickadees are in a gray dogwood. They pluck the white fruits, then land several inches away and peck at the fruit while holding it in the toes. 6:15 P.M.: four raccoons are under the bird feeder.

November 9. An eastern chipmunk gathers seeds under a bird feeder. On most witch-hazel plants, a minority of flowers still have their petals attached. In 2002, witch-hazels are still profusely flowering.

November 10. A beaver food cache includes eastern hemlock, mountain laurel, yellow birch, and striped maple on the top. Juvenile gray squirrels are clearly distinguishable from older individuals by their smaller size and lesser bulk. Massive numbers of red oak acorns cover the ground, but some red oaks still retain hundreds of mature acorns (2002). Chestnut oak acorns have germinated. Low 60s (°F): hundreds of lady beetles fly around and walk on a large rock outcrop that juts above an extensive deciduous forest.

November 11. A large crayfish crawls upstream in slow shallows along the edge of a fast-flowing river. When I pick it up, it "locks" itself into a triangle, with the large claws extending down to the tip of the tail end. When placed on the ground or in shallow water, it maintains this posture for several minutes. Highbush blueberry has many red leaves still attached. Black cherry still has many green leaves. Several cherry-faced meadowhawk dragonflies perch and fly beside a swampy pond on a sunny, mild day.

November 12. Two large, live but cold, queen bald-faced hornets are under a rotting log. An adult whirligig beetle clings to the underside of a rock that I pick up in a river. Common running clubmoss "cones" are shedding spores, but cones of nearby tree clubmoss are still tightly closed. A large flock of common grackles moves through a wooded residential area.

November 13. Large numbers of tulip-tree seeds are on the ground. Shrubby honeysuckles* and Asian bittersweet* are among the most conspicuous plants still with leaves, the yellowish leaves contrasting strongly with the dominant brown and gray of the native vegetation.

November 14. Cabbage white*, orange sulphur, and viceroy fly on a mild day at the coast. A black-capped chickadee tears apart a tulip-tree pod, discarding winged fruits; searching for insects? 9:00 P.M., 43°F, light rain: a small adult spotted salamander walks through a deciduous forest.

November 15. Black willows and highbush blueberry still have some leaves attached. Several American goldfinches eat seeds from eastern hemlock cones. 9:30 P.M., 56°F: several narrow-winged tree crickets are singing in shrubbery on the Wesleyan campus; this after some nights with temperatures in the mid-20s (°F).

November 16. Daisy fleabane is still flowering. Seven eastern bluebirds perch in a holly shrub, plucking and eating the red fruits. Milkweed pods are open and releasing seeds. A single witch-hazel plant stands out from others by still having numerous flowers.

November 17. A huge flock of common grackles plus some starlings* and red-winged blackbirds perch in my yard. A mockingbird eats Asian bittersweet* fruits. In the wind, thousands of winged fruits are helicoptering far away from tulip-trees. Sunny, low 50s (°F): a juvenile eastern rat snake (28 inches in total length) lies freshly killed on a road crossing a wooded stream course; snake is black, but the dorsal blotches are still evident; this is unusually late activity for this species. Sunny, 40s (°F): a clouded sulphur flies at Hammonasset Beach State Park. Tree clubmoss is shedding large amounts of spores.

November 18. Daisy fleabane flower petals are now wilted. Some wintergreen leaves have turned red, others are still green or partially green. Spreading dogbane pods have not yet opened.

November 19. Black-capped chickadees obtain something to eat from clusters of staghorn sumac fruits. Butter-and-eggs* and goldenrods are still flowering. A vernal pool is loaded with mostly 1-inch-long marbled salamander larvae. Another spring-fed pool contains a large-gilled spotted salamander larva, 1.3 inches snout to vent length. 8:00 A.M., 30°F: on a rocky wooded slope, three blue jays use their bills to vigorously flick leaf litter and small sticks aside, then the jays peck strongly into the exposed soil, occasionally eating something small (sunflower seeds from my feeder?). A small goldenrod, still in flower along the creek, has aphids on it. Temperatures at night have been in the low 20s. A pileated woodpecker eats grapes from a vine in a tree. Low temperature drops below 30°F for the first time this month (a warm fall).

November 20. Greenbriar still has many leaves attached. A rural road is strewn with hundreds of seeds that have been "shot" from the pods of roadside witch-hazels. 10:30 P.M., 32°F: a spanworm moth is perched outside my lighted kitchen window; I touch it and it flies away!

November 21. A mockingbird is singing. Fruits of greenbriar and Asian bittersweet* remain abundant in coastal thickets. At the suet feeder, hairy woodpeckers are consistently much warier than are downy woodpeckers.

November 22. A water strider is still active on a puddle at the edge of a lake. Common groundsel* is still flowering. Ground crickets sing continuously from forest-edge leaf litter on a sunny, mild day. Under wood and rocks in the dried-up basin of a semi-permanent kettle-pond, I find adult diving beetles, many green darner dragonfly nymphs (nearly 1.5 inches long), several planorbid snails, and several adult eastern newts, all alive and well.

November 23. A few nuts of pignut hickory are still attached to the trees; nuts are partially exposed in split-open husks. 9:00 A.M.: while skiing in a field, I see several meadow vole tunnel openings at the snow surface; some have green feces and urine visible in them; none have tracks on the snow surface outside the tunnel; 5 to 6 inches of snow on ground.

November 24. Groundsel-tree still has many leaves attached; so do some marsh-elders (mostly are leafless). Dusk, 4:26 P.M., 50°F, calm, cloudy: a large bat, probably big brown, flies slowly up and down a road bordering a stream, making several abrupt changes in direction, as if catching insects (small dipterans and moths are flying). 11:30 P.M., 42°F: caddisflies and moths appear outside a lighted window. A mockingbird sings in a coastal thicket.

November 25. Most milkweed pods are empty, but a few are still shedding seeds. Midday low tide, 30s: many common* and yellow periwinkles are active among rockweed on exposed intertidal rocks. Atlantic dogwinkles and rough periwinkles are mainly in crevices. 3:35 P.M., 60°F, calm, cloudy: narrow-winged tree crickets trill from a thicket on the Wesleyan campus.

November 26. Mild, misty rain: a wood frog calls intermittently this afternoon from a vernal pool. Hard freezes occur the past two days (around 15°F), now it is warmer (mid-40s), but crickets are silent.

November 27. 5:50 P.M., 43°F: a spanworm moth flutters at my lighted window. Beach fleas are still active under wood in high salt marsh. Early dusk: a huge flock of common grackles flies by; the flock is about 100 to 200 feet wide and about ¼ mile long.

November 28. Sunny, 45°F: an adult stonefly is active near a creek. 10:42 A.M.: a sharp-shinned hawk catches a tufted titmouse in flight near the bird feeder; the feeder area is vacant thereafter until 10:53 when black-capped chickadees arrive and resume feeding. A female mosquito flies inside my house. 9:00 P.M., 42°F: two spanworm moths and several small caddisflies perch outside a lighted window.

November 29. Mosquitolike winter crane flies appear outside my lighted kitchen window on a mild night. Horned larks pluck and eat sea-lavender fruits bordering a salt marsh.

November 30. Cattail seeds are dispersing. 4:30 P.M., 61°F, calm, cloudy, wet: a few ground crickets trill continuously from forest-edge leaf litter; so far this fall the lowest temperatures has been only in the mid-20s (and only briefly).

December

December 1. Japanese barberry* is heavily laden with fruits that birds appear to totally ignore (for now). 2:30 p.m., sunny, 65°F: several spring peepers call from the woods away from ponds; eight cherry-faced meadowhawks (dragonflies) fly and perch on leaves and low rocks in a small clearing beside a shallow pond; I easily hand-capture a firefly as it flies over a large field; ground crickets call from forest leaf litter. Midnight, 31.5°F: an inchworm moth flies to a lighted window.

December 2. Over the past few weeks, the outside part of gray squirrel ears have turned white. Three crows feed on a female ring-necked pheasant* at the edge of a rural road. The season's first "winter" stonefly (¼-inch long, not including antennae or tail filaments) appears inside my streamside house.

December 3. A large springtail walks and hops across the electric range as I make soup. At 5:25 p.m., air temperature 37°F, a spanworm moth flies outside a lighted window. A gray squirrel hulls and eats 19 sunflower seeds in one minute as it hangs upside down from a hanging tubular bird feeder. A mockingbird eats holly fruits. Tree clubmoss cones are still shedding spores. Leaves of sheep laurel in frozen soil droop vertically; some are reddish. Needle ice is 3 inches long in muddy soil.

December 4. A group of four northern flickers perches in a large smooth sumac, eating the fruits. Yellowjackets and paper wasps fly on a warm day (70°F). A woollybear caterpillar is curled up in my firewood pile; air temperature 19°F. A gray squirrel eats sumac fruits. Many winter crane flies and a small caddisfly appear at my lighted kitchen window; no ground crickets are audible in nearby woods (10:55 p.m., 47°F).

December 5. Asian bittersweet* fruits are almost all still attached. Common periwinkles* are still active in the intertidal zone of Long Island Sound (warm autumn this year). Japanese honeysuckle* still has green and red leaves attached.

December 6. Winter crane flies swarm at night outside a lighted window near a stream. A woodchuck gathers dry leaves and carries them in its mouth; it rears up awkwardly on its hind legs when stuffing the leaves into its mouth; air temperature is 39°F, but for the last few days the temperature has stayed below freezing. In a warm autumn, common chickweed* flowers abundantly along a roadside.

December 7. A flock of American goldfinches eats birch catkins.

December 8. Two gray squirrels spend the night together in a nest box on a tree; earlier, one of them chases off a third squirrel that briefly enters the box. The first tree sparrows of the season arrive at my feeder. The first significant snowfall of the season arrives, and the first flock of purple finches arrives at my bird feeder.

December 9. A blue jay picks up and ingests 25 black oil sunflower seeds and flies off. Two inches of snow yesterday, low temperature 11°F last night: last night or early this morning a shrew leaves a winding, sometimes looping, tunnel (visible as a raised ridge of snow) through a field, with openings to the surface every few inches or (on long straightaways over hard bare ground) every 2 to 3 feet.

December 10. Six adult white-footed mice huddle in a thinly roofed nest in a nest box attached to a tree.

December 11. A partial count of bats in a hibernaculum yields 492 little brown bats, 30 northern bats, and 25 eastern pipistrelles.

December 12. A woodchuck gathers dry leaves and takes them into a burrow. A large crane fly (1-inch body length) appears at a window inside the house.

December 13. Hundreds of millipedes (up to ½ inch long) and isopods are clustered under wood on frozen ground. They slowly crawl away when warmed to 35°F.

December 14. I find an average of 12.5 meadow vole tunnel openings per 100 square meters (1,080 ft^2) in a snow-covered grassy field.

December 15. A small flock of eastern bluebirds eats holly fruits. Last night was rainy, with temperatures in the upper 40s (°F); juvenile green frogs are active on a road next to a small stream. Northern mockingbid eats winterberry fruits. Tree clubmosses release clouds of spores when jostled.

December 16. A ruffed grouse perches in a fruiting holly tree at dusk; it has been there on several previous evenings. White-throated sparrows eat multiflora rose* fruits.

December 17. A red fox trail in snow leads to an old woodchuck burrow, where the fox spends the day. 7:00 to 7:30 p.m., light rain, 38°F: several winter crane flies appear outside my lighted kitchen window.

December 18. White-throated sparrows pluck fruits of multiflora rose* and Asian bittersweet*.

December 19. Common dandelion* has unweathered flowers fully open. Many newly fallen fruits of tulip-tree are scattered on the ground.

December 20. A northern flicker eats sumac fruits. An Amercian coot dives in a lake to obtain and eat aquatic plant material.

December 21. An immature yellow-bellied sapsucker eats holly fruits. In a cove along the Connecticut River are newly beaver-cut winterberry stems and a half-cut 16-inch-diameter green ash tree.

December 22. A spring peeper in the woods peeps several times on this warm (60°F), humid day. Large numbers of sanderlings feed along sandy coastal beaches.

December 23. The wind blows tufted seeds out of milkweed pods.

December 24. Purple sandpipers probe their bills among algae-covered rocks along the shore of Long Island Sound.

December 25. Ice covers half the width of the Connecticut River; low temperature last night was 10.5°F; common (2) and hooded (7) mergansers swim near the ice.

December 26. Large numbers of black birch seeds are scattered widely over the snow-covered ground.

December 27. A white-breasted nuthatch caches a small fruit into a crevice in tree bark. Poison ivy still has abundant fruits attached.

December 28. I watch a shrew partly emerge from several of a close network of tunnel openings in the snow under the bird feeder. It emerges every 15 seconds or so, and each time quickly takes one sunflower seed and immediately disappears into the tunnel. Twice the shrew plugs a tunnel opening with loose snow; once it later burrows through the plug to take another seed. Seven to eight adult white-footed mice are huddled together in a nest box I put up for them.

December 29. An adult *Allocapnia* stonefly is active. An eastern screech-owl pellet contains the remains of a short-tailed shrew.

December 30. A gray squirrel gathers dry leaves and carries them in its mouth into a nest box in a tree; it does this several times.

December 31. A flock of about a hundred American robins perches in trees and flies west along the coast. A live woollybear is curled up on top of snow; air temperature is 32°F. An immature snowy owl perches on a jetty in Long Island Sound.

Appendix A
Endangered, Threatened, and Special Concern Species in Connecticut

Many species of Connecticut wildlife have been officially listed by the Connecticut Department of Environmental Protection (DEP) as *Endangered, Threatened,* or *Special Concern* (see *http://dep.state.ct.us* for the latest list). A small number of species in Connecticut also are protected under the U.S. Endangered Species Act, applicable throughout the United States and to all U.S. citizens.

Endangered under the Connecticut Endangered Species Act of 1989 (Public Act 89-224) means any native species documented by biological research and inventory to be in danger of extirpation throughout all or a significant portion of its range within the state and to have no more than five occurrences in the state. An occurrence is an area that is currently occupied by a relatively discrete population of the species.

Threatened refers to any native species documented by biological research and inventory to be likely to become an endangered species within the foreseeable future throughout all or a significant portion of its range within the state and to have no more than nine occurrences is the state.

Special Concern indicates any native plant species or any native nonharvested wildlife species documented by scientific research and inventory to have a naturally restricted range or habitat in the state, to be at a low population level, to be in such high demand by humans that its unregulated taking would be detrimental to the conservation of its population, or has been extirpated from the state. According to DEP policy, all species that are believed to be extirpated in Connecticut are listed as *Special Concern,* as are additional species that are still present in the state but of conservation concern.

These statuses may reflect rarity, a declining population trend, or impending or ongoing threats resulting from natural or anthropogenic factors. A species that is "rare" in Connecticut generally is common in at least some other states. In Connecticut, these species generally fall into one of several nonexclusive categories: rare to locally abundant but with a small geographical distribution; restricted ecological tolerance (e.g., found only in calcareous wetlands); or wide distribution but low population density (few widely scattered individuals). Some of the invertebrates, such as the Labrador-tea tentiform leaf miner *(Phyllonorycter ledella)* and pitcher plant borer *(Papaipema appassionata),* are rare because their food plants are rare.

Some *Endangered, Threatened,* or *Special Concern* species are represented by stable populations but exist in only a few locations. Others have an insecure existence in the state and could be eliminated by small changes in land use or climate or by simple acts of vandalism. An alarmingly large number of plant and invertebrate species have not been seen in the state in many years. Many of these undoubtedly have been extirpated by habitat destruction, overcollection of naturally rare species, or natural factors, but some may still exist in as yet undetected locations, so you may be able to enjoy the thrill of "rediscovering" them. Because of the precarious status of these species, they should not be picked or collected.

To document observations of *Endangered, Threatened,* or *Special Concern* species, report the following information to the Connecticut Natural Diversity Data Base: precise location (mark the location on a topographic map or labeled sketch map); date of observation; observer's name, e-mail address, and phone number; habitat description; flowering/fruiting status (for plants); and the number of individuals or size of the plant patch you observed. If possible, take several photographs from different angles and label the photographs.

Appendix B
Conservation Supplement

Connecticut has many wildlife conservation efforts not mentioned in other parts of this book. Some of these are briefly described here.

The DEP's Wildlife Habitat Improvement Program, Wetlands Habitat and Mosquito Management Program, and other similar state programs maintain and enhance the diversity of existing wildlife habitats and species, and they provide access to public land and opportunities for wildlife-based recreational activities. Some projects are in conjunction with the USDA Farm Bill's Wildlife Habitat Incentives Program and Conservation Reserve Program, U.S. Fish and Wildlife Service, and private conservation organizations.

The state-designated Natural Area Preserves, now encompassing some 6,700 acres, protect areas with outstanding biological, scientific, educational, geological, paleontological, or scenic value. Recently designated preserves include such places as Duck Island in Westbrook, Lord Cove in Lyme and Old Lyme, Matianuck Sand Dunes in Windsor, and Merrick Brook in Scotland. Preserves have detailed management plans that ensure the protection of their unique resources.

Connecticut recently received $485,000 from the federal Commerce, Justice, and State Departments appropriations measure passed by Congress and signed by President Clinton. These funds are being used for wildlife conservation and related recreation and education projects.

A state Income Tax Check-off Program that provides funds for the Connecticut Endangered Species/Wildlife Fund is a significant source of revenue supporting research and management benefiting endangered and threatened species, Natural Area Preserves, and watchable wildlife. Contributions to this program often help the species that need it most.

The Connecticut Chapter of The Nature Conservancy (TNC) (http://nature.org/wherewework/northamerica/states/connecticut) has over the years undertaken several important wildland conservation programs in Connecticut. Recently, TNC launched a $10 million fund-raising project to provide the remaining funds needed to protect 15,370 acres of water company land in western Connecticut—the largest land conservation project in state history. The project encompasses land in more than twenty towns. Additional TNC land protection projects are underway in all of the seven areas in the state that the Conservancy has identified as Connecticut's highest quality landscapes.

Connecticut Fund for the Environment (www.cfenv.org), founded in 1978, is an important nonprofit legal champion for the environment. Working with citizen activists, other environmental groups, and elected officials, CFE uses law, science, and education as tools in their efforts to improve air and water quality, control toxic contamination, minimize the adverse impacts of highways and traffic congestion, protect public water supplies, and preserve the open space and wetlands crucial to both the state's citizens and its wildlife.

Connecticut Audubon Society (www.ctaudubon.org), an independent nonprofit founded in 1898, is the state's oldest and largest environmental education organization, dedicated to advancing people's knowledge and appreciation of Connecticut's diverse wildlife and habitats. With a comprehensive network of professional educators, seven nature facilities, and 19 wildlife sanctuaries around the state, Connecticut Audubon's environmental programs, conservation activities, and legislative advocacy annually engage over 200,000 people.

Bibliography

Space limitations allow me to list only a small fraction of the scientific publications that I consulted in preparing this book or that deal with topics discussed here. A complete bibliography likely would be by itself at least as large as the present book. So this bibliography is highly selective.

The following books, reviews, and journal articles served as significant sources of specific information or will be worthwhile further reading for those seeking additional information on the diversity of life, natural history, and conservation of Connecticut and southern New England. A leaf (⋄) indicates certain key books that I recommend as particularly informative for laypersons wanting to read more about the state's natural history. Also listed here are the sources of the previously published illustrations used in this book.

General

Connecticut Department of Environmental Protection: http://dep.state.ct.us/. This website has extensive information on animal wildlife and wildlife habitats.

⋄*Connecticut Wildlife.* Published bimonthly by the Wildlife Division, Connecticut Department of Environmental Protection.
   Essential for keeping up to date on the state's wildlife; informative, well illustrated, inexpensive.

⋄Jorgensen, N. 1978. *A Sierra Club Naturalist's Guide to Southern New England.* San Francisco: Sierra Club Books. xxi + 417 pp.
   Strength is ecological treatment of upland and wetland communities and plants.

Chapter 1. The Landscape

⋄Bell, M. 1985. *The face of Connecticut: People, geology, and the land.* State Geological and Natural History Survey of Connecticut, Bulletin 110:1–196.
   Highly readable, attractively illustrated; explains state's geological history and its effect on cultural history.

Luo, Y., S. Yang, R. J. Carley, and C. Perkins. 2002. Atmospheric deposition of nitrogen along the Connecticut coastline of Long Island Sound: A decade of measurements. *Atmospheric Environment* 36 (28): 4517–28.

Chapter 2. The Seasons

National Oceanic and Atmospheric Administration. 1985. *Climates of the states.* Two volumes. Detroit, Mich.: Gale Research Company. Volume I includes Connecticut.

Chapter 3. Coastal Waters and Wetlands: Estuarine Ecosystems

Aarset, A. V. 1982. Freezing tolerance in intertidal invertebrates. *Comparative Biochemistry and Physiology* 73A: 571–80.

⋄Andersen, T. 2002. *This fine piece of water: An environmental history of Long Island Sound.* Yale University Press, New Haven. xv + 256 pp.
   Scholarly but highly readable, discusses history of exploration and natural resource use, as well as conservation efforts.

Anderson, T. H., and G. T. Taylor. 2001. Nutrient pulses, plankton blooms, and seasonal hypoxia in western Long Island Sound. *Estuaries* 24: 228–43.

⋄Bachand, R. G. 1994. *Coastal Atlantic sea creatures: A natural history.* Norwalk, Conn.: Sea Sports Publications. viii + 184 pp.
   Informative accounts of Long Island Sound animals for a general audience.

Bertness, M. D. 1992. The ecology of a New England salt marsh. *American Scientist* 80: 260–68.

⋄Bertness, M. D. 1999. *The ecology of Atlantic shorelines.* Sunderland, Mass.: Sinauer Associates. xiv + 417 pp.
   Authoritative, up to date, accessibly written, abundantly illustrated; the best single source for coastal ecology.

Bertness, M. D., and G. H. Leonard. 1997. The role of positive interactions in communities: Lessons from intertidal habitats. *Ecology* 78: 1976–89.

Bertness, M. D., and S. M. Yeh. 1994. Cooperative and competitive interactions in the recruitment of marsh elders. *Ecology* 75: 2416–29.

Carlson, J. K., T. A. Randall, and M. E. Mroczka. 1997. Feeding habits of winter flounder *(Pleuronectes americanus)* in a habitat exposed to anthropogenic disturbance. *Journal of Northwest Atlantic Fishery Science* 21: 65:73.

Coastal Area Management Program. 1977. Long Island Sound: An atlas of natural resources. Connecticut Department of Environmental Protection, Hartford. 52 pp.

Deevey, G. B. 1956. Oceanography of Long Island Sound, 1952–1954. V. Zooplankton. *Bulletin of the Bingham Oceanographic Collection* 15: 113–55.

◆Dreyer, G. D., and W. A. Niering, eds. 1995. *Tidal marshes of Long Island Sound: Ecology, history and restoration.* Connecticut College Arboretum Bulletin No. 34.

    Highly informative, accessibly written, well illustrated.

Durbin, A. G., and E. G. Durbin. 1998. Effects of menhaden predation on plankton populations in Narragansett Bay Rhode Island. *Estuaries* 21: 449–65.

Fell, P. E., S. P. Weissbach, D. A. Jones, M. A. Fallon, J. A. Zeppieri, E. K. Faison, K. A. Lennon, K. J. Newberry, and L. K. Reddington. 1998. Does invasion of oligohaline tidal marshes by reed grass, *Phragmites australis* (Cav.) Trin. Ex Steud., affect the availability of prey resources for the mummichog, *Fundulus heteroclitus* L.? *Journal of Experimental Marine Biology and Ecology* 222: 59–77.

Gottschall, K. F., M. W. Johnson, and D. G. Simpson. 2000. *The distribution and size composition of finfish, American lobster, and long-finned squid in Long Island Sound based on the Connecticut Fisheries Division bottom trawl survey, 1984–1994.* NOAA Technical Report NMFS 148.

    Major publication focusing on fish populations (except small shoreline species).

Harwell, M. C., and R. J. Orth. 2002. Long-distance dispersal potential in a marine macrophyte. *Ecology* 83: 3319–30. Deals with eelgrass.

Kennish, M. J., ed. 2000. *Estuary restoration and maintenance: The National Estuary Program.* Boca Raton, Fl.: CRC Press, 359 pp.

    Includes discussion of the Long Island Sound Study's Comprehensive Conservation and Management Plan.

Keser, M., J. T. Swenarton, J. M. Vozarik, and J. T. Foertch. 2003. Decline in eelgrass (*Zostera marina* L.) in Long Island Sound near Millstone Point, Connecticut (USA) unrelated to thermal input. *Journal of Sea Research* 49: 11–26.

Levin, P. S., J. A. Coyer, R. Petrik, and T. P. Good. 2002. Community-wide effects of nonindigenous species on temperate rocky reefs. *Ecology* 83: 3182–93. Impacts of green fleece and an ectoproct on kelp.

Miner, R. W. 1950. *Field book of seashore life.* New York: G. P. Putnam's Sons. xv + 888 pp.

    Minimal ecological information; dated taxonomy; source of several illustrations in this book.

Morris, J. T., P. V. Sundareshwar, C. T. Nietch, B. Kjerfve, and D. R. Cahoon. 2002. Responses of coastal wetlands to rising sea level. *Ecology* 83: 2869–77.

Niering, W. A., and R. S. Warren. 1980. Vegetation patterns and processes in New England salt marshes. *BioScience* 30: 301–306.

Olmstead, N. C., ed. 1978. *Plants and animals of the estuary.* Connecticut College Arboretum Bulletin 23. 44 pp.

    Informative nontechnical accounts of major species.

Olmstead, N. C., and P. E. Fell. 1974. *Tidal marsh invertebrates of Connecticut.* Connecticut College Arboretum Bulletin 20. 36 pp.

    Informative nonetchnical accounts of major species.

Orson, R. A., R. S. Warren, and W. A. Niering. 1987. Development of a tidal marsh in a New England river valley. *Estuaries* 10 (1): 20–27.

◆Patton, Peter C., and James M. Kent. 1992. *A movable shore: The fate of the Connecticut coast.* Duke University Press, Durham, North Carolina. xiv + 143 pp.

    Essential reading for those interested in coastal shore dynamics; abundantly illustrated.

Richards, S. W., and G. A. Riley. 1967. The benthic epifauna of Long Island Sound. *Bulletin of the Bingham Oceanographic Collection* 19: 89–131.

Roman, C. T., W. A. Niering, and R. S. Warren. 1984. Salt marsh vegetation change in response to tidal restriction. *Environmental Management* 8: 141–50.

Saltonstall, K. 2002. Cryptic invasion by a non-native genotype of the common reed, *Phragmites australis*, into North America. *Proceedings of the National Academy of Sciences* 99: 2445–49.

Sanders, H. L. 1956. Oceanography of Long Island Sound, 1952–1954. X. The biology of marine bottom communities. *Bulletin of the Bingham Oceanographic Collection* 15: 345–414.

Sinicrope, T. L., P. G. Hine, R. S. Warren, and W. A. Niering. 1990. Restoration of an impounded salt marsh in New England. *Estuaries* 13: 25–30.

Tiner, R. W., Jr. 1987. *A field guide to coastal wetland plants of the northeastern United States.* Amherst: University of Massachusetts Press.

Warren, R. S., and W. A. Niering. 1993. Vegetation change on a northeast tidal marsh: Interaction of sea-level rise and marsh accretion. *Ecology* 74: 96–103.

Warren, R. W., P. E. Fell, J. L. Grimsby, E. L. Buck, G. C. Rilling, and R. A. Fertik. 2001. Rates, patterns, and impacts of *Phragmites australis* expansion and effects of experimental *Phragmites* control on vegetation, macroinverbetaces, and fish within tidelands of the lower Connecticut River. *Estuaries* 23: 90–107.

Warren, R. S., P. E. Fell, R. Rozsa, A. H. Brawley, A. C. Orsted, E. T. Olson, V. Swamy, and W. A. Niering. 2002. Salt marsh restoration in Connecticut: 20 years of science and management. *Restoration Ecology* 10:497–513.

◆Weiss, H. M. 1995. *Marine animals of southern New England and New York.* State Geological and Natural History Survey of Connecticut Bulletin 115.

    Lavishly illustrated, with minimal ecological information; the place to go to identify coastal animals, but too big to carry in the field.

Zajac, R. N., and R. B. Whitlatch. 2001. Response of macrobenthic communities to restoration efforts in a New England estuary. *Estuaries* 24: 167–83.

Chapter 4. Streams and Associated Wetlands: Riverine Ecosystems

Clausen, J. C., K. Guillard, and C. M. Sigmund. 2000. Water quality changes from riparian buffer restoration in Connecticut. *Journal of Environmental Quality* 29: 1751–61.

Craig, R. J. 1990. *Historic trends in the distribution and populations of estuarine marsh birds of the Connecticut River.* University of

Connecticut, Storrs Agricultural Experiment Station, Department of Natural Resources Management and Engineering, Research Report 83. ii + 38 pp.

Craig, R. J., and K. G. Beal. 1992. The influence of habitat variables on marsh bird communities of the Connecticut River estuary. *Wilson Bulletin* 104:295-311.

☙Dreyer, G. D., and M. Caplis, eds. 2001. *Living resources and habitats of the lower Connecticut River.* Connecticut College Arboretum Bulletin 37.

Scholarly nontechnical reading on river ecology and wildlife.

MacBroom, J. G. 1998. *The river book: The nature and management of streams in glaciated terranes.* Connecticut Department of Environmental Protection, Natural Resources Center, DEP Bulletin 28. xxii + 242 pp.

Essential reading for anyone seriously interested in river conservation in Connecticut.

☙Maloney, T., J. P. Barrett, N. Barrett, S. Gephard, G. A. Hammerson, C. H. Kimball, J. Pfeiffer, N. Proctor, and J. R. Stone. 2001. *Tidewaters of the Connecticut River: An explorer's guide to hidden coves and marshes.* Essex, Conn.: River's End Press. 123 pp.

Attractively illustrated guide to paddling routes, plants, animals, and cultural history.

Merriman, D., and L. M. Thorpe, eds. 1976. *The Connecticut River ecological study: The impact of a nuclear power plant.* American Fisheries Society Monograph No. 1. xi + 252 pp.

Metzler, K. J., and A. W. H. Damman. 1985. Vegetation patterns in the Connecticut River flood plain in relation to frequency and duration of flooding. *Le Naturaliste Canadien* 112: 535-47.

☙Metzler, K. J., and R. W. Tiner. 1992. *Wetlands of Connecticut.* State Geological and Natural History Survey of Connecticut, Report of Investigations No. 13. vi + 115 pp.

Best source of information on the vegetation of Connecticut's coastal and inland wetlands.

Chapter 5. Lakes and Ponds: Lacustrine Ecosystems

Brugam, R. B. 1978. Human disturbance and the historical development of Linsley Pond. *Ecology* 59: 19-36.

Deevey, E. S., Jr. 1951. Life in the depths of a pond. *Scientific American* 185 (4): 68-72.

Field, C. K., P. A. Siver, and A.-M. Lott. 1996. Estimating the effects of changing land use patterns on Connecticut lakes. *Journal of Environmental Quality* 24: 325-33.

Jacobs, R. P., and E. B. O'Donnell. 2002. *A fisheries guide to lakes and ponds of Connecticut including the Connecticut River and its coves.* Connecticut Department of Environmental Protection Bulletin 35. xiv + 354 pp.

Nontechnical overview of lacustrine fishes and their environments.

Murphy, B., and D. Mysling. 1993. *Small ponds in Connecticut: A guide for fish management.* Connecticut Department of Environmental Protection, Fisheries Division, DEP Bulletin 19. vi + 81 pp.

Siver, P. A., R. W. Canavan, IV, and C. K. Field. 1996. Historical changes in Connecticut lakes over a 55-year period. *Journal of Environmental Quality* 25: 334-45.

Siver, P. A., A. M. Lott, and E. Cash. 1999. Century changes in Connecticut, U.S.A. lakes as inferred from siliceous algal remains and their relationships to land-use change. *Limnology and Oceanography* 44: 1928-35.

Chapter 6. Inland Wetlands: Palustrine Ecosystems

Brooks, R. T. 2000. Annual and seasonal variation and the effects of hydroperiod on benthic macroinvertebrates of seasonal forest ("vernal") ponds in central Massachusetts, USA. *Wetlands* 20: 707-15.

Brooks, R. T., and M. Hayashi. 2002. Depth-area-volume and hydroperiod relationships of ephemeral (vernal) forest pools in southern New England. *Wetlands* 22: 247-55.

Brown, B. J., R. J. Mitchell, and S. A. Graham. 2002. Competition for pollination between an invasive species (purple loosestrife) and a native congener. *Ecology* 83: 2328-36.

Calhoun, A. J. K., and M. W. Klemens. 2002. *Best development practices: Conserving pool-breeding amphibians in residential and commercial developments in the northeastern United States.* Metropolitan Conservation Alliance, Technical Paper Series 5. 57 pp.

Essential resource for land use commissions and developers.

Cowardin, L. M., V. Carter, F. C. Golet, and E. T. LaRoe. 1979. *Classification of wetlands and deepwater habitats of the United States.* U.S. Fish and Wildlife Service, Biological Services Program, FWS/OBS-79/31. 103 pp.

Craig, R. J., and J. S. Barclay. 1992. Seasonal dynamics of bird populations in small New England wetlands. *Wilson Bulletin* 104: 148-55.

Cronk, J. K., and M. S. Fennessy. 2001. *Wetland plants: Biology and ecology.* Boca Raton, Fl.: Lewis Publishers.

Donahue, D. F. No date. *A guide to the identification and protection of vernal pool wetlands of Connecticut.* University of Connecticut Cooperative Extension System. 18 pp.

Farnsworth, E. J., and D. R. Ellis. 2001. Is purple loosestrife *(Lythrum salicaria)* an invasive threat to freshwater wetlands? Conflicting evidence from several ecological metrics. *Wetlands* 21: 199-209.

Fellman, B., ed. 1998. *Our hidden wetlands: The proceedings of a symposium on vernal pools in Connecticut 1998.* Center for Coastal and Watershed Systems, Yale University School of Forestry and Environmental Studies, and Wetland Management Section, Connecticut Department of Environmental Protection. 62 pp.

Giberson, D., and M. L. Hardwick. 1999. Pitcher plants *(Sarracenia purpurea)* in eastern Canadian peatlands: Ecology and conservation of invertebrate inquilines. In *Invertebrates in freshwater wetlands of North America: Ecology and management,* ed. D. P. Batzer, R. B. Rader, and S. A. Wissinger. 401-22. New York: John Wiley & Sons.

Golet, F. C., Y. Wang, J. S. Merrow, and W. R. DeRagon. 2001. Relationship between habitat and landscape features and the avian community of red maple swamps in southern Rhode Island. *Wilson Bulletin* 113: 217-27.

Heard, S. B. 1998. Capture rates of invertebrate prey by the pitcher plant, *Sarracenia purpurea* L. *American Midland Naturalist* 139: 78–89.

❧Johnson, C. W. 1985. *Bogs of the northeast.* Hanover, N.H., University Press of New England. xiii + 269 pp.

❧Kenney, L. P., and M. R. Burne. 2000. *A field guide to the animals of the vernal pools.* Massachusetts Division of Fisheries and Wildlife Natural Heritage and Endangered Species Program and Vernal Pool Association.

Kittresge, D. M., Jr. 1996. Protection of habitat for rare wetland fauna during timber harvesting in Massachusetts (USA). *Natural Areas Journal* 16: 310–17.

Magee, A. F. 1996. Breeding bird censuses. *Journal of Field Ornithology* 67 (Supplement): 30–31, 60, 76.

McMaster, R. T., and N. D. McMaster. 2000. Vascular flora of beaver wetlands in western Massachusetts. *Rhodora* 102: 175–97.

❧Metzler, K. J., and R. W. Tiner. 1992. *Wetlands of Connecticut.* State Geological and Natural History Survey of Connecticut, Report of Investigations No. 13. vi + 115 pp.

Mitsch, W. J., and J. G. Gosselink. 2000. *Wetlands.* Third edition. New York: John Wiley. xiii + 920 pp.

Motzkin, G. 1994. Calcareous fens of western New England and adjacent New York state. *Rhodora* 96: 44–68.

Motzkin, G., W. A. Patterson, III, and N. E. R. Drake. 1993. Fire history and vegetation dynamics of a *Chamaecyparis thyoides* wetland on Cape Cod, Massachusetts. *Journal of Ecology* 81: 391–402.

Richburg, J. A., W. A. Patterson,, III, and F. Lowenstein. 2001. Effects of road salt and *Phragmites australis* invasion on the vegetation of a western Massachusetts calcareous lake-basin fen. *Wetlands* 21: 247–55.

Semlitsch, R. D. 1998. Biological delineation of terrestrial buffer zones for pond-breeding salamanders. *Conservation Biology* 12: 1113–19.

Semlitsch, R. D., and J. R. Bodie. 1998. Are small, isolated wetlands expendable? *Conservation Biology* 12: 1129–33.

Swift, B. L., J. S. Larson, and R. M. DeGraaf. 1984. Relationship of breeding bird density and diversity to habitat variables in forested wetlands. *Wilson Bulletin* 96: 48–59.

Turtle, S. L. 2000. Embryonic survivorship of the spotted salamander *(Ambystoma maculatum)* in roadside and woodland vernal pools in southeastern New Hampshire. *Journal of Herpetology* 34: 60–67.

Whitt, M. B., H. H. Prince, and R. R. Cox, Jr. 1999. Avian use of purple loosestrife dominated habitat relative to other vegetation types in a Lake Huron wetland complex. *Wilson Bulletin* 111: 105–14.

Williams, D. D. 1987. The ecology of temporary waters. London: Croom Helm, and Portland, Ore.: Timber Press.

Chapters 4, 5, and 6. Freshwater Fauna

Smith, D. G. 1991. *Keys to the freshwater macroinvertebrates of Massachusetts.* Published by the author. iv + 236 pp.

Useful in Connecticut; indispensable for students and professionals.

———. 2001. *Pennack's freshwater invertebrates of the United States.* Fourth edition. Porifera to Crustacea. New York: John Wiley & Sons. x + 638 pp.

Detailed biological information; updates Pennack's earlier editions, but taxonomically less comprehensive.

Thorp, J. H., and A. P. Covich, eds. 2001. *Ecology and classification of North American freshwater invertebrates.* Second edition. San Diego: Academic Press. xvi + 1056 pp. Comprehensive scientific treatise.

❧Voshell, J. R., Jr. 2002. *A guide to common freshwater invertebrates of North America.* Blacksburg, Va.: McDonald & Woodward Publishing Company. xiv + 443 pp.

Extensive information, nontechnical, well illustrated, small enough for field use.

Chapter 7. Uplands: Terrestrial Ecosystems

Barnett, R. J. 1977. The effect of burial by squirrels on germination and survival of oak and hickory nuts. *American Midland Naturalist* 98: 319–30.

Beatty, S. W. 1984. Influence on microtopography and canopy species on spatial patterns of forest understory plants. *Ecology* 65: 1406–19.

Bossema, I. 1979. Jays and oaks: An eco-ethological study of a symbiosis. *Behaviour* 70: 1–117.

Christianson, L. M., G. M. Lovett, M. J. Mitchell, and P. M. Groffman. 2002. The fate of nitrogen in gypsy moth frass deposited to an oak forest. *Oecologia* 131: 444–52.

Collins, S. 1961. Benefits to understory from canopy defoliation by gypsy moth larvae. *Ecology* 42: 836–38.

Darley-Hill, S., and W. C. Johnson. 1981. Dispersal of acorns by blue jays *(Cyanocitta cristata). Oecologia* 50: 213–32.

DeGraaf, R. M. 1984. *Managing New England woodlands for wildlife that uses tree cavities.* Cooperative Extension Service, University of Massachusetts, Amherst, Publication C-171.

Dixon, M. D., W. C. Johnson, and C. S. Adkisson. 1997. Effects of weevil larvae on acorn use by blue jays. *Oecologia.* 111 (2): 201–208.

Elkinton, J. S., W. M. Healy, J. P. Buonaccorsi, G. H. Boettner, A. M. Hazzard, H. R. Smith, and A. M. Liebhold. 1996. Interactions among gypsy moths, white-footed mice, and acorns. *Ecology* 77: 2332–42.

Foster, D. R., and T. M. Zebryk. 1993. Long-term vegetation dynamics and disturbance history of a *Tsuga*-dominated forest in New England. *Ecology* 74: 982–84.

Fox, J. F. 1982. Adaptation of gray squirrel behavior to autumn germination by white oak acorns. *Evolution* 36: 800–809.

Gale, G. A., J. A. DeCecco, M. R. Marshall, W. R. McClain, and R. J. Cooper. 2001. Effects of gypsy moth defoliation on forest birds: An assessment using breeding bird census data. *Journal of Field Ornithology* 72: 291–304.

Gibson, L. P. 1964. Biology and life history of acorn-infesting weevils of the genus *Conotrachelus* (Coleoptera: Curculionidae). *Annals of the Entomological Society of America* 57: 521–26.

Hill, J. D., C. D. Canham, and D. M. Wood. 1995. Patterns and causes of resistance to tree invasion in rights-of-way. *Ecological Applications* 5: 459–70.

Jenkins, J. C., J. D. Aber, and C. D. Canham. 1999. Hemlock woolly adelgid impacts on community scructure and N cycling rates in eastern hemlock forests. *Canadian Journal of Forest Research* 29: 630–45.

Johnson, W. C., and T. Webb, III. 1989. The role of blue jays *(Cyanocitta cristata)* in the postglacial dispersal of fagaceous trees in eastern North America. *Journal of Biogeography* 16: 561–72.

Johnson, W. C., C. S. Adkisson, T. R. Crow, and M. D. Dixon. 1997. Nut caching by blue jays (*Cyanocitta cristata* L.): Implications for tree demography. *American Midland Naturalist* 138: 357–70.

Kelly, D., D. E. Hart, and R. B. Allen. 2001. Evaluating the wind pollination benefits of mast seeding. *Ecology* 82: 117–26.

Kobe, R. K. 1996. Intraspecific variation in sapling mortality and growth predicts geographic variation in forest composition. *Ecological Monographs* 66: 181–201.

Lee, C. 1985. *West Rock to the Barndoor Hills: The traprock ridges of Connecticut.* State Geological and Natural History Survey of Connecticut, Vegetation of Connecticut Natural Areas 4. 60 pp.

Liang, S. Y., and S. W. Seagle. 2002. Browsing and microhabitat effects on riparian forest woody seedling demography. *Ecology* 83: 212–27.

LoGiudice, K., and R. S. Ostfeld. 2002. Interaction between mammals and trees: Predation on mammal-dispersed seeds and the effect of ambient food. *Oecologia* 130: 420–25.

Magee, A. F. 1996. Breeding bird censuses. *Journal of Field Ornithology* 67 (supplement): 30–31, 60, 70.

McLachlan, S. M., and D. R. Bazely. 2001. Recovery patterns of understory herbs and their use as indicators of deciduous forest regeneration. *Conservation Biology* 15: 98–110.

McShea, W. J. 2000. The influence of acorn crops on annual variation in rodent and bird populations. *Ecology* 81: 228–38.

McShea, W. J., and G. Schwede. 1993. Variable acorn crops: Responses of white-tailed deer and other mast consumers. *Journal of Mammalogy* 74: 999–1006.

McShea, W. J., and W. M. Healy, eds. 2002. *Oak forest ecosystems: Ecology and management for wildlife.* Baltimore: Johns Hopkins University Press.

Moffett, M. W. 1989. Life in a nutshell. *National Geographic* 175 (6): 782–96.
    High-quality illustrations of acorn-associated invertebrates.

Motzkin, G., D. Foster, A. Allen, J. Harrod, and R. Boone. 1996. Controlling site to evaluate history: Vegetation patterns of a New England sand plain. *Ecological Monographs* 66: 345–65.

Nichols, W. F., K. T. Killingbeck, and P. V. August. 1998. The influence of geomorphological heterogeneity on biodiversity. II. A landscape perspective. *Conservation Biology* 12: 371–79.

Niemelä, P., and W. J. Mattson. 1996. Invasion of North American forests by European phytophagous insects. *BioScience* 46: 741–53.

Ostfeld, R. S., C. G. Jones, and J. O. Wolff. 1996. Of mice and mast. *BioScience* 46: 323–30.

Pyare, S., J. A. Kent, D. L. Noxon, and M. T. Murphy. 1993. Acorn preference and habitat use in eastern chipmunks. *American Midland Naturalist* 130: 173–83.

Orwig, D. A., D. R. Foster, and D. L. Mausel. 2002. Landscape patterns of hemlock decline in New England due to the introduced hemlock woolly adelgid. *Journal of Biogeography* 29: 1475–87.

Sallabanks, R., and S. P. Courtney. 1992. Frugivory, seed predation, and insect-vertebrate interactions. *Annual Review of Entomology* 37: 377–400.

Semel, B., and D. C. Andersen. 1988. Vulnerability of acorn weevils (Coleoptera: Curculionidae) and attractiveness of weevils and infested *Quercus alba* acorns to *Peromyscus leucopus* and *Blarina brevicauda*. *American Midland Naturalist* 119: 385–93.

Steele, M. A., G. Turner, P. D. Smallwood, J. O. Wolff, and J. Radillo. 2001. Cache management by small mammals: Experimental evidence for the significance of acorn-embryo excision. *Journal of Mammalogy* 82: 35–42.

Steele, M. A., L. Z. Hadj-Chikh, and J. Hazeltine. 1996. Caching and feeding decisions by *Scirus carolinensis*: Responses to weevil-infested acorns. *Journal of Mammalogy* 77: 305–14.

Steele, M. A., T. Knowles, K. Bridle, and E. L. Simms. 1993. Tannins and partial consumption of acorns: Implications for dispersal of oaks by seed predators. *American Midland Naturalist* 130: 229–38.

Tingley, M. W., D. A. Orwig, R. Field, and G. Motzkin. 2002. Avian response to removal of a forest dominant: Consequences of hemlock woolly adelgid infestations. *Journal of Biogeography* 29: 1505–16.

Weckerly, F. W., K. E. Nicholson, and R. D. Semlitsch. 1989. Experimental test of discrimination by squirrels for insect-infested and noninfested acorns. *American Midland Naturalist* 122: 412–15.

Wessels, T. 1997. *Reading the forested landscape: A natural history of New England.* Woodstock, Vt.: The Countryman Press. 199 pp.
    Helpful in learning to use vegetation characteristics to interpret forest history.

Wetherell, D. V. 1997. *Traprock ridges of Connecticut: A naturalist's guide.* State Geological and Natural History Survey of Connecticut Bulletin 25. 58 pp.

Chapter 8. Algae, Fungi, and Lichens

Bessette A. E., A. R. Bessette, and D. W. Fischer. 1997. *Mushrooms of northeastern North America.* Syracuse, N.Y.: Syracuse University Press. xiv + 582 pp.

Brodo, I. M., S. D. Sharnoff, and S. Sharnoff. 2001. *Lichens of North America.* New Haven, Conn.: Yale University Press. xxiii + 795 pp.

Taylor, S. L., and M. Villalard. 1985. *Seaweeds of the Connecticut shoreline: A wader's guide.* Third edition. Connecticut College Arboretum Bulletin 18. 36 pp.

Chapter 9. Plants

Anagnostakis, S. L., and J. S. Ward. 1996. The status of flowering dogwood in five long-term forest plots in Connecticut. *Plant Disease* 80: 1403–05.

Anderson, T. E. 1995. *The poison ivy, oak and sumac book: A short natural history and cautionary account.* Ukiah, Calif.: Acton Circle. xi + 130 pp.

Beattie, A. J., and D. C. Culver. 1981. The guild of myrmecochores in the herbaceous flora of West Virginia forests. *Ecology* 67: 572–76.

Bertin, R. I., and O. D. V. Sholes. 1993. Weather, pollination, and the phenology of *Geranium maculatum*. *American Midland Naturalist* 129: 52–66.

Brown, P. M. 1997. *Wild orchids of the northeastern United States.* Ithaca, N.Y.: Cornell University Press. x + 236 pp.

Collier, M. H., J. L. Vankat, and M. R. Hughes. 2002. Diminished plant richness and abundance below *Lonicera maackii*, an invasive shrub. *American Midland Naturalist* 147: 60–71.

Dowham, J. J. 1979. *Preliminary checklist of the vascular flora of Connecticut (growing without cultivation).* State Geological and Natural History Survey of Connecticut, Report of Investigations No. 8. 176 pp.

◆Eastman, J. 1992. *The book of forest and thicket: trees, shrubs, and wildflowers of eastern North America.* Mechanicsburg, Pa.: Stackpole Books.

This and Eastman's other books are loaded with detailed, easily digested information.

◆———. 1995. *The book of swamp and bog: Trees, shrubs, and wildflowers of eastern freshwater wetlands.* Mechanicsburg, Pa.: Stackpole Books.

◆———. 2003. *The book of field and roadside: Open-country weeds, trees, and wildflowers of eastern North America.* Mechanicsburg, Pa.: Stackpole Books.

Fergus, C. 2002. *Trees of Pennsylvania and the northeast.* Mechanicsburg, Pa.: Stackpole Books. xiii + 272 pp.

Engagingly written, highly informative.

Gould, A. M. A., and D. L. Gorchov. 2000. Effects of the exotic invasive shrub *Lonicera maackii* on the survival and fecundity of three species of native annuals. *American Midland Naturalist* 144: 36–50.

Handel, S. N., S. B. Fisch, and G. E. Schatz. 1981. Ants disperse a majority of herbs in a mesic forest community in New York State. *Bulletin of the Torrey Botanical Club* 108: 430–37.

Harder, L. D., J. D. Thomson, M. B. Cruzan, and R. S. Unnasch. 1985. Sexual reproduction and variation in floral morphology in an ephemeral vernal lily, *Erythronium americanum*. *Oecologia* 67: 286–91.

Heinrich, B. 1992. Maple sugaring by red squirrels. *Journal of Mammalogy* 73: 51–54.

Johnson, W. T., and H. H. Lyon. 1991. *Insects that feed on trees and shrubs.* Second edition. Ithaca, N.Y.: Cornell University Press. 560 pp.

Lerat, S., R. Gauci, J. H. Catford, H. Vierheilig, and L. Lapointe. 2002. ^{14}C transfer between the spring ephemeral *Erythronium americanum* and sugar maple saplings via arbuscular mycorrhizal fungi in natural stands. *Oecologia* 132: 181–87.

Loik, M. E., and P. S. Nobel. 1991. Water relations and mucopolysaccharide increases for a winter hardy cactus during acclimation to subzero temperatures. *Oecologia* 88: 340–46.

Lounsberry, A. 1900. *A guide to the trees.* New York: Frederick A. Stokes Company.

Magee, D. W., and H. E. Ahles. 1999. *Flora of the Northeast.* Amherst: University of Massachusetts Press. xxxi + 1213 pp.

Technical, detailed identification keys, county distribution maps.

Martin, A. C., H. S. Zim, and A. L. Nelson. 1951. *American wildlife and plants: A guide to wildlife food habits.* New York: Dover Publication. 500 pp.

Classic account listing plants and their consumers.

Motten, A. F. 1986. Pollination ecology of the spring wildflower community of a temperate deciduous forest. *Ecological Monographs* 56: 21–42.

Nagy, E. S., L. Strong, and L. F. Galloway. 1999. Contribution of delayed autonomous selfing to reproductive success in mountain laurel, *Kalmia latifolia* (Ericaceae). *American Midland Naturalist* 142: 39–46.

Nilsen, E. T. 1991. The relationship between freezing tolerance and thermotropic leaf movement in five *Rhododendron* species. *Oecologia* 87: 63–71.

Packer, A., and C. Clay. 2000. Soil pathogens and spatial patterns of seedling mortality in a temperate tree. *Nature* 404: 278–81.

Study of black cherry.

Peattie, D. C. 1966. *A natural history of trees of eastern and central North America.* Second edition. New York: Crown Publishers. 606 pp.

Best for lore and human uses.

Primack, R. B., and C. McCall. 1986. Gender variation in a red maple population (*Acer rubrum*; Aceraceae): a seven-year study of a "polygamodioecious" species. *American Journal of Botany* 73: 1239–48.

Roberts, K. J., and R. C. Anderson. 2001. Effect of garlic mustard [*Alliaria petiolata* (Beib. Cavara & Grande)] extracts on plants and arbuscular mycorrhizal (AM) fungi. *American Midland Naturalist* 146: 146–52.

Sakai, A. K. 1990. Sex ratios of red maple (*Acer rubrum*) populations in northern lower Michigan. *Ecology* 71: 571–80.

Schemske, D. W. 1977. Flowering phenology and seed set in *Claytonia virginica* (Portilacaceae). *Bulletin of the Torrey Botanical Club* 104: 254–52.

———. 1978. Sexual reproduction in an Illinois population of *Sanguinaria canadensis* L. *American Midland Naturalist* 100:261–68.

Seymour, R. S. 1997. Plants that warm themselves. *Scientific American* 276: (3) 104–109.

Tryon, A. F., and R. C. Moran. 1997. *The ferns and allied plants of New England.* Lincoln: Massachusetts Audubon Society. xv + 325 pp.

Focuses on identification, includes detailed distribution maps.

Wen, J. 1999. Origin and evolution of eastern Asian and eastern North American disjunct distributions in flowering plants. *Annual Review of Ecology and Systematics* 30: 421–55.

Zettler, J. A., T. P. Spira, and C. R. Allen. 2001. Yellow jackets (*Vespula* spp.) disperse *Trillium* (spp.) seeds in eastern North America. *American Midland Naturalist* 146: 444–46.

Chapter 10. Sponges through Ectoprocts

Daly, M. 2002. Taxonomy, anatomy, and histology of the lined anemone, *Edwardsiella lineata* (Verrill, 1873) (Cnidaria: Anthozoa: Edwardsiidae). *Proceedings of the Biological Society of Washington* 115: 868–77.

De Santo, E. M., and P. E. Fell. 1996. Distribution and ecology of freshwater sponges in Connecticut. *Hydrobiologia* 341: 81–89.

Hyman, L. H. 1940. *The invertebrates: Protozoa through Ctenophora.* New York: McGraw-Hill Book Company.

Chapter 11. Segmented Worms

Smith, D. G. 2001. *Pennack's freshwater invertebrates of the United States.* Fourth edition. Porifera to Crustacea. New York: John Wiley & Sons. x + 638 pp.

❧Weiss, H. M. 1995. See references for Chapter 3.

Chapter 12. Mollusks

Carlton, J. T., G. J. Vermeij, D. R. Lindberg, D. A. Carlton, and E. C. Dudley. 1991. The first historical extinction of a marine invertebrate in an ocean basin: The demise of the eelgrass limpet *Lottia alveus. Biological Bulletin* 180: 72–80.

Jonkinen, E. H. 1983. *The freshwater snails of Connecticut.* State Geological and Natural History Survey of Connecticut Bulletin 109. vii + 83 pp.

Focuses on identification and distribution, aimed at scientists.

❧Nedeau, E. J., M. A. McCollough, and B. I. Swarz. 2000. *The freshwater mussels of Maine.* Maine Department of Inland Fisheries and Wildlife, Augusta, Maine. 118 pp.

Attractively packed with information, highly readable, abundantly illustrated, useful in Connecticut.

Spelke, J. A., P. E. Fell, and L. L. Helvenston. 1995. Population structure, growth and fecundity of *Melampus bidentatus* (Say) from two regions of a tidal marsh complex in Connecticut. *Nautilus* 108: 42–47.

Strayer, D. L., L. C. Smith, and D. C. Hunter. 1998. Effects of the zebra mussel *(Dreissena polymorpha)* invasion on the macrobenthos of the freshwater tidal Hudson River. *Canadian Journal of Zoology* 76: 419–25.

Tucker, A. D., S. R. Yeomans, and J. W. Gibbons. 1997. Shell strength of mud snails *(Ilyanassa obsoleta)* may deter foraging by diamondback terrapins *(Malaclemys terrapin). American Naturalist* 138: 224–29.

Turgeon, D. D., et al. 1998. *Common and scientific names of aquatic invertebrates from the United States and Canada: Mollusks.* Second edition. American Fisheries Society Special Publication No. 26.

Wilbur, A. K., and R. S. Steneck. 1999. Polychromatic patterns of *Littorina obtusata* on *Ascophyllum nodosum:* Are snails hiding in intertidal seaweed? *Northeastern Naturalist* 6: 189–98.

Williams, J. D., M. L. Warren, Jr., K. S. Cummings, J. L. Harris, and R. J. Neves. 1993. Conservation status of freshwater mussels of the United States and Canada. *Fisheries* 18 (9): 6–22.

Chapter 13. Chelicerates

Connecticut Agricultural Experimental Station: http://www.state.ct.us/caes.

Information on Lyme disease and other tick-associated diseases, tick control, and tick bite prevention.

Craig, C. L., and G. D. Bernard. 1990. Insect attraction to ultraviolet-reflecting spider webs and web decorations. *Ecology* 71: 616–23.

Eisner, T., and S. Nowicki. 1983. Spider web protection through visual advertisement: Role of the stabilimentum. *Science* 219: 185–86.

Kaston, B. J. 1981. *Spiders of Connecticut.* Revised edition. State Geological and Natural History Survey of Connecticut Bulletin 70. 1020 pp.

Myers, J.P. 1986. Sex and gluttony on Delaware Bay. *Natural History* 95 (5): 68–76.

Importance of horseshoe crabs to shorebirds.

Schoener, T. W., and D. A. Spiller. 1992. Stabilimenta characteristics of the spider *Argiope argentata* on small islands: Support of the predator-defense hypothesis. *Behavioral Ecology and Sociobiology* 31: 309–18.

Chapter 14. Insects and Relatives

Abrahamson, W. G., and A. E. Weis. 1997. *Evolutionary ecology across three trophic levels: Goldenrods, gallmakers, and natural enemies.* Princeton: Princeton University Press.

Arnett, R. H., Jr. 2000. *American insects: A handbook of the insects of America north of Mexico. Second edition.* Boca Raton, Fl.: CRC Press.

Betten, C. 1934. The caddis flies or Trichoptera of New York state. *New York State Museum Bulletin* 292. 576 pp.

Boettner, G. H., J. S. Elkinton, and C. J. Boettner. 2000. Effects of a biological control introduction on three nontarget native species of saturniid moths. *Conservation Biology* 14: 1798–1806.

Britton, W. E., et al. 1923. *Guide to the insects of Connecticut. Part IV. The Hemiptera or sucking insects of Connecticut.* State Geological and Natural History Survey Bulletin 34.

Connecticut Agricultural Experiment Station: <www.state.ct.us/caes>.

Information on hemlock woolly adelgid and other injurious organisms.

Connecticut Butterfly Atlas Project: <http://peabody.yale.edu/other/chap>.

Connecticut Department of Public Health: <http://www.state.ct.us/dph>.

Information on mosquito-borne viruses.

Cox, R. T., and c. E. Carlton. 2003. A comment on gene introgression versus en masse cycle switching in the evolution of 13-year and 17-year life cycles in periodical cicadas. *Evolution* 57: 428–32.

❧Dunkle, S. W. 2000. *Dragonflies through binoculars: A field guide to dragonflies of North America.* New York: Oxford University Press. 266 pp.

Elliott, N. B., and W. M. Elliott. 1994. Recognition and avoidance of the predator *Phymata americana* Melin on *Solidago odora* Ait. by late season floral visitors. *American Midland Naturalist* 131: 378–80.

Ellis, D. R., D. R. Prokrym, and R. G. Adams. 1999. Exotic lady beetle survey in northeastern United States: *Hippodamia variegata* and *Propylea quatuordecimpunctata* (Coleoptera: Coccinellidae). *Entomological News* 110: 73–84.

Etheredge, J. A., S. M. Perez, O. R. Taylor, and R. Jander. 1999. Monarch butterflies (*Danaus plexippus* L.) use a magnetic compass for navigation. *Proceedings of the National Academy of Sciences* 96 (24): 13845–46.

◆Glassberg, J. 1993. *Butterflies through binoculars: A field guide to butterflies in the Boston/New York/Washington region.* New York: Oxford University Press.

    Contains much natural history information.

Goldstein, E. L. 1975. Island biogeography of ants. *Evolution* 29: 750–62.

    Survey of ants on the Thimble Islands and adjacent mainland.

Heatwole, H., D. M. Davis, and A. M. Wenner. 1964. Detection of mates and hosts by parasitic insects of the genus *Megarhyssa* (Humenoptera: Ichneumonidae). *American Midland Naturalist* 71: 374–81.

◆Heinrich, B. 1974. *In a patch of fireweed.* Cambridge: Harvard University Press.

    Engaging discussion of the lives of various organisms studied in the field.

———. 1996. *The thermal warriors: Strategies of insect survival.* Cambridge: Harvard University Press. xiv + 221 pp.

Hitchcock, S. W. 1974. *Guide to the insects of Connecticut. Part VII. The Plecoptera or stoneflies of Connecticut.* State Geological and Natural History Survey of Connecticut Bulletin 107. vi + 262 pp.

Johannsen, O. A. 1934. Aquatic Diptera. Part I. Nemocera, exclusive of Chironomidae and Ceratopongonidae. Cornell University Agricultural Experiment Station Memoirs 164: 1–70.

Johnson, W. T., and H. H. Lyon. 1991. *Insects that feed on trees and shrubs.* Second edition. Ithaca, N.Y.: Cornell University Press. 560 pp.

    Lavishly illustrated, showing insects and their effects; packed with information.

Kellogg, V. L. 1905. *American insects.* New York: Henry Holt and Company. 647 pp.

Kennedy, C. H. 1915. Notes on the life history and ecology of the dragonflies (Odonata) of Washington and Oregon. *Proceedings of the U.S. National Museum* 49: 259–345.

Kizlinski, M. L., D. A. Orwig, R. C. Cobb, and D. R. Foster. 2002. Direct and indirect ecosystem consequences of an invasive pest on forests dominated by eastern hemlock. *Journal of Biogeography* 29 (10–11): 1489–1503.

Krinsky, W. L., and P. A. Godwin. 1990. Ground beetles in canopy collections in Connecticut (Coleoptera: Carabidae). *Coleopterists Bulletin* 44: 268–70.

———. 1996. Long-horned beetles from the forest canopy in New England and New York (Coleoptera: Cerambycidae). *Coleopterists Bulletin* 50: 236–40.

Krinsky, W. L., and M. K. Oliver. 2001. *Ground beetles of Connecticut (Coleoptera: Carabidae, excluding Cincindelini): An annotated checklist.* State Geological and Natural History Survey of Connecticut Bulletin 117. 308 pp.

    Distribution maps, notes on habitat, for the specialist.

Layne, J. R., Jr., C. L. Edgar, and R. E. Medwith. 1999. Cold hardiness of the woolly bear caterpillar (*Pyrrharctica isabella* Lepidoptera: Arctiidae). *American Midland Naturalist* 141: 293–304.

Lee, R. E., Jr. 1989. Insect cold-hardiness: To freeze or not to freeze. How insects survive low temperatures. *BioScience* 39: 308–13.

Leonard, J. G., and R. T. Bell. 1999. *Northeastern tiger beetles: A field guide to tiger beetles of New England and eastern Canada.* Boca Raton, Fl.: CRC Press. xii + 176 pp.

    Detailed maps, photographs, information on habitat and life history.

Malcolm, S. B., and M. P. Zalucki, editors. 1993. Biology and conservation of the monarch butterfly. *Natural History Museum of Los Angeles County, Science Series* No. 38. 419 pp.

Marshall, D. C., J. R. Cooley, and C. Simon. 2003. Holocene climate shifts, life-cycle plasticity, and speciation in periodical cicadas: A reply to Cox and Carlton. *Evolution* 57: 433–37.

Mason, L. G. 1987. Guerrillas of the goldenrod. *Natural History* 96: 34–39.

    Life history and ecology of ambush bugs.

McClure, M. S. 1987. Biology and control of hemlock woolly adelgid. *Connecticut Agricultural Experiment Station Bulletin* 851.

McClure, M. S., and C. A. S.-J Cheah. 1998. Reshaping the ecology of invading populations of hemlock woolly adelgic, *Adelges tsugae* (Homoptera: Adelgidae), in eastern North America. *Biological Invasions* 1: 247–54.

Merritt, R. W., and K. W. Cummins, eds. 1996. *An introduction to the aquatic insects of North America.* Third edition. Dubuque, Iowa: Kendall/Hunt Publishing Company. 862 pp.

    Identification keys and information on biology and ecology for serious students and professionals.

Mosquito Management Program, Connecticut Department of Environmental Protection: <http://dep.state.ct.us/mosquito/index/as>; also see <http://dep.state.ct.us/mosquito/fact/2001mosqprot.pdf> (Mosquito Protection for People, Property and Pets).

Needham, J. G. 1920. Burrowing mayflies of our larger lakes and streams. *U.S. Bureau of Fisheries Bulletin* 36: 265–92.

Omland, K. S. 2002. Larval habitat and reintroduction site selection for *Cicindela puritana* in Connecticut. *Northeastern Naturalist* 9: 433–50.

Pergande, T. 1901. The life-history of two species of plant-lice, inhabiting both the witch-hazel and birch. *U.S.D.A. Division of Entomology Technical series Bulletin* 9.

Plowright, R. D., and T. M. Laverty. 1984. The ecology and sociobiology of bumblebees. *Annual Review of Entomology* 29: 175–99.

Shorthouse, J. D., and O. Rohfritsch, eds. 1992. *Biology of insect-induced galls.* New York: Oxford University Press. 320 pp.

Storey, K. B., and J. M. Storey. 1988. Freeze tolerance in animals. *Physiological Reviews* 62: 27–84.

———. 1990. Frozen and alive. *Scientific American* 263 (6): 92–97.

Sykes, D. S. *Tiger beetles of Connecticut:* <http://collections2.eeb.uconn.edu/collections/insects/CTBnew/ctb.htm>.

　　Detailed information on biology, life history, and conservation status; color photographs.

Szlanski, A. L., D. S. Sikes, R. Bischof, and M.Fritz. 2000. Population genetics and phylogenetics of the endangered American burying beetle, *Nicrophorus americanus* (Coleopters: Silphidae). *Annals of the Entomological Society of America* 93: 589–94.

Viereck, H. L., A. D. MacGillivray, C. T. Brues, W. M. Wheeler, and S. A. Rohwer. 1916. Guide to the insects of Connecticut. Part III. The Hymenoptera, or wasp-like insects, of Connecticut. *Geological and Natural History Survey of the State of Connecticut Bulletin* 22. 824 pp.

Wagner, D. L., and M. C. Thomas. 1999. The Odonata fauna of Connecticut. *Bulletin of American Odonatology* 5 (4): 59–85.

Wagner, D. L., D. M. Simmonds, and M. C. Thomas. 1995. Three rare gomphids from the lower Connecticut River. *Journal of the New York Entomological Society* 103: 334–36.

Weseloh, R. M. 1997. Evidence for limited dispersal of larval gypsy moth, *Lymantria dispar* L. (Lepidoptera: Lymantriidae). *Canadian Entomologist* 129: 355–61.

———. 1998. Possibility for recent origin of the gypsy moth ((Lepidoptera: Lymantriidae) fungal pathogen *Entomophaga maimaiga* (Zygomcycetes: Entomophthorales) in North America. *Environmental Entomology* 27: 171–77.

Wheeler, A. G., Jr., and E. R. Hoebeke. 1995. *Coccinella novemnotata* in northeastern North America: Historical occurrence and current status (Coleoptera: Coccinellidae). *Proceedings of the Entomological Society of Washington* 97: 701–16.

Chapter 15. Crustaceans

Ahl, R. S., and S. P. Moss. 1999. Status of the nonindigenous crab, *Hemigrapsus sanguineus*, At Greenwich Point, Connecticut. *Northeastern Naturalist* 6: 221–24.

✺Berrick, S. 1986. *Crabs of Cape Cod.* Cape Cod Museum of Natural History, Natural History Series No. 3. 76 pp.

　　Much information on biology and natural history, accessibly written, excellent illustrations.

Kunkel, B. W. 1918. *The Arthrostraca of Connecticut.* State Geological and Natural History Survey of Connecticut Bulletin 26. 261 pp.

Ledesma, M. E. and N. J. O'Connor. 2001. Habitat and diet of the non-native crab *Hemigrapsus sanguineus* in southeastern New England. *Northeastern Naturalist* 8: 63–78.

McDermott, J. J. 2001. Symbionts of the hermit crab *Pagurus longicarpus* Say, 1817 (Decapoda: Anomura): New observations from New Jersey waters and a review of all known relationships. *Proceedings of the Biological Society of Washington* 114: 624–39.

Smith, D. G. 1987. The genus *Synurella* in New England (Amphipoda, Crangonyctidae). *Crustaceana* 53: 304–306.

———. 1997. An annotated checklist of malacostracans (Crustacea) inhabiting southern New England fresh waters. *Journal of Freshwater Ecology* 12: 217–23.

Taylor, C. A., M. L. Warren, J. F. Fitzpatrick, H. H. Hobbs, R. F. Jezerinac, W. L. Pflieger, H. W. Robison. 1996. Conservation status of crayfishes of the United States and Canada. *Fisheries* 21 (4): 25–38.

Verrill, A. E., and S. I. Smith. 1874. *Report upon the invertebrate animals of Vineyard Sound and adjacent waters, with an account of the physical features of the region.* Washington, D.C.: Government Printing Office.

Chapter 16. Echinoderms and Tunicates

Coe, W. R. 1912. *Echinoderms of Connecticut.* State Geological and Natural History Survey of Connecticut Bulletin 19. 152 pp.

Chapter 17. Fishes: Lampreys, Sharks and Rays, and Bony Fishes

Bigelow, H. B., and W. C. Schroeder. 1953. *Fishes of the Gulf of Maine.* Washington, D.C.: Fishery Bulletin of the Fish and Wildlife Service Volume 53, Fishery Bulletin 74.

　　Classic account, recently updated by Collette and Klein-MacPhee (2002).

Buckley, J., and B. Kynard. 1985. Yearly movements of shortnose sturgeon in the Connecticut River. *Transactions of the American Fisheries Society* 114: 813–20.

✺Collette, B. B., and G. Klein-MacPhee, eds. 2002. *Bigelow and Schroeder's fishes of the Gulf of Maine.* Third edition. Washington, D.C.: Smithsonian Institution Press. xxxiv + 748 pp.

　　Best source of information on the biology, ecology, and human use of saltwater fishes of the northwest Atlantic.

Durbin, A. G., S. W. Nixon, and C. A. Oviatt. 1979. Effects of the spawning migration of the alewife, *Alosa pseudoharengus,* on freshwater ecosystems. *Ecology* 60: 8–17.

Gephard, S., P. Moran, and E. Garcia-Vazquez. 2000. Evidence of successful natural reproduction between brown trout and mature male Atlantic salmon parr. *Transactions of the American Fisheries Society* 129: 301–306.

Gido, K. B., and J. H. Brown. 1999. Invasion of North American drainages by alien fish species. *Freshwater Biology* 42: 387–99.

Gottschall, K. F., M. W. Johnson, and D. G. Simpson. 2000. See references for chapter 3.

Halpin, P. M.,and K. L. M. Martin. 1999. Aerial respiration in the salt marsh fish *Fundulus heteroclitus* (Fundulidae). *Copeia* 1999: 743–48.

✺Hartel, K. E., D. B. Halliwell, and A. E. Launer. 2002. *Inland fishes of Massachusetts.* Lincoln: Massachusetts Audubon Society. xiii + 328 pp.

　　Useful for natural history information for Connecticut fishes.

Kynard, B., M. Horgan, M. Kieffer, and D. Seibel. 2000. Habitats used by shortnose sturgeon in two Massachusetts rivers, with notes on estuarine Atlantic sturgeon: A hierarchical approach. *Transactions of the American Fisheries Society* 129: 487–503.

Richards, S. W. 1963. The demersal fish population of Long Island Sound. I. Species composition and relative abundance in two localities, 1956–57. *Bulletin of the Bingham Oceanographic Collection* 18: 5–30.

Root, K. V. 2002. Evaluating risks for threatened aquatic species: The shortnose sturgeon in the Connecticut River. *American Fisheries Society Symposium* 28: 45–54.

Savoy, T., and D. Pacileo. 2003. Movements and important habitats of subadult Atlantic sturgeon in Connecticut waters. *Transactions of the American Fisheries Society* 132: 1–8.

Schmidt, R. E., and W. R. Whitworth. 1979. Distribution and habitat of the swamp darter *(Etheostoma fusiforme)* in southern New England. *American Midland Naturalist* 102: 408–13.

Smith, C. L. 1985. *The inland fishes of New York State.* New York State Department of Environmental Conservation.

Useful for natural history information for Connecticut fishes.

Taubert, B. D. 1980. Reproduction of shortnose sturgeon *(Acipenser brevirostrum)* in Holyoke Pool, Connecticut River, Massachusetts. *Copeia* 1980: 114–17.

Thomson, K. S., et al. 1978. *Saltwater fishes of Connecticut.* State Geological and Natural History Survey of Connecticut Bull. 105.

Minimal natural history information, for which see Collette and Klein-MacPhee (2002).

Warshaw, S. J. 1972. Effects of alewives *(Alosa pseudoharengus)* on the zooplankton of Lake Wononskopomuc, Connecticut. *Limnology and Oceanography* 17: 816–24.

Whitworth, W. R. 1996. *Freshwater fishes of Connecticut.* Second edition. State Geological and Natural History Survey of Connecticut, Bulletin 114. 243 pp.

Limited natural history information, for which see Smith (1985) and Hartel, et al. (2002).

Chapter 18. Amphibians

Bogart, J. P., and M. W. Klemens. 1997. Hybrids and genetic interactions of mole salamanders *(Ambystoma jeffersonianum* and *A. laterale)* (Amphibia: Caudata) in New York and New England. *American Museum Novitates* 3218. 78 pp.

deMaynadier, P. G., and M. L. Hunter, Jr. 1995. The relationship between forest management and amphibian ecology: A review of the North American literature. *Environmental Reviews* 3: 230–61.

———. 1998. Effects of silvicultural edges on the distribution and abundance of amphibians in Maine. *Conservation Biology* 12: 340–52.

———. 1999. Forest canopy closure and juvenile emigration by pool-breeding amphibians in Maine. *Journal of Wildlife Management* 63: 441–50.

———. 2000. Road effects on amphibian movements in a forested landscape. *Natural Areas Journal* 20: 56–65.

Frisbie, M. P., J. P. Costanzo, and R. E. Lee, Jr. 2000. Physiological and ecological aspects of low-temperature tolerance in embryos of the wood frog, *Rana sylvatica. Canadian Journal of Zoology* 78: 1032–41.

Gibbs, J. P. 1998. Amphibian movements in response to forest edges, roads, and streambeds in southern New England. *Journal of Wildlife Management* 62: 584–89.

———. 1998. Distribution of woodland amphibians along a forest fragmentation gradient. *Landscape Ecology* 13: 263–68.

◆Klemens, M. W. 1993. *Amphibians and reptiles of Connecticut and adjacent regions.* State Geological and Natural History Survey of Connecticut Bulletin 112. xii + 318 pp.

Best single source on distribution, biology, and ecology of herpetofauna in Connecticut.

———. 2000. *Amphibians and reptiles in Connecticut: A checklist with notes on conservation status, identification, and distribution.* Connecticut Department of Environmental Protection, DEP Bulletin 32.

Essentially a less expensive condensation of the preceding publication.

Layne, J. R., Jr., J. P. Costanzo, and R. E. Lee, Jr. 1998. Freeze duration influences postfreeze survival in the frog *Rana sylvatica. Journal of Experimental Zoology* 280: 197–201.

Layne, J. R., Jr., and J. Kefauver. Freeze tolerance and postfreeze recovery in the frog *Pseudacris crucifer. Copeia* 1997: 260–64.

Paton, P., S. Stevens, and L. Longo. 2000. Seasonal phenology of amphibian breeding and recruitment at a pond in Rhode Island. *Northeastern Naturalist* 7: 255–69.

Schmid, W. D. 1982. Survival of frogs at low temperature. *Science* 215: 697–98.

Semlitsch, R. D. 2003. *Amphibian conservation.* Washington, D.C.: Smithsonian Institution. xi + 324 pp.

Scholarly but readable reviews of effects of habitat destruction and alteration, climate change, chemical contamination, disease and pathogens, invasive species, and commercial exploitation, and potential solutions, including forest management guidelines.

Storey, K. B., and J. M. Storey. 1988. Freeze tolerance in animals. *Physiological Reviews* 68: 27–84.

———. 1990. Frozen and alive. *Scientific American* 263 (6): 92–97.

Chapter 19. Reptiles

Aresco, M. J. 1996. Natural history notes: *Malaclemys terrapin terrapin* (northern diamondback terrapin). Reproduction and next predation. *Herpetological Review* 27: 77.

Burke, V. J., S. J. Morreale, and E. A. Standora. 1994. Diet of the Kemp's ridley sea turtle, *Lepidochelys kempii*, in New York waters. *Fishery Bulletin* 92: 26–32.

Burke, V. J., E. A. Standora, and S. J. Morreale. 1991. Factors affecting strandings of cold-stunned juvenile Kemp's ridley and loggerhead sea turtles in Long Island, New York. *Copeia* 1991: 1136–38.

Churchill, T. A., and K. B. Storey. 1992. Freezing and survival of the garter snake *Thamnophis sirtalis parietalis. Canadian Journal of Zoology* 70: 99–105.

Claussen, D. L., P. M. Daniel, S. Jiang, and N. A. Adams. 1991. Hibernation in the eastern box turtle, *Terrapene c. carolina. Journal of Herpetology* 25: 334–41.

Compton, B. W., J. M. Rhymer, and M. McCollough. 2002. Habitat selection by wood turtles *(Clemmys insculpta)*: An application of paired logistic regression. *Ecology* 83: 833–43.

Costanzo, J. P. 1989. Effects of humidity, temperature, and submergence behavior on survivorship and energy use in hibernating garter snakes, *Thamnophis sirtalis*. *Canadian Journal of Zoology* 67: 2486–92.

Costanzo, J. P., J. B. Iverson, M. F. Wright, and R. E. Lee, Jr. 1995. Cold hardiness and overwintering strategies of hatchlings in an assemblage of northern turtles. *Ecology* 76: 1772–85.

Crocker, C. E., R. A. Feldman, G. R. Ultsch, and D. C. Jackson. 2000. Overwintering behavior and physiology of eastern painted turtles *(Chrysemys picta picta)* in Rhode Island. *Canadian Journal of Zoology* 78: 936–42.

Garber, S. D., and J. Burger. 1995. A 20-yr study documenting the relationship between turtle decline and human recreation. *Ecological Applications* 5: 1151–62.

Graham, T. E. 1995. Habitat use and population parameters of the spotted turtle, *Clemmys guttata*, a species of special concern in Massachusetts. *Chelonian Conservation and Biology* 1 (3): 207–14.

Joyal, L. A., M. McCollough, and M. L. Hunter, Jr. 2001. Landscape ecology approaches to wetland species conservation: A case study of two turtle species in southern Maine. *Conservation Biology* 14: 1775–62.

Kiviat, E., and J. G. Barbour. 1996. Wood turtles, *Clemmys insculpta*, in the fresh-tidal Hudson River. *Canadian Field-Naturalist* 110: 341–43.

◆Klemens, M. W. 1993, 2000. See references for chapter 18.

Milam, J. C., and S. M. Melvin. 2001. Density, habitat use, movements, and conservation of spotted turtles *(Clemmys tuttata)* in Massachusetts. *Journal of Herpetology* 35: 418–27.

Morreale, S. J., A. B. Meylan, S. S. Sadove, and E. A. Standora. 1992. Annual occurrence and winter mortality of marine turtles in New York waters. *Journal of Herpetology* 26: 301–308.

Perillo, K. M. 1997. Seasonal movements and habitat preferences of spotted turtles *(Clemmys guttata)* in north central Connecticut. *Chelonian Conservation and Biology* 2 (3): 445–47.

◆Petersen, R. C., and R. W. Fritsch. 1986. *Connecticut venomous snakes: The timber rattlesnake and northern copperhead.* State Geological and Natural History Survey Bulletin 111: 1–48.

Roosenburg, W. M., W. Cresko, M. Modesitte, and M. B. Robbins. 1997. Diamondback terrapin *(Malaclemys terrapin)* mortality in crab pots. *Conservation Biology* 2: 1166–72.

Storey, J. M., and K. B. Storey. 1992. Out cold: The winter life of painted turtles. *Natural History* 101 (1): 22–25.

Storey, K. B., J. R. Layne, Jr., M. M. Cutwa, T. A. Churchill, and J. M. Storey. 1993. Freezing survival and metabolism of box turtles, *Terrapene carolina*. *Copeia* 1993: 628–34.

Tyning, T. F., and D. W. Kimball, eds. 1992. *Conservation of the timber rattlesnake in the northeast.* The timber rattlesnake in New England—A symposium. Lincoln: Massachusetts Audubon Society, . v + 40 pp.

Deals with symposia held in 1978 and 1991.

Ultsch, G. R., R. W. Hanley, and T. R. Bauman. 1985. Responses to anoxia during simulated hibernation in northern and southern painted turtles. *Ecology* 66: 388–95.

Whitelaw, D. M., and R. N. Zajac. 2002. Assessment of prey availability for terrapins in a Connecticut salt marsh. *Northeastern Naturalist* 9: 407–18.

Chapter 20. Birds

American Ornithologist's Union. 1998. *Check-list of North American birds.* Seventh edition. Washington, D.C.: American Ornithologists' Union.

Askins, R. A. 1990. *Birds of the Connecticut College Arboretum: Population changes over forty years.* Connecticut College Arboretum Bulletin No. 31. v + 43 pp.

———. 1999. History of grassland birds in eastern North America. *Studies in Avian Biology* 19: 60–71.

◆———. 2000. *Restoring North America's birds: Lessons from landscape ecology.* New Haven: Yale University Press. xiii + 320 pp.

Askins, R. A., and M. J. Philbrick. 1987. Effects of changes in regional forest abundance on the decline and recovery of a forest bird community. *Wilson Bulletin* 99: 7–21.

Askins, R. A., J. F. Lynch, and R. Greenberg. 1990. Population declines in migratory birds in eastern North America. *Current Ornithology* 7: 1–57.

Askins, R. A., M. J. Philbrick, and D. S. Sugeno. 1987. Relationship between regional abundance of forest and the composition of forest bird communities. *Biological Conservation* 39: 129–52.

Benoit, L. K., and R. A. Askins. 1999. Impact of the spread of *Phragmites* on the distribution of birds in Connecticut tidal marshes. *Wetlands* 19: 194–208.

———. 2002. Relationship between habitat area and the distribution of tidal marsh birds. *Wilson Bulletin* 114: 314–23.

◆Bevier, L. R., ed. 1994. *The atlas of breeding birds of Connecticut.* State Geological and Natural History Survey of Connecticut Bull. 113.

Best source of information on birds nesting in Connecticut; maps, species accounts, data from surveys done from 1982 to 1986.

◆Billings, G. 1990. *Birds of prey in Connecticut.* Torrington, Conn.: Rainbow Press.

Extensive detailed but nontechnical information.

Borowicz, V. A. 1988. Fruit consumption by birds in relation to fat content of pulp. *American Midland Naturalist* 119: 121–27.

Brawley, A. H., R. S. Warren, and R. A. Askins. 1998. Bird use of restoration and reference marshes within the Barn Island Wildlife Management Area, Stonington, Connecticut. *Environmental Management* 22: 625–33.

Bushman, E. S., and G. D. Therres. 1988. *Habitat management guidelines for forest interior breeding birds of coastal Maryland.* Maryland Department of Natural Resources, Forest, Park & Wildlife Service, Wildlife Technical Publications 88-1.

Chalfoun, A. D., F. R. Thompson, III, and M. J. Ratnaswamy. 2002. Nest predators and fragmentation: A review and meta-analysis. *Conservation Biology* 16: 306–18.

Connecticut Ornithological Association (www.ctbirding.org). *The Connecticut Warbler.*

Quarterly journal; detailed information on birds and birding in Connecticut, most useful for hard-core birders.

Conover, M. R., and G. G. Chasko. 1985. Nuisance Canada geese problems in the eastern United States. *Wildlife Society Bulletin* 13: 228–33.

Conover, M. R., and G. S. Kania. 1994. Impact of interspecific aggression and herbivory by mute swans on native waterfowl and aquatic vegetation in New England. *Auk* 111: 744–48.

Corser, J. D., M. Amaral, C. J. Martin, and C. C. Rimmer. 1999. Recovery of a cliff-nesting peregrine falcon, *Falco peregrinus*, population in northern New York and New England, 1984–1996. *Canadian Field-Naturalist* 113: 472–80.

Crossland, C. R., and S. P. Vander Kloet. 1996. Berry consumption by the American robin, *Turdus migratorius*, and the subsequent effect on seed germination, plant vigor, and dispersal of the lowbush blueberry, *Vaccinium angustifolium*. *Canadian Field-Naturalist* 110: 303–309.

Dearborn, D. C. 1996. Video documentation of a brown-headed cowbird nestling ejecting an indigo bunting nestling from the next. *Condor* 98: 645–49.

DeGraaf, R. M., and J. H. Rappole. 1995. *Neotropical migratory birds: Natural history, distribution, and population change.* Ithaca, N.Y.: Cornell University Press.

☙Devine, A., and D. G. Smith. 1996. *Connecticut birding guide.* Dexter, Mich.: Thomson-Shore. xiii + 569 pp.

DiQuinzio, D. A., P. W. C. Paton, and W. R. Eddleman. 2001. Site fidelity, philopatry, and survival of promiscuous saltmarsh sharp-tailed sparrows in Rhode Island. *Auk* 118: 888–99.

Elliott, P. F. 1999. Killing of host nestlings by the brown-headed cowbird. *Journal of Field Ornithology* 70: 55–57.

Gale, G. A., L. A. Hanners, and S. R. Patton. 1997. Reproductive success of worm-eating warblers in a forested landscape. *Conservation Biology* 11: 246–50.

Gervais, J. A., and N. T. Wheelwright. 1994. Winter fruit removal in four plant species in Maine. *Maine Naturalist* 2 (1): 15–24.

Greenberg, R., and S. Droege. 1999. On the decline of the rusty blackbird and the use of ornithological literature to document long-term population trends. *Conservation Biology* 13: 553–59.

Hammerson, G. A. 1988. Diet of the common barn-owl in Middlefield, Connecticut. *Connecticut Warbler* 8 (3): 60.

Hartup, B. K., J. M. Bickal, A. A. Dhondt, D. H. Ley, and G. V. Kollias. 2001. Dynamics of conjunctivitis and *Mycoplasma gallisepticum* infections in house finches. *Auk* 118: 327–33.

Heinrich, B. 1993. Kinglets' realm of cold. *Natural History* 102 (2): 4–9.

Helzer, C. J., and D. E. Jelinski. 1999. The relative importance of patch area and perimeter-area ratio to grassland breeding birds. *Ecological Applications* 9: 1448–58.

Heusmann, H. W., T. J. Early, and B. J. Nikula. 2000. Evidence of an increasing hooded merganser population in Massachusetts. *Wilson Bulletin* 112: 413–15.

Hodgman, T. P., W. G. Shriver, and P. D. Vickery. 2002. Redefining range overlap between the sharp-tailed sparrows of coastal New England. *Wilson Bulletin* 114: 38–43.

Holmes, R. T., and T. W. Sherry. 2001. Thirty-year bird population trends in an unfragmented temperate deciduous forest: Importance of habitat change. *Auk* 118: 589–609.

Kilpatrick, H. J., T. P. Husband, and C. A. Pringle. 1988. Winter roost site characteristics of eastern wild turkeys. *Journal of Wildlife Management* 52: 461–63.

Kluza, D. A., C. R. Griffin, and R. M. DeGraaf. 2000. Housing developments in rural New England: Effects on forest birds. *Animal Conservation* 3: 15–26.

Kreuger, B., and D. A. Potter. 1994. Changes in saponins and tannins in ripening holly fruits and effects of fruit consumption on nonadapted insect herbivores. *American Midland Naturalist* 132: 183–91.

Krischik, V., E. S. McCloud, and J. A. Davidson. 1989. Selective avoidance by vertebrate frugivores of green holly berries infested with a cecidomyiid fly (Diptera: Cecidomyiidae). *American Midland Naturalist* 121: 350–54.

Lopez, J. M. 2001. The impact of communication towers on Neotropical songbird populations. *Endangered Species Update.* 18: 50–54.

Martínez del Rio, C. 1990. Sugar preferences in hummingbirds: The influence of subtle chemical difference on food choice. *Condor* 92: 1022–30.

McCarty, J. P., and A. L. Secord. 1999. Reproductive ecology of tree swallows (*Tachycineta bicolor*) with high levels of PCB contamination. *Environmental Toxicology and Chemistry* 18: 1433–39.

McKechnie, A. E., and B. G. Lovegrove. 2002. Avian facultative hypothermic responses: A review. *Condor* 104: 705–24.

Merola, P. R., and G. G. Chasko. 1989. *Waterfowl in Connecticut.* Connecticut Department of Environmental Protection, Wildlife Bureau, Publication WF-4.

Meyer, G. A., and M. C. Witmer. 1998. Influence of seed processing by frugivorous birds on germination success of three North American shrubs. *American Midland Naturalist* 140: 129–39.

Moore, J. A., C. Lewis, and M. R. Anderson. 1999. Diet of a great horned owl (*Bubo virginianus*) on a small coastal Connecticut Island. *Postilla* 220: 1–10.

Norment, C. 2002. On grassland bird conservation in the Northeast. *Auk* 119: 271–79.

Norris, D. R., and B. J. M. Stutchbury. 2001. Extraterritorial movements of a forest songbird in a fragmented landscape. *Conservation Biology* 15: 729–36.

Olson, J. M., W. R. Dawson, and J. J. Camilliere. 1988. Fat from black-capped chickadees: Avian brown adipose tissue? *Condor* 90: 529–37.

Parrish, J. D. 1997. Patterns of frugivory and energetic condition in Nearctic landbirds during autumn migration. *Condor* 99: 681–97.

Place, A. R., and E. W. Stiles. 1992. Living off the fat of the land: Bayberries and yellow-rumped warblers. *Auk* 109: 334–45.

Rappole, J. H., and M. V. McDonald. 1994. Cause and effect in population declines of migratory birds. *Auk* 111: 652–60.

Rodewald, P. G., and M. C. Brittingham. 2002. Habitat use and behavior of mixed species landbird flocks during fall migration. *Wilson Bulletin* 114: 87–98.

Rosenberg, K. V., R. W. Rohrbaugh, Jr., S. E. Barker, R. S. Hames, J. D. Lowe, and A. A. Dhondt. 1999. *A land manager's guide to improving habitat for scarlet tanagers and other forest-interior birds.* Ithaca, N.Y.: Cornell Lab of Ornithology. 23 pp.

Rosgen, D., and G. Billings. 1996. *Finding birds in Connecticut: A habitat-based guide to 450 sites.* Torrington, Connecticut: Rainbow Press. vii + 640 pp.

Rosgen, D., and J. M. Zingo. 1993. The Connecticut bluebird restoration project: Successfully managing for eastern bluebird and other native cavity-nesting birds. *Connecticut Warbler* 13: 91–103.

Saarela, S., et al. 1989. Do birds possess brown adipose tissue? *Comparative Biochemistry and Physiology* 92A: 219–28.

Schmidt, K. A., and R. S. Ostfeld. 2003. Songbird populations in fluctuating environments: Predator responses to pulsed resources. *Ecology* 84: 406–15.

Schmidt, K. A., and C. J. Whelan. 1999. Effects of *Lonicera* and *Rhamnus* on songbird nest predation. *Conservation Biology* 13: 1502–1506.

Smith, D. G., and A. Devine. 1993. Winter ecology of the long-eared owl in Connecticut. *Connecticut Warbler* 13: 44–53.

Smith, D. G., and R. Gilbert. 1984. Eastern screech-owl home range and use of suburban habitats in southern Connecticut. *Journal of Field Ornithology* 55: 322–29.

Smith, D. G., T. Becker, and A. Devine. 2001. The sharp-shinned hawk in Connecticut. *Connecticut Warbler* 21: 124–29.

Smith, D. G., T. Bosakowski, and A. Devine. 1999. Nest selection by urban and rural great horned owls in the northeast. *Journal of Field Ornithology* 70: 535–42.

Stutchbury, B. J. M. 1997. Effects of female cowbird removal on reproductive success of hooded warblers. *Wilson Bulletin* 109: 74–81.

Titus, K., and M. R. Fuller. 1990. Recent trends in counts of migrant hawks from northeastern North America. *Journal of Wildlife Management* 54: 463–70.

Tsipoura, N., and J. Burger. 1999. Shorebird diet during spring migration stopover on Delaware Bay. *Condor* 101: 635–44.

Villard, M.-A. 1998. On forest-interior species, edge avoidance, area sensitivity, and dogmas in avian conservation. *Auk* 115: 801–805.

Walk, J. W., and R. E. Warner. 1999. Effects of habitat area on the occurrences of grassland birds in Illinois. *American Midland Naturalist* 141: 339–44.

Willson, M. F. 1986. Avian frugivory and seed dispersal in eastern North America. *Current Ornithology* 3: 223–79.

Willson, M. F., and C. J. Whelan. 1990. The evolution of fruit color in fleshy-fruited plants. *American Naturalist* 136: 790–809.

❧Zeranski, J. D., and T. R. Baptist. 1990. *Connecticut birds.* Hanover, N.H.: University Press of New England.

Chapter 21. Mammals

Augustine, D. J., and L. E. Frelich. 1998. Effects of white-tailed deer on populations of an understory forb in fragmented deciduous forests. *Conservation Biology* 12: 995–1004.

Bozinovic, F., and J. F. Merritt. 1992. Summer and winter thermal conductance of *Blarina brevicauda* (Mammalia: Insectivora: Soricidae) inhabiting the Appalachian Mountains. *Annals of Carnegie Museum* 61 (1): 33–37.

Bruseo, J. A., and R. E. Barry, Jr. 1995. Temporal activity of syntopic *Peromyscus* in the central Appalachians. *Journal of Mammalogy* 76: 78–82.

Buech, R. R., and D. J. Rugg. 1989. Temperature in beaver lodges and bank dens in a near-boreal environment. *Canadian Journal of Zoology* 67: 1061–66.

Burke da Silva, K., D. L. Kramer, and D. M. Weary. 1994. Context-specific alarm calls of the eastern chipmunk, *Tamias striatus*. *Canadian Journal of Zoology* 72: 1087–92.

Burke da Silva, K., C. Mahan, and J. da Silva. 2002. The trill of the chase: Eastern chipmunks call to warn kin. *Journal of Mammalogy* 83: 546–622.

Dietz, B. A., A. E. Hagerman, and G. W. Barrett. 1994. Role of condensed tannin on salivary tannin-binding proteins, bioenergetics, and nitrogen digestibility in *Microtus pennsylvanicus*. *Journal of Mammalogy* 75: 880–89.

Drickamer, L. C. 1987. Influence of time of day on captures of two species of *Peromyscus* in a New England deciduous forest. *Journal of Mammalogy* 68: 702–703.

Dumont, A., and M. Crête. 1996. The meningeal worm, *Parelaphostrongylus tenuis*, a marginal limiting factor for moose, *Alces alces,* in southern Quebec. *Canadian Field-Naturalist* 110: 413–18.

Ellingwood, M. R., and S. L. Caturano. 1988. *An evaluation of deer management options.* New England Chapter of The Wildlife Society and Northeast Deer Technical Committee. Publication DR-11. 12 pp.

Elowe, K. D., and W. E. Dodge. 1989. Factors affecting black bear reproductive success and cub survival. *Journal of Wildlife Management* 53: 962–68.

Study done in western Massachusetts.

English, E. I. 1994. Vegetational gradients and proximity to woodchuck *(Marmota monax)* burrows in an old field. *Journal of Mammalogy* 75: 775–80.

Fascione, N., L. G. L. Osborn, S. R. Kendrot, and P. C. Paquet. 2001. Canis soupus: Eastern wolf genetics and its implications for wolf recovery in the northeast United States. *Endangered Species Update* 18 (4): 159–63.

Ferron, J. 1996. How do woodchucks *(Marmota monax)* cope with harsh winter conditions? *Journal of Mammalogy* 77: 412–16.

Freeman, M. 2003. Working the bear traps in Connecticut forests. *Natural New England* (14): 6–9.

French, A. R. 2000. Interdependency of stored food and changes in body temperature during hibernation of the eastern chipmunk, *Tamias straitus*. *Journal of Mammalogy* 81: 979–85.

Fridell, R. A., and J. A. Litvaitis. 1991. Influence of resource distribution and abundance on home-range characteristics of southern flying squirrels. *Canadian Journal of Zoology* 69 2589–93.

❧Godin, A. J. 1977. *Wild mammals of New England.* Baltimore: Johns Hopkins University Press. xii + 304 pp.

Dated but still useful.

Gompper, M. E. 2002. Top carnivores in the suburbs? Ecological and conservation issues raised by colonization of northeastern North America by coyotes. *BioScience* 52: 185–90.

Gould, E., W. McShea, and T. Grand. 1993. Function of the star in the star-nosed mole, *Condylura cristata*. *Journal of Mammalogy* 74: 108–16.

Griesemer, S. J., T. K. Fuller, and R. M. DeGraaf. 1996. Denning patterns of porcupines, *Erethizon dorsatum*. *Canadian Field-Naturalist* 110: 634–37.

———. 1998. Habitat use by porcupines, *Erethizon dorsatum*, in central Massachusetts: Effects of topography and forest composition. *American Midland Naturalist* 140: 271–79.

Hammerson, G. A. 1994. Beaver *(Castor candensis)*: Ecosystem alterations, management, and monitoring. *Natural Areas Journal* 14:44–57.

Harrison, D. J., J. A. Bissonette, and J. A. Sherburne. 1989. Spatial relationships between coyotes and red foxes in eastern Maine. *Journal of Wildlife Management* 53: 181–85.

Heinrich, B. 1991. Nutcracker sweets. *Natural History* 100 (2): 4–8. Squirrel use of maple sap.

Heske, E. J. 1995. Mammalian abundances on forest-farm edges versus forest interiors in southern Illinois: Is there an edge effect? *Journal of Mammalogy* 76: 562–68.

Hoff, J. G. 1987. Status and distribution of two species of cottontail rabbits, *Sylvilagus transitionalis* and *S. floridanus*, in southeastern Massachusetts. *Canadian Field-Naturalist* 101: 88–89.

Hossler, R. J., J. B. McAninch, and J. D. Harder. 1994. Maternal denning behavior and survival of juveniles in opossums in southeastern New York. *Journal of Mammalogy* 75: 60–70.

Humphries, M. M., D. W. Thomas, C. L. Hall, J. R. Speakman, and K. L. Kramer. 2002. The energetics of autumn mast hoarding in eastern chipmunks. *Oecologia* 133: 30–37.

Kilpatrick, H. J., and P. W. Rego. 1994. Influence of season, sex, and site availability on fisher *(Martes pennanti)* rest-site selection in the central hardwood forest. *Canadian Journal of Zoology* 72: 1416–19.

Kilpatrick, H. J., S. M. Spohr, and G. G. Chasko. 1997. A controlled deer hunt on a state-owned coastal reserve in Connecticut: Controversies, strategies, and results. *Wildlife Society Bulletin* 24: 451–56.

Kilpatrick, H. J., S. M. Spohr, and K. K. Lima. 2001. Effects of population reduction on home ranges of female white-tailed deer at high densities. *Canadian Journal of Zoology* 79: 949–54.

Lehman, N., et al. 1991. Introgression of coyote mitochondrial DNA into sympatric North American gray wolf populations. *Evolution* 45: 104–19.

Litvaitis, J. A., and D. J. Harrison. 1989. Bobcat-coyote niche relationships during a period of coyote population increase. *Canadian Journal of Zoology* 67: 1180–88.

LoGiudice, K. 2003. Trophically transmitted parasites and the conservation of small populations: Raccoon roundworm and the imperiled Allegheny woodrat. *Conservation Biology* 17: 258–66.

Mahan, C. G., and R. H. Yahner. 1998. Lack of population response by eastern chipmunks *(Tamias straitus)* to forest fragmentation. *American Midland Naturalist* 140: 382–86.

———. 1999. Effects of forest fragmentation on behaviour patterns in the eastern chipmunk *(Tamias straitus)*. *Canadian Journal of Zoology* 77: 1991–97.

Maier, T. J. 2002. Long-distance movements by female white-footed mice, *Peromyscus leucopus*, in extensive mixed-wood forest. *Canadian Field-Naturalist* 116: 108–11.

Marfori, M. A., P. G. Parker, T. G. Gregg, J. G. Vendenbergh, and N. G. Solomon. 1997. Using DNA fingerprinting to estimate relatedness within social groups of pine voles. *Journal of Mammalogy* 78: 715–24.

McShea, W. J., and J. H. Rappole. 2002. Managing the abundance and diversity of breeding bird populations through manipulation of deer populations. *Conservation Biology* 14: 1161–70.

Merritt, J. F. 1986. Winter survival adaptations of the short-tailed shrew *(Blarina brevicauda)* in an Appalachian montane forest. *Journal of Mammalogy* 67: 450–64.

Merritt, J. F. 1995. Seasonal thermogenesis and changes in body mass of masked shrews, *Sorex cinereus*. *Journal of Mammalogy* 76: 1020–35.

Merritt, J. F., and D. A. Zegers. 1991. Seasonal thermogenesis and body-mass dynamics of *Clethrionomys gapperi*. *Canadian Journal of Zoology* 69: 2771–77.

Merritt, J. F., D. A. Zegers, and L. R. Rose. 2001 Seasonal thermogenesis of southern flying squirrels *(Glaucomys volans)*. *Journal of Mammalogy* 82: 51–64.

Miller, D. H., and L. L. Getz. 1969. Life-history notes on *Microtus pinetorum* in central Connecticut. *Journal of Mammalogy* 50: 777–84.

Morrison, P., and F. A. Ryser. 1962. Metabolism and body temperature in a small hibernator, the meadow jumping mouse, *Zapus hudsonius*. *Journal of Cellular Physiology* 60: 169–80.

Nowak, R. M. 2002. The original status of wolves in eastern North America. *Southeastern Naturalist* 1: 95–130.

Nupp, T. E., and R. K. Swihart. 1998. Effects of forest fragmentation on population attributes of white-footed mice and eastern chipmunks. *Journal of Mammalogy* 79: 1234–43.

Ostfeld, R. S., and C. D. Canham. 1993. Effects of meadow vole population density on tree seedling survival in old fields. *Ecology* 74: 1792–1801.

Page, L. K., R. K. Swihart, and K. R. Kazacos. 1993. Seed preferences and foraging by granivores at raccoon latrines in the transmission dynamics of raccoon roundworm *(Baylisascaris procyonis)*. *Canadian Journal of Zoology* 79: 616–22.

Pepe, T. Coyotes content with smaller ranges in resource-rich Connecticut. *Natural New England* (14): 26–27.

Pietz, P. J., and D. A. Granfors. 2000. White-tailed deer *(Odocoileus virginianus)* predation on grassland songbird nestlings. *American Midland Naturalist* 144: 419–22.

Pusenius, J., R. S. Ostfeld, and F. Keesing. 2000. Patch selection and tree-seedling predation by resident vs. immigrant meadow voles. *Ecology* 81: 2951–56.

Robbins, C. T. 1993. *Wildlife feeding and nutrition*. San Diego: Academic Press.

Roze, U. 1985. How to select, climb, and eat a tree. *Natural History* 94: (5) 62–68. Porcupine feeding behavior, written for a general audience.

Russell, F. L., D. B. Zippin, and N. L. Fowler. 2001. Effects of white-tailed deer *(Odocoileus virginianus)* on plants, plant populations, and communities: A review. *American Midland Naturalist* 146: 1–26.

Sawyer, T. G., R. L. Marchinton, and K. V. Miller. 1989. Response of female white-tailed deer to scrapes and antler rubs. *Journal of Mammalogy* 70: 431–33.

Schmitz, O. J., and T. D. Nudds. 1994. Parasite-mediated competition in deer and moose: How strong is the effect of meningeal worm on moose? *Ecological Applications* 4: 91–103.

Smith, D. F., and J. A. Litvaitis. 2000. Foraging strategies of sympatric lagomorphs: Implications for differential success in fragmented landscapes. *Canadian Journal of Zoology* 78: 2134–41. Deals with eastern and New England cottontails.

Stabb, M. A., M. E. Gartshore, and P. L. Aird. 1989. Interactions of southern flying squirrels, *Glaucomys volans*, and cavity-nesting birds. *Canadian Field-Naturalist* 103: 401–403.

Stapp, P. 1992. Energetic influences on the life history of *Glaucomys volans*. *Journal of Mammalogy* 73: 914–20.

Swihart, R. K., and P. M. Picone. 1995. Use of woodchuck burrows by small mammals in agricultural habitats. *American Midland Naturalist* 133: 360–63.

———. 1998. Selection of mature growth stages of coniferous browse in temperate forests by white-tailed deer *(Odocoileus virginianus)*. *American Midland Naturalist* 139: 269–74.

Tilghman, N. G. 1989. Impacts of white-tailed deer on forest regeneration in northwestern Pennsylvania. *Journal of Wildlife Management* 53: 524–32.

Tomasi, T. E. 1979. Echolocation by the short-tailed shrew *Blarina brevicauda*. *Journal of Mammalogy* 60: 751–59.

Vernes, K. 2001. Gliding performance of the northern flying squirrel *Glaucomys sabrinus)* in mature mixed forest of eastern Canada. *Journal of Mammalogy* 82: 1026–33.

Vispo, C., and I. D. Hume. 1995. The digestive tract and digestive function in the North American porcupine and beaver. *Canadian Journal of Zoology* 73: 967–74.

Vogt, F. D. 1981. *Survival strategies in the in the white-footed mouse,* Peromyscus leucopus: *Winter adaptation and energetics.* Ph.D. dissertation, Wesleyan University, Middletown, Conn.

Vogt, F. D., and P. Kakooza. 1993. The influence of nest sharing on the expression of daily torpor in the white-footed mouse. *Canadian Journal of Zoology* 71: 1297–1302.

◆Wilson, J. M. 2001. *Beavers in Connecticut: Their natural history and management.* Connecticut Department of Environmental Protection, Wildlife Division, Hartford. 18 pp.

Wilson, P. M., et al. 2000. DNA profiles of the eastern Canadian wolf and the red wolf provide evidence for a common evolutionary history independent of the gray wolf. *Canadian Journal of Zoology* 78: 2156–66.

Worden, K. A., and P. J. Pekins. 1995. Seasonal change in feed intake, body composition, and metabolic rate of white-tailed deer. *Canadian Journal of Zoology* 43: 452–57.

Yahner, R. H. 2001. *Fascinating mammals: conservation and ecology in the mid-eastern states.* Pittsburgh: University of Pittsburgh Press.

Zegers, D. A., and J. F. Merritt. 1988. Adaptations of *Peromyscus* for winter survival in an Appalachian montane forest. *Journal of Mammalogy* 69: 516–23.

Zervanos, S. M., and C. M. Salsbury. 2003. Seasonal body temperature fluctuations and energetic strategies in free-ranging eastern woodchucks *(Marmota monax)*. *Journal of Mammalogy* 84: 299–310.

Index

Scientific names are indexed only if no English name is available for the species. For species listed only by English name, you can find the scientific name in the text, generally in the main account (indicated in **bold**).

Acentria ephemerella, 69
Acorns, 51, **106–10**, 350, 384, 385, 386, 391, 401, 409, 410, 418, 420, 421, 422, 423, 424, 425, 426. *See also* Oaks
Adder's-tongue, northern, 56, 131, 132
Adelgid: eastern spruce gall, 268–69; hemlock woolly, 111, 119, 139, **231–32**, 349, 356; woolly, 136
Agalinus, sand-plain *(Agalinus acuta),* 119
Ailanthus, 24, 117, 158, 416
Alder(s), 48, 53, 72, 89, 97, 136, **144–45**, 176, 180, 408, 409, 410; smooth, 49, 73, **144**; speckled, 49, 54, 73, 75, **144–45**, 408, 420
Alderflies (Sialidae), 68, 83
Alewife, 46, 60, 205, **289**, 293, 305, 306
Alewife Cove, 38
Alexandrium, 34
Algae, **15–17**, **122–25**; banded red, 16, 125; brown, 16, 18, 123–24; crustose 16, 20, 124, 125; filamentous brown, 124; golden-brown, 67; green, 18, 67, 122–23; hollow green, 16; red, 16–18, 124–25
Amaranths, 252
American Heritage River, 63
Ampharete acutifrons, 21
Amphibians: conservation, 315–16; frogs and toads, 310–13; habitats, 313; reproduction, 314; salamanders, 307–10; in winter, 315
Amphipod(s), 17–18, 20, 22, 24, 26, 30, 43, 49, 50, 68, 73, 83, 84, 85, **276–77**, 282, 288, 289, 290, 294, 303, 419, 421, 425; blind tanaid, 20; caprellid, 21; digger, 24; four-eyed (ampeliscid), 21, 276; *Leptocheirus,* 288; marsh, 277; Mystic Valley, 277; *Orchestia grillus,* 29; piedmont groundwater, 277; tube-dwelling *(Corophium),* 21
Anchovy, bay, 288, 304
Ancylid, creeping, 44, 202, 203
Anemone(s), 22; **189–90**, 280; burrowing *(Ceriantheopsis americanus),* 301; clonal plumose, 189–90; comb jelly, 190; lined, 190; northern red, 190; orange-striped green (striped), 29; 190; white (ghost), 190
Angelwing, false, 209
Ant(s), 105, 145, 156, 175, 212, 230, 231, 233, **240–41**, 348, 420; acorn, 110; Allengheny mound, 241; carpenter, 241, 348; *Formica,* 260; *Myrmica lobifrons,* 80; slave-making, 241; *Tapinoma sessile,* 110
Antlion(s), **232–33**, 417, 418, 425
Apes, great, 383
Aphid(s), 105, 141, 146, 152, 172, **229–31**, 240, 265, 268, 420, 423, 427; beech blight, 147, 231; poplar petioleball, 269, 423; witch-hazel bud and leaf gall, 269, 420; woolly alder *(Paraprociphilus tesselatus),* 145, 231, 250, 420, 423, 425; woolly beech, 250, 425
Aplexa, lance, 83, 203
Arbutus, trailing, 98, 102, 135, 136, **168–69**, 175, 409, 410, 412
Aristolochia, 250
Ark, transverse, 207, 209
Arrow-arum *(Peltandra virginica),* 32, 47, 48, 50, 67, 71, 90, 91, 397, 417, 425
Arrow-grass, seaside *(Triglochin maritima),* 28
Arrowhead(s) *(Sagittaria* spp.), 31, 47, 48, 49, 50, 67, 71, 88, 89, 419
Arrowwood, northern, 33, 48, 49, 53, 72, 74, 75, 97, **174**, 366, 416, 417, 418, 421
Ash, 136, **171–72**, 176, 366, 423, 424; black, 75, 97, **171**; green, 48, 49, 53, 89, 91, 97, **171**, 184, 412, 416, 429; white, 54, 89, 97, 98, 102, **171**, 424
Askins, Robert, 37, 115
Aspen(s), 97, 136, **141–42**, 176, 250, 251, 386, 408, 410, 419, 423, 426; bigtooth, 33, 116, **141**, 410, 422; galls, 268; quaking, 116, 180, **141**, 410
Aster(s), 71, 116, **180**, 251; flat-topped white, 180, 251; heart-leaved *(Aster cordifolius),* 425; New England *(A. novae-angliae),* 72, 87, 180; salt-marsh *(A. subulatus, A. tenuifolius),* 26, 28, 29; swamp *(A. puniceus),* 54; white wood *(A. divaricatus),* 102, 422, 425
Atlantic Coastal Fisheries Cooperative Management Act, 306
Atlantic State Marine Fisheries Commission, 211, 278, 288, 290, 298, 301, 303, 305
Autumn-olive, 28, 33, 100, 113, 114, 116, **165**, 366, 408, 409, 410, 413, 423, 425
Autumn wildflowers, 178–80
Avens, purple *(Geum rivale),* 73, 75
Awningclam, Atlantic, 209
Azalea(s), 98, **170**; pink, 33, 97, 136, **170**, 246, 247, 412, 413; swamp, 74, 75, 77, 78, 79, 97, 136, **170**, 413, 416, 417
Azure, spring and complex *(Celastrina ladon* and relatives), **250**, 408, 409, 411, 412

Babesiosis, 215
Backswimmer(s), 45, 50, 68, 73, 83, 84, 85, 216, **228–29**
Bacteria, 21, 45, 49, 84, 85, 87–88, 102, 105, 196, 201, 210, 215, 263, 275, 377, 397; nitrogen-fixing, 143, 157
Baneberry, red *(Actaea rubra),* 412
Bantam Lake, 65, 133, 291, 292, 341
Barberry, Japanese, 113, 119, 120, **150**, 364, 365, 366, 394, 406, 407, 409, 412, 428
Barkhamsted Reservoir, 337
Barnacle(s), 13, 16, 17, 18, 19, 20, 21, 198, 201, **275**, 280, 301, 420
Barndoor Hills, 1
Barn Island Wildlife Management Area, 25, 33, 370
Barrel-bubble, channeled, 21
Barrett, Nels, 31, 47
Basalt, 1, 5
Baskettail, prince, 220
Bass, 46; black sea, 298; largemouth, 45, 46, 68, 70, 205, **299**, 305, 420; rock, 46, 205, **298**; smallmouth, 46, 68, **299**; striped, 14, 26, 46, 295, **297–98**, 305
Basswood, 33, 44, 53, 55, 75, 97, 98, 99, 102, 136, **164–65**, 180, 416, 425
Bats, 47, 53, 69, 105, 106, 225, **383–83**, 398, 403, 420, 425; big brown, 117, **382**, 402, 403, 427; in buildings, 382; in caves and tunnels, 382–83; eastern pipistrelle, **383**, 402, 407, 425, 428; eastern small-footed, 382; hibernacula, 404; hoary, **382**, 399; Indiana, 383; little brown, 117, **382–83**, 402, 403, 407, 425, 428; northern, **382–83**, 396, 402, 407, 425, 428; red, 115, **382**, 399, 424; silver-haired, 115, **382**, 396, 399; in trees, 382
Bayberry, 24, 28, 97, 114, **142–43**, 351, 356, 366, 405, 421
Beaches, coastal, 22–25; animals, 24–25; vegetation, 23–24

Beach flea(s), 24, 122, **277**, 425, 427
Beach-heather *(Hudsonia tomentosa)*, 24
Beach hopper(s), 24, 122, **277**
Beak-rush, capillary *(Rhynchospora capillacea)*, 73
Bear, black, 107, 109, 147, 148, 155, 157, 166, 168, 169, 171, 365, **391**, 396, 398, 402
Bearberry, 97, 111, 135, **168**, 366
Beaver, 47, 51, 55, 69, 73, 74, 77, 141, 144, 145, 146, 152, 161, 166, 172, **385–86**, 397, 398, 399, 402, 407, 408, 412, 424, 426, 429; wetlands, **53–54**, 331, 335, 340, 349, 351, 375
Bedstraw *(Galium tinctorium)*, 48, 54
Bee(s), 29, 79, 150, 151, 152, 153, 154, 155, 156, 157, 158, 159, 162, 163, 164, 166, 168, 170, 172, 173, 178, 181, 183, 228, **245–47**, 265, 413, 415, 417; anthophorid, 175; carpenter, 24, 167, 171, 173, **245–46**, 265, 409, 410, 411, 415, 419, 421; halictid, 44, 140, 164, 165, 171, 175, 177; honey-, 44, 159, 167, 169, 175, 177, 214, 228, **247**, 407, 415, 419, 421; megachilid, 175; mining, 140, 154, 165, 171, 177. *See also* Bumblebees
Beech, American, 97, 98, 101, 103, 104, 110, 126, 135, 136, **147**, 176, 180, 181, 366, 384, 386, 391, 409, 410, 411, 412, 413, 415, 419, 421, 422, 425
Beech-drops, 147, **181**, 421
Beetle(s), 29, 43, 45, 47, 68, 80, 85, 105, 151, 153, 155, 156, 159, 166, 167, 172, 212, **233–39**, 265, 348, 413, 414, 416, 418, 419; alder flea *(Macrohaltica ambiens)*, 145; American burying, 235; Asian long-horned, 238; bark, 239; bombadier, 234; borer *(see* Borers); carrion, 235; chrysomelid, 241; click, 236, 414; crawling water, 43, 50, 83; dermestid, 260; diving *(see* predaceous diving); elm bark, introduced *(Scolytus multistriatus)*, 150; elm bark, native *(Hylurgopinus rufipes)*, 150; eyed elator, 414; firefly, 236–37, 416, 422, 428; *Galerucella* spp., 92; giant root borer, 237–38, 417; goldenrod, 179; grapevine *(Pelidnota punctata)*, 419; ground, 45, 192, 234; hickory bark *(Scolytus quadrispinosus)*, 144; Japanese, 151, 236, 265, 416; lady, 145, 232, **237**, 266, 424, 426; leaf, 238; long-horned, 105, 237–38, 416; marsh *(Scirtidae* or *Helodidae)*, 83, 84, 86; May (June bug), 236; *Mordellistena*, 270; milkweed, **238**, 415, 419; minute moss (Limnebiidae), 83; predaceous diving (dytiscid), 43, 45, 50, 68, 73, 83, 84, 86, **235**, 423, 427; pselaphids (short-winged mold), 86; riffle, 43; round-headed borer, 238; rove, 235, 408; sap, 107, 108; scarab, 235–36; snout, 238, 239; tiger, 24, 45, 59, 61, 233–34; water-lily leaf, 44, 50, 68, **238**, 415, 420; water penny, 43, 236; water scavenger, 43, 50, 73, 83, 84, **235**; whirligig, 43, 50, 68, 83, 84, **234**, 420, 426; willow leaf *(Pagiodera versicolora)*, 141; wood-boring, 149
Beggar-ticks *(Bidens)*, 83; common *(B. frondosa)*, 116
Beluga, 379
Benoit, Lori, 37
Bentgrass, creeping *(Agrostis stolonifera)*, 31, 48
Berlin, 319
Bertness, Mark, 25

Betony, wood, 183
Bindweed, 49; hedge- *(Calystegia sepium)*, 24
Birch(es), 5, 89, 99, 103, 116, 126, 136, **145–47**, 152, 176, 250, 251, 269, 364, 366, 386, 422, 423, 424, 426; black, 33, 54, 75, 96, 97, 99, 100, 102, 110, 111, **145**, 411, 412, 414, 419, 422, 423, 429; bog (swamp), 73, 74, 78, 79, **146**; gray, 33, 54, 77, 78, 97, 100, 102, 114, 116, **146**; paper, 54, 78, 97, 102, **146**, 180, 406; river, 146; yellow, 54, 75, 76, 77, 78, 97, 98, 110, **145–46**, 385, 408, 411, 412, 425, 426
Bird(s): and cats, 377; climate change, 377; coastal and riverine birds, 375–77; conservation, 372–78; contaminants, 376–77; diversity, 329; exotic species, 377–78; food augmentation, 377; food and feeding, 363–67; forest fragmentation, predation, and cowbird parasitism, 372–74; grassland birds, 375; habitats, 363; Important Bird Areas Initiative, 378; migration, 369–71; reproduction, 367–69; seasonal diversity, 362–63; in winter, 371–72. *See also* Fruits
Birds-foot trefoil *(Lotus corniculatus)*, 116
Bittern: American, 331, 424; least, 32, 331
Bittersweet: American, **160**, 366, 405; Asian, 24, 33, 53, 75, 104, 113, 116, 119, **160**, 184, 394–95, 405, 408, 424, 426, 427, 428, 429
Bittium, alternate, 200
Black dash *(Euphyes conspicua)*, 73
Black-eyed susan *(Rudbeckia hirta)*, 116
Blackfish, 301
Black-haw, 174
Black knot *(Apiosporina morbosa = Dibotryon morbosum)*, 156
Black widow, 214
Blackberries, 154; fruits, 365; highbush *(Rubus allegheniensis)*, 72. *See also* Bramble(s)
Blackbird(s), 49, 370; red-winged, 30, 32, 37, 51, 73, 74, 117, 173, **360**, 367, 372, 408, 410, 419, 422, 426; rusty, 360
Bladdernut, 53, **160**, 418
Bladderworts, 54, 67, 78, **80–81**, 89, 90, 420
Blazing star, northern *(Liatris scariosa)*, 119
Block, Adriaen, 60, 64
Bloodroot, 102, 134, 175, **176–77**, 240, 410
Bloodworms, 264
Blueberries, 6, 78, 98, 111, 112, **170–71**, 250, 261, 364, 365, 367, 414, 415, 418, 421; galls, 271; highbush, 33, 72, 73, 75, 77, 79, 83, 97, 104, **171**, 366, 410, 411, 412, 413, 414, 416, 417, 419, 421, 426; lowbush, 33, 96, 97, 101, 103, 136, **170**, 411, 412, 417, 422
Bluebird, eastern, 54, 73, 75, 115, 142, 159, **354**, 355, 367, 369, 372, 377, 406, 426, 428
Bluefish, 14, 26, 36, 295, 297, **300**, 304
Blue flag *(Iris versicolor)*, 49, 71, 86, 414
Bluegill, 46, 68, 70, **299**, 305
Bluejoint *(Calamagrostis canadensis)*, 48, 54, 71, 86, 87
Bluestem, 252; little *(Schizachyrium scoparium)*, 114, 252
Bluet: big *(Enallagma durum)*, 50; familiar, 83, 219; vesper, 220

Bluff Point Coastal Reserve (State Park), 25, 33, 371
Bobcat, **393**, 399
Bobolink, 116, 119, **360**, 375
Bobwhite, northern, 152, **340**, 366; quail, 226
Boghaunter, ringed, 77, 82, 219
Bog lemming, southern, 388
Bog-rosemary, 78, 97, 135, 136, **168**
Bolleswood Natural Area, 113
Boneset *(Eupatorium perfoliatum)*, 71, 72
Bonito, Atlantic, 302
Borer: bronze birch *(Agrilus anxius)*, 146; dogwood *(Synanthedon scitula)*, 156; elder borer *(Desmocerus palliatus)*, 173; giant root, 237–38, 417; locust borer *(Megacyllene robiniae)*, 157; round-headed, 238, 416
Box elder, 53, 97, **160–61**, 366
Bracken *(Pteridium aquilinum)*, 116
Brackish marshes: fauna, 32–33; salinity, 31; vegetation patterns, 31–32
Bradley International Airport, 119, 342
Bramble(s) *(Rubus* spp.), 33, 99, 114, 116, 136, **154**
Brant, 14, 20, 22, 123, **334**, 424, 426
Brazilian elodea or giant waterweed *(Egeria densa)*, 62
Breeding Bird Survey, 373
Bridgeport, 338; Harbor, 20
Brittle star, little, 283, 284
Broker, Steve, 388, 410
Brooks. *See* Riverine ecosystems
Brown: Appalachian *(Satyrodes appalachia)*, 251; eyed *(S. eurydice)*, 251
Brownstone, 1, 61
Brown tide. *See* Long Island Sound, brown tide
Brown University, 25
Bryozoans. *See* Ectoprocts
Buckbean *(Menyanthes trifoliata)*, 73
Buckeye, common *(Junonia coenia)*, **251**, 267, 421
Buckmoth, eastern, 258
Buckthorns, **163–64**, 250, 374; alder-leaved (swamp), 73, 74, 97, 163; common, 33, 163, 366; glossy (European), 113, 119, 163, 413
Bufflehead, 14, 69, 336
Bugs, 212, **227–29**, 265; ambush, 85, **227–28**, 418, 420; giant water, 45, 50, 68, 73, 83, 85, **229**, 408, 420; leaf-footed, 227, 406; milkweed, **228**, 266; stink-, 227
Bullhead(s), 46; brown, 46, 68, **292**, 417; yellow, 291
Bulrushes, 51, 54, 67, 88; river *(Scirpus fluviatilis)*, 47, 48, 49; salt-marsh *(S. cylindricus)*, 31, 415; *S. acutus*, 73; *S. pendulus*, 73; soft-stem *(S. tabernaemontani)*, 31, 47, 49, 67, 71
Bumblebees, 44, 140, 147, 149, 154, 155, 159, 164, 165, 167, 168, 169, 170, 171, 175, 177, 179, 182, 183, 267, 409, 410, 412, 413, 414, 415, 416, 419, 420, 421, 422, 423; foraging behavior, 246; nests, 246; reproduction, 247; spring emergence, 246; winter, 247
Bunting: indigo, 115, **359**; snow, 25, 117, **359**, 426
Burbot, 294
Bur-cucumber, 53, 135, 424
Burdock *(Arctium)*, 134; common *(A. minus)*, 116

Bur-marigold (*Bidens* spp.), 31, 48, 91; nodding (*B. cernua*), 86
Bur-reed(s) (*Sparganium* spp.), 47, 48, 49, 54, 67, 71, 83, 86, 90
Bush-clover (*Lespedeza*), 116, 251
Bush-honeysuckle, northern, 136, **173**, 416
Butter-and-eggs (*Linaria vulgaris*), 24, 116, 425, 427
Buttercup, common tall (*Ranunculus acris*), 116
Butterfish, 14, 302
Butterflies, 29, 116, 119, 175, 183, **248–56**, 265. *See also* Monarch *and specific species*
Butternut, 53, 97, 98, **143**, 261, 413, 416, 417, 420
Buttonbush, 44, 48, 49, 53, 72, 73, 74, 79, 83, 89, 91, 97, 135, 136, **172–73**, 366, 416, 423

Caddisflies, 43, 45, 50, 68, 73, 83, 84, 193, **247–48**, 299, 411, 412, 413, 418, 422, 423, 424, 427, 428
Calothrix, 16
Campeloma, pointed, 203
Campion, white (*Silene latifolia*), 135
Cancer-root, one-flowered, 181
Candlewood Lake, 65
Canvasback, 14, 46, **336**, 370
Captain Islands, 4, 7
Cardinal, northern, 115, 144, 151, 154, 163, 164, 166, 329, **359**, 367, 372, 377, 405, 406, 407, 408
Cardinal-flower (*Lobelia cardinalis*), 49, 90, 347, 420, 424
Caring for Our Lakes, 70
Carnivorous plants, 79–81
Carp: common, 46, 68, 205, **290**, 305; grass, 306
Carrot, 250; wild (Queen Anne's lace, *Daucus carota*), 116, 134
Catalpa (*Catalpa*), 53
Catbird, gray, 75, 77, 106, 115, 140, 143, 150, 151, 153, 154, 157, 164, 166, 173, 329, **354**, 355, 361, 366, 373, 416, 417, 422, 425
Catbriar. *See* Greenbriar (catbriar)
Catfish, 46, 304, 305; channel, 46, 206, **292**; white, 46, 68, **291**
Cathedral Pines, 112, 113
Cats, domestic and feral, 117, 344, 347, 375, 377, 380, 388, 404, 413, 417, 418, 420
Cattails, 47, 49, 50, 51, 67, 71, 72, 74, 86, 88, 90, 91, 331, 353, 405, 425, 427; broadleaf (*Typha latifolia*), 48, 49, 54; hybrid, 32; narrowleaf (*T. angustifolia*), 31, 48, 79
Cedar: Atlantic white, **76–77**, 78, 94, 97, 135, **139**, 250; northern white, **139**, 147; red- (*see* Red-cedar)
Cedar-apple rust, **140**, 410
Centipedes, 105, **271–72**, 408
Central Valley, 1, 3
Ceratium lineatum, 13
Charles Island, 332
Charlock (*Sinapis arvensis*), 116
Chat, yellow-breasted, 115, **357**
Checkerspot: Baltimore (*Euphydryas phaeton*), **251**, 414; Harris' (*Chlosyne harrisii*), 180, **251**
Chelicerates, 211–16
Cherries, **155–56**, 250, 251, 401, 413; black, 33, 54, 97, 98, 100, 113, 116, **155–56**, 180, 250, 270, 366, 367, 409, 410, 411, 412, 413, 414, 419, 426; choke, 114, **156**, 366; fruits, 364, 365, 367; pin, 97, 100, 116, **155**, 366, 410, 411
Chester, 64
Chestnut: American, 5, 96, 97, 99, 101, **119–20**, 136, **147–48**, 391, 416, 417, 423; chestnut blight (*Cryphonectria parasitica*), 119, 120, 125, 147; water (*Trapa natans*), 62
Chickadee, black-capped, 77, 105, 106, 113, 115, 117, 121, 139, 163, 270, 329, **352**, 354, 366, 367, 369, 371, 372, 405, 406, 407, 417, 426, 427
Chickweed, common (*Stellaria media*), 116, 135, 428
Chicory (*Cichorium intybus*), 116, 418
Chink shell. *See* Lacuna, northern
Chipmunk, eastern, 106, 107, 110, 115, 143, 144, 147, 148, 155, 163, 164, 171, 174, 180, 215, 322, 373, **385**, 399, 401, 403, 406, 407, 408, 417, 423, 426
Chokeberries, 33, 74, 75, 77, 97, 114, **154–55**, 363, 413, 419; black 78, **154**, 366, 412, 420; galls on, 268; purple, 154; red, 154
Chub, creek, 45, 291
Chubsucker, creek, 45, 291
Cicada(s), **229–30**, 245, 380, 416, 417–18
Cicada killer. *See* Wasp(s), cicada killer
Cinquefoil (*Potentilla* spp.), 116
Cities, 117
Civilian Conservation Corps, 137
Cladocerans, 13, 43, 68, 83, 84, 85, **273–74**, 299
Clam(s), 34, 35, **204–10**, 295, 303, 421; amethyst gem-, 21, 210; Asian, 44, 55, **206–7**; Atlantic awning-, 209; Atlantic jackknife (razor clam), 208; Atlantic nut-, 210; Atlantic surf-, 208; dwarf surf-, 21, 210; fingernail, 43, 68, 73, 83, 85, **206**; freshwater, 204–7; pea-, **206**; saltwater, 207–10; softshell, 20, 208, 209; surf-, 36, 208. *See also specific groups and species*
Clam shrimp, 86, **273**
Clean Water Act, 33, 38
Cliff-brakes, 132
Climate, 9–10
Cloudywing: northern (*Thorybes pylades*); 251; southern (*T. bathyllus*), 251
Clover, 250; red (*Trifolium pratense*), 116, 424; white (*T. repens*), 116
Clubmosses, 130–31; common running, 102, 422, 426; running, 102, 130; tree, 102, 131, 422, 426, 427, 428
Clubtail(s), 61, 219; arrow, 45; beaverpond, 219; cobra, 219; midland, 45, 219; riverine, 45; skillet, 219
Coast, 1–2
Coastal slope, 3
Cocklebur (*Xanthium* spp.), 23; common (*X. strumarium*), 23–24, 116
Cockroaches, 117, 192, 226, 227
Cohosh, blue (*Caulophyllum thalictroides*), 102, 129
Columbine, 252; wild (*Aquilegia canadensis*), 102, 410
Comb jellies, 13, **190–91**, 297, 302, 415; Beroe's, 190–91; Leidy's (sea walnut), 190–91, 421; sea gooseberry, 191

Comma: eastern (*Polygonia comma*), 150, **251**, 408; gray (*P. progne*), 152
Community-Based Restoration Program (NOAA), 56
Compsilura concinnata, 265
Conehead, 116; robust, 224; sword-bearing, 224
Coniferous and mixed forests and woodlands: eastern hemlock, 110–11; eastern white pine, 111; fauna, 113; human influences, 112; hurricanes, 112–13; sand plain communities, 111–12
Conifers, 110–13, 133–34, 136–40
Connecticut Agricultural Experiment Stration, 215, 238, 239, 395
Connecticut Audubon Society, 433
Connecticut Botanical Society, 62
Connecticut Butterfly Atlas Project, 253
Connecticut Coastal Access Guide, 38
Connecticut Coastal Management Act, 38
Connecticut Coastal Management Program, 38
Connecticut College, 25, 37, 113; Arboretum, 115
Connecticut Department of Environmental Protection (DEP), 37, 38, 61, 62, 63, 64, 70, 278, 289, 300, 375, 395, 404, 431; Bureau of Waste Management, 69; Bureau of Water Management, 70; Inland Water Resources Division, 92; Office of Long Island Sound Programs, 39, 62; Pesticides Management Division, 69; Wetlands Habitat and Mosquito Management Program, 62, 433; Wildlife Diversity Program, 354; Wildlife Division, 339, 346, 390, 391; Wildlife Habitat Improvement Program, 433
Connecticut Department of Transportation, 339
Connecticut Endangered Species/Wildlife Fund, 433
Connecticut Federation of Lakes, 70
Connecticut Fund for the Environment, 433
Connecticut Income Tax Check-off Program, 433
Connecticut Invasive Plant Working Group, 184
Connecticut Natural Diversity Data Base, 431
Connecticut Ornithological Association, 378
Connecticut River, 12, 20, 23, 30, 31, 37, 40, 42, 44, 45, 46, 47, 48, 50, 51, 55, **56–64**, 205, 234, 286, 288, 289, 290, 291, 292, 293, 294, 295, 298, 299, 300, 303, 306, 308, 312, 331, 332, 336, 337, 340, 343, 357, 358, 361, 405, 406, 407, 408, 410, 414, 416, 417, 418, 419, 420, 423, 429; boat traffic, 60–61; bridges and canals, 60; dams and flood control, 59–60; dredging, 61; flow into Long Island Sound, 11; importance, 62–64; invasive plants, 62; map, 57; mouth, aerial photos, 58; physical and hydrological aspects, 56–59; power plant impacts, 60; restoration, 62; sewage treatment, 61; streamflow, 58; tidal wetlands map, 63
Connecticut River Estuary Regional Planning Agency (CRERPA), 64
Connecticut River Watershed Council, 63, 64
Connecticut River Watershed/Long Island Sound Invasive Plant Control Initiative Strategic Plan, 61
Connecticut Tidal Wetlands Act, 38
Conservation Reserve Program, 433

Convention on Wetlands of International Importance especially as Waterfowl Habitat, 63
Coontail. *See* Hornwort
Coot, American, 122, **341**, 429
Copepods, 13, 21, 43, 68, 83, 84, **274**, 302
Copper: American *(Lycaena phlaeas)*, 250; bog *(L. epixanthe)*, 82, 171, **250**; bronze *(L. hyllus)*, 50, **250**
Copperhead. *See* Snakes, copperhead
Coral, northern star, 189
Coral-root, 183
Coral weed, 124, 125
Cordgrass, 295; big *(Spartina cynosuroides)*, 31; freshwater, 24, 28, 32, 47, 48, 418; salt-meadow, 25, 25, 26, 27, 199, 418; smooth, 20, 25, 26, 27–28, 29, 30, 124, 200, 418, 419, 421
Cormorant(s), 15, 19, 46, 410, 419; great, 14, 47, **331**; double-crested, 14, 19, 32, 47, 53, 69, **331**, 420
Cornwall, 112, 113
Corophium, 21
Corporal, white *(Ladona exusta)*, 82
Corydalis, yellow *(Corydalis flavula)*, 175
Cotton-grass, 78; *Eriophorum viridicarinatum*, 73
Cottontail(s), 141, 146, 154, 159, 163, 165, 166, 392, 393, 398, 399, 400, 410, 413, 419; eastern, 25, 115, **383**; New England, 383
Cottonwoods, 89, 136, **142**, 176, 251, 386, 410, 412, 419, 420, 423, 424; eastern, 53, 59, 86, 97, **142**; galls, 268, 269; swamp, 53, 75, 142
Cougar, eastern, 393
Coventry, 70
Cowbird, brown-headed, 115, 357, **360–61**, 373
Coyote, 115, 117, 377, **389–90**, 393, 399, 406, 411, 416
Crab(s), 13, 20–22, 30, **277–81**, 287, 294, 295, 297, 300, 301, 304; Asian shore, 18, 21, 36, **280**, 410, 425, 426; Atlantic rock, 13, 16, 18, 21, **278**, 318, 425, 426; Atlantic sand, 281; blue, 26, 29, 44, 50, **278**, 419, 420; fiddler, 26, 29, 30, 32, 279, 423; green, 16–17, 21, 26, 29, 36, 199, **279**, 415, 421; hermit, 21, **280–81**; horseshoe (*see* Horseshoe crab); Jonah, 279; lady, 278; mole, 281; mud, 18, 29, 208, **279**, 423; oyster pea, 280; parasites on, 276; sand, 24; Say mud, 21; spider, 188, **278**, 318, 424; squatter pea, 280
Crab-apple, 33, 407, 408, 417
Crabgrass *(Digitaria sanguinalis)*, 24, 116, 424
Craig, Robert, 51
Cranberries, 78, 79, 97, 135, 136, **171**, 250, 366
Crangonyx, 267
Crappie: black, 68, 299; white, 299
Crayfish, 43, 44, 47, 50, 68, 207, **282**, 298, 300, 391, 426
Creeks. *See* Riverine ecosystems
Creeper, 205
Creeper, brown, 54, 105, 113, 144, **352–53**, 369, 372, 410
Creeper, Virginia, 24, 33, 75, 136, **164**, 259, 364, 366, 421, 424
Crickets, 192, **222–26**, 427; black-horned tree, 223, 225, 422, 423; camel, 105, 225; fall field, 223; field, 116, 224; four-spotted tree, 223; ground,

29, 45, 50, 105, 116, 223, 224, 422, 425, 426, 427, 428; house, 222, 223; narrow-winged tree, 223, 424, 426, 427; northern mole, 222, 223, 224; sand field, 223; snowy tree, 223, 418, 422; sphagnum, ground *(Neonemobius palustris)*, 82; spring field, 222, 223; tree, 105, 173, 222, 224, 421
Croaker, Atlantic, 301
Cromwell, 234, 298
Crossbill, 113, 136, 139, 367; red, **362**, 371; white-winged, 137, **362**, 371
Crow, 69, 346, 369, 406; American, 106, 113, 115, 117, 147, 329, 344, 346, **350**, 366, 372, 377, 405, 408, 409, 410, 428; fish, 351
Crowfoot *(Ranunculus* spp.), 71
Crustaceans, 15, 19, 21, **273–82**. *See also specific groups and species*
Cryptotis, 25
Cuckoo, 260, 369, 370; black-billed, 345; yellow-billed, 345
Cucumber, sea, 21, 284; hairy, 284; wild *(Echinocystis lobata)*, 53, 413
Cummings, Vicki, 319
Cunner, 301
Curculio(s): butternut *(Conotrachelus juglandis)*, 143; hickorynut *(Conotrachelus)*, 143, 144
Curlew, eskimo, 342
Currants, 97, 138, **152**
Cusk-eel, striped, 293
Cutgrass, rice *(Leersia oryzoides)*, 48, 54, 83, 86
Cyanobacteria, 16, 67

Dace, 46; blacknose, 45, 205, **291**, 417; longnose, 45, 205, **291**
Daddy-long-legs, 214
Daisy, ox-eye *(Leucanthemum vulgare)*, 116
Damselflies, 43, 44, 45, 50, 68, 73, 83, 84, 85, **218–21**, 419; broad-winged, 220; narrow-winged, 219, 220; spread-winged, 219, 220. *See also specific species*
Dandelion, common *(Taraxacum officinale)*, 116, 178, 429
Dangleberry, 77
Darner(s): *Aeshna*, 220; common green, 29, 45, 85, 219, 220, 267, 411, 412, 421, 427; fawn, 220; shadow, 220; swamp, 220
Darter(s), 305; swamp, 77, **299**; tessellated, 45, 46, **299**
Dash: long *(Polites mystic)*, 252; northern broken- *(Wallengrenia egeremet)*, 252
Dasher, blue, 216
Daylength, 10
Deciduous forest(s): acorn ecology, 106–10; air pollution, 101–2; amphibians, 105; biotic influences, 101; birds, 105–6; fire and logging, 99–101; forest floor flora and fungi, 102; invasive species, 119–20; invertebrates, 104–5; leaf fall and color change, 102–4; mammals, 106; microrelief, 98–99; reptiles, 105; shade tolerance, 98; severe weather events, 101; soil relationships, 96–98; trap-rock ridges, 101–2
Deep River, 64
Deer, white-tailed, 51, 73, 74, 77, 85, 101, 106, 107,

109, 110, 115, 117, 126, 139, 140, 141, 142, 144, 146, 147, 149, 150, 151, 152, 154, 155, 156, 157, 158, 159, 161, 163, 164, 166, 167, 169, 170, 171, 172, 173, 174, 215, 364, 390, 393, **394–95**, 397, 398, 399, 402, 406, 415, 419
Deforestation, 6
DEP. *See* Connecticut Department of Environmental Protection
Dermo, 26, 208
Destroying angel, 127
Dewberries, 53, 154; swamp, 75. *See also* Bramble(s)
Diatoms, 12–13, 17, 21, 43, 45, 67, 201, 248, 282, 310
Dickcissel, 359
Dinoflagellates, 12–13, 34
Dobsonfly, eastern, 43, 45, 50, **232**
Dock *(Rumex)*, 88, 250; curly *(R. crispus)*, 24, 50, 116, 250; great water *(R. orbiculatus)*, 50, 250
Dodder, **181–82**, 420, 422, 424
Dogbane *(Apocynum* spp.), 116, 411, 427
Dogfish: smooth, 287; spiny, 287
Dogs, domestic and feral, 38, 117, 346, 375, 377, 404, 420
Dogwinkle, Atlantic, 16, 18, **201**, 275, 427
Dogwood(s), 128, **165–66**, 250, 364, 366, 367; flowering, 97, 98, 103, 113, 114, 115, **166**, 411, 412, 421, 424, 425; galls, 270; gray, 74, 97, 114, **166**, 270, 426; silky, 49, 53, 72, 74, 75, 97, 114, **166**, 416, 423, 424
Dove: mourning, 115, 117, 329, **344–45**, 367, 372, 406, 407, 409, 410; rock *(see* Pigeon, rock)
Dowitcher, short-billed, 21, **343**, 413
Dragonflies, 43, 44, 45, 50, 67, 73, 84, 85, 86, **218–21**, 263, 410, 415; aeshnid, 83; libellulid, 83. *See also specific species*
Dragonlet, seaside, 26, 29, 219
Dragon's-mouth *(Arethusa)*, 81, 183
Drill(s), 280; Atlantic oyster, 16, 17, 18, **201**, 208, 420; thick-lip, 201
Drumlin, 2; Woodstock, 3
Duck(s), 15, 54, 61, 89, 122, 172, 209, 333, 369; Amerian black, 14, 21, 30, 32, 46, 51, 69, 73, 84, **335**; Labrador, 336; long-tailed (oldsquaw), 14, 336; ring-necked, 46, 69, **336**; ruddy, 336; wood, 51, 53, 54, 69, 73, 75, 77, 84, 153, 162, 166, 167, 172, **334–35**, 366, 369, 415, 416, 417. *See also additional species*
Duck Island, 332, 433
Duckweed(s), 67; lesser *(Lemna minor)*, 54
Dulse, 16, 124
Dunes: coastal, 22–25; animals, 24–25; vegetation, 23–24
Dunlin, 21, 30, 197, **343**, 370, 413, 426
Duskywing: columbine *(Erynnis lucilius)*, 252; dreamy *(E. icelus)*, 251; Horace's *(E. horatius)*, 252; Juvenal's *(E. juvenalis)*, 102, 149, **252**, 411; Persius *(E. persius)*, 252; sleepy duskywing *(E. brizo)*, 251; wild indigo *(E. baptisiae)*, 252; zarruco *(E. zarucco)*, 157, 252
Dusty miller *(Artemisia stelleriana)*, 23, 415
Dutchman's breeches *(Dicentra cucullaria)*, 102, 175, 240, 246, 410

Eagle, bald, 47, 53, 61, **337**, 367, 371, 376, 406, 407, 408, 410; viewings, 337
Earwigs, 24, **226**, 425
East Haddam, 59, 64
East Hampton, 70, 323
East Hartford, 61, 358
East Rock, 1
Echinoderms, 283–84
Ectoprocts (bryozoans), 17, 22, **192–93**, 208, 211, 278, 280; lacy, 192, 193
Eel, American, 32, 45, 46, **288**, 305, 413
Eelgrass *(Zostera marina)*, 15, **21–22**, 192, 197, 199, 201, 209, 276, 280, 284, 285, 294, 296, 297, 298, 301, 318, 421
Egret, 30; cattle, 329, **332**; great, 21, 26, 30, 33, 47, 51, 69, **332**, 376, 419, 421; snowy, 21, 22, 26, 30, 32, 33, 47, 51, **332**, 376, 416, 421
Ehrlichiosis, 215, 216
Elders (elderberry), 173–74; common, 74, 97, **173–74**, 366, 367, 415; elder borer beetle *(Desmocerus palliatus)*; fruits, 364, 366, 367; red-berried, **173–74**, 412
Elevation, 2, 6
Elfin: brown *(Callophrys augustinus)*, 102, 170, 171, 250, 249, **250**, 411; eastern pine *(C. niphon)*, 113, 138, **250**; frosted *(Incisalia irus)*, 250; Henry's *(C. henrici)*, 250; hoary *(I. polios)*, 168, 169
Elimia, piedmont, 44, **202**
Elliptio, eastern, 44, 205, 207
Elm(s), 53, 119, 136, **149–50**, 176, 251; American, 49, 53, 55, 75, 85, 89, 91, 97, **149**, 150, 409, 410, 412; slippery, 149
Embrace-A-Stream, 56
Emerald: brush-tipped *(Somatochlora walshii)*, 82; petite (Dorocordulia lepida), 82; *Somatochlora*, 220
Emperor (butterflies), 150; hackberry *(Asterocampa celtis)*, 150, **251**; tawny *(A. clyton)*, 251
Encephalitus: eastern equine, 35, 263; powassan, 205
Endangered species, 431
Endangered Species Act: Connecticut, 431; United States, 431
Enfield, 56; Enfield Dam, 290, 298
Entoprocts, 22
Environmental Protection Agency (EPA), 39; Long Island Sound Office, 38
Ermine, 392
Essex, 64, 288
Estuarine ecosystems: brackish marshes, 31–33; coastal forests, 33; conservation, 33–39; Long Island Sound physical environment, 11–12; muddy and sandy shores and shallows, 20–25; open-water biota, 12–15; rocky shores, 15–20; salt marshes, 25–30; stress gradients, 33
Evening-primrose *(Oenothera biennis, O. parviflora)*, 24, 116

Fairfield, 336
Fairy shrimp, 83, 84, 85, 86, **273**
Falcon, peregrine, **338–39**, 369, 371, 376, 422
Falkner Island, 4, 7, 11, 302, 344, 375, 376

Fallfish, 45, 205, 291
False Solomon's seal, star-flowered *(Smilacina stellata)*, 76
Fanwort *(Cabomba caroliniana)*, 62, 67, 418, 420
Farmington, 318
Fell, Paul, 25
Fern(s), **132–33**, 261; adder's-tongue, 131, 132; bracken, 132, 133; bulblet bladder, 132; chain-, 132, 133; Christmas, 102, 132, 133; cinnamon, 74, 75, 76, 77, 132, 133, 412, 413; cliff-brake, 132; climbing, 132; grape-, 131; hairy lip-, 132; hay-scented, 129; interrupted, 128, 413; maidenhair-, 128; marsh 28, 32, 48, 67, 71, 74, 75, 77, 83, 132; Massachusetts, 77; mountain spleenwort, 132; netted chain- *(Woodwardia areolata)*, 77; *Osmunda*, 133; ostrich, 132, 133, 261, 417; polypody, 132; royal, 48, 67, 74, 75, 77, 132, 412; rusty woodsia, 132; sensitive, 48, 53, 54, 71, 74, 75, 128, 132, 133; spinulose wood *(Dryopteris carthusiana)*, 75; Virginia chain- *(Woodwardia virginica)*, 77, 78; walking, 129, 132; wall-rue spleenwort, 132; wood, 102, 132
Fern allies, 130–32
Fescue, red *(Festuca rubra)*, 28
Fetter-bush, 73, 75, 77, 97, 136, **170**, 414
Fiddler crabs. See Crabs
Field crickets, 222
Fields: fauna, 116–17; vegetation, 115–16
Filefish, 14
Finch(es), 138, 139, 144, 366, 370; house, 115, 117, 140, 329, **361**, 362, 367, 372, 377, 411, 416; purple, 113, 151, 172, **361**, 371, 372, 428. See also Goldfinch, American
Fir, 104; balsam, 110, 366
Fire, 6, 99–101
Firebrats, 117
Fireflies. See Beetles, firefly
Firmosses, 130–31; shining, 75, 102, 131
Fisher, 106, 373, **392–93**, 396, 398, 399, 400
Fishes, 13, 14–15, 18–19, 21, 22, **286–306**; cartilaginous fishes, 287–88; conservation, 305–6; food and feeding, 304; habitat, 304; lampreys, 286; ray-finned fishes, 288–304; reproduction, 304–5; in winter, 305
Fishflies, 43, 45, 83, **232**, 412, 416, 417, 418
Flatworm(s), 43, 68, 84, 85, **191**; zebra, 281
Fleabane, common *(Erigeron philadelphicus)*, 175; daisy- *(E. annuus, E. strigosus)*, 116, 425, 426, 427
Fleas, 265–66
Flicker, northern, 115, 158, 329, **348**, 369, 428, 429
Flies (Diptera), 45, 126, 151, 152, 155, 158, 163, 164, 167, 168, 176, 245, **262–65**, 299, 413, 427; anthomyiid, 140; bee, 151, 155, 169, 175, 177, 183, 246, **265**, 419; black, 43, **263–64**, 410, 411, 414; blow-, 178; crane, 43, 50, 86, 192, **262**, 414, 428; deer, **264**, 415, 416; *Fletcherimyia fletcheri*, 80; flower, 265; gall, 150; greenhead, 26, 29, **264**; hippoboscid, 421, 424; horse, 264; hover, 265; moth (Psychodidae), 86; parasitic, 260, **265**; robber, 24, 45, **264–65**; sand (Psychodidae), 43, 50; sarcophagid, 79, 80, 225; seaweed, 24; stratiomyiid, 140; syrphid, 29, 44, 86, 140, 165, 170,

175, 177, 183, **265**, 415, 421, 423; tachinid, 183, 225, 258, **265**; winter crane, 267, 405, 407, 427, 428
Floater: alewife, 44, 205; brook, 205; eastern, 205; triangle, 44, 205
Floating-heart *(Nymphoides cordata)*, 67, 88, 420
Floodplain forests: fauna, 53; vegetation, 52–53
Flounder(s), 21, 26, 209, 282, 284, 304, 425; fourspot, 14, **303**; smallmouth, 302; summer, **303**, 305; windowpane, 14, **302**; winter, 14, 30, 35, 36, **303**, 305; yellowtail, **303**, 305
Flycatcher(s), 127, 151, 221, 363, 364, 366; Acadian, 53, 111, **349**, 373; alder, 145, 166, **349**; great crested, 105, **349**, 369, 411, 412; least, 349; olive-sided, 82, 329, **349**, 369, 419; willow, 75, 166, **349**, 368; yellow-bellied, 82, **349**
Foamflower *(Tiarella cordifolia)*, 76
Foliage, color change and leaf fall, 103–4
Forest(s): coastal, 33; coniferous and mixed, 110–13; conservation, 117–21; deciduous, 95–110; deforestation, 117–18; edge, 113–15; floodplain (see Floodplain forests); forestry practices and missing resources, 118–19; fragmentation, 403–4; hemlock, 110–11; invasive species, 119–20; management, 404; pine, 111; sand plain, 111–12; wetlands [see Forested wetland(s)]. See also specific forest types
Forest edge: fauna, 115; vegetation, 113–14
Forested wetland(s): evergreen swamps, 76–77, 94; fauna, 77–78; protection, 94; red maple swamps, 75–77; vegetation, 75–77;
Forktail, eastern, 219
Fossaria: boreal, 202; pygmy *(Fossaria parva)*, 83
Fourth Connecticut Lake, 56
Fox(es), 139, 150, 164, 166, 171, 173, 174, 365, 399, 401; gray, 106, 115, **391**, 396, 398, 399, 400, 409, 412, 428
Frankia, 145
Franklin Wildlife Management Area, 383, 396
Freshwater marshes and meadows (nontidal): fauna, 73; vegetation, 71–73
Frissel, Mount, 3
Fritillary: aphrodite *(Speyeria aphrodite)*, 251; Atlantis *(S. atlantis)*, 251; great spangled *(S. cybele)*, 249, **251**, 416, 420, 422; meadow *(Boloria bellona)*, 251; silver-bordered *(B. selene)*, 251; variegated *(Euptoieta claudia)*, 250
Frog(s), 51, 54, 59, 73, 116, 221, 292, 391; bull-, 50, 69, 73, 86, **312**, 313, 314, 315, 409, 412, 424; gray tree-, 50, 73, 74, 77, 84, 86, 105, **311**, 314, 315, 411, 412, 413, 414, 416, 417, 422; green, 46, 50, 53, 69, 73, 74, 77, 81, 310, **312**, 313, 314, 315, 409, 411, 413, 414, 415, 416, 417, 418, 422, 428; mink, 315; pickerel, 50, 69, 73, 81, 115, **312**, 314, 315, 409, 412, 414, 418, 419; northern leopard, 50, 81, 86, **312–13**, 314, 315, 316, 415, 417; wood, 77, 83, 85, 93, 105, 310, **313**, 314, 315, 407, 408, 409, 411, 413, 415, 416, 417, 420, 423, 427
Fruits: and birds, 367–67; color, 365; competition for, 367; and foliar flagging, 365; function, 364–65; quality, 366; seasonality, 365–66; size, 364

Index | 455

Fungi, 45, 49, 84, 85, 102–3, 105, **125–27**, 142, 146, 157, 214, 275, 384, 385, 387, 388, 398, 422, 423; anthracnose, 153, 166; *Apioplagiostoma populi*, 141; of aspen, 348; black cherry, 156; cedar-apple rust, 140; *Cercospora kalmiae*, 170; chestnut blight, 147–48; coral, 419; Dutch elm disease, 119, 125, 150; and fruits, 366; of gypsy moth, 259–60; heart-rot, 147; hen-of-the-woods (*Gifola frondosa*), 424; mycorrhizal, 106, 125–26, 171, 385; *Nectria coccinea faginata*, 147; pine blister rust (Cronartium ribicola), 138, 152; powdery mildew, 150; *Sirococcus*, 143; sooty mold, 231; viscid violet cort (*Cortinarius iodes*), 419

Gadwall, 46, **335**, 370
Gale, sweet, 74, 78, 97, **143**
Galls, **267–71**; adelgid, 268–69; aphid, 269; beetles in, 270; blueberry, 271; birch, 268, 269; cherry, 268, 270; chokeberry, 268; cynipid wasps, 270–71, 423, 424, 425; definition of, 267; formation, 268; grape, 420; hackberry, 268, 269, 424; hazelnut, 270; hickory, 268, 414; homopteran, 268–69; maple, 268; midge, 269–70; mites, 268; moth, 270, 271; oak apple, 413, 421, 423; poplar, 269; psyllid, 269; rose, 271; spiraea, 270; spruce, 268–69; willow, 269–70; witch-hazel, 268–69
Gammarus, 277
Gannet, northern, 14–15, **331**
Gardner Lake, 300
Garlic mustard (*Alliaria petiolata*), 114, 184, 410
Gateway: committee, 64; conservation zone, 64
Gentian, fringed (*Gentianopsis crinita*), 73
Geology, 1
Gerardia, 251; seaside (*Agalinus maritima*), 26, 28
Germander, seaside (*Teucrium canadense*), 24
Ginger, wild (*Asarum canadense*), 102, 175, 240
Ginseng (*Panax*), 129; dwarf (*P. trifolius*), 102
Glacial features, 1–4; Lake Hitchcock, 1, 3, 4; Lake Middletown, 4. See also Kettles
Glasswort (*Salicornia* spp.), 26, 28, 423, 424
Glassywing, little (*Pompeius verna*), 252
Glastonbury, 62, 323
Glider: spot-winged (*Pantala hymenaea*), 86, 219, 220; wandering (*P. flavescens*), 86, 219, 220
Globeflower, spreading (*Trollius laxus*), 73
Gnat(s): fungus (Mycetophilidae), 152; wood (Anisopodidae), 86
Gnatcatcher, blue-gray, 77, 105, 127, **354**, 368
Goby, naked, 302
Golden aster, sickle-leaved (*Pityopsis falcata*), 119
Golden club (*Orontium aquaticum*), 47, 48
Goldeneye, common, 14, 46, **336**, 370
Golden-heather (*Hudsonia ericoides*), 24
Golden-plover, American, 117, **341**
Goldenrod (*Solidago* spp.), 54, 72, 116, 134, 157, 179, 180, 227–28, 241, 243, 417, 420, 423, 424, 425, 427; beetles, 179; galls, 270, 271; seaside (*S. sempervirens*) 23, 24, 28, 29, 412, 421, 426
Goldfinch, American, 73, 74, 115, 144, 145, 146, 150, 166, 355, **362**, 368, 372, 408, 409, 410, 417, 418, 426, 428

Goldthread (*Coptis trifolia*), 75, 76, 77, 78
Goose (geese), 15, 122, 333, 369; Canada, 14, 32, 46, 50, 51, 69, 73, 117, **333–34**, 378, 411, 412, 413, 414, 417; snow, 333
Gooseberries, 138, 152
Goosefish, 294
Goosefoot (*Chenopodium* spp.), 24, 29
Goshawk, northern, **338**, 367, 371
Goshen Hills, 3
Grackle: boat-tailed, 360; common, 30, 32, 51, 73, 117, 173, 329, **360**, 372, 425, 426, 427
Grape(s), 49, 97, 114, 136, **164**, 259, 348, 364, 366, 420, 423, 427; fox, 53, 75, **164**, 420, 424; New England, 164; riverbank, 53, 75, **164**; seeds, 365; summer, **164**, 422, 423
Grass(es), 71, 72, 96, 116, 251, 252, 397; American beach (*Ammophila breviligulata*), 22, 23, 419, 421; barnyard (*Echinochloa crus-galli*), 24; Bermuda, 252; blue-eyed (*Sisyrinchium*), 415; bluejoint (see Bluejoint); bluestem (see Bluestem); manna (*Glyceria* spp.), 67; Muhly (*Muhlenbergia glomerata*), 73; *Panicum* (panic), 86, 252; rattlesnake (*Glyceria canadensis*), 54; reed canary (*Phalaris arundinacea*), 48, 49, 54, 71; switch- (*Panicum virgatum*), 24, 26, 28, 32
Grasshopper(s), 49, 50, 192, **222–26**, 265, 390; crackler, 223; red-legged, 225–26
Grassland(s): conservation, 119; fauna, 116–17; sand plain, 119; vegetation, 115–16
Grass-of-Parnassus (*Parnassia glauca*), 73
Grass-pink (*Calapogon tuberosus*), 81, 86, 183, 416, 417
Great Captain Island, 332
Great Island, 370
Great Meadow (Essex), 31
Great Meadows (Stratford), 337, 360
Grebe(s), 15, 369; horned, 14, **330**; pied billed, 14, 47, 69, 73, **330**; red-necked, 330
Greenbriar (catbriar), 33, 75, 113, 136, **140**, 408, 411, 413, 427; common, 140; fruits, 366
Green dragon (*Arisaema dracontium*), 53
Green fleece, 16, 20, **122–23**
Greenwich, 11, 332, 338, 370; Cove, 20
Grinnellia americana, 122
Griswold Point, 20, 23, 24, 56, 370
Grosbeak, 150, 153, 368; blue, 359; evening, 144, 161, 163, 329, **362**, 371, 372; pine, **361**, 371; rose-breasted, 105, 115, 164, **359**, 366, 369, 412, 417
Groton, 371, 395
Groundhog. See Woodchuck
Ground-ivy (*Glechoma hederacea*), 116
Groundnut (*Apios americana*), 48, 53
Groundsel, common (*Senecio vulgaris*), 427
Groundsel-tree, 26, 28, 136, **175**, 422, 424, 427
Grouse, ruffed, 105, 107, 140, 141, 144, 146, 147, 148, 149, 152, 153, 154, 156, 157, 158, 159, 161, 163, 166, 168, 169, 170, 171, 173, 174, **339**, 366, 369, 377, 409, 411, 414, 428
Grubby, 14, 19, 21, **297**, 305, 423
Gruner, Hank, 328
Guilford, 202
Gull(s), 15, 19, 377; Bonaparte's, 14, **343**, 369; great black-backed, 14, 19, 25, 30, 32, 47, 51, **343**, 421, 426; herring, 14–15, 19, 25, 30, 32, 47, 51, 200, 209, **343**, 344, 408, 419, 421, 424, 425, 426; laughing, 14, 25, **343**, 419, 421; ring-billed, 14, 24, 25, 47, 69, 117, 197, **343–44**, 406, 424, 426
Gum: black (tupelo), 33, 75, 77, 86–87, 89, 91, 97, 103, 129, **166–67**, 366, 421; var. *biflora* (swamp tupelo), 89; sweet, 75, 89, 103
Gunnel, rock, 301–2

Hackberry, **150**, 250, 251, 424; fruits, 366; nipple gall psyllid, 269, 424
Haddam, 42, 44, 62, 64, 288, 336, 405, 406, 408
Haddam Meadows State Park, 292
Haddam Neck, 60
Haddock, 284
Hairstreak: Acadian (*Satyrium acadica*), 141, **250**; banded (*S. calanus*), 149, **250**, 415; coral (*S. titus*), 155, **250**; Edwards' (*S. edwardsii*), 102, **250**; Hessel's (*Callophrys hesseli*), 77, 139, **250**; gray (*Strymon melinus*), **250**; hickory (*Satyrium caryaevorum*), 143, **250**; Juniper/olive (*Callophrys gryneus*), 102, 140, **250**; oak (*Satyrium favonius*), **250**; red-banded (*Calycopis cecrops*), 159; striped (*Satyrium liparops*), 155, **250**; white M hairstreak (*Parrhasius m-album*), 149, **250**
Hake: red, 14, **294**; silver, 14, **293**; spotted, **294**, 305
Hamburg Cove, 43
Hammonasset Beach State Park, 11, 25, 38, 351, 370, 383, 393, 427
Hanging Hills, 1, 3
Hardhack, 54, **156**, 417
Hare(s): European, 384; snowshoe, 115, 141, 146, 163, **383–84**, 393, 399
Harrier, northern, 30, **337**, 370, 371, 422
Hartford, 56, 60, 62, 64, 338, 350
Harvester (*Feniseca tarquinius*), 145, **250**
Harvestmen, 214
Hawk(s), 110, 368, 387, 401; broad-winged, 77, **338**, 369, 370, 411, 415, 422; Cooper's, **338**, 370, 371, 422; red-shouldered, 77, **338**, 367, 370, 373; red-tailed, 113, 115, 117, 119, **338**, 346, 367, 370, 371; sharp-shinned, **337–38**, 370, 371, 372, 422, 427
Hawkweed (*Hieracium* spp.), 116
Hawthorns, 33, 114, **156–57**, 364, 366
Hazelnut(s), 33, 97, 136, 144, 176, 408, 409, 410, 416; American, **144**, 366; beaked, **144**, 366, 416, 418; galls, 270
Heal-all (*Prunella vulgaris*), 116
Heather, beach-, 24; golden-, 24
Heath hen, 339
Hedge-bindweed (*Calystegia sepium*), 24
Hedgehog, the, 1
Heinrich, Bernd, 267
Hellebore, false (*Veratrum viride*), 75, 409
Hellgrammites, 45, 232
Hemlock, eastern, 5, 75, 76, 77, 78, 89, 97, 98, 99, 100, 101, 102, 104, **110–11**, 113, 135, 136, **138–39**, 146, 182, 231–32, 346, 349, 357, 366, 406, 409, 419, 420, 423, 426

456 | Index

Hemp: salt-marsh (*Amaranthus cannabinus*), 31, 32, 48; water (*A. tuberculatus*), 48
Hepatica (*Hepatica* spp.), 102, 175, 408, 410
Heritage Lake, 70
Heron(s), 30, 46, 49, 295; great blue, 21, 30, 32, 47, 51, 53, 54, 69, **331–32**, 410, 421, 424; green, 21, 32, 47, 51, 53, 69, **332**; little blue, 33, 329, **332**, 376; tricolored, 332
Herrings, 32, 46, 288–90, 304, 306; Atlantic, 14, **290**; blueback, 46, 60, **289**, 305, 306
Hickories, 5, 98, 99, 100, 102, 103, 136, **143–44**, 176, 250, 261, 384, 401, 413, 419, 424; bitternut, 33, 53, 55, 97, **143**, 419; galls, 268; mockernut, 33, 97, **143**; pignut, 33, 97, **143**, 414, 417, 424, 425, 427; shagbark, 97, 101, **143**, 419
Higganum, 407; Creek, 414
Highlands: Housatonic, 3; Hudson, 3; Northwest, 2, 3, 137, 348, 349, 352, 354, 356, 357, 359, 362, 393, 414
Hildenbrandia, 20
Hoary edge (*Achalarus lyciades*), 251
Hogchoker, 302
Hog-peanut (*Amphicarpaea bracteata*), 53
Hollow green weeds (*Enteromorpha* spp.), 16, 17, 19, 23, 122, **123**
Holly, 364, 365, 366, 367, 405, 406, 415, 426, 428, 429; American, 111, 250; mountain-, 74, 76, 77, 78, 97, **160**, 412
Homopterans, 50, **229–32**
Honeysuckle(s), 49, **173**, 364, 374, 408, 409, 426; Japanese, 113, 114, 119, **173**, 405, 412, 428; Morrow's, 28, 33, 72, 113, 119, **173**, 413; tartarian, 113–14, 119, 173
Hopeville Pond State Park, 112
Hop-hornbean, eastern, 44, 97, 98, 102, **144**, 410
Hornbeam, American, 33, 75, 76, 97, 98, 135, 136, **144**, 145, 176, 410, 419, 423
Hornet, bald-faced, 167, **242**, 244, 265, 267, 419, 425, 426
Horntail wasp, 240
Hornworm: tobacco, 258–59; tomato, 258–59
Hornwort (coontail; *Ceratophyllum demersum*), 32, 43, 48, 67, 86, 88, 89, 90, 91
Hornworts, 129
Horsehair worms, 192
Horse Island, 240
Horse-nettle (*Solanum carolinense*), 116
Horseshoe crab, 21, 197, **211**, 370, 376, 415
Horsetails, **130**, 397; field, 130; water (*Equisetum fluviatile*), 48, 130
Horseweed (*Conyza canadensis*), 116, 135
Huckleberries, 6, 33, 96, 111, 112, 364, 365, 414, 415, 418, 429; black, 97, 101, **169**, 417, 422; dwarf, 78
Humans, 383; impact of, 6
Hummingbird, ruby-throated, 77,127, 163, **347**, 368, 417, 420
Hurricanes, 9, 10, 12, 59, 99, 101, 113, 231
Hyallela azteca, 277
Hydroids, hydrozoans, 17, 21, 22, **189**, 208, 211, 278, 280, 303
Hyperia galba, 276
Hypoxia. *See* Long Island Sound, hypoxia

Ibis, glossy, 26, 30, 329, **333**
Ice Age (Pleistocene), 2
Ichneumon, giant, 239–40
Important Bird Areas Initiative, 378
Income Tax Check-off Program, 433
Indian pipe, 102, **182–83**, 415, 418, 420, 421, 422, 425
Indigo: false, 24, 48, 49, 53, 89, **157**, 251, 415; wild, 250, 252
Inkberry (*Ilex glabra*), **77**, 366
Inland wetlands. *See* Wetlands, inland
Insects, **217–71**; flightless relatives, 217–18; galls, 267–71; in winter, 266–67
Interstate Fisheries Management Program, 305
Invasive Species Executive Order, 185
Irish moss, 15, 16, 18, 20, **124–25**
Ironweed, New York (*Vernonia noveboracensis*), 67, 71, 72, 87
Ironwood. *See* Hornbeam, American
Isopods, 17, 22, 26, 30, 43, 45, 50, 73, 83, 84, 85, 105, 132, **275–76**, 406, 408, 414, 428; *Caecidotea*, 84; *Philoscia vittata*, 29
Ivy, poison. *See* Poison ivy

Jack(s), 304; crevalle, 14, 300; yellow, 14, 300
Jack-in-the-pulpit (*Arisaema triphyllum*), 75, 87, 412
Jay, blue, 77, 98, 99, 105, 107, 108, 110, 113, 115, 117, 121, 136, 144, 147, 148, 256, 329, **350**, 368, 372, 373, 405, 407, 409, 415, 417, 422, 423, 427, 428
Jellyfish, **188–89**, 297, 302, 318; freshwater, **189**, 420; lion's-mane, 13, **188**, 426; moon, 13, **188**, 415; parasites on, 276; sea nettle, 13, **188**
Jewelweed, spotted (*Impatiens capensis*), 54, 71, 75, 87, 420
Jewelwing, ebony (*Calopteryx maculata*), 219
Jingle, common, 207, 209
Joe-pye weed (*Eupatorium* spp.), 54, 72, 422; spotted (*E. maculatum*), 54
Jokinen, Eileen, 202
Junco, dark-eyed, 82, 115, 139, **359**, 372, 405, 411, 423, 424
Juneberry, 157
Juniper, common, 97, 100, 114, **139**, 366

Katydids, 192, **222–25**, 245, 246, 422; bush, 105, 224, 421, 423; conehead, 116, 224, 418; meadow, 29, 50, 116, 224, 423, 426; oblong-winged, 418; round-headed, 224; true, 105, 223, 224, 417, 418, 419, 420, 421, 422, 424, 425
Ked, deer (Hippoboscidae), 424
Keeney Cove, 410
Kelp, 16, 18, 122, **123–24**
Kensington Hatchery, 293
Kent, 42, 70, 356
Kestrel, American, **338**, 369, 370, 371, 422
Kettles, 76, 78; sand plain, 86
Killdeer, 47, 117, **341**, 416
Killifish, 14, 304, 305; banded, 32, 46, 50, 205, **295**; rainwater, 296; striped, 19, 21, 26, 30, **295–96**, 421, 423
King, John, 322
Kingbird, eastern, 75, 173, 221, **349**, 369, 370, 416
Kingfish, northern, 301

Kingfisher, belted, 32, 46, 47, 53, 69, **347**, 412
Kinglet: golden-crowned, 82, 139, **353–54**; ruby-crowned, **354**, 410
Knapweed (*Centaurea*), 420
Knot, red, 342
Knotted wrack, 15, 16, **124**
Knotweed, Japanese (*Polygonum cuspidatum*), 114
Kokanee, 68, **292–93**

Labrador-tea, 78, 97, 135, 136, **170**
Labrador-tea tentiform leaf miner (*Phyllonorycter ledella*), 170, 431
Lacewings, 232, 407
Lacuna, northern, 200
Lacustrine ecosystems. *See* Lakes and ponds
Ladies'-tresses, 81, 183; nodding (*Spiranthes cernua*), 87, 183; shining (*S. lucida*), 73
Lady: American (*Vanessa virginiensis*), 102, 249, **251**, 267, 415, 421; painted (*V. cardui*), 251, 412
Ladybugs. *See* Beetles, lady
Lady's-slippers, 183; large yellow (*Cypripedium pubescens*), 73; pink, 98, 184, 413; showy (*C. reginae*), 73
Lady's thumb (*Polygonum persicaria*), 87
Lakes and ponds: amphibians, 68–69; birds, 69; conservation, 69–70; fishes, 68; fish kills, 70; invasive species, 70; invertebrates, 68; mammals, 69; physical characteristics, 65–67; pollution, 69–70; reptiles, 69; vegetation, 67
Lamb's-quarters (*Chenopodium album*), 23, 116, 252
Lamprey: American brook, 286; sea, 32, 46, **286–87**, 304, 414
Larch, 113, 135; American, 78, 97, **136–37**; European, 112, **136**, **137**
Lark, horned, 24, 30, 116, 117, **351**, 367, 427
Laurel(s), 101, 136; bog, 78, 97, 135, **170**; great, 75, 76, 77, 135, 136, **170**, 180–81; mountain, 76, 77, 96, 97, 98, 135, **169–70**, 182, 405, 407, 412, 414, 416, 420, 423, 424, 426; sheep, 74, 78, 97, 111, 135, 136, **169**, 428
Laver (*Porphyra* spp.), 16, 17
Leatherback, 318
Leatherleaf, 73, 74, 78, 79, 97, 135, **168–69**, 172, 412, 413
Leeches, 43, 68, 83, 85, **195–96**, 415
Legumes, 251
Leitoscoloplos, 21
Lemming, southern bog, 82, 171, **388**, 399
Lettuce (*Lactuca* spp.), 24; wild (*Lactuca canadensis*), 116
Leucothoe, swamp. 170
Lichen(s), 16, 96, 98, 101, 102, **127**, 422; beard (*Usnea*), 356
Lighthouse Point (New Haven), 338, 370, 371
Lightning bugs. *See* Beetles, firefly
Lilaeopsis (*Lilaeopsis chinensis*), 31, 416
Lily, trout, **178**, 410
Lily-of-the-valley, wild (*Maianthemum canadense*), 77, 111, 410, 413
Limestone, 3–4, 72–73, 74, 75
Limnodrilus hoffmeisteri, 44

Index | 457

Limpets: eelgrass (bowl), 199; freshwater, 203; plant, 199; saltwater, 22, 199
Linsley Pond, 69
Lion, mountain, 393
Listen to the Sound 2000, 39
Little yellow (*Eurema lisa*), 250
Liverworts, 76, 77, 129
Lizard(s), 320
Lizardfish, inshore, 293
Lobelia, water (*Lobelia dortmanna*), 67
Lobster, American, 18, 21, 35, 36, 209, **277–78**, 287; gill worm, 195
Locust: black, 53, 116, 119, **157–58**, 251, 413, 414; sand, 24, 225, 226
Loggerhead, 318
Long Island Sound, algae, 16–17; algal blooms, 34–35; birds, 14–15; brown tide, 34–35; conservation, 33–39; depth, 11; fisheries, 36; fishes, 14; glacial features, 7; geography, 12; hypoxia, 34; invasive species, 36; mammals, 15; marsh and cove degradation, 37–38; muddy/sandy shores and shallows, 20–25; nitrogen, 34; open-water macroinvertebrates, 13–14; pathogens, 35–36; physical environment, 11–12; phytoplankton, 12–13; public access, 38; rocky shores, 15–20; salinity, 11; salt marshes, 25–30; satellite image, 5; sediment disposal, 61; shoreline modifications, 36–37; temperature, 11; tides, 11–12; toxins, 35; trash, 36; 2003 Agreement, 39; water quality, 55; zooplankton, 13
Long Island Sound Estuary Program, 33–34; Comprehensive Conservation and Management Plan, 33–34
Long Island Sound Fund, 38
Longspur, Lapland, 117, **359**
Lookdown, 300
Loon, 15, 369; common, 14, 69, **330**; red-throated, 14, **330**
Loosestrife: fringed (*Lysimachia ciliata*), 49; purple (*Lythrum salicaria*), 32, 48, 49, 62, 71, 87, 88, 91, 92, 419; tufted (*Lysimachia thyrsiflora*), 75, 76; yellow (*Lysimachia terrestris*), 54
Lopseed, 129
Lord Cove, 32, 433
Lumpfish, 297
Lyme, 64, 215
Lyme disease, 215
Lymnaea, mimic, 202
Lynde Point, 56
Lynx, 393

Mackerel, 304; Atlantic, 302; Spanish, 14, 302, 305
Madison, 4, 11, 370
Magee, Andrew, 74, 106, 113
Maleberry, 73, 75, 77, 78, 97, 136, **170**
Mallard, 14, 30, 46, 50, 51, 69, 73, 75, 84, 117, **335**, 370, 411
Mallow family, 252
Mammals, 15, 379–404; body mass, 401; communal nesting, 399; conservation, 403–4; diversity, 379; fat, 403; food and feeding, 396, 399–401; fur, 401; habitats, 396; heat production, 402; hibernation, 402–3; nest insulation, 399; reduced activity, 401; reproduction, 398–99; subnivean activity, 401; torpor, 401; in winter, 399–403
Managing Urban Deer in Connecticut, 395
Manatee, West Indian, 379
Mansfield Hollow Reservoir, 292
Mantid(s), 29, 50; European, 226; Chinese, 226
Mantis, praying. *See* Mantid(s)
Maple(s), 5, 99, 103, 136, 176, 424, 425; Norway, 33, **161**, 411, 413; mountain, 97; red, 33, 48, 49, 53, 54, 72, 74, **75–76**, 78, 83, 86, 89, 96, 97, 100, 101, **161–62**, 346, 385, 405, 407, 408, 409, 410, 413, 414, 415, 423, 424; silver, 48, 49, 52, 53, 97, **162**, 410, 411, 412, 413; striped, 97, 102, **161**, 412, 426; sugar, 55, 97, 98, 102, 135, **162–63**, 180, 385, 410, 411, 413, 416
Marble, 1, 4, 42
Marble valleys, 3, 42
Maritime Center (Norwalk), 38
Marsh cress (*Rorippa palustris*), 86
Marsh-elder, 26, 28, 97, 136, **175**, 421, 426
Marshes: brackish (*see* Brackish marshes); freshwater (*see* Freshwater marshes and meadows; Riverine and floodplain marshes); salt (*see* Salt marshes); tidal freshwater (*see* Tidal freshwater marshes)
Marshmallow (*Althaea officinalis*), 28, 87
Marsh-marigold (*Caltha palustris*), 75, 90, 409
Marsh St. John's-wort (*Triadenum virginicum*), 77
Marten, American, 392
Martin, purple, **351**, 369, 416
Matianuck Sand Dunes, 433
Mattabassett District Treatment Plant, 61
Mattituck Sill, 7
Mayflies, 43, 50, 68, 73, 83, 85, **218**, 304, 410, 411, 412, 413, 418
McMaster, 54
Meadow-beauty (*Rhexia virginica*), 67
Meadowhawks (*Sympetrum*), 220; cherry-faced (*S. internum*), 220, 426, 428; yellow-legged, 220
Meadowlark, eastern, 116, 117, 119, **360**, 375
Meadow-rue: early (*Thalictrum dioicum*), 175; tall (*T. polygamum*), 87, 415
Meadows, wet. *See* Freshwater marshes and meadows
Meadow-sweet, 54, **156**, 270, 416
Mediomastus ambiseta, 21
Melampus, eastern, 26, 29, 30, **199–200**, 421
Menhaden, Atlantic, 14, 26, 32, 46, **290**, 298, 304, 421, 423
Merganser, 15, 333; common, 46, 69, **335**, 369, 370, 405, 418, 429; hooded, 46, 54, 69, 73, 75, 77, 167, **335**, 367, 369, 429; red-breasted, 14, 19, **336**, 370, 408
Meriden, 319
Merlin, **338**, 370, 371, 422, 424
Mermaid's hair (*Cladophora* spp.), 16, 123
Mermaid-weed (*Proserpinaca palustris*), 32, 71
Merrick Brook, 433
Metalmark, northern (*Calephelis borealis*), 250
Metzler, Ken, 31, 47, 72, 75, 76, 79

Middletown, 12, 51, 56, 61, 349, 350, 389
Middletown Water Pollution Control Plant, 61
Midge(s), 44, 68, 73, 83, 84, 85, 141, 216, 264, 292, 293, 299, 411, 414, 415, 417; biting (punkies, Ceratopogonidae), 43, 50, 86; chironomid, 43, 80, 84, 86, **264**; gall (Cecidomyiidae), 144, 156, 164, 166, 180, **269–70**; holly berry, 367; *Metriocnemus knabi*, 80; phantom, 73, 83, 84, **262**, 263; willow gall, 266, 269–70; *Youngomyia umbellicola*, 173
Migration, bird: autumn, 370–71; boreal seed-eaters, 371; examples of, 369; spring, 369–70
Milford, 333; Harbor, 332; Point, 23, 25, 370, 375
Milkweeds (*Asclepias*), 116, 228, 238, 251, 255–56, 415, 416, 419, 423, 426, 427, 429; swamp (*A. incarnata*), 71, 73
Millipedes, 45, 105, 132, **271–72**, 408, 418, 428
Mink, 33, 46, 51, 69, 73, 75, **392**, 399, 417, 418
Minnow, 304, 305; bluntnose, 205, 291; cutlips, 290; fathead, 291; sheepshead, 21, 25, 26, 30, **296**, 413, 421–22
Mistletoe, dwarf, 137, **182**
Mite(s), 29, 105, 110, 132, 154, 156, 171, **214–16**, 235; *Arrenurus*, 85; eriophyid, 146, 150, 268; gall, 144, 155, **268**, 413; on honeybees, 247; slime, 80; water, 45, 68, 73, 83, 84, 85, **216**
Miterwort (*Mitella nuda*), 75, 76; false, 129
Mockingbird, northern, 24, 115, 117, 139, 140, 154, 164, **354–55**, 366, 367, 406, 426, 427, 428
Mohawk State Forest, 78
Moles, 105, 380, **381–82**, 399; eastern, 115, 117, **381**, 406, 407, 408; hairy-tailed, 382; star-nosed, 51, 69, 73, 75, 78, 82, 346, **381–82**
Mollusks, 13, 15, 17, 20–21, **197–210**
Monarch, 24, 251, **253–56**, 414, 421, 423, 424; caterpillar diet, 255–56; distribution, 253; eggs, 255; larvae, 255; migration, 253–54, 267; navigation, 254; pupae, 255; reproduction, 254–55; toxins, 255–56; uniqueness, 253; winter, 254
Monostoma pulchrum, 122
Moonfish, Atlantic, 14, 300
Moonseed, 129
Moonworts, 131
Moorhen, common, 329, **340–41**
Moose, 73, **395–96**, 398
Moraine(s), 4, 76; Harbor Hill, 4, 7; Madison, 7; Old Saybrook, 7; Ronkonkoma, 4, 7
Morse Point, 344
Mosquitoes, 35, 37, 43, 83, 183, 216, **262–63**, 415, 425, 427; *Aedes*, 84; *Anopheles*, 85; salt marsh, 26, 29; tree-hole (*Aedes triseriatus*), 86
Mosses, 75, 76, 96, 98, 102, **129**; sphagnum (*see* Sphagnum)
Moth(s), 47, 141, 148, 152, 158, 183, 228, 248–49, **256–61**, 411, 414, 418, 420, 425, 427; Abbot's sphinx, 259; acorn, 107; *Acrobasis comptoniella*, 142; ailanthus silk- (*Samia cynthia*), 158; ailanthus webworm (*Atteva punctella*), 158, 416, 417; arched hooktip (masked birch caterpillar, *Drepana arcuata*), 411; azalea sphinx, 259, 413; borer, 80, 133, 261, 431, 417; Carolina sphinx, 258; cecropia, 166, 173, 258, 413; common oak (*Phoberia atomaris*), 409; conservation,

261–62; copper underwing (*Amphipyra pyramidoides*), 418; dagger, 261; dart, 409; dot-and-dash swordgrass (*Xylena curvimacula*), 407; eastern buck-, 113, 258; filbert, 107, 108; five-spotted hawk, 258; flower, 180; gypsy, 109, 111, 119–21, 138, 139, 145, 146, 147, 149, 151, 155, 258, **259–60**, 265, 345, 408, 415, 417; hawk, 258–59; hummingbird sphinx (*Hemaris thysbe*), 414; imperial, 258; inchworm (geometer, spanworm), 147, 257, 267, 407, 408, 424, 427, 428; Isabella tiger, 260, 415; lappet, 257, 411; laurel sphinx (*Sphinx kalmiae*), 420; linden looper (*Erannis tiliaria*), 164; luna, 258, 413; Morrison's sallow, 261; oakworm, 149; owlet, 163, **260–61**, 267, 407, 408, 410; Pandora sphinx (*Eumorpha pandorus*), 164; pitcher plant, 80, 431; promethea (*Callosamia promethea*), 151; pyralid, 44–45, 68; regal, 258; royal, 258; silkworm, 258; silver-spotted ghost, 145; skiff, 424; slug caterpillar, 256; sphinx, 173, 258–59; sycamore tussock, 153; tent caterpillar (*see* Tent caterpillars); tiger, 260; underwing, 143, 261, 418, 419, 421; white-marked tussock (*Orgyia leucostigma*), 136, 166; woollybear, 260, 415; yellow underwing, 261; yellow-banded (*Catocala cerogama*), 164

Mountain-ash, American, **153**, 366

Mountain: Avon, 1; Bear, 3; Beseck, 1; Canaan, 3; Higby, 1; Lamentation, 1; Manitook, 1; Onion, 1; Penwood, 1; Talcott, 1; Totket, 1; West Suffield, 1

Mourning cloak (*Nymphalis antiopa*), 102, 141, 249, **251**, 266, 408, 412, 415

Mouse, 126, 138, 152, 154, 155, 164, 180, 337, 390, 393; deer, 107, **387**, 396, 398, 399; house, 117, 388; jumping, 78, **388–89**, 398, 402, 403; white-footed, 53, 78, 106, 107, 108, 109, 110, 115, 143, 144, 148, 151, 155, 171, 172, 174, 215, 260, 322, 354, 373, **386–87**, 396, 398, 399, 400, 401, 402, 403, 406, 410, 414, 415, 420, 422, 423, 428, 429

MSX, 36, 208

Mud dauber. *See* Wasps, mud dauber

Muddy and sandy shores and shallows: aquatic predators, 21; burrowers, 20; eelgrass meadows, 21–22; epifauna, 20–21; oxygen, 20; sandy beaches and dunes, 22–25; tidal and shallow subtidal flats, 20

Mud flats. *See* Muddy and sandy shores and shallows

Mud-plantain (*Heteranthera dubia*), 32

Mudpuppy, 46, **307**, 313, 314, 315

Mudsnail: eastern, 26, 29, **199**, 280, 320, 414; threeline, 18, 21, **199**, 280

Mudwort (*Limosella australis*), 31, 416

Mulberries, 53, **150**, 364, 365, 413, 415, 416, 417

Mulberry wing (*Poanes massasoit*), 73, **252**

Mullein, common (*Verbascum thapsus*), 24, 116, 134

Mullet: striped, 294, 305; white, 295

Mumford Cove, 395

Mummichog, 19, 21, 25, 26, 30, 32, 46, **295**, 413, 415, 421

Mushrooms. *See* Fungi

Muskrat, 30, 33, 37, 47, 51, 69, 73, 75, 206, **388**, 399

Mussel(s), 34, 35; associated crabs, 280; blue, 16, 17, 18, 19, 20, 21, 201, 207, **209**, 301, 420, 425; dwarf wedge-, 44, 205; freshwater, 43, 59, 68, **204–6**, 216, 288, 388; lamp-, 44, 205, 206; ribbed-, 26, 30, 207, **209**, 342; zebra, 55, 70, 205, 207. *See also* Creeper; Elliptio; Floater; Pearlshell; Tidewater mucket

Mustard, 250, 424; field, 24

Mystic, 333

Naiad (*Najas* spp.), 43, 48, 54, 67

Nannyberry, 49, 74, 75, **174**, 175, 413, 418

Narragansett Bay, 34

National Audubon Society, 39, 378

National Estuary Program, 33

National Fish and Wildlife Foundation, 62

National Marine Fisheries Service, 278

National Oceanic and Atmospheric Administration, 56

National Water-Quality Assessment Program, 55

National Wetlands Inventory, 40, 71

Natural Area Preserves, 433

Nature Conservancy, The, 37, 62, 63, 64, 79, 184, 341, 375, 433

Needlefish, Atlantic, 295

Nells Island, 23

Nematodes, 43, 85, **191–92**, 274, 387

Nettle, 251; false (*Boehmeria cylindrica*), 48, 53, 71; stinging (*Urtica dioica*), 116; wood- (*Laportea canadensis*), 53

New Hartford, 392

New Haven, 337, 370, 420; Harbor, 35, 123, 290

New Preston, 335

Newt, eastern, 69, 81, 84, 86, 105, **310**, 313, 314, 315, 322, 409, 411, 412, 413, 415, 417, 418, 422, 427

Niering, William, 25

Nighthawk, common, **347**, 363, 368, 369, 370, 420

Night-heron: black-crowned, 30, 33, **332**, 344, 376, 419; yellow-crowned, 33, 329, **332**

Nightshade, climbing (*Solanum dulcamara*), 24, 417

Norfolk, 9

North American Bluebird Society, 354

North Branford, 69, 309

Norwalk, 9

Norwalk Islands, 2, 6, 7, 12, 332, 333

Nutclam, Atlantic, 210

Nuthatch, 138, 146, 372; red-breasted, 113, **352**, 369, 371, 372; white-breasted, 105, 115, **352**, 353, 369, 372, 406, 424, 429

Oaks, 5, 98, 99, 100, 102, 103, 104, 126, 136, **148–49**, 176, 181, 182, 250, 252, 261, 346, 366, 384, 407, 412, 413, 415, 419, 422, 423, 424, 425; acorn ecology, 106–10; black, 33, 97, 101, 106 107, 111, **149**, 412, 421, 422, 423, 424; black/red oak group, 106–10; bur (mossy-cup), **148**; chestnut, 96, 97, 101, 102, 106, 107, 111, **148**, 259, 271, 419, 420, 421, 422, 423, 424, 426; dwarf chestnut, 97, 111, 135, **148**; galls, 270–71; northern red, 96, 98, 102, 104, 106, 107, 110, 111, **149**, 179, 180, 385, 408, 410, 412, 413, 418, 420, 421, 422, 423, 425, 426; pin, 49, 53, 75, 97, 106, **149**, 413, 421, 422; post, 111, **148**; scarlet, 33, 97, 106, 107, 111, **149**, 410, 421, 422, 423; scrub, 97, 106, 111, 112, 135, **149**, 251, 421; swamp white, 53, 75, 97, 98, **148**, **423**; white, 33, 97, 100, 101, 106, 107, 110, 111, **148**, 259, 271, 385, 413, 421, 422, 423, 424, 425; white oak group, 106–10; yellow (chinquapin), 148

Oil Pollution Act, 61

Old fields: conservation, 119; fauna, 115; vegetation, 113–14

Old Lyme, 56, 64, 370, 381

Old Saybrook, 11, 12, 56, 64

Opossum, 19, 25, 53, 115, 150, 373, 375, **379–80**, 393, 396, 398, 399, 406, 407, 410, 411

Orach (*Atriplex*), 29; seabeach (*Atriplex pentandra*), 23

Orangetip, falcate (*Anthocharis midea*), 102, 250, 410, 412

Orchids, 78, 91, 183–84; fringed, 81; peatland, 81; white-fringed, 183; yellow-fringed, 183

Oriole(s), 157, 368, 370; Baltimore, 115, 121, 150, **361**, 366, 369, 373, 412, 414, 417; orchard, 361, 369

Osier, red, 49, 73, 74, 75, 97, **166**

Osprey, 15, 30, 32, 47, 51, 53, 69, **336–37**, 370, 371, 376, 408, 422

Ostracods, 43, 50, 83, 84, 85, **274–75**

Otter, river, 33, 46, 47, 69, **391–92**, 399, 414, 415

Outwash delta: Bridgeport, 7; New Haven, 7

Ovenbird, 77, 105, 106, 113, 115, **357**, 361, 369, 373, 411

Owl(s), 167, 367, 380, 401; barn, 117, **345–46**, 367, 371, 375, 381, 387, 410, 415; barred, 77, 105, **346**, 367, 373; great horned, 77, 113, 115, **346**, 367, 368, 385, 406, 409, 420; long-eared, 329, **346**, 367, 371; migration, 371; northern saw-whet, **346**, 367, 369, 371; short-eared, 30, **346**, 407; snowy, **346**, 429

Oyster, eastern, 34, 35, 36, 123, 201, **207–8**; associated crabs, 279, 280, 423

Oystercatcher, American, 19, 209, 329, **341**, 375, 421, 426

Pachaug State Forest, 77

Palustrine ecosystems. *See* Wetlands, inland

Pannes, salt marsh, 26, 28

Parakeet: Carolina, 345; monk, 345

Paralytic shellfish poisoning, 34

Paramoeba, 36

Parasitic plants, 181–83

Parsnip, water (*Sium suave*), 48

Partridge-berry, 77, 98, 102, 129, **173**, 366, 415

Parula, northern, 356

Pea, 250; beach (*Lathyrus japonicus*), 24, 414

Pearl crescent (*Phyciodes tharos*), 180, **251**

Pearlshell, eastern, 205, 206

Pearly-eye, northern (*Enodia anthedon*), 251

Peatlands: bogs, 78; carnivorous plants, 79–81; fauna, 81–82; fens, 79; orchids, 81

Pectinatella magnifica, 43, 44, 193

Peeper, spring, 50, 73, 77, 81, 84, 105, 115, **311–12**, 314, 315, 408, 409, 410, 411, 412, 416, 417, 424, 428, 429

Pelecinid, American, 240
Pepperbush, sweet, 33, 72, 74, 75, 77, 97, 136, **167**, 418, 419
Pepper-grass (*Lepidium virginicum*), 24; field (*L. campestre*), 135
Perch: white, 46, 68, 205, **297–98**, 305; yellow, 46, 68, 205, **289–90**, 305, 409, 410
Periwinkle(s), 280; common, 16, 17, 18, 20, 21, 36, **198–99**, 200, 408, 410, 420, 424, 426, 427, 428; rough, 16, 17, 18, 21, 29, **199**, 427; yellow, 16, 17, 18, 124, **199**, 275, 427
Petaltail, gray, 221
Pheasant, ring-necked, **339**, 372, 428
Phoebe, eastern, 115, **349**, 367, 368, 408, 409, 412, 422
Photoperiod, 10
Phragmites. See Reed, common
Phylloxera, Phylloxeridae, 144, 268
Physa: pewter, 202; vernal, 83, 203
Physiography, 3
Phytoplankton, 12–14, 30, 43, 67, 68, 73, 290
Pickerel: chain, 45, 68, 70, 292; redfin, 292, 417
Pickerelweed (*Pontederia cordata*), 44, 45, 47, 48, 49, 50, 67, 71, 90, 411, 416, 419
Pigeon: passenger, 147; **345**, 367; rock, 117, **344**, 367, 372, 410
Pigweed (*Amaranthus* spp.), 116
Pike, northern, 46, 68, **292**, 305
Pimpernel: false (*Linernia dubia*), 47; water (*Samolus valerandi*), 31
Pine, 126, 136, 176, 250, 424; eastern white, 75, 76, 77, 78, 89, 97, 98, 100, 104, 110, 111, 112, 113, **138**, 179, 346, 357, 410, 412; pitch, 78, 89, 97, 98, 104, 111–12, **138**, 357, 366; red, **137–38**, 366
Pinesap, **182**, 423
Pinkweed (*Polygonum pensylvanicum*), 116
Pintail, northern, 336
Pinweed, beach (*Lechea maritima*), 24
Pinxter-flower, 170
Pipefish, 305; northern, **296–97**, 423
Pipewort (*Eriocaulon* spp.), 67, 416
Pipistrelle, eastern, 382, 383
Pipit, American, 117, **355**
Pipsissewa (*Chimaphila umbellata*), 102
Pitcher plant, 72, 77, 78, **79–80**, 414; associated insects, 80, 261, 263, 431
Pitcher plant borer, 80, 261
Pitcher plant mosquito, 80
Pitcher plant moth, 80
Plant(s), 128–85; annual, 134; biennial, 134; biogeography, 128–29; carnivorous, 79–81; defenses, 396–97; flowering, 134–35; importance, 128; invasive, 61, 184–85; nonvascular, 129; parasitic, 181–82; perennial, 134; vascular, 130–35; in winter, 180–81
Plantain, 251; downy rattlesnake- (*Goodyera pubescens*), 417, 419; English (*Plantago lanceolata*), 116, 251; seaside (*P. maritima*), 26, 28
Planthoppers (Homoptera), 418
Plover, 25; black-bellied, 21, 117, **341**, 413, 419, 421, 426; piping, 24, 341, 375, 376, 408, 412, 413, 414, 416, 417; semipalmated, 19, 21, **341**, 413, 418, 419, 421

Plums, **155**, 250, 365; beach, 24, **155**, 412, 419
Poaching, 404
Pocotopaug, Lake, 70
Pogonia, rose, 81, 183, 416, 417
Pogonia, whorled, 183
Poison-hemlock (*Conium maculatum*), 116
Poison ivy, 24, 28, 33, 53, 72, 113, 136, **158–59**, 364, 366, 424, 426, 429
Pokeweed, 116, 134, 166, 421, 425
Pollock, 294, 305
Pollution: air, 6, 8; noise, 38; water, 6, 55
Polychaetes. *See* Worm(s), polychaete
Polysiphonia, 16, 124, 125
Pomperaug Valley, 3
Pond-lily. *See* Water-lily
Ponds. *See* Lakes and ponds
Pondweed: curly, 32, 62; Hill's (*Potamogeton hillii*), 67; horned (*Zannichellia palustris*), 32, 43; *Potamogeton* spp., 32, 43, 48, 54, 67, 86, 90, 418; ribbonleaf, 32; slender (*P. strictifolius*), 67
Pool(s): vernal [*see* Vernal pool(s)]; sand plain, 86; tree-hole, 86
Poplar(s), **141–42**, 251, 261; galls, 269
Population, human, 1
Porcupine, North American, 137, 138, 141, 161, 163, 172, **389**, 392, 396, 397, 398
Porgy, 300
Porpoise, harbor, 15, **389**
Portland, 61, 234, 323, 349
Portuguese man-of-war, 189
Pout, ocean, 301, 305
Prairie-chicken, greater, 339
Praying mantis. *See* Mantid(s)
Precipitation, 9–10
Prickly-pear, 111, 136, 165, 180
Privet (*Ligustrum vulgare*, and possibly other species), 33, 119
Project Oceanology (Avery Point), 38
Protozoa, 21, 85, 278
Psyllid, hackberry nipple gall, 150, 269
Puffer, northern, 284, 303, **304**
Pumpkinseed, 45, 68, 70, 205, **299**, 305
Punkies. *See* Midge(s), biting
Pureka, Chris, 151
Purple-top (tall red top, *Tridens flavus*), 252
Purslane: marsh (*Ludwigia palustris*), 54; water (*Elatine americana*), 47, 48
Pyralidae, 44–45

Quahog: false, 209; northern, 20, 207, **208**
Quaker Ridge, 338, 370
Question mark (*Polygonia interrogationis*), 150, **251**, 267
Quillworts, 131
Quinebaug lowlands, 3
Quonnipaug, Lake, 202

Rabbits, 141, 144, 150, 151, 152, 154, 155, 157, 158, 161, 164, 172, 390, 392, 397, 398. *See also* Cottontail(s); Hare(s)
Rabies, 382, 391, 393
Raccoon, 19, 25, 30, 33, 69, 73, 75, 84, 85, 139, 140, 150, 153, 155, 162, 164, 166, 167, 171, 173, 320, 365,

366, 373, 375, 376, 387, **391**, 392, 393, 396, 399, 400, 407, 408, 410, 424, 426
Race, the, 11, 12
Racer. *See* Snake(s), racer
Radish, wild, 135
Ragweed (*Ambrosia* spp.), 179, 252; common (*A. artemisiifolia*), 116; giant (*A. trifida*), 116, 135
Ragwort, round-leaved (*Senecio obovatus*), 250
Rails, 50; black, 340; clapper, 26, 30, 32, **340**, 368; king, 32, 51, 329, **340**; Virginia, 30, 32, 51, **340**, 418
Rainfall, 9–10
Ralfsia, 20
Rams-horn: marsh, 202; thicklip, 83, 203; two-ridge, 202
Range: Bolton, 3; Mohegan, 3; Tolland, 3
Raspberries, 154, 364, 366. *See also* Bramble(s)
Rat, Norway, 19, 117, 346, **388**, 410
Rattlesnake, timber, 73, 74, 105, **323–24**, 326, 380, 410, 415, 419, 420, 421, 423, 424; collecting, vandalism, and incidental mortality, 328; habitat loss, degradation, and fragmentation, 327
Rattlesnake-plantain (*Goodyera*), 183
Raven: common, 69, **351**, 367
Red admiral (*Vanessa atalanta*), 251
Red-cedar, 24, 33, 97, 101, 102, 114, 115, **139–40**, 250, 410, 411, 413; fruits, 366
Redhead (*Aythya americana*), 46
Red maple swamps. *See* Forested wetland(s)
Redpoll, common, 145, 146, **362**, 371, 372, 406
Red-spotted admiral (*Limenitis arthemis*), 155, **251**
Redstart, American, 106, 115, **357**, 364
Reed, common (*Phragmites australis*), 26, 28, 31, 32, 37, 47, 48, 49, 50, 62, 71, 88, 92, 114, 252, 331, 351, 419
Reptile(s): anoxia tolerance, 326–27; conservation, 327–28; freeze-tolerance, 326; habitats, 324; hibernation, 325–26; lizards, 320; reproduction and activity, 324–25; snakes, 320–24; stored fat, 326; supercooling, 326; turtles, 317–20; winter metabolic adjustments, 326
Rhododendrons. *See* Laurel(s), great
Rhodora, 78, 136, **170**, 412
Ribbed-mussel. *See* Mussel(s), ribbed-
Ribbon worms, 191
Rice, wild (*Zizania palustris*), 31, 47, 48, 50, 51, 86, 90, 91, 135, 252, 417, 418, 419, 422, 424
Ridley, Kemp's, 318
Ringlet, common (*Coenonympha tullia*), 251
River(s): Bantam, 43; Branford, 31; Byram, 291; East, 25, 31; Eight Mile, 292; Farmington, 40, 153, 289, 290, 293, 312, 359, 418; flora, fauna (*see* Riverine ecosystems); Hammonasset, 12, 21, 344; Hockanum, 40; Housatonic, 12, 20, 23, 31, 40, 42, 55, 288, 289, 290, 291, 294, 300, 312, 332, 333, 352, 357; Mystic, 21; Natchaug, 41; Naugatuck, 40, 55, 293, 333; Niantic, 21; Norwalk, 56; Quinebaug, 40, 299; Quinnipiac, 12, 31, 40, 42, 337; Saugatuck, 20; Salmon, 40, 43, 286, 292, 293, 416; Scantic, 41; Shepaug, 40; Shetucket, 41, 293; Thames, 12, 21, 31, 40, 42, 55, 61, 288, 289, 293, 345; West, 31. *See also* Connecticut River

460 | Index

River conservation: dams, 54–55; land development, 55; non-native species, 55; pollution, 55; protection, 56
Riverine and floodplain marshes: amphibians, 50; birds, 50–51; energy flow, 50; fishes, 50; invertebrates, 50; mammals, 51; nontidal river marshes, 49; tidal freshwater marshes, 47–49; reptiles, 50; vegetation, 47–49
Riverine ecosystems: beaver-influenced wetlands, 53–54; conservation, 54–56; dams, 55; drainage basins, map of, 42; floodplain forests, 51–53; effects of land development, 55; major rivers, map, 41; non-native species, 55–56; open waters and aquatic beds, 43–47; pollution, 55; protection, 56; riverine and floodplain marshes, 47–51. *See also* Connecticut River
Riverweed (*Podostemum ceratophyllum*), 43
Robin, American, 75, 106, 113, 115, 117, 140, 150, 154, 159, 164, 173, 329, **354**, 361, 366, 367, 405, 406, 407, 408, 409, 413, 416, 417, 425, 429
Rock-cresses (*Arabis* spp.), 102, 250, 410
Rockling, fourbeard, 294
Rockrose, bushy (*Helianthemum dumosum*), 119
Rockweed, 15, 16, 17, 18, 19, 27, 30, 122, **124**, 420, 427
Rocky Hill, 4
Rocky intertidal shores, 15–20; algae, 16–17; amphipods, 17; barnacles, 17; birds, 19; crabs, 18; fishes, 18–19; grazers, 17–18; isopods, 17; lichens, 16; low-gradient, 19–20; mammals, 19; mussels, 17; predatory snails, 18; sea stars, 18; storms, 19; zonation, 16–17
Rocky Mountain spotted fever, 215, 216
Rogers Lake, 300
Rose(s), 33, 136, **153–54**; fruits, 364, 366; galls, 271; multiflora, 33, 100, 113, 114, 116, 119, **153–54**, 366, 406, 408, 409, 410, 428, 429; northeastern (shining), 153; pasture, 97, 153; salt-spray, 24, **153**, 154; swamp, 48, 49, 53, 72, 74, 90, 97, 135, **153**, 416; wild, 153
Roseland Lake, 202
Rose-mallow (*Hibiscus moscheutos*), 31, 32, 419
Rotifers, 80, 85
Roxbury, 403
Rozsa, Ron, 31, 47
Rudderfish, banded, 300
Rue anemone (*Thalictrum thalictroides*), 102
Rushes, 71; black (black grass), 25, 26, 27–28; *Juncus*, 54; soft (*J. effusus*), 47, 54, 71

Saddlebags (*Tramea* spp.), 221; black, 29, 421
Salamander(s), 321; blue-spotted, 77, 105, **308**, 314, 315; four-toed, 77, 81, **309**, 313, 314, 315, 425; Jefferson, 84, 85, 93, **308**, 314, 315; marbled, 83–84, 85, 93, 105, **308**, 314, 315, 409, 413, 421, 422, 423, 427; northern dusky, 46, **308**, 309, 313, 314, 315, 418; northern slimy, 105, **310**, 313, 314, 315; redback, 77, 84, 105, **309–10**, 313, 314, 315, 409, 418, 425; spotted, 81, 83, 85, 93, 105, **308**, 314, 315, 407, 409, 411, 412, 413, 417, 422, 426, 427; spring, 42, **309**, 313, 314, 315; two-lined, 46, **309**, 313, 314, 315, 420. *See also* Mudpuppy
Salisbury, 202, 207, 391

Sallow, Morrison's (*Eupsilia morrisoni*), 261
Salmon, 305; Atlantic, 205, **293**, 305, 306; sockeye, 292–93
Salmon Cove, 416, 418
Saltgrass. *See* Spikegrass
Salt marshes, 23; birds, 30; fauna, 26, 29–30; fishes, 30; history and changes, 25; interactions, 26; invertebrates, 29–30; mammals, 30; marshes to visit, 25; plant responses to salty soils, 29; reptiles, 30; stress gradients, 33; vegetation patterns, 25–28
Saltonstall, Lake, 300
Saltonstall Ridge, 1
Sand crab, Atlantic, 281
Sand dollars, 284
Sanderling, 24, 197, **342–43**, 406, 413, 421, 424, 426, 429
Sand lance, 15, 305; inshore, 302; offshore, 302
Sandflat, 302
Sandpipers, 19, 24, 369; least, 19, 21, 47, 51, 69, 117, **343**, 413, 419; pectoral, 117, **343**; purple, 19, **343**, 429; semipalmated, 19, 21, **343**, 370, 413, 419, 421; solitary, 69, **342**; spotted, 19, 47, 51, 53, **342**, 419; upland, 116, 329, **342**, 375; western, 343; white-rumped, 343
Sand plains, 111–12
Sandwort, seabeach (*Honkenya peploides*), 23, 24
Sandy Point, 344, 370
Sapsucker, yellow-bellied, 106, 137, 141, 144, 146, 151, **348**, 369, 406, 429
Sarsaparilla (*Aralia*), 77, 129; wild, 102, 412, 417
Sassafras, 33, 97, 100, 113, 116, 136, **151–52**, 250, 257, 366, 410, 413, 416, 425
Satellite image(s), 1–2, 5
Saugatuck Reservoir, 300
Sawbriar, 140
Sawflies (Hymenoptera), 141
Sawfly, larch (*Pristiphora erichsonii*), 136
Saxifrage: early (*Saxifraga virginiensis*), 102; swamp (*S. pensylvanica*), 75
Scad: bigeye, 300; mackerel, 300; rough, 14, 300
Scale insects, 138; beech (*Cryptococcus fagisuga*), 147
Scallops, 34, 35; bay, 22, 208, **209**; sea, 208
Scaup: greater, 14, 46, **336**; lesser, 46, **336**
Scorpionflies, common, **262**, 415
Scoter: black, 336; surf, 336; white-winged, 14, 336
Scouring-rush, 130
Screech-owl, eastern, 73, 75, 77, 105, 163, **346**, 367, 369, 407, 408, 420, 429
Scrippsiella trochoidea, 13
Sculpins, 46, 305; longhorn, **297**, 305; shorthorn, **297**, 305; slimy, 45, 205, **297**
Scup, 14, **300–301**
Sea-blight (*Suaeda* spp.), 24
Sea gooseberry, 191
Sea Grant, Connecticut, New York, 278
Sea grape, 285
Seahorse, 305; lined, 296
Seal: gray, 394; harbor, 15, 47, 297, **393–94**, 399, 408, 409
Sea-lavender (*Limonium carolinianum*), 26, 28, 29, 30, 420, 421, 427

Sea lettuce, 16, 17, 19–20, 27, 122, **123**, 336
Sea level, 2, 4–5, 23, 25
Sea nettle, 13
Sea raven, 187, 297, 305
Sea-rocket (*Cakile edentula*), 23, 418
Sea spiders, 211–12
Sea squirt(s), 201, 276; Pacific rough, 285. *See also* Tunicates
Sea stars, 13, 21, 208, 209, **283–84**; common, 21, **283**; little brittle star, 283; northern, 283
Seaweeds, 16–19
Sedges, 49, 71, 72, 75, 96, 251, 252; *Carex*, 54, 73, 75, 77, 78, 82, 83, 86, 102; *C. aquatilis*, 79; *C. crinita*, 54; *C. flava*, 79; *C. interior*, 79; *C. intumescens*, 414; *C. lacustris*, 79; *C. lasiocarpa*, 79; *C. leptalea*, 79; *C. silicea*, 24; *C. stipata*, 54; false hop (*C. lupuliformis*), 83; *Scirpus*, 50; three-way (*Dulichium arundinaceum*), 47, 67, 74; tussock (*C. stricta*), 32, 48, 54, 71, 72, 74, 75, 76, 88, 252; woodland (*C. pensylvanica*), 98, 102, 410; woolly (*S. cyperinus*), 54, 67, 71
Seedbox (*Ludwigia alternifolia*), 48, 405
Selden Cove, 43
Selden Creek, 47
Senna(s), 250
Serviceberry, 157
Sessions Woods Wildlife Management Area, 383, 391
Shad, 46, 304; American, 30, 32, 46, 60, **289–90**, 305, 306; gizzard, 290; hickory, 46, **289**, 305
Shadbush, 33, 74, 77, 97, **157**, 364, 366, 411, 412, 415, 424
Shade Swamp Sanctuary, 318
Shadowdragons (*Neurocordulia* spp.), 220; umber 219
Shamrock, water, 43, 86, 133
Shark(s), 304, 305; sandbar, 287
Shepaug Bald Eagle Viewing Area, 337
Shepherd's-purse (*Capsella bursa-pastoris*), 116
Sherwood Island State Park, 25
Shieldback, American, 224
Shiner: bridle, 290; common, 45, 205, **290**; golden, 45, 46, 68, 205, **290**; mimic, 291; spottail, 30, 46, **291**, 416
Shipworm, common (naval), 210
Shoveler, northern, 370
Shrew(s), 25, 30, 53, 82, 105, 106, 115, 260, 393, 399, 417, 428, 429; least, 25, **381**; masked, 25, **380**, 398, 402; short-tailed, 25, 78, 107, 337, 346, **380–81**, 399, 401, 402, 405, 429; smoky, 25, **380**; water, 47, 77–78, **380**
Shrike, loggerhead, 349–50; northern, 349
Shrimp(s), 13, 22, 287, 288, 289, 290, 293, 294, 300, 302; clam (*see* Clam shrimp); fairy (*see* Fairy shrimp); grass (shore), 26, 29, **281–82**, 298, 415; mantis, 282; opossum (*Neomysis americana*) 21, 282; parasites on, 276; sand, 21, 26, 29, **281–82**, 303; seed, 274–75; seven-spine bay, 282; skeleton, 21
Shrub swamp: fauna, 74–75; vegetation, 73–74
Silverfish, 117
Silver Lake, 319

Silver-rod (*Solidago bicolor*), 422
Silverside, 30, 32, 298, 421, 422; Atlantic, 21, 26, **294–95**, 423; inland, 295
Simsbury, 153
Siskin, pine, 113, 139, 145, 146, **362**, 367, 371, 372
Skate(s), 304; little, 14, 21, **287–88**; winter, 21, **288**
Skeletonema costatum, 13
Skimmer: black, 329, **344**, 375, 417; slaty (*Libellula incesta*), 219; twelve-spotted (*L. pulchella*), 218
Skimmer dragonflies, 219
Skink, five-lined, **320**, 325, 326
Skipper, 414; Arctic (*Carterocephalus palaemon*), 252; broad-winged (*Poanes viator*), 73, 252; cobweb (*Hesperia metea*), 252; common checkered- (*Pyrgus communis*), 252; crossline (*Polites origenes*), 252; Delaware (*Anatrytone logan*), 252; dion (*Euphyes dion*), 82, 252; dun (*E. vestris*), 252; dusted (*Atrytonopsis hianna*), 252; European (*Thymelicus lineola*), 252, 414; fiery (*Hylephila phyleus*), 252; Hobomok (*Poanes hobomok*), 252; Indian (*Hesperia sassacus*), 252; least (*Ancyloxypha numitor*), 252; Leonard's (*Hesperia leonardus*), 252; long-tailed (*Urbanus proteus*), 251; ocola (*Panoquina ocola*), 252; Peck's (*Polites peckius*), 252; pepper and salt (*Amblyscirtes hegon*), 252; silver-spotted (*Epargyreus clarus*), 157, 251; swarthy (*Nastra lherminier*), 252; tawny-edged (*Polites themistocles*), 252; Zabulon (*Poanes zabulon*), 252
Skunk, striped, 19, 25, 115, 150, 375, **393**, 399, 407, 421
Skunk cabbage, 71, 74, 75, 76, 77, 86, 90, 129, **176**, 397, 409, 410, 411
Sleeping Giant, 1
Slugs, 105, 126, **204**, 422, 425
Smartweed, 86; water (*Polygonum amphibium, P. punctatum*), 32, 48, 49, 54, 67, 71
Smelt(s): American, 305; rainbow, **292**, 305
Smooth cord weed, 17
Snail(s), 20–22, 80, 126, 192, **197–203**, 214, 280, 295, 401, 425; dove-, 201, 280; freshwater, 43, 45, 50, 68, 73, 84, 85, **201–3**; golden amber- (*Succinea wilsoni*), 29; horn, 200; hydrobiids, 203; marsh pond-, 202; mystery-, 44, 196, 202–3, 207; moon-, **200**, 280, 343; orb (planorbid), 86, 203, 427; pond, 203; pouch, 203; prosobranch, 202–3; pulmonates, 202–3; round-mouthed, 202; salt marsh (*see* Melampus, eastern); saltwater, **197–201**, 209; slipper-, 21, 24, **197–98**, 281, 406, 424; terrestrial, 105, 203–4; turret, 202; Virginia river, 44; woodland pond-, 202. *See also additional species*
Snailfur (hydroids), 189
Snake(s), 387, 388; brown, **321**, 324, 325, 326; common garter, 30, 73, 81, 105, 115, **322**, 325, 326, 413, 415, 418, 419, 420, 421, 422, 423, 425; copperhead, **323**, 325, 326, 380, 417; eastern hognose, **322**, 325, 326, 415, 420; eastern rat, 105, 322, **323**, 324, 325, 326, 373, 417, 427; eastern ribbon, 73, 74, **322**, 324, 325, 326, 409, 410, 411, 412, 414, 415, 416, 418, 422, 424; milk, 105, 115, **322**, 325, 326, 373; northern water, 46, 50, 69, 73, 74, 81, **322**, 324, 325, 326, 411, 417, 420; racer, 105, 115, **321**, 324, 325, 326, 373, 411, 412, 413, 424; redbelly, 81, **321**, 325, 326; ringneck, 105, **321**, 325, 326, 413, 414, 415, 416, 424; smooth green, 116, **321**, 325, 326; worm, **321**, 325, 326. *See also* Rattlesnake, timber
Snakeflies, 232
Sneezeweed (*Helenium autumnale*), 48
Snipe, Wilson's, 73, 342
Snout, American (*Libytheana carinenta*), 150, **250**
Snow, 9–10
Snowberry, 366; creeping (*Gaultheria hispidula*), 77, 78
Snowflea, **217**, 267
Soapwort (*Saponaria officinalis*), 24, 116
Solomon's-seal, 366
Sootywing, common (*Pholisora catullus*), 252
Sora, 51, 340, 413, 422
Sorex, 25
Southbury, 335
Southwest Hills, 3
Spadefoot, eastern, 86, **310–11**, 314, 315, 316
Spanworms. *See* Moth(s), inchworm
Sparrows, 117, 119, 152, 180, 366, 370; American tree, 115, **358**, 369, 372, 409, 428; chipping, 115, 138, 140, 329, **358**, 366, 369, 411; field, 115, 119, **358**, 372, 411; fox, **359**, 369, 372; grasshopper, 117, 329, **358**, 375; Henslow's, 358; house, 117, 329, 351, 354, 362, 372; Nelson's sharp-tailed, 358; saltmarsh sharp-tailed, 26, 30, 32, 37, **359**; savannah, 358; seaside, 26, 30, 32, 37, **359**, 368; song, 24, 30, 32, 50, 73, 75, 115, 140, 154, 329, **359**, 367, 372; swamp, 30, 32, 37, 50, 73, 74, **359**; vesper, 329, **358**; white-crowned, 359; white-throated, 82, 115, 159, **359**, 366, 372, 405, 411, 423, 424, 428, 429
Spearscale (*Atriplex patula*), 23
Special Concern species, 431
Spendelow, Jeff, 344
Sphagnum, 74, 76, 77, 78, 79, 82, 309, 388
Sphinx moths, 258–59. *See also* Moth(s)
Spicebush, 53, 75, 76, 77, 97, 98, 135, **151–52**, 250, 364, 366, 367, 409, 410, 418, 419, 420, 421, 422
Spider(s), 29, 105, **212–14**, 245, 263, 266, 408, 425; black-and-yellow garden, 212–13; black widow, 214; crab, **213**, 214, 420, 423; filmy dome, **213**, 420; fishing, 50, 212; jumping, **213**, 414; marbled, 422; orb-web, 212; sea (*see* Sea Spiders); wolf, 213
Spikegrass (*Distichlis spicata*), 27–28, 199
Spikemosses, 130
Spikerush (*Eleocharis* spp.), 31, 32, 48, 54; *Eleocharis acicularis*, 86
Spiketail, 220; tiger, 219, 221
Spindle-tree, winged [*Evonymus (Euonymus) alata*], 119, 395
Spiraea, 72, 73, 97, 136, **156**. *See also* Hardhack; Meadow-sweet; Steeplebush
Spit, sand, 22, 23, 24, 25
Spittlebugs, 138
Spleenwort: mountain, 132; wall-rue, 132
Sponge(s), 17, 22, **186–87**, 208, 276, 297; boring, **186–87**, 208; freshwater, 43, 73, 83, **187**, 232;

halichondria, **186**, 421; haliclona, **187**, 421; on crabs, 278; red beard, 186
Spongillaflies, 43, 232
Sport Fish Restoration Program, 306
Spot, 21, **301**, 305
Spreadwing: amber-winged (*Lestes eurinus*), 82; emerald (*L. dryas*), 83, 220; slender (*L. rectangularis*), 83
Spring beauty, 102, **177**
Springtails, 29, 68, 83, 105, 110, **217–18**, 267, 428
Spring wildflowers, 175–78
Spruce, 104, 136; black, 76, 77, 78, 97, **137**, 182; galls, 268–69; Norway, 112, 346; red, 76, 110, **137**
Spurge: cypress (*Euphorbia cyparissias*), 114, 116; seaside (*Chamaesyce polygonifolia*), 23; spotted (*C. maculata*), 116
Squantz Pond, 300
Squawroot, 181
Squids, **210**, 289, 290, 293, 294, 300, 302, 303
Squirrel(s), 98, 99, 105, 107, 126, 136, 138, 139, 143, 144, 147, 148, 149, 150, 151, 157, 158, 161, 162, 163, 164, 166, 167, 390, 399, 401, 402, 417, 424; flying, 78, 106, 107, 108, 143, 373, **385**, 396, 399, 401, 403, 415, 425; gray, 53, 78, 106, 107, 108, 109, 115, 117, 143, 144, 150, 159, 161, 166, 270, 346, 373, **384**, 392, 393, 396, 398, 400, 402, 403, 407, 416, 417, 418, 419, 420, 421, 422, 423, 426, 428, 429; red, 113, 136, 137, 163, **384–85**, 396, 402
Star-flower (*Trientalis borealis*), 76, 77, 110, 412
Stargrass, water (*Heteranthera dubia*), 43, 48, 419
Starling, European, 115, 117, 142, 329, 348, 351, 354, **355**, 362, 366, 369, 372, 377, 410, 412, 413, 416, 426
Steblospio benedicti, 21
Steeplebush, 136, **156**
Stewart B. McKinney National Wildlife Refuge, 8
Stickleback, 14, 305; blackspotted, 296; fourspine, 21, 30, **296**; ninespine, 296; threespine, 205, **296**
St. John's-wort, dwarf (*Hypericum mutilum*), 86
Stoneflies, 43, 45, 50, 54, 55, 83, **221–22**, 406, 407, 413, 418, 427, 428, 429; *Allocapnia recta*, 221; *Brachytera*, 221; *Taeniopteryx*, 221
Stone walls, 7
Stonington, 12, 101, 370
Storm-petrel, Wilson's, 331
Storms, 19
Stratford, 337, 360
Strawberry: fruits, 364, 365; wild (*Fragaria virginiana*), 116
Stream orders, 40
Streams. *See* Riverine ecosystems
Sturgeon, 304, 305; Atlantic, 46, **288**, 305; shortnose, 46, 206, **288**, 305, 306
Suburban yards, 117
Suckers, 46, 68, 304, 304; longnose, 291; white, 45, 205, **291**
Sugarloaf, the, 1
Sulphur: clouded (*Colias philodice*), **250**, 425, 427; cloudless (*Phoebis sennae*), **250**; orange (*C. eurytheme*), **250**, 421, 424, 426
Sumacs, 116, 136, 159, 364, 366, 405, 429; poison,

72, 74, 75, 77, 78, 97, **159**; shining, 24, 97, 103, 114, **159**, 419, 420, 424; smooth, 33, 114, **159**, 415, 419, 428; staghorn, 114, **159**, 415, 419, 427
Sundews, 77, 79, 80; round-leaved, 78
Sunfish, 46, 304, 305; banded, 298; green, 299; redbreast, 298, 299
Sunflower family, 251
Swallow, 363; bank, 30, 32, 47, 51, 53, 69, **351**, 369, 417; barn, 30, 32, 47, 51, **352**, 369, 416; cliff, 352; northern rough-winged, 47, 53, **351**; tree, 24, 30, 32, 37, 47, 51, 54, 73, 75, 143, **351**, 354, 369, 376–77, 408, 409, 419, 420, 421
Swallowtail(s), 250; black swallowtail (*Papilio polyxenes*), **250**, 412; eastern tiger (*P. glaucus*), 102, 151, 155, 172, 238, **250**, 411, 412, 413, 416; pipevine (*Battus philenor*), 250; spicebush swallowtail (*P. troilus*), 15, 151, **250**
Swamp(s), evergreen, forested, red maple. *See* Forested wetland(s)
Swamp, shrub. *See* Shrub swamp
Swamp-buttercup, northern (*Ranunculus hispidus*), 75
Swan(s), 15, 333; mute, 14, 15, 19, 21, 30, 32, 46, 50, 51, 69, 122, 123, **334**, 377–78, 390, 416, 418; tundra, 334
Sweet-fern, 97, 111, 114, 116, 136, **142**, 414
Sweetflag (*Acorus americanus*), 32, 47, 48, 49, 89
Swift, chimney, 117, 153, **347**, 363, 369, 370, 411, 412, 422
Swimmer's itch, 191, 199
Sycamore, American, 53, 89, 91, 97, **153**, 408, 412, 424
Sylvio O. Conte National Fish and Wildlife Refuge, 62, 63
Synapta: pink, 284; white, 284

Taconic Plateau, 3
Tailed-blue, eastern (*Everes comyntas*), 250
Talus, 5
Tanager, scarlet, 105, 106, 113, 115, **358**, 366, 368, 370, 373
Tanaid, blind, 20
Tardigrades (water bears), 85, 132
Tautog, 14, 18, 209, **301**, 304
Teal, 51, **335**, 370
Tearthumb, 71; arrow-leaved tearthumb (*Polygonum sagittatum*), 48; halberd-leaved (*P. arifolium*), 48
Temperature, 9–10
Tench, 291
Tent caterpillars, 146, 147, 156, **257**, 266, 410, 411, 412, 413
Termites, **227**, 414
Terns, 15, 19, 25, 30, 302; common, 14, 26, 30, **344**, 376, 415, 416, 418, 419; Forster's, 344; least, 14–15, 24, 26, 30, 296, **344**, 375, 413, 416; roseate, 14, 329, **344**, 375, 376, 415
Terrapin, diamondback, 26, 30, 32, **320**, 324, 325, 326, 417
Terrestrial ecosystems, 95–121; coniferous and mixed forests and woodlands, 110–13; deciduous forests, 95–110; forest edges and shrubby old fields, 113–15; fields and grasslands, 115–17; suburban yards, 117; cities, 117; conservation, 117–21
Thalassiosira nordenskioldii, 13
Thimble Islands, 2, 6, 7, 240
Thistle, 251, 362, 420; bull, 116, 134; Canada (*Cirsium arvense*), 116; Russian, 23, 24; swamp (*C. muticum*), 75
Thomas, Michael, 221
Thompsonville, 56, 58
Thrasher, brown, 115, **355**, 366
Threatened species, 431
Three-square (*Scirpus americanus, S. pungens*), 28, 31, 32, 42, 47, 48, 67, 408, 414
Thrush, 139, 140, 150, 151, 154, 157, 164, 166, 169, 173, 174, 182, 366, 367; hermit, 105, 113, **354**, 369, 407; Swainson's, **354**; wood, 105, 106, 113, 115, 121, 170, **354**, 369, 373, 411, 416, 421
Ticks, 214–16; black-legged (deer) **215–16**, 415; dog, 215, 216
Tickseed, lance-leaved (*Coreopsis lanceolata*), 24
Tick-trefoil, 251
Tidal and shallow subtidal flats: aquatic predators, 21; birds, 21; epifauna and burrowing invertebrates, 20–21; sand-cobble mosiaics, 21
Tidal freshwater marshes: fauna, 50–51; vegetation, 47–49
Tides, 11–12
Tidewater mucket (*Leptodea ochracea*), 44, 206
Timothy, 252, 417
Tiner, Ralph, 72, 75, 76, 79
Titmouse, tufted, 77, 105, 106, 115, 117, **352**, 367, 369, 372, 377, 405, 406, 407, 424, 427
Toad(s), 59, 322; American, 50, 77, 86, 105, 115, **311**, 314, 315, 316, 409, 410, 411, 412, 414, 415, 417; Fowler's, **311**, 314, 315, 417
Toadfish, oyster, 284, **294**, 305
Toadflax (*Linaria*), 251
Tobacco, Indian (*Lobelia inflata*), 425
Tolland Range, 3
Tomcod, Atlantic, **294**, 305
Toothworts, 250
Topography, 6
Tortoiseshell: Compton (*Nymphalis vaualbum*), 102, **251**, 408; Milbert's (*N. milberti*), 251
Towhee, eastern, 106, 115, 121, **358**, 366, 369, 372, 377, 413, 414
Tracy, Tara, 133
Trap-rock ridge(s), 1, 4, 5, 101–2
Trawl samples, 14
Tree-hole pools, 86
Treehoppers, 240–41
Trichocorixa, 29
Trillium (*Trillium* spp.), 129, 175, 243; red (*T. erectum*), 102
Trout, 46, 54, 205, 221, 304, 305; brook, 45, 68, 113, 221, **293**, 305; brown, 45, 68, **293**; rainbow, 68, **292**
Trout Unlimited, 56
Tufted red weed, 16
Tularemia, 215
Tulip-tree, 97, 98, 104, 129, 136, **150–51**, 250, 257, 366, 405, 407, 410, 411, 413, 414, 420, 424, 425, 426, 427, 429
Tulip-tree beauty, 89, 151, **257**, 412, 414, 421
Tulip-tree silkmoth (*Callosamia angulifera*), 151
Tumblebug, eastern, 236
Tunicates, 17, 22, 208, 278, **284–85**, 302; golden star, 285; Pacific colonial, 285
Tunny, little, 302
Tupelo. *See* Gum, black
Turkey, wild, 105, 107, 109, 113, 117, 144, 147, 148, 149, 150, 151, 152, 166, 168, 169, 172, 226, **339–40**, 369, 372, 377, 390, 408, 416, 425
Turnstone, ruddy, 19, 24, 197, **342**, 370, 406, 419
Turtle(s), 54, 73, **317–20**, 414; bog, 82, 319, 324, 325, 326, 327, 328, 391; collecting, vandalism, and incidental mortality, 328; diamondback terrapin, 320, 326; eastern box, 84, 105, 115, 116, 320, 324, 325, 326, 415, 416; green, 318; habitat loss, degradation, and fragmentation, 327; Kemp's ridley, 318; leatherback, 318; loggerhead, 318; painted, 32, 46, 50, 53, 69, 73, 74, 81, 84, 86, **319**, 324, 325, 326, 327, 409, 413, 416; snapping, 30, 32, 46, 50, 68, 73, 81, **317–18**, 324, 325, 326, 410, 416, 420; spotted, 69, 73, 74, 81, 84, **319**, 324, 325, 326, 409, 411, 413, 415, 418; stinkpot (common musk), 69, 73, **318**, 325, 326, 420; wood, 46, **319**, 324, 325, 326, 327, 328
Turtlehead (*Chelone glabra*), 75, 251
Twayblade, lily-leaved, 183
Twig-rush (*Cladium mariscoides*), 67, 73, 79
Twin Lakes, 70; East Twin Lake, 202, 207, 292–93; West Twin Lake, 207

Ulothrix, 16
University of Connecticut Department of Plant Science, 62
Uplands: western, 1; eastern, 1
Upper Bolton Lake, 70
Urchins: sea, 17, 18, **284**, 304; green, 284; purple, 284
U.S. Army Corps of Engineers, 59, 61
U.S. Coast Guard, 61
U.S. Department of Agriculture: Animal and Plant Health Inspection Service Cooperative, 62; Farm Bill, 433; Conservation Reserve Program, 433; Wildlife Habitat Incentives Program, 433
U.S. Environmental Protection Agency, 61
U.S. Fish and Wildlife Service, 63, 64; Partners for Wildlife Program, 62
U.S. Geological Survey, 55
U.S. Navy, 61

Veery, 75, 77, 105, 106, 113, 115, **354**, 369, 413
Vernal pool(s), 102, 418, 423, 425; amphibians, **83–84**, 316, 408, 411, 413, 415, 416, 417, 421, 422, 427; colonization, 85; dry phase, 85; invertebrates, 83; other temporary pools, 85–86; protection, 93–94; recognizing dry basins, 83; seasonal changes, 84–85; sources of water and physical conditions, 82
Vernon, 70
Verrucaria erichsenii, 16
Vervain, blue (*Verbena hastata*), 71
Vetch, crown, 252

Viburnum(s), 174, 250, 364, 366, 367, 408, 416; maple-leaved, 97, 103, **174**, 366, 413, 414, 416, 417, 420, 425
Viceroy (*Limenitis archippus*), 141, 180, **251**, 256, 426
Violet(s) (*Viola* spp.), 102, 116, 175, 250, 251; northern white (*V. macloskeyi*), 75
Vireo, 366, 368, 369, 370; blue-headed, 82, 113, **350**; red-eyed, 77, 105, 106, 113, 115, **350**, 361, 364, 366, 373; warbling, 53, 77, **350**, 411; white-eyed, 115, **350**; yellow-throated, 77, 105, **350**, 411
Virginia river snail, 202
Vole(s), 82, 136, 154, 156, 171, 337, 346, 390, 392, 393, 399, 401, 402; meadow, 30, 51, 114, 117, 119, 346, **387–88**, 390, 397, 398, 400, 406, 410, 417, 418, 427, 428; red-backed, 78, 106, **387**; woodland, 106, 115, **388**, 398, 407
Voluntown, 77
Vulture: black, **333**, 367; turkey, **333**, 370, 406, 422

Wagner, David, 221
Walkingsticks, 170, **222**, 420, 422, 423, 424
Walleye, 300
Waramaug, Lake, 70, 335
Warbler(s), 366, 368, 369, 370; bay-breasted, 357; black-and-white, 77, 105, 113, 115, **357**, 369, 410, 411; blackburnian, 111, 113, 139, **356**; blackpoll, 357; black-throated blue, **356**, 421; black-throated green, 82, 111, 113, 139, 146, **356**, 411; blue-winged, 106, 115, **355**; Brewster's, 356; Canada, 77, 113, **357**, 369; Cape May, 357; cerulean, 53, 105, **357**; chestnut-sided, 106, 115, **356**; Connecticut, 369; golden-winged, 115, **355–56**; hooded, 53, 77, 115, 170, **357**, 373, 411; Kentucky, 53, **357**; magnolia, 113, **356**; Nashville, 82, 115, **356**; palm, 357; pine, 113, 138, **357**, 369; prairie, 115, 119, **357**; Tennessee, 357; Wilson's, 357; worm-eating, 105, 170, **357**, 373, 411; yellow, 74, 115, 166, **356**, 361; yellow-rumped, 82, 113, 115, 139, 143, **356**, 405, 424; yellow-throated, 356–57
Warren, 70
Warren, R. Scott, 25
Washington, 70
Wasp(s), 24, 29, 152, 159, 163, 169, 212, 228, **239–45**, 411, 415, 421, 428; American pelecinid, **240**, 420; Chalcididae, 139, 261, 270; cicada killer, **245**, 418, 421; common sand, 245; digger, 245; gall (cynipid), 149, 154, 171, 240, **270–71**, 413, 419, 423, 424, 425; ichneumon, **239–40**, 271, 409, 421; horntail, 240; mud dauber, 245; paper, 159, 228, 241, **242**, 244, 408, 421; potter, 173; sphecid, 245, 246; spider, 173, **245**, 416; spiny rose gall (*D. bicolor*), 271. *See also* Hornet(s), bald-faced; Yellowjacket(s)
Water boatmen, 29, 50, 68, 73, 83, 84, 85, 216, 227, **229**
Water-celery (wild-celery, tapegrass; *Vallisneria americana*), 32, 43, 45, 46, 48, 67, 86, 90, 418, 419, 420
Watercress (*Nasturtium officinale*), 86
Water-horehound (*Lycopus* spp.), 71, 75
Water-lily, 44, 45, 48, 68, 86; bullhead-lily (*Nuphar variegatum*); white (*Nymphaea odorata*), 43, 67, 90, 413; yellow (spatterdock; *Nuphar* spp.), 43, 47, 50, 54, 67, 86, 90, 412
Water measurers (Hydrometridae), 50
Water-milfoil (*Myriophyllum* spp.), 43, 48, 88, 90; Eurasian (*M. spicatum*), 32, 62, 69, 70, 86
Water millet (*Echinochloa walteri*), 48
Water-plantain (*Alisma* spp.), 48, 86, 90
Waterscorpions, 45, 50, 73, 83, 216, **229**
Water-shield (*Brasenia schreberi*), 54, 67
Water-starwort (*Callitriche* spp.), 43
Water striders, 43, 45, 50, 68, 83, 84, 85, 227, **228**, 229, 427; broad-shouldered, 85
Waterthrush: Louisiana, 53, 77, **357**, 373, 411, 416; northern, 77, 82, **357**
Waterweed (*Elodea* spp.), 32, 43, 48, 67, 90, 418, 419, 420
Water-willow (swamp loosestrife, *Decodon verticillatus*), 67, 71, 74, 78, 89, 90
Waxwing, cedar, 73, 75, 115, 139, 150, 153, 157, 166, 174, 182, **355**, 366, 367, 408, 409, 416, 425
Wayfaring-tree (*Viburnum lantana*), 405
Weakfish, 14, **301**
Weasel(s), 106, 115, 399; long-tailed, 392; short-tailed, 392
Weevil(s), 238–39, 423; acorn, **107–9**, 148, **239**; flower-feeding (*Nanophyes marmoratus*), 92; northern pine, 138, **239**; root-mining (*Hylobius transversovittatus*), 92; short-snouted, 107. *See also* Curculios
Westbrook, 332, 333
West Hartford, 350
West Haven, 336, 344, 370
West Hill Pond, 292
Wetlands, inland: bogs, 78; conservation, 91–94; fens, 79; forested wetlands, 75–78; freshwater marshes and meadows, 71–73; habitat protection, 93–94; indicators, 86–87; invasive plants, 92–93; plant reproduction in water, 89–91; red maple swamps, 75–76; regulations, 91–92; shrub swamps, 73–75; soils and plant responses, 87–89; values, 91; vernal pools, 82–86
West Nile virus, 35, 263
West Rock, 1
Whale, long-finned pilot, 389
Wharton Brook State Park, 112
Wheeler Wildlife Management Area, 25
Whelk(s), 280, 281; channeled, 200, 201; knobbed, 200, 201
Whimbrel, 342
Whip-poor-will, **347**, 373
White(s) (butterflies): cabbage (*Pieris rapa*), **250**, 409, 410, 421, 426; West Virginia (*P. virginianus*), 250
Whiteface: crimson-ringed (*Leucorrhinia glacialis*), 82; frosted (*L. frigida*), 82; Hudsonian (*L. hudsonica*), 82
Whitefish, round, 293
White-fringed orchid (*Platanthera blephariglottis*), 183
Whitelip, 203
White Memorial Foundation, 360
Whitetail, common, 219
Widgeon-grass (ditch-grass; *Ruppia maritima*), 32
Wigeon, American, 46, **336**, 370
Wild calla (*Calla palustris*), 78
Wildlife Habitat Incentives Program, 433
Wild-raisin, northern, 74, 75, **174**
Willet, 21, 26, 37, 329, **342**, 413, 419, 421
Willimantic, 55
Willimantic Basin, 3
Willow(s), 48, 49, 72, 74, 90, 97, 126, 136, **140–41**, 180, 250, 251, 261, 349, 386, 423; autumn, 73; beaked gall midge, 269–70, 423; black, 53, 89, 91, 104, 140–41; cone gall midge, 269, 270; hoary, 73; pussy, 73, **140–41**; silky, 54
Windham Hills, 3
Windowpane, 302
Windsor, 359; Locks, 56, 60
Wineberry, 33, 154
Winterberries, 33, 49, 54, 72, 75, 78, **159–60**, 366, 405, 406, 407, 408, 415, 416, 420, 421, 422, 423, 428, 429; common, 74, 77, 97, **159**; smooth, 74, 77, 97, **159**
Winter-cress, common, 135
Wintergreen, 98, 102, 135, 136, **169**, 366, 410, 422, 423, 427; spotted, 102, 136, **167–68**, 416, 417
Witch-hazel, 33, 97, 98, 136, 146, **152–53**, 179, 366, 415, 418, 419, 420, 422, 423, 424, 425, 426, 427; bud gall aphid, 269; leaf gall aphid, 268, 269
Witch-hobble, 110, **174**, 412
Wolf, gray, 390
Wononscopomuc, Lake, 292–93
Wood anemone (*Anemone quinquefolia*), 102, 177
Woodchuck, 115, 117, **384**, 390, 393, 396, 400, 402, 403, 407, 408, 411, 417, 428
Woodcock, American, 145, **342**, 367, 369, 412, 414
Woodlice, 132. *See also* Isopods
Wood-nymph, common (*Cercyonis pegala*), 251
Woodpecker(s), 54, 73, 75, 107, 138, 142, 144, 147, 150, 151, 157, 159, 164, 166, 349, 352, 366, 367, 372, 385; downy, 77, 105, 106, 113, 115, 270, **348**, 369, 372, 406, 415, 423, 427; hairy, 77, 113, **348**, 369, 372, 373, 427; pileated, 77, 118, 241, **348**, 369, 373, 409, 427; red-bellied, 53, 77, 105, 106, 311, **348**, 369, 372, 377, 405, 411, 417, 425; red-headed, **348**, 369
Wood-pewee, eastern, 105, 106, 115, 349, 369
Woodrat, Allegheny, 387
Wood-satyr, little (*Megisto cymela*), 251
Woodsia, rusty, 132
Woodstock, 202
Wool-grass. *See* Sedges, woolly
Woollybear, **260**, 266, 405, 407, 424, 425, 426
Worm(s), 24, 85, 235, 287, 301, 302, 303, 333, 421; brain-, 395; carnation, 195; clam, 29, 194; coiled, 194–95; cone, 21, 195; earth-, 45, 105, 192, **195**, 214, 415, 418, 421; glow-, 237, 420; horsehair, 53, 192; lobster gill, 195; oligochaete, 43, 44, 45, 68, 83, 84, **195**; painted, 21; peanut, 276; polychaete, 13, 20, 21, 22, 24, 26, **194–95**, 208, 276, 280, 281, 282, 288, 292, 294, 295, 297, 301, 303, 416; ribbon, 21, **191**; wire-, 236. *See also* Flatworm(s)

Wren: Carolina, 53, 77, 115, 353, 367, 372, 377, 405, 406, 412, 417; house, 115, 121, **353**, 354, 369; marsh, 30, 32, 37, 50, 51, 73, 75, **353**, 368; sedge, 73, **353**, 413; winter, 53, **353**, 369

Xerces Society, 262

Yarrow, common *(Achillea millefolium)*, 116
Yellow cress: creeping *(Rorippa sylvestris)*, 86; *R. amphibia*, 87
Yellow-eyed grass *(Xyris* spp.*)*, 67

Yellow iris *(Iris pseudacornus)*, 71, 86, 414
Yellowjackets, 159, 167, 175, **242–44**, 265, 419, 420, 421, 425, 426, 428; colony demise, 243–44; eggs, 243; food and foraging, 243; larvae, 243; males, 243; nest usurpers, 244; nests, 242–43; queens, 243; sex determination, 243; stings and venom, 244; winter, 242
Yellowlegs, 30; lesser, 21, 69, 342, 419, 421; greater, 21, 69, 342, 413, 419, 421, 426
Yellowthroat, common, 30, 32, 50, 73, 74, 77, 106, 329, **357**, 361

Yew, 406; American *(Taxus canadensis)*, 110
Yoldia, file, 21, 209

Zebra mussel, 207
Zepko, George, 223, 415
Zoar, Lake, 335
Zooplankton, 13–14, 35, 43, 60, 68, 73, 189, 190, 200, 288, 289, 290, 291, 292, 293, 295, 299, 300, 301, 302, 308